AF602069

Birkhäuser

Trends in Mathematics

Trends in Mathematics is a series devoted to the publication of volumes arising from conferences and lecture series focusing on a particular topic from any area of mathematics. Its aim is to make current developments available to the community as rapidly as possible without compromise to quality and to archive these for reference.

Proposals for volumes can be submitted using the Online Book Project Submission Form at our website www.birkhauser-science.com.

Material submitted for publication must be screened and prepared as follows:

All contributions should undergo a reviewing process similar to that carried out by journals and be checked for correct use of language which, as a rule, is English. Articles without proofs, or which do not contain any significantly new results, should be rejected. High quality survey papers, however, are welcome.

We expect the organizers to deliver manuscripts in a form that is essentially ready for direct reproduction. Any version of TEX is acceptable, but the entire collection of files must be in one particular dialect of TEX and unified according to simple instructions available from Birkhäuser.

Furthermore, in order to guarantee the timely appearance of the proceedings it is essential that the final version of the entire material be submitted no later than one year after the conference.

More information about this series at http://www.springer.com/series/4961

Gerard Buskes • Marcel de Jeu •
Peter Dodds • Anton Schep • Fedor Sukochev •
Jan van Neerven • Anthony Wickstead
Editors

Positivity and Noncommutative Analysis

Festschrift in Honour of Ben de Pagter on the Occasion of his 65th Birthday

Editors
Gerard Buskes
Department of Mathematics
University of Mississippi
Oxford, MS, USA

Marcel de Jeu
Mathematical Institute
Leiden University
Leiden, The Netherlands

Department of Mathematics
and Applied Mathematics
University of Pretoria
Pretoria, South Africa

Peter Dodds
College of Science and Engineering
Flinders University
Adelaide, Australia

Anton Schep
Department of Mathematics
University of South Carolina
Columbia, SC, USA

Fedor Sukochev
School of Mathematics and Statistics
University of New South Wales
Sydney, Australia

Jan van Neerven
Delft Institute of Applied Mathematics
Delft University of Technology
Delft, The Netherlands

Anthony Wickstead
Mathematical Sciences Research Centre
Queen's University Belfast
Belfast, UK

ISSN 2297-0215 ISSN 2297-024X (electronic)
Trends in Mathematics
ISBN 978-3-030-10849-6 ISBN 978-3-030-10850-2 (eBook)
https://doi.org/10.1007/978-3-030-10850-2

Mathematics Subject Classification (2010): 46-XX, 58J42, 46L51, 47-XX, 35-XX

This book is published under the imprint Birkhäuser, www.birkhauser-science.com, by the registered company Springer Nature Switzerland AG.
The registered company address is: Gewerbestrasse 11, 6330 Cham, Switzerland

This Festschrift is dedicated to Ben de Pagter
on the occasion of his 65th birthday

Preface

This volume is dedicated to Ben de Pagter on the occasion of his 65th birthday in November 2018. It originates in the workshop 'Positivity and Noncommutative Analysis' that was held in Delft from 26 to 28 September 2018 to celebrate Ben's fundamental contributions as a mathematician as well as his personal leadership.

The papers in this work cover a spectrum of topics in positivity, noncommutative analysis, semigroups, analysis in Banach spaces, partial differential equations, measure theory, harmonic analysis, operator theory, special functions, and topology. The range of topics not only reflects Ben's wide mathematical interests, but also the extent of the circle of researchers, some of them former colleagues in Delft, who have expressed their appreciation of Ben by contributing to this volume.

The editors gratefully acknowledge the financial support of Delft Institute of Applied Mathematics and Birkhäuser that made the workshop possible. They also thank Sarah Goob, Sabrina Hoecklin, and Kathleen Moriarty for the pleasant relationship with the publisher while preparing this volume.

Finally, the editors want to acknowledge their personal admiration for Ben as a mathematician and their appreciation for him as a person. His mathematical formation took inspiration from the role played by partial order in algebra and analysis, an area which these days lies at the heart of the field known as Positivity. His research contributions in this area have been profound, and his ongoing role as Editor-in-Chief of the journal *Positivity* bears his personal imprint. Positivity is also a hallmark of his personality. It is always a pleasure to meet Ben and to work with him, and not only because of the mathematics.

Gerard Buskes
Marcel de Jeu
Peter Dodds
Anton Schep
Fedor Sukochev
Jan van Neerven
Anthony Wickstead

Ben de Pagter: Curriculum Vitae

Ben de Pagter, born on 10 November 1953 in The Hague, The Netherlands, started his studies in mathematics at Leiden University in 1972. He obtained his bachelor's degree in 1974 and his master's degree (cum laude) in 1977. In September 1975, Ben became a teaching assistant at the Mathematical Institute in Leiden, and this was followed by an appointment as a doctoral assistant in 1977. The latter position enabled him to pursue his doctoral studies under the supervision of Prof. A.C. Zaanen and Dr. C.B. Huijsmans, resulting in the dissertation 'f-Algebras and orthomorphisms', which he defended in Leiden in 1981. He then moved to a postdoctoral position at California Institute of Technology which he held until 1983, when he became an assistant professor at Delft University of Technology. In 1986, he received a five-year Christiaan and Constantijn Huygens Research Fellowship from the Nederlandse Organisatie voor Zuiver-Wetenschappelijk Onderzoek, nowadays called the Netherlands Organisation for Scientific Research. In 1987, he held a Research Fellowship at The Flinders University of South Australia in Adelaide that was provided by the Australian Research Grant Scheme, nowadays called the Australian Research Council. He was appointed full professor to the Chair of Analysis at Delft Institute of Applied Mathematics in 2005. His research areas include the theory of Banach lattices and positive operators and noncommutative integration theory.

Ben has been Editor-in-Chief of the journal *Positivity* since 2008.

Ben is a co-author of several books. The monograph *One-Parameter Semigroups of Operators*, written jointly with Philippe Clément, Henk Heijmans, Sigurd Angenent, and Hans van Duijn, was published in 1987, and the textbook *An Invitation to Functional Analysis*, written jointly with Arnoud van Rooij, appeared in 2013. Currently, he is working, jointly with Peter Dodds and Fedor Sukochev, on a two-volume standard work on noncommutative integration.

Ben has (co-)supervised seven PhD students: Jan van Neerven, Marc Uiterdijk, Henrico Witvliet, Sjoerd Dirksen, Marten Wortel, Jan Rozendaal, and Nikita Moriakov.

Ben has held several administrative positions in Delft. He was Director of Education at Delft Institute of Applied Mathematics from 2001 to 2007 and its Chairman from 2008 to 2016.

Contents

2-Local Automorphisms on AW^*-Algebras

Shavkat Ayupov, Karimbergen Kudaybergenov, and Turabay Kalandarov

With deep respect, we dedicate this paper to the 65th anniversary of Professor Ben de Pagter

Abstract The paper is devoted to 2-local automorphisms on AW^*-algebras. Using the technique of matrix algebras over a unital Banach algebra we prove that any 2-local automorphism on an arbitrary AW^*-algebra without finite type I direct summands is a global automorphism.

Keywords AW^*-Algebra · Matrix algebra · Automorphism · 2-Local automorphism

1 Introduction and the Main Theorem

In 1990, Kadison [9] and Larson and Sourour [11] independently introduced the concept of a local derivation. A linear map $\Delta : \mathcal{A} \to \mathcal{M}$ is called a *local derivation* if for every $x \in \mathcal{A}$ there exists a derivation D_x (depending on x) such that $\Delta(x) = D_x(x)$. It is natural to consider under which conditions local derivations automatically become derivations. Many partial results have been done in this problem. In [9] Kadison shows that every norm-continuous local derivation from a von Neumann algebra M into a dual M-bimodule is a derivation. In [8] Johnson

Sh. Ayupov (✉)
V.I.Romanovskiy Institute of Mathematics, Uzbekistan Academy of Sciences, Tashkent, Uzbekistan

National University of Uzbekistan, Tashkent, Uzbekistan
e-mail: ayupovshavkat@gmail.com; ayupovshavkat52@gmail.com; sh_ayupov@mail.ru

K. Kudaybergenov · T. Kalandarov
Ch. Abdirov 1, Department of Mathematics, Karakalpak State University, Nukus, Uzbekistan
e-mail: karim2006@mail.ru; turaboy_kts@mail.ru

G. Buskes et al. (eds.), *Positivity and Noncommutative Analysis*,
Trends in Mathematics, https://doi.org/10.1007/978-3-030-10850-2_1

extends Kadison's result and proves every local derivation from a C^*-algebra $\mathcal{A}$ into any Banach $\mathcal{A}$-bimodule is a derivation.

In 1997, Šemrl [12] initiated the study of so-called 2-local derivations and 2-local automorphisms on algebras. Namely, he described such maps on the algebra $B(H)$ of all bounded linear operators on an infinite dimensional separable Hilbert space H.

In the above notations, a map $\Delta : \mathcal{A} \to \mathcal{A}$ (not necessarily linear) is called a *2-local automorphism* if, for every $x, y \in \mathcal{A}$, there exists an automorphism $\Phi_{x,y} : \mathcal{A} \to \mathcal{A}$ such that $\Phi_{x,y}(x) = \Delta(x)$ and $\Phi_{x,y}(y) = \Delta(y)$.

Afterwards local derivations and 2-local derivations have been investigated by many authors on different algebras and many results have been obtained in [1–3, 9, 10, 12].

In [7] it was established that every 2-local $*$-homomorphism from a von Neumann algebra into a C^*-algebra is a linear $*$-homomorphism. These authors also proved that every 2-local Jordan $*$-homomorphism from a JBW*-algebra into a JB*-algebra is a Jordan *-homomorphism.

In the present paper we extend the result obtained in [1] for 2-local derivations on AW^*-algebras to the case of 2-local automorphisms on AW^*-algebras.

If $\Delta : \mathcal{A} \to \mathcal{A}$ is a 2-local automorphism, then from the definition it easily follows that Δ is homogenous. At the same time,

$$\Delta(x^2) = \Phi_{x,x^2}(x^2) = \Phi_{x,x^2}(x)\Phi_{x,x^2}(x) = \Delta(x)^2$$

for each $x \in \mathcal{A}$. This means that additive (and hence, linear) 2-local automorphism is a Jordan automorphism.

The following theorem is the main result of this paper.

Theorem 1.1 *Let M be an arbitrary AW^*-algebra without finite type I direct summands. Then any 2-local automorphism Δ on M is an automorphism.*

The proof of this theorem is based on representations of AW^*-algebras as matrix algebras over a unital Banach algebra with the following two properties:

(J): *for any Jordan automorphism Φ on $\mathcal{A}$ there exists a decomposition $\mathcal{A} = \mathcal{A}_1 \oplus \mathcal{A}_2$ such that*

$$x \in \mathcal{A} \mapsto p_1(\Phi(x)) \in \mathcal{A}_1$$

is a homomorphism and

$$x \in \mathcal{A} \mapsto p_2(\Phi(x)) \in \mathcal{A}_2$$

is an anti-homomorphism, where p_i is a projection from $\mathcal{A}$ onto $\mathcal{A}_i$, $i = 1, 2$

(M): *There exist elements $x, y \in \mathcal{A}$ such that $xy = 0$ and $yx \neq 0$.*

Remark 1.2 Note that if an algebra $\mathcal{A}$ contains a subalgebra isomorphic to the matrix algebra $M_2(\mathbb{C})$, then it satisfies the condition **(M)**. Indeed, for matrices $x = \begin{pmatrix} 0 & 1 \\ 0 & 0 \end{pmatrix}$ and $y = \begin{pmatrix} 1 & 0 \\ 0 & 0 \end{pmatrix}$, we have $xy = 0$ and $yx \neq 0$.

2 The Proof of the Main Result

The key tool for the proof of Theorem 1.1 is the following.

Theorem 2.1 *Let $\mathcal{A}$ be a unital Banach algebra with the properties* **(J)** *and* **(M)** *and let $M_{2^n}(\mathcal{A})$ be the algebra of all $2^n \times 2^n$-matrices over $\mathcal{A}$, where $n \geq 2$. Then any 2-local automorphism Δ on $M_{2^n}(\mathcal{A})$ is an automorphism.*

The proof of Theorem 2.1 consists of two steps. In the first step we shall show additivity of Δ on the subalgebra of diagonal matrices from $M_n(\mathcal{A})$.

Let $\{e_{i,j}\}_{i,j=1}^n$ be the system of matrix units in $M_n(\mathcal{A})$. For $x \in M_n(\mathcal{A})$ by $x_{i,j}$ we denote the (i,j)-entry of x, where $1 \leq i, j \leq n$. We shall, if necessary, identify this element with the matrix from $M_n(\mathcal{A})$ whose (i,j)-entry is $x_{i,j}$, other entries are zero, i.e. $x_{i,j} = e_{i,i} x e_{j,j}$.

Each element $x \in M_n(\mathcal{A})$ has the form

$$x = \sum_{i,j=1}^{n} x_{ij} e_{ij}, \ x_{ij} \in \mathcal{A}, i, j \in \overline{1, n}.$$

Let $\psi : \mathcal{A} \to \mathcal{A}$ be an automorphism. Setting

$$\overline{\psi}(x) = \sum_{i,j=1}^{n} \psi(x_{ij}) e_{ij}, \ x_{ij} \in \mathcal{A}, i, j \in \overline{1, n} \tag{2.1}$$

we obtain a well-defined linear operator $\overline{\psi}$ on $M_n(\mathcal{A})$. Moreover $\overline{\psi}$ is an automorphism.

For an invertible element $a \in M_n(\mathcal{A})$ set

$$\Phi_a(x) = axa^{-1}, \ x \in M_n(\mathcal{A}).$$

Then Φ_a is an automorphism and it is called a spatial automorphism.

It is known [4, Corollary 3.14] that every automorphism Φ on $M_n(\mathcal{A})$ can be represented as a product

$$\Phi = \Phi_a \circ \overline{\psi}, \tag{2.2}$$

where Φ_a is a spatial automorphism implemented by an invertible element $a \in M_n(\mathcal{A})$, while $\overline{\psi}$ is the automorphism of the form (2.1) generated by an automorphism ψ on $\mathcal{A}$.

Consider the following two matrices:

$$u = \sum_{i=1}^{n} \frac{1}{2^i} e_{i,i}, \ v = \sum_{i=2}^{n} e_{i-1,i}. \tag{2.3}$$

It is easy to see that an element $x \in M_n(\mathcal{A})$ commutes with u if and only if it is diagonal, and if an element $a \in M_n(\mathcal{A})$ commutes with v, then a is of the form

$$a = \begin{pmatrix} a_1 & a_2 & a_3 & \dots & a_n \\ 0 & a_1 & a_2 & \dots & a_{n-1} \\ 0 & 0 & a_1 & \dots & a_{n-2} \\ \vdots & \vdots & \vdots & \vdots & \vdots \\ 0 & 0 & \dots & a_1 & a_2 \\ 0 & 0 & \dots & 0 & a_1 \end{pmatrix}. \tag{2.4}$$

Further in Lemmata 2.2–2.5 we assume that $n \geq 2$.

Lemma 2.2 *For every 2-local automorphism Δ on $M_n(\mathcal{A})$ there exists an automorphism Φ such that $\Delta|_{sp\{e_{i,j}\}_{i,j=1}^n} = \Phi|_{sp\{e_{i,j}\}_{i,j=1}^n}$, where $sp\{e_{i,j}\}_{i,j=1}^n$ is the linear span of the set $\{e_{i,j}\}_{i,j=1}^n$.*

Proof Take an automorphism $\Phi_{u,v}$ on $M_n(\mathcal{A})$ such that

$$\Delta(u) = \Phi_{u,v}(u), \ \Delta(v) = \Phi_{u,v}(v),$$

where u, v are the elements from (2.3). Replacing Δ by $\Phi_{u,v}^{-1} \circ \Delta$, if necessary, we can assume that $\Delta(u) = u$, $\Delta(v) = v$.

Let $i, j \in \overline{1,n}$. Take an automorphism $\Phi = \Phi_a \circ \overline{\psi}$ of the form (2.2) such that

$$\Delta(e_{i,j}) = a\overline{\psi}(e_{ij})a^{-1}, \ \Delta(u) = a\overline{\psi}(u)a^{-1}.$$

Since $\Delta(u) = u$ and $\overline{\psi}(u) = u$, it follows that $[a, u] = 0$, and therefore a has a diagonal form, i.e. $a = \sum_{s=1}^{n} a_s e_{s,s}$, $a_s \in \mathcal{A}$, $s \in \overline{1,n}$.

In the same way, but starting with the element v instead of u, we obtain

$$\Delta(e_{i,j}) = be_{i,j}b^{-1},$$

where b has the form (2.4), depending on $e_{i,j}$. So

$$\Delta(e_{i,j}) = ae_{i,j}a^{-1} = be_{i,j}b^{-1}.$$

Since

$$ae_{i,j}a^{-1} = a_i a_j^{-1} e_{i,j}$$

and

$$[be_{i,j}b^{-1}]_{i,j} = 1,$$

it follows that $\Delta(e_{i,j}) = e_{i,j}$.

Now let us take a matrix $x = \sum_{i,j=1}^{n} \lambda_{i,j} e_{i,j} \in M_n(\mathbb{C})$. Then

$$\begin{aligned} e_{j,i}\Delta(x)e_{j,i} &= \Delta(e_{j,i})\Delta(x)\Delta(e_{j,i}) = \Phi_{e_{j,i},x}(e_{j,i})\Phi_{e_{i,j},x}(x)\Phi_{e_{j,i},x}(e_{j,i}) = \\ &= \Phi_{e_{j,i},x}(e_{j,i}xe_{j,i}) = \Phi_{e_{j,i},x}(\lambda_{i,j}e_{j,i}) = \\ &= \lambda_{i,j}\Phi_{e_{j,i},x}(e_{j,i}) = \lambda_{i,j}e_{j,i}, \end{aligned}$$

i.e. $e_{i,i}\Delta(x)e_{j,j} = \lambda_{i,j}e_{i,j}$ for all $i, j \in \overline{1, n}$. This means that $\Delta(x) = x$. The proof is complete. □

Further in Lemmata 2.3–2.8 we assume that Δ is a 2-local automorphism on $M_n(\mathcal{A})$ such that $\Delta|_{\mathrm{sp}\{e_{i,j}\}_{i,j=1}^n} = id|_{\mathrm{sp}\{e_{i,j}\}_{i,j=1}^n}$.

Let $\Delta_{i,j}$ be the restriction of Δ onto $\mathcal{A}_{i,j} = e_{i,i}M_n(\mathcal{A})e_{j,j}$, where $1 \le i, j \le n$.

Lemma 2.3 *$\Delta_{i,j}$ maps $\mathcal{A}_{i,j}$ into itself.*

Proof Let us show that

$$\Delta_{i,j}(x) = e_{i,i}\Delta(x)e_{j,j} \tag{2.5}$$

for all $x \in \mathcal{A}_{i,j}$.

Take $x = x_{i,j} \in \mathcal{A}_{i,j}$, and consider an automorphism $\Phi = \Phi_a \circ \overline{\psi}$ of the form (2.2) such that

$$\Delta(x) = a\overline{\psi}(x)a^{-1}, \ \Delta(u) = a\overline{\psi}(u)a^{-1},$$

where u is the element from (2.3). Since $\Delta(u) = u$ and $\overline{\psi}(u) = u$, it follows that $[a, u] = 0$, and therefore a has a diagonal form. Then $\Delta(x) = a_i\psi(x_{ij})a_j^{-1}e_{ij}$. This means that $\Delta(x) \in \mathcal{A}_{i,j}$. The proof is complete. □

Lemma 2.4 *Let $x = \sum_{i=1}^{n} x_{i,i}$ be a diagonal matrix. Then*

$$e_{k,k}\Delta(x)e_{k,k} = \Delta(x_{k,k}) \tag{2.6}$$

for all $k \in \overline{1, n}$.

Proof Take an automorphism Φ of the form (2.2) such that

$$\Delta(x) = a\overline{\psi}(x)a^{-1} \text{ and } \Delta(x_{k,k}) = a\overline{\psi}(x_{kk})a^{-1}.$$

If necessary, replacing $x_{k,k}$ by $\lambda e + x_{k,k}$ ($|\lambda| > ||x_{k,k}||$) we can assume that $x_{k,k}$ is invertible. Using the equality (2.5), we obtain that $\Delta(x_{k,k}) \in \mathcal{A}_{k,k}$. Since $\Delta(x_{k,k})a = a\overline{\psi}(x_{kk})$,

$$0 = (\Delta(x_{k,k})a)_{k,i} = x_{k,k}a_{k,i},$$
$$0 = (a\overline{\psi}(x_{k,k}))_{i,k} = a_{i,k}\overline{\psi}(x_{k,k})$$

for all $i \neq k$. Since $x_{k,k}$ and $\overline{\psi}(x_{k,k})$ are invertible, we have that $a_{i,k} = a_{k,i} = 0$ for all $i \neq k$. Further

$$\Delta(x_{k,k}) = e_{k,k}\Delta(x_{k,k})e_{k,k} = e_{k,k}a\overline{\psi}(x_{k,k})a^{-1}e_{k,k} = a_{k,k}\overline{\psi}(x_{k,k})a_{k,k}^{-1}.$$

Since x is a diagonal matrix and $a_{i,k} = a_{k,i} = 0$ for all $i \neq k$. we get

$$e_{k,k}\Delta(x)e_{k,k} = e_{k,k}a\overline{\psi}(x)a^{-1}e_{k,k} = a_{k,k}\overline{\psi}(x_{k,k})a_{k,k}^{-1}.$$

Thus $e_{k,k}\Delta(x)e_{k,k} = \Delta(x_{k,k})$. The proof is complete. □

Lemma 2.5 *Let $x = x_{i,i} \in \mathcal{A}_{i,i}$. Then*

$$e_{j,i}\Delta(x)e_{i,j} = \Delta(e_{j,i}xe_{i,j}) \tag{2.7}$$

for every $j \in \{1, \cdots, n\}$.

Proof The case when $i = j$ has been already proved (see Lemma 2.4).

Suppose that $i \neq j$. For an arbitrary element $x = x_{i,i} \in \mathcal{A}_{i,i}$, consider $y = x + e_{j,i}xe_{i,j} \in \mathcal{A}_{i,i} + \mathcal{A}_{j,j}$. Take an automorphism Φ of the form (2.2) such that

$$\Delta(y) = a\overline{\psi}(y)a^{-1} \text{ and } \Delta(v) = a\overline{\psi}(v)a^{-1},$$

where v is the element from (2.3). Since $\Delta(v) = v$ and $\overline{\delta}(v) = v$, it follows that a has the form (2.4). By Lemma 2.4 we obtain that

$$e_{j,i}\Delta(x)e_{i,j} = e_{j,i}e_{i,i}\Delta(y)e_{i,i}e_{i,j} = a_1\overline{\psi}(y)a_1^{-1}e_{j,j},$$
$$\Delta(e_{j,i}xe_{i,j}) = e_{j,j}\Delta(y)e_{j,j} = a_1\overline{\psi}(x)a_1^{-1}e_{j,j}.$$

The proof is complete. □

Further in Lemmata 2.6–2.11 we assume that $n \geq 3$.

Lemma 2.6 $\Delta_{i,i}$ *is additive for all $i \in \overline{1,n}$.*

Proof Let $i \in \overline{1, n}$. Since $n \geq 3$, we can take different numbers k, s such that $(k - i)(s - i) \neq 0$.

For arbitrary $x, y \in \mathcal{A}_{i,i}$ consider the diagonal element $z \in \mathcal{A}_{i,i} + \mathcal{A}_{k,k} + \mathcal{A}_{s,s}$ such that $z_{i,i} = x + y$, $z_{k,k} = x$, $z_{s,s} = y$. Take an automorphism Φ of the form (2.2) such that

$$\Delta(z) = a\overline{\psi}(z)a^{-1} \text{ and } \Delta(v) = a\overline{\psi}(v)a^{-1},$$

where v is the element from (2.3). Since $\Delta(v) = v$ and $\overline{\delta}(v) = v$, it follows that a has the form (2.4). Using Lemmata 2.4 and 2.5 we obtain that

$$\begin{aligned}
\Delta_{i,i}(x + y) &\overset{(2.6)}{=} e_{i,i}\Delta(z)e_{i,i} = a_1\overline{\psi}(x + y)a_1^{-1}e_{i,i},\\
\Delta_{i,i}(x) &\overset{(2.7)}{=} e_{i,k}\Delta(e_{k,i}xe_{i,k})e_{k,i} \overset{(2.6)}{=} e_{i,k}e_{k,k}\Delta(z)e_{k,k}e_{k,i} =\\
&= a_1\overline{\psi}(x)a_1^{-1}e_{i,i},\\
\Delta_{i,i}(y) &\overset{(2.7)}{=} e_{i,s}\Delta(e_{s,i}ye_{i,s})e_{s,i} \overset{(2.6)}{=} e_{i,s}e_{s,s}\Delta(z)e_{s,s}e_{s,i} =\\
&= a_1\overline{\psi}(y)a_1^{-1}e_{i,i}.
\end{aligned}$$

Hence

$$\Delta_{i,i}(x + y) = \Delta_{i,i}(x) + \Delta_{i,i}(y).$$

The proof is complete. □

As it was mentioned in the beginning of the section any additive 2-local automorphism is a Jordan automorphism. Since $\mathcal{A}_{i,i} \cong \mathcal{A}$ has the property **(J)**, by Lemma 2.6 there exists a decomposition $\mathcal{A} = \mathcal{A}_1 \oplus \mathcal{A}_2$ such that

$$x \in \mathcal{A} \mapsto p_1(\Delta_{i,i}(x)) \in \mathcal{A}_1$$

is a homomorphism and

$$x \in \mathcal{A} \mapsto p_2(\Delta_{i,i}(x)) \in \mathcal{A}_2$$

is an anti-homomorphism.

Suppose that $p_2 \neq 0$. By the condition **(M)** we can find elements $x, y \in \mathcal{A}$ such that $xy = 0$ and $yx \neq 0$. Then

$$0 = p_2(\Delta_{i,i}(xy)) = p_2(\Delta_{i,i}(y))p_2(\Delta_{i,i}(x)).$$

On the other hand,

$$\Delta_{i,i}(y)\Delta_{i,i}(x) = \Phi_{x,y}(y)\Phi_{x,y}(x) = \Phi_{x,y}(yx) \neq 0.$$

From this contradiction we obtain that $p_2 = 0$. So, we have the following

Lemma 2.7 $\Delta_{i,i}$ *is an automorphism for all* $i \in \overline{1,n}$.

Denote by $\mathcal{D}_n(\mathcal{A})$ the set of all diagonal matrices from $M_n(\mathcal{A})$, i.e. the set of all matrices of the following form

$$x = \begin{pmatrix} x_1 & 0 & 0 & \dots & 0 \\ 0 & x_2 & 0 & \dots & 0 \\ \vdots & \vdots & \vdots & \vdots & \vdots \\ 0 & 0 & \dots & x_{n-1} & 0 \\ 0 & 0 & \dots & 0 & x_n \end{pmatrix}.$$

Let us consider an operator $\overline{\Delta_{1,1}}$ of the form (2.1). By Lemmata 2.4 and 2.5 we obtain that

Lemma 2.8 $\Delta|_{\mathcal{D}_n(\mathcal{A})} = \overline{\Delta_{1,1}}|_{\mathcal{D}_n(\mathcal{A})}$ *and* $\overline{\Delta_{1,1}}|_{sp\{e_{i,j}\}_{i,j=1}^n} = id|_{sp\{e_{i,j}\}_{i,j=1}^n}$.

Now we are in position to pass to the second step of our proof. In this step we show that if a 2-local automorphism Δ satisfies the following conditions

$$\Delta|_{\mathcal{D}_n(\mathcal{A})} \equiv id|_{\mathcal{D}_n(\mathcal{A})} \text{ and } \Delta|_{\mathrm{sp}\{e_{i,j}\}_{i,j=1}^n} \equiv id|_{\mathrm{sp}\{e_{i,j}\}_{i,j=1}^n},$$

then it is the identical map.

In following five Lemmata 2.9–2.13 we shall consider 2-local automorphisms which satisfy the latter equalities.

We denote by e the unit of the algebra $\mathcal{A}$.

Lemma 2.9 *Let* $x \in M_n(\mathcal{A})$. *Then* $\Delta(x)_{k,k} = x_{k,k}$ *for all* $k \in \overline{1,n}$.

Proof Let $x \in M_n(\mathcal{A})$, and fix $k \in \overline{1,n}$. Since Δ is homogeneous, we can assume that $\|x_{k,k}\| < 1$, where $\|\cdot\|$ is the norm on $\mathcal{A}$. Take a diagonal element y in $M_n(\mathcal{A})$ with $y_{k,k} = e + x_{k,k}$ and $y_{i,i} = 0$ otherwise. Since $\|x_{k,k}\| < 1$, it follows that $e + x_{k,k}$ is invertible in $\mathcal{A}$. Take an automorphism Φ of the form (2.2) such that

$$\Delta(x) = a\overline{\psi}(x)a^{-1} \text{ and } \Delta(y) = a\overline{\psi}(y)a^{-1}.$$

Since $y \in \mathcal{D}_n(\mathcal{A})$ we have that $y = \Delta(y) = a\overline{\psi}(y)a^{-1}$, and therefore

$$0 = \Delta(y)_{i,k} = a_{i,k}(e + x_{k,k}),$$
$$0 = \Delta(y)_{k,i} = -(e + x_{k,k})a_{k,i}$$

for all $i \neq k$. Thus

$$a_{i,k} = a_{k,i} = 0$$

for all $i \neq k$. The above equalities imply that

$$\Delta(x)_{k,k} = \Delta(y)_{k,k} = x_{k,k}.$$

The proof is complete. □

Lemma 2.10 *Let x be a matrix with $x_{k,s} = \lambda e$. Then $\Delta(x)_{k,s} = \lambda e$.*

Proof We have

$$e_{s,k}\Delta(x)e_{s,k} = \Delta(e_{s,k})\Delta(x)\Delta(e_{s,k}) = \Phi_{e_{s,k},x}(e_{s,k})\Phi_{e_{s,k},x}(x)\Phi_{e_{s,k},x}(e_{s,k}) =$$
$$= \Phi_{e_{s,k},x}(e_{s,k}xe_{s,k}) = \Phi_{e_{s,k},x}(\lambda e_{s,k}) = \lambda\Delta(e_{s,k}) = \lambda e_{s,k}.$$

Thus

$$e_{k,k}\Delta(x)e_{s,s} = e_{k,s}e_{s,k}\Delta(x)e_{s,k}e_{k,s} = \lambda e_{k,s}.$$

This means that $\Delta(x)_{k,s} = \lambda e$. The proof is complete. □

Lemma 2.11 *Let k, s be numbers such that $k \neq s$ and let x be a matrix with $x_{k,s} = \lambda e$, $\lambda \neq 0$. Then $\Delta(x)_{s,k} = x_{s,k}$.*

Proof Take a diagonal element y such that $y_{k,k} = x_{s,k}$ and $y_{i,i} = \lambda_i e$ otherwise, where λ_i $(i \neq k)$ are distinct numbers with $|\lambda_i| > \|x_{s,k}\|$. Take an automorphism Φ such that

$$\Delta(x) = \Phi(x) \text{ and } \Delta(y) = \Phi(y).$$

Then $ya = a\overline{\psi}(y)$, and therefore

$$0 = (ya - a\overline{\psi}(y))_{ij} = \lambda_j a_{i,j} - \lambda_i a_{i,j} = a_{i,j}(\lambda_j - \lambda_i) \text{ for } (i-j)(i-k)(j-k) \neq 0,$$
$$0 = (ya - a\overline{\psi}(y))_{i,k} = a_{i,k}\overline{\psi}(y_{k,k}) - \lambda_i a_{i,k} = a_{i,k}(\overline{\psi}(x_{s,k}) - \lambda_i) \text{ for } i \neq k,$$
$$0 = (ya - a\overline{\psi}(y))_{k,j} = a_{k,j}\lambda_j - \overline{\psi}(y_{kk})a_{kj} = (\lambda_j - \overline{\psi}(x_{s,k}))a_{k,j} \text{ for } j \neq k.$$

Thus $a_{i,j} = 0$ for all $i \neq j$, i.e. a is a diagonal element. Since

$$\lambda e = \Delta(x)_{ks} = a_{kk}\lambda e a_{ss}^{-1},$$

it follows that $a_{k,k} = a_{s,s}$. Finally,

$$\Delta(x)_{s,k} = a_{s,s}\overline{\psi}(x_{s,k})a_{k,k}^{-1} =$$
$$= a_{k,k}\overline{\psi}(y_{k,k})a_{k,k}^{-1} = \Delta(y)_{k,k} = x_{s,k}.$$

The proof is complete. □

In the next two lemmata we assume that Δ is a 2-local automorphism on $M_2(\mathcal{A})$.

Lemma 2.12 *Let* $x = \begin{pmatrix} x_{1,1} & \lambda e \\ x_{2,1} & x_{2,2} \end{pmatrix}$ *and* $y = \begin{pmatrix} x_{1,1} & x_{1,2} \\ x_{2,1} & x_{2,2} \end{pmatrix}$, *where* $|\lambda| > ||\Delta(y)_{1,2}||$. *Then* $\Delta(x)_{2,1} = \Delta(y)_{2,1}$.

Proof Take an automorphism Φ such that

$$\Delta(x) = \Phi(x) \text{ and } \Delta(y) = \Phi(y).$$

Then

$$\begin{pmatrix} 0 & \lambda e - \Delta(y)_{1,2} \\ (\Delta(x) - \Delta(y))_{2,1} & 0 \end{pmatrix} \begin{pmatrix} a_{1,1} & a_{1,2} \\ a_{2,1} & a_{2,2} \end{pmatrix} = \begin{pmatrix} a_{1,1} & a_{1,2} \\ a_{2,1} & a_{2,2} \end{pmatrix} \begin{pmatrix} 0 & \lambda e - x_{1,2} \\ 0 & 0 \end{pmatrix}.$$

Thus

$$\begin{cases} (\lambda e - \Delta(y)_{1,2})a_{2,1} = 0, \\ (\Delta(x)_{2,1} - \Delta(y)_{2,1})a_{1,1} = 0. \end{cases}$$

Since $|\lambda| > ||\Delta(y)_{1,2}||$, it follows that $\lambda e - \Delta(y)_{1,2}$ is invertible in $\mathcal{A}$, and therefore the first equality implies that $a_{2,1} = 0$. Thus $a_{1,1}$ is invertible and the second equality gives us $\Delta(x)_{2,1} = \Delta(y)_{2,1}$. The proof is complete. □

Lemma 2.13 $\Delta = id$.

Proof Let $x \in M_2(\mathcal{A})$. By Lemma 2.9 we have that $\Delta(x)_{k,k} = x_{k,k}$ for $k = 1, 2$.

Let now $k \neq s$. Take a matrix y with $y_{s,k} = \lambda e$ and $y_{i,j} = x_{i,j}$ otherwise. By Lemma 2.11 we have that $\Delta(y)_{k,s} = x_{k,s}$. Further Lemma 2.12 implies that

$$\Delta(x)_{k,s} = \Delta(y)_{k,s} = x_{k,s}.$$

Thus $\Delta(x)_{k,s} = \Delta(y)_{k,s} = x_{k,s}$ for all $k, s = 1, 2$, and therefore $\Delta(x) = x$. The proof is complete. □

Now we are in position to prove Theorem 2.1.

Proof of Theorem 2.1 Let Δ be a 2-local automorphism on $M_{2^n}(\mathcal{A})$, where $n \geq 2$. By Lemma 2.2 there exists an automorphism Φ_1 on $M_{2^n}(\mathcal{A})$ such that $\Delta|_{\mathrm{sp}\{e_{i,j}\}_{i,j=1}^{2^n}} = \Phi_1|_{\mathrm{sp}\{e_{i,j}\}_{i,j=1}^{2^n}}$. Replacing, if necessary, Δ by $\Phi_1^{-1} \circ \Delta$, we may assume that Δ is identical on $\mathrm{sp}\{e_{i,j}\}_{i,j=1}^{2^n}$. Further, by Lemma 2.8 there exists an automorphism Φ_2 on $M_{2^n}(\mathcal{A})$ such that $\Delta|_{\mathcal{D}_{2^n}} = \Phi_2|_{\mathcal{D}_{2^n}}$. Now replacing Δ by $\Phi_2^{-1} \circ \Delta$, we can assume that Δ acts as the identity on $\mathcal{D}_{2^n}$. So, we can assume that

$$\Delta|_{\mathrm{sp}\{e_{i,j}\}_{i,j=1}^{2^n}} \equiv id|_{\mathrm{sp}\{e_{i,j}\}_{i,j=1}^{2^n}} \text{ and } \Delta|_{\mathcal{D}_{2^n}} \equiv id|_{\mathcal{D}_{2^n}}.$$

Let us show that $\Delta \equiv id$. We proceed by induction on n.

Let $n = 2$. We identify the algebra $M_4(\mathcal{A})$ with the algebra of 2×2-matrices $M_2(\mathcal{B})$, over $\mathcal{B} = M_2(\mathcal{A})$.

Let $\{e_{i,j}\}_{i,j=1}^4$ be a system of matrix units in $M_4(\mathcal{A})$. Then

$$p_{1,1} = e_{1,1} + e_{2,2},\ p_{2,2} = e_{3,3} + e_{4,4},\ p_{1,2} = e_{1,3} + e_{2,4},\ p_{2,1} = e_{3,1} + e_{4,2}$$

is the system of matrix units in $M_2(\mathcal{B})$. Since $\Delta|_{\mathrm{sp}\{e_{i,j}\}_{i,j=1}^4} \equiv id|_{\mathrm{sp}\{e_{i,j}\}_{i,j=1}^4}$, it follows that $\Delta|_{\mathrm{sp}\{p_{i,j}\}_{i,j=1}^2} \equiv id|_{\mathrm{sp}\{p_{i,j}\}_{i,j=1}^2}$.

Take an arbitrary element $x \in p_{1,1}M_2(\mathcal{B})p_{1,1} \equiv \mathcal{B}$. Choose an automorphism Φ on $M_2(\mathcal{B})$ such that

$$\Delta(x) = \Phi(x),\ \Delta(p_{1,1}) = \Phi(p_{1,1}).$$

Since $\Delta(p_{1,1}) = p_{1,1}$, we obtain that

$$p_{1,1}\Delta(x)p_{1,1} = p_{1,1}\Phi(x)p_{1,1} = \Delta(x).$$

This means that the restriction $\Delta_{1,1}$ of Δ onto $p_{1,1}M_2(\mathcal{B})p_{1,1} \equiv \mathcal{B}$ maps $\mathcal{B} = M_2(\mathcal{A})$ into itself.

If $\mathcal{D}_4$ is the subalgebra of diagonal matrices from $M_4(\mathcal{A})$, then $p_{1,1}\mathcal{D}_4 p_{1,1}$ is the subalgebra of diagonal matrices in the algebra $M_2(\mathcal{A})$. Since $\Delta|_{\mathcal{D}_4} \equiv id|_{\mathcal{D}_4}$, it follows that $\Delta_{1,1}$ acts identically on diagonal matrices from $M_2(\mathcal{A})$. So,

$$\Delta_{1,1}|_{\mathrm{sp}\{e_{i,j}\}_{i,j=1}^2} \equiv id|_{\mathrm{sp}\{e_{i,j}\}_{i,j=1}^2} \text{ and } \Delta_{1,1}|_{p_{1,1}\mathcal{D}_4 p_{1,1}} \equiv id|_{p_{1,1}\mathcal{D}_4 p_{1,1}}.$$

By Lemma 2.13 it follows that $\Delta_{1,1} \equiv id$.

Let $\mathcal{D}_2$ be the set of diagonal matrices from $M_2(\mathcal{B})$. Since

$$\mathcal{D}_2 = \begin{pmatrix} \mathcal{B} & 0 \\ 0 & \mathcal{B} \end{pmatrix}$$

and $\Delta_{1,1} = id$, Lemma 2.4 implies that $\Delta|_{\mathcal{D}_2} \equiv id|_{\mathcal{D}_2}$. Hence, Δ is a 2-local derivation on $M_2(\mathcal{B})$ such that

$$\Delta|_{\mathrm{sp}\{p_{i,j}\}_{i,j=1}^2} \equiv id|_{\mathrm{sp}\{p_{i,j}\}_{i,j=1}^2} \text{ and } \Delta|_{\mathcal{D}_2} \equiv id|_{\mathcal{D}_2}.$$

Again by Lemma 2.13 it follows that $\Delta \equiv id$.

Now assume that the assertion of the Theorem is true for $n - 1$.

Considering the algebra $M_{2^n}(\mathcal{A})$ as the algebra of 2×2-matrices $M_2(\mathcal{B})$ over $\mathcal{B} = M_{2^{n-1}}(\mathcal{A})$ and repeating the above arguments we obtain that $\Delta \equiv id$. The proof is complete. □

Now we apply Theorem 2.1 to the proof of our main result which describes 2-local automorphism on AW^*-algebras.

First note that by Stormer [13, Theorem 3.3] (see also [6, Theorem 3.2.3]) any C^*-algebra, in particular, AW^*-algebra, has the property **(J)**.

Proof of Theorem 1.1 Let M be an arbitrary AW^*-algebra without finite type I direct summands. Then there exist mutually orthogonal central projections z_1, z_2, z_3 in M such that $M = z_1M \oplus z_2M \oplus z_3M$, where z_1M, z_2M, z_3M are algebras of types I_∞, II and III, respectively. Then the halving Lemma [5, P. 120, Theorem 1] applied to each summand implies that the unit z_i of the algebra z_iM, $(i = 1, 2, 3)$ can be represented as a sum of mutually equivalent orthogonal projections $e_1^{(i)}, e_2^{(i)}, e_3^{(i)}, e_4^{(i)}$ from z_iM. Set $e_k = \sum\limits_{i=1}^{3} e_k^{(i)}$, $k = 1, 2, 3, 4$. Then the map $x \mapsto \sum\limits_{i,j=1}^{4} e_ixe_j$ defines an isomorphism between the algebra M and the matrix algebra $M_4(\mathcal{A})$, where $\mathcal{A} = e_{1,1}Me_{1,1}$. Moreover, the algebra $\mathcal{A}$ has the properties **(J)** and **(M)** (see Remark 1.2 after the definition of property **(M)**). Therefore Theorem 2.1 implies that any 2-local automorphism on M is an automorphism. The proof is complete. □

Acknowledgements The authors are indebted to the reviewer for useful remarks.

References

1. Sh.A. Ayupov, K.K. Kudaybergenov, 2-Local derivations on matrix algebras over semi-prime Banach algebras and on AW^*-algebras. J. Phys. Conf. Ser. **697**, 1–10 (2016)
2. Sh.A. Ayupov, K.K. Kudaybergenov, Derivations, local and 2-local derivations on algebras of measurable operators, in *Topics in Functional Analysis and Algebra, Contemporary Mathematics*, vol. 672 (American Mathematical Society, Providence, 2016), pp. 51–72
3. Sh.A. Ayupov, K.K. Kudaybergenov, A.M. Peralta, A survey on local and 2-local derivations on C*- and von Neumann algebras, in *Topics in Functional Analysis and Algebra, Contemporary Mathematics*, vol. 672 (American Mathematical Society, Providence, 2016), pp. 73–126
4. G.M. Benkart, J.M. Osborn, Derivations and automorphisms of nonassociative matrix algebras. Trans. Am. Math. Soc. **263**(2), 411–430 (1981)
5. S. Berberian, *Bear *-rings*, 2nd edn. 2011 (Springer, New York, 1972)
6. O. Brattelli, D. Robinson, *Operator Algebras and Quantum Statistical Mechanics,* 2nd edn. (Springer, Berlin, 2002)
7. M.J. Burgos, F.J. Fernandez Polo, J.J. Garces, A.M. Peralta, A Kowalski-Slodkowski theorem for 2-local $*$-homomorphisms on von Neumann algebras. Rev. Ser. A Mat. **109**(2), 551–568 (2015)
8. B.E. Johnson, Local derivations on C^*-algebras are derivations. Trans. Am. Math. Soc. **353**(1), 313–325 (2001)
9. R.V. Kadison, Local derivations. J. Algebra **130**, 494–509 (1990)
10. S.O. Kim, J.S. Kim, Local automorphisms and derivations on $\mathbb{M}_n$. Proc. Am. Math. Soc. **132**(5), 1389–1392 (2004)
11. D.R. Larson, A.R. Sourour, Local derivations and local automorphisms of $B(X)$, in *Operator Theory: Operator Algebras and Applications, Part 2 (Durham, NH, 1988)*, Proc. Sympos. Pure Math. 51, Part 2 (American Mathematical Society, Providence, 1990), pp. 187–194

12. P. Šemrl, Local automorphisms and derivations on $B(H)$. Proc. Am. Math. Soc. **125**, 2677–2680 (1997)
13. E. Stormer, On the Jordan structure of C^*-algebras. Trans. Am. Math. Soc. **120**, 438–447 (1965)

Orthosymmetric Archimedean-Valued Vector Lattices

Mohamed Amine Ben Amor, Karim Boulabiar, and Jamel Jaber

We would like to dedicate this paper to our friend Professor Ben de Pagter, whose works strongly influenced our research. He is, undoubtedly, the most cited author in our articles

Abstract We introduce and study the notion of orthosymmetric Archimedean-valued vector lattices as a generalization of finite-dimensional Euclidean inner spaces. A special attention has been paid to linear operators on these spaces.

Keywords $\mathbb{V}$-valued orthosymmetric product · Adjoint operators · Orthomorphisms

1 Introduction and First Properties

We take it for granted that the reader is familiar with the elementary theory of vector lattices (i.e., Riesz spaces) and positive operators. For terminology, notation, and properties not explained or proved, we refer to the standard texts [15, 19].

Let $\mathbb{V}$ be an Archimedean vector lattice. Following Buskes and van Rooij in [7], we call a $\mathbb{V}$*-valued orthosymmetric product* on a vector lattice L any function that takes each ordered pair (f, g) of elements of L to a vector $\langle f, g\rangle$ of $\mathbb{V}$ and has the following two properties.

(1) (*Positivity*) $\langle f, g\rangle \in \mathbb{V}^{+}$ for all $f, g \in L^{+}$.
(2) (*Orthosymmetry*) $\langle f, g\rangle = 0$ in $\mathbb{V}$ for all $f, g \in L$ with $f \wedge g = 0$.

M. A. Ben Amor · K. Boulabiar (✉) · J. Jaber
GOSAEF, Laboratoire de Recherche LATAO, Département de Mathématiques, Faculté des Sciences de Tunis, Université de Tunis El Manar, El Manar, Tunisia
e-mail: mohamedamine.benamor@ipest.rnu.tn; Karim.boulabiar@ipest.rnu.tn; jamel.jaber@free.fr

G. Buskes et al. (eds.), *Positivity and Noncommutative Analysis*, Trends in Mathematics, https://doi.org/10.1007/978-3-030-10850-2_2

By an *orthosymmetric* $\mathbb{V}$*-valued vector lattice* (or, just an *orthosymmetric space* if no confusion can arise) we mean a vector lattice L along with a $\mathbb{V}$-valued orthosymmetric product on L. As the next example shows, the classical Euclidean spaces fit within the framework of orthosymmetric spaces.

Example 1.1 As usual, $\mathbb{R}$ denotes the Archimedean vector lattice of all real numbers. Pick $n \in \mathbb{N} = \{1, 2, \ldots\}$ and suppose that the vector space $\mathbb{R}^n$ is equipped with its usual structure of Euclidean space. In particular,

$$\langle f, g \rangle = \sum_{k=1}^{n} \langle f, e_k \rangle \langle g, e_k \rangle \text{ for all } f, g \in \mathbb{R}^n,$$

where $(e_1, \ldots, e_n)$ is the canonical (orthogonal) basis of $\mathbb{R}^n$. Simultaneously, $\mathbb{R}^n$ is a vector lattice with respect to the coordinatewise ordering. That is,

$$f \geq 0 \text{ in } \mathbb{R}^n \text{ if and only if } \langle f, e_k \rangle \in \mathbb{R}^+ \text{ for all } k \in \{1, \ldots, n\}.$$

Consequently, if $f, g \geq 0$ in $\mathbb{R}^n$, then $\langle f, g \rangle \in \mathbb{R}^+$. Furthermore, let $f, g \in \mathbb{R}^n$ such that $f \wedge g = 0$. Whence,

$$\min \{\langle f, e_k \rangle, \langle g, e_k \rangle\} = 0 \text{ for all } k \in \{1, 2, \ldots, n\}.$$

Therefore, $\langle f, g \rangle = 0$ meaning that the inner product on $\mathbb{R}^n$ is an $\mathbb{R}$-valued orthosymmetric product. Thus, the Euclidean space $\mathbb{R}^n$ is an orthosymmetric $\mathbb{R}$-valued vector lattice.

Beginning with the next lines, we shall impose the blanket assumption that all orthosymmetric spaces under consideration are orthosymmetric $\mathbb{V}$-valued vector lattices, where $\mathbb{V}$ is a fixed Archimedean vector lattice (unless otherwise stated explicitly).

The following property is useful for later purposes.

Lemma 1.2 *Let L be an orthosymmetric space. Then,*

$$\langle f, f \rangle = \langle |f|, |f| \rangle \in \mathbb{V}^+ \textit{ for all } f \in L.$$

Proof If $f \in L$, then $f^+ \wedge f^- = 0$. It follows that

$$\left\langle f^+, f^- \right\rangle = \left\langle f^-, f^+ \right\rangle = 0.$$

Hence,

$$\begin{aligned}\langle f, f \rangle &= \left\langle f^+ - f^-, f^+ - f^- \right\rangle = \left\langle f^+, f^+ \right\rangle + \left\langle f^-, f^- \right\rangle \\ &= \left\langle f^+ + f^-, f^+ + f^- \right\rangle = \langle |f|, |f| \rangle \in \mathbb{V}^+.\end{aligned}$$

This is the desired result. □

At first sight, it might seem that it is easy to establish the following remarkable property of orthosymmetric spaces. However, all proofs that can be found in the literature are quite involved and far from being trivial (see, e.g., Corollary 2 in [7] and Theorem 3.8.14 in [18]). By the way, it should be pointed out that this property is based on the fact that $\mathbb{V}$ is Archimedean.

Lemma 1.3 *Let L be an orthosymmetric space. Then,*

$$\langle f, g\rangle = \langle g, f\rangle \text{ for all } f, g \in L.$$

Roughly speaking, any $\mathbb{V}$-valued orthosymmetric product on a vector lattice is symmetric (a multidimensional version of Lemma 1.3 can be found in [5]). We emphasize that results in Lemmas 1.2 and 1.3 could be used below without further mention.

Before proceeding our investigation, we note that our next terminology comes in part from the Inner Product Theory linguistic repertoire (see [4]). Let L be an orthosymmetric space. An element f in L is said to be *neutral* if $\langle f, f\rangle = 0$. Obviously, the zero vector is neutral. The set of all neutral elements in L is called the *neutral part* of L and is denoted by L^0. Namely,

$$L^0 = \{f \in L : \langle f, f\rangle = 0\}.$$

The neutral part of L has a nice characterization.

Lemma 1.4 *Let L be an orthosymmetric space. Then,*

$$L^0 = \{f \in L : \langle f, g\rangle = 0 \text{ for all } g \in L\}.$$

Proof Obviously, if $f \in L$ with $\langle f, g\rangle = 0$ for all $g \in L$, then $\langle f, f\rangle = 0$ so $f \in L^0$. Conversely, let $f \in L^0$ and pick $g \in L$. Choose $n \in \mathbb{N}$ and observe that

$$0 \leq \langle g - nf, g - nf\rangle = \langle g, g\rangle - 2n\langle f, g\rangle + n^2\langle f, f\rangle = \langle g, g\rangle - 2n\langle f, g\rangle.$$

Therefore,

$$2n\langle f, g\rangle \leq \langle g, g\rangle \quad \text{for all } n \in \mathbb{N}.$$

Replacing f by $-f$ in the above inequality, we obtain

$$-2n\langle f, g\rangle \leq \langle g, g\rangle \quad \text{for all } n \in \mathbb{N}.$$

But then $\langle f, g\rangle = 0$ because $\mathbb{V}$ is Archimedean. The proof is complete. □

An interesting lattice-ordered property of the neutral part of an orthosymmetric space is obtained as a consequence of previous lemma.

Theorem 1.5 *The neutral part L^0 of an orthosymmetric space L is an ideal in L.*

Proof Let r be a real number and $f, g \in L^0$. Then,

$$\langle f + rg, f + rg\rangle = \langle f, f\rangle + 2r\langle f, g\rangle + r^2\langle g, g\rangle = 0$$

(where we use Lemma 1.4). It follows that $f + rg \in L^0$ and so L^0 is a vector subspace of L.

Secondly, let $f \in L^0$ and observe that

$$\langle |f|, |f|\rangle = \langle f, f\rangle = 0.$$

Hence, $|f| \in L^0$ and thus L^0 is a vector sublattice of L.

Finally, let $f, g \in L$ such that $0 \leq f \leq g$ and $g \in L^0$. Whence,

$$0 \leq \langle f, f\rangle \leq \langle f, g\rangle \leq \langle g, g\rangle = 0.$$

This yields that L^0 is a solid in L and finishes the proof. □

An orthosymmetric space need not be Archimedean as it is shown in the next example.

Example 1.6 Assume that the Euclidean plan $\mathbb{R}^2$ is endowed with its lexicographic ordering. Hence, $\mathbb{R}^2$ is a non-Archimedean vector lattice. Put

$$\langle f, g\rangle = x_1 y_1 \text{ for all } f = (x_1, x_2), g = (y_1, y_2) \text{ in } \mathbb{R}^2.$$

Since $\mathbb{R}^2$ is totally ordered (i.e., linearly ordered), this formula defines an $\mathbb{R}$-valued orthosymmetric product on $\mathbb{R}^2$. This means that $\mathbb{R}^2$ is a non-Archimedean orthosymmetric $\mathbb{R}$-valued vector lattice.

The orthosymmetric space L is said to be *definite* if its neutral part is trivial. That is to say, L is definite if and only if

$$\langle f, f\rangle = 0 \text{ in } \mathbb{V} \text{ implies } f = 0 \text{ in } L.$$

Definite orthosymmetric spaces have a better behavior as explained in the following.

Proposition 1.7 *Any definite orthosymmetric space is Archimedean.*

Proof Let L be a definite orthosymmetric space and choose $f, g \in L^+$ with

$$0 \leq nf \leq g \text{ for all } n \in \mathbb{N}.$$

Pick $n \in \mathbb{N}$ and observe that $g - nf \in L^+$. So,

$$0 \leq \langle g - nf, f\rangle = \langle g, f\rangle - n\langle f, f\rangle.$$

Hence,

$$0 \leq n \langle f, f \rangle \leq \langle g, f \rangle \text{ for all } n \in \mathbb{N}.$$

Since $\mathbb{V}$ is Archimedean, we get $\langle f, f \rangle = 0$. But then $f = 0$ because L is definite. □

By Theorem 1.5, the neutral part L^0 is an ideal in L. Hence, we may consider the quotient vector lattice L/L^0 (see Chapter 9 in [15]). The equivalence class (i.e., the residue class)

$$f + L^0 = \left\{ f + g : g \in L^0 \right\}$$

of a vector $f \in L$ is denoted by $[f]$. It turns out that L/L^0 is automatically equipped with a structure of an orthosymmetric space.

Theorem 1.8 *Let L be an orthosymmetric space. Then L/L^0 is a definite orthosymmetric space with respect to the $\mathbb{V}$-valued orthosymmetric product given by*

$$\langle [f], [g] \rangle = \langle f, g \rangle \textit{ for all } f, g \in L.$$

Proof First of all, let's prove that the function that takes each ordered pair $([f], [g])$ of elements of L/L^0 to a vector $\langle [f], [g] \rangle$ of $\mathbb{V}$ and given

$$\langle [f], [g] \rangle = \langle f, g \rangle \text{ for all } f, g \in L \qquad (*)$$

is well-defined. Indeed, choose $f, g \in L$ and $h, k \in L^0$. By Lemma 1.4, we have

$$\langle f, k \rangle = \langle h, g \rangle = \langle h, k \rangle = 0.$$

So,

$$\langle f + h, g + k \rangle = \langle f, g \rangle + \langle f, k \rangle + \langle h, g \rangle + \langle h, k \rangle = \langle f, g \rangle .$$

We derive that the function given by $(*)$ is well-defined (its bilinearity is obvious). Now, let $f, g \in L$ such that $[f], [g]$ are positive in L/L^0. Hence, there exist $h, k \in L^0$ such that $h \leq f$ and $k \leq g$. Whence, $0 \leq f - h$ and $0 \leq g - k$ from which it follows that

$$0 \leq \langle f - h, g - k \rangle = \langle f, g \rangle - \langle f, k \rangle - \langle h, g \rangle + \langle h, k \rangle = \langle f, g \rangle = \langle [f], [g] \rangle .$$

Moreover, if $[f] \wedge [g] = [0]$ in L/L^0, then

$$[f \wedge g] = [f] \wedge [g] = [0] .$$

This means that $f \wedge g \in L^0$. This together with Lemma 1.4 yields quickly that

$$\langle [f], [g] \rangle = \langle f, g \rangle = \langle f - f \wedge g, g - f \wedge g \rangle .$$

But then $\langle [f], [g] \rangle = 0$ because

$$(f - f \wedge g) \wedge (g - f \wedge g) = 0.$$

Accordingly, L/L^0 is an orthosymmetric space. It remains to show that L/L^0 is definite. To see this, let $f \in L$ such that $\langle [f], [f] \rangle = 0$. Hence, $\langle f, f \rangle = 0$ from which it follows that $f \in L^0$, so $[f] = [0]$. This completes the proof. □

Taking into account Proposition 1.7 and Theorem 1.8, we infer directly that the quotient vector lattice L/L^0 is Archimedean and so L^0 is a uniformly closed ideal in the vector lattice L (see Theorem 60.2 in [15]).

2 Multiplication Operators in Orthosymmetric Product Spaces

Let M be an ordered vector space. A vector subspace V of M is called a *lattice-subspace* of M if V is a vector lattice with respect to the ordering inherited from M (see Definition 5.58 in [1]). On the other hand, in general, we cannot talk about V being a vector sublattice of M as the latter space need not be a vector lattice. In order to cope with this terminology problem, we call after Abramovich and Wickstead in [2] the lattice-subspace V of M a *generalized vector sublattice* of M if the supremum in M of each $v, w \in V$ exists and coincides with their supremum in V. Hence, if M turns out to be a vector lattice, then the word 'generalized' becomes superfluous. Moreover, it is trivial that every generalized vector sublattice of M is a lattice-subspace. Nevertheless, the converse is not true as we can see in the example provided in [1, Page 229].

We start this section with the following general lemma which is presumably well-known, though we have not been able to locate a precise reference for it. As usual, the kernel and the range of any linear operator T are denoted by $\ker T$ and $\operatorname{Im} T$, respectively.

Lemma 2.1 *Let L be a vector lattice and M be an ordered vector space. Suppose that $T : L \to M$ is a positive operator such that $f^- \in \ker T$ for all $f \in L$ with $Tf \in M^+$. Then* $\operatorname{Im} T$ *is a lattice-subspace of M and T is a lattice homomorphism from L onto* $\operatorname{Im} T$.

Proof Observe first that $\ker T$ is a vector sublattice of L. Indeed, choose f in $\ker T$. So, $0 = Tf \geq 0$ and thus f^- is in $\ker T$. It follows that f^+ is also in $\ker T$ yielding that $|f| \in \ker T$. This means that $\ker T$ is a vector sublattice of L, as claimed. Now, let $f, g \in L$. Since T is positive,

$$T(f \vee g) \geq Tf \text{ and } T(f \vee g) \geq Tg.$$

This means that $T(f \vee g)$ is an upper bound of $\{Tf, Tg\}$ in $\operatorname{Im} T$. Let $v \in \operatorname{Im} T$ another upper bound of $\{Tf, Tg\}$ in $\operatorname{Im} T$. There exists $h \in L$ such that $v = Th$. Since

$$Th = v \geq Tf \text{ and } Th = v \geq Tg,$$

we get

$$T(h - f) \geq 0 \text{ and } T(h - g) \geq 0.$$

Therefore,

$$(h - f)^{-} \in \ker T \text{ and } (h - g)^{-} \in \ker T.$$

But then

$$(h - (f \vee g))^{-} = (h - f)^{-} \vee (h - g)^{-} \in \ker T$$

because $\ker T$ is a vector sublattice of M. It follows that

$$T(h - ((f \vee g))) = T\left(f - ((f \vee g))^{+}\right) \geq 0$$

and so

$$v = Th \geq T(f \vee g).$$

We derive that $T(f \vee g)$ is a supremum of $\{Tf, Tg\}$ in $\operatorname{Im} T$. That is,

$$T(f \vee g) = Tf \vee Tg \text{ in } \operatorname{Im} T$$

and the proof is complete. □

Lemma 2.1 does not hold without the condition (ii). An example in this direction is given next.

Example 2.2 Let $L = M = C[0, \pi]$, where $C[0, \pi]$ is the Archimedean vector lattice of all real-valued continuous functions on the real interval $[0, \pi]$. Define $T : L \to M$ by putting

$$(Tf)(x) = \int_0^x f(y)\,\mathrm{d}y \text{ for all } f \in L \text{ and } x \in [0, \pi].$$

Obviously, T is a positive operator. Moreover, T is one-to-one and so $\ker T = \{0\}$ is an ideal in L. However, if $f \in L$ is defined by

$$f(x) = 2\sin x \cos x \text{ for all } x \in [0, \pi],$$

then

$$(Tf)(x) = (\sin x)^2 \text{ for all } x \in [0, \pi].$$

Furthermore,

$$f^-(x) = \begin{cases} 0 \text{ if } x \in [0, \pi/2] \\ -2\sin x \cos x \text{ if } x \in [\pi/2, \pi]. \end{cases}$$

Thus,

$$\left(T\left(f^-\right)\right)(\pi/2) = -\int_{\pi/2}^{\pi} 2\sin x \cos x \mathrm{d}x = 1 \neq 0.$$

Hence, $f^- \notin \ker T$. Observe now that

$$\operatorname{Im} T = \{f \in L : f(0) = 0 \text{ and } f \text{ continuously differentiable}\}$$

is not a lattice-subspace of M.

Hence, Example 2.2 proves that the condition (ii) is not redundant in Lemma 2.1. In spite of that, the following observation deserves particular attention.

Remark 2.3 Let L be a vector lattice, M be an ordered vector space, and $T : L \to M$ be a positive operator such that $\ker T$ is a vector sublattice of L. We derive quickly that $\ker T$ is an ideal in L. Hence, we may speak about the vector lattice $L/\ker T$ (see [15, Chapter 9]). In this situation, it is not hard to see that an ordering can be defined on $\operatorname{Im} T$ by putting

$$Tf \preccurlyeq Tg \text{ in } \operatorname{Im} T \text{ if and only if } g - f \geq h \text{ for some } h \in \ker T.$$

This ordering makes $\operatorname{Im} T$ into a vector lattice which is lattice isomorphic with $L/\ker T$. The lattice operations in $\operatorname{Im} T$ are given pointwise as follows:

$$Tf \curlyvee Tg = T(f \vee g) \text{ and } Tf \curlywedge Tg = T(f \wedge g) \text{ for all } f, g \in L.$$

However, $\operatorname{Im} T$ need not be, in general, a lattice-subspace of M. As a matter of fact, *Example 2.2* illustrates this situation. Indeed, we have observed already that

$$\operatorname{Im} T = \{f \in C[0, \pi] : f(0) = 0 \text{ and } f \text{ continuously differentiable}\}$$

is not a lattice-subspace of $C[0,\pi]$. Nevertheless, $\operatorname{Im} T$ is a vector lattice with respect to the 'new' ordering defined by

$$f \preccurlyeq g \text{ if and only if } f' \leq g',$$

where f' and g' denote the derivative of f and g, respectively. Moreover, the lattice operations in this vector lattice are given by

$$(f \curlyvee g)(x) = \int_0^x (f' \vee g')(y)\,\mathrm{d}y \text{ and } (f \curlywedge g)(x) = \int_0^x (f' \wedge g')(y)\,\mathrm{d}y$$

for all $f, g \in \operatorname{Im} T$ and $x \in [0,\pi]$.

Henceforth, L stands for an orthosymmetric space (i.e. an orthosymmetric $\mathbb{V}$-valued vector lattice) and $\mathcal{L}^+(L,\mathbb{V})$ denotes the set of all positive operators from L into $\mathbb{V}$. It could be helpful to recall that an operator $T : L \to \mathbb{V}$ is said to be *regular* if $T = R - S$ for some $R, S \in \mathcal{L}^+(L,\mathbb{V})$. The set $\mathcal{L}_r(L,\mathbb{V})$ of all regular operators from L into $\mathbb{V}$ is an Archimedean ordered vector space with respect to the pointwise addition, scalar multiplication, and ordering. By the way, $\mathcal{L}^+(L,,\mathbb{V})$ is the positive cone of $\mathcal{L}_r(L,\mathbb{V})$.

At this point, let $f \in L$ and define a map $\Phi_f : L \to \mathbb{V}$ by putting

$$\Phi_f g = \langle f, g\rangle \text{ for all } g \in L.$$

Obviously, Φ_f is a linear operator, which is referred to as a *multiplication operator* on the orthosymmetric space L. Further elementary (but very useful) properties of such operators are gathered next.

Lemma 2.4 *Let L be an orthosymmetric space and $f \in L$. Then the following hold:*

(i) $\Phi_f \in \mathcal{L}^+(L,\mathbb{V})$ *whenever* $f \in L^+$.
(ii) $\Phi_f = \Phi_{f^+} - \Phi_{f^-} \in \mathcal{L}_r(L,\mathbb{V})$.
(iii) $\Phi_f = 0$ *if and only if* $f \in L^0$.
(iv) $\Phi_f \in \mathcal{L}^+(L,\mathbb{V})$ *if and only if* $f^- \in L^0$.

Proof

(i) This follows immediately from the positivity of the $\mathbb{V}$-valued orthosymmetric product on L.
(ii) If $f, g \in L$, then

$$\begin{aligned}\Phi_f g &= \langle f, g\rangle = \langle f^+ - f^-, g\rangle = \langle f^+, g\rangle - \langle f^-, g\rangle \\ &= \Phi_{f^+} g - \Phi_{f^-} g = \left(\Phi_{f^+} - \Phi_{f^-}\right) g.\end{aligned}$$

Thus, $\Phi_f = \Phi_{f^+} - \Phi_{f^-}$. Moreover, $\Phi_{f^+}, \Phi_{f^-} \in \mathcal{L}^+(L,\mathbb{V})$ as $f^+, f^- \in L^+$ (where we use (i)). It follows that Φ_f is regular.

(iii) This is a direct consequence of Lemma 1.4.
(iv) If $f^- \in L^0$ then, by (iii), $\Phi_{f^-} = 0$. Using (ii) then (i), we get $\Phi_f = \Phi_{f^+} \in \mathcal{L}^+(L, \mathbb{V})$. Conversely, suppose that $\Phi_f \in \mathcal{L}^+(L, \mathbb{V})$. Then,

$$\begin{aligned} 0 \leq \Phi_f(f^-) &= \Phi_{f^+}(f^-) - \Phi_{f^-}(f^-) \\ &= \langle f^+, f^- \rangle - \langle f^-, f^- \rangle = -\langle f^-, f^- \rangle \leq 0. \end{aligned}$$

We derive that $\langle f^-, f^- \rangle = 0$ so $f^- \in L^0$, as required. □

As we shall see in what follows, it turns out that the set

$$\mathcal{M}(L, \mathbb{V}) = \{\Phi_f : f \in L\}$$

of all multiplication operators on the orthosymmetric space L enjoys a very interesting lattice-ordered structure.

Theorem 2.5 *Let L be an orthosymmetric space. Then $\mathcal{M}(L, \mathbb{V})$ is a generalized vector sublattice of $\mathcal{L}_r(L, \mathbb{V})$ and the map $\Phi : L \to \mathcal{M}(L, \mathbb{V})$ defined by*

$$\Phi f = \Phi_f \text{ for all } f \in L$$

is a surjective lattice homomorphism with $\ker \Phi = L^0$. *In particular, the vector lattice L/L^0 and $\mathcal{M}(L, \mathbb{V})$ are lattice isomorphic.*

Proof We have seen in Lemma 2.4 (ii) that $\mathcal{M}(L, \mathbb{V})$ is contained in $\mathcal{L}_r(L, \mathbb{V})$. Moreover, it is a simple exercise to check that $\mathcal{M}(L, \mathbb{V})$ is a vector subspace of $\mathcal{L}_r(L, \mathbb{V})$. Also, we may check directly that Φ is a linear operator and, by Lemma 2.4 (i), Φ is positive. Furthermore, Lemma 2.4 (iii) yields that $\ker \Phi = L^0$. In particular, $\ker \Phi$ is a vector sublattice of L (where we use Theorem 1.5). Moreover, Lemma 2.4 (iv) shows that $f^- \in \ker \Phi$ whenever $f \in L$ and $0 \leq \Phi f \in \mathcal{M}(L, \mathbb{V})$. Consequently, since $\operatorname{Im} \Phi = \mathcal{M}(L, \mathbb{V})$, all conditions of Lemma 2.1 are fulfilled. So, $\mathcal{M}(L, \mathbb{V})$ is a lattice-subspace of $\mathcal{L}_r(L, \mathbb{V})$ and Φ is a lattice homomorphism from L onto $\mathcal{M}(L, \mathbb{V})$. In particular, if $f \in L$, then the absolute value of Φ_f in $\mathcal{M}(L, \mathbb{V})$ equals $\Phi_{|f|}$. At this point, we claim that $\mathcal{M}(L, \mathbb{V})$ is a generalized vector sublattice of $\mathcal{L}_r(L, \mathbb{V})$. To this end, we shall prove that for each $f \in L$ and $g \in L^+$ the set

$$A(f, g) = \{|\Phi_f h| : h \in L \text{ and } |h| \leq g\}$$

has $\Phi_{|f|} g$ as a supremum in $\mathbb{V}$. The Riesz-Kantorovich formula (see, e.g., Theorem 1.14 in [3]) will thus give the desired result.

So, let $f \in L$ and $g \in L^+$. If $h \in L$ with $|h| \leq g$, then

$$|\Phi_f h| = |\langle f, h \rangle| \leq \langle |f|, |h| \rangle \leq \langle |f|, g \rangle = \Phi_{|f|} g.$$

That is, $\Phi_{|f|} g$ is an upper bound of $A(f, g)$ in $\mathbb{V}$. We now proceed into three steps.

Step 1 Assume that L is Dedekind-complete. In particular, any band is a projection band. Let $P_{|f|}$ denote the order projection on the principal band in L generated by $|f|$. In particular, we have

$$\left|P_{|f|}g\right| \leq g.$$

We derive, via an elementary calculation, that

$$\Phi_{|f|}g = \langle |f|, g\rangle = \left\langle f, P_{|f|}g\right\rangle = \left|\left\langle f, P_{|f|}g\right\rangle\right| = \left|\Phi_f\left(P_{|f|}g\right)\right| \in A(f, g).$$

It follows that the set $A(f, g)$ has a supremum in $\mathbb{V}$ and

$$\Phi_{|f|}g = \sup A(f, g) = \sup\left\{\left|\Phi_f h\right| : h \in L \text{ and } |h| \leq g\right\}.$$

This completes the first step.

Step 2 Suppose that L is Archimedean. Let L^δ and $\mathbb{V}^\delta$ denote the Dedekind-completions of L and $\mathbb{V}$, respectively. There exists a $\mathbb{V}^\delta$-valued orthosymmetric product on L^δ which extends the L's (see Theorem 4.1 in [6]). The image under this product of each ordered pair (f, g) of elements in L^δ is denoted by $\langle f, g\rangle^\delta$. Furthermore, we define $\Phi_f^\delta : L^\delta \to \mathbb{V}^\delta$ by putting

$$\Phi_f^\delta g = \langle f, g\rangle^\delta \text{ for all } g \in L^\delta.$$

Let $T \in \mathcal{L}_r(L, \mathbb{V})$ such that

$$T \geq \pm\Phi_f \text{ in } \mathcal{L}_r(L, \mathbb{V}).$$

In particular, T is positive. Choose a positive extension $T^\delta \in \mathcal{L}_r\left(L^\delta, \mathbb{V}^\delta\right)$ of T (see, e.g., Corollary 1.5.9 in [16]). By the first case (as L^δ is Dedekind-complete), we get

$$T^\delta \geq \Phi_{|f|} \text{ in } \mathcal{L}_r\left(L^\delta, \mathbb{V}^\delta\right).$$

Thus, if $g \in L^+$, then

$$Tg = T^\delta g \geq \Phi_{|f|}^\delta g = \langle |f|, g\rangle^\delta = \langle |f|, g\rangle = \Phi_{|f|}g.$$

This yields that

$$T \geq \Phi_{|f|} \text{ in } \mathcal{L}_r(L, \mathbb{V})$$

and so

$$\left|\Phi_f\right| = \Phi_{|f|} \text{ in } \mathcal{L}_r(L, \mathbb{V}).$$

We focus next on the general case.

Step 3 Here we do not assume any extra condition on L. Set

$$A([f],[g]) = \left\{ \left|\Phi_f h\right| : h \in L \text{ and } |[h]| \leq [g] \text{ in } L/L^0 \right\}.$$

We claim that

$$A(f,g) = A([f],[g]).$$

The inclusion

$$A(f,g) \subset A([f],[g])$$

being obvious, we prove the converse inclusion. Hence, let $h \in L$ such that $|[h]| \leq [g]$ in L/L^0. From Theorem 59.1 in [15], it follows that there $k \in L^0$ such that $|k| \leq |h|$ and $|h-k| \leq |g|$. But then

$$\left|\Phi_f h\right| = \left|\Phi_f (h-k)\right| \in A(f,g)$$

and the required inclusion follows. Theorem 1.8 together with the Archimedean case leads to

$$\begin{aligned}\Phi_{|f|} g &= \langle |f|, g\rangle = \langle [|f|],[g]\rangle = \langle |[f]|,[g]\rangle \\ &= \Phi_{|[f]|}[g] = \sup A([f],[g]) = \sup A(f,g).\end{aligned}$$

This completes the proof of theorem. □

In what follows we shall discuss an example which illustrates Theorem 2.5.

Example 2.6 Let $C(\mathbb{N})$ be the Archimedean vector lattice of all sequences of real numbers. The vector sublattice of $C(\mathbb{N})$ of all bounded sequences is denoted by $C^*(\mathbb{N})$ (here, we follow notations from *[10]*). Define $T \in \mathcal{L}_r(C^*(\mathbb{N}), C(\mathbb{N}))$ by

$$(Tf)(n) = \begin{cases} f(n) - f(n+1) & \text{if } n \in \mathbb{N} \text{ is odd} \\ \\ 0 \text{ if } n \in \mathbb{N} \text{ is even} \end{cases} \qquad \text{for all } f \in C^*(\mathbb{N}).$$

Since $C(\mathbb{N})$ is Dedekind-complete, $\mathcal{L}_r(C^*(\mathbb{N}), C(\mathbb{N}))$ is a vector lattice and thus T has an absolute value $|T|$. We intend to calculate $|T|$. Consider $u, v \in C^*(\mathbb{N})$ with

$$u = \begin{cases} 1 \text{ if } n \text{ is odd} \\ \\ 0 \text{ if } n \text{ is even} \end{cases} \quad \text{and } v = \begin{cases} 0 \text{ if } n \text{ is odd} \\ \\ 1 \text{ if } n \text{ is even.} \end{cases}$$

Moreover, define the shift operator $S \in \mathcal{L}^+ (C^* (\mathbb{N}), C (\mathbb{N}))$ by putting

$$(Sf)(n) = f(n+1) \text{ for all } f \in C^* (\mathbb{N}) \text{ and } n \in \mathbb{N}.$$

Also, if $f, g \in C^* (\mathbb{N})$ then fg is defined by

$$(fg)(n) = f(n) g(n) \text{ for all } f \in C^* (\mathbb{N}) \text{ and } n \in \mathbb{N}.$$

Then, it is readily checked that the formula

$$\langle f, g \rangle = ufg + S(vfg) \text{ for all } f, g \in L.$$

makes $C^* (\mathbb{N})$ into an orthosymmetric $C (\mathbb{N})$-valued vector lattice. An easy calculation yields that

$$Tf = \langle u - v, f \rangle = \Phi_{u-v} f \text{ for all } f \in L.$$

By Theorem 2.5, we derive that T has an absolute value in $\mathcal{L}_r (C^* (\mathbb{N}), C (\mathbb{N}))$ which is given by

$$|T| = |\Phi_{u-v}| = \Phi_{|u-v|} = \Phi_{u+v}.$$

That is,

$$(|T| f)(n) = \begin{cases} f(n) + f(n+1) \text{ if } n \in \mathbb{N} \text{ is odd} \\ \\ 0 \text{ if } n \in \mathbb{N} \text{ is even} \end{cases} \quad \text{for all } f \in C^* (\mathbb{N}).$$

By the way, let $C (\mathbb{N}^*)$ be the vector sublattice of $C (\mathbb{N})$ of all convergent sequences (where $\mathbb{N}^*$ denote the point-one compactification of $\mathbb{N}$). Also, T leaves $C (\mathbb{N}^*)$ invariant and thus T can be considered as an element of $\mathcal{L}_r (C (\mathbb{N}^*), C (\mathbb{N}^*))$. Using an example by Kaplan in [14] (see also Example 1.17 in [3]), it turns out that T has no absolute value in $\mathcal{L}_r (C (\mathbb{N}^*), C (\mathbb{N}^*))$.

The following consequence of Theorem 2.5 is straightforward but worth talking about. First recall that the orthosymmetric space is definite if L^0 is trivial.

Corollary 2.7 *Let L be a definite orthosymmetric space. The map $T : L \to \mathcal{M} (L, \mathbb{V})$ defined by*

$$\Phi f = \Phi_f \textit{ for all } f \in L$$

is a lattice isomorphism.

Roughly speaking, any definite orthosymmetric space L can be embedded in $\mathcal{L}_r (L, \mathbb{V})$ as a generalized vector sublattice. In particular, if $\mathbb{V}$ is in addition

Dedekind-complete, then L has a vector sublattice copy in the Dedekind-complete $\mathcal{L}_r(L, \mathbb{V})$. For instance, any definite orthosymmetric $\mathbb{R}$-valued vector lattice can be considered as a vector sublattice of its order dual.

3 Adjoint Operators on Orthosymmetric Spaces

Also in this section, *all given orthosymmetric spaces are orthosymmetric* $\mathbb{V}$*-valued vector lattices*. Let L, M be two orthosymmetric spaces. The ordered vector space of all (linear) operators from L into M is denoted by $\mathcal{L}_{\mathrm{r}}(L, M)$ and by $\mathcal{L}_{\mathrm{r}}(L)$ if $L = M$. Recall that we call a *positive orthomorphism* on L any positive operator $T \in \mathcal{L}_{\mathrm{r}}(L)$ for which

$$f \wedge Tg = 0 \text{ for all } f, g \in L \text{ with } f \wedge g = 0.$$

An operator $T \in \mathcal{L}_{\mathrm{r}}(L)$ is called an *orthomorphism* if $T = R - S$ for some positive orthomorphisms R, S on L. The set of all orthomorphisms on L is denoted by $\mathrm{Orth}(L)$. For orthomorphisms, the reader can consult the Thesis [8] or Chapter 20 in [19] (further results can be found in [9, 11–13]).

It turns out that orthomorphisms on orthosymmetric spaces have an interesting property.

Theorem 3.1 *Let L be an orthosymmetric space. Then*

$$\langle f, Tg \rangle = \langle Tf, g \rangle \text{ for all } T \in \mathrm{Orth}(L) \text{ and } f, g \in L.$$

Proof Obviously, we can prove the formula only for positive orthomorphisms. Hence, let T be a positive orthomorphism on L and define a bilinear map which assigns the vector

$$\langle f, g \rangle_T = \langle f, Tg \rangle \in \mathbb{V}$$

to each ordered pair $(f, g) \in L \times L$. Clearly, the above formula defines a new $\mathbb{V}$-valued orthosymmetric product on L. In view of Lemma 1.3, we conclude that if $f, g \in L$ then

$$\langle f, Tg \rangle = \langle f, g \rangle_T = \langle g, f \rangle_T = \langle g, Tf \rangle = \langle Tf, g \rangle$$

and the theorem follows. □

Theorem 3.1 motivates us to introduce the following concept. We say that $T \in \mathcal{L}_{\mathrm{r}}(L, M)$ has an *adjoint* in $\mathcal{L}_{\mathrm{r}}(M, L)$ if there is $S \in \mathcal{L}_{\mathrm{r}}(M, L)$ for which

$$\langle Tf, g \rangle = \langle f, Sg \rangle \text{ for all } f \in L \text{ and } g \in M.$$

The subset of $\mathcal{L}_r(M, L)$ of all adjoints of $T \in \mathcal{L}_r(L, M)$ is denoted by $\operatorname{adj}(T)$. The operator $T \in \mathcal{L}_r(L)$ is said to be *selfadjoint* if $T \in \operatorname{adj}(T)$. It follows from Theorem 3.1 that any orthomorphism on L is selfadjoint. Of course, the converse is not valid, i.e., a selfadjoint operator in $\mathcal{L}_r(L)$ need not be an orthomorphism. Consider the orthosymmetric space $\mathbb{R}^2$ as defined in Example 1.1. The operator $T \in \mathcal{L}_r(\mathbb{R}^2)$ given by the 2×2 matrix

$$T = \begin{pmatrix} 1 & 2 \\ 2 & 0 \end{pmatrix}$$

is selfadjoint but fails to be an orthomorphism. Recall by the way that any orthomorphism on the vector lattice $\mathbb{R}^n$ ($n \in \mathbb{N}$) is given by a diagonal matrix as shown in [19, Exercice 141.7]. Next, we shall show *via* an example that $\operatorname{adj}(T)$ can be empty.

Example 3.2 Put $L = \mathbb{V} = C[-1, 1]$, where $C[-1, 1]$ is the Archimedean vector lattice of all real-valued continuous functions on the real interval $[0, 1]$. It is not hard to see that L is an orthosymmetric space under the $\mathbb{V}$-valued orthosymmetric product given by

$$\langle f, g \rangle = fg \text{ for all } f, g \in L.$$

Now, define $T \in \mathcal{L}_r(L)$ by

$$(Tf)(x) = x \int_{-1}^{1} f(s)\, ds \text{ for all } f \in L \text{ and } x \in [-1, 1].$$

Suppose that $\operatorname{adj}(T)$ contains R. Let $f \in L$ and put

$$g(x) = x \text{ for all } x \in [-1, 1].$$

Hence, $Tg = 0$ and so

$$x(Rf)(x) = \langle Rf, g \rangle = \langle f, Tg \rangle = 0 \text{ for all } x \in [-1, 1].$$

It follows that $R = 0$ and thus

$$Tf = \langle Tf, \mathbf{1} \rangle = \langle f, R\mathbf{1} \rangle = 0 \text{ for all } f \in L$$

(here, $\mathbf{1}$ is the constant function whose is the constant 1). This is an obvious absurdity and thus $\operatorname{adj}(T)$ is empty.

Next, we provide an example in which we shall see that $\operatorname{adj}(T)$ may contain more than one element.

Example 3.3 Again, we put $L = \mathbb{V} = C[-1, 1]$. For each $f, g \in L$ we set

$$\langle f, g\rangle (x) = \begin{cases} x \int_{-1}^{0} (fg)(s)\, \mathrm{d}s & \text{if } x \in [0, 1] \\ 0 \quad \text{if } x \in [-1, 0]. \end{cases}$$

It is an easy task to check that this formula makes L into an orthosymmetric $\mathbb{V}$-valued vector lattice. Define $T \in \mathcal{L}_{\mathrm{r}}(L)$ by

$$(Tf)(x) = x \int_{-1}^{0} f(s)\, \mathrm{d}s \text{ for all } f \in L \text{ and } x \in [-1, 1].$$

Moreover, consider $R, S \in \mathcal{L}_{\mathrm{r}}(L)$ such that

$$(Rf)(x) = \int_{-1}^{0} sf(s)\, \mathrm{d}s \text{ and } Sf = fw + Rf \text{ for all } f \in L \text{ and } x \in [-1, 1],$$

where

$$w(x) = 0 \text{ if } x \in [-1, 0] \text{ and } w(x) = x \text{ if } x \in [0, 1].$$

A direct calculation reveals that

$$\langle f, Tg\rangle = \langle Rf, g\rangle = \langle Sf, g\rangle \text{ for all } f, g \in L.$$

Hence, $R, S \in \mathrm{adj}(T)$ and $R \neq S$.

The fact that an operator $T \in \mathcal{L}_{\mathrm{r}}(L, M)$ could have more than one adjoint is rather inconvenient. In order to get around the problem and thus make our theory more or less reasonable, *we shall assume from now on that the codomain orthosymmetric space M is definite*. Recall here that the orthosymmetric space M is definite if 0 is the only neutral element in M. This means that

$$M^0 = \{g \in M : \langle g, g\rangle = 0\} = \{0\}.$$

As we shall prove next, an operator on a definite orthosymmetric space has at most one adjoint.

Proposition 3.4 *Let L, M be orthosymmetric spaces with M definite. If $T \in \mathcal{L}_{\mathrm{r}}(L, M)$ then* $\mathrm{adj}(T)$ *has at most one point.*

Proof Suppose that $R, S \in \mathrm{adj}(T)$. Let $f \in L$ and $g \in M$. Observe that

$$\begin{aligned} \langle Rf - Sf, g\rangle &= \langle Rf, g\rangle - \langle Sf, g\rangle \\ &= \langle f, Tg\rangle - \langle f, Tg\rangle = 0. \end{aligned}$$

Since g are arbitrary in M and M is definite, we get

$$Rf - Sf = 0 \text{ for all } f \in L.$$

This means that $R = S$ and we are done. □

As it would be expected, if adj (T) is non-empty, then its unique element is called the *adjoint* of T and denoted by T^*. In this situation, we have

$$T^* \in \mathcal{L}_r(M, L) \text{ and } \langle Tf, g\rangle = \left\langle f, T^* g\right\rangle \text{ for all } f \in L \text{ and } g \in M.$$

A first property of adjoint operators is given below.

Proposition 3.5 *Let L, M be orthosymmetric spaces with M definite and $T \in \mathcal{L}_r(L, M)$ such that $T^* \in \mathcal{L}_r(M, L)$ exists. If T is positive, then so is T^*.*

Proof Let $f \in L$ and $g \in M$. Keeping the same notation as previously used in Sect. 2, we can write

$$\Phi_{T^*g} f = \left\langle f, T^* g\right\rangle = \langle Tf, g\rangle = \Phi_g (Tf) = \left(\Phi_g \circ T\right) g.$$

Hence,

$$\Phi_{T^*g} = \Phi_g \circ T \text{ for all } g \in L.$$

Now, assume that T is positive and let $g \in M^+$. We claim that $T^* g \in M^+$. Since $g \in M^+$, the operator Φ_g is positive (see Lemma 2.4 (i)). Hence, $\Phi_g \circ T$ is positive and so is Φ_{T^*g}. But then

$$\left(T^* g\right)^- \in M^0 = \{0\}$$

as M is definite (where we use Lemma 2.4 (iv)). It follows that $T^* g \in M^+$ which leads to the desired result. □

Now, we shall focus on lattice homomorphisms from L into M. Recall that $T \in \mathcal{L}_r(L, M)$ is a *lattice homomorphism* if

$$|Tf| = T\,|f| \quad \text{for all } f \in L.$$

Next, we obtain a (quite amazing) characterization of lattice homomorphisms in terms of adjoint.

Theorem 3.6 *Let L, M be orthosymmetric spaces with M definite and $T \in \mathcal{L}_r(L, M)$ such that T is positive and $T^* \in \mathcal{L}_r(M, L)$ exists. Then T is a lattice homomorphism if and only if $T^*T \in$ Orth (L).*

Proof Assume that T is a lattice homomorphism. Since T and T^* are positive (see Proposition 3.5), so is T^*T. Let $f, g \in L$ with $f \wedge g = 0$. From $Tf \wedge Tg = 0$ it follows that $\langle Tf, Tg\rangle = 0$. We get

$$\begin{aligned} 0 &\leq \left\langle \left(T^*T\right) f \wedge g, \left(T^*T\right) f \wedge g\right\rangle \leq \left\langle \left(T^*T\right) f, g\right\rangle \\ &= \left\langle T^*Tf, g\right\rangle = \langle Tf, Tg\rangle = 0. \end{aligned}$$

Since M is definite, we derive that $(T^*T)\, f \wedge g = 0$ so $T^*T \in \operatorname{Orth}(L)$.

Conversely, suppose that $T^*T \in \operatorname{Orth}(L)$ and pick $f, g \in L$ with $f \wedge g = 0$. Hence, $(T^*T)\, f \wedge g = 0$ because T^*T is a positive orthomorphism. Therefore,

$$\langle Tf, Tg\rangle = \left\langle \left(T^*T\right) f, g\right\rangle = 0.$$

Consequently,

$$0 \leq \langle Tf \wedge Tg, Tf \wedge Tg\rangle \leq \langle Tf, Tg\rangle = 0.$$

We derive that $Tf \wedge Tg = 0$ as M is definite. This yields that T is a lattice homomorphism and completes the proof. □

A quite curious consequences of Theorem 3.6 is discussed next. Let T be a positive $n \times n$ matrix ($n \in \mathbb{N}$) such that T^*T is diagonal. Then each row of T contains at most one nonzero (positive) entry. Indeed, since T^*T is diagonal, T^*T is an orthomorphism on $\mathbb{R}^n$. But then T is a lattice homomorphism (where we use Theorem 3.6) and the result follows (see [1] or [17]).

We proceed to a question which arises naturally. Namely, is the adjoint of a lattice homomorphism on orthosymmetric spaces again a lattice homomorphism? The following simple example proves that this is not true in general.

Example 3.7 Let $L = \mathbb{R}^3$ be equipped with its structure of definite orthosymmetric $\mathbb{R}$-valued vector lattice as explained in *Example 1.1*. Consider

$$T = \begin{pmatrix} 0 & 0 & 0 \\ 0 & 1 & 0 \\ 0 & 1 & 0 \end{pmatrix} \in \mathcal{L}_r(L)$$

and observe that T is a lattice homomorphism on L. However,

$$T^* = \begin{pmatrix} 0 & 0 & 0 \\ 0 & 1 & 1 \\ 0 & 0 & 0 \end{pmatrix}$$

is not a lattice homomorphism.

The situation improves if T is onto as we prove in the following.

Corollary 3.8 *Let L, M be orthosymmetric spaces with M definite and $T \in \mathcal{L}_r(T)$ be a surjective lattice homomorphism such that $T^* \in \mathcal{L}_r(M, L)$ exists. Then T^* is again a lattice homomorphism.*

Proof Let $g \in M$ and $f \in L$ such that $g = Tf$. By Theorem 3.6, the operator T^*T is an orthomorphism. Thus,

$$T^* |g| = T^* |Tf| = T^*T |f| = |T^*Tf| = |T^*g|.$$

This means that T^* is a lattice homomorphism, as desired. □

The following shows that only normal lattice homomorphisms have adjoints. First, recall that a lattice homomorphism $T \in \mathcal{L}_r(L, M)$ is said to be *normal* if its kernel $\ker(T)$ is a band of L.

Corollary 3.9 *Let L, M be orthosymmetric spaces with M definite and $T \in \mathcal{L}_r(L, M)$ be a lattice homomorphism such that $T^* \in \mathcal{L}_r(M, L)$ exists. Then T is normal.*

Proof We prove that $\ker(T)$ is band. We claim that $\ker(T^*T) = \ker(T)$. Obviously, $\ker(T^*T)$ contains $\ker(T)$. Conversely, if $f \in \ker(T^*T)$ then

$$0 \leq \langle Tf, Tf\rangle = \langle T^*Tf, f\rangle = 0.$$

It follows that $Tf = 0$ because M is definite. We derive that $\ker(T^*T) = \ker(T)$, as required. On the other hand, Theorem 3.6 guaranties that T^*T is an orthomorphism on L. Accordingly, $\ker(T^*T)$ is a band of L and so is $\ker(T)$. This yields that T is normal and completes the proof. □

The last part of the paper deals with interval preserving operators. For any positive element f in a vector lattice L, we set

$$[0, f] = \{g \in M : 0 \leq g \leq f\}.$$

A positive operator $T : M \to L$ is said to be *interval preserving* if

$$T[0, f] = [0, Tf] \text{ for all } f \in M^+,$$

A sufficient condition for a positive operator acting on orthosymmetric spaces to be a lattice homomorphism is to have an interval preserving adjoint.

Theorem 3.10 *Let L, M be orthosymmetric spaces with M definite and $T \in \mathcal{L}_r(L, M)$ be positive such that T^* exists. If T^* is interval preserving, then T is a lattice homomorphism.*

Proof We choose $f \in L$ and we claim that $T\left(f^{+}\right) = (Tf)^{+}$. To this end, observe that if $0 \leq h \in M^{+}$ then

$$\begin{aligned}\Phi_{(Tf)^{+}}h = \Phi_{Tf}^{+}h &= \sup\{\langle Tf, g\rangle : 0 \leq g \leq h \text{ in } M\} \\ &= \sup\left\{\left\langle f, T^{*}g\right\rangle : 0 \leq g \leq h \text{ in } M\right\} \\ &\leq \sup\left\{\langle f, u\rangle : 0 \leq u \leq T^{*}h \text{ in } L\right\}.\end{aligned}$$

On the other hand, if $0 \leq u \leq T^{*}h$ in L, then there exists $g \in [0, h]$ such that $u = T^{*}g$. Accordingly,

$$\begin{aligned}\Phi_{(Tf)^{+}}h &= \sup\left\{\langle f, u\rangle : 0 \leq u \leq T^{*}h \text{ in } L\right\} \\ &= \Phi_{f}^{+}T^{*}h = \Phi_{f^{+}}T^{*}h = \Phi_{T(f^{+})}h.\end{aligned}$$

It follows that $\Phi_{(Tf)^{+}} = \Phi_{T(f^{+})}$ and so $(Tf)^{+} = T\left(f^{+}\right)$ because M is definite. This ends the proof of the theorem. □

The converse of the previous theorem fails as it can be seen in the following example, which is the last item of this paper.

Example 3.11 Let $L = M = C[0, 1]$ be equipped with any structure of definite symmetric space and define $T \in \text{Orth}(L)$ by

$$(Tf)(x) = xf(x) \text{ for all } f \in L \text{ and } x \in [0, 1].$$

Since $T \in \text{Orth}(L)$, we derive that T^{*} exists and $T = T^{*}$ (see Theorem 3.1). Of course, T is a lattice homomorphism. However, T is not interval preserving. Indeed, if $g \in L$ is defined by

$$g(x) = x\left|\sin\frac{1}{x}\right| \text{ if } x \in \left]0, 1\right] \text{ and } g(0) = 0,$$

then $0 \leq g \leq T\mathbf{1}$ but there is not $f \in L$ such that $g = Tf$.

References

1. Y.A. Abramovich, C.D. Aliprantis, *An Invitation to Operator Theory*. Graduate Studies Math. 50 (American Mathematical Society, Providence, 2002)
2. Y.A. Abramovich, A.W. Wickstead, Recent results on the order structure of compact operators. Irish Math. Soc. Bull. **32**, 32–45 (1994)
3. C.D. Aliprantis, O. Burkinshaw, *Positive Operators* (Springer, Dordrecht, 2006)
4. J. Bognár, *Indefinite Inner Product Spaces* (Springer, Berlin, 1974)
5. K. Boulabiar, Some aspects of Riesz bimorphisms. Indag. Math. **13**, 419–432 (2002)

6. G. Buskes, A.G. Kusraev, Representation and extension of orthoregular bilinear operators. Vladikavkaz. Mat. Zh. **9**, 16–29 (2007)
7. G. Buskes, A. van Rooij, Almost f-algebras: commutativity and the Cauchy-Schwarz inequality. Positivity **4**, 227–231 (2000)
8. B. de Pagter, f-algebras and orthomorphism. Ph.D. Thesis, Leiden, 1981
9. B. de Pagter, The space of extended orthomorphisms in a Riesz space. Pacific J. Math. **112**, 193–210 (1984)
10. L. Gillman, M. Jerison, *Rings of Continuous Functions* (Springer, Berlin, 1976)
11. C.B. Huijsmans, B. de Pagter, Ideal theory in f-algebras. Amer. Math. Soc. **269**, 225–245 (1982)
12. C.B. Huijsmans, B. de Pagter, The order bidual of a lattice-ordered algebra. J. Funct. Anal. **59**, 41–64 (1984)
13. C.B. Huijsmans, B. de Pagter, Averaging operators and positive contractive projections. J. Math. Anal. Appl. **113**, 163–184 (1986)
14. S. Kaplan, An example in the space of bounded operators from $C(X)$ to $C(Y)$. Proc. Am. Math. Soc. **38**, 595–597 (1973)
15. W.A.J. Luxemburg, A.C. Zaanen, *Riesz Spaces I* (North-Holland, Amsterdam, 1971)
16. P. Meyer-Neiberg, *Banach Lattices* (Springer, Berlin, 1991)
17. H.H. Schaefer, *Banach Lattices and Positive Operators* (Springer, Berlin, 1974)
18. S. Steinberg, *Lattice-Ordered Rings and Modules* (Springer, Dordrecht, 2010)
19. A.C. Zaanen, *Riesz Spaces II* (North-Holland, Amsterdam, 1983)

Arens Extensions for Polynomials and the Woodbury–Schep Formula

Gerard Buskes and Stephan Roberts

Dedicated to the 65th birthday of Professor Ben de Pagter

Abstract Among other results, we provide a formula for the order continuous component of a positive s-homogeneous polynomial on a vector lattice with Dedekind complete range.

Keywords Vector lattice · Polynomial · Order continuity · Arens adjoint

1 Introduction and Preliminaries

The theory of polynomials on vector lattices and Banach lattices has been rapidly developing over the last decade. Surprisingly, order continuity of such polynomials has not been investigated before. We define and derive the order continuity of polynomials on vector lattices via the order continuity of multilinear maps on vector lattices. Our main results are Theorems 2.8 and 3.5. They extend corresponding results for linear maps on vector lattices to polynomials on vector lattices. These extensions are relatively straightforward and they are essential to a sequel to this note that will recast the connection between orthosymmetric maps and orthogonally additive polynomials via polynomial valuations.

We use $E, E_1, \ldots, E_s$, and F to denote Archimedean vector lattices. A net $(x_\alpha)_{\alpha\in A}$ in E *converges in order* to x if there exists a net $(y_\beta)_{\beta\in B}$ in E such that $y_\beta \downarrow 0$ and for each β there exists an index α_0 for which $|x - x_\alpha| \leq y_\beta$ for all $\alpha \geq \alpha_0$ (see [1]). We write $x_\alpha \to x$ to denote x_α converges in order to x. Let $T\colon E_1 \times \ldots \times E_s \to F$ be an s-linear map. We call T *separately order continuous*

G. Buskes (✉) · S. Roberts
Department of Mathematics, University of Mississippi, Oxford, MS, USA
e-mail: mmbuskes@olemiss.edu; scrober2@olemiss.edu

G. Buskes et al. (eds.), *Positivity and Noncommutative Analysis*,
Trends in Mathematics, https://doi.org/10.1007/978-3-030-10850-2_3

if for each $j \in \{1, \ldots, s\}$ and $x_i \in E_i$ $(i \in \{1, \ldots, s\} \setminus \{j\})$ the linear map

$$x \mapsto T(x_1, \ldots, x, \ldots, x_s) \; (x \in E_j)$$

is order continuous. Let A_i be a directed set and let $x_i \colon A_i \to E_i$ be a net $(i \in \{1, \ldots, s\})$. Define $(x_1, \ldots, x_s) : A_1 \times \ldots \times A_s \to E_1 \times \ldots \times E_s$ by $(x_1, \ldots, x_s)(\alpha_1, \ldots, \alpha_s) = (x_1(\alpha_1), \ldots, x_s(\alpha_s))$ for all $(\alpha_1, \ldots, \alpha_s) \in A_1 \times \ldots \times A_s$. When $A_1 \times \ldots \times A_s$ is directed coordinate-wise, $(x_1, \ldots, x_s)$ defines a net, which we denote by $(x_{1,\alpha_1}, \ldots, x_{s,\alpha_s})_{(\alpha_1,\ldots,\alpha_s) \in A_1 \times \ldots \times A_s}$. We call T *order continuous* if

$$T(x_{1,\alpha_1}, \ldots, x_{s,\alpha_s}) \to T(x_1, \ldots, x_s)$$

in F whenever $(x_{1,\alpha_1}, \ldots, x_{s,\alpha_s})_{(\alpha_1,\ldots,\alpha_s) \in A_1 \times \ldots \times A_s}$ is a net in $E_1 \times \ldots \times E_s$ such that $(x_{1,\alpha_1}, \ldots, x_{s,\alpha_s}) \to (x_1, \ldots, x_s)$.

For $x \in E^+$, a *partition* of x is a finite sequence of elements of E^+ whose sum is equal to x. The set of all partitions of x is denoted by Πx. We write a to abbreviate the partition $(a_1, \ldots, a_n)$ of x. An s-linear map $T \colon E_1 \times \ldots \times E_s \to F$ is *of order bounded variation* if

$$\left\{ \sum_{k_1,\ldots,k_s} |T(a_{1,k_1}, \ldots, a_{s,k_s})| : a_1 \in \Pi x_1, \ldots, a_s \in \Pi x_s \right\} \; (x_i \in E_i^+, i \in \{1, \ldots, s\})$$

is order bounded. The set of all s-linear maps of order bounded variation from $E_1 \times \ldots \times E_s$ to F is denoted by $\mathcal{L}^{obv}(E_1, \ldots, E_s; F)$. For $s \geq 2$, we denote the set of all symmetric maps in $\mathcal{L}^{obv}(E_1, \ldots, E_s; F)$ by $\mathcal{L}^{obv}_{sym}(E_1, \ldots, E_s; F)$. The set of all separately order continuous s-linear maps in $\mathcal{L}^{obv}(E_1, \ldots, E_s; F)$ is denoted by $\mathcal{L}_n(E_1, \ldots, E_s; F)$. If F is Dedekind complete, then $\mathcal{L}^{obv}(E_1, \ldots, E_s; F)$ is a Dedekind complete vector lattice (Theorem 1.1 of [4]) and $\mathcal{L}^{obv}_{sym}(E_1, \ldots, E_s; F)$ is a Dedekind complete vector sublattice of $\mathcal{L}^{obv}(E_1, \ldots, E_s; F)$ (Lemma 2.15 of [7]).

Let F be Dedekind complete. In this note, we characterize $\mathcal{L}_n(E_1, \ldots, E_s; F)$ as a band in $\mathcal{L}^{obv}(E_1, \ldots, E_s; F)$ (Theorem 2.3). This extends a result of Nakano (Theorem 3.4 of [8]). In addition, we define order continuous polynomials of order bounded variation and show that they form a band in the vector lattice of s-homogeneous polynomials of order bounded variation (Theorem 2.7). The order continuous component T_n of a positive linear map $T \colon E \to F$ is given by

$$T_n(x) = \inf \left\{ \sup_{\alpha} T(x_\alpha) : 0 \leq x_\alpha \uparrow x \right\} \; (x \in E^+).$$

This formula originates in [10] by Woodbury and appears in the above form in [9] by Schep. We extend this Woodbury–Schep formula to positive s-homogeneous polynomials (Theorem 2.8).

The order dual of E is denoted by $E^{\sim}$ and the order continuous dual of E is denoted by $E^{\sim}_n$. The *Arens adjoint* of an s-linear map $T: E_1 \times \cdots \times E_s \to F$ is the map $T^*: F^{\sim} \times E_1 \times \ldots \times E_{s-1} \to E^{\sim}_s$ that is defined by

$$T^*(f, x_1, \ldots, x_{s-1})(x_s) = f(T(x_1, \ldots, x_{s-1}, x_s))$$

for all $f \in F^{\sim}, x_1 \in E_1, \ldots, x_s \in E_s$. Let $T^{[1]*} := T^*$ and inductively define $T^{[k]*}$ by $T^{[k]*} = (T^{[k-1]*})^*$ $(k \geq 2)$. We prove that the restriction of $T^{[s+1]*}$ to $(E^{\sim}_1)^{\sim}_n \times \ldots \times (E^{\sim}_s)^{\sim}_n$ is separately order continuous if T is of order bounded variation (Theorem 3.4), which extends Theorem 2.1(ii) of [3]. We also prove that if $E = E_1 = .. = E_s$ and if T is of order bounded variation and symmetric then $T^{[s+1]*}$ restricted to $(E^{\sim})^{\sim}_n \times \ldots \times (E^{\sim})^{\sim}_n$ is symmetric (Theorem 3.4). This extends Theorem 2.1 (iv) of [3]. Finally, we draw conclusions for the polynomial that is associated with $T^{[s+1]*}$ (Theorem 3.5).

2 Order Continuous Maps of Order Bounded Variation

An s-homogeneous polynomial $P: E \to F$ is *of order bounded variation* if

$$\left\{\sum_k |P(a_k)| : a \in \Pi x\right\} \quad (x \in E^+)$$

is order bounded. We denote the set of all s-homogeneous polynomials of order bounded variation from E to F by $\mathcal{P}^{obv}({}^sE; F)$. Define $\Delta: E \to E^s$ by $\Delta(x) = (x, \ldots, x)$ for all $x \in E$. For an s-homogeneous polynomial $P: E \to F$, the unique symmetric s-linear map $T: E \times \ldots \times E \to F$ such that $P = T \circ \Delta$ is denoted by $\check{P}$.

Proposition 2.1 *Let $P: E \to F$ be an s-homogeneous polynomial. Then the following are equivalent.*

1. *P is of order bounded variation.*
2. *$\check{P}$ is of order bounded variation.*

Proof We need only prove that **1** implies **2**. Suppose that $P: E \to F$ is an s-homogeneous polynomial of order bounded variation. Let $x_i \in E^+$ and $a_i \in \Pi x_i$ $(i \in \{1, \ldots, s\})$. By the Mazur–Orlicz polarization formula,

$$\begin{aligned}\sum_{k_1,\ldots,k_s} |\check{P}(a_{1,k_1}, \ldots, a_{s,k_s})| &= \frac{1}{s!} \sum_{k_1,\ldots,k_s} \left| \sum_{\delta_i=0,1} (-1)^{s-\sum \delta_i} P\left(\sum_{i=1}^{s} \delta_i a_{i,k_i}\right) \right| \\ &\leq \frac{1}{s!} \sum_{k_1,\ldots,k_s} \sum_{\delta_i=0,1} \left| P\left(\sum_{i=1}^{s} \delta_i a_{i,k_i}\right) \right|.\end{aligned}$$

Fix $\delta_1, \ldots, \delta_s \in \{0, 1\}$. Note that $\delta_i a_i \in \Pi \delta_i x_i$ $(i \in \{1, \ldots, s\})$ implies $\sum_{i=1}^{s} \delta_i a_i \in \Pi \sum_{i=1}^{s} \delta_i x_i$. Since P is of order bounded variation, there exists some $y_{\delta_1,\ldots,\delta_s} \in F$ such that $\sum_{k_1,\ldots,k_s} \left| P\left(\sum_{i=1}^{s} \delta_i a_{i,k_i}\right) \right| \leq y_{\delta_1,\ldots,\delta_s}$. Hence

$$\frac{1}{s!} \sum_{k_1,\ldots,k_s} \sum_{\delta_i=0,1} \left| P\left(\sum_{i=1}^{s} \delta_i a_{i,k_i}\right) \right| = \frac{1}{s!} \sum_{\delta_i=0,1} \sum_{k_1,\ldots,k_s} \left| P\left(\sum_{i=1}^{s} \delta_i a_{i,k_i}\right) \right|$$
$$\leq \frac{1}{s!} \sum_{\delta_i=0,1} y_{\delta_1,\ldots,\delta_s} \in F.$$

Then

$$\left\{ \sum_{k_1,\ldots,k_s} |\check{P}(a_{1,k_1}, \ldots, a_{s,k_s})| : a_1 \in \Pi x_1, \ldots, a_s \in \Pi x_s \right\}$$

is order bounded. Therefore, $\check{P}$ is of order bounded variation. □

The next theorem follows from Lemma 2.16 of [7] and Proposition 2.1.

Theorem 2.2 *Let F be Dedekind complete. The following hold.*

1. *$P \in \mathcal{P}^{obv}({}^{s}E; F)$ if and only if $\check{P}$ is regular.*
2. *$\mathcal{P}^{obv}({}^{s}E; F)$ is a Dedekind complete vector lattice.*

Let $s \geq 2$. Define $K : \mathcal{L}^{obv}(E_1, \ldots, E_s; F) \to \mathcal{L}^{b}(E_s; \mathcal{L}^{obv}(E_1, \ldots, E_{s-1}; F))$ by

$$K(T)(x_s)(x_1, \ldots, x_{s-1}) = T(x_1, \ldots, x_s)$$

for all $T \in \mathcal{L}^{obv}(E_1, \ldots, E_s; F)$ and $x_i \in E_i$ $(i \in \{1, \ldots, s\})$.

Theorem 2.3 *Let F be Dedekind complete. Define K as above. The following hold.*

1. *$K(\mathcal{L}_n(E_1, \ldots, E_s; F)) = \mathcal{L}_n(E_1, \ldots, E_{s-1}; \mathcal{L}_n(E_s; F))$ $(s \geq 2)$.*
2. *$\mathcal{L}_n(E_1, \ldots, E_s; F)$ is a band in $\mathcal{L}^{obv}(E_1, \ldots, E_s; F)$.*
3. *If $E = E_1 = \ldots = E_s$, then $\mathcal{L}_{sym,n}({}^{s}E; F)$ is a band in $\mathcal{L}^{obv}_{sym}({}^{s}E; F)$ $(s \geq 2)$.*

Proof

1. Let $T \in \mathcal{L}_n(E_1, \ldots, E_s; F)$ and let $x_i \in E_i$ $(i \in \{1, \ldots, s\})$. If $(x_{s,\alpha_s})_{\alpha_s \in A_s}$ is a net in E_s such that $x_{s,\alpha_s} \to x_s$ then it follows from the order continuity in the last argument of T that

$$K(T)(x_{s,\alpha_s})(x_1, \ldots, x_{s-1}) \to K(T)(x_s)(x_1, \ldots, x_{s-1}).$$

Hence $K(T) \in \mathcal{L}_n(E_s; \mathcal{L}^{obv}(E_1, \ldots, E_{s-1}; F))$. If $1 \leq i \leq s-1$, then by the separate order continuity of T we have that

$$K(T)(x_s)(x_1, \ldots, x_{i,\alpha_i}, \ldots, x_{s-1}) \to K(T)(x_s)(x_1, \ldots, x_i, \ldots, x_{s-1}).$$

Thus

$$K(\mathcal{L}_n(E_1, \ldots, E_s; F)) \subseteq \mathcal{L}_n(E_s; \mathcal{L}_n(E_1, \ldots, E_{s-1}; F)).$$

Since the reverse inclusion is straightforward it follows that

$$K(\mathcal{L}_n(E_1, \ldots, E_s; F)) = \mathcal{L}_n(E_1, \ldots, E_{s-1}; \mathcal{L}_n(E_s; F)).$$

2. We proceed by induction on s. It is known that $\mathcal{L}_n(E_1; F)$ is a band in $\mathcal{L}^b(E_1; F)$ (Theorem 1.57 of [2]). Assume $\mathcal{L}_n(E_1, \ldots, E_s; F)$ is a band in $\mathcal{L}^{obv}(E_1, \ldots, E_s; F)$ for some $s \geq 1$. Note that $\mathcal{L}_n(E_{s+1}; \mathcal{L}_n(E_1, \ldots, E_s; F))$ is a band in $\mathcal{L}^b(E_{s+1}; \mathcal{L}_n(E_1, \ldots, E_s; F))$ and by assumption $\mathcal{L}^b(E_{s+1}; \mathcal{L}_n(E_1, \ldots, E_s; F))$ is a band in $\mathcal{L}^b(E_{s+1}; \mathcal{L}^{obv}(E_1, \ldots, E_s; F))$. Thus $\mathcal{L}_n(E_{s+1}; \mathcal{L}_n(E_1, \ldots, E_s; F))$ is a band in $\mathcal{L}^b(E_{s+1}; \mathcal{L}^{obv}(E_1, \ldots, E_s; F))$. Since K is a lattice isomorphism (see Corollary 3.1 of [4]), it follows from part **1** that $\mathcal{L}_n(E_1, \ldots, E_{s+1}; F)$ is a band in $\mathcal{L}^{obv}(E_1, \ldots, E_{s+1}; F)$.
3. Lemma 2.15 of [7] implies $\mathcal{L}^{obv}_{sym}({}^sE; F)$ is a Dedekind complete vector lattice. It follows that $\mathcal{L}_{sym,n}({}^sE; F)$ is a band in $\mathcal{L}^{obv}_{sym}({}^sE; F)$.

□

By Theorem 2.3 we have that

$$\mathcal{L}^{obv}(E_1, \ldots, E_s; F) = \mathcal{L}_n(E_1, \ldots, E_s; F) \oplus \mathcal{L}_n(E_1, \ldots, E_s; F)^d$$

whenever F is Dedekind complete. For $T \in \mathcal{L}^{obv}(E_1, \ldots, E_s; F)$, the order continuous component of T is denoted by T_n. In the next theorem we generalize the Woodbury–Schep formula to positive s-linear maps.

Theorem 2.4 *Let F be Dedekind complete and let $T\colon E_1 \times \ldots \times E_s \to F$ be a positive s-linear map. Then*

$$T_n(x_1, \ldots, x_s) = \inf \left\{ \sup_{(\alpha_1, \ldots, \alpha_s)} T(x_{1,\alpha_1}, \ldots, x_{s,\alpha_s}) : 0 \leq x_{1,\alpha_1} \uparrow x_1, \ldots, 0 \leq x_{s,\alpha_s} \uparrow x_s \right\}$$

for all $x_i \in E_i^+$ ($i \in \{1, \ldots, s\}$).

Proof Define $S\colon E_1^+ \times \ldots \times E_s^+ \to F^+$ by

$$S(x_1, \ldots, x_s) = \inf \left\{ \sup_{(\alpha_1, \ldots, \alpha_s)} T(x_{1,\alpha_1}, \ldots, x_{s,\alpha_s}) : 0 \leq x_{1,\alpha_1} \uparrow x_1, \ldots, 0 \leq x_{s,\alpha_s} \uparrow x_s \right\}$$

for all $x_i \in E_i^+$ $(i \in \{1, \ldots, s\})$. We first prove that S is separately additive. Fix $j \in \{1, \ldots, s\}$. Let $y, z \in E_j^+$ and let $(x_{\alpha_j})_{\alpha_j \in A_j}$ be a net in E_j^+ such that $x_{\alpha_j} \uparrow (y+z)$. For each $\alpha_j \in A_j$, let $y_{\alpha_j} := x_{\alpha_j} \wedge y$ and let $z_{\alpha_j} := x_{\alpha_j} - y_{\alpha_j}$. Note that $y_{\alpha_j} \uparrow y$ and that $z_{\alpha_j} \uparrow z$. For nets $(x_{i,\alpha_i})_{\alpha_i \in A_i}$ in E_i^+ such that $x_{i,\alpha_i} \uparrow x_i$ $(i \in \{1, \ldots s\} \setminus \{j\})$ we have that

$$\begin{aligned} & S(x_1, \ldots, y, \ldots, x_s) + S(x_1, \ldots, z, \ldots, x_s) \\ \leq & \sup_{(\alpha_1, \ldots, \alpha_s)} T(x_{1,\alpha_1}, \ldots, y_{\alpha_j}, \ldots, x_{s,\alpha_s}) + \sup_{(\alpha_1, \ldots, \alpha_s)} T(x_{1,\alpha_1}, \ldots, z_{\alpha_j}, \ldots, x_{s,\alpha_s}) \\ = & \sup_{(\alpha_1, \ldots, \alpha_s)} T(x_{1,\alpha_1}, \ldots, x_{\alpha_j}, \ldots, x_{s,\alpha_s}), \end{aligned}$$

where the equality follows from Lemma 21.3 of [11]. Then

$$S(x_1, \ldots, y, \ldots, x_s) + S(x_1, \ldots, z, \ldots, x_s) \leq S(x_1, \ldots, y+z, \ldots, x_s).$$

Similarly,

$$S(x_1, \ldots, y+z, \ldots, x_s) \leq S(x_1, \ldots, y, \ldots, x_s) + S(x_1, \ldots, z, \ldots, x_s) \leq S(x_1, \ldots, y+z, \ldots, x_s).$$

It follows from Theorem 2.3 of [7] that S extends to a positive s-linear map $E_1 \times \ldots \times E_s \to F$ which we also denote by S.

It remains to prove that $S = T_n$. Suppose $x_\beta \downarrow 0$ in E_j $(j \in \{1, \ldots, s\})$. Define $K(T)\colon E_1 \times \ldots \times E_{j-1} \times E_{j+1} \times \ldots \times E_s \to \mathcal{L}^b(E_j; F)$ by

$$K(T)(x_1, \ldots, x_{j-1}, x_{j+1}, \ldots, x_s)(x_j) = T(x_1, \ldots, x_s)(x_i \in E_i, i \in \{1, \ldots, s\}).$$

Then the Woodbury–Schep formula for linear maps implies

$$\begin{aligned} S(x_1, \ldots, x_\beta, \ldots, x_s) &\leq \inf \left\{ \sup_{\alpha_j} T(x_1, \ldots, x_{j,\alpha_j}, \ldots, x_s) : 0 \leq x_{j,\alpha_j} \uparrow x_\beta \right\} \\ &\leq \inf \left\{ \sup_{\alpha_j} K(T)(x_1, \ldots, x_{j-1}, x_{j+1}, \ldots, x_s)(x_{j,\alpha_j}) : 0 \leq x_{j,\alpha_j} \uparrow x_\beta \right\} \\ &= (K(T)(x_1, \ldots, x_{j-1}, x_{j+1}, \ldots, x_s))_n(x_\beta) \downarrow 0 \end{aligned}$$

for $x_i \in E_i^+$ $(i \in \{1,\ldots,s\} \setminus \{j\})$. Hence S is separately order continuous. Since

$$\begin{aligned} T_n(x_1,\ldots,x_s) =& \inf\left\{ \sup_{(\alpha_1,\ldots,\alpha_s)} T_n(x_{1,\alpha_1},\ldots,x_{s,\alpha_s}) : 0 \le x_{1,\alpha_1} \uparrow x_1,\ldots,0 \le x_{s,\alpha_s} \uparrow x_s \right\} \\ \le& \inf\left\{ \sup_{(\alpha_1,\ldots,\alpha_s)} T(x_{1,\alpha_1},\ldots,x_{s,\alpha_s}) : 0 \le x_{1,\alpha_1} \uparrow x_1,\ldots,0 \le x_{s,\alpha_s} \uparrow x_s \right\} \\ =& S(x_1,\ldots,x_s), \end{aligned}$$

it follows that $T_n = S$. □

Our final goal for this section is to prove the Woodbury–Schep formula for positive s-homogeneous polynomials (Theorem 2.8). An s-homogeneous polynomial $P\colon E \to F$ is called *order continuous* if $P(x_\alpha) \to P(x)$ whenever $x_\alpha \to x$. We denote the set of all order continuous s-homogeneous polynomials of order bounded variation from E to F by $\mathcal{P}_n({}^sE; F)$. We use the following pair of lemmas to prove that $\mathcal{P}_n({}^sE; F)$ is a vector lattice when F is Dedekind complete.

Lemma 2.5 *Let $P\colon E \to F$ be an s-homogeneous polynomial. The following are equivalent.*

1. *P is order continuous.*
2. *$\check{P}$ is order continuous.*

Proof We need only prove that **1** implies **2**. Suppose P is order continuous. Let $(x_{i,\alpha_i})_{\alpha_i \in A_i}$ be a net such that $x_{i,\alpha_i} \to x_i$ in E $(i \in \{1,\ldots s\})$. The Mazur–Orlicz polarization formula and the order continuity of P imply

$$\check{P}(x_{1,\alpha_1},\ldots,x_{s,\alpha_s}) =$$
$$= \frac{1}{s!}\sum_{\delta_i=0,1}(-1)^{s-\sum\delta_i} P\left(\sum_{i=1}^{s}\delta_i x_{i,\alpha_i}\right) \to \frac{1}{s!}\sum_{\delta_i=0,1}(-1)^{s-\sum\delta_i} P_T\left(\sum_{i=1}^{s}\delta_i x_i\right) = T(x_1,\ldots,x_s).$$

□

The following lemma extends Proposition 5.4.2 of [5].

Lemma 2.6 *Let $T\colon E_1 \times \ldots \times E_s \to F$ be a positive s-linear map. The following are equivalent.*

1. *T is order continuous.*
2. *T is separately order continuous.*

Proof We need only prove that **2** implies **1**. Suppose that T is separately order continuous. We induct on s to prove that T is order continuous. Trivially T is order continuous if $s = 1$. Assume that the lemma holds for some natural number $s \ge 1$. Let $(x_{1,\alpha_1},\ldots,x_{s+1,\alpha_{s+1}})_{(\alpha_1,\ldots,\alpha_{s+1})\in A_1\times\ldots\times A_{s+1}}$ be a net such that $(x_{1,\alpha_1},\ldots,x_{s+1,\alpha_{s+1}}) \to (x_1,\ldots,x_{s+1})$ in $E_1\times\ldots\times E_{s+1}$. For each $i \in \{1,\ldots,s\}$

there exists a net $(y_{i,\delta_i})_{\delta_i \in D_i}$ such that $y_{i,\delta_i} \downarrow 0$ and such that for any $\delta_{i_0} \in D_i$ there exists an $\alpha_{i_0} \in A_i$ for which $|x_{i,\alpha_i}| \leq y_{i,\delta_{i_0}} + |x| := z_{\delta_{i,0}}$ whenever $\alpha_i \geq \alpha_{i_0}$. For a fixed $(\delta_{1_0}, \ldots, \delta_{s0}) \in D_1 \times \ldots \times D_s$ we have that

$$\begin{aligned}
&|T(x_{1,\alpha_1}, \ldots, x_{s,\alpha_s}, x_{s+1,\alpha_{s+1}}) - T(x_1, \ldots, x_{s+1})| \leq \\
&\leq \left|T(x_{1,\alpha_1}, \ldots, x_{s,\alpha_s}, |x_{s+1,\alpha_{s+1}} - x_{s+1}|)\right| + \left|T(x_{1,\alpha_1}, \ldots, x_{s,\alpha_s}, x_{s+1}) - T(x_1, \ldots, x_{s+1})\right| \\
&\leq \left|T(z_{\delta_{1,0}}, \ldots, z_{\delta_{s,0}}, |x_{s+1,\alpha_{s+1}} - x_{s+1}|)\right| + \left|T(x_{1,\alpha_1}, \ldots, x_{s,\alpha_s}, x_{s+1}) - T(x_1, \ldots, x_{s+1})\right|
\end{aligned}$$

for all $\alpha_i \geq \alpha_{i_0}$ $(i \in \{1, \ldots, s\})$ and $\alpha_{s+1} \in A_{s+1}$. It follows from the order continuity of T in its last argument and the induction hypothesis that $|T(x_{1,\alpha_1}, \ldots, x_{s+1,\alpha_{s+1}}) - T(x_1, \ldots, x_{s+1})| \to 0$. □

Theorem 2.7 *If F is Dedekind complete, then $\mathcal{P}_n({}^sE; F)$ is a band in $\mathcal{P}^{obv}({}^sE; F)$.*

Proof By Theorem 2.3, we have that $\mathcal{L}_{sym,n}({}^sE; F)$ is a band in $\mathcal{L}^{obv}_{sym}({}^sE; F)$. Lemmas 2.5 and 2.6 imply that $\mathcal{P}_n({}^sE; F)$ is the image of $\mathcal{L}_{sym,n}({}^sE; F)$ under the lattice isomorphism $\check{P} \mapsto P$. Therefore, $\mathcal{P}_n({}^sE; F)$ is a band in $\mathcal{P}^{obv}({}^sE; F)$. □

Let F be a Dedekind complete vector lattice. It follows from Theorem 2.7 that

$$\mathcal{P}^{obv}({}^sE; F) = \mathcal{P}_n({}^sE; F) \oplus \mathcal{P}_n({}^sE; F)^d.$$

For $P \in \mathcal{P}^{obv}({}^sE; F)$, we denote the order continuous component of P by P_n.

Theorem 2.8 *Let F be a Dedekind complete vector lattice. If $P\colon E \to F$ is a positive s-homogeneous polynomial, then*

$$P_n(x) = \inf\left\{\sup_\alpha P(x_\alpha) : 0 \leq x_\alpha \uparrow x\right\}.$$

Proof Let $x \in E^+$ and let $(x_{i,\alpha_i})_{\alpha_i \in A_i}$ be a net in E^+ such that $x_{i,\alpha_i} \uparrow x$ $(i \in \{1, \ldots, s\})$. For each $\alpha = (\alpha_1, \ldots, \alpha_s) \in A := A_1 \times \ldots \times A_s$, let $z_\alpha := x_{1,\alpha_1} \wedge \ldots \wedge x_{s,\alpha_s}$. Since $z_\alpha \uparrow x$ we have that

$$\inf\left\{\sup_\alpha P(x_\alpha) : 0 \leq x_\alpha \uparrow x\right\} \leq \sup_\alpha P(z_\alpha) = \sup_{(\alpha_1,\ldots,\alpha_s)} P(x_{1,\alpha_1} \wedge \ldots \wedge x_{s,\alpha_s}).$$

The positivity of $\check{P}$ implies

$$\sup_{(\alpha_1,\ldots,\alpha_s)} P(x_{1,\alpha_1} \wedge \ldots \wedge x_{s,\alpha_s}) \leq \sup_{(\alpha_1,\ldots,\alpha_s)} \check{P}(x_{1,\alpha_1}, \ldots, x_{s,\alpha_s}).$$

Thus

$$\begin{aligned}&\inf\left\{\sup_{\alpha} P(x_\alpha) : 0 \leq x_\alpha \uparrow x\right\} \leq \\ &\leq \inf\left\{\sup_{(\alpha_1,\ldots,\alpha_s)} \check{P}(x_{1,\alpha_1},\ldots,x_{s,\alpha_s}) : 0 \leq x_{1,\alpha_1} \uparrow x,\ldots,0 \leq x_{s,\alpha_s} \uparrow x\right\} \\ &= \check{P}_n(x,\ldots,x).\end{aligned}$$

On the other hand,

$$\check{P}_n(x,\ldots,x) \leq \inf\left\{\sup_{\alpha} \check{P}(x_\alpha,\ldots,x_\alpha) : 0 \leq x_\alpha \uparrow x\right\} = \inf\left\{\sup_{\alpha} P(x_\alpha) : 0 \leq x_\alpha \uparrow x\right\}.$$

Hence

$$\check{P}_n(x,\ldots,x) = \inf\left\{\sup_{\alpha} P(x_\alpha) : 0 \leq x_\alpha \uparrow x\right\}.$$

It remains to show that $\check{P}_n \circ \Delta = P_n$. Lemma 2.5 implies $\widecheck{P_n} = (\widecheck{P_n})_n \leq \check{P}_n$. Thus $P_n \leq \check{P}_n \circ \Delta$. On the other hand, $\check{P}_n \circ \Delta$ is an order continuous s-homogeneous polynomial such that $\check{P}_n \circ \Delta \leq P$. This implies $\check{P}_n \circ \Delta \leq P_n$. Therefore, $\check{P}_n \circ \Delta = P_n$. □

3 Arens Extensions

We collect some preliminary results before we prove the main theorems of this section. The proofs of the next two of lemmas are similar, though for a higher number of variables, to the proofs of Theorem 2.1(i) of [3] and Theorem 1.76 of [2], respectively.

Lemma 3.1 *If $T\colon E_1 \times \ldots \times E_s \to F$ is an s-linear map of order bounded variation, then $T^*\colon F^\sim \times E_1 \times \ldots \times E_{s-1} \to E_s^\sim$ is of order bounded variation.*

Lemma 3.2 *Let F be Dedekind complete. If $T\colon E_1 \times \ldots \times E_s \to F$ is an s-linear map of order bounded variation then $|T^*|(f, x_1,\ldots,x_{s-1}) = |T|^*(f, x_1,\ldots,x_{s-1})$ $(f \in F_n^\sim, x_i \in E_i,\ i \in \{1,\ldots,s-1\})$.*

We prove that the ith adjoint of an s-linear map of order bounded variation is order continuous in argument i $(1 \leq i \leq s)$.

Lemma 3.3 *Let F be Dedekind complete. Let $1 \leq i \leq s$. If $T\colon E_1 \times \ldots \times E_s \to F$ is an s-linear map of order bounded variation and if $f_\alpha \to f$ in $(F_n^\sim)$ then*

$$T^{[i]*}(\psi_{s-i+2},\ldots,\psi_s, f_\alpha, x_1,\ldots,x_{s-i}) \to T^{[i]*}(\psi_{s-i+2},\ldots,\psi_s, f, x_1,\ldots,x_{s-i})$$

for all $\psi_j \in (E_j^\sim)_n^\sim$ $(j \in \{s-i+2, \ldots, s\})$ *and* $x_j \in E_j$ $(j \in \{1, \ldots, s-i\})$.

Proof Let $(f_\alpha)_{\alpha\in A}$ be a net in $F_n^\sim$ such that $f_\alpha \downarrow 0$. Note that

$$|T^*(f_\alpha, x_1, \ldots, x_{s-1})|(x_s) = \sup\left\{\sum_k \left|f_\alpha\big(T(x_1, \ldots, x_{s-1}, a_k)\big)\right| : a \in \Pi x_s\right\}$$
$$\leq f_\alpha\left(|T|(x_1, \ldots, x_{s-1}, x_s)\right) \downarrow 0$$

for all $x_j \in E_j^+$ $(j \in \{1, \ldots, s\})$. Assume that the lemma holds for some natural number $i \in \{1, .., s-1\}$. It follows from Lemma 3.2 and our assumption that

$$\left|T^{[i+1]*}(\psi_{s-i+1}, \psi_{s-i+2}, \ldots, \psi_s, f_\alpha, x_1, \ldots, x_{s-i-1})\right|(x_{s-i})$$
$$\leq \left|T^{[i+1]*}\right|(\psi_{s-i+1}, \psi_{s-i+2}, \ldots, \psi_s, f_\alpha, x_1, \ldots, x_{s-i-1})(x_{s-i})$$
$$= |T|^{[i+1]*}(\psi_{s-i+1}, \psi_{s-i+2}, \ldots, \psi_s, f_\alpha, x_1, \ldots, x_{s-i-1})(x_{s-i})$$
$$= \psi_{s-i+1}\big(|T|^{[i]*}(\psi_{s-i+2}, \ldots, \psi_s, f_\alpha, x_1, \ldots, x_{s-i-1}, x_{s-i})\big) \downarrow 0$$

for all $\psi_j \in (E_j^\sim)_n^{\sim+}$ $(j \in \{s-i+1, \ldots, s\})$ and $x_j \in E_j^+$ $(j \in \{1, \ldots, s-i\})$. □

For $x \in E$, define $\hat{x} \in (E^\sim)_n^\sim$ by $\hat{x}(f) = f(x)$ $(f \in E^\sim)$. We define $\hat{E}$ to be $\{\hat{x} : x \in E\}$. For a subset S of $(E^\sim)_n^\sim$ define

$$\mathcal{I}S = \{x \in (E^\sim)_n^\sim : x_\alpha \uparrow x \text{ for some net } (x_\alpha)_{\alpha\in A} \text{ in } S\}$$

and

$$\mathcal{D}S = \{x \in (E^\sim)_n^\sim : x_\alpha \downarrow x \text{ for some net } (x_\alpha)_{\alpha\in A} \text{ in } S\}.$$

Theorem 3.4 *Let* $T : E_1 \times \ldots \times E_s \to F$ *be an s-linear map of order bounded variation. The following hold.*

1. $T^{[s+1]*}$ *is separately order continuous on* $(E_1^\sim)_n^\sim \times \ldots \times (E_s^\sim)_n^\sim$.
2. *If* $E = E_1 = \ldots = E_s$ *and if* $T : E \times \ldots \times E \to F$ *is symmetric then* $T^{[s+1]*}$ *is symmetric on* $(E^\sim)_n^\sim \times \ldots \times (E^\sim)_n^\sim$.

Proof

1. Fix $j \in \{1, \ldots, s\}$. Let $(\psi_\alpha)_{\alpha\in A}$ be a net in $(E_j^\sim)_n^\sim$ such that $\psi_\alpha \downarrow 0$. Lemma 3.2 implies

$$\left|T^{[s+1]*}(\psi_1, \ldots, \psi_\alpha, \ldots, \psi_s)\right| \leq \left|T^{[s+1-j]*}\right|^{[j]*}(\psi_1, \ldots, \psi_\alpha, \ldots, \psi_s)$$

for $\psi_i \in (E_i^\sim)_n^{\sim+}$ $(i \in \{1, \ldots, s\} \setminus \{j\})$. It follows from Lemma 3.3 that

$$\left|T^{[s+1-j]*}\right|^{[j]*}(\psi_1, \ldots, \psi_\alpha, \ldots, \psi_s) \downarrow 0.$$

Then $T^{[s+1]*}$ is separately order continuous.

2. Let $\psi_1, \ldots, \psi_s \in (E^\sim)_n^\sim$. Fix $i, j \in \{1, \ldots, s\}$ with $i \neq j$. It will suffice to prove that

$$T^{[s+1]*}(\psi_1, \ldots, \psi_i, \ldots, \psi_j, \ldots, \psi_s) = T^{[s+1]*}(\psi_1, \ldots, \psi_j, \ldots, \psi_i, \ldots, \psi_s).$$

Suppose that $\psi_i \in \mathcal{I}\hat{E}$ $(i \in \{1, \ldots, s\})$. By definition, there exists a net $(\hat{x}_{i,\alpha_i})_{\alpha_i \in A_i}$ in $\hat{E}$ such that $\hat{x}_{i,\alpha_i} \uparrow \psi_i$ $(i \in \{1, \ldots, s\})$. Then by the symmetry of T, part (1) of this theorem, and Lemma 2.6, we have that

$$T^{[s+1]*}(\hat{x}_{1,\alpha_1}, \ldots, \hat{x}_{i,\alpha_i}, \ldots, \hat{x}_{j,\alpha_j}, \ldots, \hat{x}_{s,\alpha_s}) =$$
$$= T^{[s+1]*}(\hat{x}_{1,\alpha_1}, \ldots, \hat{x}_{j,\alpha_j}, \ldots, \hat{x}_{i,\alpha_i}, \ldots, \hat{x}_{s,\alpha_s}) \to T^{[s+1]*}(\psi_1, \ldots, \psi_j, \ldots, \psi_i, \ldots, \psi_s).$$

Then

$$T^{[s+1]*}(\psi_1, \ldots, \psi_i, \ldots, \psi_j, \ldots, \psi_s) = T^{[s+1]*}(\psi_1, \ldots, \psi_j, \ldots, \psi_i, \ldots, \psi_s).$$

Similarly,

$$T^{[s+1]*}(\psi_1, \ldots, \psi_i, \ldots, \psi_j, \ldots, \psi_s) = T^{[s+1]*}(\psi_1, \ldots, \psi_j, \ldots, \psi_i, \ldots, \psi_s)$$

for $\psi_i \in \mathcal{DI}\hat{E}(i \in \{1, \ldots, s\})$. By repeating this action, once up and once down, we obtain

$$T^{[s+1]*}(\psi_1, \ldots, \psi_i, \ldots, \psi_j, \ldots, \psi_s) = T^{[s+1]*}(\psi_1, \ldots, \psi_j, \ldots, \psi_i, \ldots, \psi_s)$$

for $\psi_i \in \mathcal{DIDI}(\hat{E})$ $(i \in \{1, \ldots, s\})$. Since $\mathcal{DIDI}(\hat{E}) = (E^\sim)_n^\sim$ (see [6]), it follows that $T^{[s+1]*}$ is symmetric on $(E^\sim)_n^\sim \times \ldots \times (E^\sim)_n^\sim$. □

For an s-homogeneous polynomial $P\colon E \to F$, we define $\bar{P}\colon E^{\sim\sim} \to F^{\sim\sim}$ by $\bar{P}(\psi) = \check{P}^{[s+1]*}(\psi, \ldots, \psi)$ for all $\psi \in E^{\sim\sim}$. For readability, we denote the unique symmetric s-linear map $T\colon E^{\sim\sim} \times \ldots \times E^{\sim\sim} \to F^{\sim\sim}$ for which $\bar{P} = T \circ \Delta$ by $\bar{P}^\vee$. The following theorem implies $\bar{P}$ is the canonical extension of P to $(E^\sim)_n^\sim$.

Theorem 3.5 *Let $P\colon E \to F$ be an s-homogeneous polynomial of order bounded variation. The following hold.*

1. *$\bar{P}$ is of order bounded variation.*
2. *$\bar{P}$ is order continuous on $(E^\sim)_n^\sim$.*
3. *$\bar{P}^\vee = \check{P}^{[s+1]*}$ on $(E^\sim)_n^\sim \times \ldots \times (E^\sim)_n^\sim$.*

Proof

1. Proposition 2.1 and Lemma 3.1 imply $\check{P}^{[s+1]*}$ is of order bounded variation. Therefore, $\bar{P}$ is of order bounded variation.
2. Theorem 3.4 implies that $\check{P}^{[s+1]*}$ restricted to $(E^{\sim})_n^{\sim} \times \ldots \times (E^{\sim})_n^{\sim}$ is separately order continuous. It follows from Lemma 2.6 that $\bar{P}$ is order continuous on $(E^{\sim})_n^{\sim}$.
3. By Theorem 3.4 we have that $\check{P}^{[s+1]*}$ is symmetric on $(E^{\sim})_n^{\sim} \times \ldots \times (E^{\sim})_n^{\sim}$. It follows from the uniqueness of $\bar{P}^{\vee}$ that $\bar{P}^{\vee} = \check{P}^{[s+1]*}$ is symmetric on $(E^{\sim})_n^{\sim} \times \ldots \times (E^{\sim})_n^{\sim}$.

□

References

1. Y. Abramovich, G. Sirotkin, On order convergence of nets. Positivity **9**(3), 287–292 (2005)
2. C.D. Aliprantis, O. Burkinshaw, *Positive Operators* (Springer, Dordrecht, 2006)
3. K. Boulabiar, G. Buskes, R. Page, On some properties of bilinear maps of order bounded variation. Positivity **9**(3), 401–414 (2005)
4. G. Buskes, A. van Rooij, Bounded variation and tensor products of Banach lattices. Positivity **7**(1–2), 47–59 (2003)
5. R. Cristescu, *Ordered Vector Spaces and Linear Operators* (Abacus Press, Tunbridge, 1976)
6. D.H. Fremlin, Abstract Köthe spaces. I. Proc. Camb. Philol. Soc. **63**, 653–660, 681 (1967)
7. J. Loane, Polynomials on Riesz spaces. Ph.D. Thesis, National University of Ireland, Galway, 2007
8. H. Nakano, Product spaces of semi-ordered linear spaces. J. Fac. Sci. Hokkaido Univ. Ser. I **12**, 163–210 (1953)
9. A.R. Schep, Order continuous components of operators and measures. Indag. Math. **40**(1), 110–117 (1978)
10. M.A. Woodbury, A decomposition theorem for finitely additive set functions (preliminary report). Bull. Am. Math. Soc. **56**, 171–172 (1950)
11. A.C. Zaanen, *Introduction to Operator Theory in Riesz Spaces* (Springer, Berlin, 1997)

On the Endpoints of De Leeuw Restriction Theorems

Martijn Caspers

Dedicated to Ben de Pagter's 65th birthday

Abstract We prove a De Leeuw restriction theorem for Fourier multipliers on certain quasi-normed spaces. The proof is based on methods that were recently used in order to resolve problems on perturbations of commutators.

Keywords De Leeuw theorems · Fourier multipliers

1 Introduction

In 1965 Karel de Leeuw proved three fundamental theorems on L_p-multipliers [12]: his restriction theorem, lattice approximation theorem and compactification theorem. The strongest De Leeuw theorem is the compactification theorem, which shows that boundedness of an L_p-Fourier multiplier does not depend on the topology. More precisely, let $m \in L_\infty(\mathbb{R}^n)$ be continuous and let $m_d \in \ell_\infty(\mathbb{R}^n_{\text{disc}})$ be equal to m but then on $\mathbb{R}^n$ with the discrete topology. Then, the Fourier multipliers $T_m : L_p(\mathbb{R}^n) \to L_p(\mathbb{R}^n)$ and $T_{m_d} : L_p(\mathbb{R}^n_{\text{Bohr}}) \to L_p(\mathbb{R}^n_{\text{Bohr}})$ have the same norm. Here $\mathbb{R}^n_{\text{Bohr}}$ is the Bohr compactification of $\mathbb{R}^n$ which can be viewed as the Pontrjagin dual group of $\mathbb{R}^n_{\text{disc}}$. In order to show this, De Leeuw proved that for the discrete subgroup $\mathbb{Z}^n$ of $\mathbb{R}^n$ and $m \in L_\infty(\mathbb{R}^n)$, the symbol $m_{\mathbb{Z}^n} = m|_{\mathbb{Z}^n}$ gives rise to a Fourier multiplier with norms related by

$$\| T_{m|_{\mathbb{Z}^n}} : L_p(\mathbb{T}^n) \to L_p(\mathbb{T}^n) \| \leq \| T_m : L_p(\mathbb{R}^n) \to L_p(\mathbb{R}^n) \|, \qquad 1 \leq p < \infty. \tag{1.1}$$

M. Caspers (✉)
TU Delft, EWI/DIAM, Delft, The Netherlands
e-mail: m.p.t.caspers@tudelft.nl

G. Buskes et al. (eds.), *Positivity and Noncommutative Analysis*,
Trends in Mathematics, https://doi.org/10.1007/978-3-030-10850-2_4

The inequality (1.1) is known as De Leeuw's restriction theorem. Shortly after [12] De Leeuw theorems have been obtained for arbitrary locally compact Abelian groups by Saeki [29]. These theorems are widely applied in harmonic analysis and the analysis of singular integrals.

Recent developments in non-commutative integration motivated an extension of De Leeuw theorems in various contexts (see, e.g., [6, 8]). One of these motivations comes from the theory of perturbations of commutators. A central question in this theory (going back at least to Krein [22]) asks the following. Let $f : \mathbb{R} \to \mathbb{R}$ be Lipschitz. Is there an absolute constant C_{abs} such that for every $x \in B(H)$ and $A \in B(H)$ self-adjoint we have a commutator estimate,

$$\| [f(A), x] \| \leq C_{abs} \| [A, x] \|. \tag{1.2}$$

The question has a long history and is proven to be false [10, 17–19, 21], unless further differentiability conditions are imposed on f [1, 25] or if the uniform is replaced by the Schatten $\mathcal{S}_p$-norm [5, 27]. Important relations (in fact equivalences) between commutator estimates and non-commutative Lipschitz functions were obtained in [15, 26] and the problem was recast into the language of double operator integrals [13].

Nazarov and Peller [23] were the first ones to obtain a weak (1, 1) estimate for commutators (this would be the optimal solution to estimates of the form (1.2)). They showed that if $[A, x]$ is in $\mathcal{S}_1$ and has rank 1, then $[f(A), x] \in L_{1,\infty}$ and moreover we have the estimate

$$\| [f(A), x] \|_{1,\infty} \leq C_{abs} \| [A, x] \|_1.$$

The question whether the rank 1 condition could be removed remained open for quite some time. The question was resolved in [8] (see also [7] for the absolute value map). The proof is based on two important ingredients: (1) Parcet's semi-commutative Calderón–Zygmund theorem [2, 24], (2) a De Leeuw theorem for $L_{1,\infty}$-spaces which was implicitly proved in [8].

The aim of this text is to prove this De Leeuw theorem more explicitly and in a more general context of symmetric spaces (with conditions), see [14]. More precisely, we show that for $\mathbb{Z}^n$ as a discrete subgroup of $\mathbb{R}^n$ and $\| \; \|_\star$ a suitable norm on both $L_\infty(\mathbb{R}^n)$ and $L_\infty(\mathbb{T}^n)$ (to be made precise) we get for $m \in L_\infty(\mathbb{R}^n)$ smooth that,

$$\| T_m : L_1(\mathbb{T}^n) \to L_\star(\mathbb{T}^n) \| \preceq \| T_m : L_1(\mathbb{R}^n) \to L_\star(\mathbb{R}^n) \|.$$

We also prove the analogous statement in the completely bounded setting. Natural examples of such norms $\| \; \|_\star$ are weak L_1-norms and $M_{1,\infty}$-norms. We state some open questions in this direction in Sect. 4.

2 Preliminaries and Notation

2.1 General Notation

For von Neumann algebra theory we refer to [30]. We write $\prec$ for an inequality that holds up to some constant and $\prec_n$ for an inequality that holds up to a constant only depending on the dimension n. Let $B_r \in \mathbb{R}^n$ be the ball with radius r. For a function $f : \mathbb{R}^n \to \mathbb{C}$ and $s \in \mathbb{R}^n$ we set $f_s(t) = f(t-s)$. The circle $\mathbb{T}$ is identified with the unit circle in $\mathbb{C}$ and is equipped with the Haar measure with total mass 2π. We set harmonic functions $e_\gamma(z) = z^\gamma$, $\gamma \in \mathbb{Z}$, $z \in \mathbb{T}$. We say that a function $\widehat{f} = \sum_{\gamma \in F} c_\gamma e_\gamma$, $c_\gamma \in \mathbb{C}$ on $\mathbb{T}$ has finite frequency support if F is finite.

2.2 Symmetric Spaces

Let $\mathcal{N}$ be a semi-finite von Neumann algebra with faithful, tracial weight $\tau : \mathcal{N}^+ \to [0,\infty]$. Let $T(\mathcal{N})$ be the space of closed, densely defined operators that are affiliated with $\mathcal{N}$. A closed densely defined operator x is in $T(\mathcal{N})$ if for the polar decomposition $x = u|x|$ we have that $u \in \mathcal{N}$ and all spectral projections $E_{[0,\lambda]}(|x|)$, $\lambda > 0$ are in $\mathcal{N}$. Let $T_\tau(\mathcal{N})$ be the subspace of τ-measurable operators, namely all $x \in \mathcal{N}$ for which there exists a τ-finite projection $p \in \mathcal{N}$ such that xp is bounded. $T_\tau(\mathcal{N})$ forms a $*$-algebra with respect to the strong sum and strong product (closures of sums and products).

For $x \in T_\tau(\mathcal{N})$ we define the decreasing rearrangement function,

$$\mu_t(x) = \inf\{\|xp\| \mid p \in \mathcal{N} \text{ projection with } \tau(1-p) \leq t\}.$$

We have $\mu_t(x) = \mu_t(|x|)$. In this paper we consider quasi-norms $\| \ \|_\star$ where $\star$ is to be specified. Our prime examples are

$$L_{1,\infty}(\mathcal{N}) = \{x \in T_\tau(\mathcal{N}) \mid \|x\|_{1,\infty} < \infty\},$$

where

$$\|x\|_{1,\infty} = \sup_{t>0} t\mu_t(x), \qquad x \in T_\tau(\mathcal{N}).$$

And further,

$$M_{1,\infty}(\mathcal{N}) = \{x \in T_\tau(\mathcal{N}) \mid \|x\|_{M_{1,\infty}} < \infty\},$$

where

$$\|x\|_{M_{1,\infty}} = \sup_{t>0} \frac{1}{\log(1+t)} \int_0^t \mu_s(x) ds.$$

We have relations

$$\|x\|_{M_{1,\infty}} \leq \|x\|_{1,\infty} \leq \|x\|_1, \qquad x \in T_\tau(\mathcal{N}).$$

In general, we take $\| \ \|_\star : T_\tau(\mathcal{N}) \to [0, \infty]$ and set

$$L_\star(\mathcal{N}) = \{x \in T_\tau(\mathcal{N}) \mid \|x\|_\star < \infty\},$$

which is assumed to be a vector space. Moreover, we restrict $\| \ \|_\star$ to $L_\star(\mathcal{N})$ and assume it has the following properties:

Assumptions on $\| \ \|_\star$

1. $\| \ \|_\star$ is a (non-degenerate) quasi-norm. In particular, there is a constant $K \geq 1$ such that we have the quasi-triangle inequality,

$$\|x + y\|_\star \leq K(\|x\|_\star + \|y\|_\star), \qquad \forall x, y \in L_\star(\mathcal{N}).$$

2. $\| \ \|_\star$ satisfies the following dilation invariance property. There exists a constant $K > 0$ such that

$$K^{-1}\lambda\|x\|_\star \leq \|y\|_\star \leq K\lambda\|x\|_\star.$$

   for any $\lambda > 0$ and for any $x, y \in L_\star(\mathcal{N})$ such that $\mu_t(x) = \mu_{\lambda t}(y)$.
3. $L_\star(\mathcal{N})$ is complete with respect to $\| \ \|_\star$.

Property (2) implies that (up to a constant) the quasi-norm $\| \ \|_\star$ is rearrangement invariant. Below we need to make a stronger assumption on the norm to relate norms of different spaces.

It is clear that the $L_{1,\infty}$-quasi-norm satisfies these properties (see also [28, Lemma 1.4]). Further, so does the $M_{1,\infty}$-quasi norm by the following lemma:

Lemma 2.1 $\| \ \|_{M_{1,\infty}}$ *satisfies properties* (1), (2) *and* (3) *above.*

Proof By [4, Theorem 2.1] (see also [3, Theorem 4.5]) the $M_{1,\infty}$-quasi norm on $T_\tau(\mathcal{N})$ is equivalent to the norm,

$$\|x\|_{\mathcal{Z}_1} = \sup_{p>1}(p-1)\|x\|_p, \qquad x \in \mathcal{N}.$$

Suppose that x and y are elements of $\mathcal{N}$ and suppose there exists $\lambda > 0$ such that for the decreasing rearrangements for every $t > 0$ we have $\mu_t(x) = \mu_{\lambda t}(y)$. Then because the $\| \ \|_p$-norm has the property that $\|x\|_p = \lambda^{1/p}\|y\|_p$ we find that $\|x\|_{\mathcal{Z}_1} = \|y\|_{\mathcal{Z}_1}$. By equivalence of $\mathcal{Z}_1$- and $M_{1,\infty}$-norms this shows that there exists an absolute constant $K > 0$ such that

$$K^{-1}\lambda\|x\|_{M_{1,\infty}} \leq \|y\|_{M_{1,\infty}} \leq K\lambda\|x\|_{M_{1,\infty}}.$$

Completeness of $\mathcal{Z}_1$ (and hence $M_{1,\infty}$) follows from completeness of L_p-spaces and the quasi-norm property may be derived from [28, Lemma 1.4]. □

Remark 2.2 Suppose that for all $x \in \mathcal{N}$ we have that $\|x\|_p \leq \|x\|_\star$ for some $1 \leq p \leq \infty$. Then $p = 1$ because otherwise (2) would be violated. This shows that the spaces to which our De Leeuw theorems apply can be viewed as end-point spaces at $p = 1$.

2.3 *Fourier Multipliers*

Let G be a locally compact Abelian group. We shall only be concerned with $\mathsf{G} = \mathbb{R}^n$ and $\mathsf{G} = \mathbb{T}^n$. Let $\widehat{\mathsf{G}}$ be the Pontrjagin dual group of characters. So $\widehat{\mathbb{R}^n} = \mathbb{R}^n$ and $\widehat{\mathbb{Z}^n} = \mathbb{T}^n$. Further, let $\mathcal{F}$ be the Fourier transform $L_2(\mathsf{G}) \to L_2(\widehat{\mathsf{G}})$. Consider a function $m \in L_\infty(\mathsf{G})$ and set $T_m : L_2(\mathsf{G}) \to L_2(\mathsf{G})$ by $T_m = \mathcal{F}^{-1} \circ m \circ \mathcal{F}$ where we view m as a multiplication operator. Concretely, $T_m = m(\nabla)$ where $\nabla = -i(\frac{\partial}{\partial \xi_1}, \ldots, \frac{\partial}{\partial \xi_n})$ is the gradient operator on either $\mathsf{G} = \mathbb{R}^n$ or $\mathsf{G} = \mathbb{T}^n$.

If the operator T_m extends to a bounded map $L_p(\mathsf{G}) \to L_p(\mathsf{G})$ for some $1 \leq p < \infty$, then we call m an L_p-Fourier multiplier or briefly an L_p-multiplier. We have the translational behaviour of an L_p-multiplier $m \in L_\infty(\mathbb{R}^n)$:

$$T_m(f e_s) = T_{m_s}(f) e_s, \qquad s \in \mathbb{R}^n, f \in L_p(\mathbb{R}^n), \tag{2.1}$$

with $m_s(t) = m(t - s)$. Similarly, we will say that m is an $L_1 - L_\star$ multiplier if for every $f \in L_1 \cap L_2$ we have that $T_m(f) \in L_\star$ and further there exists a constant $K > 0$ such that $\|T_m(f)\|_\star \leq K\|f\|_1$. The infimum over such $K > 0$ is then the norm $\|T_m : L_1 \to L_\star\|$.

Remark 2.3 Multipliers $L_1 \to L_{1,\infty}$ are extensively studied in the context of Calderón–Zygmund theory [20]. The (weaker) multipliers $L_1 \to M_{1,\infty}$ naturally occur in problems finding the best constants of certain commutative and non-commutative estimates. See [5, Corollary 5.6] for such an application.

3 De Leeuw Restriction Theorems at the Endpoints

3.1 *Symmetric Quasi-Norms*

Let $\mathcal{M}$ be a semi-finite von Neumann algebra. In this section we presume that $\| \ \|_\star$ is a quasi-norm on either $L_\star(\mathcal{M} \otimes L_\infty(\mathbb{R}^n))$ and $L_\star(\mathcal{M} \otimes L_\infty(\mathbb{T}^n))$ satisfying the properties (1) and (2). We will not distinguish in notation to view $\| \ \|_\star$ on either

$L_\star(\mathcal{M} \otimes L_\infty(\mathbb{R}^n))$ or $L_\star(\mathcal{M} \otimes L_\infty(\mathbb{T}^n))$. However, we will impose the following natural condition relating quasi-norms of different spaces:

Assumption If $x \in L_\star(\mathcal{M} \otimes L_\infty(\mathbb{R}^n))$ and $y \in L_\star(\mathcal{M} \otimes L_\infty(\mathbb{T}^n))$ have decreasing rearrangements such that $\mu_t(x) \leq \mu_t(y)$, then $\|x\|_\star \leq \|y\|_\star$. In particular, the quasi-norms on the spaces $L_\star(\mathcal{M} \otimes L_\infty(\mathbb{R}^n))$ and $L_\star(\mathcal{M} \otimes L_\infty(\mathbb{T}^n))$ themselves are rearrangement invariant (i.e. only depend on the decreasing rearrangement).

For functions $f_i : \mathbb{R}^{n_i} \to \mathbb{C}$ we write $f_1 \otimes f_2 : \mathbb{R}^{n_1+n_2} \to \mathbb{C}$ for $(f_1 \otimes f_2)(s_1, s_2) = f_1(s_1) f_2(s_2)$. Let $G_k^{(n)} : \mathbb{R}^n \to \mathbb{C}$ be the L_1-normalized Gaussian function set by

$$G_k(\xi) = \frac{1}{k\sqrt{\pi}} \exp(-|\xi|^2/k^2), \qquad \xi \in \mathbb{R}.$$

and then $G_k^{(n)} = G_k^{\otimes n}$. For $f : \mathbb{T}^n \to \mathcal{M}$ set the periodization $\mathrm{per}(f) : \mathbb{R}^n \to \mathcal{M}$ by $f(x) = f(x \bmod 2\pi)$ where we identify $\mathbb{T}^n$ with the block interval $[0, 2\pi)^n$ and the mod 2π is taken coordinate-wise. Then set

$$\Phi_k(f)(t) = \mathrm{per}(f)(t) G_k^{(n)}(t), \qquad k \in \mathbb{N}_{\geq 1}, t \in \mathbb{R}^n.$$

We say that an element $x \in \mathcal{M} \otimes L_\infty(\mathbb{T}^n)$ has finite frequency support if $x = \sum_{\gamma \in F} x_\gamma \otimes e_\gamma$ with F finite.

Lemma 3.1 *For $x \in L_{1,\infty}(\mathcal{M} \otimes L_\infty(\mathbb{T}^n))$ we have that $\Phi_k(x) \in L_{1,\infty}(\mathcal{M} \otimes L_\infty(\mathbb{R}^n))$. Moreover, $\|x\|_{1,\infty} \prec_n \|\Phi_k(x)\|_{1,\infty}$.*

Proof As $\Phi_k(|x|) = |\Phi_k(x)|$ we may assume without loss of generality that $x \geq 0$ by considering $|x|$ instead. Set constants $c_n = \pi^{n/2} e^{-1}$. We estimate

$$c_n k^{-n} \sum_{l \in \mathbb{Z}^n, |l| \leq k-1} \chi_{2\pi l + [0,2\pi)^n} \leq G_k^{(n)},$$

so that

$$c_n k^{-n} \sum_{l \in \mathbb{Z}^n, |l| \leq k-1} \chi_{2\pi l + [0,2\pi)^n} \mathrm{per}(x) \leq G_k^{(n)} \mathrm{per}(x).$$

Taking $\| \ \|_\star$-quasi-norms we find

$$c_n k^{-n} \| \sum_{l \in \mathbb{Z}^n, |l| \leq k-1} \chi_{2\pi l + [0,2\pi)^n} \mathrm{per}(x) \|_\star \leq \|G_k^{(n)} \mathrm{per}(x)\|_\star = \|\Phi_k(x)\|_\star. \qquad (3.1)$$

Set

$$\lambda_n = \#\{l \in \mathbb{Z}^n \mid |l| \leq k-1\}.$$

Note that $\lambda_n \approx (k-1)^n$. Then,

$$\mu_t(c_n k^{-n} \sum_{l \in \mathbb{Z}^n, |l| \leq k-1} \chi_{2\pi l + [0,2\pi)^n} \mathrm{per}(x)) = c_n k^{-n} \mu_{\lambda_n^{-1} t}(x).$$

Therefore, by our assumptions on $\| \ \|_\star$ the left-hand side of (3.1) is up to a constant equal to

$$c_n k^{-n} \lambda_n \|x\|_\star \approx c_n k^{-n} (k-1)^n \|x\|_\star.$$

As $k^{-n}(k-1)^n \to 1$ if $k \to \infty$ this concludes that for all k we have $c_n \|x\|_{1,\infty} \prec \|\Phi_k(x)\|_{1,\infty}$. This concludes the lemma. □

Lemma 3.2 *For $x \in L_1(\mathcal{M} \otimes L_\infty(\mathbb{T}^n))$ we have that $\Phi_k(x) \in L_1(\mathcal{M} \otimes L_\infty(\mathbb{R}^n))$. Moreover, $\lim_k \|\Phi_k(x)\|_1 = \|x\|_1$.*

Proof Without loss of generality we may assume that x is positive by replacing it with $|x|$. For $l \in \mathbb{Z}^n$ we let again $A_l = 2\pi l + [0, 2\pi)^n \subseteq \mathbb{R}^n$. We set the functions G_k^+ and G_k^- on $\mathbb{Z}^n$ by

$$G_k^+(l) = \max_{s \in A_l} G_k^{(n)}(s), \qquad G_k^-(l) = \min_{s \in A_l} G_k^{(n)}(s).$$

We estimate

$$\sum_{l \in \mathbb{Z}^n} \mathrm{per}(x) \chi_{A_l} G_k^-(l) \leq \Phi_k(x) \leq \sum_{l \in \mathbb{Z}^n} \mathrm{per}(x) \chi_{A_l} G_k^+(l). \tag{3.2}$$

And so the same estimates hold after taking the L_1-norm. We further get

$$\| \sum_{l \in \mathbb{Z}^n} \mathrm{per}(x) \chi_{A_l} G_k^\pm(l) \|_1 = \| \bigoplus_{l \in \mathbb{Z}^n} G_k^\pm(l) x \|_{L_1(\oplus_{l \in \mathbb{Z}^n} \mathcal{M} \otimes L_\infty(\mathbb{T}^n))} = \|x\|_1 \sum_{l \in \mathbb{Z}^n} G_k^\pm(l).$$

Then as we have more and more refined approximations,

$$\sum_{l \in \mathbb{Z}^n} G_k^\pm(l) = \frac{1}{k^n} \sum_{l \in \mathbb{Z}^n} \max_{s \in A_l} G_1^\pm\left(\frac{s}{k}\right)$$

has limit $\|G_1^{(n)}\|_1 = 1$ for both $\pm$ either $+$ or $-$. We conclude from (3.2) that

$$\|x\|_1 \leq \lim_k \|\Phi_k(x)\|_1 \leq \|x\|_1.$$

□

We define the Fourier algebra (see also [16]) as

$$\mathcal{A}(\mathbb{R}^n) = \mathcal{F}(L_1(\widehat{\mathbb{R}^n})).$$

So $\mathcal{A}(\mathbb{R}^n)$ consists of all functions in $C_0(\mathbb{R}^n)$ whose (distributional) Fourier transform lies in $L_1(\mathbb{R}^n)$. Every $m \in \mathcal{A}(\mathbb{R}^n)$ is in particular an $L_1 \to L_1$ Fourier multiplier and $T_m(f) = \widehat{m} * f$ with $\widehat{m} = \mathcal{F}(m)$. Moreover, such multipliers are completely bounded and therefore for any (semifinite) von Neumann algebra $\mathcal{M}$ we obtain a bounded map $(\mathrm{id}_{\mathcal{M}} \otimes T_m) : L_1(\mathcal{M} \otimes L_\infty(\mathbb{R}^n)) \to L_1(\mathcal{M} \otimes L_\infty(\mathbb{R}^n))$, see also [11] for these results as well as far-reaching generalizations. For $m \in C_0(\mathbb{R}^n)$ we write $m_d \in c_0(\mathbb{Z}^n)$ for its discretization $m_d = m|_{\mathbb{Z}^n}$ (restriction to $\mathbb{Z}^n$).

Proposition 3.3 *Let $m \in \mathcal{A}(\mathbb{R}^n)$ and let $x \in L_\infty(\mathbb{T}^n, \mathcal{M}) \simeq \mathcal{M} \otimes L_\infty(\mathbb{T}^n)$ have finite frequency support. We have that*

$$\|\Phi_k((id_{\mathcal{M}} \otimes T_{m_d})(x)) - (id_{\mathcal{M}} \otimes T_m)(\Phi_k(x))\|_1 \to 0.$$

Proof By the triangle inequality for $\|\cdot\|_1$ and the translation behaviour (2.1) of Fourier multipliers we may assume without loss of generality that $x = x_0 \otimes e_0$, $x_0 \in \mathcal{M}$. In fact, this shows that we may assume that $\mathcal{M} = \mathbb{C}$ and $x = e_0$. Suppose that $\widehat{m} = \mathcal{F}(m)$ is positive and $\|\widehat{m}\|_1 = 1$. Each multiplier in $\mathcal{A}(\mathbb{R}^n)$ may be written as a linear combination of four positive multipliers in $\mathcal{A}(\mathbb{R}^n)$. It follows that $m(0) = \|\widehat{m}\|_1 = 1$. So it remains to prove that

$$\|G_k^{(n)} - T_m(G_k^{(n)})\|_1 \to 0.$$

Let $\varepsilon > 0$ and let $B_r \subseteq \mathbb{R}^n$ be a ball such that $\|\widehat{m}\chi_{B_r} - \widehat{m}\|_1 \le \varepsilon$. Set $\widehat{m}_r = \widehat{m}\chi_{B_r}$. Then,

$$\|(\widehat{m} - \widehat{m}_r) * G_k^{(n)}\|_1 \le \|\widehat{m} - \widehat{m}_r\|_1 \|G_k^{(n)}\|_1 \le \varepsilon.$$

Further, we have

$$T_{m_r}(G_k^{(n)}) = \widehat{m}_r * G_k^{(n)},$$

and

$$\min_{|s| \le r} G_k^{(n)}(t+s) \le (\widehat{m}_r * G_k^{(n)})(t) \le \max_{|s| \le r} G_k^{(n)}(t+s).$$

We therefore get, with G_k' the gradient of G_k,

$$G_k^{(n)}(t) - r \max_{|s| \le r} |G_k^{(n)'}(t+s)| \le (\widehat{m}_r * G_k^{(n)})(t) \le G_k^{(n)}(t) + r \max_{|s| \le r} |G_k^{(n)'}(t+s)|.$$

We find

$$|G_k^{(n)} - \widehat{m}_r * G_k^{(n)}| \le r \max_{|s| \le r} |G_k^{(n)'}(\,\cdot\, + s)|.$$

Further, the right-hand side of this inequality converges to 0 in the $\|\cdot\|_1$-norm as $k \to \infty$. In all we conclude that for $\varepsilon > 0$ for k large we have

$$\begin{aligned}\|G_k^{(n)} - T_m(G_k^{(n)})\|_1 \leq& \|G_k^{(n)} - T_{m_r}(G_k^{(n)})\|_1 + \|(T_m - T_{m_r})(G_k^{(n)})\|_1 \\ \leq& \|G_k^{(n)} - \widehat{m}_r * G_k^{(n)}\|_1 + \|\widehat{m} - \widehat{m}_r\|_1 \leq 2\varepsilon.\end{aligned}$$

□

Now we arrive at the following De Leeuw restriction theorem.

Theorem 3.4 *Let $m \in \mathcal{A}(\mathbb{R}^n)$ and let $m_d = m|_{\mathbb{Z}^n} \in c_0(\mathbb{Z}^n)$ be its restriction. If $id_{\mathcal{M}} \otimes T_m$ is bounded as a multiplier $L_1(\mathcal{M} \otimes L_\infty(\mathbb{R}^n)) \to L_\star(\mathcal{M} \otimes L_\infty(\mathbb{R}^n))$. Then also $id_{\mathcal{M}} \otimes T_{m_d}$ is bounded as a multiplier $L_1(\mathcal{M} \otimes L_\infty(\mathbb{T}^n)) \to L_\star(\mathcal{M} \otimes L_\infty(\mathbb{T}^n))$. Further,*

$$\begin{aligned}&\|id_{\mathcal{M}} \otimes T_{m_d} : L_1(\mathcal{M} \otimes L_\infty(\mathbb{T}^n)) \to L_{1,\infty}(\mathcal{M} \otimes L_\infty(\mathbb{T}^n))\| \\ &\quad \prec \|id_{\mathcal{M}} \otimes T_m : L_1(\mathcal{M} \otimes L_\infty(\mathbb{R}^n)) \to L_{1,\infty}(\mathcal{M} \otimes L_\infty(\mathbb{R}^n))\|.\end{aligned}$$

Proof Let $x \in L_1(\mathcal{M} \otimes \mathbb{T}^n)$ with finite frequency support. We get by Lemma 3.1, Proposition 3.3 and Lemma 3.2 respectively that

$$\begin{aligned}\|(\mathrm{id}_{\mathcal{M}} \otimes T_{m_d})(x)\|_{1,\infty} \prec& \limsup_k \|\Phi_k((\mathrm{id}_{\mathcal{M}} \otimes T_{m_d})(x))\|_{1,\infty} \\ \prec& \limsup_k \|(\mathrm{id}_{\mathcal{M}} \otimes T_m)(\Phi_k(x))\|_{1,\infty} \\ &+ \quad \|\Phi_k((\mathrm{id}_{\mathcal{M}} \otimes T_{m_d})(x)) - (\mathrm{id}_{\mathcal{M}} \otimes T_m)(\Phi_k(x))\|_{1,\infty} \\ \leq& \limsup_k \|(\mathrm{id}_{\mathcal{M}} \otimes T_m)(\Phi_k(x))\|_{1,\infty} \\ &+ \quad \|\Phi_k((\mathrm{id}_{\mathcal{M}} \otimes T_{m_d})(x)) - (\mathrm{id}_{\mathcal{M}} \otimes T_m)(\Phi_k(x))\|_1 \\ =& \limsup_k \|(\mathrm{id}_{\mathcal{M}} \otimes T_m)(\Phi_k(x))\|_{1,\infty} \\ \leq& \|(\mathrm{id} \otimes T_m) : L_1 \to L_{1,\infty}\| \limsup_k \|\Phi_k(x)\|_1 \\ =& \|T_m : L_1 \to L_{1,\infty}\| \|x\|_1.\end{aligned} \tag{3.3}$$

This concludes our proof from the following density argument. The elements in $L_1(\mathcal{M} \otimes L_\infty(\mathbb{T}^n))$ with finite frequency support are dense in $L_1(\mathcal{M} \otimes L_\infty(\mathbb{T}^n))$ (this follows directly from Fejér's theorem [9, Lemma A.2] for example). Then let $x \in L_1(\mathcal{M} \otimes \mathbb{T}^n)$ and let $x_n \in L_1(\mathcal{M} \otimes \mathbb{T}^n)$ be such that $x_n \to x$ and x_n having finite frequency support. Then by (3.3) we see that $(\mathrm{id}_{\mathcal{M}} \otimes T_{m_d})(x_n)$ is Cauchy and hence converges within $L_\star(\mathcal{M} \otimes L_\infty(\mathbb{T}^n))$ which is assumed to be complete. Say that the limit is $T(x)$. If x is also in $L_2(\mathcal{M} \otimes \mathbb{T}^n)$, then also the approximating sequence x_n

can be taken in $L_2(\mathcal{M} \otimes \mathbb{T}^n)$ [9, Remark A.1]. Then $T(x) = (\mathrm{id}_{\mathcal{M}} \otimes T_m)(x)$. So the multiplier is an extension of the L_2-multiplier. □

We extend the result to smooth multipliers.

Theorem 3.5 *Let $m \in L_\infty(\mathbb{R}^n \backslash \{0\})$ be smooth on $\mathbb{R}^n \backslash \{0\}$ and let $m_d = m|_{\mathbb{Z}^n} \in \ell^\infty(\mathbb{Z}^n)$ be its restriction. If $id_{\mathcal{M}} \otimes T_m$ is bounded as a multiplier $L_1(\mathcal{M} \otimes L_\infty(\mathbb{R}^n)) \to L_\star(\mathcal{M} \otimes L_\infty(\mathbb{R}^n))$, then also $id_{\mathcal{M}} \otimes T_{m_d}$ is bounded as a multiplier $L_1(\mathcal{M} \otimes L_\infty(\mathbb{T}^n)) \to L_\star(\mathcal{M} \otimes L_\infty(\mathbb{T}^n))$. Moreover,*

$$\begin{aligned} &\|id_{\mathcal{M}} \otimes T_{m_d} : L_1(\mathcal{M} \otimes L_\infty(\mathbb{T}^n)) \to L_\star(\mathcal{M} \otimes L_\infty(\mathbb{T}^n))\| \\ &\prec \|id_{\mathcal{M}} \otimes T_m : L_1(\mathcal{M} \otimes L_\infty(\mathbb{R}^n)) \to L_\star(\mathcal{M} \otimes L_\infty(\mathbb{R}^n))\|. \end{aligned}$$

Proof Let φ_k be Schwartz functions such that $\varphi_k(0) = 0$, such that for all $s \in B_k \backslash B_{\frac{1}{k}}$ we have $\varphi_k(s) = 1$ and finally such that $\|\widehat{\varphi}_k\|_1$ is uniformly bounded in k, say by a constant A. Such functions exist as was justified in [8, Footnote 5]. We have that

$$\begin{aligned} \|T_{m\varphi_k} : L_1 \to L_\star\| \leq& \|T_{\varphi_k} : L_1 \to L_1\| \|T_m : L_1 \to L_\star\| \\ \leq& \|\widehat{\varphi}_k\| \|T_m : L_1 \to L_\star\|. \end{aligned}$$

By construction $m\varphi_k$ is Schwartz. Let $x \in L_1(\mathcal{M} \otimes L_\infty(\mathbb{T}^n))$ with finite frequency support. Let $\mathcal{E}_0$ be the conditional expectation of $L_\infty(\mathbb{T}^n)$ onto $\mathbb{C}1_{\mathbb{T}^n}$. It extends to a complete contraction of $L_1(\mathbb{T}^n)$ onto $\mathbb{C}1_{\mathbb{T}}$. Set $\mathcal{E}_0(x) = x_0$ and $x_1 = x - x_0$. So,

$$\|(\mathrm{id}_{\mathcal{M}} \otimes T_{m_d})(x)\|_\star \leq \|(\mathrm{id}_{\mathcal{M}} \otimes T_{m_d})(x_0)\|_\star + \|(\mathrm{id}_{\mathcal{M}} \otimes T_{m_d})(x_1)\|_\star.$$

We have that

$$\|(\mathrm{id}_{\mathcal{M}} \otimes T_{m_d})(x_0)\|_\star \prec |m_d(0)| \|x_0\|_1 \leq |m_d(0)| \|x\|_1.$$

Further, we get for k large such that the frequency support of x_1 lies in B_k that

$$\begin{aligned} \|(\mathrm{id}_{\mathcal{M}} \otimes T_{m_d})(x_1)\|_\star =& \|(\mathrm{id}_{\mathcal{M}} \otimes T_{m_d\varphi_k})(x_1)\|_\star \\ \prec& \|(\mathrm{id}_{\mathcal{M}} \otimes T_{m\varphi_k}) : L_1 \to L_\star\| \|x_1\|_1 \\ \prec& \|\widehat{\varphi}_k\| \|(\mathrm{id}_{\mathcal{M}} \otimes T_m) : L_1 \to L_\star\| \|x_1\|_1 \\ \leq& A \|(\mathrm{id}_{\mathcal{M}} \otimes T_m) : L_1 \to L_\star\| \|x\|_1. \end{aligned}$$

We conclude by density of the functions with finite frequency support in $L_1(\mathbb{R}^n)$, just as in the proof of Theorem 3.4. □

We single out the commutative case.

Corollary 3.6 *Let $m \in L_\infty(\mathbb{R}^n\backslash\{0\})$ be smooth on $\mathbb{R}^n\backslash\{0\}$ and let $m_d = m|_{\mathbb{Z}^n} \in \ell^\infty(\mathbb{Z}^n)$ be its restriction. If T_m is $L_1 \to L_\star$-bounded, then T_{m_d} is $L_1 \to L_\star$-bounded. Moreover,*

$$\|T_{m_d} : L_1(\mathbb{T}^n) \to L_\star(\mathbb{T}^n)\| \prec \|T_m : L_1(\mathbb{R}^n) \to L_\star(\mathbb{R}^n)\|.$$

4 Open Questions

We conclude this paper with a couple of questions which we believe are interesting.

Question Does a De Leeuw theorem hold for general multipliers $m \in L_\infty(\mathbb{R}^n)$ of weak type $(1, 1)$? That is, can one drop additional (smoothness) assumptions as in Theorems 3.4 and 3.5.

Question The classical de Leeuw theorem, the constant in Theorem 3.4 that is incorporated in the symbol $\prec$ is in fact 1, i.e. one gets a true inequality $\leq$. We do not know if this is true in the weak type $(1, 1)$ case.

Question Naturally the question arises if the lattice approximation and the compactification theorem of De Leeuw hold at the endpoints.

Question In [6] a non-commutative De Leeuw restriction theorem was proved for multipliers T_m acting on $L_p(\widehat{\mathsf{G}})$. Here G is a locally compact group and $L_p(\widehat{\mathsf{G}})$ is the non-commutative L_p-space of its group von Neumann algebra. If Γ is a discrete subgroup of G and G has small almost invariant neighbourhoods with respect to Γ (see [6] for the precise definition), then a De Leeuw theorem holds for L_p-spaces. The weak $(1, 1)$ restriction theorem for any class of multipliers on such non-commutative spaces is open. For example, is there a De Leeuw restriction theorem for weak $(1, 1)$ type multipliers for the Heisenberg group?

References

1. M.S. Birman, M.Z. Solomyak, *Double Stieltjes Operator Integrals (Russian)* (Probl. Math. Phys., Izdat. Leningrad. Univ., Leningrad, 1966), pp. 33–67. English translation in: Topics in Mathematical Physics, vol. 1 (Consultants Bureau Plenum Publishing Corporation, New York, 1967), pp. 25–54
2. L. Cadilhac, Weak boundedness of Calderón–Zygmund operators on noncommutative L_1-spaces. J. Funct. Anal. **274**(3), 769–796 (2018)
3. A. Carey, A. Rennie, A. Sedaev, F. Sukochev, The Dixmier trace and asymptotics of zeta functions. J. Funct. Anal. **249**(2), 253–283 (2007)
4. A. Carey, V. Gayral, A. Rennie, F. Sukochev, Integration on locally compact noncommutative spaces. J. Funct. Anal. **263**(2), 383–414 (2012)
5. M. Caspers, S. Montgomery-Smith, D. Potapov, F. Sukochev, The best constants for operator Lipschitz functions on Schatten classes. J. Funct. Anal. **267**(10), 3557–3579 (2014)

6. M. Caspers, J. Parcet, M. Perrin, E. Ricard, Noncommutative de Leeuw theorems. Forum Math. Sigma **3**, e21 (2015)
7. M. Caspers, D. Potapov, F. Sukochev, D. Zanin, Weak type estimates for the absolute value mapping. J. Operator Theory **73**(2), 361–384 (2015)
8. M. Caspers, D. Potapov, F. Sukochev, D. Zanin, Weak type commutator and Lipschitz estimates: resolution of the Nazarov-Peller conjecture. arXiv preprint, arXiv:1506.00778
9. M. Caspers, F. Sukochev, D. Zanin, Weak type operator Lipschitz and commutator estimates for commuting tuples. arXiv preprint, arXiv:1703.03089
10. E. Davies, Lipschitz continuity of functions of operators in the Schatten classes. J. Lond. Math. Soc. **37**, 148–157 (1988)
11. J. de Cannière, U. Haagerup, Multipliers of the Fourier algebras of some simple Lie groups and their discrete subgroups. Am. J. Math. **107**(2), 455–500 (1985)
12. K. de Leeuw, On L_p multipliers. Ann. Math. (2) **81**, 364–379 (1965)
13. B. de Pagter, H. Witvliet, F.A. Sukochev, Double operator integrals. J. Funct. Anal. **192**(1), 52–111 (2002)
14. P.G. Dodds, T. Dodds, B. de Pagter, Noncommutative Banach function spaces. Math. Z. **201**(4), 583–597 (1989)
15. P.G. Dodds, B. de Pagter, F.A. Sukochev, Lipschitz continuity of the absolute value and Riesz projections in symmetric operator spaces. J. Funct. Anal. **148**(1), 28–69 (1997)
16. P. Eymard, L'algèbre de Fourier d'un groupe localement compact. Bull. Soc. Math. France **92**, 181–236 (1964)
17. Y. Farforovskaya, An estimate of the nearness of the spectral decompositions of self-adjoint operators in the Kantorovich–Rubinstein metric. Vestnik Leningrad. Univ. **22**(19), 155–156 (1967)
18. Y. Farforovskaya, The connection of the Kantorovich–Rubinstein metric for spectral resolutions of selfadjoint operators with functions of operators. Vestnik Leningrad. Univ. **23**(19), 94–97 (1968)
19. Y. Farforovskaya, An example of a Lipschitz function of self-adjoint operators with nonnuclear difference under a nuclear perturbation. Zap. Nauchn. Sem. Leningrad. Otdel. Mat. Inst. Steklov. **30**, 146–153 (1972)
20. L. Grafakos, *Classical and Modern Fourier Analysis* (Pearson Education, Upper Saddle River, 2004), pp. xii+931
21. T. Kato, Continuity of the map $S \mapsto |S|$ for linear operators. Proc. Jpn. Acad. **49**, 157–160 (1973)
22. M. Krein, Some new studies in the theory of perturbations of self-adjoint operators, in *First Math. Summer School, Part I (Russian), Izdat. "Naukova Dumka", Kiev* (1964), pp. 103–187
23. P. Nazarov, V. Peller, Lipschitz functions of perturbed operators. C. R. Math. Acad. Sci. Paris **347**(15–16), 857–862 (2009)
24. J. Parcet, Pseudo-localization of singular integrals and noncommutative Calderón–Zygmund theory. J. Funct. Anal. **256**(2), 509–593 (2009)
25. V. Peller, Hankel operators in the theory of perturbations of unitary and selfadjoint operators. Funktsional. Anal. i Prilozhen. **19**(2), 37–51, 96 (1985)
26. D. Potapov, F. Sukochev, Lipschitz and commutator estimates in symmetric operator spaces. J. Operator Theory **59**(1), 211–234 (2008)
27. D. Potapov, F. Sukochev, Operator-Lipschitz functions in Schatten-von Neumann classes. Acta Math. **207**(2), 375–389 (2011)
28. N. Randrianantoanina, A weak type inequality for non-commutative martingales and applications. Proc. Lond. Math. Soc. (3) **91**(2), 509–542 (2005)
29. S. Saeki, Translation invariant operators on groups. Tohoku Math. J. (2) **22**, 409–419 (1970)
30. M. Takesaki, *Theory of Operator Algebras. I* (Springer, Berlin, 2002), pp. xx+415

Lebesgue Topologies and Mixed Topologies

Jurie Conradie

To Ben de Pagter on the occasion of his sixty-fifth birthday

Abstract Lebesgue topologies on Riesz spaces are topologies that are order continuous: every net that is order convergent to 0 converges to 0 in the topology. In this paper we give an overview of various types of Lebesgue topologies, replacing order convergent nets by nets that are unboundedly order convergent and bounded in some sense. In the case of order boundedness we obtain Lebesgue topologies as usually defined. In normed Riesz spaces, norm-boundedness can be used to define another class of Lebesgue topologies, the uniformly Lebesgue topologies. The relationship between these two types of Lebesgue topologies and the corresponding pre-Lebesgue topologies is investigated, as well as the role of unbounded Lebesgue topologies. We identify the finest uniformly Lebesgue topologies on a Banach lattice, and show that it may be regarded as a mixed topology, and that the coarsest Hausdorff Lebesgue topology also enters into the picture.

Keywords Lebesgue topologies · Unbounded order convergence · Mixed topologies

The classical Lebesgue dominated convergence theorem establishes a link between two types of convergence for sequences of measurable functions: almost everywhere convergence, and convergence with respect to the L_1-norm. For sequences dominated by an integrable function, convergence in the norm follows from almost everywhere convergence. It is possible to consider this type of implication in the abstract setting of normed Riesz spaces (or normed vector lattices). In this setting dominated almost everywhere convergence is replaced by order convergence, and

---

J. Conradie (✉)
Department of Mathematics and Applied Mathematics, University of Cape Town, Cape Town, South Africa
e-mail: jurie.conradie@uct.ac.za

G. Buskes et al. (eds.), *Positivity and Noncommutative Analysis*,
Trends in Mathematics, https://doi.org/10.1007/978-3-030-10850-2_5

norms for which such convergence implies norm convergence are called *order continuous*. There is no need to confine such a generalization to norms. Linear topologies on Riesz spaces that are in some sense compatible with the lattice structure and are order continuous have been studied under a variety of names. Following Fremlin ([11], see also [2]), we shall call them *Lebesgue topologies*.

In Riesz spaces of measurable functions, almost everywhere convergence is, in general, not equivalent to order convergence, but rather to *unbounded order convergence*. The abstract equivalent, in Riesz spaces, of dominated almost everywhere convergence is therefore "order bounded unbounded order convergence". It is from this perspective that we look at Lebesgue topologies in this paper. In a recent paper [14], a related form of convergence in Banach lattices, where order boundedness of the sequence (or net) is replaced by norm-boundedness, was considered in an investigation of a duality theory for unbounded order convergence. Topologies that are continuous with respect to this type of convergence are closely related to the *uniform Lebesgue topologies* introduced in [17] (and also studied in [6]).

The aim of this paper is to give a general framework into which these different types of Lebesgue topologies can be fitted. This is achieved by introducing the notion of order convergence with respect to a bornology, and using such order convergences to define corresponding Lebesgue topologies. To every such Lebesgue topology there corresponds a pre-Lebesgue topology. In the definition of pre-Lebesgue topologies the unboundedly order convergent nets are replaced by disjoint sequences. Locally convex topologies of this kind have equicontinuous sets that are order precompact. This fact, and the relationship between Lebesgue and pre-Lebesgue topologies, makes it possible to characterize Lebesgue topologies that are the finest of their kind as mixed topologies.

The first section of this paper is devoted to an overview of various types of order convergence, and the corresponding order continuous linear functionals. In the second section we look at the (usual) Lebesgue topologies obtained when we consider the bornology of order bounded sets, while the third is devoted to a more in-depth study of the uniformly Lebesgue topologies that result from using the bornology of norm-bounded sets in a normed Riesz space. The final section introduces mixed topologies into the picture, and we show that the characterization of the finest uniformly Lebesgue topology as a mixed topology helps in answering questions about the possibility of topologizing norm-bounded unbounded order convergence.

The terminology for Riesz spaces (vector lattices) and topologies on these will be that of [11] and [2], to which we also refer the reader for any unexplained notions. Some of the more important notions and results are summarized below.

All Riesz spaces will be assumed to be Archimedean. The positive cone of a Riesz space E will be denoted by E_+. An *order interval* in E is a set of the form $[x, y] = \{z \in E : x \leq z \leq y\}$, for $x, y \in E$. A subset A of E is *order bounded* if it is contained in an order interval, and it is *solid* if $x \in E$, $y \in A$ and $|x| \leq |y|$ implies $x \in A$; a solid linear subspace of E is called an i*deal*. A Riesz subspace of E is a linear subspace which is also a sublattice. A Riesz subspace F of E is order dense in E if for every $0 < x \in E$ there is a $y \in F$ such that $0 < y \leq x$. A set A in

E is *order closed* if whenever (x_i) is a net in A and (x_i) order converges to $x \in E$, also $x \in A$. An order closed ideal is called a *band*.

The Dedekind completion of E will be denoted by E^δ, and its universal completion by E^u.

1 Order Convergences

In this section we look at a number of different forms of order convergence in Riesz spaces that have appeared in the literature, as well as at the corresponding order continuous linear functionals.

Definition 1.1

(a) A filter $\mathcal{F}$ on E converges in order to x, denoted $\mathcal{F} \to x$, if $\mathcal{F}$ contains a family of order intervals with intersection $\{x\}$.

(b) A net $(x_\alpha)_{\alpha \in A}$ converges in order to x in E (denoted $x_\alpha \xrightarrow{o} x$) if there is a net $(y_\beta)_{\beta \in B}$ in E^+ such that $y_\beta \downarrow 0$ and for every $\beta \in B$ there is an $\alpha_\beta \in A$ such that $|x_\alpha - x| \leq y_\beta$ for every $\alpha \geq \alpha_\beta$. This is equivalent to the order convergence of the filter $\mathcal{F}$ with base $\{z : |z - x| \leq y_\beta\} : \beta \in B\}$.

Order convergence of filters is an example of a convergence structure, in the sense of [4].

In the literature, order convergence for a net is often defined differently: A net $(x_\alpha)_{\alpha \in A}$ converges in order to x in E if there is a net $(y_\alpha)_{\alpha \in A}$ in E^+ such that $y_\alpha \downarrow 0$ and $|x_\alpha - x| \leq y_\alpha$ for every $\alpha \in A$. This definition can lead to anomalies, as pointed out in [1]. These can be avoided if the definition is amended by requiring that there exists an $\alpha_0 \in A$ such that $|x_\alpha - x| \leq y_\alpha$ for every $\alpha \geq \alpha_0$ (and not for every $\alpha \in A$). We denote convergence in the sense of this modified definition by $x_\alpha \xrightarrow{o_1} x$. The relationship between these two definitions is given by the following result.

Proposition 1.2 ([1], Proposition 1.5) *For a net $(x_\alpha)_{\alpha \in A}$ in a Riesz space E, the following are equivalent:*

(a) $x_\alpha \xrightarrow{o} x$ in E;
(b) $x_\alpha \xrightarrow{o_1} x$ in E^δ.

In particular, these two definitions coincide for Dedekind-complete Riesz spaces. In the rest of this paper we use order convergence in the sense of Definition 1.1. We note that an order convergent net need, in general, not be order bounded.

Definition 1.3 A net $(x_\alpha)_{\alpha \in A}$ *converges unboundedly in order* to x in E if for every $u \in E_+$, $|x_\alpha - x| \wedge u \xrightarrow{o} 0$. We denote this by $x_\alpha \xrightarrow{uo} x$ and say (x_α) is uo-convergent to x.

The filter convergence associated with unbounded order convergence is also a convergence structure, see also [11], 23E.

It is easy to check that for a net $(x_\alpha)_{\alpha\in A}$ in E, $x_\alpha \xrightarrow{o} 0$ if and only if there is an $\alpha_0 \in A$ such that $\{x_\alpha : \alpha \geq \alpha_0\}$ is order bounded in E and $x_\alpha \xrightarrow{uo} x$. Thus order convergence implies unbounded order convergence.

The following examples illustrate the difference between the two types of convergence. Let (X, Σ, μ) be a semi-finite measure space. In the spaces $L_p(X, \Sigma, \mu)$, for $0 < p \leq \infty$, a sequence (f_n) converges in order to f if and only if (f_n) is order bounded and converges μ-almost everywhere to f, while (f_n) converges unboundedly to f if and only if it converges μ-almost everywhere to f. In the space $L_0(X, \Sigma, \mu)$ of (equivalence classes of) measurable functions, the two modes of convergence coincide, and both are equivalent to μ-almost everywhere convergence. There is a sense in which this is typical of the general situation, as the following result shows. Recall that the universal completion of $L_p(X, \Sigma, \mu)$ is $L_0(X, \Sigma, \mu)$.

Proposition 1.4 *A sequence (x_n) in a Riesz space E is unboundedly order convergent to 0 in E if and only if it is order convergent to 0 in the universal completion E^u of E.*

We shall repeatedly use the following property of disjoint sequences.

Proposition 1.5 ([13], Corollary 3.6) *If (x_n) is a disjoint sequence in a Riesz space E, then $x_n \xrightarrow{uo} 0$.*

It is possible to use unbounded order convergence as the primary form of order convergence and then to introduce a family of order convergences of which the usual order convergence is the prime example. We need the following definition.

Definition 1.6 A family $\mathcal{B}$ of subsets of a vector space E is a *vector bornology* on E if

(a) $\bigcup\{B : B \in \mathcal{B}\} = E$;
(b) $B \in \mathcal{B}, C \subseteq B \Rightarrow C \in \mathcal{B}$;
(c) $B_1, B_2 \in \mathcal{B} \Rightarrow B_1 + B_2 \in \mathcal{B}$;
(d) $B \in \mathcal{B}, 0 \neq \lambda \in \mathbb{R} \Rightarrow \lambda B \in \mathcal{B}$;
(e) $\mathcal{B}$ is closed under the formation of balanced hulls.

A subclass $\mathcal{B}_0$ of $\mathcal{B}$ is a basis for $\mathcal{B}$ if for every $B \in \mathcal{B}$, there is a $B_0 \in \mathcal{B}_0$ such that $B \subseteq B_0$. A vector bornology is *countable* if it has a countable basis. It is *solid* (in the case where E is a Riesz space) if it has a basis consisting of solid sets.

The two examples of solid bornologies that we will be primarily concerned with are the bornology of order bounded sets in a Riesz space, and the bornology of norm-bounded sets in a normed Riesz space. The latter is a countable bornology.

Definition 1.7 Let $\mathcal{B}$ be a solid bornology in a Riesz space E. A net $(x_\alpha)_{\alpha\in A}$ in E is $\mathcal{B}$-order convergent to x in E if $\{x_\alpha : \alpha \geq \alpha_0\} \in \mathcal{B}$ for some $\alpha_0 \in A$ and $x_\alpha \xrightarrow{uo} x$.

As an example, if $\mathcal{B}$ is the bornology of order bounded sets in E, (x_α) is $\mathcal{B}$-order convergent to x if and only if $x_\alpha \xrightarrow{o} x$. We recover unbounded order convergence by considering the bornology of all subsets of E.

We note that the filter convergence corresponding to $\mathcal{B}$-order convergence is an example of what is called *specified sets convergence* in [4], p. 3 and p. 86.

We finish this section with a look at order continuous linear functionals on E. Recall that a linear functional $f : E \to \mathbb{R}$ is called *order bounded* if it maps order bounded sets in E to order bounded sets in $\mathbb{R}$. We denote the set of all order bounded functionals on E by $E^\sim$. A linear functional f on E is *order continuous* if $x_\alpha \xrightarrow{o} 0$ in E implies $f(x_\alpha) \to 0$ in $\mathbb{R}$. A linear functional f on E is called *unboundedly order continuous* if $x_\alpha \xrightarrow{uo} 0$ in E implies $f(x_\alpha) \to 0$ in $\mathbb{R}$. The existence of non-trivial functionals of this type depends on the existence of atoms in E, as the following result shows.

Theorem 1.8 ([14], Proposition 2.2) *If f is a non-zero unbounded order continuous linear functional on E, then it is a finite linear combination of coordinate functionals of atoms of E.*

Definition 1.9 Let $\mathcal{B}$ be a solid bornology on a Riesz space E. A linear functional f on E is $\mathcal{B}$-order continuous if $f(x_\alpha) \to 0$ in $\mathbb{R}$ whenever (x_α) is $\mathcal{B}$-order convergent to 0 in E.

Example

(a) If $\mathcal{B}$ is the bornology of order bounded sets, the $\mathcal{B}$-order continuous linear functionals are the usual order continuous functionals. The vector space of all such functionals on E will be denoted by $E^\times$; it is known that $E^\times \subseteq E^\sim$.
(b) If $\mathcal{B}$ is the bornology of norm-bounded sets in a normed Riesz space, the $\mathcal{B}$-order continuous linear functionals are the *boundedly uo-continuous functionals* of [14], Definition 2.1.

The boundedly uo-continuous functionals were characterized in [14], Theorem 2.3. We will return to this topic in Sect. 3.

2 Lebesgue Topologies

A vector topology τ on a Riesz space E is a *locally solid topology* if it has a neighbourhood base at 0 consisting of solid sets. A locally solid topology is generated by a family of Riesz pseudo-norms. A function $p : E \to [0, \infty)$ is a *Riesz pseudo-norm* if for all $x, y \in E$

(a) $p(x + y) \leq p(x) + p(y)$,
(b) $|x| \leq |y|$ implies $p(x) \leq p(y)$,
(c) $p(\lambda x) \to 0$ as $\lambda \to 0$ in $\mathbb{R}$.

The space of all τ-continuous linear functionals on E will be denoted by $(E, \tau)'$, or, if there is no danger of confusion, by E'. For a locally solid topology τ on E, $(E, \tau)'$ is an ideal in $E^\sim$.

A *Fatou pseudo-norm* is a Riesz pseudo-norm p for which $x_\alpha \uparrow x$ in E^+ implies $p(x_\alpha) \to p(x)$. A locally solid topology generated by a family of Fatou pseudo-norms, or equivalently one with a neighbourhood base at 0 consisting of order closed sets, is a *Fatou topology*.

Definition 2.1 A locally solid topology τ on E is

(a) *pre-Lebesgue* if for every disjoint order bounded sequence (x_n) in E, $x_n \xrightarrow{\tau} 0$;
(b) *unbounded Lebesgue* if $x_\alpha \xrightarrow{uo} 0$ implies that $x_\alpha \xrightarrow{\tau} 0$;
(c) *Lebesgue* if for every order bounded net (x_α) such that $x_\alpha \xrightarrow{uo} 0$ we have that $x_\alpha \xrightarrow{\tau} 0$.

It now follows from the fact that $x_\alpha \xrightarrow{o} 0$ if and only if (x_α) is eventually order bounded and unboundedly order convergent to 0 that the locally solid topology τ is Lebesgue if and only if $x_\alpha \xrightarrow{o} 0$ implies $x_\alpha \xrightarrow{\tau} 0$. This in its turn is equivalent to requiring that $x_\alpha \downarrow 0$ implies that $x_\alpha \xrightarrow{\tau} 0$, the definition of a Lebesgue topology that is more common in the literature (see, e.g., [2], Definition 3.1).

There is an characterization of pre-Lebesgue topologies (sometimes used as a definition) that we will return to later.

Theorem 2.2 ([2], Theorem 3.22) *A locally solid topology τ on a Riesz space E is pre-Lebesgue if and only if every increasing order bounded sequence (x_n) in E_+ is a τ-Cauchy sequence.*

It is immediately clear from earlier remarks that every unbounded Lebesgue topology is Lebesgue, and that every Lebesgue topology is pre-Lebesgue. A pre-Lebesgue topology need not be Lebesgue, however [2, Example 3.2].

Example A Lebesgue topology need not be an unbounded Lebesgue topology. As an example, the usual supremum norm induces a Lebesgue topology on the sequence space c_0, but this topology is not an unbounded Lebesgue topology. To see this let e_n denotes the element of c_0 with nth term 1 and all other terms 0, then (e_n) converges unboundedly to 0 in c_0, but does not converge to 0 in the norm of c_0.

The relationship between Lebesgue and pre-Lebesgue topologies is clarified further by the following results.

Theorem 2.3 ([2], Theorems 3.24 and 4.8) *Let τ be a pre-Lebesgue topology on a Riesz space E. Then τ is a Lebesgue topology if either*

(a) τ is a Fatou topology, or
(b) E is τ-complete.

To clarify the relationship between Lebesgue and unbounded Lebesgue topologies, we need the notion of an unbounded topology, introduced by Taylor in [19].

Definition 2.4 Let τ be a locally solid topology on the Riesz space E, generated by the family $\mathcal{P}$ of Riesz pseudo-norms. The locally solid topology $u\tau$ on E is generated by the family $\{p_u : u \in E_+\}$ of Riesz pseudo-norms, where for each $u \in E_+$, p_u is defined by

$$p_u(x) = p(|x| \wedge u), \qquad x \in E.$$

We shall refer to $u\tau$ as the *unbounded topology associated with* τ. It is clear that $u\tau$ is coarser than τ; we say that a locally solid topology is *unbounded* if $\tau = u\tau$.

Theorem 2.5 ([19], Theorem 5.9) *A locally solid topology on a Riesz space is an unbounded Lebesgue topology if and only if it is both a Lebesgue topology and an unbounded topology.*

To give an example of an unbounded Lebesgue topology, we need the following result of Amemiya and Mori.

Theorem 2.6 ([2], Theorem 4.22) *Any two Hausdorff Lebesgue topologies on a Riesz space E induce the same topology on the order intervals of E.*

If τ_1 and τ_2 are Hausdorff Lebesgue topologies on E, then this theorem implies that $u\tau_1 = u\tau_2$. Since $u\tau$ is coarser than τ for any Hausdorff Lebesgue topology τ on E, we can conclude that if E admits a Hausdorff Lebesgue topology τ, there is a coarsest Hausdorff Lebesgue topology on E, which we shall denote by τ_m. If $\mathcal{P}$ is a family of defining Riesz pseudo-norms for τ, the topology τ_m is just the topology $u\tau$, and is defined by the pseudo-norms $\{p_u : p \in \mathcal{P}, u \in E_+\}$. It suffices, in fact, that E contains an order dense ideal F which admits a Hausdorff Lebesgue topology τ_0, and in this case the topology τ_m is defined by the family of pseudo-norms $\{p_u : p \in \mathcal{P}_0, u \in F_+\}$, where $\mathcal{P}_0$ is a family of defining pseudo-norms for τ_0.

The topology τ_m is Hausdorff, unbounded, and Lebesgue, and the only such topology on E. It need not be a locally convex topology.

If (X, Σ, μ) is a semi-finite measure space, the topology τ_m can be defined on $E = L_0(X, \Sigma, \mu)$, and it equals the topology of convergence in measure on subsets of X of finite measure. A detailed investigation of the topology τ_m can be found in [7].

If we confine ourselves to Hausdorff topologies, we can characterize Lebesgue topologies in terms of the topology τ_m.

Proposition 2.7 *Suppose a Riesz space E admits a Hausdorff Lebesgue topology. Then a Hausdorff locally solid topology τ on E is Lebesgue if and only if for every order bounded net (x_α) in E, $x_\alpha \xrightarrow{\tau_m} 0$ implies $x_\alpha \xrightarrow{\tau} 0$.*

Proof Since E admits a Hausdorff Lebesgue topology, the topology τ_m can be defined on E. If τ is a Hausdorff Lebesgue topology on E, τ and τ_m coincide on order bounded sets, and therefore $x_\alpha \xrightarrow{\tau_m} 0$ implies $x_\alpha \xrightarrow{\tau} 0$. Conversely, if τ

satisfies the stated condition and (x_α) is an order bounded net such that $x_\alpha \xrightarrow{uo} 0$, then $x_\alpha \xrightarrow{\tau_m} 0$ and so by assumption $x_\alpha \xrightarrow{\tau} 0$. □

Locally convex pre-Lebesgue topologies can be characterized in terms of their equicontinuous sets. To do this, we need to recall the notions of order precompactness and absolute weak topologies.

If τ is a locally solid topology on a Riesz space E, a subset A of E is *τ-quasi-order precompact* (respectively, *τ-order precompact*) if for every solid τ-neighbourhood U of 0 in E there is a positive x_U in E (respectively, the ideal generated by A in E) such that $A \subset [-x_U, x_U]+U$, or equivalently, $(|x|-x_U)^+ = |x| - |x| \wedge x_U \in U$ for every $x \in A$. If τ is a pre-Lebesgue topology, the τ-order precompact and quasi-order precompact sets coincide. Clearly every order interval is order precompact. More information about order precompactness can be found in [10].

Order precompact sets approximate order intervals, and in the light of Theorem 2.6 the following result does not come as a surprise.

Proposition 2.8 ([7], Corollary 7.2, [16], Theorem 2.5) *Let τ be a Hausdorff Lebesgue topology on a Riesz space E and A a τ-order precompact set in E. Then the topologies τ and τ_m coincide on A.*

The following sufficient condition for order precompactness is contained in the proof of [3], Theorem 4 and will be needed in the next section.

Theorem 2.9 *Let τ_1 and τ_2 be locally solid locally convex topologies on the Riesz space E, with τ_2 coarser than τ_1 and E τ_1-complete. Suppose that K is a solid τ_1-bounded subset of E such that for every disjoint sequence (x_n) in K, $x_n \xrightarrow{\tau_2} 0$. Then K is τ_2-order precompact.*

For a vector subspace F of $E^\sim$, the *absolute weak topology* $|\sigma|(E, F)$ is the locally convex topology generated by the seminorms $\{p_f : f \in F\}$, where $p_f(x) = |f|(|x|)$. The topological dual of E under $|\sigma|(E, F)$ equals the order ideal I_F generated in $E^\sim$ by F. The absolute weak topology $|\sigma|(F, E)$ is defined similarly; it is generated by the seminorms $\{p_x : x \in E\}$, where $p_x(f) = |f|(|x|)$. The topology $|\sigma|(E, F)$ (respectively, $|\sigma|(F, E)$) can also be described as the topology of uniform convergence on the order intervals in F (respectively, the order intervals in E).

The following results show how order precompactness can be used in the characterization of locally convex Lebesgue and pre-Lebesgue topologies.

Theorem 2.10 ([3], Theorems 2.7 and 2.8) *Let E be a Riesz space and τ a locally convex locally solid topology on E, with dual E'.*

1. *τ is a pre-Lebesgue topology if and only if every τ-equicontinuous subset of $E' = (E, \tau)'$ is $|\sigma|(E', E)$-order precompact.*
2. *$\beta(E', E)$ is a Lebesgue topology if and only if every τ-bounded subset of E is $|\sigma|(E, E')$-order precompact.*

Since any order interval is order precompact, it is clear that for any order ideal F of $E^{\sim}$, $|\sigma|(E, F)$ is a pre-Lebesgue topology.

Proposition 2.11 ([2], Theorem 3.12) *Let τ be a locally convex locally solid topology on a Riesz space E and $E' = (E, \tau)'$. Then τ is a Lebesgue topology if and only if $E' \subseteq E^{\times}$.*

For future reference we note the following obvious corollary.

Corollary 2.12 *If E is a Riesz space and $f \in E^{\sim}$, then the following are equivalent:*

(a) $f \in E^{\times}$.
(b) For every order bounded net (x_α) in E such that $x_\alpha \xrightarrow{uo} 0$, $f(x_\alpha) \to 0$.
(c) There is a Lebesgue topology τ on E such that f is τ-continuous.

If E is a Riesz space for which $E^{\times} \neq E^{\sim}$ (for example $E = \ell_\infty$), then $|\sigma|(E, E^{\sim})$ provides another example of a pre-Lebesgue topology which is not Lebesgue.

3 Uniformly Lebesgue Topologies

The definitions given in the previous section for pre-Lebesgue and Lebesgue topologies are both formulated as conditions on order bounded nets (or sequences). Replacing the bornology of order bounded sets by a larger bornology will result in a smaller class of topologies. This is the motivation for the following definitions.

Definition 3.1 Let $\mathcal{B}$ be a solid vector bornology on a Riesz space E. A locally solid topology τ on E is

(a) $\mathcal{B}$-pre-Lebesgue if every disjoint sequence (x_n) in E such that $\{x_n : n \in \mathbb{N}\} \in \mathcal{B}$ is τ-convergent to 0;
(b) $\mathcal{B}$-Lebesgue if for every net (x_α) in E such that $\{x_\alpha : \alpha \in A\} \in \mathcal{B}$ and $x_\alpha \xrightarrow{uo} 0$, we have $x_\alpha \xrightarrow{\tau} 0$.

It is immediately clear that if $\mathcal{B}$ and $\mathcal{C}$ are both solid vector bornologies on E and $\mathcal{B} \subseteq \mathcal{C}$, then a $\mathcal{C}$-pre-Lebesgue (respectively, $\mathcal{C}$-Lebesgue) topology is $\mathcal{B}$-pre-Lebesgue (respectively, $\mathcal{B}$-Lebesgue).

In the case where $\mathcal{B}$ is the bornology of solid order bounded sets in E, we recover the definitions of pre-Lebesgue and Lebesgue topologies in Definition 2.1. If $\mathcal{B}$ is the bornology of all solid subsets of E, the $\mathcal{B}$-Lebesgue topologies are the unbounded Lebesgue topologies.

If E is a normed Riesz space and $\mathcal{B}$ is the bornology of solid norm-bounded sets in E, we shall call the $\mathcal{B}$-pre-Lebesgue and $\mathcal{B}$-Lebesgue topologies *uniformly pre-Lebesgue* and *uniformly Lebesgue*, respectively. The terminology, in a slightly different context, is due to Nowak ([17], see also [6]).

Clearly every uniformly pre-Lebesgue (respectively, Lebesgue) topology is pre-Lebesgue (respectively, Lebesgue). It follows immediately from Proposition 1.5 that every unbounded Lebesgue topology is a uniformly Lebesgue topology, and that every uniformly Lebesgue topology is uniformly pre-Lebesgue.

The following duality result will play an important role in the rest of the paper, and enables us to give examples of uniformly pre-Lebesgue topologies.

Theorem 3.2 *Let E be a Riesz space and F an ideal in $E^{\sim}$. Let $\mathcal{K}$ be a bornology of solid relatively $\sigma(E, F)$-compact subsets of E, and $\mathcal{L}$ a family of solid relatively $\sigma(F, E)$-compact subsets of F. Denote by $\tau_{\mathcal{L}}$ the topology on E of uniform convergence on the sets in $\mathcal{L}$, and by $\tau_{\mathcal{K}}$ the topology on F of uniform convergence on the sets in $\mathcal{K}$. Consider the following statements:*

(a) Every $K \in \mathcal{K}$ is $\tau_{\mathcal{L}}$-order precompact, and every $L \in \mathcal{L}$ is $|\sigma|(F, E)$-order precompact.
(b) Every $L \in \mathcal{L}$ is $\tau_{\mathcal{K}}$-order precompact, and every $K \in \mathcal{K}$ is $|\sigma|(E, F)$-order precompact.
(c) For every $L \in \mathcal{L}$ and for every disjoint sequence (f_n) in L, $f_n \xrightarrow{\tau_{\mathcal{K}}} 0$.
(d) For every $K \in \mathcal{K}$ and for every disjoint sequence (x_n) in K, $x_n \xrightarrow{\tau_{\mathcal{L}}} 0$.

Then (a) and (b) are equivalent, (a) implies (d) and (b) implies (c). If there is a locally convex locally solid topology τ_1 on E finer than $\tau_{\mathcal{L}}$ for which (E, τ_1) is complete and each $K \in \mathcal{K}$ is τ_1-bounded, then (d) implies (a). Similarly, if there is a locally convex locally solid topology τ_2 on F finer than $\tau_{\mathcal{K}}$ for which (E, τ_2) is complete and each $L \in \mathcal{L}$ is τ_2-bounded, then (c) implies (b).

Proof

(a) $\Leftrightarrow$ (b): The equivalence of (a) and (b), with order precompactness replaced by quasi-order precompactness, is given in [5], Theorem 3.1. We show that under the stated conditions quasi-order precompactness in (a) and (b) is equivalent to order compactness. For (a), note that since $|\sigma|(F, E) = |\sigma|(E', E)$ is a Lebesgue topology [2, Theorem 6.4], $|\sigma|(F, E)$-quasi-order precompact sets are $|\sigma|(F, E)$-order precompact. The fact that every $L \in \mathcal{L}$ is $|\sigma|(F, E)$-order precompact means that $\tau_{\mathcal{L}}$ is a pre-Lebesgue topology (Theorem 2.10), and so the $\tau_{\mathcal{L}}$-quasi-order precompact sets are in fact order precompact. A similar argument using the fact that $|\sigma|(E, E')$ is a pre-Lebesgue topology shows that quasi-order precompactness may be replaced by order precompactness in (b).

(b) $\Rightarrow$ (c) and (a) $\Rightarrow$ (d): The same argument as in [6], Theorem 3.1 can be used here.

(d) $\Rightarrow$ (a): It follows at once from Theorem 2.9 that every $K \in \mathcal{K}$ is $\tau_{\mathcal{L}}$ order precompact. Since $(E, \tau_{\mathcal{L}})' = F$ and every $L \in \mathcal{L}$ is $\tau_{\mathcal{L}}$-equicontinuous, it follows from Theorem 2.10 that every L is $|\sigma|(F, E)$-order precompact.

(c) $\Rightarrow$ (b): The proof is similar to that for the previous implication.

□

If we take $\mathcal{K}$ to be the bornology of solid norm-bounded sets in a Banach lattice E, we obtain several characterizations of uniformly pre-Lebesgue topologies. We denote the largest ideal in E' (respectively $E^{\times}$) on which the norm of E' induces a Lebesgue topology by E'_a (respectively, $E^{\times}_a$).

Corollary 3.3 *Let E be a Banach lattice with dual E', and τ a locally solid locally convex topology on E, coarser than the norm topology, and suppose that $F = (E, \tau)'$ is a norm-closed ideal of E'. Then the following are equivalent:*

(a) τ is a pre-Lebesgue topology, and every solid norm-bounded set in E is τ-order precompact.
(b) $F \subseteq E'_a$ and every τ-equicontinuous set in F is order precompact for the norm of F.
(c) For every τ-equicontinuous set L in F and every disjoint sequence (f_n) in L, (f_n) converges to 0 in norm.
(d) τ is a uniformly pre-Lebesgue topology.

Proof Let $\mathcal{K}$ be the bornology of solid norm-bounded sets in E, and $\mathcal{L}$ the family of τ-equicontinuous sets in F. Then $\tau_{\mathcal{K}}$ is the topology on F induced by the norm of E', and $\tau_{\mathcal{L}}$ is the topology τ. Since E' is norm-complete and F is norm-closed, F is norm-complete as well. Let τ_1 be the norm topology on E and τ_2 the norm topology on F. Then E is τ_1-complete and every $K \in \mathcal{K}$ is τ_1-bounded; F is τ_2-complete and every $L \in \mathcal{L}$ is τ_2-bounded. Since $(E, \tau)' = F$, the sets in $\mathcal{L}$ are relatively $\sigma(F, E)$-compact. The sets in $\mathcal{K}$ need not be relatively $\sigma(E, F)$-compact, since $(E', \tau_{\mathcal{K}})' = E''$. However, we can still use Theorem 3.2 to deduce that (a), (b) and (c) are equivalent and that (a) implies (d); from Theorem 2.10 it follows that the $|\sigma|(F, E)$-order precompactness of the sets in $\mathcal{L}$ ensures that τ is a pre-Lebesgue topology, and the $|\sigma|(E, F)$-order precompactness of the sets in $\mathcal{K}$ ensures that $\tau_{\mathcal{K}} = \beta(F, E)$ is a Lebesgue topology and hence $F \subseteq E'_a$. For the implication (d) $\Rightarrow$ (a) we note that as in the proof of the same implication in Theorem 3.2 we can still deduce that every solid norm-bounded set in E is τ-order precompact. Since every uniformly pre-Lebesgue topology is a pre-Lebesgue topology, it follows that (d) implies (a). □

Corollary 3.4 *Let E be a Banach lattice and τ a locally convex locally solid topology on E with dual F, and suppose F is a band in E'_a. Then τ is uniformly pre-Lebesgue if and only if every τ-equicontinuous set in F is order precompact for the norm on F. In particular, if F a band in E'_a, $|\sigma|(E, F)$ is a uniformly pre-Lebesgue topology.*

Proof It suffices to note that since the norm on E' induces a Fatou topology [11, 23N(h)], bands in E' are closed [2, Theorem 4.20]. The result then follows from the equivalence of (b) and (d) in Corollary 3.3. □

Example A locally convex locally solid topology τ on a Banach lattice E for which $(E, \tau)' \subset E'_a$ need not be a uniformly pre-Lebesgue topology. Let $E = L_2[0, 1]$ and τ the usual norm topology of E. Then $E' = L_2[0, 1] = E'_a$. The norm topology on E cannot be uniformly pre-Lebesgue, for if it were, then it would follow from the equivalence of (a) and (d) in Corollary 3.3 that the closed unit ball of E must be order precompact for the norm. But then E must be finite-dimensional, by Grobler [15], Lemma 4.4, a contradiction.

Theorem 2.2 gives an alternative characterization of pre-Lebesgue topologies that can also be generalized, as before by replacing order boundedness by norm-boundedness. We recall that a locally solid topology τ on a Riesz space E is called a *Levi topology* if every increasing τ-bounded net in E has an upper bound in E.

Proposition 3.5 *Let E be a Banach lattice for which the norm topology is a Levi topology, and let τ be a locally solid topology on E. Then every norm-bounded increasing sequence in E is a τ-Cauchy sequence if and only if τ is a pre-Lebesgue topology.*

Proof Suppose τ is a pre-Lebesgue topology, and let (x_n) be an increasing norm-bounded sequence. Since the norm topology is a Levi topology, there is an $x \in E$ such that $x_n \leq x$ for all $n \in \mathbb{N}$. It follows from the fact that τ is pre-Lebesgue and Theorem 2.2 that (x_n) is a τ-Cauchy sequence. The converse follows easily from the fact that order bounded sets are norm-bounded, and Theorem 2.2. □

There is a large class of Banach lattices for which the condition in the theorem is satisfied: the norm topology on Banach function spaces with a Fatou norm is a Levi topology [11, 65E]. It follows that a pre-Lebesgue topology τ on such a space has the property that every increasing norm-bounded sequence is a τ-Cauchy sequence, but τ need not be a uniformly pre-Lebesgue topology in the sense of Definition 3.1, as is clear from the equivalence of (a) and (d) in Corollary 3.3.

We now turn to uniformly Lebesgue topologies, and start of by showing that as long as we restrict ourselves to Hausdorff topologies, we can obtain a characterization similar to the one in Proposition 2.7.

Proposition 3.6 *Let τ be a Hausdorff Lebesgue topology on a Banach lattice E.*

(a) If the norm-bounded sets in E are τ-order precompact, then τ is uniformly Lebesgue, and the converse holds if τ is locally convex.
(b) If τ is locally convex, it is a uniformly Lebesgue topology if and only if for every norm-bounded net (x_α) in E such that $x_\alpha \xrightarrow{\tau_m} 0$, we have $x_\alpha \xrightarrow{\tau} 0$.

Proof Note that since τ is assumed to be a Hausdorff Lebesgue topology, the topology τ_m can be defined on E.

(a) If the norm-bounded sets are τ-order precompact, the topologies τ and τ_m coincide on norm-bounded sets (Proposition 2.8). Hence if (x_α) is a norm-bounded net such that $x_\alpha \xrightarrow{uo} 0$, then also $x_\alpha \xrightarrow{\tau_m} 0$ (since τ_m is an unbounded Lebesgue topology), and therefore $x_\alpha \xrightarrow{\tau} 0$. This shows that τ is a uniformly Lebesgue topology. In the case where τ is locally convex and uniformly Lebesgue, it is uniformly pre-Lebesgue, and therefore by the equivalence of (a) and (d) in Corollary 3.3 norm-bounded sets are τ-order precompact.

(b) If τ is locally convex and uniformly Lebesgue, norm-bounded sets are τ-order precompact by (a), and then τ and τ_m coincide on norm-bounded sets. Conversely, if the condition is satisfied and (x_α) is a norm-bounded net such $x_\alpha \xrightarrow{uo} 0$, then $x_\alpha \xrightarrow{\tau_m} 0$ and hence, by assumption $x_\alpha \xrightarrow{\tau} 0$.

□

Proposition 3.7 *If τ is a uniformly Lebesgue topology on a Banach lattice E, $(E,\tau)' \subseteq E_a^\times$.*

Proof The topology τ is Lebesgue, and therefore $(E,\tau)' \subseteq E^\times$. Since τ is also uniformly pre Lebesgue, $(E,\tau)' \subseteq E_a'$ (by Corollary 3.3), and we can therefore deduce that $(E,\tau)' \subseteq E^\times \cap E_a' = E_a^\times$. □

A locally solid topology on E for which $(E,\tau)' \subseteq E_a^\times$ need not be uniformly Lebesgue. The example following Corollary 3.4 suffices to show this (since in the example $E_a' = E_a^\times = E^\times$).

Corollary 3.8 *For a Hausdorff locally convex uniformly pre-Lebesgue topology τ on a Banach lattice E the following are equivalent:*

(a) τ is a uniformly Lebesgue topology;
(b) τ is a Lebesgue topology;
(c) τ is a Fatou topology.

Proof The equivalence of (b) and (c) follows from Theorem 2.3, and that of (a) and (b) from Proposition 3.6 and the equivalence of (a) and (d) in Corollary 3.3. □

If E is a Banach lattice such that $E_a^\times$ separates the points of E, then it follows from Corollary 3.8, Proposition 2.11 and Corollary 3.4 that $|\sigma|(E, E_a^\times)$ is a uniformly Lebesgue topology.

If E is a Banach lattice such that $E^\times$ is a proper subspace of E', and $E' = E_a'$, then $\tau = |\sigma|(E, E_a')$ is an example of a uniformly pre-Lebesgue topology which is not a uniformly Lebesgue topology (since $(E,\tau)' = E_a' \nsubseteq E\times_a$).

We can now also give an example of a uniformly Lebesgue topology which is not an unbounded Lebesgue topology. Let $E = L_\infty[0,1]$, then $E^\times = E_a^\times = L_1[0,1]$, and by the comments above, $|\sigma|(E, E_a^\times) = |\sigma|(L_\infty[0,1], L_1[0,1])$ is a uniformly pre-Lebesgue topology. It is not an unbounded Lebesgue topology; however, it is easily checked that if we put $f_n = n\chi_{[0,\frac{1}{n}]}$, then $f_n \xrightarrow{uo} 0$, but (f_n) does not converge to 0 for $|\sigma|(L_\infty[0,1], L_1[0,1])$.

The following two results are analogues of Corollary 2.12 for uniform topologies. The second is a generalization of Theorem 2.3 in [13].

Proposition 3.9 *Let E be a Banach lattice and $f \in E'$. Then the following statements are equivalent:*

(a) For every norm-bounded disjoint sequence (x_n) in E, $f(x_n) \to 0$.
(b) There is a locally convex uniformly pre-Lebesgue topology such that f is τ-continuous.
(c) $f \in E_a'$.

Proof

(a) $\Rightarrow$ (b): It follows from the Riesz–Kantorovich formula that for every norm-bounded disjoint sequence (x_n) in E, $|f|(|x_n|) \to 0$. Let τ be the locally solid locally convex topology on E defined by the seminorm $p_f(x) = |f|(|x|)$. Then (a) shows that τ is uniformly pre-Lebesgue, and since $|f(x)| \leq |f|(|x|)$ for all $x \in E$, f is τ-continuous.

(b) $\Rightarrow$ (a): If τ is uniformly pre-Lebesgue, a norm-bounded disjoint sequence is τ-convergent to 0, and the τ-continuity of f ensures that $f(x_n) \to 0$.

(b) $\Rightarrow$ (c): This follows from (d) $\Rightarrow$ (b) in Corollary 3.3.

(c) $\Rightarrow$ (b): If $f \in E'_a$, f is τ-continuous for the locally convex locally solid topology $\tau = |\sigma|(E, E'_a)$.

□

Proposition 3.10 *Let E be a Banach lattice and $f \in E^\times$. Then the following statements are equivalent:*

(a) For every norm-bounded net (x_α) in E such that $x_\alpha \xrightarrow{uo} 0$, $f(x_\alpha) \to 0$.
(b) For every norm-bounded disjoint sequence (x_n) in E, $f(x_n) \to 0$.
(c) There is a locally convex uniformly Lebesgue topology such that f is τ-continuous.
(d) $f \in E_a^\times$.

Proof

(a) $\Rightarrow$ (b): Use the fact that disjoint sequences are uo-convergent to 0.

(b) $\Rightarrow$ (c): The proof is similar to that of the implication (a) $\Rightarrow$ (b) in Proposition 3.9. Since $f \in E^\times$, the topology τ is now Lebesgue, and uniformly pre-Lebesgue by (b), and therefore uniformly Lebesgue, by Corollary 3.8.

(c) $\Rightarrow$ (a): For a norm-bounded net (x_α) in E such that $x_\alpha \xrightarrow{uo} 0$, $x_\alpha \xrightarrow{\tau} 0$ and therefore $f(x_\alpha \to 0$.

(c) $\Rightarrow$ (d): This is Proposition 3.7.

(d) $\Rightarrow$ (c): Let $\tau = |\sigma|(E, E_a^\times)$.

□

The last result of this section characterizes locally convex uniformly Lebesgue topologies in terms of their equicontinuous sets.

Proposition 3.11 *A locally solid locally convex topology τ on a Banach lattice E is uniformly Lebesgue if and only if $F = (E, \tau)' \subseteq E_a^\times$ and every τ-equicontinuous subset of F is order precompact for the norm induced on F by the norm of E'.*

Proof If τ is uniformly Lebesgue, $F = (E, \tau)' \subseteq E_a^\times$ (Proposition 3.7), and τ is uniformly pre-Lebesgue. Hence every τ-equicontinuous subset of F is norm-order precompact (Corollary 3.3). Conversely, if $(E, \tau)' \subseteq E_a^\times \subseteq E^\times$, then τ is Lebesgue, and if every τ-equicontinuous subset of $(E, \tau)'$ is norm-order precompact, τ is also uniformly pre-Lebesgue (Corollary 3.3). Then τ is uniformly Lebesgue, by Corollary 3.8. □

4 Mixed Topologies

The Lebesgue topologies we have considered are characterized by the fact that unbounded order convergence restricted to sets that are bounded (in some sense) implies convergence in the topology. This raises a natural question: Are the convergences we have looked at topological? In general, order convergence (where the nets or sequences are required to be order bounded) on a Riesz space E is not *topological*, that is, there need not be a topology on E such that a net (or sequence) in E is convergent with respect to the topology if and only if it is unbounded order convergent on order bounded sets [9].

In this section we investigate the question for norm-bounded unbounded order convergence, that is, given a Banach lattice E, we look for a locally solid topology on E for which the nets that are convergent in the topology are exactly the norm-bounded uo-convergent nets. It is clear that we can confine our search to uniformly Lebesgue topologies, and since limits of uo-convergent nets are unique, we may restrict attention to Hausdorff topologies. Since uniformly Lebesgue topologies are Lebesgue, we shall further assume that E admits a Hausdorff Lebesgue topology, and therefore that the topology τ_m is defined on E. We shall, for the moment, also confine our search to locally convex topologies.

It follows from Propositions 3.7 and 3.11 that the finest locally convex uniformly Lebesgue topology on E is the topology of uniform convergence on all the solid subsets of $E_a^\times$ that are order precompact with respect to the norm induced by the norm of E'. To ensure that this topology is Hausdorff, we assume that $E_a^\times$ separates the points of E. If norm-bounded uo-convergence can be topologized, this topology appears to be the most likely candidate. To investigate it further, we have to consider the notion of a mixed topology, an example of the generalized inductive limit topologies first introduced by Garling in [12]. We give the essential background, and refer the reader to [8] for further details.

A *mixed space* is a triple $(E, \mathcal{B}, \tau)$ with E a vector space, $\mathcal{B}$ a vector bornology on E, and τ a linear topology on E such that every $B \in \mathcal{B}$ is τ-bounded. If $\mathcal{B}$ has a basis consisting of τ-closed sets, the mixed space is called *normal*. We introduce two *mixed topologies*:

(a) γ_τ is the finest locally convex topology coinciding with τ on the sets in $\mathcal{B}$;
(b) γ'_τ is the finest vector topology coinciding with τ on the sets in $\mathcal{B}$.

If τ is locally convex, it can be shown that $\gamma'_\tau = \gamma_\tau$. Denote by β the finest locally convex topology on E for which every $B \in \mathcal{B}$ is bounded.

Let E^b denote the space of all linear functionals on E which are bounded on the sets in $\mathcal{B}$, equipped with the topology of uniform convergence on the sets in $\mathcal{B}$.

Theorem 4.1 *If $(E, \mathcal{B}, \tau)$ is a normal mixed space and τ locally convex, then*

(a) $\tau \leq \gamma_\tau \leq \beta$;
(b) $(E, \gamma_\tau)'$ *is the closure of* $(E, \tau)'$ *in* E^b;

(c) *a subset* A *of* $(E, \gamma_\tau)'$ *is* γ_τ*-equicontinuous iff for every* $\epsilon > 0$ *and every* τ*-closed* B *in a basis for* $\mathcal{B}$*, there is a* τ*-equicontinuous set* $A(\epsilon, B)$ *such that* $A \subseteq A(\epsilon, B) + \epsilon B^\circ$*, where the polar* B° *is taken in* E^b*.*

Theorem 4.2 *Let* $(E, \mathcal{B}, \tau)$ *be a normal mixed space, with* $\mathcal{B}$ *a countable bornology and* τ *a locally convex topology. Then*

1. $\gamma'_\tau = \gamma_\tau$*;*
2. *a sequence* (x_n) *in* E *is* γ_τ*-convergent to* $x \in E$ *if and only if* $\{x_n : n \in \mathbb{N}\} \in \mathcal{B}$ *and* $x_n \xrightarrow{\tau} x$*.*

Proposition 4.3 *For any mixed space* $(E, \mathcal{B}, \tau)$*, a linear functional* f *on* E *is* γ_τ*-continuous iff the restriction of* f *to each* $B \in \mathcal{B}$ *is* τ*-continuous.*

We now specialize to the case where E is a Banach lattice with its norm a Fatou norm. Let $\mathcal{B}$ be the countable bornology of norm-bounded sets in E. If τ is a Hausdorff locally solid topology on E for which every norm-bounded set is τ-bounded (in particular, if τ is coarser than the norm topology), $(E, \mathcal{B}, \tau)$ is a mixed space. If in addition τ is a Fatou topology, $(E, \mathcal{B}, \tau)$ is a normal mixed space; the norm-closed balls in E are order closed (since the norm topology is Fatou), and therefore τ-closed. In the case where $E_a^\times$ separates the points of E, we can take τ to be the locally convex Hausdorff Lebesgue (and therefore Fatou) topology $|\sigma|(E, E_a^\times)$. The topology β is now the norm topology on E, $E^b = E^\sim = E'$, the norm-dual of E, and the topology of uniform convergence on the sets in $\mathcal{B}$ is the norm topology of E'. Since $E_a^\times$ is a band in $E^\sim$, $(E, |\sigma|(E, E_a^\times))' = E_a^\times = (E, \gamma'_{|\sigma|(E,E_a^\times)})'$ and the $\gamma'_{|\sigma|(E,E_a^\times)}$-equicontinuous sets in $E_a^\times$ are the sets that are order precompact in the norm of E', by Theorem 4.1(b) and (c). Since $\mathcal{B}$ is a countable bornology we can apply Theorem 4.2 to get:

Theorem 4.4 *If* E *is a Banach lattice with a Fatou norm,* $E_a^\times$ *separates the points of* E *and* $\mathcal{B}$ *is the bornology of norm-bounded sets in* E*, then* $(E, \mathcal{B}, |\sigma|(E, E_a^\times))$ *is a normal mixed space. The mixed topologies* $\gamma_{|\sigma|(E,E_a^\times)}$ *and* $\gamma'_{|\sigma|(E,E_a^\times)}$ *are both equal to the topology on* E *of uniform convergence on the norm-order precompact subsets of* $E_a^\times$*, the finest uniformly Lebesgue topology on* E*. A sequence* (x_n) *in* E *is* $\gamma_{|\sigma|(E,E_a^\times)}$*-convergent to* x *if and only if it is norm-bounded and* $|\sigma|(E, E_a^\times)$*-convergent to* x*.*

Since $\gamma_{|\sigma|(E,E_a^\times)}$ is a locally convex Hausdorff uniformly Lebesgue topology, it follows from Proposition 3.6 that the norm-bounded sets in E are $\gamma_{|\sigma|(E,E_a^\times)}$-order precompact. Proposition 2.8 then shows that the topologies $\gamma_{|\sigma|(E,E_a^\times)}$ and τ_m coincide on the norm-bounded sets. This gives:

Corollary 4.5 *Let* E *and* $\mathcal{B}$ *be as in the theorem. Then a sequence* (x_n) *in* E *is* $\gamma_{|\sigma|(E,E_a^\times)}$*-convergent to* x *if and only if it is norm-bounded and* τ_m*-convergent to* x*.*

It follows from the corollary that the norm-bounded uo-convergent sequences in E will coincide with the $\gamma_{|\sigma|(E,E_a^\times)}$-convergent sequences if and only if the norm-bounded uo-convergent sequences and τ_m-convergent sequences coincide. But one of the standard examples of a norm-bounded (and also order bounded) sequence in $L_\infty 1[0, 1]$ which converges in measure but not almost everywhere [18] shows that this is not possible in general. It is not difficult to see, using the results in [9] and the proof of Proposition 3.5 in [7] that it is possible in the case where E is an atomic lattice.

Proposition 4.3 now allows us to add another equivalence to the list in Proposition 3.7.

Proposition 4.6 *Let E be a Banach lattice with a Fatou norm and such that $E_a^\times$ separates the points of E. Then $f \in E_a^\times$ if and only if f is τ_m-continuous when restricted to norm-bounded sets.*

It is also possible to characterize the finest Lebesgue topology on a Riesz space E for which $E^\times$ separates the points of E in a similar way as a mixed topology, by using the bornology of order bounded sets. In this case the finest Lebesgue topology is the absolute Mackey topology $|\tau|(E, E^\times)$ (the topology of uniform convergence on the solid $\sigma(E^\times, E)$-compact sets in $E^\times$). Since the bornology in this case is in general not countable, it is no longer possible to use Theorem 4.2 to characterize the convergent sequences of this topology. These and related results can be found in [8].

Acknowledgement The author would like to thank the University of Cape Town for partial financial support.

References

1. Y. Abramovich, G. Sirotkin, On order convergence of nets. Positivity **9**, 287–292 (2005)
2. C.D. Aliprantis, O. Burkinshaw, *Locally Solid Riesz Spaces with Applications to Economics*, 2nd edn. (AMS, Providence, 2003)
3. C.D. Aliprantis, O. Burkinshaw, M. Duhoux, Compactness properties of abstract kernel operators. Pac. J. Math. **100**(1), 1–22 (1982)
4. R. Beattie, H.-P. Butzmann, *Convergence Structures and Applications to Functional Analysis* (Kluwer, Dordrecht, 2002)
5. J.J. Conradie, Duality results for order precompact sets in locally solid Riesz spaces. Indag. Math. (N.S.) **2**, 19–28 (1991)
6. J.J. Conradie, Generalized precompactness and mixed topologies. Collect. Math. **44**, 164–172 (1993)
7. J.J. Conradie, The coarsest Hausdorff Lebesgue topology. Quaest. Math. **28**, 287–304 (2005)
8. J. Conradie, Mackey topologies and mixed topologies in Riesz spaces. Positivity **10**, 591–606 (2006)
9. Y.A. Dabboorasad, E.Y. Emelyanov, M.A.A. Marabeh, Order Convergence in infinite-dimensional vector lattices is not topological. Preprint, arXiv:1705.09883 [math.FA]
10. M. Duhoux, Order precompactness in topological Riesz spaces. J. Lond. Math. Soc. (2) **23**, 509–522 (1981)

11. D.H. Fremlin, *Topological Riesz Spaces and Measure Theory* (Cambridge University Press, Cambridge, 1974)
12. D.J.H. Garling, A generalized form of inductive limit topology for vector spaces. Proc. Lond. Math. Soc. **14**, 1–28 (1964)
13. N. Gao, V.G. Troitsky, F. Xanthos, Uo-convergence and its applications to Cesáro means in Banach lattices. Israel J. Math. **220**, 649–689 (2017)
14. N. Gao, D. Leung, F. Xanthos, Duality for unbounded order convergence and applications. Positivity **22**, 711–725 (2018)
15. J.J. Grobler, Indices for Banach function spaces. Math. Z. 145, 99–109 (1975)
16. J.J. Grobler, P. van Eldik, Lebesgue-type convergence theorems in Banach lattices with applications to compact operators. Indag. Math. **41**, 425–437 (1979)
17. M. Nowak, Mixed topology on normed function spaces I. Bull. Pol. Ac. Math. **36**, 251–262 (1988)
18. E.T. Ordman, Convergence almost everywhere is not topological. Am. Math. Monthly **2**, 182–183 (1966)
19. M.A. Taylor, Unbounded topologies and uo-convergence in locally solid vector lattices. Preprint, arXiv:1706.01575 [math.FA]

Lattice Homomorphisms in Harmonic Analysis

H. Garth Dales and Marcel de Jeu

Dedicated to Ben de Pagter on the occasion of his 65th birthday

Abstract Let S be a non-empty, closed subspace of a locally compact group G that is a subsemigroup of G. Suppose that X, Y, and Z are Banach lattices that are vector sublattices of the order dual $\mathrm{C_c}(S, \mathbb{R})^\sim$ of the real-valued, continuous functions with compact support on S, and where Z is Dedekind complete. Suppose that $* : X \times Y \to Z$ is a positive bilinear map such that $\operatorname{supp}(x * y) \subseteq \operatorname{supp} x \cdot \operatorname{supp} y$ for all $x \in X^+$ and $y \in Y^+$ with compact support. We show that, under mild conditions, the canonically associated map from X into the vector lattice of regular operators from Y into Z is then a lattice homomorphism. Applications of this result are given in the context of convolutions, answering questions previously posed in the literature.

As a preparation, we show that the order dual of the continuous, compactly supported functions on a closed subspace of a locally compact space can be canonically viewed as an order ideal of the order dual of the continuous, compactly supported functions on the larger space.

As another preparation, we show that L^p-spaces and Banach lattices of measures on a locally compact space can be embedded as vector sublattices of the order dual of the continuous, compactly supported functions on that space.

Keywords Locally compact group · Convolution · Banach lattice · Lattice homomorphism · Locally compact space · Order dual · Radon measure

---

H. G. Dales
Department of Mathematics and Statistics, University of Lancaster, Lancaster, UK
e-mail: g.dales@lancaster.ac.uk

M. de Jeu (✉)
Mathematical Institute, Leiden University, Leiden, The Netherlands

Department of Mathematics and Applied Mathematics, University of Pretoria, Pretoria, South Africa
e-mail: mdejeu@math.leidenuniv.nl

G. Buskes et al. (eds.), *Positivity and Noncommutative Analysis*,
Trends in Mathematics, https://doi.org/10.1007/978-3-030-10850-2_6

1 Introduction and Overview

Let G be a locally compact group with (real) measure algebra $\mathrm{M}(G, \mathbb{R})$. Then $\mathrm{M}(G, \mathbb{R})$ is not only a Banach algebra with convolution as multiplication, but also a Banach lattice. The left regular representation π of $\mathrm{M}(G, \mathbb{R})$ is easily seen to take its values in the algebra of regular operators $\mathrm{L_r}(\mathrm{M}(G, \mathbb{R}))$ on $\mathrm{M}(G, \mathbb{R})$, so that we actually have an algebra homomorphism $\pi : \mathrm{M}(G, \mathbb{R}) \to \mathrm{L_r}(\mathrm{M}(G, \mathbb{R}))$. Furthermore, $\mathrm{M}(G, \mathbb{R})$ is Dedekind complete, so that $\mathrm{L_r}(\mathrm{M}(G, \mathbb{R}))$ is a vector lattice again. Hence it is meaningful to wonder whether the left regular representation $\pi : \mathrm{M}(G, \mathbb{R}) \to \mathrm{L_r}(\mathrm{M}(G, \mathbb{R}))$ is not only an algebra homomorphism, but also a lattice homomorphism. This question was raised during a workshop on ordered Banach algebras at the Lorentz Center in Leiden in 2014, and it occurs in Wickstead's list of open problems based on those that were posed during this workshop, see [52].

The natural approach to this question is to start with one of the Riesz–Kantorovich formulae as a basis to determine whether π is a lattice homomorphism, and to use the explicit formula for the convolution of two measures while doing so. Then the expressions become complicated very quickly, and an answer has not been obtained along these lines so far.

Nevertheless, the answer to the question is known: the left regular representation $\pi : \mathrm{M}(G, \mathbb{R}) \to \mathrm{L_r}(\mathrm{M}(G, \mathbb{R}))$ is indeed a lattice homomorphism. The first proof of this, as obtained by the present authors, is surprisingly simple. It uses just a little more than the fact that the support of the convolution of two measures with compact support is contained in the products of the support, combined with the general fact that the modulus on a vector lattice is additive on finite sums of mutually disjoint elements. The Riesz–Kantorovich formulae and the explicit expression for the convolution of two measures are not needed.

A closer look at the proof showed that, in fact, it does not really use that the objects involved are measures. Essentially the same proof establishes that, for $1 \leq p < \infty$, the natural action of $\mathrm{L}^1(G, \mathbb{R})$ on $\mathrm{L}^p(G, \mathbb{R})$ by convolution gives a lattice homomorphism from $\mathrm{L}^1(G, \mathbb{R})$ into the regular operators $\mathrm{L_r}(\mathrm{L}^p(G, \mathbb{R}))$ on $\mathrm{L}^p(G, \mathbb{R})$. In fact, under mild conditions, it shows that, 'whenever' a Banach lattice X on G convolves a Banach lattice Y on G into a Dedekind complete Banach lattice Z on G, then the natural map from X into the regular operators from Y into Z is a lattice homomorphism. A still closer look showed that it is not even necessary that the action of X on Y be given by convolution. As long as it is a positive map that satisfies the property for supports mentioned above, essentially the same proof as for $\mathrm{M}(G, \mathbb{R})$ shows that the natural map from X into the regular operators from Y into Z is still a lattice homomorphism. As a rule of thumb, this is 'always' true for convolution-like positive bilinear maps. Exaggerating a little, one could say that the main problem with the original question for $\mathrm{M}(G, \mathbb{R})$ is that there is too much information that obscures the underlying picture.

Above, we have spoken loosely about 'essentially the same proof' and 'Banach lattices on G'. It is evidently desirable to be able to make this precise, and then—hopefully—give the 'essential' proof of one central theorem that clarifies the mathematical backbone of the situation, and that specialises to various practical cases of interest. This is, indeed, possible. As will become apparent, the order dual $C_c(G, \mathbb{R})^\sim$ of the continuous functions with compact support on G can act as a large vector lattice that—this is true in a more general context of locally compact spaces—contains various familiar Banach lattices as vector sublattices. It is in this framework that such a central theorem can, indeed, be established 'once and for all'. The ensuing result, which is the group case of Theorem 10.3, below, is the heart of this article.

There are many examples of Banach algebras on a locally compact *semi*group S, provided with a convolution-like product, that are also Dedekind complete Banach lattices. Again, one can ask whether the left regular representation of these algebras is a lattice homomorphism. More generally again, if a Banach lattice X on S 'convolves' a Banach lattice Y on S into a Banach lattice Z on S, where Z is Dedekind complete, is the canonically associated map from X into the regular operators from Y into Z then a lattice homomorphism? Unfortunately, the proof of the general theorem as given for groups is then no longer valid. Results can still be obtained, however, when one supposes that S is actually a closed subset of a locally compact group G. It is then possible to reduce the problem for S to the problem for G, where the answer is known. For this, one merely needs to be able to view Banach lattices that are sublattices of $C_c(S, \mathbb{R})^\sim$ as Banach lattices that are sublattices of $C_c(G, \mathbb{R})^\sim$. This is indeed possible, since—this is a special case of a general result for closed subspaces of locally compact spaces—it can be shown that one can canonically embed $C_c(S, \mathbb{R})^\sim$ as a vector sublattice of $C_c(G, \mathbb{R})^\sim$, with supports being preserved under the embedding. It is thus that the group case of our main result, Theorem 10.3, below, can actually be used to establish a similar result for semigroups that are closed subsets of locally compact groups. In the end, the original result for locally compact groups (where the actual key proof can be given) is then a special case of Theorem 10.3. This final result is described in the abstract of this article.

It may have become obvious from the above discussion that the present article is at the interface of the fields of positivity, abstract harmonic analysis, and Banach algebras. It is, perhaps, not yet very common to be familiar with the basic notions of these three disciplines together. It is for this reason that we have decided to explain the necessary terms and to review the necessary results from each of these fields in an attempt to make this article accessible to all readers, regardless of their background. We also hope that, by doing this, we shall facilitate further research at the junction of these disciplines.

This article is organised as follows.

Section 2 contains basic notions and results for vector lattices and Banach lattices.

Banach lattices can be complexified to yield complex Banach lattices; this is the topic of Sect. 3.

Section 4 covers the basic notions of Banach algebras and Banach lattice algebras, and introduces complex Banach lattice algebras.

Section 5 is concerned with locally compact spaces, and notably with the order dual $\mathrm{C_c}(X,\mathbb{R})^\sim$ of the continuous, compactly supported functions on a locally compact space X. As will be explained in that section, this order dual is Bourbaki's space of Radon measures on X as in [12].

Section 6 shows how the order dual $\mathrm{C_c}(Y,\mathbb{R})^\sim$ for a closed subspace Y of a locally compact space X can be embedded into $\mathrm{C_c}(X,\mathbb{R})^\sim$ as an order ideal. The reader whose interest lies in groups and not in semigroups can omit this section in its entirety. We are not aware of a reference for the results in this section, which may also find applications elsewhere.

Let X be a locally compact space. As explained above in the context where X is a locally compact group, it is necessary to embed various familiar Banach lattices on X as vector sublattice of $\mathrm{C_c}(X,\mathbb{R})^\sim$. This is done in Sect. 7. We are not aware of earlier results in this direction, where the role of $\mathrm{C_c}(X,\mathbb{R})^\sim$ is not dissimilar to that of the space of distributions on an open subset of $\mathbb{R}^d$ in the sense of Schwartz.

Section 8 contains the necessary material on locally compact groups and on Banach lattices and Banach lattice algebras on such groups.

Section 9 is of a similar nature as Sect. 8, but now for semigroups. Taken together, Sects. 8 and 9 contain a good stockpile of Banach lattice algebras. Some of them are semisimple, while others are radical—this does not seem to influence the order properties of the left regular representations. We hope that these examples can also serve as test cases for further study of Banach lattice algebras in general.

Section 10 contains our key results. This section is the core of the present article and the other sections are, in a sense, merely auxiliary. The reader may actually wish to have a look at this section, and notably at the proof for the group case of Theorem 10.3, before reading other sections.

In Sect. 11, all is put together. The general results from Sect. 10, combined with the embedding results from Sect. 7, are now easily combined to yield that various canonical maps are actually lattice homomorphism. The left regular representation of $\mathrm{M}(G,\mathbb{R})$ is one of them. We also include in this section a list of cases where it is known whether the left regular representation of a Dedekind complete Banach lattice algebra is a lattice homomorphism or not.

Section 12 discusses the relation between one of the results in Sect. 11 and earlier work by Arendt, Brainerd and Edwards, and Gilbert. This leads to questions for further research, on which we hope to be able to report in the future.

We conclude this section by introducing a few conventions and notations.

The vector spaces and algebras in this article are all over the real field, $\mathbb{R}$, unless stated otherwise. This is the canonical convention in the field of positivity. On the other hand, the canonical convention in the context of Banach algebras and abstract harmonic analysis is that the base field be the complex field, $\mathbb{C}$. There seems to be no natural way to reconcile these two conventions where these disciplines meet. In view of the prominent role of ordering in the present article, we have chosen to consistently side with the convention in positivity. Readers from a different background are, therefore, cautioned to realise that a Banach algebra is a real Banach algebra, and that, e.g., the measure algebra of a locally compact group consists of the *real* signed regular Borel measures on the group. We apologise for the mental dissonance that such consequences of our efforts to be precise and consistent will almost inevitably cause. In a further attempt to prevent misunderstanding as much as possible, we have included the field in the notation for concrete spaces. The group algebra of a locally compact group is denoted by $\mathrm{L}^1(G, \mathbb{R})$, for example.

We shall let $\mathbb{F}$ denote the choice for either $\mathbb{R}$ or $\mathbb{C}$ when results are valid in both cases.

Algebras are always linear and associative. An algebra need not have an identity element. An algebra homomorphism between two unital algebras need not map the identity element to the identity element.

Topological spaces are always supposed to be Hausdorff, unless stated otherwise.

Let X be a topological space. Then we let $\mathrm{C}(X, \mathbb{R})$ denote the real-valued, continuous functions on X, we let $\mathrm{C_b}(X, \mathbb{R})$ denote the real-valued, bounded, continuous functions on X, we let $\mathrm{C_0}(X, \mathbb{R})$ denote the real-valued, continuous functions on X that vanish at infinity, and we let $\mathrm{C_c}(X, \mathbb{R})$ denote the real-valued, continuous functions on X with compact support. Their complex counterparts $\mathrm{C}(X, \mathbb{C})$, $\mathrm{C_b}(X, \mathbb{C})$, $\mathrm{C_0}(X, \mathbb{C})$, and $\mathrm{C_c}(X, \mathbb{C})$ are similarly defined. The Borel σ-algebra of X is the σ-algebra of subsets of X that is generated by the open subsets of X.

Let S be a non-empty set. Then $\|f\|_\infty$ denotes the uniform norm of a bounded, real- or complex-valued function f on S. Sometimes we shall write $\|f\|_{\infty,S}$ if confusion could arise otherwise.

Let E and F be normed spaces over $\mathbb{F}$. Then $\mathrm{B}(E, F)$ denotes the bounded linear operators from E into F. We shall write $\mathrm{B}(E)$ for $\mathrm{B}(E, E)$.

The identity element of a group G is denoted by e_G.

Semigroups need not have identity elements.

Let S be a semigroup, and suppose that A_1 and A_2 are non-empty subsets of S. Then we set $A_1 \cdot A_2 := \{a_1 a_2 : a_1 \in A_1,\, a_2 \in A_2\}$.

2 Vector Lattices and Banach Lattices

In this section, we shall cover some basic material on vector and Banach lattices. The details can be found in introductory books such as [22, 56]. More advanced general references are [1, 2, 4, 5, 34, 35, 46, 54, 55].

Suppose that E is a partially ordered vector space, i.e., a vector space that is supplied with a partial ordering such that $x + z \geq y + z$ for all $z \in E$ whenever $x, y \in E$ are such that $x \geq y$, and such that $\alpha x \geq 0$ whenever $x \geq 0$ in E and $\alpha \geq 0$ in $\mathbb{R}$. The subset of positive elements of E is then a cone, and it is denoted by E^+.

A *vector lattice* or *Riesz space* is a partially ordered vector space E such that every two elements x, y of E have a least upper bound in E; this supremum of the set $\{x, y\}$ is denoted by $x \vee y$. The infimum of $\{x, y\}$ then also exists; it is denoted by $x \wedge y$. For $x \in E$, we define its *modulus* $|x|$ as $|x| := x \vee (-x)$, its *positive part* x^+ as $x^+ := x \vee 0$, and its *negative part* x^- as $x^- := (-x) \vee 0$. Then $x^+, x^- \in E^+$, $x = x^+ - x^-$, and $|x| = x^+ + x^-$.

Let E be a vector lattice. Two elements x and y of E are *disjoint* if $|x| \wedge |y| = 0$; this is denoted by $x \perp y$. When this is the case, then $|x + y| = |x| + |y|$. This latter property lies at the heart of the results in this article, and can be found in [34, Theorem 14.4(i)] and [56, Theorem 8.2(i)], for example.

Let $x \in E$. Then $x^+ \perp x^-$. Suppose that $x = y_1 - y_2$ with $y_1, y_2 \in E^+$. Then $y_1 \geq x^+$ and $y_2 \geq x^-$. Suppose, further, that $y_1 \perp y_2$. Then $y_1 = x^+$ and $y_2 = x^-$.

Let E be a vector lattice, and let F be a linear subspace of E. Then F is a *vector sublattice of* E if $x \vee y \in F$ whenever $x, y \in F$; then also $x \wedge y \in F$ whenever $x, y \in F$, and $|x| \in F$ whenever $x \in F$.

Let E be a vector lattice, and let F be a vector sublattice of E. Then F is an *order ideal of* E if $x \in F$ whenever $x, y \in E$ are such that $|x| \leq |y|$ and $y \in F$.

An *order interval* in a vector lattice E is a subset of the form

$$\{x \in E : a \leq x \leq b\}$$

for some $a \leq b$ in E. A subset of E is *order bounded* if it is contained in an order interval.

A vector lattice E is *Dedekind complete* or *order complete* if every non-empty subset of E that is bounded above in E has a supremum in E.

Example 2.1 Let X be a non-empty, topological space. Then $\mathrm{C}(X, \mathbb{R})$, $\mathrm{C_b}(X, \mathbb{R})$, $\mathrm{C_0}(X, \mathbb{R})$, and $\mathrm{C_c}(X, \mathbb{R})$ are vector lattices when supplied with the pointwise ordering.

Let X be a non-empty, compact space. Then $\mathrm{C}(X, \mathbb{R})$ is Dedekind complete if and only if X is extremely disconnected (some sources write 'extremally disconnected'), i.e., if and only if the closure of every open subset of X is open. This result is due to Nakano, see [18, Proposition 4.2.9], [19, Theorem 2.3.3], or [22, Theorem 12.16], for example. The Stone–Čech compactification $\beta\mathbb{N}$ of the natural numbers $\mathbb{N}$ is an example of a compact, extremely disconnected space.

Example 2.2 Let X be a non-empty set, let $\mathfrak{B}$ be a σ-algebra of subsets of X, and let $\mu : \mathfrak{B} \to [0, \infty]$ be a measure on $\mathfrak{B}$. For $1 \leq p \leq \infty$, we supply $\mathrm{L}^p(X, \mathfrak{B}, \mu, \mathbb{R})$ with the pointwise μ-almost everywhere partial ordering. Then $\mathrm{L}^p(X, \mathfrak{B}, \mathbb{R})$ is a vector lattice. For $1 \leq p < \infty$, it is Dedekind complete. For $p = \infty$, it is Dedekind complete if μ is localisable, i.e., if every measurable subset of X of infinite measure has a measurable subset of finite, strictly positive measure and the measure algebra of X is order complete. In particular, $\mathrm{L}^\infty(X, \mathfrak{B}, \mu)$ is Dedekind complete when μ is σ-finite. We refer to [34, pp. 126–127] and [26, Definition 211G, Theorem 211L, and Theorem 243H] for proofs.

An example, taken from [49], where $\mathrm{L}^\infty(X, \mathfrak{B}, \mu)$ is not Dedekind complete, is as follows. Let X be an uncountable set, and let $\mathfrak{B}$ be the σ-algebra of all subsets A of X such that either A or $X \setminus A$ is uncountable. Let μ be the counting measure on $\mathfrak{B}$. Take a subset U of X such that both U and $X \setminus U$ are uncountable, and set

$$S := \{1_A : A \subset U \text{ and } A \text{ is countable}\}.$$

The S is a subset of $\mathrm{L}^\infty(X, \mathfrak{B}, \mu)$ that is bounded above, but S has no supremum in $\mathrm{L}^\infty(X, \mathfrak{B}, \mu)$. Hence $\mathrm{L}^\infty(X, \mathfrak{B}, \mu)$ is not Dedekind complete.

Example 2.3 Let X be a non-empty set, and let $\mathfrak{B}$ be a σ-algebra of subsets of X. We let $\mathrm{M}(X, \mathfrak{B}, \mathbb{R})$ be the vector space of all signed measures $\mu : \mathfrak{B} \to \mathbb{R}$. We introduce a partial ordering on $\mathrm{M}(X, \mathfrak{B}, \mathbb{R})$ by setting $\mu \geq \nu$ whenever $\mu, \nu \in \mathrm{M}(X, \mathfrak{B}, \mathbb{R})$ are such that $\mu(A) \geq \nu(A)$ for all $A \in \mathfrak{B}$. Then $\mathrm{M}(X, \mathfrak{B}, \mathbb{R})$ is a Dedekind complete vector lattice, see [56, p. 187]. For $\mu, \nu \in \mathrm{M}(X, \mathfrak{B}, \mathbb{R})$, the supremum $\mu \vee \nu$ of μ and ν is given by the formula

$$(\mu \vee \nu)(A) = \sup\{\, \mu(B) + \nu(A \setminus B) : B \in \mathfrak{B},\ B \subseteq A \,\} \tag{2.1}$$

for $A \in \mathfrak{B}$. The formula for the infimum is similar, and, for $\mu \in \mathrm{M}(X, \mathfrak{B}, \mathbb{R})$, we have

$$|\mu|(A) = \sup\left\{ \sum_{i=1}^{n} |\mu(B_i)| : B_1, \ldots, B_n \in \mathfrak{B} \text{ form a disjoint partition of } A \right\} \tag{2.2}$$

for $A \in \mathfrak{B}$. That is, $|\mu|$ is the usual total variation measure of μ.

Suppose that E and F are vector lattices and that $T : E \to F$ is a linear operator. Then T is *order bounded* if T maps order bounded subsets of E to order bounded subsets of F. Equivalently, T should map order intervals in E into order intervals in F. The order bounded linear operators from E into F form a vector space that is denoted by $\mathrm{L_b}(E, F)$. We shall write $\mathrm{L_b}(E)$ for $\mathrm{L_b}(E, E)$.

Let $S, T : E \to F$ be order bounded linear operators. Then we say that $S \geq T$ if $Sx \geq Tx$ for all $x \in E^+$. This introduces a partially ordering on $\mathrm{L_b}(E, F)$. The *regular operators* from E into F are the elements of the subspace $\mathrm{L_r}(E, F)$ of $\mathrm{L_b}(E, F)$ that is spanned by the positive linear operators from E into F. Thus the regular operators from E into F are the linear operators T from E into F that can be written as $T = S_1 - S_2$, where $S_1, S_2 \in \mathrm{L_b}(E, F)$ are both positive. We shall write $\mathrm{L_r}(E)$ for $\mathrm{L_r}(E, E)$.

It is not generally true that the partially ordered vector spaces $\mathrm{L_b}(E, F)$ or $\mathrm{L_r}(E, F)$ are again vector lattices, but there is a sufficient condition on the codomain for this to be the case. We have the following, see [5, Theorem 1.18] or [56, Theorem 20.4], for example.

Theorem 2.4 *Let E and F be vector lattices such that F is Dedekind complete. Then the spaces $\mathrm{L_b}(E, F)$ and $\mathrm{L_r}(E, F)$ coincide. Moreover, $\mathrm{L_r}(E, F)$ is a Dedekind complete vector lattice, where the lattice operations are given by*

$$|T|(x) = \sup\{\, |Ty| : |y| \leq x \,\}, \tag{2.3}$$

$$[S \vee T](x) = \sup\{\, Sy + Tx : y, z \in E^+,\ y + z = x \,\}, \text{ and} \tag{2.4}$$

$$[S \wedge T](x) = \inf\{\, Sy + Tx : y, z \in E^+,\ y + z = x \,\} \tag{2.5}$$

for all $S, T \in \mathrm{L_r}(E, F)$ and $x \in E^+$.

The formulae in the above theorem are the *Riesz–Kantorovich formulae*.

Applying the theorem with $F = \mathbb{R}$, we see that the order bounded linear functionals on E coincide with the regular ones, and that they form a vector lattice. This vector lattice is denoted by $E^\sim$, and it is called the *order dual of* E. Of course, for $\varphi \in E^\sim$, we have $\varphi \geq 0$ if and only if $\langle \varphi, x\rangle \geq 0$ for all $x \in E^+$.

Suppose that E and F are vector lattices. A linear operator $T : E \to F$ is a *lattice homomorphism* if $T(x \vee y) = Tx \vee Ty$ for all $x, y \in E$. This is equivalent to requiring that $T(x \wedge y) = Tx \wedge Ty$ for all $x, y \in E$, and also equivalent to requiring that $|Tx| = T|x|$ for all $x \in E$. Lattice homomorphisms are positive linear operators.

A linear operator $T : E \to F$ is *interval preserving* if it is positive and such that $T([0, x]) = [0, Tx]$ for all $x \in E^+$. The positivity of T already implies that $T([0, x]) \subseteq [0, Tx]$; the point is that equality should hold.

Let $T : E \to F$ be an order bounded linear operator. Then its *order adjoint* $T^\sim : F^\sim \to E^\sim$ is defined by setting

$$\left\langle T^\sim\varphi, x\right\rangle := \langle \varphi, Tx\rangle$$

for $x \in E$ and $\varphi \in E^\sim$. In Sect. 5, we shall use the following two results, see [5, Theorems 2.19 and 2.20].

Proposition 2.5 *Let $T : E \to F$ be an interval preserving linear operator between the vector lattices E and F. Then $T^\sim : F^\sim \to E^\sim$ is a lattice homomorphism.*

Proposition 2.6 *Let $T : E \to F$ be a positive linear operator between the vector lattices E and F, where F is such that $F^\sim$ separates the points of F. Then T is a lattice homomorphism if and only if $T^\sim : F^\sim \to E^\sim$ is interval preserving.*

Let E be a vector lattice. Then a norm $\|\cdot\|$ on E is a *lattice norm* if $\|x\| \leq \|y\|$ whenever x and y in E are such that $|x| \leq |y|$.

Definition 2.7 A Banach space $(E, \|\cdot\|)$ for which E is a vector lattice and $\|\cdot\|$ is a lattice norm is a *Banach lattice*.

Example 2.8 Let X be a topological space. Then the vector lattices $\mathrm{C_b}(X, \mathbb{R})$ and $\mathrm{C_0}(X, \mathbb{R})$ from Example 2.1 are Banach lattices when supplied with the uniform norm $\|\cdot\|_\infty$.

Example 2.9 Let X be a non-empty set, let $\mathfrak{B}$ be a σ-algebra of subsets of X, and let $\mu : \mathfrak{B} \to [0, \infty]$ be a measure on $\mathfrak{B}$. Then the vector lattices $\mathrm{L}^p(X, \mathfrak{B}, \mu, \mathbb{R})$ from Example 2.2 are Banach lattices when supplied with the usual p-norm $\|\cdot\|_p$.

Example 2.10 Let X be a non-empty set, and let $\mathfrak{B}$ be a σ-algebra of subsets of X. Then the vector lattice $\mathrm{M}(X, \mathfrak{B}, \mathbb{R})$ of real-valued measures on $\mathfrak{B}$ from Example 2.3 is a Banach lattice when supplied with the norm $\|\cdot\|$ that is obtained by setting

$$\|\mu\| := |\mu|(X). \tag{2.6}$$

Let E be a Banach lattice. Then E has an order dual $E^\sim$ as a vector lattice, as well as a topological dual E' as a Banach space. It is a fundamental fact that $E^\sim = E'$, see [5, Corollary 4.4] or [56, Theorem 25.8(iii)], for example.

Suppose that E is a Banach lattice, that F is a normed vector lattice, and that the map $T : E \to F$ is an order bounded linear operator. Then E is automatically continuous, see [5, Theorem 4.3], for example. In the sequel we shall repeatedly use the special case that a positive linear operator from a Banach lattice into a normed vector lattice is automatically continuous.

Let E and F be Banach lattices, where F is Dedekind complete. Then we know from Theorem 2.4 that $\mathrm{L_r}(E, F)$ is a Dedekind complete vector lattice. It can be supplied with the operator norm, but this is not generally a lattice norm. One can, however, define the *regular norm* $\|\cdot\|_{\mathrm{r}}$ on $\mathrm{L_r}(E, F)$ by setting

$$\|T\|_{\mathrm{r}} := \| |T| \|$$

for $T \in \mathrm{L_r}(E, F)$. The regular norm is a lattice norm on $\mathrm{L_r}(E, F)$, and $\mathrm{L_r}(E, F)$ is then a Dedekind Banach lattice, see [5, Theorem 4.74], for example.

3 Complex Banach Lattices

In abstract harmonic analysis, Banach spaces and Banach algebras are almost always over the complex numbers. It is for this reason that we include the following material on complex Banach lattices. Details can be found in [1, Section 3.2], [35, Section 2.2], or [46, Section 2.11], for example.

Let E be a Banach lattice. Then its complexified vector space $E_{\mathbb{C}}$ can be supplied with a modulus $|\cdot|_{\mathbb{C}} : E_{\mathbb{C}} \to E$. The definition of $|\cdot|_{\mathbb{C}}$ is analogous to one of the possible descriptions of the modulus of a complex number, as follows. For $x, y \in E$, the supremum

$$\sup\{\,\mathrm{Re}(\mathrm{e}^{\mathrm{i}\theta}(x + \mathrm{i}y)) : 0 \leq \theta \leq 2\pi\,\} = \sup\{\,x\cos\theta + y\sin\theta : 0 \leq \theta \leq 2\pi\,\}$$

can be shown to exist in E, and we define this supremum to be the modulus $|x + \mathrm{i}y|_{\mathbb{C}}$ of the element $x + \mathrm{i}y$ of $E_{\mathbb{C}}$. Then $|\cdot|_{\mathbb{C}}$ extends the modulus $|\cdot|$ on E. Take $z \in E_{\mathbb{C}}$. Then $|z|_{\mathbb{C}} = 0$ if and only if $z = 0$. Furthermore, $|\alpha z|_{\mathbb{C}} = |\alpha||z|_{\mathbb{C}}$ for all $\alpha \in \mathbb{C}$ and $z \in E_{\mathbb{C}}$, and $|w + z|_{\mathbb{C}} \leq |w|_{\mathbb{C}} + |z|_{\mathbb{C}}$ for all $w, z \in E_{\mathbb{C}}$.

Set $\|z\|_{\mathbb{C}} := \| |z|_{\mathbb{C}} \|$ for $z \in E_{\mathbb{C}}$. Then $\|\cdot\|_{\mathbb{C}}$ is a norm on $E_{\mathbb{C}}$ that extends the norm on E, and $(E_{\mathbb{C}}, \|\cdot\|_{\mathbb{C}})$ is a complex Banach space that is called a *complex Banach lattice*. As a topological vector space, $E_{\mathbb{C}}$ is $\mathbb{R}$-linearly homeomorphic to the Cartesian product $E \times E$. One of the things to remember is that the non-zero complex Banach lattices are not lattices: they do have a modulus, but there is no role for a partial ordering on $E_{\mathbb{C}}$ as a whole.

Example 3.1 Let X be a topological space. Then the complexifications of the Banach lattice $\mathrm{C_b}(X, \mathbb{R})$, respectively, $\mathrm{C_0}(X, \mathbb{R})$, from Example 2.8 can be identified with the Banach space $\mathrm{C_b}(X, \mathbb{C})$, respectively, $\mathrm{C_0}(X, \mathbb{C})$, with the usual pointwise complex modulus and with the uniform norm $\|\cdot\|_\infty$.

Example 3.2 Let X be a non-empty set, let $\mathfrak{B}$ be a σ-algebra of subsets of X, and let $\mu : \mathfrak{B} \to [0, \infty]$ be a measure on $\mathfrak{B}$. Then the complexifications of the Banach lattices $\mathrm{L}^p(X, \mathfrak{B}, \mu, \mathbb{R})$ from Example 2.9 can be identified with the Banach spaces $\mathrm{L}^p(X, \mathfrak{B}, \mu, \mathbb{C})$, with the usual pointwise μ-almost everywhere complex modulus and with the usual p-norm $\|\cdot\|_p$.

Example 3.3 Let X be a non-empty set, and let $\mathfrak{B}$ be a σ-algebra of subsets of X. Then the complexification of the Banach lattice $\mathrm{M}(X, \mathfrak{B}, \mathbb{R})$ of real-valued measures on $\mathfrak{B}$ from Example 2.10 can be identified with the Banach space $\mathrm{M}(X, \mathfrak{B}, \mathbb{C})$ of complex-valued measures on $\mathfrak{B}$, where the modulus, respectively, the norm, is again given by Eq. (2.2), respectively, Eq. (2.6).

Let E and F be Banach lattices, and let $T : E \to F$ be a bounded linear operator. Then its complex-linear extension $T_\mathbb{C} : (E_\mathbb{C}, \|\cdot\|_\mathbb{C}) \to (F_\mathbb{C}, \|\cdot\|_\mathbb{C})$ is a bounded linear operator, and $\|T\| \le \|T_\mathbb{C}\| \le 2\|T\|$. If $T \ge 0$, then $\|T_\mathbb{C}\| = \|T\|$.

Let $E_\mathbb{C}$ and $F_\mathbb{C}$ be complex Banach lattices. Then every complex-linear operator $T : E_\mathbb{C} \to F_\mathbb{C}$ has a unique expression as $T = S_1 + \mathrm{i}S_2$, where $S_1, S_2 : E \to F$ are real-linear operators, and

$$(S_1 + \mathrm{i}S_2)(x + \mathrm{i}y) = (Sx - Ty) + \mathrm{i}(Sy + Tx)$$

for $x, y \in E$. Then T is *order bounded* (respectively, *regular*) if both S_1 and S_2 are order bounded (respectively, regular). The complex vector space of all order bounded (respectively, regular) complex-linear operators from $E_\mathbb{C}$ into $F_\mathbb{C}$ is denoted by $\mathrm{L_b}(E_\mathbb{C}, F_\mathbb{C})$ (respectively, $\mathrm{L_r}(E_\mathbb{C}, F_\mathbb{C})$). Then $\mathrm{L_r}(E_\mathbb{C}, F_\mathbb{C}) \subseteq \mathrm{L_b}(E_\mathbb{C}, F_\mathbb{C}) \subseteq \mathrm{B}(E_\mathbb{C}, F_\mathbb{C})$. A complex-linear operator $T : E_\mathbb{C} \to F_\mathbb{C}$ is *positive* if $T(E^+) \subseteq F^+$; this implies that $T(E) \subseteq F$. For such positive T, we have $|Tz|_\mathbb{C} \le T(|z|_\mathbb{C})$ for $z \in E_\mathbb{C}$. A complex-linear operator $T : E_\mathbb{C} \to F_\mathbb{C}$ is a *complex lattice homomorphism* if $|Tz|_\mathbb{C} = T(|z|_\mathbb{C})$ for all $z \in E_\mathbb{C}$. This is the case if and only if T leaves E invariant and the restricted map $T\,|_E : E \to E$ is a lattice homomorphism, see [45, p. 136].

Let E and F be Banach lattices, where F is Dedekind complete. Then the space $(\mathrm{L_r}(E, F), \|\cdot\|_\mathrm{r})$ is a Dedekind complete Banach lattice, so that we can consider the complex Banach lattice $([\mathrm{L_r}(E, F)]_\mathbb{C}, \|\cdot\|_{\mathrm{r},\mathbb{C}})$. For $T \in [\mathrm{L_r}(E, F)]_\mathbb{C}$, we have, by definition, that

$$\|T\|_{\mathrm{r},\mathbb{C}} = \||T|_\mathbb{C}\|_\mathrm{r} = \||T|_\mathbb{C}\|,$$

and then the norm $\|\cdot\|_{\mathrm{r},\mathbb{C}}$ on $[\mathrm{L_r}(E, F)]_\mathbb{C}$ extends the norm $\|\cdot\|_\mathrm{r}$ on $\mathrm{L_r}(E, F)$. It is clear from the definitions that $\mathrm{L_r}(E_\mathbb{C}, F_\mathbb{C})$ and $[\mathrm{L_r}(E, F)]_\mathbb{C}$ can be identified as complex vector spaces. Let $T \in \mathrm{L_r}(E_\mathbb{C}, F_\mathbb{C})$. Then, viewing T as an element of $[\mathrm{L_r}(E, F)]_{\mathbb{C}}$, so that $|T|_{\mathbb{C}}$ is defined in $[\mathrm{L_r}(E, F)]_{\mathbb{C}}$, and viewing $|T|_{\mathbb{C}}$ as an element of $\mathrm{L_r}(E_{\mathbb{C}}, F_{\mathbb{C}})$ again, we have

$$|T|_{\mathbb{C}}x = \sup\left\{ |Tz|_{\mathbb{C}} : z \in E_{\mathbb{C}},\ |z|_{\mathbb{C}} \leq x \right\}$$

for all $x \in E^+$, and

$$|Tz|_{\mathbb{C}} \leq |T|_{\mathbb{C}}|z|_{\mathbb{C}} \tag{3.7}$$

for all $z \subset E_{\mathbb{C}}$.

Let $(E, \|\cdot\|)$ be a Banach lattice with dual Banach lattice $(E', \|\cdot\|')$. It follows from Eq. (3.7) that the norm dual of the complex Banach lattice $(E_{\mathbb{C}}, \|\cdot\|_{\mathbb{C}})$ is canonically isometrically isomorphic as a complex Banach space to the complex Banach lattice $\left((E')_{\mathbb{C}}, (\|\cdot\|')_{\mathbb{C}}\right)$. In particular, analogously to the case of real scalars, the norm dual of a complex Banach lattice is again a complex Banach lattice.

4 Banach Algebras and Banach Lattice Algebras

In this section, we shall review some material about Banach algebras, Banach lattice algebras, and their complex versions.

A *Banach algebra* (respectively, a *complex Banach algebra*) is a pair $(A, \|\cdot\|)$, where A is an algebra (respectively, a complex algebra) with a norm $\|\cdot\|$ such that $(A, \|\cdot\|)$ is a Banach space (respectively, a complex Banach space) and

$$\|a_1 a_2\| \leq \|a_1\| \|a_2\|$$

for $a_1, a_2 \in A$. An identity element, if present, need not have norm 1. A net $(a_i)_{i \in I}$ in A is an *approximate identity* if $\lim_i a_i a = \lim_i a a_i = a$ for all $a \in A$. If, in addition, $\|a_i\| \leq 1$ for all $i \in I$, then the approximate identity $(a_i)_{i \in I}$ is *contractive*.

Let A and B be Banach algebras. Then a map $\pi : A \to B$ is a *Banach algebra homomorphism* if it is a continuous algebra homomorphism. The notion of a *complex Banach algebra homomorphism* between two complex Banach algebras is similarly defined.

For an introduction to the theory of complex Banach algebras, see [6], for example; a more substantial account is given in [18]. As long as one does not move into topics where working over the complex field is manifestly essential—the latter actually constitute most of the theory—several of the (more basic) results about complex Banach algebras are obviously also true for Banach algebras.

Canonical examples of Banach algebras are $\mathrm{B}(E)$, where E is a Banach space, and $\mathrm{C_b}(X, \mathbb{R})$ and $\mathrm{C_0}(X, \mathbb{R})$, where X is a topological space and where the norm on both algebras is the supremum-norm $\|\cdot\|_\infty$. Examples of complex Banach algebras are obtained likewise.

In Sect. 8, we shall give examples of Banach algebras and complex Banach algebras on locally compact groups that involve convolution.

Let A be a complex algebra. A proper left ideal I in A is *modular* if there exists $u \in A$ with $a - au \in I$ for all $a \in A$. The family of modular left ideals in A (if non-empty) has maximal members, and the (*Jacobson*) *radical of* A is the intersection of the maximal modular left ideals of A [18, Section 1.5]; it is denoted by $\operatorname{rad} A$, where we set $\operatorname{rad} A := A$ when A has no maximal modular left ideals. In fact, $\operatorname{rad} A$ is a (two-sided) ideal in A. The complex algebra A is *semisimple* when $\operatorname{rad} A = \{0\}$ and *radical* when $\operatorname{rad} A = A$.

Let A be a complex Banach algebra. Then $\operatorname{rad} A$ is closed in A, and $A/\operatorname{rad} A$ is a semisimple complex Banach algebra. An element $a \in A$ is *quasi-nilpotent* if $\lim_{n\to\infty} \|a^n\|^{1/n} = 0$. Each quasi-nilpotent element belongs to $\operatorname{rad} A$, and $\operatorname{rad} A$ is equal to the set of quasi-nilpotent elements in the special case that A is commutative.

Banach lattice algebras combine the structures of Banach lattices and of Banach algebras. Their definition in the present article is as follows.

Definition 4.1 Let A be a Banach lattice that is also a Banach algebra such that the product of two positive elements is again positive. Then A is a *Banach lattice algebra.*

We note that the norm on a Banach lattice algebra is compatible with both the order and product.

There are further remarks concerning the definition of a Banach lattice algebra, in particular involving the role of an identity, in [51]. In the present article, we leave this unspecified: the algebra need not be unital, nor need an identity element, if present, be positive.

As compared to the general theory of Banach algebras or operator algebras the theory of Banach lattice algebras is largely undeveloped. We refer to [51, 52] for a survey and for open problems. Problems 6 and 7 in [52] are resolved by Corollary 11.4 and Theorem 11.1, respectively, in the present article.

Let A be a Banach lattice algebra, and take $a_1, a_2 \in A$. By splitting each of a_1 and a_2 into their positive and negative parts, it follows easily that $|a_1 a_2| \leq |a_1||a_2|$. This holds, in fact, in every so-called *Riesz algebra*, i.e., in every vector lattice that is an algebra with the property that the product of two positive elements is again positive.

Example 4.2 Let X be a topological space. Then $\mathrm{C_b}(X, \mathbb{R})$ and $\mathrm{C_0}(X, \mathbb{R})$, with the uniform norm and pointwise ordering, are Banach lattice algebras.

Example 4.3 Let E be a Dedekind complete Banach lattice. Then $\mathrm{L_r}(E)$ is a Dedekind complete Banach lattice and also an algebra. It is, in fact, a Riesz algebra. Since then $|T_1 T_2| \leq |T_1||T_2|$ for $T_1, T_2 \in \mathrm{L_r}(E)$, it follows that the regular norm $\|\cdot\|_{\mathrm{r}}$ is submultiplicative on $\mathrm{L_r}(E)$. Hence $(\mathrm{L_r}(E), \|\cdot\|_{\mathrm{r}})$ is a Dedekind complete Banach lattice algebra.

In Sect. 8, we shall define the group algebra and the measure algebra of a locally compact group. These Banach algebras are Banach lattice algebras.

Definition 4.4 Let A and B be Banach lattice algebras. Then a map $\pi : A \to B$ is a *Banach lattice algebra homomorphism* if π is a Banach algebra homomorphism as well as a lattice homomorphism.

Banach algebra homomorphisms are supposed to be continuous. However, since Banach lattice algebra homomorphisms are, in particular, positive linear maps between Banach lattices, their continuity is, in fact, already automatic.

Definition 4.5 Let A be a Banach lattice algebra, and let E be a Dedekind complete Banach lattice. Suppose that $\pi : A \to \mathrm{L_r}(E)$ is a Banach lattice algebra homomorphism. Then π is a *Banach lattice algebra representation of A on E*.

Let A be a Banach algebra. Then the *left regular representation of* A is the map $\pi : A \to \mathrm{B}(A)$ that is obtained by setting $\pi(a_1)a_2 := a_1 a_2$ for $a_1, a_2 \in A$. The left regular representation of a complex Banach algebra is similarly defined.

Let A be a Dedekind complete Banach lattice algebra. Since $A = A^+ - A^-$, it follows that the left regular representation π of A is, in fact, a positive algebra homomorphism $\pi : A \to \mathrm{L_r}(A) \subseteq \mathrm{B}(A)$ from A into the regular operators on A. Since A is a Dedekind complete Banach lattice, it is a meaningful question whether the left regular representation π of A as a Banach algebra is, in fact, a Banach lattice algebra representation of A on itself. That is, is the map $\pi : A \to \mathrm{L_r}(A)$ a lattice homomorphism? This question is raised in [52, Problem 1]. In Remark 11.9, below, we summarise what is known to us.

We shall now introduce complex Banach lattice algebras.

Let A be a Banach lattice algebra with norm $\|\cdot\|$. Applying the general procedure for the complexification of a Banach lattice, one obtains the complex Banach lattice $(A_\mathbb{C}, \|\cdot\|_\mathbb{C})$. Furthermore, $A_\mathbb{C}$ is also a complex algebra. It is a non-trivial fact that $|z_1 z_2|_\mathbb{C} \leq |z_1|_\mathbb{C} |z_2|_\mathbb{C}$ for all $z_1, z_2 \in A_\mathbb{C}$. We refer to [7, Lemma 1.5] or [47, Satz 1.1] for a proof of this result, which was later generalised to arbitrary Archimedean relatively uniformly complete Riesz algebras in [31]. The submultiplicativity of the lattice norm $\|\cdot\|$ on A then immediately implies that $\|z_1 z_2\|_\mathbb{C} \leq \|z_1\|_\mathbb{C} \|z_2\|_\mathbb{C}$ for $z_1, z_2 \in A_\mathbb{C}$. Hence the complex Banach space $(A_\mathbb{C}, \|\cdot\|_\mathbb{C})$ is also a complex Banach algebra. The complex Banach space $(A_\mathbb{C}, \|\cdot\|_\mathbb{C})$, with its structures of a complex Banach lattice and of a complex Banach algebra, is a *complex Banach lattice algebra*.

Example 4.6 Let X be a topological space. Complexification of the Banach lattice algebra $(\mathrm{C_0}(X, \mathbb{R}), \|\cdot\|_\infty)$, respectively, $(\mathrm{C_b}(X, \mathbb{R}), \|\cdot\|_\infty)$, yields the complex Banach lattice algebra $(\mathrm{C_0}(X, \mathbb{C}), \|\cdot\|_\infty)$, respectively, $(\mathrm{C_b}(X, \mathbb{C}), \|\cdot\|_\infty)$.

Example 4.7 Let E be a Dedekind complete Banach lattice. Then $(\mathrm{L_r}(E), \|\cdot\|_\mathrm{r})$ is a Banach lattice algebra, and complexification yields the complex Banach lattice algebra $(\mathrm{L_r}(E_\mathbb{C}), \|\cdot\|_{\mathrm{r},\mathbb{C}})$.

As we shall see later in Sect. 8, the complex group algebra (respectively, the complex measure algebra) of a locally compact group can be identified, as a complex algebra, with the complexification of the group algebra (respectively, the measure algebra) of the group. It is not difficult to see that the usual norms on these two complex Banach algebras coincide with the norms they obtain as complexifications of the pertinent Banach lattice algebras. Hence the complex group algebra and the complex measure algebra of a locally compact group, with the usual norm, are both complex Banach lattice algebras.

Remark 4.8 It is possible to complexify arbitrary Banach algebras. Indeed, suppose that A is a Banach algebra. Then the algebraic complexification $A_{\mathbb{C}}$ can be given a norm $\|\cdot\|_{\mathbb{C}}$ such that $(A_{\mathbb{C}}, \|\cdot\|_{\mathbb{C}})$ is a complex Banach algebra and the natural embedding $a \mapsto (a, 0)$ from A into $A_{\mathbb{C}}$ is an isometry. Furthermore, all norms on $A_{\mathbb{C}}$ with this property are equivalent. We refer to [42, Theorem 1.3.2] for these results.

There is, in fact, an explicit construction of such a norm in [42]. It would be interesting to investigate whether, for the complexifications of the Banach lattice algebras in the present article, this particular norm in [42] coincides with the norm as found above via the complexification of Banach lattices. If this were even true for general Banach lattice algebras, then this would yield an alternative proof of the submultiplicativity of the norm found via the complexifications of Banach lattices that would not need the results in [7, Lemma 1.5], [31], or [47, Satz 1.1] referred to above.

Definition 4.9 Let A and B be Banach lattice algebras. Then a map $\pi : A_{\mathbb{C}} \to B_{\mathbb{C}}$ is a *complex Banach lattice algebra homomorphism* if π is a complex Banach algebra homomorphism as well as a complex lattice homomorphism.

Let A and B be Banach lattice algebras. Then a map $\pi : A_{\mathbb{C}} \to B_{\mathbb{C}}$ is a complex Banach lattice algebra homomorphism if and only if π maps A into B and the restricted map $\pi\,|_A\colon A \to B$ is a Banach lattice algebra homomorphism.

A complex Banach lattice homomorphism is automatically continuous.

Definition 4.10 Let A be a Banach lattice algebra, and let E be a Dedekind complete Banach lattice. Suppose that $\pi : A_{\mathbb{C}} \to \mathrm{L_r}(E_{\mathbb{C}})$ is a complex Banach lattice algebra homomorphism. Then π is a *complex Banach lattice algebra representation of A on $E_{\mathbb{C}}$*.

Let A be a Banach lattice algebra, and let E be a Dedekind complete Banach lattice. Then, by combining Definitions 4.4, 4.5, 4.9, and 4.10, we see that a complex algebra homomorphism $\pi : A_{\mathbb{C}} \to \mathrm{L_r}(E_{\mathbb{C}})$ is a complex Banach lattice algebra representation of $A_{\mathbb{C}}$ on $E_{\mathbb{C}}$ if and only if π maps A into $\mathrm{L_r}(E)$ and the restricted map $\pi\,|_A\colon A \to \mathrm{L_r}(E)$ is a Banach lattice algebra representation of A on E.

Let A be a Dedekind complete Banach lattice algebra. Then the left regular representation π of the complex Banach algebra $E_{\mathbb{C}}$ is a positive algebra homomorphism $\pi : E_{\mathbb{C}} :\to \mathrm{L_r}(E_{\mathbb{C}})$. The left regular representation of $A_{\mathbb{C}}$ is a complex Banach lattice algebra representation of $A_{\mathbb{C}}$ on itself if and only if the left regular representation of A is a Banach lattice algebra representation of A on itself.

We mention the following. Let A be a complex Banach algebra. Suppose that ${}^* : A \to A$ is a conjugate-linear map such that $(a^*)^* = a$ for $a \in A$, $(a_1a_2)^* = a_2^*a_1^*$ for $a_1, a_2 \in A$, and $\|a^*\| = \|a\|$ for $a \in A$. Then the map * is an *involution on* A, and

A is a *complex Banach* **-algebra*. For complex Banach *-algebras, see [38, 39], for example. The theory of *-representations of complex Banach *-algebras on complex Hilbert spaces is well developed.

In our context, one can consider complex Banach lattice algebras that are also complex Banach *-algebras. Examples are $C_0(X, \mathbb{C})$ and $C_b(X, \mathbb{C})$ for a topological space X, provided with complex conjugation as involution. The complex group algebra and the complex measure algebra of a locally compact group are other natural examples of complex Banach lattice *-algebras. However, there does not seem to be a natural role for the involution in the representation theory of Banach lattice *-algebras. The reason is that the complex Banach lattice algebra $L_r(E_{\mathbb{C}})$, where E is a Dedekind complete Banach lattice, does not have a natural involution. It has a natural conjugation, but this preserves the order of the factors in a product of linear operators rather than reverses it.

5 Locally Compact Spaces

In this section, we shall let X denote a non-empty, locally compact space. As for all topological spaces in this article, X is supposed to be Hausdorff.

We shall be concerned with the order dual $C_c(X, \mathbb{R})^\sim$ of $C_c(X, \mathbb{R})$. As explained in Sect. 1, the role of $C_c(X, \mathbb{R})^\sim$ in the present article is to be present as a large vector lattice that contains various familiar vector lattices as sublattices, see Theorems 7.5 and 7.9, below, for example.

The first step to be taken is to observe that $C_c(X, \mathbb{R})^\sim$ is equal to the space of real Radon measures on X in the sense of Bourbaki [12]. This will make a few (not too deep) known results for these Radon measures and their supports available. For this, we shall briefly recall the definition of Bourbaki's Radon measures on X.

As usual, for a real- or complex-valued function f on X, the *support of* f, denoted by $\operatorname{supp} f$, is the closure of the set consisting of those $x \in X$ such that $f(x) \neq 0$.

For each non-empty subset S of X, we let $C_c(X, \mathbb{R}; S)$ denote the set of those $f \in C_c(X, \mathbb{R})$ such that $\operatorname{supp} f \subseteq S$. Let K be a non-empty, compact subset of X. With the uniform norm, $C_c(X, \mathbb{R}; K)$ is a (possibly zero) Banach space. The space $C_c(X, \mathbb{R})$ is the union of the spaces $C_c(X, \mathbb{R}; K)$ as K runs over all non-empty, compact subsets of X. Consider the family $\mathcal{N}$ of all absorbing, symmetric, convex subsets V of $C_c(X, \mathbb{R})$ such that $V \cap C_c(X, \mathbb{R}; K)$ is a neighbourhood of 0 in $C_c(X, \mathbb{R}; K)$ for each non-empty, compact subset K of X. According to [11, II, § 4, No. 4, Proposition 5], $\mathcal{N}$ is a local base at 0 for a locally convex vector space topology $\mathcal{T}$ on $C_c(X, \mathbb{R})$. Furthermore, a linear map from $C_c(X, \mathbb{R})$ into a locally convex space is continuous with respect to $\mathcal{T}$ if and only if its restriction to $C_c(X, \mathbb{R}; K)$ is continuous for each non-empty, compact subset K of X, and $\mathcal{T}$ is the only locally convex topology on $C_c(X, \mathbb{R})$ with this property. The topology $\mathcal{T}$ is also the strongest locally convex topology on $C_c(X, \mathbb{R})$ such that the inclusion map from $C_c(X, \mathbb{R}; K)$ into $C_c(X, \mathbb{R})$ is continuous for each non-empty, compact subset K of X. The topology $\mathcal{T}$ on $C_c(X, \mathbb{R})$ is called the *direct limit* or *inductive limit* of the topologies on the spaces $C_c(X, \mathbb{R}; K)$ for non-empty, compact subsets K of X.

A real *Radon measure on* X in the sense of Bourbaki is a real-valued linear functional on $C_c(X, \mathbb{R})$ that is continuous with respect to the topology $\mathcal{T}$ specified above, see [12, III, § 1, No. 3, Definition 2]. In [12], Bourbaki uses the notation $\mathcal{M}(X; \mathbf{R})$ for the space of real Radon measures on X.

An alternative description of $C_c(X, \mathbb{R})^\sim$ is given by the following result. It can already be found in the literature as [12, paragraph preceding III, § 1, No. 5, Theorem 3], but we thought it worthwhile to make it explicit and also to include the easy proof, as we wish to combine some of the available results on Bourbaki's Radon measures with their lattice structure, which is not as prominent in Bourbaki as we shall need it.

Proposition 5.1 *Let* X *be a non-empty, locally compact space. Then* $C_c(X, \mathbb{R})^\sim$ *is the space* $\mathcal{M}(X; \mathbf{R})$ *of real Radon measures on* X *in the sense of Bourbaki.*

Proof Suppose that $\varphi : C_c(X, \mathbb{R}) \to \mathbb{R}$ is a Radon measure in the sense of Bourbaki. Let $S \subseteq C_c(X, \mathbb{R})$ be an order bounded subset. Then there exists $g \in C_c(X, \mathbb{R})$ such that $|f| \leq g$ for all $f \in S$. This implies that S is a uniformly bounded subset of $C_c(X, \mathbb{R}; \operatorname{supp} g)$. Since the restriction of φ to $C_c(X, \mathbb{R}; \operatorname{supp} g)$ is continuous, $\varphi(A)$ is a bounded, and then also an order bounded, subset of $\mathbb{R}$. Hence $\varphi \in C_c(X, \mathbb{R})^\sim$.

Conversely, suppose that $\varphi \in C_c(X, \mathbb{R})^\sim$. Let K be a non-empty, compact subset of X. Then the restriction of φ to $C_c(X, \mathbb{R}; K)$ is a regular linear functional. Since $C_c(X, \mathbb{R}; K)$ is a Banach lattice, this restriction is continuous. Hence φ is a Radon measure in the sense of Bourbaki. □

Let X be a non-empty, locally compact space. The above proposition makes it slightly easier to see that a linear functional on $C_c(X, \mathbb{R})$ is a Radon measure. Indeed, it will usually be obvious that it is regular if this be, in fact, the case, whereas seeing that it is continuous on each subspace $C_c(X, \mathbb{R}; K)$ could be (marginally) more complicated.

It is now also possible to make contact with measure theory in the other, perhaps more usual, sense of the word. In order to do so, we recall that a positive measure $\mu : \mathcal{B} \to [0, \infty]$ on the Borel σ-algebra $\mathcal{B}$ of X is:

(1) a *Borel measure* if $\mu(K) < \infty$ for all compact subsets K of X;
(2) *outer regular on* $A \in \mathcal{B}$ if $\mu(A) = \inf\{\, \mu(V) : V \text{ open and } A \subseteq V \,\}$;
(3) *inner regular on* $A \in \mathcal{B}$ if $\mu(A) = \sup\{\, \mu(K) : K \text{ compact and } K \subseteq A \,\}$.

Using the terminology in [3, p. 352], μ is a *positive regular Borel measure on* X if it is a positive Borel measure that is outer regular on all $A \in \mathcal{B}$ and inner regular on all open subsets of X. The measure μ is *finite* if $\mu(X) < \infty$.

The nomenclature is not uniform in the literature; sometimes the inner regularity on all elements of $\mathcal{B}$ rather than just on the open subsets is incorporated in the definition of a regular Borel measure, as in [24, p. 212]. In [24, p. 212], our positive regular Borel measures are called Radon measures. In view of the possibility of confusion with Bourbaki's terminology, we prefer to speak of positive regular Borel measures in the present article.

We shall now review a number of properties of regular Borel measures on X. Details can be found in [3], for example; this reference puts more emphasis on the lattice structure than several other sources.

The set of positive regular Borel measures on X is a cone that is denoted by $\mathrm{M_r}(X, \mathfrak{B}, \overline{\mathbb{R}^+})$. Its subcone consisting of the finite positive regular Borel measures on X is denoted by $\mathrm{M_r}(X, \mathfrak{B}, \mathbb{R}^+)$. By definition, the real-linear span of $\mathrm{M_r}(X, \mathfrak{B}, \mathbb{R}^+)$ is the vector space $\mathrm{M_r}(X, \mathfrak{B}, \mathbb{R})$ of *real regular Borel measures on* X. The vector space $\mathrm{M_r}(X, \mathfrak{B}, \mathbb{R})$ is, in fact, a Dedekind complete Banach sublattice of the Banach lattice $\mathrm{M}(X, \mathfrak{B}, \mathbb{R})$ from Example 2.10. The supremum of two elements is given by Eq. (2.1), the modulus by Eq. (2.6), and the norm by Eq. (2.6).

Let $\varphi \in \mathrm{C_c}(X, \mathbb{R})^\sim$. After splitting φ into its positive and negative parts, the Riesz representation theorem for positive functionals on $\mathrm{C_c}(X, \mathbb{R})$ implies that there exist $\mu^+, \mu^- \in \mathrm{M_r}(X, \mathfrak{B}, \overline{\mathbb{R}^+})$ such that

$$\langle \varphi, f \rangle = \int_X f \, \mathrm{d}\mu^+ - \int_X f \, \mathrm{d}\mu^- \tag{5.8}$$

for all $f \in \mathrm{C_c}(X, \mathbb{R})$. If $\varphi \geq 0$, then one can take $\mu^- = 0$, and in this case μ^+ is uniquely determined.

Let $\varphi \in \mathrm{C_c}(X, \mathbb{R})^\sim$, and suppose that φ is a continuous linear functional on $(\mathrm{C_c}(X, \mathbb{R}), \|\cdot\|_\infty)$; equivalently, one can suppose that φ is the restriction to $\mathrm{C_c}(X, \mathbb{R})$ of a continuous linear functional on $(\mathrm{C_0}(X, \mathbb{R}), \|\cdot\|_\infty)$. Then μ^+ and μ^- in Eq. (5.8) can both be taken to be elements of $\mathrm{M_r}(X, \mathfrak{B}, \mathbb{R}^+)$. Conversely, if $\mu^+, \mu^- \in \mathrm{M_r}(X, \mathfrak{B}, \mathbb{R}^+)$, then the right-hand side of Eq. (5.8) defines a continuous linear functional φ on $(\mathrm{C_0}(X, \mathbb{R}), \|\cdot\|_\infty)$. In this way, an isometric isomorphism of Banach lattices between the norm (or order) dual of the Banach lattice $(\mathrm{C_0}(X, \mathbb{R}), \|\cdot\|_\infty)$ and the Banach lattice $(\mathrm{M_r}(X, \mathfrak{B}, \mathbb{R}), \|\cdot\|)$ is obtained, see [3, Theorem 38.7], for example.

Remark 5.2 The measures μ^+ and μ^- in Eq. (5.8) can be infinite simultaneously, so that it is meaningless to say that φ is represented by the measure $\mu^+ - \mu^-$ because the latter cannot generally be properly defined. This is where Bourbaki's terminology for Radon 'measures' conflicts with that in measure theory in the sense of Lebesgue and Caratheodory.

Let X be a non-empty, locally compact space. The Riesz representation theorem provides a means to define the product of a bounded Borel measurable function on X and an element of $\mathrm{C_c}(X, \mathbb{R})^\sim$. We shall now explain this.

Let $\varphi \in \mathrm{C_c}(X, \mathbb{R})^\sim$. Suppose that U is a non-empty, open, and relatively compact subset of X. Since $\mathrm{C_c}(X, \mathbb{R}; U) \subseteq \mathrm{C_c}(X, \mathbb{R}; \overline{U})$, the restriction of φ to $\mathrm{C_c}(X, \mathbb{R}; U)$ is continuous when $\mathrm{C_c}(X, \mathbb{R}; U)$ is supplied with the uniform norm.

Therefore, there exists a unique finite regular Borel measure μ on U such that

$$\langle h, \varphi\rangle = \int_U h \, \mathrm{d}\mu_U$$

for all $h \in \mathrm{C_c}(X, \mathbb{R}; U)$.

Suppose that V is an open and relatively compact subset of X with $V \supseteq U$. Then it is a consequence of [25, Section 7.2, Exercise 7] and the uniqueness part of the Riesz representation theorem that μ_U equals the restriction of μ_V to U. Consequently, suppose that U and V are two non-empty, open, and relatively compact subsets of X such that $U \cap V \neq \emptyset$. Then the restrictions of μ_U and μ_V to $U \cap V$ are identical.

Let $g : X \to \mathbb{R}$ be a bounded Borel measurable function on X. Suppose that $f \in \mathrm{C_c}(X, \mathbb{R})$, and choose an open and relatively compact neighbourhood U of $\operatorname{supp} f$ in X. Since fg is zero outside U, it follows from the above that the integral

$$\int_U fg \, \mathrm{d}\mu_U$$

does not depend on the choice of U. Hence we can set

$$\langle g\varphi, f\rangle := \int_U fg \, \mathrm{d}\mu_U$$

as a well-defined element of $\mathbb{R}$, thus obtaining a map $g\varphi : \mathrm{C_c}(X, \mathbb{R}) \to \mathbb{R}$. It is then routine to verify that $g\varphi \in \mathrm{C_c}(X, \mathbb{R})^\sim$, and that $g\varphi$ depends bilinearly on the bounded Borel measurable function g on X and the element φ of $\mathrm{C_c}(X, \mathbb{R})^\sim$. The element $g\varphi$ of $\mathrm{C_c}(X, \mathbb{R})^\sim$ is the *product of* g *and* φ.

Although we shall not need this, let us note that, more generally, a similar argument that is based on local applications of the Riesz representation theorem can be employed to define the product $g\varphi$ of a Borel measurable function g on X that is locally integrable (in the canonical sense) with respect to $|\varphi|$ for a given $\varphi \in \mathrm{C_c}(X, \mathbb{R})^\sim$. It is possible to avoid the Riesz representation theorem in defining such products, see [12, V, § 5. No. 2], but the definition using the Riesz representation theorem may be a little more transparent.

Following Bourbaki (see [12, III, § 2, Nos. 1 and 2]), we shall now introduce the supports of elements of $\mathrm{C_c}(X, \mathbb{R})^\sim$.

Let U be non-empty, open subset of X. An element φ of $\mathrm{C_c}(X, \mathbb{R})^\sim$ *vanishes on* U if $\langle \varphi, f\rangle = 0$ for all $f \in \mathrm{C_c}(X, \mathbb{R}; U)$. By definition, φ vanishes on the empty set. A partition of unity argument shows that φ vanishes on the open subset $\mathcal{U}$ of X that is the union of all open subsets of X on which φ vanishes. The closed subset $X \setminus \mathcal{U}$ of X is called the *support of* φ; it is denoted by $\operatorname{supp} \varphi$. Thus a point x in X is in the support of φ if and only if, for every open neighbourhood U of x, there exists $f \in \mathrm{C_c}(X, \mathbb{R}; U)$ such that $\langle \varphi, f\rangle \neq 0$.

Let $\varphi \in \mathrm{C_c}(X,\mathbb{R})^{\sim}$. Then $\operatorname{supp}\varphi = \operatorname{supp}|\varphi| = \operatorname{supp}\varphi^+ \cup \operatorname{supp}\varphi^-$, see [12, III, § 2, No. 2, Propositions 2].

Let $\varphi_1, \varphi_2 \in \mathrm{C_c}(X,\mathbb{R})^{\sim}$. Then $\operatorname{supp}(\varphi_1+\varphi_2) \subseteq \operatorname{supp}\varphi_1 \cup \operatorname{supp}\varphi_2$, and if $|\varphi_1| \leq |\varphi_2|$, then $\operatorname{supp}\varphi_1 \subseteq \operatorname{supp}\varphi_2$, see [12, III, § 2, No. 2, Propositions 3 and 4]. Consequently, if S is an arbitrary subset of X, then the subset of $\mathrm{C_c}(X,\mathbb{R})^{\sim}$ consisting of all elements φ of $\mathrm{C_c}(X,\mathbb{R})^{\sim}$ such that $\operatorname{supp}\varphi \subseteq S$ is an order ideal of $\mathrm{C_c}(X,\mathbb{R})^{\sim}$.

Let $\varphi \in \mathrm{C_c}(X,\mathbb{R})^{\sim}$. It can happen that $\operatorname{supp}\varphi^+ = \operatorname{supp}\varphi^- = X$, see [12, V, Exercises, § 5, Exerc. 4]. Hence the disjointness of two elements of $\mathrm{C_c}(X,\mathbb{R})^{\sim}$ does not imply that their supports are disjoint subsets of X. The following result shows that the converse implication *does* hold.

Lemma 5.3 *Let X be a non-empty, locally compact space. Let $\varphi_1, \varphi_2 \in \mathrm{C_c}(X,\mathbb{R})^{\sim}$ be such that* $\operatorname{supp}\varphi_1$ *and* $\operatorname{supp}\varphi_2$ *are disjoint subsets of X. Then φ_1 and φ_2 are disjoint elements of* $\mathrm{C_c}(X,\mathbb{R})^{\sim}$. *Consequently,* $|\varphi_1+\varphi_2| = |\varphi_1| + |\varphi_2|$.

Proof Using the fact that $\operatorname{supp}\varphi = \operatorname{supp}|\varphi|$ for $\varphi \in \mathrm{C_c}(X,\mathbb{R})^{\sim}$, we may suppose that $\varphi_1, \varphi_2 \in (\mathrm{C_c}(X,\mathbb{R})^{\sim})^+$.

Then Eq. (2.5) yields that, for $f \in \mathrm{C_c}(X,\mathbb{R})^+$, we have

$$(\varphi_1 \wedge \varphi_2)(f) = \inf\{\, \varphi_1(f_1) + \varphi_2(f_2) : f_1, f_2 \in \mathrm{C_c}(X,\mathbb{R})^+,\ f_1 + f_2 = f \,\}.$$

Since $\operatorname{supp}\varphi_1$ and $\operatorname{supp}\varphi_2$ are disjoint, we have

$$\operatorname{supp} f \subseteq X = (X \setminus \operatorname{supp}\varphi_1) \cup (X \setminus \operatorname{supp}\varphi_2)\,.$$

We can then find continuous functions $g_1, g_2 : X \to [0,1]$ with compact support such that $g_1 + g_2 = 1$, $\operatorname{supp} g_1 \subseteq X \setminus \operatorname{supp}\varphi_1$, and $\operatorname{supp} g_2 \subseteq X \setminus \operatorname{supp}\varphi_2$. For the resulting decomposition $f = g_1 f + g_2 f$, we have $\varphi_1(g_1 f) = \varphi_2(g_2 f) = 0$, and this shows that $(\varphi_1 \wedge \varphi_2)(f) \leq 0$. Since obviously $(\varphi_1 \wedge \varphi_2)(f) \geq 0$, we see that $(\varphi_1 \wedge \varphi_2)(f) = 0$. Hence $\varphi_1 \wedge \varphi_2 = 0$.

Now that we have established that φ_1 and φ_2 are disjoint, the final statement follows from the general principle in vector lattices that the modulus is additive on the sum of two (in fact, of finitely many) mutually disjoint elements. □

Remark 5.4 Lemma 5.3, with its elementary proof, is also a consequence of the technically considerably more demanding [12, V, § 5, No. 7, Proposition 13], where a necessary and sufficient condition for two elements of $\mathrm{C_c}(X,\mathbb{R})^{\sim}$ to be disjoint—Bourbaki calls such elements *alien* (to each other)—is given. The reader may wish to consult [12, IV, § 2, No. 2, Proposition 5 and IV, § 5, No. 2, Definition 3] to see that an element φ of $\mathrm{C_c}(X,\mathbb{R})^{\sim}$ is *concentrated on* $\operatorname{supp}\varphi$ in the sense of [12, V, § 5, No. 7, Definition 4], after which it is immediate from [12, V, § 5, No. 7, Proposition 13] that the disjointness of the supports of two elements of $\mathrm{C_c}(X,\mathbb{R})^{\sim}$ implies their disjointness in the vector lattice $\mathrm{C_c}(X,\mathbb{R})^{\sim}$.

The relevance of the following result will become clear in the proof of Theorem 7.5, below.

Lemma 5.5 *Let X be a non-empty, locally compact space, and let $f \in C_c(X, \mathbb{R})$. Take $\varepsilon > 0$. Then there exist $g^+, g^- \in C_c(X, \mathbb{R})$ such that:*

(1) $0 \le g^+ \le f^+$ *and* $0 \le g^- \le f^-$*;*
(2) $0 \le f^+ - g^+ \le \varepsilon 1_X$ *and* $0 \le f^- - g^- \le \varepsilon 1_X$*;*
(3) $\operatorname{supp} g^+ \cap \operatorname{supp} g^- = \emptyset$.

Proof If $f^+ = 0$, then we can take $g^+ = 0$ and $g^- = f^-$; if $f^- = 0$, then we can take $g^+ = f^+$ and $g^- = 0$. Hence we may suppose that there exists $\delta > 0$ such that $\{ x \in X : f^+(x) > \delta \}$ and $\{ x \in X : f^-(x) > \delta \}$ are both non-empty subsets of X. It is then sufficient to prove the result for all ε such that $0 < \varepsilon < \delta$. For such a fixed ε, set

$$g^+ := (f^+ \vee \varepsilon 1_X) - \varepsilon 1_X .$$

Then $g^+ \in C(X, \mathbb{R})$, $0 \le g^+ \le f^+$, and $0 \le f^+ - g^+ \le \varepsilon 1_X$; we see that $g^+ \in C_c(X, \mathbb{R})$. Likewise, we set

$$g^- := (f^- \vee \varepsilon 1_X) - \varepsilon 1_X ,$$

and then $g^- \in C_c(X, \mathbb{R})$, $0 \le g^- \le f^-$, and $0 \le f^- - g^- \le \varepsilon 1_X$.

Let $x \in X$. If

$$x \in \operatorname{supp} g^+ = \overline{\{ x \in X : g^+(x) \ne 0 \}} \subseteq \overline{\{ x \in X : f^+(x) > \varepsilon \}},$$

then the continuity of f^+ implies that $f^+(x) \ge \varepsilon$. Hence $f(x) \ge \varepsilon$. Likewise, if $x \in \operatorname{supp} g^-$, then $f^-(x) \ge \varepsilon$, which implies that $f(x) \le -\varepsilon$. Since $\varepsilon > 0$, this shows that $\operatorname{supp} g^+ \cap \operatorname{supp} g^- = \emptyset$. □

6 Closed Subspaces of Locally Compact Spaces

Let X be a non-empty, locally compact space, and let Y be a non-empty, closed subspace of X. Then Y is again a locally compact space. We shall now prove that $C_c(Y, \mathbb{R})^\sim$ can be canonically viewed as the order ideal of $C_c(X, \mathbb{R})^\sim$ that consists of those elements of $C_c(X, \mathbb{R})^\sim$ with support contained in Y. The reader who is interested in Banach lattices on groups, but not on semigroups, can omit this section in its entirety.

We are not aware of references for the results in this section, which may find applications elsewhere.

Let Y be a non-empty, closed subset of a locally compact space X. Then we define the *restriction map* $R_Y : C_c(X, \mathbb{R}) \to C_c(Y, \mathbb{R})$ by setting $R_Y f := f \mid_Y$ for $f \in C_c(X, \mathbb{R})$. As we shall see, the order adjoint

$$R_Y^\sim : C_c(Y, \mathbb{R})^\sim \to C_c(X, \mathbb{R})^\sim$$

of R_Y is injective, and the image of $\mathrm{C_c}(Y, \mathbb{R})^\sim$ under $R_Y^\sim$ is the order ideal of $\mathrm{C_c}(X, \mathbb{R})^\sim$ that consists of those elements of $\mathrm{C_c}(X, \mathbb{R})^\sim$ with support contained in Y.

We shall require two preparatory results. The first one is a slight strengthening of a version of Tietze's extension theorem [44, Theorem 20.4], on which it is also based.

Proposition 6.1 *Let X be a non-empty, locally compact space, let Y be a non-empty, closed subspace of X, and let $f \in \mathrm{C_c}(Y, \mathbb{R})$. Then there exists $F \in \mathrm{C_c}(X, \mathbb{R})$ such that $R_Y F = f$ and $\|F\|_{\infty,X} = \|f\|_{\infty,Y}$. If $f \geq 0$, then it can be arranged that also $F \geq 0$.*

Proof Let $f \in \mathrm{C_c}(Y, \mathbb{R})$. Take a relatively compact open neighbourhood U of $\operatorname{supp} f$ in X. Since $\overline{U} \cap Y$ is a compact subset of $\overline{U}$, Tietze's extension theorem shows that there exists an element g of $\mathrm{C}(\overline{U}, \mathbb{R})$ such that $g \mid \overline{U} \cap Y = f \mid \overline{U} \cap Y$ as well as $\|g\|_{\infty,\overline{U}} = \|f\|_{\infty,\overline{U}\cap Y} = \|f\|_{\infty,Y}$. By a version of Urysohn's lemma [44, Theorem 2.12], there exists $h \in \mathrm{C}(\overline{U}, \mathbb{R})$ such that $h(\overline{U}) \subseteq [0, 1]$, $h(y) = 1$ for $y \in \operatorname{supp} f$, and $\operatorname{supp} h \subseteq U$.

Set $F := gh$, so that $F \in \mathrm{C}(\overline{U}, \mathbb{R})$ and $\operatorname{supp} F \subseteq U$. We extend F to be an element of $\mathrm{C_c}(X, \mathbb{R})$ by setting $F(x) := 0$ for $x \in X \setminus \overline{U}$. Then we have $\|F\|_{\infty,X} \leq \|g\|_{\infty,\overline{U}} = \|f\|_{\infty,Y}$.

For $y \in \operatorname{supp} f$, we have $F(y) = g(y)h(y) = g(y) = f(y)$; this also shows that $\|F\|_{\infty,X} \geq \|f\|_{\infty,Y}$. For $y \in (\overline{U} \cap Y) \setminus \operatorname{supp} f$, we have $F(y) = 0 = f(y)$ because $g(y) = f(y) = 0$. For $y \in Y \setminus \overline{U}$, we have $F(y) = 0 = f(y)$ because F vanishes on $X \setminus \overline{U}$. We conclude that $R_Y F = f$ and that $\|F\|_{\infty,X} = \|f\|_{\infty,Y}$.

If $f \geq 0$, then replacing F by F^+ shows that we can also arrange that $F \geq 0$. □

Corollary 6.2 *Let X be a non-empty, locally compact space, and let Y be a non-empty, closed subspace of X. Then $R_Y : \mathrm{C_c}(X, \mathbb{R}) \to \mathrm{C_c}(Y, \mathbb{R})$ is a continuous, interval preserving, and surjective lattice homomorphism.*

Proof The map R_Y is clearly a lattice homomorphism, and it is immediate from the properties of the topologies of $\mathrm{C_c}(X, \mathbb{R})$ and $\mathrm{C_c}(Y, \mathbb{R})$ that R_Y is continuous. The surjectivity follows from Proposition 6.1.

It remains to show that the positive linear operator $R_Y : \mathrm{C_c}(X, \mathbb{R}) \to \mathrm{C_c}(Y, \mathbb{R})$ is interval preserving. For this, take $F \in \mathrm{C_c}(X, \mathbb{R})^+$, and suppose that $g \in \mathrm{C_c}(Y, \mathbb{R})$ is such that $0 \leq g \leq R_Y F$. By Proposition 6.1, there exists $G \in \mathrm{C_c}(X, \mathbb{R})^+$ such that $R_Y G = g$. Then $0 \leq F \wedge G \leq F$ and $R_Y(F \wedge G) = R_Y F \wedge R_Y G = R_Y F \wedge g = g$. Thus $R_Y([0, F]) = [0, R_Y F]$, as required. □

Theorem 6.3 *Let X be a non-empty, locally compact space, and let Y be a non-empty, closed subspace of X. Then $R_Y^\sim : \mathrm{C_c}(Y, \mathbb{R})^\sim \to \mathrm{C_c}(X, \mathbb{R})^\sim$ is a weak*-continuous, injective, and interval preserving lattice homomorphism.*

Furthermore, $\operatorname{supp} \varphi = \operatorname{supp} R_Y^\sim \varphi$ *for all* $\varphi \in \mathrm{C_c}(Y, \mathbb{R})^\sim$.

The image of $\mathrm{C_c}(Y, \mathbb{R})^\sim$ *under* $R_Y^\sim$ *is the order ideal of* $\mathrm{C_c}(X, \mathbb{R})^\sim$ *that consists of all elements* Φ *of* $\mathrm{C_c}(X, \mathbb{R})^\sim$ *such that* $\operatorname{supp} \Phi \subseteq Y$.

Suppose that g is a bounded Borel measurable function on Y. Extend g to a Borel measurable function $\widetilde{g}$ on X by setting $\widetilde{g}(x) := 0$ for $x \in X \setminus Y$. Then $R_Y^{\sim}(g\varphi) = \widetilde{g}R_Y^{\sim}\varphi$ for all $\varphi \in C_c(Y, \mathbb{R})^{\sim}$.

Proof In view of Corollary 6.2 and Propositions 2.5 and 2.6, it is clear that $R_Y^{\sim}$, which is obviously weak*-continuous, is an injective and interval preserving lattice homomorphism.

We turn to the second statement.

Let $\varphi \in C_c(Y, \mathbb{R})^{\sim}$. Let $x \in X$, and suppose that $x \notin \operatorname{supp}\varphi$. Since Y is a closed subset of X, $\operatorname{supp}\varphi$ is a closed subset of X. Hence there exists an open neighbourhood U of x in X such that $U \cap \operatorname{supp}\varphi = \emptyset$. Let $f \in C_c(X, \mathbb{R})$ be such that $\operatorname{supp} f \subseteq U$. If $R_Y f = 0$, then certainly $\langle R_Y^{\sim}\varphi, f\rangle = \langle \varphi, R_Y f\rangle = 0$. If $R_Y f \neq 0$, then $R_Y f$ is an element of $C_c(Y, \mathbb{R})$ such that $\operatorname{supp} R_Y f \subseteq U \cap Y$. Since $U \cap Y$ is then a non-empty, open subset of Y that is disjoint from $\operatorname{supp}\varphi$, we have $\langle \varphi, R_Y f\rangle = 0$. Hence $\langle R_Y^{\sim}\varphi, f\rangle = 0$. We conclude that $R_Y^{\sim}\varphi$ vanishes on U, and hence $x \notin \operatorname{supp} R_Y^{\sim}\varphi$. It follows that $\operatorname{supp}\varphi \supseteq \operatorname{supp} R_Y^{\sim}\varphi$.

For the reverse inclusion, take $x \in Y$, and suppose that $x \in \operatorname{supp}\varphi$. Let U be an open neighbourhood of x in X. Take $f \in C_c(Y, \mathbb{R})$ such that $\operatorname{supp} f \subseteq U \cap Y$ and $\langle \varphi, f\rangle \neq 0$. By Proposition 6.1, there exists $F \in C_c(X, \mathbb{R})$ such that $R_Y F = f$, and Urysohn's lemma furnishes $G \in C_c(X, \mathbb{R})$ such that $G = 1$ on $\operatorname{supp} f$ and $\operatorname{supp} G \subseteq U$. Set $H := FG$. Then $H \in C_c(X, \mathbb{R})$, $\operatorname{supp} H \subseteq U$, and $R_Y H = f$. We then conclude from $\langle R_Y^{\sim}\varphi, H\rangle = \langle \varphi, R_Y H\rangle = \langle \varphi, f\rangle \neq 0$ that $R_Y^{\sim}\varphi$ does not vanish on U. Hence $x \in \operatorname{supp} R_Y^{\sim}\varphi$. This shows that $\operatorname{supp}\varphi \subseteq \operatorname{supp} R_Y^{\sim}\varphi$.

We turn to the statement on the range of $R_Y^{\sim}$.

From what we have already established, it is clear that the support of $R_Y^{\sim}\varphi$ is contained in Y for all $\varphi \in C_c(Y, \mathbb{R})^{\sim}$. Conversely, suppose that $\Phi \in C_c(X, \mathbb{R})^{\sim}$ is such that $\operatorname{supp}\Phi \subseteq Y$. We shall establish the existence of a $\varphi \in C_c(Y, \mathbb{R})^{\sim}$ such that $R_Y^{\sim}\varphi = \Phi$, as follows. Let $f \in C_c(Y, \mathbb{R})$. Using Proposition 6.1, we choose $F \in C_c(X, \mathbb{R})$ such that $R_Y F = f$, and we define $\varphi : C_c(Y, \mathbb{R}) \to \mathbb{R}$ by setting $\langle \varphi, f\rangle := \langle \Phi, F\rangle$. We shall show that this is well defined. For this, it is clearly sufficient to show that $\langle \Phi, F\rangle = 0$ whenever $F \in C_c(X, \mathbb{R})$ is such that $R_Y F = 0$. Fix such an F, and choose an open and relatively compact neighbourhood U of $\operatorname{supp} F$ in X. Then there exists a constant $M \geq 0$ such that $|\langle \Phi, G\rangle| \leq M\|G\|_{\infty,X}$ for all $G \in C_c(X, \mathbb{R}; \overline{U})$. Let $\varepsilon > 0$ be fixed, and set $V_\varepsilon := \{x \in X : |F(x)| < \varepsilon\}$. Since $R_Y F = 0$, V_ε is an open neighbourhood of Y in X; in particular, V_ε is an open neighbourhood of $Y \cap \operatorname{supp} F$ in X. Take an open and relatively compact subset W_ε of X such that $Y \cap \operatorname{supp} F \subseteq W_\varepsilon \subseteq \overline{W_\varepsilon} \subseteq V_\varepsilon$, and take $G_\varepsilon \in C_c(X, \mathbb{R})$ such that $0 \leq G_\varepsilon \leq 1$, $G_\varepsilon = 1$ on $\overline{W_\varepsilon}$, and $\operatorname{supp} G_\varepsilon \subseteq V_\varepsilon$.

Let $x \in X$, and suppose that $(FG_\varepsilon - F)(x) \neq 0$. Then certainly $G_\varepsilon(x) \neq 1$, so that $x \notin \overline{W_\varepsilon}$. In particular, $x \notin W_\varepsilon$. We conclude that $\operatorname{supp}(FG_\varepsilon - F) \subseteq X \setminus W_\varepsilon$. Evidently, $\operatorname{supp}(FG_\varepsilon - F) \subseteq \operatorname{supp} F$, so $\operatorname{supp}(FG_\varepsilon - F) \subseteq (X \setminus W_\varepsilon) \cap \operatorname{supp} F$. Hence

$$\begin{aligned}\operatorname{supp}(FG_\varepsilon - F) \cap \operatorname{supp}\Phi &\subseteq \operatorname{supp}(FG_\varepsilon - F) \cap Y \\ &\subseteq (X \setminus W_\varepsilon) \cap Y \cap \operatorname{supp} F = \emptyset,\end{aligned}$$

since $Y \cap \operatorname{supp} F \subseteq W_\varepsilon$. It follows from this that $\langle \Phi, F \rangle = \langle \Phi, FG_\varepsilon \rangle$. Since, in addition, $FG_\varepsilon \in C_c(X, \mathbb{R}; \overline{U})$ and $\|FG_\varepsilon\|_{\infty, X} \leq \varepsilon$, we have $|\langle \Phi, FG_\varepsilon \rangle| \leq \varepsilon M_{\overline{U}}$.

We thus see that $|\langle \Phi, F \rangle| \leq \varepsilon M_{\overline{U}}$ for all $\varepsilon > 0$. Hence $\langle \Phi, F \rangle = 0$. This establishes our claim.

Now that we know that the map $\varphi : C_c(Y, \mathbb{R}) \to \mathbb{R}$ is well defined, it is immediate that it is linear. Combining the facts that a positive $f \in C_c(Y, \mathbb{R})$ has a positive extension, as asserted by Proposition 6.1, and that $\Phi = \Phi^+ - \Phi^-$ in $C_c(X, \mathbb{R})^\sim$, it is easy to see that $\varphi \in C_c(Y, \mathbb{R})^\sim$. Finally, for $F \in C_c(Y, \mathbb{R})$, we have, using the definition of φ, that $\langle R_Y^\sim \varphi, F \rangle = \langle \varphi, R_Y F \rangle = \langle \Phi, F \rangle$. Hence $R_Y^\sim \varphi = \Phi$.

We have now shown that the image of $C_c(Y, \mathbb{R})^\sim$ under $R_Y^\sim$ is the subset of $C_c(X, \mathbb{R})^\sim$ that consists of all elements Φ of $C_c(X, \mathbb{R})^\sim$ such that $\operatorname{supp} \Phi \subseteq Y$. Since such a subset of $C_c(X, \mathbb{R})^\sim$ is an order ideal of $C_c(X, \mathbb{R})^\sim$ for an arbitrary subset Y of X, the proof of the statement on the range of $R_Y^\sim$ is complete.

We turn to the final statement.

Let g be a bounded Borel measurable function on Y, and let $\varphi \in C_c(Y, \mathbb{R})^\sim$. Suppose that $f \in C_c(X, \mathbb{R})$. Choose a non-empty, open, relatively compact neighbourhood U of $\operatorname{supp} f$ in X; we may suppose that $U \cap Y \neq \emptyset$. Then $U \cap Y$ is a non-empty, open, relatively compact neighbourhood of $\operatorname{supp}(R_Y f)$ in Y. There exists a unique regular Borel measure μ on $U \cap Y$ such that

$$\langle \varphi, h \rangle = \int_{U \cap Y} h \, d\mu \tag{6.9}$$

for all $h \in C_c(Y, \mathbb{R}; U \cap Y)$. Suppose that A is an arbitrary Borel subset of U, and set $\widetilde{\mu}(A) := \mu(A \cap (U \cap Y))$. The fact that $U \cap Y$ is closed in U implies that this defines a regular Borel measure $\widetilde{\mu}$ on U. It is easily seen that

$$\int_U k \, d\widetilde{\mu} = \int_{U \cap Y} R_{U \cap Y} k \, d\mu \tag{6.10}$$

for all bounded Borel measurable functions k on U.

On the other hand, there exists a unique regular Borel measure ν on U such that

$$\langle R_Y^\sim \varphi, k \rangle = \int_U k \, d\nu \tag{6.11}$$

for all $k \in C_c(X, \mathbb{R}; U)$.

Combining Eqs. (6.9)–(6.11), we see that, for $k \in C_c(X, \mathbb{R}; U)$, we have

$$\int_U k \, d\nu = \langle R_Y^\sim \varphi, k \rangle = \langle \varphi, R_Y k \rangle = \int_{U \cap Y} R_{U \cap Y} k \, d\mu = \int_U k \, d\widetilde{\mu}.$$

It follows that $\nu = \widetilde{\mu}$.

Using the definitions of $\widetilde{g}R_Y^{\sim}\varphi$ and $R_Y^{\sim}(g\varphi)$, we see that this implies that

$$\langle \widetilde{g}R_Y^{\sim}\varphi, f\rangle = \int_U \widetilde{g} f \,\mathrm{d}\nu = \int_U \widetilde{g} f \,\mathrm{d}\widetilde{\mu} = \int_{U\cap Y} g R_Y f \,\mathrm{d}\mu = \langle g\varphi, R_Y f\rangle = \langle R_Y^{\sim}(g\varphi), f\rangle.$$

Hence $\widetilde{g}R_Y^{\sim}\varphi = R_Y^{\sim}(g\varphi)$. □

We are not aware of earlier results in the vein of Theorem 6.3. Bourbaki introduces restrictions of his Radon measures in [12, III, § 2, No. 1 and IV, § 5, No. 7], but does not seem to consider what are essentially extensions as in Theorem 6.3.

7 Embedding Familiar Vector Lattices into $\mathrm{C_c}(X, \mathbb{R})^{\sim}$

In this section, X is a non-empty, locally compact space. We shall see how various familiar vector lattices can be embedded into $\mathrm{C_c}(X, \mathbb{R})^{\sim}$.

Let $\mu \in \mathrm{M_r}(X, \mathfrak{B}, \overline{\mathbb{R}^+})$ be a positive regular Borel measure on X. Suppose that $g : X \to \mathbb{R}$ is Borel measurable. Then g is *locally integrable with respect to* μ, or *locally* μ*-integrable* if

$$\int_K |g(x)| \,\mathrm{d}\mu < \infty$$

for every compact subset K of X. We shall identify two locally μ-integrable functions g_1 and g_2 that are locally μ-almost everywhere equal, i.e., which are such that

$$\mu(\{ x \in K : g_1(x) \neq g_2(x) \}) = 0$$

for all compact subsets K of X. The equivalence classes of locally μ-integrable functions on X form a vector lattice when the vector space operations and ordering are defined pointwise locally almost everywhere using representatives of equivalence classes. The vector lattice of equivalence classes thus obtained is denoted by $\mathrm{L}^{1,\mathrm{loc}}(X, \mathfrak{B}, \mu, \mathbb{R})$.

We shall shortly show that there exists a canonical lattice isomorphism Φ from $\mathrm{L}^{1,\mathrm{loc}}(X, \mathfrak{B}, \mu, \mathbb{R})$ into $\mathrm{C_c}(X, \mathbb{R})^{\sim}$, see Proposition 7.2, below. The spaces $\mathrm{L}^p(X, \mathfrak{B}, \mu, \mathbb{R})$ for $1 \leq p < \infty$ are sublattices of $\mathrm{L}^{1,\mathrm{loc}}(X, \mathfrak{B}, \mu, \mathbb{R})$, see Lemma 7.4, below. For $1 \leq p < \infty$, the restrictions of Φ to these sublattices will, therefore, yield embeddings of the vector lattices $\mathrm{L}^p(X, \mathfrak{B}, \mu, \mathbb{R})$ as vector sublattices of $\mathrm{C_c}(X, \mathbb{R})^{\sim}$, see Theorem 7.5, below.

We shall need the following auxiliary result, which can be found as [44, Corollary to Lusin's Theorem 2.24], for example.

Proposition 7.1 *Let X be a non-empty, locally compact space, let γ be a bounded Borel measurable function on X, let $\mu \in \mathrm{M_r}(X, \mathfrak{B}, \overline{\mathbb{R}^+})$, and let $A \in \mathfrak{B}$ be such that $\mu(A) < \infty$. Suppose that γ vanishes outside A and that $\|\gamma\|_\infty \leq 1$. Then there exists a sequence (γ_n) in $\mathrm{C_c}(X, \mathbb{R})$ such that $\|\gamma_n\|_\infty \leq 1$ for all $n \geq 1$, and $\gamma(x) = \lim_{n\to\infty} \gamma_n(x)$ for μ-almost all x in X.*

Proposition 7.2 *Let X be a non-empty, locally compact space, and suppose that $\mu \in \mathrm{M_r}(X, \mathfrak{B}, \overline{\mathbb{R}^+})$. For $g \in \mathrm{L}^{1,\mathrm{loc}}(X, \mathfrak{B}, \mu, \mathbb{R})$, set*

$$\langle \varphi_g, f \rangle := \int_X fg \,\mathrm{d}\mu$$

for $f \in \mathrm{C_c}(X, \mathbb{R})$. Then $\varphi_g \in \mathrm{C_c}(X, \mathbb{R})^\sim$, and the map $\Phi : g \mapsto \varphi_g$ defines an injective lattice homomorphism $\Phi : \mathrm{L}^{1,\mathrm{loc}}(X, \mathfrak{B}, \mu, \mathbb{R}) \to \mathrm{C_c}(X, \mathbb{R})^\sim$. Suppose that h is a bounded Borel measurable function on X. Then $\varphi_{hg} = h\varphi_g$. Furthermore, $\operatorname{supp} \varphi_g \subseteq \operatorname{supp} g$ for $g \in \mathrm{C_c}(X, \mathbb{R})$.

Proof Let $g \in \mathrm{L}^{1,\mathrm{loc}}(X, \mathfrak{B}, \mu, \mathbb{R})$. It is clear that $\varphi_g \in \mathrm{C_c}(X, \mathbb{R})^\sim$. We shall first prove that Φ is a lattice homomorphism by showing that $|\varphi_g| = \varphi_{|g|}$. For this, we apply Eq. (2.5) to see that

$$\begin{aligned}\langle |\varphi_g|, f \rangle &= \sup \{ \langle \varphi_g, h \rangle : h \in \mathrm{C_c}(X, \mathbb{R}),\ |h| \leq f \} \\ &= \sup \left\{ \int_X hg \,\mathrm{d}\mu : h \in \mathrm{C_c}(X, \mathbb{R}),\ |h| \leq f \right\}\end{aligned} \tag{7.12}$$

for $f \in \mathrm{C_c}(X, \mathbb{R})^+$.

Fix $f \in \mathrm{C_c}(X, \mathbb{R})^+$, and take $h \in \mathrm{C_c}(X, \mathbb{R})$ with $|h| \leq f$. Then

$$\int_X hg \,\mathrm{d}\mu \leq \left| \int_X hg \,\mathrm{d}\mu \right| \leq \int_X |h||g| \,\mathrm{d}\mu \leq \int_X f|g| \,\mathrm{d}\mu = \langle \varphi_{|g|}, f \rangle.$$

This shows that

$$\sup \left\{ \int_X hg \,\mathrm{d}\mu : h \in \mathrm{C_c}(X, \mathbb{R}),\ |h| \leq f \right\} \leq \langle \varphi_{|g|}, f \rangle. \tag{7.13}$$

For the reverse inequality, we define $\gamma : X \to \mathbb{R}$ by

$$\gamma(x) = \begin{cases} 0 & \text{if } x \notin \operatorname{supp} f, \\ \operatorname{sgn}(g) & \text{if } x \in \operatorname{supp} f. \end{cases}$$

Since $\operatorname{supp} f$ is compact, it has finite μ-measure, so that Proposition 7.1 yields a sequence (γ_n) in $\mathrm{C_c}(X, \mathbb{R})$ such that $\|\gamma_n\|_\infty \leq 1$ for all $n \geq 1$, and $\gamma_n(x) \to \gamma(x)$ for μ-almost all x in X. Note that $\gamma_n f \in \mathrm{C_c}(X, \mathbb{R})$, that $|\gamma_n f| \leq f$ for all $n \geq 1$, and that

$$\lim_{n\to\infty}\int_X(\gamma_n f)g\,\mathrm{d}\mu = \lim_{n\to\infty}\int_X \chi_{\operatorname{supp} f} f\gamma_n g\,\mathrm{d}\mu = \int_X \chi_{\operatorname{supp} f} f|g|\,\mathrm{d}\mu$$
$$= \int_X f|g|\,\mathrm{d}\mu = \langle f, \varphi_{|g|}\rangle.$$

Here the dominated convergence theorem was applied in the second step, and this is valid since $\chi_{\operatorname{supp} f} f|g|$ is integrable. We thus see that

$$\sup\left\{\int_X hg\,\mathrm{d}\mu : h \in \mathrm{C_c}(X, \mathbb{R}),\ |h| \le f\right\} \ge \langle \varphi_{|g|}, f\rangle. \tag{7.14}$$

Combining Eqs. (7.13) and (7.14), we obtain

$$\sup\left\{\int_X hg\,\mathrm{d}\mu : h \in \mathrm{C_c}(X, \mathbb{R}),\ |h| \le f\right\} = \langle \varphi_{|g|}, f\rangle,$$

and then Eq. (7.12) shows that $\langle |\varphi_g|, f\rangle = \langle \varphi_{|g|}, f\rangle$. Hence $|\varphi_g| = \varphi_{|g|}$.

It is now easy to prove that Φ is injective. Indeed, let $g \in \mathrm{L}^{1,\mathrm{loc}}(X, \mathfrak{B}, \mu, \mathbb{R})$ be such that $\varphi_g = 0$. Then also $\varphi_{|g|} = |\varphi_g| = 0$. Suppose that K is a compact subset of X, and take $f \in \mathrm{C_c}(X, \mathbb{R})^+$ such that $f = 1$ on K. Then

$$\int_K |g|\,\mathrm{d}\mu \le \int_X f|g|\,\mathrm{d}\mu = \langle \varphi_{|g|}, f\rangle = 0.$$

Hence g is locally μ-almost everywhere equal to zero, as required.

The statements on the multiplication by bounded Borel measurable functions and on supports are clear. □

Remark 7.3 Proposition 7.2 also follows from [12, V, § 5, No. 2, Corollary to Proposition 2]. Bourbaki's approach is different from ours. It does not use the dominated convergence theorem, for example, as there are no integrals present at all.

Lemma 7.4 *Let X be a non-empty, locally compact space, let $\mu \in \mathrm{M_r}(X, \mathfrak{B}, \overline{\mathbb{R}^+})$, and let $1 \le p < \infty$. Then $\mathrm{L}^p(X, \mathfrak{B}, \mu, \mathbb{R})$ is a vector sublattice of $\mathrm{L}^{1,\mathrm{loc}}(X, \mathfrak{B}, \mu)$.*

Proof If a measurable function is μ-almost everywhere equal zero, then it is clearly locally μ-almost everywhere equal to zero. Furthermore, Hölder's inequality implies that every p-integrable measurable function is locally integrable. Hence there exists a canonical lattice homomorphism from $\mathrm{L}^p(X, \mathfrak{B}, \mu, \mathbb{R})$ into $\mathrm{L}^{1,\mathrm{loc}}(X, \mathfrak{B}, \mu)$. We need to show that this homomorphism is injective. To this end, suppose that g is a measurable function on X such that

$$\int_X |g|^p\,\mathrm{d}\mu < \infty$$

and

$$\int_K |g| \, d\mu = 0$$

for every compact subset K of X. For $n = 1, 2, \ldots$, set

$$A_n := \{ x \in X : |g(x)| \geq 1/n \}.$$

Then $\mu(A_n) < \infty$ and

$$\{ x \in X : g(x) \neq 0 \} = \bigcup_{n=1}^{\infty} A_n. \tag{7.15}$$

Take $n \geq 1$. Since $\mu(A_n) < \infty$, [24, Proposition 7.5] shows that

$$\mu(A_n) = \sup \{ \mu(K) : K \text{ compact and } K \subseteq A_n \}. \tag{7.16}$$

Suppose that K is a compact subset of X such that $K \subseteq A_n$. Then

$$\frac{1}{n}\mu(K) = \int_K \frac{1}{n} \, d\mu \leq \int_K |g| \, d\mu = 0.$$

Hence Eq. (7.16) shows that $\mu(A_n) = 0$, and then Eq. (7.15) implies that g is μ-almost everywhere equal to zero. □

We can now establish our embedding theorem for L^p-spaces.

Theorem 7.5 *Let X be a non-empty, locally compact space, take p with $1 \leq p < \infty$, and let $\mu \in M_r(X, \mathfrak{B}, \overline{\mathbb{R}^+})$. For $g \in L^p(X, \mathfrak{B}, \mu, \mathbb{R})$, set*

$$\langle \varphi_g, f \rangle := \int_X fg \, d\mu$$

for $f \in C_c(X, \mathbb{R})$. Then $\varphi_g \in C_c(X, \mathbb{R})^\sim$, and the map $\Phi : g \mapsto \varphi_g$ defines an injective lattice homomorphism $\Phi : L^p(X, \mathfrak{B}, \mu, \mathbb{R}) \to C_c(X, \mathbb{R})^\sim$. Suppose that h is a bounded Borel measurable function on X. Then $\varphi_{hg} = h\varphi_g$.

For $g \in L^p(X, \mathfrak{B}, \mu, \mathbb{R})$, set $\|\varphi_g\| := \|g\|_p$, thus making $\Phi(L^p(X, \mathfrak{B}, \mu, \mathbb{R}))$ into a Dedekind complete Banach lattice. Then the set

$$\{ \varphi_g : g \in C_c(X, \mathbb{R}), \ \operatorname{supp} g^+ \cap \operatorname{supp} g^- = \emptyset \}$$

is a dense subset of the Banach lattice $\Phi(\mathrm{L}^p(X, \mathfrak{B}, \mu, \mathbb{R}))$. *Consequently, the set*

$$\left\{ \varphi \in \Phi(\mathrm{L}^p(X, \mathfrak{B}, \mu, \mathbb{R})) : \operatorname{supp}\varphi \textit{ is compact and } \operatorname{supp}\varphi^+ \cap \operatorname{supp}\varphi^- = \emptyset \right\}$$

is a dense subset of the Banach lattice $\Phi(\mathrm{L}^p(X, \mathfrak{B}, \mu, \mathbb{R}))$.

Proof It is clear from Proposition 7.2 and Lemma 7.4 that Φ is an injective lattice homomorphism that is compatible with multiplication by bounded Borel measurable functions. We establish the remaining statements.

It is obvious that $\mathrm{L}^p(X, \mathfrak{B}, \mu, \mathbb{R})$ is a Dedekind complete Banach lattice when the norm is transported via the lattice isomorphism Φ.

We turn to the density statements. Let $h \in C_c(X, \mathbb{R})$, and let $\varepsilon > 0$. It follows from Lemma 5.5 that there exists $g \in C_c(X, \mathbb{R})$ such that $\operatorname{supp} g \subseteq \operatorname{supp} h$, $\|h - g\|_\infty < \varepsilon$, and $\operatorname{supp} g^+ \cap \operatorname{supp} g^- = \emptyset$. Since then $\|h - g\|_p \leq \varepsilon\mu(\operatorname{supp} h)^{1/p}$ and since $C_c(X, \mathbb{R})$ is dense in $\mathrm{L}^p(X, \mathfrak{B}, \mu, \mathbb{R})$, it follows that

$$\left\{ g \in C_c(X, \mathbb{R}) : \operatorname{supp} g^+ \cap \operatorname{supp} g^- = \emptyset \right\}$$

is a dense subset of $\mathrm{L}^p(X, \mathfrak{B}, \mu, \mathbb{R})$. Applying the isometry Φ, we see that

$$\left\{ \varphi_g : g \in C_c(X, \mathbb{R}),\ \operatorname{supp} g^+ \cap \operatorname{supp} g^- = \emptyset \right\}$$

is a dense subset of the Banach lattice $\Phi(\mathrm{L}^p(X, \mathfrak{B}, \mu, \mathbb{R}))$.

Suppose that $g \in C_c(X, \mathbb{R})$ is such that $\operatorname{supp} g^+ \cap \operatorname{supp} g^- = \emptyset$. It follows from the inclusions $\operatorname{supp} \varphi_{g^+} \subseteq \operatorname{supp} g^+$ and $\operatorname{supp} \varphi_{g^-} \subseteq \operatorname{supp} g^-$ that we also have $\operatorname{supp} \varphi_{g^+} \cap \operatorname{supp} \varphi_{g^-} = \emptyset$. Hence Lemma 5.3 shows that φ_{g^+} and φ_{g^-} are disjoint elements of $C_c(X, \mathbb{R})^\sim$, and this implies that the equality $\varphi_g = \varphi_{g^+} - \varphi_{g^-}$ gives the decomposition of φ_g in $C_c(X, \mathbb{R})^\sim$ into its positive and negative part φ_g^+ and φ_g^-, respectively. The final density statement is now clear. □

We shall now show that $M_r(X, \mathfrak{B}, \mathbb{R})$ can also be embedded as a vector sublattice of $C_c(X, \mathbb{R})^\sim$. For this, we shall use the following auxiliary result. It is a slightly rephrased version of [44, Theorem 6.12], which is a consequence of the Radon–Nikodým theorem.

Proposition 7.6 *Let μ be a finite, real-valued measure on a σ-algebra of subsets of a set X. Then there is a measurable function γ on X such that $|\gamma(x)| = 1$ for all $x \in X$ and $\gamma \,\mathrm{d}\mu = \mathrm{d}|\mu|$.*

Proposition 7.7 *Let X be a non-empty, locally compact space. For a finite, real-valued measure $\mu \in M_r(X, \mathfrak{B}, \mathbb{R})$, set*

$$\langle \varphi_\mu, f \rangle := \int_X f \,\mathrm{d}\mu$$

for $f \in C_c(X, \mathbb{R})$. Then $\varphi_\mu \in C_c(X, \mathbb{R})^\sim$, and the map $\Phi : \mu \mapsto \varphi_\mu$ defines an injective lattice homomorphism $\Phi : M_r(X, \mathfrak{B}, \mathbb{R}) \to C_c(X, \mathbb{R})^\sim$.

Proof Let $\mu \in \mathrm{M_r}(X, \mathfrak{B}, \mathbb{R})$. We shall prove that $\varphi_{|\mu|} = |\varphi_\mu|$. The proof for this is quite similar to the proof of Proposition 7.2. Again we apply Eq. (2.5) to see that

$$\begin{aligned}\left\langle |\varphi_\mu|, f \right\rangle &= \sup\left\{ \left\langle \varphi_\mu, h \right\rangle : h \in \mathrm{C_c}(X, \mathbb{R}),\ |h| \leq f \right\} \\ &= \sup\left\{ \int_X h \,\mathrm{d}\mu : h \in \mathrm{C_c}(X, \mathbb{R}),\ |h| \leq f \right\}\end{aligned} \tag{7.17}$$

for $f \in \mathrm{C_c}(X, \mathbb{R})^+$.

Fix $f \in \mathrm{C_c}(X, \mathbb{R})^+$. If $h \in \mathrm{C_c}(X, \mathbb{R})$ and $|h| \leq f$, then

$$\int_X h \,\mathrm{d}\mu \leq \left| \int_X h \,\mathrm{d}\mu \right| \leq \int_X |h| \,\mathrm{d}|\mu| \leq \int_X f \,\mathrm{d}|\mu| = \left\langle \varphi_{|\mu|}, f \right\rangle.$$

This shows that

$$\sup\left\{ \int_X h \,\mathrm{d}\mu : h \in \mathrm{C_c}(X, \mathbb{R}),\ |h| \leq f \right\} \leq \left\langle \varphi_{|\mu|}, f \right\rangle. \tag{7.18}$$

For the reverse inequality, we use the unimodular measurable function γ such that $\gamma \,\mathrm{d}\mu = \mathrm{d}|\mu|$ that is supplied by Proposition 7.6. Since $|\mu|$ is a finite measure, Proposition 7.1 yields a sequence (γ_n) in $\mathrm{C_c}(X, \mathbb{R})$ such that $\|\gamma_n\|_\infty \leq 1$ for all $n \geq 1$, and $\gamma_n(x) \to \gamma(x)$ for $|\mu|$-almost all x in X. Note that $\gamma_n f \in \mathrm{C_c}(X, \mathbb{R})$ and $|\gamma_n f| \leq f$ for all $n \geq 1$, and that, by the dominated convergence theorem,

$$\lim_{n\to\infty} \int_X (\gamma_n f) \,\mathrm{d}\mu = \int_X f\gamma \,\mathrm{d}\mu = \int_X f \,\mathrm{d}|\mu| = \left\langle f, \varphi_{|\mu|} \right\rangle.$$

We thus see that

$$\sup\left\{ \int_X h \,\mathrm{d}\mu : h \in \mathrm{C_c}(X, \mathbb{R}),\ |h| \leq f \right\} \geq \left\langle \varphi_{|\mu|}, f \right\rangle. \tag{7.19}$$

Combining Eqs. (7.18) and (7.19), we obtain that

$$\sup\left\{ \int_X h \,\mathrm{d}\mu : h \in \mathrm{C_c}(X, \mathbb{R}),\ |h| \leq f \right\} = \left\langle \varphi_{|\mu|}, f \right\rangle,$$

and then Eq. (7.17) shows that $\left\langle |\varphi_\mu|, f \right\rangle = \left\langle \varphi_{|\mu|}, f \right\rangle$. Hence $|\varphi_\mu| = \varphi_{|\mu|}$.

It follows that Φ is a lattice homomorphism.

Suppose that $\varphi_\mu = 0$. We need to show that $\mu = 0$. Since also $\varphi_{|\mu|} = |\varphi_\mu| = 0$, we may suppose that $\mu \geq 0$. Let V be a non-empty, open subset of V. One of the explicit formulas in the Riesz representation theorem (see [24, Theorem 7.2]) shows that

$$\mu(V) = \sup\left\{ \int_X f \,\mathrm{d}\mu : f \in \mathrm{C_c}(X, \mathbb{R}),\ \operatorname{supp} f \subseteq V,\ 0 \leq f \leq 1_X \right\}.$$

Since all integrals in the set on the right-hand side are zero by assumption, μ vanishes on all open subsets of X. The outer regularity of μ at all Borel subsets of X then implies that $\mu = 0$. □

Remark 7.8

(1) An alternative proof of Proposition 7.7 goes as follows. It is generally true that the norm dual E' of a normed vector lattice E is a vector sublattice of the order dual $E^{\sim}$ of E, see [3, Theorem 30.8]. Since $(\mathrm{C_c}(X, \mathbb{R}), \|\cdot\|_\infty)'$ is (isometrically) lattice isomorphic to $\mathrm{M_r}(X, \mathfrak{B}, \mathbb{R})$, it is now immediate that the map Φ in Proposition 7.7 is an injective lattice homomorphism.

This alternative approach uses the vector lattice part of the Riesz representation theorem, whereas our earlier proof does not.

(2) There does not seem to be a result in the vein of Proposition 7.7 in [12]; presumably this is because the space $\mathrm{M_r}(X, \mathfrak{B}, \mathbb{R})$, which consists of measures in the sense of Caratheodory and Lebesgue, simply does not exist for Bourbaki.

We can now establish the following analogue of Theorem 7.5.

Theorem 7.9 *Let X be a non-empty, locally compact space. For $\mu \in \mathrm{M_r}(X, \mathfrak{B}, \mathbb{R})$, set*

$$\langle \varphi_\mu, f \rangle := \int_X f \,\mathrm{d}\mu$$

for $f \in \mathrm{C_c}(X, \mathbb{R})$. Then $\varphi_\mu \in \mathrm{C_c}(X, \mathbb{R})^{\sim}$, and the map $\Phi : \mu \mapsto \varphi_\mu$ defines an injective lattice homomorphism $\Phi : \mathrm{M_r}(X, \mathfrak{B}, \mathbb{R}) \to \mathrm{C_c}(X, \mathbb{R})^{\sim}$. Suppose that h is a bounded Borel measurable function on X. Then $\varphi_{h\mu} = h\varphi_\mu$.

For $\mu \in \mathrm{M_r}(X, \mathfrak{B}, \mathbb{R})$, set $\|\varphi_\mu\| := \|\mu\|$, thus making $\Phi(\mathrm{M_r}(X, \mathfrak{B}, \mathbb{R}))$ into a Dedekind complete Banach lattice. Then the set

$$\left\{ \varphi \in \Phi(\mathrm{M_r}(X, \mathfrak{B}, \mathbb{R})) : \operatorname{supp} \varphi \textit{ is compact and } \operatorname{supp} \varphi^+ \cap \operatorname{supp} \varphi^- = \emptyset \right\}$$

is a dense subset of the Banach lattice $\Phi(\mathrm{M_r}(X, \mathfrak{B}, \mathbb{R}))$.

Proof It is clear that Φ is compatible with the multiplication by bounded Borel measurable functions. In view of Proposition 7.7, it is then only the density statement that requires proof. Let $\mu \in \mathrm{M_r}(X, \mathfrak{B}, \mathbb{R})$, and let $\mu = \mu^+ - \mu^-$ be its decomposition into its positive and negative parts. There exists a partition of X into disjoint Borel measurable subsets X^+ and X^- of X such that $\mu^+(X^-) = 0$, $\mu^-(X^+) = 0$, $\mu^+(A^+) \geq 0$ for every Borel subset A^+ of X^+, and $\mu^-(A^-) \geq 0$ for every Borel subset A^- of X^-, see [44, Theorem 6.14]. Let $\varepsilon > 0$. Since μ^+, being finite, is inner regular at all Borel subsets of X (see [24, Proposition 7.5]), there exists a compact subset K^+ of X^+ such that $0 \leq \mu^+(X^+) - \mu^+(K^+) < \varepsilon/2$. Likewise, there exists a compact subset K^- of X^- with the property that $0 \leq \mu^-(X^-) - \mu^-(K^-) < \varepsilon/2$. For $A \in \mathfrak{B}$, we set $\nu^+(A) := \mu^+(A \cap K^+)$ and $\nu^-(A) := \mu^-(A \cap K^-)$, thus defining positive measures ν^+, ν^- on $\mathfrak{B}$. Since

μ^+ and μ^- are *finite* positive regular Borel measures, [25, Section 7.2, Exercise 7] shows that $\nu^+, \nu^- \in \mathrm{M_r}(X, \mathfrak{B}, \mathbb{R})$. Set $\nu := \nu^+ - \nu^-$. Then $\nu \in \mathrm{M_r}(X, \mathfrak{B}, \mathbb{R})$ and $\|\mu - \nu\| < \varepsilon$. Furthermore, $\operatorname{supp} \varphi_\nu \subseteq K^+ \cup K^-$ is a compact subset of X.

Since $\operatorname{supp} \varphi_{\nu^+} \subseteq K^+$, $\operatorname{supp} \varphi_{\nu^-} \subseteq K^-$, and $K^+ \cap K^- = \emptyset$, it follows that $\operatorname{supp} \varphi_{\nu^+} \cap \operatorname{supp} \varphi_{\nu^-} = \emptyset$. Hence Lemma 5.3 shows that φ_{ν^+} and φ_{ν^-} are disjoint elements of $\mathrm{C_c}(X, \mathbb{R})^\sim$, and this implies that the equality $\varphi_\nu = \varphi_{\nu^+} - \varphi_{\nu^-}$ gives the decomposition of φ_ν in $\mathrm{C_c}(X, \mathbb{R})^\sim$ into its positive and negative part φ_ν^+ and φ_ν^-, respectively. Since $\|\varphi_\mu - \varphi_\nu\| = \|\mu - \nu\| < \varepsilon$ by definition, the proof of the theorem is complete. □

8 Locally Compact Groups

In this section, we shall review some material on locally compact groups and on Banach lattice and Banach lattice algebras on such groups. In particular, we shall describe various well-known Banach algebras that are studied within harmonic analysis. For details, see [13, 25, 30, 43], and also [18, Sections 3.3 and 4.5], for example.

A group that is also a locally compact space is a *locally compact group* whenever the group operations are continuous.

Let G be a locally compact group. As for general locally compact spaces, the Borel σ-algebra of G will be denoted by $\mathfrak{B}$. Conforming to the customary notation in abstract harmonic analysis as much as possible, we shall write $\mathrm{M}(G, \mathbb{R})$ for $\mathrm{M_r}(G, \mathfrak{B}, \mathbb{R})$ and $\mathrm{M}(G, \mathbb{C})$ for $\mathrm{M_r}(G, \mathfrak{B}, \mathbb{C})$. There exists a non-zero, positive regular Borel measure m_G on G such that $m_G(s \cdot A) = m_G(A)$ for all $s \in G$ and all Borel subsets A of G. Such a measure is a *(left) Haar measure on* G; it is unique up to a non-zero positive multiplicative constant. We shall write $\mathrm{L}^p(G, \mathbb{R})$ for $\mathrm{L}^p(G, \mathfrak{B}, m_G, \mathbb{R})$ and $\mathrm{L}^p(G, \mathbb{C})$ for $\mathrm{L}^p(G, \mathfrak{B}, m_G, \mathbb{C})$.

Let G be a locally compact group, and let m_G be a Haar measure on G. Then

$$\int_G f(as) \, \mathrm{d}m_G(s) = \int_G f(s) \, \mathrm{d}m_G(s)$$

for all $f \in \mathrm{L}^1(G, \mathbb{C})$ and $a \in G$. When G is abelian, the left Haar measure is trivially also right invariant, but this is not generally the case. There exists a continuous group homomorphism $\Delta : G \to (0, \infty)$ such that

$$\int_G f(sa) \, \mathrm{d}m_G(s) = \Delta(a^{-1}) \int_G f(s) \, \mathrm{d}m_G(s)$$

for all $f \in \mathrm{L}^1(G, \mathbb{C})$ and $a \in G$; some authors write $\Delta(a)$ where we use $\Delta(a^{-1})$. The homomorphism Δ is the *modular function of* G. It is easy to see that $\Delta(s) = 1$ for all $s \in G$ when G is compact or discrete, so that the left Haar measure is then also right invariant.

Let G be a locally compact group. We recall from the general theory for locally compact spaces that the Banach lattice $\mathrm{M}(G, \mathbb{R})$ is isometrically lattice isomorphic to the Banach lattice $\mathrm{C}_0(G, \mathbb{R})'$. By combining this isomorphism with the group structure of the underlying locally compact space G, a multiplication on $\mathrm{M}(G, \mathbb{R})$ can be introduced such that it becomes a Banach lattice algebra. Take $\mu, \nu \in \mathrm{M}(G, \mathbb{R})$. Then the *convolution product* $\mu \star \nu$ *of* μ *and* ν is defined by

$$\langle \mu \star \nu, f \rangle := \int_G \int_G f(st)\, \mathrm{d}\mu(s)\, \mathrm{d}\nu(t) \tag{8.20}$$

for all $f \in \mathrm{C}_0(G, \mathbb{R})$. With this multiplication, $\mathrm{M}(G, \mathbb{R})$ is a Banach lattice algebra. The unit mass at e_G is denoted by δ_{e_G}; it is the identity element of $\mathrm{M}(G, \mathbb{R})$. One can describe $\mu \star \nu$ at the level of the Borel subsets of G by

$$(\mu \star \nu)(A) = \int_G \nu(s^{-1} \cdot A)\, \mathrm{d}\mu(s) = \int_G \mu(A \cdot s^{-1})\, \mathrm{d}\nu(s) \tag{8.21}$$

for $A \in \mathfrak{B}$.

The following basic result is very well known. Since it is essential to the results in Sect. 10, we nevertheless include the proof.

Proposition 8.1 *Let G be a locally compact group, and take $\mu, \nu \in \mathrm{M}(G, \mathbb{R})$ with compact support. Then* $\operatorname{supp}(\mu \star \nu) \subseteq \operatorname{supp} \mu \cdot \operatorname{supp} \nu$.

Proof We may suppose that $\operatorname{supp} \mu \cdot \operatorname{supp} \nu \neq G$. Then $G \setminus (\operatorname{supp} \mu \cdot \operatorname{supp} \nu)$ is a non-empty, open subset of G. Take $f \in \mathrm{C}_c(G, \mathbb{R}; G \setminus (\operatorname{supp} \mu \cdot \operatorname{supp} \nu))$. Then it is immediate from Eq. (8.20) that $\langle \mu \star \nu, f \rangle = 0$. Hence $\mu \star \nu$ vanishes on $G \setminus (\operatorname{supp} \mu \cdot \operatorname{supp} \nu)$. The result follows. □

The complex Banach lattice $\mathrm{M}(G, \mathbb{C})$ is the complexification of the Banach lattice $\mathrm{M}(G, \mathbb{R})$. Since $\mathrm{M}(G, \mathbb{R})$ is, in fact, a Banach lattice algebra, $\mathrm{M}(G, \mathbb{C})$ is a complex Banach lattice algebra. It is then easily checked that the obvious complex analogues of Eqs. (8.20) and (8.21) hold.

Take $\mu \in \mathrm{M}(G, \mathbb{C})$. Set $\mu^*(A) = \overline{\mu(A^{-1})}$ for each Borel subset A of G. Then $\mu \mapsto \mu^*$ is an involution on $\mathrm{M}(G, \mathbb{C})$.

The following theorem is basic, see [18, Section 3.3].

Theorem 8.2 *Let G be a locally compact group. Then $\mathrm{M}(G, \mathbb{R})$ is a Dedekind complete, unital Banach lattice algebra, and $\mathrm{M}(G, \mathbb{C})$ is a unital, semisimple, complex Banach lattice *-algebra. The identity element of both algebras is δ_{e_G}.*

The commutativity of $\mathrm{M}(G, \mathbb{R})$ and that of $\mathrm{M}(G, \mathbb{C})$ are both equivalent to the group G being abelian.

Remark 8.3 In the literature on abstract harmonic analysis, the complex Banach lattice algebra $\mathrm{M}(G, \mathbb{C})$ is usually denoted by $\mathrm{M}(G)$, and it is called the measure algebra of G, without a reference to the complex field. It is then studied as a complex

Banach *-algebra. We, on the other hand, concentrate on the lattice properties of $\mathrm{M}(G, \mathbb{R})$.

Let G be a locally compact group with left Haar measure m_G. The subspace of $\mathrm{M}(G, \mathbb{R})$ consisting of all elements that are absolutely continuous with respect to m_G is a Banach sublattice of $\mathrm{M}(G, \mathbb{R})$; it is also an algebra ideal and an order ideal of $\mathrm{M}(G, \mathbb{R})$. This Banach sublattice is isometrically lattice isomorphic to $\mathrm{L}^1(G, \mathbb{R})$ by using the Radon–Nikodým theorem: each $f \in \mathrm{L}^1(G, \mathbb{R})$ corresponds to the measure $f\, \mathrm{d}m_G$ in $\mathrm{M}(G, \mathbb{R})$. This identification provides $\mathrm{L}^1(G, \mathbb{R})$ with a product; the convolution product of f and g in $\mathrm{L}^1(G, \mathbb{R})$ is then given by the formulae

$$(f \star g)(t) = \int_G f(s)g(s^{-1}t)\, \mathrm{d}m_G(s) = \int_G f(ts)g(s^{-1})\, \mathrm{d}m_G(s) \tag{8.22}$$

for m_G-almost all $t \in G$.

Similar remarks apply to $\mathrm{M}(G, \mathbb{C})$ and $\mathrm{L}^1(G, \mathbb{C})$, with the additional feature that the subspace of $\mathrm{M}(G, \mathbb{C})$ consisting of all elements that are absolutely continuous with respect to m_G is now an algebra *-ideal. The identification of $\mathrm{L}^1(G, \mathbb{C})$ with this subspace then provides $\mathrm{L}^1(G, \mathbb{C})$ with an involution, denoted by * again. For $f \in \mathrm{L}^1(G, \mathbb{C})$, the involution is given by

$$f^*(s) = \overline{f(s^{-1})}\Delta(s^{-1})$$

for m_G-almost $s \in G$.

We then have the following result.

Theorem 8.4 *Let G be a locally compact group. Then $\mathrm{L}^1(G, \mathbb{R})$ is a Dedekind complete Banach lattice algebra which is a closed algebra ideal and an order ideal of $\mathrm{M}(G, \mathbb{R})$, and $\mathrm{L}^1(G, \mathbb{C})$ is a semisimple, complex Banach lattice *-algebra which is a closed algebra *-ideal of $\mathrm{M}(G, \mathbb{C})$.*

The commutativity of $\mathrm{L}^1(G, \mathbb{R})$ and that of $\mathrm{L}^1(G, \mathbb{C})$ are both equivalent to the group G being abelian. Both algebras have a positive contractive approximate identity, and both are unital if and only if G is discrete. In the latter case, $\mathrm{L}^1(G, \mathbb{R}) = \mathrm{M}(G, \mathbb{R})$ and $\mathrm{L}^1(G, \mathbb{C}) = \mathrm{M}(G, \mathbb{C})$. It is then customary to write $\ell^1(G, \mathbb{R})$ and $\ell^1(G, \mathbb{C})$ for the coinciding convolution algebras over the respective fields.

We remark that the space $\mathrm{L}^1(G, \mathbb{R})$ is not just an order ideal of $\mathrm{M}(G, \mathbb{R})$, but that it is, in fact, a so-called *band* of $\mathrm{M}(G, \mathbb{R})$. More precisely, it is the band that is generated by m_G. We have not defined what a band is in the present article, and we shall not pursue this matter further.

Remark 8.5 In the literature on abstract harmonic analysis, the complex Banach lattice algebra $\mathrm{L}^1(G, \mathbb{C})$ is usually denoted by $\mathrm{L}^1(G)$, and it is called the group algebra of G, without a reference to the complex field. It is then studied as a complex Banach *-algebra, whereas we concentrate on the lattice properties of $\mathrm{L}^1(G, \mathbb{R})$.

Let G be a locally compact group, and take p with $1 \leq p < \infty$. Now take $\mu \in \mathrm{M}(G, \mathbb{F})$ and $g \in \mathrm{L}^p(G, \mathbb{F})$, and define

$$(\mu \star_p g)(s) := \int_G g(t^{-1}s)\, \mathrm{d}\mu(t) \tag{8.23}$$

and

$$(g \star_p \mu)(s) := \int_G g(st^{-1})\Delta_G^{1/p}(t^{-1})\, \mathrm{d}\mu(t) \tag{8.24}$$

for those $s \in G$ for which these integrals exist; this can be shown to be m_G-almost everywhere the case.

Now take $f \in \mathrm{L}^1(G, \mathbb{F})$ and $g \in \mathrm{L}^p(G, \mathbb{F})$. Identifying f and $f\,\mathrm{d}m_G$, Eqs. (8.23) and (8.24) specialise to

$$(f \star_p g)(s) := \int_G f(t)g(t^{-1}s)(t)\, \mathrm{d}m_G(t) \tag{8.25}$$

and

$$(g \star_p f)(s) := \int_G g(st^{-1})f(t)\Delta_G^{1/p}(t^{-1})\, \mathrm{d}m_G(t) \tag{8.26}$$

for m_G-almost all $s \in G$.

The following theorem is contained in [18, Section 3.3], see also [30, (20.19)].

Theorem 8.6 *Let G be a locally compact group, and take p with $1 \leq p < \infty$. Take $\mu \in \mathrm{M}(G, \mathbb{F})$ and $g \in \mathrm{L}^p(G, \mathbb{F})$. Then the functions $\mu \star_p g$ and $g \star_p \mu$ belong to $\mathrm{L}^p(G, \mathbb{F})$, and we have $\|\mu \star_p g\|_p \leq \|\mu\|\|g\|_p$ and $\|f \star_p \mu\|_p \leq \|\mu\|\|f\|_p$.*

The following is now clear.

Corollary 8.7 *Let G be a locally compact group, and take p with $1 \leq p < \infty$.*

For $\mu \in \mathrm{M}(G, \mathbb{F})$, define $\pi_\mu : \mathrm{L}^p(G, \mathbb{F}) \to \mathrm{L}^p(G, \mathbb{F})$ by setting

$$\pi_\mu(g) := \mu \star_p g$$

for $g \in \mathrm{L}^p(G, \mathbb{F})$. Then $\pi_\mu \in \mathrm{L_r}(\mathrm{L}^p(G, \mathbb{F}))$, and the map $\pi : \mu \mapsto \pi_\mu$ defines a positive Banach algebra homomorphism $\pi : \mathrm{M}(G, \mathbb{F}) \to \mathrm{L_r}(\mathrm{L}^p(G, \mathbb{F}))$.

For $f \in \mathrm{L}^1(G, \mathbb{F})$, define $\pi_f : \mathrm{L}^p(G, \mathbb{F}) \to \mathrm{L}^p(G, \mathbb{F})$ by setting

$$\pi_f(g) := f \star_p g$$

for $g \in \mathrm{L}^p(G, \mathbb{F})$. *Then* $\pi_f \in \mathrm{L_r}(\mathrm{L}^p(G, \mathbb{F}))$, *and the map* $\pi : f \mapsto \pi_f$ *defines a positive Banach algebra homomorphism* $\pi : \mathrm{L}^1(G, \mathbb{F}) \to \mathrm{L_r}(\mathrm{L}^p(G, \mathbb{F}))$.

Similarly, one can define a map $g \mapsto g \star_p \mu$, respectively, $g \mapsto g \star_p f$. Then the resulting map from $\mathrm{M}(G, \mathbb{F})$, respectively, $\mathrm{L}^1(G, \mathbb{F})$ into $\mathrm{L_r}(\mathrm{L}^p(G, \mathbb{F}))$ has the same properties as its left-sided analogue, save that is an anti-homomorphism.

We shall see later that the two Banach algebra homomorphisms π in Corollary 8.7 are both Banach lattice algebra homomorphisms, see Theorem 11.2 and Corollary 11.4, below.

9 Locally Compact Semigroups

In this section, we shall collect some material on Banach lattice algebras on locally compact semigroups. It is for these algebras that we shall benefit from the results in Sect. 6 by using them in the proof of our main result, Theorem 10.3, below.

Definition 9.1 Let S be a locally compact semigroup. A *weight on* S is a continuous function $\omega : S \to (0, \infty)$ such that

$$\omega(st) \leq \omega(s)\omega(t)$$

for all $s, t \in S$.

Let G be a locally compact group, and let S be a closed subspace of G that is a subsemigroup of G. Suppose that ω is a weight on S, and consider the subset $\mathrm{M}(S, \omega, \mathbb{R})$ of $\mathrm{M}(S, \mathbb{R})$ consisting of all elements μ of $\mathrm{M}(S, \mathbb{R})$ such that

$$\int_S \omega(t)\, \mathrm{d}|\mu|(t) < \infty.$$

Then $\mathrm{M}(S, \omega, \mathbb{R})$ is a Dedekind complete vector sublattice of $\mathrm{M}(S, \mathbb{R})$. (It is, in fact, even a band in $\mathrm{M}(S, \mathbb{R})$.) Since $\mathrm{M}(S, \mathbb{R})$ can be embedded as a sublattice of $\mathrm{C_c}(S, \mathbb{R})^\sim$ by Theorem 7.9, this is also the case for the sublattice $\mathrm{M}(S, \omega, \mathbb{R})$ of $\mathrm{M}(S, \mathbb{R})$. Since, furthermore, $\mathrm{C_c}(S, \mathbb{R})^\sim$ can be embedded as a sublattice of $\mathrm{C_c}(G, \mathbb{R})^\sim$ by Theorem 6.3, we see that $\mathrm{M}(S, \omega, \mathbb{R})$ can be embedded as a sublattice of $\mathrm{C_c}(G, \mathbb{R})^\sim$. The embedded copy is easily checked to be a subalgebra of $\mathrm{M}(G, \mathbb{R})$, and hence the embedding of $\mathrm{M}(S, \omega, \mathbb{R})$ into $\mathrm{C_c}(G, \mathbb{R})^\sim$ provides $\mathrm{M}(S, \omega, \mathbb{R})$ with a (convolution) product.

We introduce a norm $\|\cdot\|_\omega$ on $\mathrm{M}(S, \omega, \mathbb{R})$ by setting

$$\|\mu\|_\omega := \int_S \omega(t)\, \mathrm{d}|\mu|(t)$$

for $\mu \in \mathrm{M}(S, \omega, \mathbb{R})$. Then $(\mathrm{M}(S, \omega, \mathbb{R}), \|\cdot\|_\omega)$ is a Banach algebra. The algebra $(\mathrm{M}(S, \omega, \mathbb{R}), \|\cdot\|_\omega)$ is called a *Beurling algebra*. It is a Dedekind complete Banach lattice algebra.

We then have the following companion result of Theorem 7.9.

Theorem 9.2 *Let G be a locally compact group, and let S be a closed subspace of G that is a subsemigroup of G. Suppose that ω is a weight on S. For each $\mu \in \mathrm{M}(S, \omega, \mathbb{R})$, set*

$$\langle \varphi_\mu, f \rangle := \int_S f \, \mathrm{d}\mu$$

for $f \in \mathrm{C_c}(G, \mathbb{R})$. Then the map $\Phi : \mu \mapsto \varphi_\mu$ defines an injective lattice homomorphism $\Phi : \mathrm{M}(S, \omega, \mathbb{R}) \to \mathrm{C_c}(G, \mathbb{R})^\sim$. Suppose that h is a bounded Borel measurable function on S, and extend h to a Borel measurable function $\widetilde{h}$ on G by setting $\widetilde{h}(t) := 0$ for $t \in G \setminus S$. Then $\varphi_{h\mu} = \widetilde{h}\varphi_\mu$.

For $\mu \in \mathrm{M}(S, \omega, \mathbb{R})$, set $\|\varphi_\mu\| := \|\mu\|_\omega$, thus making $\Phi(\mathrm{M}(S, \omega, \mathbb{R}))$ into a Dedekind complete Banach lattice. Then the set

$$\left\{ \varphi \in \Phi(\mathrm{M}(S, \omega, \mathbb{R})) : \operatorname{supp} \varphi \textit{ is compact and } \operatorname{supp} \varphi^+ \cap \operatorname{supp} \varphi^- = \emptyset \right\}$$

is a dense subset of the Banach lattice $\Phi(\mathrm{M}(S, \omega, \mathbb{R}))$.

Proof In view of the above, all is clear except the density statement. For this, let $\mu \in \mathrm{M}(S, \omega, \mathbb{R})$. Since ω is strictly positive and continuous, the measure $\omega \, \mathrm{d}|\mu|$ is a positive regular Borel measure on S, see [24, Section 7.2, Exercise 9]. An easy modification of the argument in the proof of Theorem 7.9 then shows that the subset

$$\{ \mu \in \mathrm{M}(S, \omega, \mathbb{R}) : \operatorname{supp} \mu \text{ is compact and } \operatorname{supp} \mu^+ \cap \operatorname{supp} \mu^- = \emptyset \}$$

is a dense subset of $\mathrm{M}(S, \omega, \mathbb{R})$. As in the proof of Theorem 7.9, the density statement for the embedded copy $\Phi(\mathrm{M}(S, \omega, \mathbb{R}))$ of $\mathrm{M}(S, \omega, \mathbb{R})$ is then immediate. □

Let G be a locally compact group, and let S be a closed subspace of G that is a subsemigroup of G. Suppose that ω is a weight on S. It is obvious how to define the complex analogue $\mathrm{M}(S, \omega, \mathbb{C})$ of $\mathrm{M}(S, \omega, \mathbb{R})$. Then $\mathrm{M}(S, \omega, \mathbb{C})$ is the complexification of $\mathrm{M}(S, \omega, \mathbb{R})$; hence $\mathrm{M}(S, \omega, \mathbb{C})$ is a complex Banach lattice algebra.

Let S be a semigroup, supplied with the discrete topology, and let ω be a weight on S. Instead of considering real-valued measures on S as above, we now consider ℓ^1-spaces for weighted counting measures, as follows. Let $\ell^1(S, \omega, \mathbb{R})$ consist of the functions $f : S \to \mathbb{R}$ such that

$$\sum_{s \in S} |f(s)| \, \omega(s) < \infty. \tag{9.27}$$

We introduce a norm $\|\cdot\|_\omega$ on $\ell^1(S, \omega, \mathbb{R})$ by setting

$$\|f\|_\omega := \sum_{s \in S} |f(s)|\,\omega(s) \tag{9.28}$$

for $f \in \ell^1(S, \omega)$. Then $(\ell^1(S, \omega, \mathbb{R}), \|\cdot\|_\omega)$ is a Banach space. For $s \in S$, we let δ_s denote the characteristic function of the subset $\{s\}$ of S. Then there is a unique continuous product on $\ell^1(S, \omega, \mathbb{R})$ such that $\delta_{s_1} \star \delta_{s_2} = \delta_{s_1 s_2}$ for $s_1, s_2 \in S$. When supplied with the pointwise ordering, the weighted ℓ^1-space $(\ell^1(S, \omega, \mathbb{R}), \|\cdot\|_\omega)$ is then a Dedekind complete Banach lattice algebra, which is also called a *Beurling algebra*.

Let S be a semigroup, supplied with the discrete topology, and let ω be a weight on S. It is obvious how to use Eqs. (9.27) and (9.28) to define the complex analogue $\ell^1(S, \omega, \mathbb{C})$ of $\ell^1(S, \omega, \mathbb{R})$. Then $\ell^1(S, \omega, \mathbb{C})$ is the complexification of $\ell^1(S, \omega, \mathbb{R})$. Hence $\ell^1(S, \omega, \mathbb{C})$ is a complex Banach lattice algebra.

Let G be a group, supplied with the discrete topology, and let ω be a weight on G. Then it is a notorious open question whether the Beurling algebra $\ell^1(G, \omega, \mathbb{C})$ is always semisimple. It is proved in [21, Theorem 7.13] that this is the case whenever G is a maximally almost periodic group and ω is an arbitrary weight on G, and also whenever G is an arbitrary group and ω is a symmetric weight on G, in the sense that $\omega(s^{-1}) = \omega(s)$ for $s \in G$.

For semigroups, however, it is known that such Beurling algebras need not be semisimple. They can even be radical, as we shall now indicate.

Let S be a semigroup, supplied with the discrete topology, and let ω be a weight on S. For $s \in S$, the element δ_s of the Beurling algebra $\ell^1(S, \omega, \mathbb{C})$ is obviously quasi-nilpotent if and only if

$$\lim_{n \to \infty} \omega(s^n)^{1/n} = 0.$$

It is shown in [18, Example 2.3.13(ii)] that $\ell^1(S, \omega, \mathbb{C})$ is a radical Banach algebra whenever δ_s is quasi-nilpotent for all $s \in S$ and $\omega(st) = \omega(ts)$ for all $s, t \in S$. For example, take $S = \mathbb{Z}^+$ and set $\omega(n) := \exp(-n^2)$ for $n \in \mathbb{Z}^+$, or take S to be the free semigroup on two generators and set $\omega(w) = \exp(-|w|^2)$ for a word w in S, where $|w|$ is the length of the word w. Then in both cases $\ell^1(S, \omega, \mathbb{C})$ is a radical Banach algebra.

For a study of the algebras $\ell^1(S, \omega, \mathbb{C})$ when S is a subsemigroup of $\mathbb{R}$, see [20]. In the case where $S = \mathbb{Z}^+$, the algebras $\ell^1(\mathbb{Z}^+, \omega, \mathbb{C})$ are examples of *Banach algebras of power series*; for a study of these algebras, see [10, 18].

We shall now consider continuous analogues of the Beurling algebras $\ell^1(S, \omega, \mathbb{R})$ and $\ell^1(S, \omega, \mathbb{C})$ above.

Consider the unital additive semigroup $\mathbb{R}^+ := [0, \infty)$. Suppose that ω is a weight on $\mathbb{R}^+$. Then we define $\mathrm{L}^1(\mathbb{R}^+, \omega, \mathbb{R})$ to be the vector space of measurable functions f on $\mathbb{R}^+$ such that

$$\int_{\mathbb{R}^+} |f(t)|\,\omega(t)\,\mathrm{d}t < \infty, \tag{9.29}$$

and we introduce a norm $\|\cdot\|_\omega$ on $\mathrm{L}^1(\mathbb{R}^+,\omega,\mathbb{R})$ by setting

$$\|f\|_\omega := \int_{\mathbb{R}^+} |f(t)|\,\omega(t)\,\mathrm{d}t \tag{9.30}$$

for $f \in \mathrm{L}^1(\mathbb{R}^+,\omega,\mathbb{R})$. For $f, g \in \mathrm{L}^1(\mathbb{R}^+,\omega,\mathbb{R})$, we set

$$(f \star g)(s) := \int_{[0,s]} f(t)g(s-t)\,\mathrm{d}t \tag{9.31}$$

for all $s \in \mathbb{R}^+$ for which the integral exists, which can be shown to be the case almost everywhere. With this (convolution) product, $(\mathrm{L}^1(\mathbb{R}^+,\omega,\mathbb{R}), \|\cdot\|_\omega)$ is a Dedekind complete Banach lattice algebra that is again an example of a Beurling algebra.

It is obvious how to use Eqs. (9.29)–(9.31) to define the complex analogue $\mathrm{L}^1(\mathbb{R}^+,\omega,\mathbb{C})$ of $\mathrm{L}^1(\mathbb{R}^+,\omega,\mathbb{R})$. Then $\mathrm{L}^1(\mathbb{R}^+,\omega,\mathbb{C})$ is the complexification of $\mathrm{L}^1(\mathbb{R}^+,\omega,\mathbb{R})$. Hence $\mathrm{L}^1(\mathbb{R}^+,\omega,\mathbb{C})$ is a complex Banach lattice algebra.

The Beurling algebras $\mathrm{L}^1(\mathbb{R}^+,\omega,\mathbb{C})$ are studied in [9, 18], for example. It can be shown that $\rho_\omega := \lim_{t\to\infty}\omega(t)^{1/t}$ always exists, that $\mathrm{L}^1(\mathbb{R}^+,\omega,\mathbb{C})$ is semisimple if $\rho_\omega > 0$, and that $\mathrm{L}^1(\mathbb{R}^+,\omega,\mathbb{C})$ is radical if $\rho_\omega = 0$. For example, the weight $\omega : t \mapsto \exp(-t^2)$ gives a radical Beurling algebra on $\mathbb{R}^+$.

Let ω be a weight on $\mathbb{R}^+$. Then it follows from Titchmarsh's convolution theorem [18, Theorem 4.7.22] that the Beurling algebras $\mathrm{L}^1(\mathbb{R}^+,\omega,\mathbb{R})$ and $\mathrm{L}^1(\mathbb{R}^+,\omega,\mathbb{C})$ are integral domains.

10 Main Theorem

In this section, we shall establish our main result, Theorem 10.3, below, in the context of non-empty, closed semigroups in locally compact groups, as well as a related, easier, result in the context of discrete semigroups.

We start with the following preparatory result.

Lemma 10.1 *Let G be a locally compact group, and let K_1 and K_2 be non-empty, disjoint, compact subsets of G. Then there exists an open neighbourhood U of e_G such that K_1U and K_2U are disjoint subsets of G.*

Proof Since G is locally compact, there exists open neighbourhoods W_1 and W_2 of K_1 and K_2, respectively, such that $\overline{W_1}$ and $\overline{W_2}$ are disjoint. It is easy to see that, for $i = 1, 2$, there is an open neighbourhood U_i of e_G with $K_iU_i \subseteq W_i$.

Set $U := U_1 \cap U_2$. Then U is an open neighbourhood of e_G and we also see that $K_1 U \cap K_2 U \subseteq K_1 U_1 \cap K_2 U_2 \subseteq W_1 \cap W_2 = \emptyset$. □

For the ease of formulation, we introduce the following terminology.

Definition 10.2 Let $\varphi \in \mathrm{C_c}(X, \mathbb{R})^{\sim}$. Then *the support of* φ *is separated*, or φ *has separated support*, if the supports of φ^+ and φ^- are disjoint subsets of X.

We now come to our main result, Theorem 10.3. It employs a notation $\circledast$ for a bilinear map $\circledast$ that suggests convolution, without requiring that this actually be the case. The reason is that we also want to cover situations where, for example, μ is a positive measure on the Borel σ-algebra of a locally compact group G and a function f_1 acts on a function f_2 via the formula

$$f_1 \circledast f_2(s) = \int_G f_1(t) f_2(t^{-1}s)\, \mathrm{d}\mu(t) \tag{10.32}$$

for μ-almost all $s \in G$. Unless the measure μ is a left Haar measure on G, this is not an actual convolution, but obviously it still satisfies the relation

$$\{s \in G : (f_1 \circledast f_2)(s) \neq 0\} \subseteq \{s \in G : f_1(s) \neq 0\} \cdot \{s \in G : f_2(s) \neq 0\},$$

which is akin to the inclusion relation in the crucial first clause of the hypotheses in Theorem 10.3. Such bilinear maps occur in [36, 37], for example, and Theorem 10.3 is likely to be applicable in such contexts.

Theorem 10.3 *Let G be a locally compact group, and let S be a non-empty, closed subspace of G that is a subsemigroup of G. Let X, Y, and Z be vector sublattices of* $\mathrm{C_c}(S, \mathbb{R})^{\sim}$ *that are Banach lattices, and where Z is Dedekind complete.*

Suppose that $\circledast : X \times Y \to Z$ is a bilinear map such that $x \circledast y \in Z^+$ whenever $x \in X^+$ and $y \in Y^+$. Define the positive linear map $\pi : X \to \mathrm{L_r}(Y, Z)$ by $\pi_x(y) := x \circledast y$ for $x \in X$ and $y \in Y$.

Suppose that the following conditions are satisfied:

(1) $\mathrm{supp}\,(x \circledast y) \subseteq \mathrm{supp}\,x \cdot \mathrm{supp}\,y$ *for all $x \in X^+$ and $y \in Y^+$ with compact support;*
(2) *the elements of X with compact, separated support are dense in X;*
(3) *the elements of Y^+ with compact support are dense in Y^+;*
(4) *$\chi_A y$ is an element of Y again, whenever $y \in Y^+$ has compact support and A is a Borel subset of* $\mathrm{supp}\,y$.

Then π is a lattice homomorphism.

For the sake of clarity, we recall that semigroups are not supposed to be unital.

Proof We start with the case where $S = G$. We are to prove that $|\pi_x| = \pi_{|x|}$ for all $x \in X$.

Recalling that positive linear maps between Banach lattices are continuous, that $\mathrm{L_r}(Y, Z)$ is a Banach lattice in the regular norm, and that the modulus is continuous on Banach lattices, we see that the maps $x \mapsto |\pi_x|$ and $x \mapsto \pi_{|x|}$ are both continuous maps from X into $\mathrm{L_r}(Y, Z)$. By density, it is thus sufficient to prove that $|\pi_x| = \pi_{|x|}$ for all elements x of X with separated, compact support.

For this, we need to show that $|\pi_x|(y) = \pi_{|x|}(y)$ for all $y \in Y$. It is sufficient to establish this for all $y \in Y^+$. By the continuity of the regular operators $|\pi_x|$ and $\pi_{|x|}$ on the Banach lattice Y, it is, by density, sufficient to prove that $|\pi_x|(y) = \pi_{|x|}(y)$ for all elements of Y^+ with compact support.

All in all, we see that it is sufficient to demonstrate that $|\pi_x|(y) = \pi_{|x|}(y)$, whenever x is an element of X with separated, compact support and y is an element of Y^+ with compact support.

In order to do so, we fix an element x of X with compact, separated support, and we let $x = x^+ - x^-$ be the decomposition of x into its disjoint positive and negative parts. The supports of x^+ and x^- are disjoint, compact subsets of X. Using Lemma 10.1, we can then choose and fix a relatively compact open neighbourhood U of the e_G such that $(\operatorname{supp} x^+)U \cap (\operatorname{supp} x^-)U = \emptyset$.

We shall now first consider the special case in which the support of $y \in Y^+$ is not only compact, but where it is also 'sufficiently small'. To be precise, suppose that y is an element of Y^+ with compact support such that $\operatorname{supp} y \subseteq Us$ for some $s \in G$. We shall show that then $|\pi_x|(y) = \pi_{|x|}(y)$.

First, since π is positive, it is automatic that $|\pi_x| \leq \pi_{|x|}$, so that we have $|\pi_x|(y) \leq \pi_{|x|}(y)$.

Second, for the reverse inequality, we notice that certainly $|\pi_x|(y) \geq \pm\pi_x(y)$, so that $|\pi_x|(y) \geq |\pi_x(y)|$. Since

$$\operatorname{supp}(x^+ \circledast y) \subseteq \operatorname{supp} x^+ \cdot \operatorname{supp} y \subseteq \operatorname{supp} x^+ \cdot Us$$

and

$$\operatorname{supp}(x^- \circledast y) \subseteq \operatorname{supp} x^- \cdot \operatorname{supp} y \subseteq \operatorname{supp} x^- \cdot Us,$$

the supports of $x^+ \circledast y$ and $x^- \circledast y$ are still disjoint subsets of X. Lemma 5.3, therefore, implies that

$$\begin{aligned} |\pi_x(y)| &= |x^+ \circledast y - x^- \circledast y| = |x^+ \circledast y| + |-x^- \circledast y| \\ &= x^+ \circledast y + x^- \circledast y = |x| \circledast y = \pi_{|x|}(y). \end{aligned}$$

We conclude that $|\pi_x|(y) \geq \pi_{|x|}(y)$. We have established that $|\pi_x|(y) = \pi_{|x|}(y)$ in the special case where $y \in Y^+$ is such that $\operatorname{supp} y$ is contained in Us for some $g \in G$.

Now suppose that y is an arbitrary element of Y^+ with compact support. Choose an open neighbourhood V of e_G such that $\overline{V} \subseteq U$. Then $\operatorname{supp} y$ is contained in a union of finitely many right translates of V. Since $\chi_A y$ is still in Y for all Borel subsets A of supp y, it is then easy to see that y is a finite sum of elements of Y^+, each of which is supported in a right translate of $\overline{V}$, hence in a right translate of U. By linearity, it follows from the result as established for the special case that $|\pi_x|(y) = \pi_{|x|}(y)$.

We have now established the theorem in the case where $S = G$.

Next, we turn to the case of a general closed subspace S of G that is a subsemigroup of G. The problem with the above proof in this case is that translates of open subsets need not be open again. Even if S is unital, the proof of Lemma 10.1 breaks down, as does the argument in the final paragraph for the group case.

In order to circumvent this, we use Theorem 6.3 to embed $\mathrm{C_c}(S, \mathbb{R})^\sim$ as a vector sublattice of $\mathrm{C_c}(G, \mathbb{R})^\sim$. By restriction, this global embedding yields embeddings of X, Y, and Z as vector sublattices $X^\#$, $Y^\#$, and $Z^\#$ of $\mathrm{C_c}(G, \mathbb{R})^\sim$. By transporting the norms, these vector sublattices $X^\#$, $Y^\#$, and $Z^\#$ then become Banach lattices. The bilinear map $\circledast : X \times Y \to Z$ yields a bilinear map $\circledast^\# : X^\# \times Y^\# \to Z^\#$. Since Theorem 6.3 also states that supports are preserved under the embedding of $\mathrm{C_c}(S, \mathbb{R})^\sim$ into $\mathrm{C_c}(G, \mathbb{R})^\sim$, it is then immediate that the hypotheses in the theorem are satisfied for the sublattices $X^\#$, $Y^\#$, and $Z^\#$ of $\mathrm{C_c}(G, \mathbb{R})^\sim$ and the bilinear map $\circledast^\#$. We can now apply the result for the group case to these data. By transport of structure in the reverse direction, the result for the semigroup case then follows. □

During the above proof, it was indicated why the argument for the case of groups cannot in general be directly applied to the case of arbitrary semigroups. A closer inspection, however, shows that the argument for the case of groups *is* valid in the case of a semigroup that is discrete and cancellative. We recall that a semigroup S is *cancellative* if the maps $s \mapsto st$ and $s \mapsto ts$ from S to S are both injective for each $t \in S$. We shall now indicate the ingredients for the proof in this case.

Suppose that S is a cancellative semigroup, supplied with the discrete topology. Then $\mathrm{C_c}(S, \mathbb{R})$ consists of the real-valued functions with finite support, and $\mathrm{C_c}(S, \mathbb{R})^\sim$ can be identified as a vector lattice with the real-valued functions on S. Consequently, φ has separated support for all $\varphi \in \mathrm{C_c}(S, \mathbb{R})^\sim$. Suppose that $x \in \mathrm{C_c}(S, \mathbb{R})^\sim$, and let $x = x^+ - x^-$ be the decomposition of x into its disjoint positive and negative parts. Then $\operatorname{supp} x^+ \cdot s$ and $\operatorname{supp} x^- \cdot s$ are disjoint subsets of S for all $s \in S$, due to the fact that S is cancellative. Finally, if $y \in \mathrm{C_c}(S, \mathbb{R})^\sim$ has compact support, then y is a finite sum of elements of $\mathrm{C_c}(S, \mathbb{R})^\sim$, each of which is supported in a subset $\{s\}$ of S for some $s \in S$.

After these preliminary remarks, the reader will have no difficulty verifying the following result along the lines of the proof of Theorem 10.3.

Theorem 10.4 *Let S be a cancellative semigroup, supplied with the discrete topology. Let X, Y, and Z be vector sublattices of $\mathrm{C_c}(S, \mathbb{R})^\sim$ that are Banach lattices, and where Z is Dedekind complete.*

Suppose that $\circledast : X \times Y \to Z$ is a bilinear map such that $x \circledast y \in Z^+$ whenever $x \in X^+$ and $y \in Y^+$. Define the positive linear map $\pi : X \to \mathrm{L_r}(Y, Z)$ by $\pi_x(y) := x \circledast y$ for $x \in X$ and $y \in Y$.

Suppose that the following conditions are satisfied:

(1) $\mathrm{supp}\,(x \circledast y) \subseteq \mathrm{supp}\,x \cdot \mathrm{supp}\,y$ *for arbitrary* $x \in X^+$ *and for all* $y \in Y^+$ *with finite support;*
(2) *the elements of* Y^+ *with finite support are dense in* Y^+*;*
(3) $\chi_{\{s\}}y$ *is an element of* Y *again, whenever* $y \in Y^+$ *has finite support and* $s \in S$.

Then π *is a lattice homomorphism.*

Naturally, Theorem 10.4 follows from Theorem 10.3 for all semigroups that are subsemigroups of groups. It is known that every abelian cancellative semigroup is a subsemigroup of a group, in which case the enveloping group can even be taken to be of the same cardinality as S, see [18, Proposition 1.2.10]. In general, however, a unital cancellative semigroup is not necessarily a subsemigroup of any group; necessary and sufficient conditions for this, and examples where the conditions fail, are given in [16, Chapter 10]. This shows that Theorem 10.4 has value independent of Theorem 10.3.

11 Lattice Homomorphisms in Harmonic Analysis

All material is now in place to show that a number of natural positive maps in harmonic analysis are, in fact, lattice homomorphisms. For each of them, all that needs to be done is merely to establish that Theorem 10.3 or Theorem 10.4 is applicable in the relevant context. In view of the general nature of these two results, it seems not unlikely that they can have future applications to cases that are not covered in the present section.

Our first result answers the original question mentioned in Sect. 1, which is [52, Problem 7].

Theorem 11.1 *Let* G *be a locally compact group. Then the left regular representation* $\pi : \mathrm{M}(G, \mathbb{F}) \to \mathrm{L_r}(\mathrm{M}(G, \mathbb{F}))$ *is an isometric Banach lattice algebra homomorphism.*

Proof Since $\mathrm{M}(G, \mathbb{F})$ is a unital Banach algebra in which the identity element has norm one, it is clear that π is an isometric Banach algebra homomorphism. It remains to be shown that π is a lattice homomorphism.

For this, we start with the case where $\mathbb{F} = \mathbb{R}$. Then Theorem 7.9 shows that $\mathrm{M}(G, \mathbb{R})$ can be embedded as a sublattice of $\mathrm{C_c}(G, \mathbb{R})^\sim$ via a map $\mu \mapsto \varphi_\mu$ for $\mu \in \mathrm{M}(G, \mathbb{R})$, and that, after transport of the norm, the embedded sublattice is a Dedekind complete Banach lattice X such that its elements with compact, separated support are dense. We now resort to Theorem 10.3, where we take Y and Z to be equal to X. Then the clauses (2) and (3) of the hypotheses of Theorem 10.3 are satisfied, and it is easy to see that clause (4) of these hypotheses is also satisfied. Furthermore, Proposition 8.1 shows that clause (1) of the hypothesis of

Theorem 10.3 is also satisfied when we set

$$\varphi_\mu \circledast \varphi_\nu := \varphi_{\mu \star \nu}$$

for $\mu, \nu \in \mathrm{M}(G, \mathbb{R})$. An appeal to Theorem 10.3 concludes the proof for the case where $\mathbb{F} = \mathbb{R}$.

As explained earlier, the complex case follows from the real case on general grounds because the complex Banach lattice algebra $\mathrm{M}(G, \mathbb{C})$ is the complexification of the Banach lattice algebra $\mathrm{M}(G, \mathbb{R})$. □

Our next result concerns the action of $\mathrm{M}(G, \mathbb{F})$ on $\mathrm{L}^p(G, \mathbb{F})$ for $1 < p < \infty$ as defined in Eq. (8.23). As announced in the beginning of this section, the proof is quite similar to that of Theorem 11.1.

Theorem 11.2 *Let G be a locally compact group, and take p with $1 \leq p < \infty$. For $\mu \in \mathrm{M}(G, \mathbb{F})$ and $g \in \mathrm{L}^p(G, \mathbb{F})$, define*

$$(\pi_\mu g)(s) := \int_G g(t^{-1}s)\,\mathrm{d}\mu(t)$$

for those m_G-almost $s \in G$ for which the integral exists. Then $\pi_\mu g \in \mathrm{L}^p(G, \mathbb{F})$ for all $\mu \in \mathrm{M}(G, \mathbb{F})$ and $g \in \mathrm{L}^p(G, \mathbb{F})$, π_μ is a regular operator on $\mathrm{L}^p(G, \mathbb{F})$ for all $\mu \in \mathrm{M}(G, \mathbb{F})$, and the map $\mu \mapsto \pi_\mu$ is an injective Banach lattice algebra homomorphism $\pi : \mathrm{M}(G, \mathbb{F}) \to \mathrm{L_r}(\mathrm{L}^p(G, \mathbb{F}))$.

Proof All statements in the theorem are well known, except the one that states that π is a lattice homomorphism. The proof for this has a general outline that is similar to that of Theorem 11.1.

Again, we start with the case where $\mathbb{F} = \mathbb{R}$. In this case, Theorem 7.9 shows again that $\mathrm{M}(G, \mathbb{R})$ can be embedded as a sublattice of $\mathrm{C_c}(G, \mathbb{R})^\sim$ by means of a map $\mu \mapsto \varphi_\mu$ for $\mu \in \mathrm{M}(G, \mathbb{R})$ and that, after transport of the norm, the embedded sublattice is a Banach lattice X such that its elements with compact, separated support are dense. Furthermore, Theorem 7.5 shows that $\mathrm{L}^p(G, \mathbb{R})$ can be embedded as a sublattice of $\mathrm{C_c}(G, \mathbb{R})^\sim$ via a map $g \to \varphi_g$ and that, after transport of the norm, the embedded sublattice is a Dedekind complete Banach lattice Y such that its elements with compact support are dense. We now resort to Theorem 10.3, where we take Z to be equal to Y, and where we set

$$\varphi_\mu \circledast \varphi_g := \varphi_{\pi_\mu(g)}$$

for $\mu \in \mathrm{M}(G, \mathbb{R})$ and $g \in \mathrm{L}^p(G, \mathbb{R})$. Then the hypotheses of Theorem 10.3 are satisfied, and an application of this theorem concludes the proof for the case where $\mathbb{F} = \mathbb{R}$.

As explained earlier, the complex case follows from the real case on general grounds because the complex Banach lattice $\mathrm{M}(G, \mathbb{C})$ is the complexification of the Banach lattice $\mathrm{M}(G, \mathbb{R})$ and the complex Banach lattice $\mathrm{L}^p(G, \mathbb{C})$ is the complexification of the Banach lattice $\mathrm{L}^p(G, \mathbb{R})$. □

Remark 11.3 The action of $\mathrm{M}(G, \mathbb{F})$ on $\mathrm{L}^p(G, \mathbb{F})$ by convolution was previously studied by Arendt, using earlier results by Brainerd and Edwards (see [14]) and Gilbert (see [27]) on convolutions. In [8], Arendt showed that the map π in Theorem 11.2 is a lattice homomorphism when $p = 1$, and also when $1 < p < \infty$ and G is amenable. As Theorem 11.2 shows, for $1 < p < \infty$, the assumption that G be amenable is redundant.

We shall discuss Arendt's approach in more detail in Sect. 12.

Since $\mathrm{L}^p(G, \mathbb{F})$ is a Banach lattice subalgebra of $\mathrm{M}(G, \mathbb{F})$, we have the following consequence of Theorem 11.2, where the action of $\mathrm{L}^1(G, \mathbb{F})$ on $\mathrm{L}^p(G, \mathbb{F})$ is now given by Eq. (8.25). It solves [52, Problem 6]. It can also be found in earlier work by Kok (see [33, Example 7.6]), where it was obtained via a different approach.

Corollary 11.4 *Let G be a locally compact group, and take p with $1 \leq p < \infty$. For $f \in \mathrm{L}^1(G, \mathbb{F})$ and $g \in \mathrm{L}^p(G, \mathbb{F})$, define*

$$(\pi_f g)(s) := \int_G f(t) g(t^{-1}s)\, \mathrm{d}m_G(t)$$

for those m_G-almost $s \in G$ for which the integral exists. Then $\pi_f g \in \mathrm{L}^p(G, \mathbb{F})$ for all $f \in \mathrm{L}^1(G, \mathbb{F})$ and $g \in \mathrm{L}^p(G, \mathbb{F})$, π_f is a regular operator on $\mathrm{L}^p(G, \mathbb{F})$ for all $f \in \mathrm{L}^p(G, \mathbb{F})$, and the map $f \mapsto \pi_f$ is an injective Banach lattice algebra homomorphism $\pi : \mathrm{L}^1(G, \mathbb{F}) \to \mathrm{L_r}(\mathrm{L}^p(G, \mathbb{F}))$.

Remark 11.5 In Corollary 11.4, in the case where $p = 1$, the left regular representation $\pi : \mathrm{L}^1(G, \mathbb{F}) \to \mathrm{L_r}(\mathrm{L}^1(G, \mathbb{F}))$ is an isometric Banach algebra lattice homomorphism. The fact that π is an isometry is an immediate consequence of the fact that $\mathrm{L}^1(G, \mathbb{R})$ is a Banach algebra with a positive contractive approximate identity. In view of Theorem 12.1, below, we refrain from making a statement on the isometric nature of π if $1 < p < \infty$.

It is, of course, also possible to prove Corollary 11.4 directly from the central result Theorem 10.3, by using Theorem 7.5 to obtain an embedded copy X of $\mathrm{L}^1(G, \mathbb{R})$ and an embedded copy Y of $\mathrm{L}^p(G, \mathbb{R})$ in $\mathrm{C_c}(G, \mathbb{R})^\sim$, and taking Z to be equal to Y.

Now we turn to the case of semigroups, where we shall benefit from the results in Sect. 6 via their role in the proof of Theorem 10.3.

Theorem 11.6 *Let G be a locally compact group, and let S be a closed subspace of G that is a subsemigroup of G. Suppose that ω is a weight on S. Then the left regular representation $\pi : \mathrm{M}(S, \omega, \mathbb{F}) \to \mathrm{L_r}(\mathrm{M}(S, \omega, \mathbb{F}))$ of the Beurling algebra $\mathrm{M}(S, \omega, \mathbb{F})$ is a Banach lattice algebra homomorphism.*

Proof We again start with the case where $\mathbb{F} = \mathbb{R}$. Then Theorem 9.2 shows that $\mathrm{M}(S, \omega, \mathbb{R})$ can be embedded as a vector sublattice of $\mathrm{C_c}(S, \mathbb{R})^\sim$ and that, after transport of the norm, the embedded copy becomes a Dedekind complete Banach lattice such that its elements with compact, separated support are dense. Completely analogously to the proof of Theorem 11.1, Theorem 10.3 then shows that the present theorem holds for $\mathbb{F} = \mathbb{R}$. As earlier, it then also holds for $\mathbb{F} = \mathbb{C}$. □

In a similar vein, we have the following.

Theorem 11.7 *Suppose that ω is a weight on $\mathbb{R}^+$. Then the left regular representation $\pi : \mathrm{L}^1(\mathbb{R}^+, \omega, \mathbb{F}) \to \mathrm{L_r}(\mathrm{L}^1(\mathbb{R}^+, \omega, \mathbb{F}))$ of $\mathrm{L}^1(\mathbb{R}^+, \omega, \mathbb{F})$ is a Banach lattice algebra homomorphism.*

Proof We start with $\mathbb{F} = \mathbb{R}$. Since ω is strictly positive and continuous, [24, Exercise 7.2.9] yields that the measure $\omega\,\mathrm{d}x$ is a regular Borel measure on $\mathbb{R}^+$. Hence Theorem 7.5 shows that $\mathrm{L}^1(\mathbb{R}^+, \omega, \mathbb{R})$ can be embedded as a vector sublattice of $\mathrm{C_c}(\mathbb{R}^+, \mathbb{R})^\sim$. An application of Theorem 10.3 then shows that the present theorem holds for $\mathbb{F} = \mathbb{R}$. As earlier, it then also holds for $\mathbb{F} = \mathbb{C}$. □

We conclude the results in this section with an application of Theorem 10.4. The proof will be rather obvious by now, and is left to the reader.

Theorem 11.8 *Let S be a cancellative semigroup, and let ω be a weight on S. Then the left regular representation $\pi : \ell^1(S, \omega, \mathbb{F}) \to \mathrm{L_r}(\ell^1(S, \omega, \mathbb{F}))$ of $\ell^1(S, \omega, \mathbb{F})$ is an injective Banach lattice algebra homomorphism.*

We recall that $\ell^1(S, \omega, \mathbb{F})$ can be a radical Banach algebra, so that our theorems on left regular representations being Banach lattice algebra homomorphisms are not restricted to the semisimple case.

Remark 11.9 Let us collect what we know about the left regular representation of a Dedekind complete Banach lattice algebra A being a Banach lattice algebra homomorphism from A into $\mathrm{L_r}(A)$ or not.

On the positive side, we have the following.

Theorem 11.10 *The left regular representation is a (real or complex) Banach lattice algebra homomorphism from A into $\mathrm{L_r}(A)$ in the following cases:*

(1) *A is the measure algebra $\mathrm{M}(G, \mathbb{F})$ of a locally compact group G;*
(2) *A is the group algebra $\mathrm{L}^1(G, \mathbb{F})$ of a locally compact group G;*
(3) *A is a Beurling algebra $\mathrm{M}(S, \omega, \mathbb{F})$, where S is a closed subspace of a locally compact group G that is a subsemigroup of G, and where ω is a weight on S;*
(4) *$A = \mathrm{L}^1(\mathbb{R}^+, \omega, \mathbb{F})$, where ω is a weight on $\mathbb{R}^+$;*
(5) *$A = \ell^1(S, \omega, \mathbb{F})$, where S is a cancellative semigroup and ω is a weight on S;*
(6) *$A = \mathrm{L_r}(E)$ for a Dedekind complete Banach lattice E.*

The first five of these results can be found in the present section. The sixth one follows from a result of Synnatzschke's on two-sided multiplication operators, see [48, Satz 3.1]. For further results on two-sided multiplication operators on and between vector lattices of regular operators we refer to [15, 50].

One could argue that Theorems 10.3 and 10.4 indicate that the left regular representation will be a Banach lattice algebra homomorphism for very many Dedekind complete Banach lattice algebras on groups or semigroups, whenever the multiplication is akin to a convolution. We are, in fact, not aware of a Dedekind Banach lattice algebra A in harmonic analysis where the left regular representation of A is not a lattice homomorphism from A into $\mathrm{L_r}(A)$.

On the negative side, there exist uncountably many two-dimensional, mutually non-isomorphic, commutative Banach lattice algebras with a positive identity element of norm one that have no faithful, finite-dimensional Banach lattice algebra representations at all, see [53]. In particular, their left regular representations are *not* lattice homomorphisms.

It is unclear if there is an 'underlying' property that distinguishes the above Banach lattice algebras on the positive side from those on the negative side. Such a property, and preferably one that is easily verified or falsified in a given case, would be desirable. This question is posed in [52, Problem 1], together with various refinements of it.

12 Further Questions in Ordered Harmonic Analysis

The previous sections were centred around a convolution-like bilinear map from two Banach lattices on a locally compact (semi)group to a third. There do not seem to be too many results available with the same flavour of 'ordered harmonic analysis', i.e., results that are in the area where harmonic analysis and positivity meet. In this section, we shall discuss results by Arendt, Brainerd and Edwards, and Gilbert that are at this interface and that are related to our results in Sect. 11. Our exposition is based on [8, Section 3], to which the reader is referred for details and additional material. As we shall see, this discussion leads to natural research questions in ordered harmonic analysis. We hope to be able to report on these questions in the future.

Let G be a locally compact group, and let $1 \leq p < \infty$. Then G acts (not generally isometrically) on $\mathrm{L}^p(G, \mathbb{C})$ via the formula

$$\rho_t f(s) := f(st)$$

for $s, t \in G$ and $f \in \mathrm{L}^p(G, \mathbb{C})$. We shall be interested in operators on $\mathrm{L}^p(G, \mathbb{C})$ that commute with all ρ_s. To this end, we set

$$\mathrm{CV}_p(G, \mathbb{C}) := \{ T \in \mathrm{B}(\mathrm{L}^p(G, \mathbb{C})) : T \circ \rho_s = \rho_s \circ T \text{ for all } s \in G \}$$

and

$$\mathrm{CV}_{p,\mathrm{r}}(G, \mathbb{C}) := \{ T \in \mathrm{L_r}(\mathrm{L}^p(G, \mathbb{C})) : T \circ \rho_s = \rho_s \circ T \text{ for all } s \in G \}.$$

Then $\mathrm{CV}_{p,\mathrm{r}}(G, \mathbb{C})$ is a complex Banach lattice subalgebra of $\mathrm{L_r}(\mathrm{L}^p(G, \mathbb{C}))$. There is an easy proof of this fact, as follows. For $t \in G$, the map $T \mapsto \rho_{t^{-1}} \circ T \circ \rho_t$ from $\mathrm{L_r}(\mathrm{L}^p(G, \mathbb{C}))$ into itself is an algebra automorphism of $\mathrm{L_r}(\mathrm{L}^p(G, \mathbb{C}))$. It is a positive map, and since its inverse is clearly also positive—it is the map $T \mapsto \rho_t \circ T \circ \rho_{t^{-1}}$—it is a complex Banach algebra lattice automorphism. Hence its fixed point set, which is the commutant of ρ_t in $\mathrm{L_r}(\mathrm{L}^p(G, \mathbb{C}))$, is a complex Banach lattice subalgebra of $\mathrm{L_r}(\mathrm{L}^p(G, \mathbb{C}))$. Since $\mathrm{CV}_{p,\mathrm{r}}(G, \mathbb{C})$ is the intersection of these commutants as t ranges over G, the space $\mathrm{CV}_{p,\mathrm{r}}(G, \mathbb{C})$ is indeed a complex Banach lattice subalgebra of $\mathrm{L_r}(\mathrm{L}^p(G, \mathbb{C}))$. This argument is due to Arendt, see [8, Proof of Proposition 3.3].

There is an easy way to obtain elements of $\mathrm{CV}_{p,\mathrm{r}}(G, \mathbb{C})$ from elements of $\mathrm{M}(G, \mathbb{C})$. Take $\mu \in \mathrm{M}(G, \mathbb{C})$, and set (we repeat Eq. (8.23) for convenience)

$$(\mu \star_p f)(s) := \int_G f(t^{-1}s) \, \mathrm{d}\mu(t) \tag{12.33}$$

for $f \in \mathrm{L}^p(G, \mathbb{C})$ and $s \in G$. It is easily checked that the (left) convolution operator π_μ on $\mathrm{L}^p(G, \mathbb{C})$ that is thus defined commutes with all right translations. Obviously, if μ is positive, then π_μ is a positive element of $\mathrm{CV}_{p,\mathrm{r}}(G, \mathbb{C})$. Conversely, if T is a positive element of $\mathrm{CV}_{p,\mathrm{r}}(G, \mathbb{C})$, then, according to [14], there exists a positive regular Borel measure μ on G such that Tf equals $\mu \star_p f$ as in Eq. (12.33) for all $f \in \mathrm{C_c}(G, \mathbb{C})$ and $s \in G$. Note that we do not write that $T = \pi_\mu$ because this representation theorem by Brainerd and Edwards does not assert that μ is a *bounded* measure. When $p = 1$ this is always the case, but for $1 < p < \infty$ this is related to whether or not G is amenable. The following result is due to Gilbert, see [27, Theorem A] and also [23, Theorem 18.3.6], [28, Theorem 17], [29, Theorem 2.2.1], [40], and [41, Definition 8.3.1, Theorems 8.3.2, and Theorem 8.3.18] for the equivalence of various characterisations of amenable locally compact groups.

Theorem 12.1 *Let G be a locally compact group, and let $1 < p < \infty$. Then the following are equivalent:*

(1) *G is amenable;*
(2) *$\|\pi_\mu\| = \|\mu\|$ for all $\mu \in \mathrm{M}(G, \mathbb{C})^+$;*
(3) *Whenever μ is a positive regular Borel measure on G such that $\mu \star_p f$, as defined in Eq. (12.33), is in $\mathrm{L}^p(G, \mathbb{C})$ for all $f \in \mathrm{C_c}(G, \mathbb{C})$, and such that there exists a $c \geq 0$ such that $\|\mu \star_p f\|_p \leq c\|f\|_p$ for all $f \in \mathrm{C_c}(G, \mathbb{C})$, then $\mu \in \mathrm{M}(G, \mathbb{C})^+$.*

Suppose that $p = 1$ or that $1 < p < \infty$ and that G is amenable. Combining Theorem 12.1 with the representation theorem by Brainerd and Edwards, we see that the natural map $\mu \mapsto \pi_\mu$ defines a bipositive complex algebra isomorphism

$\pi : \mathrm{M}(G, \mathbb{C}) \to \mathrm{CV}_{p,\mathrm{r}}(G, \mathbb{C})$ between $\mathrm{M}(G, \mathbb{C})$ and $\mathrm{CV}_{p,\mathrm{r}}(G, \mathbb{C})$. Since we know that $\mathrm{CV}_{p,\mathrm{r}}(G, \mathbb{C})$ is a complex Banach lattice algebra, this bipositive vector space isomorphism π is a complex Banach lattice algebra isomorphism. Since we also know that $\mathrm{CV}_{p,\mathrm{r}}(G, \mathbb{C})$ is a complex Banach lattice subalgebra of $\mathrm{L_r}(\mathrm{L}^p(G, \mathbb{C}))$, we see that the map $\pi : \mathrm{M}(G, \mathbb{C}) \to \mathrm{L_r}(\mathrm{L}^p(G, \mathbb{C}))$ is a complex Banach lattice algebra homomorphism. This fact is a part of the statement of [8, Proposition 3.3].

Remark 12.2 Allowing ourselves a somewhat imprecise notation, we know from the above that $\mathrm{CV}_{1,\mathrm{r}}(G, \mathbb{C}) = \mathrm{M}(G, \mathbb{C})$ and, in addition, that, for $1 < p < \infty$, $\mathrm{CV}_{p,\mathrm{r}}(G, \mathbb{C}) = \mathrm{M}(G, \mathbb{C})$ whenever G is amenable. Since all bounded operators on an L^1-space are regular (see [32]), we see that $\mathrm{CV}_1(G, \mathbb{C}) = \mathrm{M}(G, \mathbb{C})$. This is a form of Wendel's theorem, see [18, Theorem 3.3.40], for example.

As we know from Theorem 11.2, the map $\pi : \mathrm{M}(G, \mathbb{C}) \to \mathrm{L_r}(\mathrm{L}^p(G, \mathbb{C}))$ is a complex Banach lattice homomorphism for all p such that $1 \leq p < \infty$. The amenability of G is not relevant for this. As long as one is interested only in π being a lattice homomorphism or not, the results in [8] are, therefore, not yet optimal. Comparing the machinery needed, including [14] and [27], for the approach in [8] on the one hand, with the proof of Theorem 11.2 as based on Theorem 10.3 on the other hand, one could also argue that—as long as one is interested only in π being a lattice homomorphism or not—the approach in [8] is more complicated than necessary.

Nevertheless, the approach in [8] raises a few natural questions, triggered by the description of $\mathrm{CV}_{p,\mathrm{r}}(G, \mathbb{C})$ that it uses. For example, is there a more general underlying phenomenon that explains what is so special about $p = 1$, which is the only case where the amenability of G does *not* play a role in the description of the regular operators on $\mathrm{L}^p(G, \mathbb{C})$ that commute with all right translations? A way to investigate this would be to consider a general Banach function space E on G that is invariant under left and right translations. Under reasonable hypotheses, at least the bounded measures will act on E via left convolutions. Is there then a representation theorem as in [14] again, stating that a positive operator on E that commutes with all right translations is a left convolution with a (possibly unbounded) positive regular Borel measure? What are the properties of E that determine whether the amenability of G is relevant or not for such a measure to be automatically bounded, as in Gilbert's work in [27]?

Let us return to the spaces $\mathrm{L}^p(G, \mathbb{C})$. Clearly, $\mathrm{CV}_{p,\mathrm{r}}(G, \mathbb{C}) \subseteq \mathrm{CV}_p(G, \mathbb{C})$ for all $1 \leq p < \infty$. Can the inclusions be proper? For $p = 1$, all bounded operators on $\mathrm{L}^1(G, \mathbb{C})$ are regular, as was already mentioned above, so in this case equality is automatic. For $1 < p < \infty$, we have the following partial answer.

Theorem 12.3 *Let G be an infinite, amenable, locally compact group, and take p with $1 < p < \infty$. Then $\mathrm{CV}_{p,\mathrm{r}}(G, \mathbb{C}) \subsetneq \mathrm{CV}_p(G, \mathbb{C})$, and so there are bounded operators on $\mathrm{L}^p(G, \mathbb{C})$ that commute with all right translations, but are not regular.*

Proof Assume, to the contrary, that $\mathrm{CV}_{p,\mathrm{r}}(G, \mathbb{C}) = \mathrm{CV}_p(G, \mathbb{C})$ for some p such that $1 < p < \infty$. Then $\mathrm{CV}_p(G, \mathbb{C}) = \mathrm{M}(G, \mathbb{C})$. We conclude from this that

$\mathrm{CV}_p(G, \mathbb{C}) = \mathrm{M}(G, \mathbb{C}) \subseteq \mathrm{CV}_q(G, \mathbb{C})$ for all q such that $1 < q < \infty$. This, however, contradicts [17, Theorem 2]. □

This result leads to a few further questions.

First, is the analogue of Theorem 12.3 true for more general Banach function spaces E on amenable groups that are invariant under left and right translations? To be more specific: for a translation invariant Banach function space E on a amenable group, is it true that, whenever there are bounded operators on E that are not regular, there are also bounded operators on E that are not regular and that commute with all right translations?

Second, is the amenability of G a necessary condition in Theorem 12.3 for the inclusion to be proper? Put more generally: for a translation invariant Banach function space E on a locally compact group, is it true that, whenever there are bounded operators on E that are not regular, there are also bounded operators on E that are not regular and that commute with all right translations?

Acknowledgements The results in this article were obtained in part when the first author held the Kloosterman Chair in Leiden in October 2017, and when the second author visited Lancaster University in October 2018. The financial support by the Mathematical Institute of Leiden University and the London Mathematical Society is gratefully acknowledged.

References

1. Y.A. Abramovich, C.D. Aliprantis, *An Invitation to Operator Theory*. Graduate Studies in Mathematics, vol. 50 (American Mathematical Society, Providence, 2002)
2. Y.A. Abramovich, C.D. Aliprantis, *Problems in Operator Theory*. Graduate Studies in Mathematics, vol. 51 (American Mathematical Society, Providence, 2002)
3. C.D. Aliprantis, O. Burkinshaw, *Principles of Real Analysis*, 3rd edn. (Academic, San Diego, 1998)
4. C.D. Aliprantis, O. Burkinshaw, *Locally Solid Riesz Spaces with Applications to Economics*. Mathematical Surveys and Monographs, 2nd edn., vol. 105 (American Mathematical Society, Providence, 2003)
5. C.D. Aliprantis, O. Burkinshaw, *Positive Operators* (Springer, Dordrecht, 2006). Reprint of the 1985 original
6. G.R. Allan, *Introduction to Banach Spaces and Algebras*. Oxford Graduate Texts in Mathematics, vol. 20 (Oxford University Press, Oxford, 2011)
7. W. Arendt, Über das Specktrum regulärer Operatoren. Ph.D. Thesis, Tübingen, 1979
8. W. Arendt, On the o-spectrum of regular operators and the spectrum of measures. Math. Z. **178**, 271–287 (1981)
9. W.G. Bade, H.G. Dales, Norms and ideals in radical convolution algebras. J. Funct. Anal. **41**, 77–109 (1981)
10. W.G. Bade, H.G. Dales, Discontinuous derivations from algebras of power series. Proc. Lond. Math. Soc. (3) **59**, 133–152 (1989)
11. N. Bourbaki, *Topological Vector Spaces*. Elements of Mathematics, Chapters 1–5 (Springer, Berlin, 1987)
12. N. Bourbaki, *Integration. I. Chapters 1–6*. Elements of Mathematics (Springer, Berlin, 2004)
13. N. Bourbaki, *Integration. II. Chapters 7–9*. Elements of Mathematics (Springer, Berlin, 2004)

14. B. Brainerd, R.E. Edwards, Linear operators which commute with translations. I. Representation theorems. J. Aust. Math. Soc. **6**, 289–327 (1966)
15. J.X. Chen, A.R. Schep, Two-sided multiplication operators on the space of regular operators. Proc. Am. Math. Soc. **144**, 2495–2501 (2016)
16. A.H. Clifford, G.B. Preston, *The Algebraic Theory of Semigroups. Vol. II*. Mathematical Surveys, vol. 7 (American Mathematical Society, Providence, 1967)
17. M.G. Cowling, J.J.F. Fournier, Inclusions and noninclusion of spaces of convolution operators. Trans. Am. Math. Soc. **221**, 59–95 (1976)
18. H.G. Dales, *Banach Algebras and Automatic Continuity*. London Mathematical Society Monographs, vol. 24 (Clarendon Press, Oxford, 2000)
19. H.G. Dales, F.K. Dashiell Jr., A.T.-M. Lau, D. Strauss, *Banach Spaces of Continuous Functions as Dual Spaces*. CMS Books in Mathematics/Ouvrages de Mathématiques de la SMC (Springer, New York, 2016)
20. H.G. Dales, H.V. Dedania, Weighted convolution algebras on subsemigroups of the real line. Dissertationes Math. **459** (2009)
21. H.G. Dales, A.T.-M. Lau, The second duals of Beurling algebras. Mem. Am. Math. Soc. **177** (836) (2005)
22. E. de Jonge, A.C.M. van Rooij, *Introduction to Riesz Spaces*. Mathematical Centre Tracts, vol. 78 (Mathematisch Centrum, Amsterdam, 1977)
23. J. Dixmier, *C^*-algebras*. North-Holland Mathematical Library, vol. 15 (North-Holland, Amsterdam/New York/Oxford, 1977)
24. G.B. Folland, *Real Analysis. Modern Techniques and Their Applications*. Pure and Applied Mathematics, 2nd edn. (Wiley, New York, 1999)
25. G.B. Folland, *A Course in Abstract Harmonic Analysis*. Textbooks in Mathematics, 2nd edn. (CRC Press, Boca Raton, 2016)
26. D.H. Fremlin, *Measure Theory. Vol. 2. Broad Foundations* (Torres Fremlin, Colchester, 2003). Corrected second printing of the 2001 original
27. J.E. Gilbert, Convolution operators on $\mathrm{L}^p(G)$ and properties of locally compact groups. Pac. J. Math. **24**, 257–268 (1968)
28. R. Godement, Les fonctions de type positif et la théorie des groupes. Trans. Am. Math. Soc. **63**, 1–84 (1948)
29. F.P. Greenleaf, *Invariant Means on Topological Groups and Their Applications*. Van Nostrand Mathematical Studies, vol. 16 (Van Nostrand Reinhold Co., New York/Toronto/London, 1969)
30. E. Hewitt, K.A. Ross, *Abstract Harmonic Analysis. Vol. I. Structure of Topological Groups, Integration Theory, Group Representations*. Grundlehren der Mathematischen Wissenschaften, 2nd edn., vol. 115 (Springer, Berlin/New York, 1979)
31. C.B. Huijsmans, An inequality in complex Riesz algebras. Studia Sci. Math. Hungar. **20**, 29–32 (1985)
32. L.V. Kantorovich, B.Z. Vulikh, Sur la représentation des opérations linéaires. Compos. Math. **5**, 119–165 (1937)
33. D. Kok, Lattice algebra representations of $\mathrm{L}^1(G)$ on translation invariant Banach function spaces. Master's thesis, Leiden University, 2016. Online at http://www.math.leidenuniv.nl/en/theses/667/
34. W.A.J. Luxemburg, A.C. Zaanen, *Riesz Spaces*, vol. I (North-Holland, Amsterdam/London; American Elsevier Publishing Co., New York, 1971)
35. P. Meyer-Nieberg, *Banach Lattices*. Universitext (Springer, Berlin, 1991)
36. S. Öztop, E. Samei, Twisted Orlicz algebras, I. Studia Math. **236**, 271–296 (2017)
37. S. Öztop, E. Samei, Twisted Orlicz algebras, II (2017). Preprint. Online at https://arxiv.org/pdf/1704.02350.pdf
38. Th.W. Palmer, *Banach Algebras and the General Theory of *-Algebras. Vol. I. Algebras and Banach Algebras*. Encyclopedia of Mathematics and its Applications, vol. 49 (Cambridge University Press, Cambridge, 1994)

39. Th.W. Palmer, *Banach Algebras and the General Theory of *-Algebras. Vol. 2. *-algebras*. Encyclopedia of Mathematics and its Applications, vol. 79 (Cambridge University Press, Cambridge, 2001)
40. H. Reiter, On some properties of locally compact groups. Nederl. Akad. Wetensch. Proc. Ser. A 68=Indag. Math. **27**, 697–701 (1965)
41. H. Reiter, J.D. Stegeman, *Classical Harmonic Analysis and Locally Compact Groups*. London Mathematical Society Monographs, vol. 22 (Clarendon Press, Oxford, 2000)
42. C.E. Rickart, *General Theory of Banach Algebras*. The University Series in Higher Mathematics (D. van Nostrand Co., Inc., Princeton/Toronto/London/New York, 1960)
43. W. Rudin, *Fourier Analysis on Groups*. Interscience Tracts in Pure and Applied Mathematics, vol. 12 (Interscience Publishers, New York/London, 1962)
44. W. Rudin, *Real and Complex Analysis*, 3rd edn. (McGraw-Hill Book Co., New York, 1987)
45. H.H. Schaefer, Some spectral properties of positive linear operators. Pac. J. Math. **10**, 1009–1019 (1960)
46. H.H. Schaefer, *Banach Lattices and Positive Operators*. Die Grundlehren der mathematischen Wissenschaften, vol. 215 (Springer, New York/Heidelberg, 1974)
47. E. Scheffold, Über komplexe Banachverbandsalgebren. J. Funct. Anal. **37**, 382–400 (1980)
48. J. Synnatzschke, Über einige verbandstheoretische Eigenschaften der Multiplikation von Operatoren in Vektorverbänden. Math. Nachr. **95**, 273–292 (1980)
49. V. Troitsky, *Introduction to Vector and Banach Lattices* (Book in preparation, 2017)
50. A.W. Wickstead, Norms of basic elementary operators on algebras of regular operators. Proc. Am. Math. Soc. **143**, 5275–5280 (2015)
51. A.W. Wickstead, Banach lattice algebras: some questions, but very few answers. Positivity **21**, 803–815 (2017)
52. A.W. Wickstead, Ordered Banach algebras and multi-norms: some open problems. Positivity **21**, 817–823 (2017)
53. A.W. Wickstead, Two dimensional unital Riesz algebras, their representations and norms. Positivity **21**, 787–801 (2017)
54. W. Wnuk, *Banach Lattices with Order Continuous Norms* (Polish Scientific Publishers PWN, Warsaw, 1999)
55. A.C. Zaanen, *Riesz Spaces. II*. North-Holland Mathematical Library, vol. 30 (North-Holland, Amsterdam, 1983)
56. A.C. Zaanen, *Introduction to Operator Theory in Riesz Spaces* (Springer, Berlin, 1997)

Noncommutative Boyd Interpolation Theorems Revisited

Peter Dodds and Theresa Dodds

To Ben de Pagter with profound respect and admiration

Abstract If E is a fully symmetric space on $(0, \infty)$, we show that the corresponding noncommutative space $E(\tau)$ of τ-measurable operators is an interpolation space for the noncommutative pair $(L_1(\tau), L_q(\tau))$ provided $1 \leq q_E < q$, where q_E is the upper Boyd index.

Keywords Measurable operators · Noncommutative symmetric spaces · Operators of weak type · Noncommutative interpolation

1 Introduction

One of the central results in the theory of interpolation of symmetric Banach function spaces is the Boyd interpolation theorem which finds its roots in the seminal work of Calderón [5]. For a given symmetric function space E on $[0, \infty)$, Boyd [4] introduced two indices p_E, q_E known, respectively, as the *lower* and *upper* Boyd indices of E. Based on weak type estimates for a linear integral operator introduced by Calderón, Boyd showed that every sublinear operator on E which is simultaneously of weak types (p, p) and (q, q) is bounded on E if and only if $1 \leq p < p_E \leq q_E < q < \infty$. More recently, an exact extension of Boyd's theorem to the setting of symmetric Banach spaces of τ-measurable operators has been established by Dirksen [7]. Appendix A, following Boyd's original approach. In addition, Dirksen established an extension of the Boyd theorem in the extreme case of weak type (p, p) and strong type (∞, ∞) interpolation in case that $1 \leq p < p_E$. In the commutative setting, such a theorem had been observed earlier

P. Dodds (✉) · T. Dodds
College of Science and Engineering, Flinders University, Adelaide, Australia
e-mail: peter@csem.flinders.edu.au; peter.dodds@flinders.edu.au; theresa@csem.flinders.edu.au

G. Buskes et al. (eds.), *Positivity and Noncommutative Analysis*,
Trends in Mathematics, https://doi.org/10.1007/978-3-030-10850-2_7

by Maligranda [17] Theorem 4.6. Subsequently, in response to a question of E.M. Semenov, Astashkin and Maligranda [1] showed that, for symmetric spaces on $[0, 1]$, the Boyd theorem continues to hold for the other extreme case of strong type $(1, 1)$ and weak-type (q, q) interpolation under the additional hypothesis that E is an interpolation space for the pair (L_1, L_∞), or equivalently, that E is fully symmetric, provided that $1 \leq q_E < q$.

The principal new result of this paper is to give an exact extension of the theorem of Astashkin and Maligranda to the noncommutative setting, thus complementing the results of Dirksen. In particular, we will show (see Theorem 6.1) that if E is a fully symmetric space on the semi-axis $[0, \infty)$ and if T is a map of strong type $(1, 1)$ and weak type (q, q), then T is bounded on the noncommutative space $E(\tau)$ provided E is fully symmetric and $1 \leq q_E < q < \infty$. As noted in [1], the assumption that E be fully symmetric cannot be omitted. Our approach follows the basic ideas of [1] and is based on estimates derived from the relevant Holmstedt formula in the noncommutative setting. We take this opportunity to present a self-contained account, not only of the extension of the Astashkin-Maligranda theorem but also of Dirksen's theorems so as to systematically exhibit the full range of noncommutative Boyd theorems.

It is a pleasure to thank Sergei Astashkin and Sjoerd Dirksen for several helpful discussions.

2 Preliminaries

In this section, we collect some of the basic facts and notation that will be used in this paper. We denote by $\mathcal{M}$ a semifinite von Neumann algebra on the Hilbert space $\mathcal{H}$, with a fixed faithful and normal semifinite trace τ. For standard facts concerning von Neumann algebras, we refer to [20]. The identity in $\mathcal{M}$ is denoted by $\mathbf{1}$. A linear operator $x : \mathcal{D}(x) \to \mathcal{H}$, with domain $\mathcal{D}(x) \subseteq \mathcal{H}$, is said to be *affiliated with* $\mathcal{M}$ if $ux = xu$ for all unitary u in the commutant $\mathcal{M}'$ of $\mathcal{M}$.

If x is a self-adjoint operator in $\mathcal{H}$ affiliated with $\mathcal{M}$, then the spectral projection $\chi_B(x)$ is an element of $\mathcal{M}$ for any Borel set $B \subseteq \mathbb{R}$. Here $\chi_\cdot$ denotes the usual indicator function. The closed and densely defined operator x affiliated with $\mathcal{M}$ is called *τ-measurable* if and only if there exists a number $s \geq 0$ such that

$$\tau\left(\chi_{(s,\infty)}(|x|)\right) < \infty.$$

The collection of all τ-measurable operators is denoted by $S(\tau)$. With the sum and product defined as the respective closures of the algebraic sum and product, $S(\tau)$ is a *-algebra. For $x \in S(\tau)$, the *generalised singular value function* $\mu(\cdot; x) = \mu(\cdot; |x|)$ is defined by

$$\mu(t; x) = \inf\{s \geq 0 : \tau(\chi_{(s,\infty)}(|x|)) \leq t\}, \quad t \geq 0. \tag{2.1}$$

It follows directly that the singular value function $\mu(x)$ is a decreasing, right-continuous function on the positive half-line $[0, \infty)$. Moreover, $\mu(uxv) \leq \|u\| \|v\| \mu(x)$ for all $u, v \in \mathcal{M}$, $x \in S(\tau)$ and

$$\mu(f(x)) = f(\mu(x))$$

whenever $0 \leq x \in S(\tau)$ and f is an increasing continuous function on $[0, \infty)$ which satisfies $f(0) = 0$. Further, for all $x, y \in S(\tau)$ and $s, t > 0$,

$$\mu(s + t; x + y) \leq \mu(s; x) + \mu(t; y).$$

If m denotes Lebesgue measure on the semiaxis $[0, \infty)$, we denote by $L_0(m)$ the class of all finite a.e. extended complex-valued measurable functions on $[0, \infty)$. We will consider $L_\infty(m)$ as an Abelian von Neumann algebra acting via multiplication on the Hilbert space $L_2(m)$, with trace given by integration with respect to Lebesgue measure m. In this case, $S(m)$ consists of all measurable functions on $[0, \infty)$ which are bounded on a set of finite measure and if $f \in S(m)$, then the generalised singular value function $\mu(f)$ is precisely the classical decreasing rearrangement of $|f|$, which is usually denoted by f^*.

If $(\mathcal{N}, \sigma)$ is a semifinite von Neumann algebra, if $x \in S(\tau)$ and $y \in S(\sigma)$, then x is said to be *submajorised* by y (in the sense of Hardy, Littlewood and Polya) if and only if

$$\int_0^t \mu(s; x) ds \leq \int_0^t \mu(s; y) ds$$

for all $t \geq 0$. We write $x \prec\prec y$, or equivalently, $\mu(x) \prec\prec \mu(y)$.

Now suppose that $E \subseteq S(\tau)$ is a normed space with norm $\|x\|_E$. The normed space E is said to be *symmetric* if whenever $x \in E$ and $y \in S(\tau)$ satisfy $\mu(y) \leq \mu(x)$, it follows that $y \in E$ and $\|y\|_E \leq \|x\|_E$. If the normed space E is symmetric, then E is said to be *strongly symmetric* if $\|y\|_E \leq \|x\|_E$ whenever $x, y \in E$ satisfy $y \prec\prec x$ and *fully symmetric* if whenever $x \in E$ and $y \in S(\tau)$ satisfy $\mu(y) \prec\prec \mu(x)$, it follows that $y \in E$ and $\|y\|_E \leq \|x\|_E$.

If $E \subseteq S(m)$ is a symmetric Banach space, the corresponding noncommutative space $(E(\tau), \|\cdot\|_E)$ is defined by setting

$$E(\tau) = \{x \in S(\tau) : \mu(x) \in E\}, \quad \|x\|_{E(\tau)} = \|\mu(x)\|_E.$$

If $E \subseteq S(m)$ is a strongly symmetric Banach space on $[0, \infty)$, then the noncommutative version $E(\tau)$ is also a strongly symmetric Banach space and if E is fully symmetric, then so is $E(\tau)$.

For further details, we refer to the survey article [8] and to the papers [9, 11, 12]. For further information on generalised singular values, we refer to [14] and to [15] for more details on symmetric spaces on the semi-axis $[0, \infty)$.

3 Interpolation Preliminaries

3.1 Lorentz $L_{p,q}$ Spaces, $1 \leq p, q \leq \infty$

It will be useful to gather here the basic definitions of the Lorentz $L_{p,q}$ spaces. More detailed properties and their proofs may be found in [2, 3].

Consider the interval $(0, \infty)$ equipped with Lebesgue measure m. For $f \in L_0(m)$, let $Mf : (0, \infty) \to [0, \infty]$ be defined by setting

$$(Mf)(t) = \frac{1}{t} \int_0^t \mu(s, f) ds, \quad t > 0.$$

Given $1 \leq p, q \leq \infty$, for each $f \in L_0(m)$, the quantity $\|f\|_{(p,q)} \in [0, \infty]$ is defined by

$$\|f\|_{(p,q)} = \left(\int_0^\infty (t^{1/p}(Mf)(t))^q \frac{dt}{t} \right)^{1/q}, \quad 1 \leq q < \infty,$$

and

$$\|f\|_{(p,\infty)} = \sup_{t>0} t^{1/p}(Mf)(t).$$

The Lorentz spaces $L_{p,q}(m)$ for $1 \leq p, q \leq \infty$ are defined by setting

$$L_{(p,q)}(\tau) = \{f \in L_0(m) : \|f\|_{(p,q)} < \infty\}.$$

If $1 < p < \infty$ and $1 \leq q \leq \infty$, the functional $\|\cdot\|_{(p,q)}$ is a norm on $L_{p,q}$ and $\left(L_{p,q}, \|\cdot\|_{(p,q)}\right)$ is a Banach space which is easily seen to be fully symmetric. If $1 \leq p < \infty$ and $1 \leq q \leq r \leq \infty$, then there exists a constant $C = C_{p,q,r}$ such that

$$\|f\|_{(p,r)} \leq C\|f\|_{(p,q)}, \quad f \in L_0(m). \tag{3.2}$$

In particular, $L_{p,q} \subseteq L_{p,r}$. It is convenient to recall [2] that if the quantity $\|f\|_{p,q} \in [0, \infty]$, $1 \leq p, q \leq \infty$ is defined by setting

$$\|f\|_{p,q} = \left(\int_0^\infty (t^{1/p}\mu(t; f))^q \frac{dt}{t} \right)^{1/q}, \quad 1 \leq q < \infty,$$

and

$$\|f\|_{p,\infty} = \sup_{t>0} t^{1/p}\mu(t; f),$$

then the functionals $\|\cdot\|_{p,q}$ are quasinorms on the corresponding Lorentz space $L_{p,q}$ that are equivalent to the norms $\|\cdot\|_{(p,q)}$.

3.2 *Real Method of Interpolation*

It is convenient now to recall briefly those elements of the K-method of real interpolation due to Peetre [3, Chapter 3] which shall be needed here. If (X_0, X_1) is a Banach couple, and if $x \in X_0 + X_1$, define for every $t > 0$,

$$K(t, x; X_0, X_1) \equiv K(t, x) = \inf\{\|x_0\|_{X_0} + t\|x_1\|_{X_1} : x = x_0 + x_1, x_i \in X_i, i = 0, 1\}.$$

For $0 < \theta < 1$, $1 \leq q \leq \infty$ or for $0 \leq \theta \leq 1$, $q = \infty$, the Banach space $[X_0, X_1]_{\theta,q;K}$ consists of all elements $x \in X_0 + X_1$ for which the functional

$$\|x\|_{\theta,q;K} = \begin{cases} \left(\int_0^\infty (t^{-\theta} K(t,x))^q \frac{dt}{t}\right)^{\frac{1}{q}}, & 0 < \theta < 1,\ 1 \leq q < \infty; \\ \sup_{t>0} t^{-\theta} K(t;x), & 0 \leq \theta \leq 1,\ q = \infty \end{cases} \tag{3.3}$$

is finite.

The functor $[\,.\,,\,.\,]_{\theta,q;K}$ is an exact interpolation functor of exponent θ [3, Theorem 3.1.2]. An immediate consequence of [11] Theorem 3.2 is that the real interpolation method lifts immediately from the commutative to the noncommutative setting.

Corollary 3.1 *If* E_0, E_1 *are fully symmetric spaces on* $[0,\infty)$, *then*

$$[E_0(\tau), E_1(\tau)]_{\theta,q:K} = [E_0, E_1]_{\theta,q:K}(\tau).$$

Note that, in the case of the Banach couple $(L_1(\tau), \mathcal{M})$, the K-functional is given by the explicit formula

$$K(t;x) = \int_0^t \mu(x)dm, \quad x \in L_1(\tau) + \mathcal{M}.$$

See [12] Proposition 2.6. This permits ready identification of the space $[L_1(\tau), \mathcal{M}]_{\theta,q;K}$ as a noncommutative Lorentz space.

Proposition 3.2

(a). *If* $0 < \theta < 1$ *and* $1 \leq q \leq \infty$, *then*

$$[L_1(\tau), \mathcal{M}]_{\theta,q;K} = L_{p,q}(\tau), \quad 1/p = 1 - \theta.$$

(b). *If $\theta = 0$ and $q = \infty$, then*

$$[L_1(\tau), \mathcal{M}]_{0,\infty;K} = L_1(\tau).$$

Proof

(a). If $x \in L_1(\tau) + \mathcal{M}$, then, using (3.3),

$$\begin{aligned}
\|x\|_{\theta,q;K} &= \left(\int_0^\infty (t^{-\theta} K(t,x))^q \frac{dt}{t}\right)^{\frac{1}{q}} \\
&= \left(\int_0^\infty (t^{-\theta} K(t,\mu(x)))^q \frac{dt}{t}\right)^{\frac{1}{q}} \\
&= \left(\int_0^\infty (t^{1-\theta} M(\mu(x))(t))^q \frac{dt}{t}\right)^{\frac{1}{q}} \\
&= \|\mu(x)\|_{L_{p,q}} = \|x\|_{L_{p,q}(\tau)},
\end{aligned}$$

where $1/p = 1 - \theta$.

(b). If $\theta = 0$ and $q = \infty$, then

$$\begin{aligned}
\|x\|_{0,\infty;K} &= \sup_{t>0} K(t;x) = \sup_{t>0} \int_0^t \mu(s;x)ds \\
&= \int_0^\infty \mu(s;x)ds = \|x\|_{L_1(\tau)}.
\end{aligned}$$

□

3.3 Holmstedt Formula

Suppose that $\mathbf{X} = (X_0, X_1)$ is a Banach couple and that

$$E_i = (X_0, X_1)_{\theta_i,q_i;K}, \quad 0 \le \theta_i < 1, \quad 1 \le q_i \le \infty, \quad i = 1, 2.$$

For all $x \in X_0 + X_1$, the *Holmstedt formula* expresses the K-functional $K(t, x; E_0, E_1)$ in terms of the K-functional $K(t, x; X_0, X_1)$ via the formula

$$\begin{aligned}
K(t, x; E_0, E_1) \sim &\left(\int_0^{t^{1/\eta}} \left(s^{-\theta_0} K(s, x; X_0, X_1)\right)^{q_0} \frac{ds}{s}\right)^{1/q_0} \\
&+ t \left(\int_{t^{1/\eta}}^\infty \left(s^{-\theta_1} K(s, x; X_0, X_1)\right)^{q_1} \frac{ds}{s}\right)^{1/q_1}
\end{aligned} \tag{3.4}$$

where

$$\eta = \theta_1 - \theta_0, \quad 0 \leq \theta_0 < \theta_1 < 1, \quad 1 \leq q_0, q_1 < \infty.$$

Here $a \sim b$ means that there exist constants $0 < c_1, c_2$ such that $c_1 a \leq b \leq c_2 a$, and the right-hand side has the natural interpretation in the case that $q_0 = \infty$ or $q_1 = \infty$. See [3] Theorem 3.6.1.

We shall need the following Hardy-type inequality.

Lemma 3.3 *Suppose that $p > 1$ and set $1/p' = 1 - 1/p$. If $x \in L_1(\tau) + \mathcal{M}$, then*

$$\begin{aligned} t\int_{t^{p'}}^{\infty} s^{-(1-\frac{1}{p})}\mu(s;x)ds &\leq t\int_{t^{p'}}^{\infty}\left(s^{-\left(1-\frac{1}{p}\right)}\int_0^s \mu(x)dm\right)\frac{ds}{s} \\ &\leq \frac{1}{1-\frac{1}{p}}\int_0^{t^{p'}}\mu(s;x)ds + \frac{t}{1-\frac{1}{p}}\int_{t^{p'}}^{\infty}\mu(s;x)s^{-(1-\frac{1}{p})}ds. \end{aligned}$$

Proof Using the fact that $s \to \mu(s;x)$ is decreasing, the first inequality follows by observing that

$$\mu(s;x) \leq \frac{1}{s}\int_0^s \mu(x)dm, \quad s > 0.$$

The second inequality follows directly via integration by parts. □

Suppose now that $1 < p < \infty$ and that $q = 1$, and observe that, using Proposition 3.2,

$$L_{p,1}(\tau) = [L_1(\tau), \mathcal{M}]_{1-\frac{1}{p},1;K},$$

and

$$L_1(\tau) = [L_1(\tau), \mathcal{M}]_{0,\infty;K}$$

Applying Holmstedt's formula to the Banach couple $(L_1(\tau), \mathcal{M})$ with

$$p_0 = 0,\ p_1 = p,\ q_0 = \infty,\ q_1 = 1,\ \theta_0 = 0,\ \eta = \theta_1 = 1 - 1/p,$$

and using the Hardy-type inequality given in Lemma 3.3, one obtains, for all $x \in L_1(\tau) + L_{p,1}(\tau)$,

$$\begin{aligned} K(t,x;L_1(\tau),L_{p,1}(\tau)) &= K(t,x;L_{0,\infty}(\tau),L_{p,1}(\tau)) \\ &= K(t,x,[L_1(\tau),\mathcal{M}]_{0,\infty;K},[L_1(\tau),\mathcal{M}]_{1-\frac{1}{p},1;K}) \end{aligned}$$

$$\sim \sup_{0 \le s \le t^{p'}} \int_0^s \mu(x)dm + t \int_{t^{p'}}^\infty \left(s^{-(1-1/p)} \int_0^s \mu(x)dm \right) \frac{ds}{s}$$

$$\sim \int_0^{t^{p'}} \mu(s;x)ds + t \int_{t^{p'}}^\infty s^{-(1-\frac{1}{p})} \mu(s;x)ds)) \tag{3.5}$$

where $1/p' = 1 - 1/p$. Further, using Proposition 3.2, observe that

$$L_{p,\infty}(\tau) = [L_1(\tau), \mathcal{M}]_{1-\frac{1}{p},\infty;K}.$$

Again applying Holmstedt's formula, we obtain

$$K(t,x; L_1(\tau), L_{p,\infty}(\tau)) \sim \int_0^{t^{p'}} \mu(s;x)ds + \sup_{s \ge t^{p'}} s^{-(1-\frac{1}{p})} \int_0^s \mu(x)dm. \tag{3.6}$$

4 Operators of Weak Type and the Calderón Operator

For $1 \le p < \infty$, the map $T : L_{p,1}(\sigma) \to S(\tau)$ is said to be of weak type (p,q), $1 \le q \le \infty$, if T maps $L_{p,1}(\sigma)$ continuously into $L_{q,\infty}(\tau)$ and there exists a constant M such that

$$\mu(t; Tx) \le M t^{-\frac{1}{q}} \|x\|_{L_{p,1}(\sigma)}, \quad t > 0, \tag{4.7}$$

for all $x \in L_{p,1}(\sigma)$. T will be said to be of weak type (∞, q) if and only if there exists a constant M such that

$$\mu(t; Tx) \le M t^{-1/q} \|x\|_{\mathcal{N}}, \quad t > 0, \tag{4.8}$$

for all $x \in \mathcal{N}$.

Remark 4.1 Note that, if $p < \infty$, and if T is a continuous *linear* mapping from $L_p(\sigma)$ to $L_q(\tau)$, then T is of weak type (p,q) since $L_{p,1}(\sigma)$ is continuously embedded in $L_p(\sigma)$ and $L_q(\tau)$ embeds continuously into $L_{q,\infty}(\tau)$.

Let ω be the closed segment in the unit square with end points $(\frac{1}{p_1}, \frac{1}{q_1})$, $(\frac{1}{p_2}, \frac{1}{q_2})$ where $1 \le p_1 < p_2 \le \infty$, $1 \le q_1, q_2 \le \infty$, $q_1 \ne q_2$.

If α denotes the slope of the segment ω, that is,

$$\alpha = \frac{\frac{1}{q_1} - \frac{1}{q_2}}{\frac{1}{p_1} - \frac{1}{p_2}} \tag{4.9}$$

then the Calderón operator $S(\omega)$ is defined on functions on $(0, \infty)$ by setting

$$
\begin{aligned}
(S(\omega)f)(t) &= t^{-\frac{1}{q_1}}\frac{1}{p_1}\int_0^{t^\alpha} s^{\frac{1}{p_1}} f(s)\frac{ds}{s} + t^{-\frac{1}{q_2}}\frac{1}{p_2}\int_{t^\alpha}^\infty s^{\frac{1}{p_2}} f(s)\frac{ds}{s}, \\
&= \int_0^{t^\alpha} f(s)d\left(\frac{s^{\frac{1}{p_1}}}{t^{\frac{1}{q_1}}}\right) + \int_{t^\alpha}^\infty f(s)d\left(\frac{s^{\frac{1}{p_2}}}{t^{\frac{1}{q_2}}}\right),
\end{aligned} \tag{4.10}
$$

if $1 \le p_2 < \infty$ and by setting

$$
(S(\omega)f)(t) = \int_0^{t^\alpha} f(s)d\left(\frac{s^{\frac{1}{p_1}}}{t^{\frac{1}{q_1}}}\right), \tag{4.11}
$$

if $p_2 = \infty$.

Lemma 4.2 *Suppose that $x \in L_1(\tau) + \mathcal{M}$, and that $t > 0$.*

(a). *If $1 \le p_2 < \infty$ and if $S(\omega)(\mu(x))(t) < \infty$, then*

$$
(|x| - \mu(t^\alpha; x)\mathbf{1})^+ \in L_{p_1,1}(\tau), \quad \text{and} \quad |x| \wedge \mu(t^\alpha; x)\mathbf{1} \in L_{p_2,1}(\tau). \tag{4.12}
$$

In particular, $x \in L_{p_1,1}(\tau) + L_{p_2,1}(\tau)$.

(b). *If $p_2 = \infty$ and if $S(\omega)(\mu(x))(t) < \infty$, then*

$$
(|x| - \mu(t^\alpha; x)\mathbf{1})^+ \in L_{p_1,1}(\tau), \quad \text{and} \quad |x| \wedge \mu(t^\alpha; x)\mathbf{1} \in \mathcal{M}. \tag{4.13}
$$

In particular, $x \in L_{p_1,1}(\tau) + \mathcal{N}$.

(c). *If $1 < p_2 < \infty$ and $\int_t^\infty s^{\frac{1}{p_2}-1}\mu(s; x)ds < \infty$, then*

$$
(|x| - \mu(t^\alpha; x)\mathbf{1})^+ \in L_1(\tau), \quad \text{and} \quad |x| \wedge \mu(t^\alpha; x)\mathbf{1} \in L_{p_2,1}(\tau). \tag{4.14}
$$

In particular, $x \in L_1(\tau) + L_{p_2,1}(\tau)$.

Proof

(a). Suppose that $1 \le p_2 < \infty$ and observe that

$$
\begin{aligned}
&t^{-\frac{1}{q_1}}\frac{1}{p_1}\int_0^\infty (\mu(s; x) - \mu(t^\alpha; x))^+ s^{\frac{1}{p_1}}\frac{ds}{s} + t^{-\frac{1}{q_2}}\frac{1}{p_2}\int_0^\infty \mu(s; x) \wedge \mu(t^\alpha; x) s^{\frac{1}{p_2}}\frac{ds}{s} \\
&= t^{-\frac{1}{q_1}}\frac{1}{p_1}\int_0^{t^\alpha} (\mu(s; x) - \mu(t^\alpha; x)) s^{\frac{1}{p_1}}\frac{ds}{s} + t^{-\frac{1}{q_2}}\frac{1}{p_2}\int_0^{t^\alpha} \mu(t^\alpha; x) s^{\frac{1}{p_2}}\frac{ds}{s} \\
&\quad + t^{-\frac{1}{q_2}}\frac{1}{p_2}\int_{t^\alpha}^\infty \mu(s; x) s^{\frac{1}{p_2}}\frac{ds}{s}
\end{aligned}
$$

$$= t^{-\frac{1}{q_1}} \frac{1}{p_1} \int_0^{t^\alpha} \mu(s;x) s^{\frac{1}{p_1}} \frac{ds}{s} + \mu(t^\alpha;x) \left(t^{\frac{\alpha}{p_2} - \frac{1}{q_2}} - t^{\frac{\alpha}{p_1} - \frac{1}{q_1}} \right)$$
$$+ t^{-\frac{1}{q_2}} \frac{1}{p_2} \int_{t^\alpha}^\infty \mu(s;x) s^{\frac{1}{p_2}} \frac{ds}{s}$$
$$= t^{-\frac{1}{q_1}} \frac{1}{p_1} \int_0^{t^\alpha} \mu(s;x) s^{\frac{1}{p_1}} \frac{ds}{s} + t^{-\frac{1}{q_2}} \frac{1}{p_2} \int_{t^\alpha}^\infty \mu(s;x) s^{\frac{1}{p_2}} \frac{ds}{s}$$
$$= S(\omega)(\mu(x))(t) < \infty.$$

Using [14] Lemma 2.5 (iv), observe that

$$\mu((|x| - \mu(t^\alpha;x)\mathbf{1})^+) = (\mu(x) - \mu(t^\alpha;x))^+, \quad \mu(|x| \wedge t^\alpha \mathbf{1}) = \mu(x) \wedge t^\alpha$$

and from this it follows that the assertion (4.12) holds. Finally, using the spectral theorem,

$$|x| = \int_0^\infty \lambda de^{|x|}(\lambda) = \int_0^\infty ((\lambda - \mu(t^\alpha, x)^+ + \lambda \wedge \mu(t^\alpha : x)) de^{|x|}(\lambda)$$
$$= (|x| - \mu(t^\alpha;x)\mathbf{1})^+ + |x| \wedge t^\alpha \mathbf{1} \in L_{p_1,1}(\tau) + L_{p_2,1}(\tau).$$

That $x \in L_{p_1,1}(\tau) + L_{p_2,1}(\tau)$ follows immediately from the polar decomposition.

(b). If $p_2 = \infty$, then

$$t^{-\frac{1}{q_1}} \frac{1}{p_1} \int_0^\infty \left(\mu(s;x) - \mu(t^\alpha;x)\right)^+ s^{\frac{1}{p_1}} \frac{ds}{s} \le t^{-\frac{1}{q_1}} \frac{1}{p_1} \int_0^{t^\alpha} \mu(s;x) s^{\frac{1}{p_1}} \frac{ds}{s}$$
$$= S(\omega)(\mu(x))(t) < \infty.$$

This implies that $(|x| - \mu(t^\alpha;x)\mathbf{1})^+ \in L_{p_1,1}(\tau)$ and since it is clear that $|x| \wedge \mu(t^\alpha;x)\mathbf{1} \in \mathcal{M}$, it follows that $x \in L_{p_1,1}(\tau) + \mathcal{M}$.

(c). Now suppose that $\int_t^\infty s^{\frac{1}{p_2}-1} \mu(s;x) ds < \infty$. The same calculation as in (a) with $p_1 = 1$ yields

$$t^{-\frac{1}{q_1}} \int_0^\infty \left(\mu(s;x) - \mu(t^\alpha;x)\right)^+ ds + t^{-\frac{1}{q_2}} \frac{1}{p_2} \int_0^\infty \mu(s;x) \wedge \mu(t^\alpha;x) s^{\frac{1}{p_2}} \frac{ds}{s}$$
$$= t^{-\frac{1}{q_1}} \int_0^{t^\alpha} \mu(s;x) ds + t^{-\frac{1}{q_2}} \frac{1}{p_2} \int_{t^\alpha}^\infty \mu(s;x) s^{\frac{1}{p_2}} \frac{ds}{s} < \infty,$$

using the fact that $\int_0^{t^\alpha} \mu(s; x)ds < \infty$ since $x \in L_1(\tau) + \mathcal{M}$. Consequently,

$$(|x| - \mu(t^m; x)\mathbf{1})^+ \in L_1(\tau) \quad \text{and} \quad |x| \wedge \mu(t^\alpha; x)\mathbf{1} \in L_{p_2,1}(\tau)$$

and it follows as above that $x \in L_1(\tau) + L_{p_2,1}(\tau)$.

□

Following [7], suppose that $D \subseteq S(\sigma)$ is a convex subset. A map $T : D \to S(\tau)$ is called *midpoint subconvex* if there exist partial isometries $v, w \in \mathcal{M}$ such that

$$|T(\frac{1}{2}x + \frac{1}{2}y)| < \frac{1}{2}v^*|Tx|v + \frac{1}{2}w^*|Ty|w \tag{4.15}$$

for all $x, y \in D$. As follows from [14] Lemma 4.3, for any $x_1, x_2 \in S(\tau)$, there exist partial isometries $v, w \in \mathcal{M}$ such that

$$|x_1 + x_1| \leq v^*|x_1|v + w^*|x_2|w$$

and from this, it is clear that any (sub)linear map is midpoint subconvex.

We now indicate how to extend Theorem 8 of [5] for sublinear operators by a simple modification of the proof given in [5]. See also [10] Proposition 5.1, and [7], Theorem A1.

Theorem 4.3 *Suppose that* $1 \leq p_1 < p_2 \leq \infty$, $1 \leq q_1, q_2 \leq \infty$ *and that* $q_1 \neq q_2$. *If* $T : L_{p_1,1}(\sigma) + L_{p_2,1}(\sigma) \longrightarrow S(\tau)$ *is a midpoint subconvex map which is simultaneously of weak types* (p_i, q_i), $i = 1, 2$ *with corresponding weak-type norms* M_i, $i = 1, 2$, *and if* $t > 0$ *satisfies*

$$S(\omega)(\mu(x))(t) < \infty,$$

then

$$\mu(t; Tx) \leq c(p_i, q_i) \max\{M_1.M_2\} S(\omega)(\mu(x))(t)$$

whenever $x \in L_{p_1,1}(\sigma) + L_{p_2,1}(\sigma)$ *in the case that* $p_2 < \infty$, *and* $x \in L_{p_1,1}(\sigma) + \mathcal{N}$ *in the case that* $p_2 = \infty$.

Proof It will suffice to prove the theorem only in the case that $p_2 < \infty$ as the proof in the case that $p_2 = \infty$ is similar. Let $x \in L_{p_1,1}(\sigma) + L_{p_2,1}(\sigma)$, and let $\beta = \mu_{t^\alpha}(x)$ where α is the slope of the segment ω. Let $x = u|x|$ be the polar decomposition of x and write $|x| = (|x| - \beta\mathbf{1})^+ + |x| \wedge \beta\mathbf{1}$. Suppose now that $p_1 < p_2$. If $t > 0$ satisfies $S(\omega)(\mu(x))(t) < \infty$, then it follows from Lemma 4.2(a) that $(|x| - \beta\mathbf{1})^+ \in L_{p_1,1}(\sigma)$, and $|x| \wedge \beta\mathbf{1} \in L_{p_2,1}(\sigma)$. Now observe that

$$\begin{aligned}\mu(t; Tx) &= \mu(t; T(u(|x| - \beta\mathbf{1})^+ + u(|x| \wedge \beta\mathbf{1}))) \\ &= \mu(t; T\left(\frac{1}{2}2u(|x| - \beta\mathbf{1})^+ + \frac{1}{2}2u|x| \wedge \beta\mathbf{1}\right))\end{aligned}$$

$$
\begin{aligned}
&\leq \frac{1}{2}\mu\left(t; v^*|T(2u(|x|-\beta\mathbf{1})^+|v + w^*|T(2u|x| \wedge \beta\mathbf{1})|w\right) \\
&\leq \frac{1}{2}\mu(\frac{t}{2}; T(2u(|x|-\beta\mathbf{1})^+)) + \frac{1}{2}\mu(\frac{t}{2}; T(2u(|x| \wedge \beta\mathbf{1}))) \\
&\leq \max_{i=1,2}\{M_i 2^{\frac{1}{q_i}}\}(t^{-\frac{1}{q_1}}\|\mu(u(|x|-\beta\mathbf{1})^+)\|_{p_1,1} + t^{-\frac{1}{q_2}}\|\mu(u(|x| \wedge \beta\mathbf{1}))\|_{p_2,1}) \\
&\leq \max_{i=1,2}\{p_i 2^{\frac{1}{q_i}} M_i\}(\frac{1}{p_1}t^{-\frac{1}{q_1}}\|(\mu(x)-\beta)^+\|_{p_1,1} + \frac{1}{p_2}t^{-\frac{1}{q_2}}\|\mu(x) \wedge \beta)\|_{p_2,1}) \\
&= \max_{i=1,2}\{p_i 2^{\frac{1}{q_i}} M_i\}\Bigg(\int_0^{t^\alpha}(\mu(x)-\mu(t^\alpha; x))d(\frac{s^{1/p_1}}{t^{1/q_1}}) + \\
&+ \int_{t^\alpha}^\infty \mu(x)d(\frac{s^{1/p_2}}{t^{1/q_2}}) + \mu(t^\alpha; x)\int_0^{t^\alpha} d(\frac{s^{1/p_2}}{t^{1/q_2}})\Bigg) \\
&= \max_{i=1,2}\{p_i 2^{\frac{1}{q_i}} M_i\}\left(\int_0^{t^\alpha}\mu(x)d(\frac{s^{1/p_1}}{t^{1/q_1}}) + \int_{t^\alpha}^\infty \mu(x)d(\frac{s^{1/p_2}}{t^{1/q_2}})\right) + \\
&+ \max_{i=1,2}\{p_i 2^{\frac{1}{q_i}} M_i\}\left(\mu(t^\alpha; x)[t^{\alpha/p_2-1/q_2} - t^{\alpha/p_1-1/q_1}]\right) \\
&= \max_{i=1,2}\{p_i 2^{\frac{1}{q_i}} M_i\} S(\omega)(\mu(x))(t),
\end{aligned}
$$

using the fact that $\alpha/p_2 - 1/q_2 = \alpha/p_1 - 1/q_1$. □

5 Boyd's Interpolation Theorem

For any measurable function f on the interval $(0, a)$, the dilation $D_s f$ of f is defined by setting

$$(D_s f)(t) = f(at)\chi_{(0,a)}(st), \quad t \in (0, a).$$

Suppose now that E is a symmetric space on $(0, a)$. Define the *lower Boyd index* p_E of E by setting

$$p_E = \sup\left\{p > 0 : \exists c > 0\ \forall 0 < s \leq 1\ \|D_s f\|_E \leq c s^{-\frac{1}{p}}\|f\|_E\right\},$$

and the *upper Boyd index* q_E of E by setting

$$q_E = \inf\left\{q > 0 : \exists c > 0\ \forall s \geq 1\ \|D_s f\|_E \leq c s^{-\frac{1}{q}}\|f\|_E\right\}.$$

If $1 \le p < q \le \infty$, we specialise the Calderón operator to $S(m)_+$ by setting $q_1 = p_1 = p$, $q_2 = p_2 = q$ to obtain

$$S_{p,q}(f)(t) = t^{-\frac{1}{p}}\frac{1}{p}\int_0^t s^{\frac{1}{p}} f(s)\frac{ds}{s} + t^{-\frac{1}{q}}\frac{1}{q}\int_t^\infty s^{\frac{1}{q}} f(s)\frac{ds}{s}, \quad t > 0,\ 0 \le f \in S(m),$$

if $1 \le p < q < \infty$, and

$$S_{p,\infty} f(t) = t^{-\frac{1}{p}}\int_0^t s^{\frac{1}{p}} f(s)\frac{ds}{s}, \quad t > 0,\ 0 \le f \in S(m),$$

if $q = \infty$.

To proceed further, again suppose that $1 \le p < q < \infty$ and define the operators P_p, Q_q on $S(m)$ by setting

$$P_p(f)(t) = t^{-\frac{1}{p}}\int_0^t s^{\frac{1}{p}} f(s)\frac{ds}{s} = \int_0^1 u^{\frac{1}{p}-1} f(ut)du;$$

$$Q_q f(t) = t^{-\frac{1}{q}}\int_t^\infty s^{\frac{1}{q}} f(s)\frac{ds}{s} = \int_1^\infty u^{\frac{1}{q}-1} f(ut)du$$

and note that

$$S_{p,q} = P_p + Q_q. \tag{5.16}$$

The theorem which follows is due to Boyd [4]. A direct proof has been given by Montgomery-Smith [18].

Theorem 5.1 *Let $E \subseteq S(m)$ be a symmetric space and suppose that $1 \le p < q < \infty$.*

(i). *The operator P_p is bounded on E if and only if $p < p_E$.*
(ii). *The operator Q_q is bounded on E if and only if $q_E < q$.*

Corollary 5.2 *Let $E \subseteq S(m)$ be a symmetric space on $(0, a)$.*

(i). *If $1 \le p < q < \infty$, then $S_{p,q}$ is bounded on E if and only if*

$$1 \le p < p_E \le q_E < q.$$

(ii). *If $1 \le p < \infty$, then $S_{p,\infty}$ is bounded on E if and only if $p < p_E$.*

Corollary 5.2 together with Lemma 4.2 now yields the following consequence.

Corollary 5.3 *Let $E \subseteq S(m)$ be a strongly symmetric space and suppose that $x \in E(\tau)$.*

(i). *If* $1 \le p < p_E \le q_E < q < \infty$, *then* $S_{p,q}(\mu(x)) < \infty$ *a.e. and* $E(\tau) \subseteq L_{p,1}(\tau) + L_{q,1}(\tau)$.
(ii). *If* $1 \le p < p_E$, *then* $S_{p,\infty}(\mu(x)) < \infty$ *a.e. and* $E(\tau) \subseteq L_{p,1}(\tau) + \mathcal{M}$.

We now prove the noncommutative Boyd Theorem.

Theorem 5.4 (Weak Type (p, p) and Weak Type (q, q) Interpolation) *Let* $E \subseteq S(m)$ *be strongly symmetric. Suppose that* $1 \le p < q \le \infty$ *and that* $T : L_{p,1}(\sigma) + L_{q,1}(\sigma) \to S(\tau)$ *is a midpoint subconvex map which is simultaneously of weak types* (p, p) *and* (q, q)*, that is,*

$$\mu(t; Tx) \le C_r t^{-\frac{1}{r}} \|x\|_{L_{r,1}(\sigma)}, \quad t > 0, \quad x \in L_{r,1}(\sigma),\ r = p, q. \tag{5.17}$$

If $1 \le p < p_E \le q_E < q < \infty$, *then* $E(\sigma) \subseteq L_{p,1}(\sigma) + L_{q,1}(\sigma)$ *and, for all* $x \in E(\sigma)$,

$$\|T(x)\|_{E(\tau)} \le c(p, q) \max\{C_p, C_q\} \|S_{p,q}\|_{E\to E} \|x\|_{E(\sigma)}.$$

Proof Suppose that $x \in E(\sigma)$, or equivalently, $\mu(x) \in E$. Since $1 \le p < p_E \le q_E < \infty$, it follows from Corollary 5.3 that $S_{p,q}(\mu(x)) < \infty$ a.e. and that $E(\sigma) \subseteq L_{p,1}(\sigma) + L_{q,1}(\sigma)$. By Lemma 4.2 and Theorem 4.3, it follows that

$$\mu(Tx) \le c(p, q) \max\{C_p, C_q\} S_{p,q}(\mu(x)) \quad \text{a.e.} \quad .$$

Since $S_{p,q}(\mu(x)) \in E$, it follows that $\mu(Tx) \in E$ and so

$$\begin{aligned} \|Tx\|_{E(\tau)} = \|\mu(Tx)\|_E &\le c(p, q) \max\{C_p, C_q\} \|S_{p,q}\|_{E\to E} \|\mu(x)\|_E \\ &= c(p, q) \max\{C_p, C_q\} \|S_{p,q}\|_{E\to E} \|x\|_{E(\sigma)}. \end{aligned}$$

□

Via Remark 4.1, Theorem 5.4 has the following consequence.

Corollary 5.5 *Let* $E \subseteq S(m)$ *be strongly symmetric space on* $(0, a)$. *If* $1 \le p < p_E \le q_E < q < \infty$, *then the pair* $(E(\sigma), E(\tau))$ *is an interpolation pair for the pair* $(L_p(\sigma), L_q(\sigma))$, $(L_{p,\infty}(\tau), L_{q,\infty}(\tau))$, *with interpolation constant depending only on* p, q, E.

Proof It need only be shown that $E(\tau)$ is intermediate for the Banach couple $(L_p(\tau), L_q(\tau))$. From Theorem 5.4 it follows that $E(\tau) \subseteq L_p(\tau) + L_q(\tau)$ and so it remains to be shown that $L_p(\tau) \cap L_q(\tau) \subseteq E(\tau)$. This follows immediately from its commutative specialisation given in [16] Proposition 2.b.3 and the fact that $L_p(\tau) \cap L_q(\tau) = (L_p \cap L_q)(\tau)$ as follows from [11] Theorem 3.2. The same arguments apply verbatim with $E(\tau)$ replaced by $E(\sigma)$. □

Theorem 5.6 (Weak Type (p, p) and Strong Type (∞, ∞) Interpolation) *Let $E \subseteq S(m)$ be strongly symmetric. Suppose that $1 \leq p < \infty$ and that $T : L_{p,1}(\sigma) + \mathcal{N} \to S(\tau)$ is a midpoint subconvex map which is simultaneously of weak type (p, p) and strong type (∞, ∞), that is,*

$$\|Tx\|_{\mathcal{M}} \leq C_\infty \|x\|_{\mathcal{N}}, \quad \|Ty\|_{L_{p,\infty}(\tau)} \leq C_p \|y\|_{L_{p,1}(\sigma)},$$

for all $x \in L_{p,1}(\sigma)$, $y \in \mathcal{N}$, respectively. If $1 \leq p < p_E$, then $E(\sigma) \subseteq L_{p,1}(\sigma) + \mathcal{N}$ and, for all $x \in E(\sigma)$,

$$\|T(x)\|_{E(\tau)} \leq c(p) \max\{C_p, C_\infty\} \|S_{p,\infty}\|_{E \to E} \|x\|_{E(\sigma)}.$$

Proof The proof follows by exactly the same argument given in the proof of Theorem 5.4, using Corollary 5.2 (ii), Lemma 4.2 (b) and Theorem 4.3. □

Again using Remark 4.1, we obtain the following consequence.

Corollary 5.7 *Let $E \subseteq S(m)$ be a strongly symmetric space. If $1 \leq p < p_E$, then the pair $(E(\sigma), E(\tau))$ is an interpolation pair for the pair $((L_p(\sigma), \mathcal{N}), (L_p(\tau), \mathcal{M}))$, with interpolation constant depending only on p and E.*

Proof As in the proof of Corollary 5.5, it suffices to show that $E(\tau)$ is intermediate for the pair $(L_p(\tau), \mathcal{M})$. From Theorem 5.6, it follows that $E(\tau) \subseteq L_p(\tau) + \mathcal{M}$. That $L_p(\tau) \cap \mathcal{M} \subseteq E(\tau)$ is proved in [6] Lemma 5.15 by adapting the relevant part of the proof of [16] Proposition 2.b.3. □

Finally, let us remark that Theorems 5.4, and 5.6 are due to Dirksen [6, 7].

6 Strong Type (1, 1) and Weak Type (q, q) Interpolation, $1 < q < \infty$

The method of the previous sections based on the Calderón inequality of Theorem 5.4 fails in the extreme case that $p_E = 1$. See Theorem 5.1 (i). In the case that $1 \leq q_E < q < \infty$, the use of the Calderón inequality will be replaced by showing that the operator T is submajorised pointwise by the operator Q_q (6.20). This idea is due to Astashkin and Maligranda [1] and the estimate will be based on considerations concerning the Holmstedt formulae (3.5).

Some initial preparation is needed. Assume that $1 < q < \infty$ and that $T : L_1(\sigma) + L_{q,1}(\sigma) \to S(\tau)$ is midpoint subconvex and satisfies

$$\|Tx\|_{L_1(\tau)} \leq C_1 \|x\|_{L_1(\sigma)}, \quad \|Ty\|_{L_{q,\infty}(\tau)} \leq C_q \|y\|_{L_{q,1}(\sigma)}.$$

for all $x \in L_1(\sigma)$, $y \in L_{q,1}(\sigma)$.

Now suppose that $x \in L_1(\sigma) + L_{q,1}(\sigma)$ and that $x = x_1 + x_2$, with $x_1 \in L_1(\sigma)$ and $x_2 \in L_{q,1}(\sigma)$. By midpoint subconvexity, there exist partial isometries $v, w \in \mathcal{M}$ such that

$$|T(x)| = |T(\frac{1}{2}2x_1 + \frac{1}{2}2x_2)| \leq \frac{1}{2}v^*|T(2x_1)|v + \frac{1}{2}w^*|T(2x_2)|w.$$

Now observe that, for all $t > 0$,

$$\begin{aligned} K(t, Tx; L_1(\tau), L_{q,\infty}(\tau)) &= \|Tx\|_{L_1(\tau)+tL_{q,\infty}(\tau)} \\ &\leq \frac{1}{2}\|T(2x_1)\|_{L_1(\tau)+tL_{q,\infty}(\tau)} + \frac{1}{2}\|T(2x_2)\|_{L_1(\tau)+tL_{q,\infty}(\tau)} \\ &\leq \frac{1}{2}\|T(2x_1)\|_{L_1(\tau)} + \frac{t}{2}\|T(2x_2)\|_{L_{q,\infty}(\tau)} \\ &\leq C_1\|x\|_{L_1(\sigma)} + tC_q\|x_2\|_{L_{q,1}(\sigma)} \\ &\leq \max\{C_1, C_q\}(\|x_1\|_{L_1(\sigma)} + t\|x_2\|_{L_{q,1}(\sigma)}). \end{aligned}$$

Consequently, taking the infimum over all decompositions $x = x_1 + x_2$ with $x_1 \in L_1(\sigma)$ and $x_2 \in L_{q,1}(\sigma)$, it follows that

$$K(t, Tx; L_1(\tau), L_{q,\infty}(\tau)) \leq CK(t, x; L_1(\sigma), L_{q,1}(\sigma)),$$

for some constant $C > 0$.

Using now the Holmstedt formulae given in (3.6), (3.5) yields the estimate

$$\begin{aligned} \int_0^{t^{q'}} \mu(s; Tx)ds &\leq C'K(t, Tx; L_1(\tau), L_{q,\infty}(\tau)) \\ &\leq C_2K(t, x; L_1(\sigma), L_{q,1}(\sigma)) \\ &\leq C_2'\left(\int_0^{t^{q'}} \mu(s; x)ds + t\int_{t^{q'}}^{\infty} s^{\frac{1}{q}-1}\mu(s; x)ds\right). \end{aligned}$$

Replacing t by $t^{\frac{1}{q'}} = t^{1-\frac{1}{q}}$ now gives

$$\int_0^t \mu(s; Tx)ds \leq C_2\left(\int_0^t \mu(s; x)ds + t^{1-\frac{1}{q}}\int_t^{\infty} s^{\frac{1}{q}-1}\mu(s; x)ds\right), \tag{6.18}$$

for all $x \in L_1(\sigma) + L_{1,q}(\sigma),\ t > 0$.

Using Fubini's theorem, observe that

$$\begin{aligned}\int_0^t Q_q(\mu(x))(s) &= \int_0^t u^{-\frac{1}{q}}\left(\int_u^\infty s^{\frac{1}{q}-1}\mu(s;x)ds\right)du\\ &= \int_0^t s^{\frac{1}{q}-1}\left(\int_0^s u^{-\frac{1}{q}}du\right)\mu(s;x)ds + \int_t^\infty s^{\frac{1}{q}-1}\left(\int_0^t u^{-\frac{1}{q}}du\right)\mu(s;x)ds\\ &= q'\left(\int_0^t \mu(s;x)ds + t^{1-\frac{1}{q}}\int_t^\infty s^{\frac{1}{q}-1}\mu(s;x)ds\right),\quad t>0.\end{aligned}\tag{6.19}$$

Comparing this with (6.18), it follows that there exists a constant C_3 such that

$$Tx \prec\prec \frac{C_3}{q'}Q_q(\mu(x)).\tag{6.20}$$

It is now possible to prove the following complement to Theorem 5.6. It is a noncommutative extension of the first statement of [1] Theorem 3.

Theorem 6.1 *Let $E \subseteq S(m)$ be strongly symmetric. Suppose that $1 < q < \infty$ and that $T : L_1(\sigma) + L_{q,1}(\sigma) \to S(\tau)$ be a midpoint subconvex map which is simultaneously of strong type $(1, 1)$ and weak type (q, q), that is,*

$$\|Tx\|_{L_1(\tau)} \le C_1\|x\|_{L_1(\sigma)},\qquad \|Ty\|_{L_{q,\infty}(\tau)} \le C_q\|y\|_{L_{q,1}(\sigma)}$$

for all $x \in L_1(\sigma)$ and $y \in L_{q,1}(\sigma)$. If $1 \le q_E < q < \infty$ then $E(\sigma) \subseteq L_1(\sigma) + L_{q,1}(\sigma)$, and if E is fully symmetric, then $T(E(\sigma)) \subseteq E(\tau)$ and there exists a constant $C(q)$ such that

$$\|Tx\|_{E(\tau)} \le C(q)\|x\|_{E(\sigma)}.$$

Proof Suppose that $x \in E(\sigma)$ or, equivalently, that $\mu(x) \in E$, and suppose that $1 \le q_E < q < \infty$. By Theorem 5.1, $Q_q(\mu(x)) \in E$. In particular, $Q_q(\mu(x))(t) < \infty$ almost everywhere. It is clear that $Q_q(\mu(x))$ is decreasing and so

$$\int_1^\infty s^{\frac{1}{q}-1}\mu(s;x)ds = Q_q(1) < \infty.$$

By Lemma 4.2 (c), it follows that $x \in L_1(\sigma) + L_{q,1}(\sigma)$ and so $E(\sigma) \subseteq L_1(\sigma) + L_{q,1}(\sigma)$. The submajorisation inequality (6.20) and the assumption that E is fully symmetric now imply that $Tx \in E(\tau)$ and that

$$\|Tx\|_{E(\tau)} \le \frac{C_4}{q'}\|Q_q\|\|x\|_{E(\sigma)},$$

and this completes the proof of the theorem. □

The lemma which follows is well-known. However, we include a proof for lack of convenient reference.

Lemma 6.2 *If* $0 \leq \phi, \psi$ *are increasing concave functions on* $[0, 1)$ *with* $\phi(0) = \psi(0) = 0$ *for which*

$$0 \leq \phi(t) \leq \psi(t), \quad 0 \leq t \leq 1, \tag{6.21}$$

then

$$\int_0^1 \mu(t; x) d\phi(t) \leq \int_0^1 \mu(t; x) d\psi(t), \quad x \in S[0, 1). \tag{6.22}$$

Proof Observe first that, if $s = \sum_{k=1}^n \alpha_k \chi_{A_k}$ is a simple function on $[0, 1)$, with decreasing rearrangement $\mu(s) = \sum_{k=1}^n \alpha_k \chi_{[0,t_k)}$ with $0 < \alpha_k \in \mathbb{R}$ and $0 < t_1 < \ldots < t_n$, then it follows directly from (6.21) that

$$\begin{aligned}\int_0^1 \mu(t; s) d\phi(t) &= \phi(0+) \sum_{k=1}^n \alpha_k + \sum_{k=1}^n \alpha_k \phi(t_k) \\ &\leq \psi(0+) \sum_{k=1}^n \alpha_k + \sum_{k=1}^n \alpha_k \psi(t_k) = \int_0^1 \mu(t; s) d\psi(t).\end{aligned}$$

Now suppose that $x \in L_\infty$. For each $n = 1, 2, \ldots$, there exists a simple function s_n such that $\|x - s_n\|_\infty < 1/n$. Using the Lorentz-Shimogaki inequality [15] Theorem II.3.1, it follows that $\|\mu(x) - \mu(s_n)\|_\infty < 1/n, \ n = 1, 2, \ldots$ and it follows by a simple argument that the inequality (6.22) holds whenever $x \in L_\infty$. Now suppose that $x \in S[0, 1)$. It is clear that we may suppose that

$$\int_0^1 \mu(t; x) d\psi(t) < \infty$$

or else there is nothing to be proved. Observing that $\mu(x) \wedge n \uparrow_n \mu(x)$, and that $\mu(x) \wedge n \in L_\infty, n \geq 1$, it follows from the monotone convergence theorem that

$$\begin{aligned}\int_0^1 \mu(t; x) d\phi(t) &= \sup_n \int_0^1 \mu(t; x) \wedge n \ d\phi(t) \\ &\leq \sup_n \int_0^1 \mu(t; x) \wedge n \ d\psi(t) = \int_0^1 \mu(t; x) d\psi(t)\end{aligned}$$

and this suffices to prove the lemma. □

Before proceeding, some further comments are necessary.

Suppose that $\psi : [0, \infty) \to [0, \infty)$ is an increasing concave function such that $\psi(0) = 0$. The Lorentz space Λ_ψ is defined to be the collection of all $f \in S(m)$ such that

$$\|x\|_{\Lambda_\psi} = \int_0^\infty \mu(t; f) d\psi(t) < \infty.$$

See, for example, [2, 15]. Note that the improper Stieltjes integral preceding may be written as

$$\|f\|_{\Lambda_\psi} = \psi(0+) + \int_0^\infty \mu(t; f)\psi'(t)dt.$$

We recall that, if E is a symmetric space on $[0, \infty)$, then the fundamental function φ_E is defined by setting

$$\varphi_E(t) = \|\chi_{[0,t)}\|_E, \quad t \geq 0.$$

If ψ_E denotes the least concave majorant of the quasiconcave function φ_E, then it is shown in [19] Proposition 12.1.3 that the Lorentz space $\Lambda^0_{\psi_E}$ embeds continuously with norm one into E and that, in general, the space $\Lambda^0_{\psi_E}$ cannot be replaced by Λ_{ψ_E}. Here $\Lambda^0_{\psi_E}$ is the closure in Λ_{ψ_E} of $L_1 \cap L_\infty$ with respect to $\|\cdot\|_{\Lambda_{\psi_E}}$. However, in the case that E is a symmetric space on $[0, 1)$, then it may be shown [13] that $\Lambda^0_{\psi_E} = \Lambda_{\psi_E}$. Consequently, in this case, we have the continuous norm one embedding

$$\Lambda_{\psi_E} \subseteq E. \tag{6.23}$$

In what follows, if E is a symmetric space on $[0, \infty)$, then $E[0, 1)$ will denote the symmetric space on $[0, 1)$ defined by setting

$$E[0, 1) = \{x \in S[0, 1) : \mu(x) \in E\}, \quad \|x\|_{E[0,1)} = \|x\|_E.$$

We shall need the following embedding which is stated in [1]. We thank Sergei Astashkin for the proof and his permission to include it here.

Lemma 6.3 *If E is a symmetric space on $[0, \infty)$, then $L_1 \cap L_{q,\infty} \subseteq E$ whenever $1 \leq q_E < q$.*

Proof Let φ_E denote the fundamental function of E. Since $1 \leq q_E < q$, it follows that, for every r such that $q_E < r < q$, there is a constant $C > 0$ such that, for all $0 < t \leq 1$, we have

$$\varphi_E(t) = \|\chi_{[0,t)}\|_E = \|D_{\frac{1}{t}} \chi_{[0,1)}\|_E \leq C_1 \|D_{\frac{1}{t}}\| \leq C t^{1/r},$$

and consequently

$$\psi_E(t) \leq Ct^{1/r}, \quad 0 \leq t \leq 1, \tag{6.24}$$

where ψ is the least concave majorant of φ. In particular, $\psi_E(0) = 0$. Using Lemma 6.2, it follows from the estimate (6.24) and the fact that $r < q$ that

$$\begin{aligned}\int_0^1 t^{-\frac{1}{q}} d\psi_E(t) &\leq C \int_0^1 t^{-\frac{1}{q}} dt^{\frac{1}{r}} \\ &= \frac{C}{r} \int_0^1 t^{-1/q+1/r-1} dt < \infty.\end{aligned}$$

Consequently, $t^{-\frac{1}{q}} \chi_{[0,1)} \in \Lambda_{\psi_E}[0,1)$. By the embedding (6.23), it now follows that $t^{-\frac{1}{q}} \chi_{[0,1)} \in E$ and

$$\|t^{-\frac{1}{q}} \chi_{[0,1)}\|_E \leq \|t^{-\frac{1}{q}} \chi_{[0,1)}\|_{\Lambda_{\psi_E}}. \tag{6.25}$$

Now suppose that $0 \leq x \in L_1 \cap L_{q,\infty}$. We set

$$\mu(x) = \mu(x)\chi_{[0,1)} + \mu(x)\chi_{[1,\infty)}. \tag{6.26}$$

Observe first that we have the pointwise estimate

$$\mu(t;x)\chi_{(0,1)}(t) \leq (\sup_{0<t\leq 1} \mu(t;x)t^{1/q})t^{-1/q} = \|x\|_{L_{q,\infty}} t^{-1/q}, \quad 0 < t \leq 1.$$

Since $t^{-1/q}\chi_{[0,1)}(t) \in E$, it follows that $\mu(x)\chi_{[0,1)} \in E$ and

$$\|\mu(x)\chi_{0,1}\|_E \leq \|x\|_{L_{q,\infty}} \|t^{-\frac{1}{q}}\|_E. \tag{6.27}$$

Noting that $\mu(x)\chi_{[1,\infty)} \leq \mu(1;x)$, it follows that $\mu(x)\chi_{[1,\infty)} \in L_1 \cap L_\infty \subseteq E$. Moreover,

$$\begin{aligned}\|\mu(x)\chi_{[1,\infty)}\|_E &\leq \|\mu(x)\chi_{[1,\infty)}\|_{L_1 \cap L_\infty} = \max(\mu(1;x), \|\mu(x)\chi_{[1,\infty)}\|_{L_1}) \\ &\leq \max(\|\mu(x)\chi_{[0,1)}\|_{L_{q,\infty}}, \|\mu(x)\chi_{[1,\infty)}\|_{L_1}) \leq \|x\|_{L_{q,\infty} \cap L_1},\end{aligned}$$

and from (6.26) and (6.27), it follows that $\mu(x) \in E$, and hence also $x \in E$, and

$$\|x\|_E = \|\mu(x)\|_E \leq (\|t^{-\frac{1}{q}}\|_E + 1)\|x\|_{L_1 \cap L_{q,\infty}},$$

and this completes the proof of the Lemma. □

We may now state the following consequence of Theorem 6.1 which is an exact noncommutative extension of [1] Theorem 3.

Corollary 6.4 *Let* $E \subseteq S(m)$ *be a fully symmetric space. If* $1 \leq q_E < q$, *then the pair* $(E(\sigma), E(\tau))$ *is an interpolation pair for the pair* $(L_1(\sigma), L_q(\sigma))$, $(L_1(\tau), L_q(\tau))$ *with interpolation constant depending only on* q *and* E.

Proof It follows from Theorem 6.1 that $E(\sigma) \subseteq L_1(\sigma) + L_q(\sigma)$ since $L_{q,1}(\sigma) \subseteq L_q(\sigma)$, with the same statement holding for $E(\tau)$. Since $1 \leq q_E < q$, the embedding $L_1 \cap L_{q,\infty} \subseteq E$ follows from Lemma 6.3. Since $L_q \subseteq L_{q,\infty}$, it follows also that $L_1 \cap L_q \subseteq E$. The desired inclusion

$$L_1(\sigma) \cap L_q(\sigma) \subseteq E(\sigma)$$

follows by applying [11] Theorem 3.2, with a similar statement for $E(\tau)$. The corollary now follows from Theorem 6.1 since any mapping of strong types $(1, 1)$ and (q, q) is necessarily of strong type $(1, 1)$ and weak type (q, q). □

It is worth noting that the assumption in Theorem 6.1 that E is fully symmetric cannot be omitted. An explicit counter-example is given in [1].

References

1. S.V. Astashkin, L. Maligranda, Interpolation between L_1 and L_p, $1 < p < \infty$. Proc. Amer. Math. Soc. **132**, 2929–2938 (2004)
2. C. Bennett, R. Sharpley, *Interpolation of Operators* (Academic, Cambridge, 1988)
3. J. Bergh, J. Löfström, *Interpolation Spaces*, vol. 223 (Springer, Berlin, 1976)
4. D.W. Boyd, Indices of function spaces and their relationship to interpolation. Canad. J. Math. **21**, 1245–1254 (1969)
5. A.P. Calderón, Spaces between L^1 and L^∞ and the theorem of Marcinkiewicz. Studia Math. **26**, 273–299 (1966)
6. S. Dirksen, Noncommutative and vector-valued Rosenthal inequalities. Thesis, Delft University of Technology, 2011
7. S. Dirksen, Noncommutative Boyd interpolation theorems. Trans. Am. Math. Soc. **367**, 4079–4110 (2015)
8. P.G. Dodds, B. dePagter, Normed Köthe spaces: a non-commutative viewpoint. Indag. Math. **25**, 206–249 (2014)
9. P.G. Dodds, T.K. Dodds, B. de Pagter, Non-commutative Banach function spaces. Math. Z. **201**, 583–597 (1989)
10. P.G. Dodds, T.K. Dodds, B. de Pagter, Remarks on non-commutative interpolation. Proc. CMA (ANU) **24**, 58–78 (1989)
11. P.G. Dodds, T.K. Dodds, B. de Pagter, Fully symmetric operator spaces. Integr. Equ. Oper. Theory **15**, 942–972 (1992)
12. P.G. Dodds, T.K. Dodds, B. de Pagter, Noncommutative Köthe duality. Trans. Am. Math. Soc. **339**, 717–750 (1993)
13. P.G. Dodds, B. dePagter, F.S. Sukochev, Noncommutative integration (forthcoming)
14. T. Fack, H. Kosaki, Generalized s-numbers of τ-measurable operators. Pac. J. Math. **123**, 269–300 (1986)

15. S.G. Krein, J.I. Petunin, E.M. Semenov, Interpolation of linear operators, translations of mathematical monographs. Amer. Math. Soc. **54**, 375 (1982)
16. J. Lindenstrauss, L. Tzafriri, *Classical Banach Spaces II* (Springer, Berlin, 1979)
17. L. Maligranda, A generalization of the Shimogaki interpolation theorem. Studia Math. **71**, 69–83 (1981)
18. S. Montgomery-Smith, *The Hardy Operator and Boyd Indices, Interaction Between Functional Analysis, Harmonic Analysis and Probability (Columbia, MO, 1994)*. Lecture Notes in Pure and Applied Math., vol. 175 (Dekker, New York, 1996), pp. 359–364
19. B.Z.A. Rubstein, G.Y. Grabarnik, M.A. Muratov, Y.S. Pashkova, *Foundations of Symmetric Spaces of Measurable Functions* (Springer, Basel, 2016)
20. M. Takesaki, *Theory of Operator Algebras I* (Springer, New York, 1979)

Strict Singularity: A Lattice Approach

Julio Flores, Francisco L. Hernández, and Pedro Tradacete

Dedicated to Ben de Pagter on the occasion of his 65th birthday

Abstract Given a Banach lattice E and a Banach space Y we say that a bounded linear operator $T : E \to Y$ is lattice strictly singular (disjointly strictly singular) if it fails to be invertible on any infinite-dimensional sublattice of E (on the span of any pairwise disjoint sequence in E). This is a survey on the existing answers up to the present day to the following questions: Is every lattice strictly singular operator also disjointly strictly singular? Do lattice strictly singular operators have a vector space structure?

Keywords Disjointly strictly singular operator · Lattice strictly singular operator · Banach lattice · Unconditional basic sequence

---

J. Flores (✉)
Dpto. Matemática Aplicada, Ciencia e Ingeniería de los Materiales y Tecnología Electrónica, ESCET, Universidad Rey Juan Carlos, Móstoles, Madrid, Spain
e-mail: julio.flores@urjc.es

F. L. Hernández
IMI and Dpto. de Análisis Matemático y Matemática Aplicada, Universidad Complutense, Madrid, Spain
e-mail: pacoh@mat.ucm.es

P. Tradacete
Instituto de Ciencias Matemáticas (CSIC-UAM-UC3M-UCM), Consejo Superior de Investigaciones Científicas, Madrid, Spain
e-mail: pedro.tradacete@icmat.es

G. Buskes et al. (eds.), *Positivity and Noncommutative Analysis*,
Trends in Mathematics, https://doi.org/10.1007/978-3-030-10850-2_8

1 Introduction

Recall that a linear operator between Banach spaces $T : Z \to X$ is *strictly singular* (SS) if it is not an isomorphism when restricted to any infinite-dimensional (closed) subspace. This well-known class was introduced by Kato [14] in connection with the perturbation theory of Fredholm operators. These operators form a closed two-sided operator ideal which contains that of compact operators, and have become a relevant tool in the modern theory of Banach spaces due to their role in connection with hereditarily indecomposable spaces.

When the domain Z is in particular a Banach lattice it is natural to expect that the additional lattice structure, which interacts with the underlying topology, can exhibit some features of this class of operators. The characterizations of strict singularity given in [5] strongly expose the interplay mentioned above, showing that strict singularity is in many cases "composed" by two simpler properties. Loosely speaking, for a large class of Banach lattices, if an operator is invertible on some infinite-dimensional subspace, then it must be invertible on a subspace which is either isomorphic to ℓ_2 or spanned by a pairwise disjoint sequence. Clearly, here is where the lattice structure comes in. Hence, when checking for strict singularity, there are basically two types of subspaces to look at, one of them being clearly defined in terms of the lattice structure. Consequently, operators which fail to be invertible on subspaces spanned by pairwise disjoint sequences appear naturally when dealing with strictly singular operators defined on a Banach lattice.

Given a Banach lattice E and a Banach space X, a bounded operator $T : E \to X$ will be called *disjointly strictly singular* (DSS) if there is no infinite sequence $(x_n)_n$ of pairwise disjoint vectors in E such that the restriction of T to its closed linear span $[x_n]$ is an isomorphism. This class was first introduced by the second author and Rodríguez-Salinas [11] in connection with ℓ_p complemented subspaces of Orlicz function spaces.

It is important to observe that although every infinite-dimensional sublattice of a Banach lattice E contains a sequence of disjoint vectors (which can be chosen to be positive), the linear span of a sequence of non-positive disjoint vectors will not be in general a sublattice of E. This remark justifies the reason to consider the class of operators which fail to be invertible on infinite-dimensional sublattices of E.

Namely, an operator $T : E \to X$ will be called *lattice strictly singular* (LSS) if for every infinite-dimensional (closed) sublattice $H \subset E$ the restriction $T|_H$ is not an isomorphism. Notice that the class of DSS operators is a left ideal [10] (but not a right ideal). Similarly, the composition (from the left) of an LSS operator with a bounded operator is clearly LSS. Notice also from the previous remark that $T : E \to X$ is LSS if and only if there is no infinite sequence $(x_n)_n$ of *positive* pairwise disjoint vectors in E such that the restriction of T to the span $[x_n]$ is an isomorphism. Hence every DSS operator is in particular LSS. Thus, we have the following inclusions:

$$SS \subset DSS \subset LSS$$

In general, the converse of the first inclusion is not true. For example, the inclusion $i : L_p(\mu) \to L_q(\mu)$ for $p > q$—μ a finite measure—is disjointly strictly singular but not strictly singular: Indeed, it cannot be an isomorphism on the span of any disjoint sequence in $L_p(\mu)$, as these subspaces are isomorphic to ℓ_p; but, by Khintchine's inequality, i is an isomorphism on the subspace generated by the Rademacher functions (which is isomorphic to ℓ_2).

Naturally, the following question arises:

(Q1) Is every LSS operator also DSS?

An additional motivation for this question stems from the following observation in [9]: If an operator $T : E \to X$ is invertible on a subspace isomorphic to c_0 generated by a disjoint sequence, then T is also invertible in a sublattice isomorphic to c_0. In particular, this implies the following preliminary result: For every compact Hausdorff space K and every Banach space X, an operator $T : C(K) \to X$ is LSS if and only if it is DSS.

This paper constitutes a survey with some partial answers to this question. The presentation will rather follow a chronological order. The reader is referred to [2, 17, 18] for any unexplained notions in Banach lattice theory.

2 Positive and Regular Operators

In [4], the first two authors considered the problem of domination for positive DSS operators. A closer look to the proof of [4, Prop. 2.4] actually renders the following:

Proposition 2.1 *Let E and F be Banach lattices, with F order continuous. A positive operator $T : E \to F$ is DSS if and only if it is LSS.*

All technicalities aside, the argument is very simple. Assume that T is not DSS; then, for some sequence (x_n) of pairwise disjoint (not necessarily positive) normalized vectors and some $\alpha > 0$, we have

$$\|\sum_n a_n T x_n\| \geq \alpha \|\sum_n a_n x_n\|$$

for every scalars (a_n). From this estimate, we would like to get something like

$$\begin{aligned} \|\sum_n a_n T|x_n|\,\| &\underset{(*)}{\geq} \beta \|\sum_n |a_n|\, T|x_n|\,\| \\ &\underset{(**)}{\geq} \beta \|\sum_a a_n T x_n\| \end{aligned}$$

$$\geq \alpha \| \sum_n a_n x_n \|$$

$$\underset{(***)}{=} \alpha \| \sum_n a_n |x_n| \|$$

as this would imply that T is not LSS.

Remark 2.2 Notice that in the previous chain of inequalities, $(*)$ follows provided $(T|x_n|)_n$ is unconditional; also, $(**)$ follows as T is positive, while $(***)$ is a consequence of the fact that (x_n) are pairwise disjoint.

We would like to emphasize that these two ingredients, unconditionality and positivity, are at the core of the argument. In what follows, it will become clear that unconditionality plays in fact a much more critical role than positivity.

In an attempt to extend the above argument at least to the class of regular operators (i.e., differences of positive operators) one might argue as follows. Assume first that T is regular and LSS and that $|T|$ is also LSS. In that case, under the assumptions of Proposition 2.1, $|T|$ would also be DSS. Thus, both $T^+ and T^-$ would be DSS according to the domination theorem given in [4]. But then T would also be DSS since DSS operators have a vector space structure. Unfortunately, even for compact T (which in particular implies T LSS), it is not true in general that $|T|$ is LSS (this was observed in [3, Example 4.1]); still, what becomes evident is that the vector space structure of the class of DSS operators plays some role in the argument.

At this point, one might be tempted to take the vector space structure of LSS operators for granted. However, the proof that the sum of two DSS operators is also DSS (see [10]), which in turn follows Kato's proof for the corresponding statement about strictly singular operators, is based on the fact that one can find subspaces spanned by certain block sequences of suitable sequences where the restriction of both the operators is compact (see [16]). This argument evidences an insurmountable obstacle when dealing with LSS operators. The reason is that a normalized block sequence of a normalized pairwise disjoint sequence (x_n) of positive elements is still disjoint but its span may not be a sublattice unless all block coefficients are positive.

Proposition 2.3 *Let E and F be Banach lattices such that E' and F are order continuous and let $T : E \to F$ be a regular operator. The following statements are equivalent:*

(i) T is LSS.
(ii) Every infinite-dimensional sublattice $H \subset E$ has some infinite-dimensional sublattice $R \subset H$ such that the restriction $T|_R$ is compact.

If T is moreover, positive, the above statements are also equivalent to

(iii) T is DSS.

The proof of the proposition above (see [3, 4]) involves the classical Kadeč–Pełczyński's Dichotomy [13]: For an order continuous Banach lattice E with weak unit, let $i : E \hookrightarrow L_1(\mu)$ denote the inclusion given by Lindenstrauss and Tzafriri [17, Theorem 1.b.14]; given any sequence (x_n) in E, either

1. There is $C > 0$ such that $\|ix_n\|_1 \leq \|x_n\|_E \leq C\|ix_n\|_1$, for every $n \in \mathbb{N}$; or
2. There exist $(n_k) \subset \mathbb{N}$ and a pairwise disjoint sequence $(y_k) \subset E$ such that $\|x_{n_k} - y_k\|_E \to 0$.

Examples provided in [3, Examples 2.6 and 2.7] show that neither the order continuity of E' nor F can be avoided in the previous result. Also from (ii), it clearly follows that all regular LSS operators between E and F form a vector space provided that both E' and F have order continuous norms. The vector space structure of LSS operators can also be derived from that of DSS operators in other cases as in the following [3, Prop. 4.2]:

Theorem 2.4 *Let $T : E \to F$ be a regular LSS operator between Banach lattices. If any of the following conditions hold, then T is DSS:*

1. *E is an L-space and F is a KB-space.*
2. *$E = L_p(\mu)$ and $F = L_q(\mu)$, $1 \leq p \leq \infty$, $1 \leq q < \infty$.*

The previous discussion naturally leads to the somehow unexpected question:

(Q2) Does the LSS operator class have a vector space structure?

Note that there is an evident connection between questions (Q1) and (Q2) insofar a positive answer to (Q1) immediately provides a positive answer to (Q2). What is less evident is that the converse implication actually holds. We will postpone this discussion to Sect. 5.

3 LSS and DSS Classes Coincide at the Local Level

We begin this section by noting that the regularity of the operators will no longer be required.

Let us recall the following finite-dimensional version of strictly singular operators introduced by Mityagin and Pełczyński in [19], the so-called super-strictly singular operators. Given X and Y Banach spaces, an operator $T : X \to Y$ is not super-strictly singular if there exist a sequence $(E_n)_n$ of subspaces of E, where each E_n is of dimension n, and a constant $C > 0$ such that $\|Tx\| \geq C\|x\|$ for every $x \in \bigcup_{n=1}^{\infty} E_n$. Of course, T is super-strictly singular when the above does not occur. Obviously, every super-strictly singular operator is strictly singular. Note that an operator $T : X \to Y$ is super-strictly singular if and only if for every free ultrafilter $\mathcal{U}$, the extension to ultrapowers $T_{\mathcal{U}} : X_{\mathcal{U}} \to Y_{\mathcal{U}}$ is strictly singular.

In [6], the first two authors together with Y. Raynaud considered the super-versions of DSS and LSS operators, respectively, as the lattice counterpart of the super-strictly singular operators:

Definition 3.1 Let E be a Banach lattice, X be a Banach space and $T : E \to X$ a bounded operator. We say that $T : E \to X$ is *super-DSS* (respectively, *super-LSS*) if for every $\varepsilon > 0$ there exists $N \in \mathbb{N}$ such that for each sequence $(x_n)_{n=1}^N$ of disjoint elements in E (respectively, positive disjoint), there is $x \in [x_n]_{n=1}^N$ with $\|Tx\| < \varepsilon\|x\|$.

Notice that T is not super-disjointly strictly singular (respectively, super-lattice strictly singular) when there is a sequence $(E_n)_n$ of subspaces of E, with each E_n being the linear span of n pairwise disjoint (respectively, positive pairwise disjoint) vectors in E, and a constant $C > 0$ such that $\|Tx\| \geq C\|x\|$ for every $x \in \bigcup_{n=1}^{\infty} E_n$.

It is clear that every super-DSS operator is DSS, and that every super-LSS operator is in turn LSS. The pertinence of considering these classes of operators in connection to question (Q1) lies in the following result given in [6, Prop. 3.4] that mimics what happens to super-SS operators. Precisely, an operator $T : E \to X$ is super-DSS (respectively, super-LSS) if and only if for every free ultrafilter $\mathcal{U}$, the extension $T_{\mathcal{U}} : E_{\mathcal{U}} \to X_{\mathcal{U}}$ is DSS (respectively, LSS).

Thus, it becomes natural to consider question (Q1) at the super level and ask whether T super-DSS and T super-LSS are equivalent conditions. Indeed, note that a positive answer to (Q1) immediately means that T super-DSS and T super-LSS are equivalent notions too. From the opposite side, if T super-LSS does not imply T super-DSS, then the answer to our (Q1) must be negative. But in [6], the following was proved:

Proposition 3.2 *Let E be a Banach lattice and Y be a Banach space. An operator $T : E \to Y$ is super-DSS if and only if T is super-LSS.*

Note that this result implies that every "ultraoperator" $T_{\mathcal{U}}$ stemming from the bounded operator $T : E \to Y$ is LSS if and only it is DSS.

The proof of Proposition 3.2 involves several ingredients. Especially relevant is the well-known theorem of Krivine of finite representability of ℓ_p, $1 \leq p \leq \infty$, in an arbitrary normalized basic sequence, which basically allows us to assume that, in the ultrapower, this sequence is equivalent to the unit basis of ℓ_p. Equally relevant is Brunel–Sucheston's result regarding unconditionality in spreading models, which brings Remark 2.2 up.

The same proof actually extends for an LSS operator $T : E \to Y$ when E and Y are stable Banach spaces (in the sense of [15]) and Y has an unconditional basis. Indeed, when the spaces involved are stable, the pairwise disjoint sequence (x_n), on whose span the operator T is invertible, can be taken to be equivalent to the unit basis of ℓ_p for some $1 \leq p < \infty$. From here we can proceed as in the proof of the previous proposition to get the following:

Proposition 3.3 *Let E be a stable Banach lattice and Y be a stable Banach space with an unconditional finite-dimensional decomposition. A bounded operator $T : E \to Y$ is LSS if and only if it is DSS.*

In the next section, we will show that a further step can be taken removing the hypothesis of stability on both spaces. In contrast, unconditionality will remain an issue of capital importance as anticipated in Remark 2.2.

4 More General Results

This section collects the most general answers known to the authors to questions (Q1) and (Q2). It contains some of the material presented in [7] by the first and third authors together with López-Abad.

Let us start by introducing some terminology. Let E be a Banach lattice, Y a Banach space and X be a Banach space with an unconditional basis (hence, an atomic Banach lattice). An operator $T : E \to Y$ is X-DSS (respectively, X-LSS) if it is never an isomorphism when restricted to a subspace of E generated by a pairwise disjoint sequence (respectively, a sublattice of E) which is isomorphic to X.

As mentioned in the introduction, one of the motivations to our problem refers to the coincidence between c_0-DSS and c_0-LSS operators. Interestingly enough, an analogous situation holds for ℓ_1 as is collected in the next result (see [7, Proposition 2.1]).

Proposition 4.1 *Let E be a Banach lattice, X a Banach space and $T : E \to X$ a bounded operator.*

1. *T is c_0-DSS if and only if it is c_0-LSS.*
2. *T is ℓ_1-DSS if and only if it is ℓ_1-LSS.*

Interestingly, this result shows that from the point of view of convexity, or lattice indexes [17], the two extreme cases (1 and ∞) behave well in connection to our problem. We remark also that no conditions whatsoever are required on the spaces or the operator, which makes this a fairly general result.

At this point, a natural question arises: For $1 < p < \infty$, is every ℓ_p-LSS operator also ℓ_p-DSS? This is partly motivated by the proof of Proposition 3.2 in connection with Krivine's theorem. No answer to this question is known to the authors.

Notice that Proposition 4.1 together with James theorem (cf. [16, Theorem 1.c.12]) yields the following corollary (notice that here (x_n) is unconditional):

Corollary 4.2 *Let E be a Banach lattice, X a Banach space and $T : E \to X$ an LSS operator. Suppose there exists a disjoint sequence (x_n) in E such that the restriction $T|_{[x_n]}$ is an isomorphism. Then $[x_n]$ is reflexive.*

The following lemma [7, Lemma 2.3] has a simple proof based on a general argument of perturbation for basic sequences [16, Proposition 1.a.9], but lies at the very core of our main result. It refers to the behaviour of blocks with positive coefficients built on a disjoint sequence on whose span T is invertible. As customary notation $(x_n) \sim (y_n)$ means that (x_n) and (y_n) are equivalent basic sequences, that is for some constant $C \geq 1$, and any scalars (a_n), one has

$$\frac{1}{C}\|\sum_n a_n x_n\| \leq \|\sum_n a_n y_n\| \leq C\|\sum_n a_n x_n\|.$$

Lemma 4.3 *Let E be a Banach lattice, X a Banach space, and $T : E \to X$ an LSS operator. Suppose that (x_n) is a normalized sequence of disjoint vectors in E such that $(Tx_n) \sim (x_n)$. Then every normalized block sequence of (x_n) with positive coefficients $a_j \geq 0$*

$$y_n = \sum_{j \in A_n} a_j x_j$$

satisfies

$$(y_n) \sim (\sum_{j \in A_n} a_j x_j^+) \sim (\sum_{j \in A_n} a_j x_j^-).$$

Lemma 4.3 comes in handy in the following argument. Assume that $T : E \to X$ is LSS but not DSS; thus, there exists a pairwise disjoint normalized sequence (x_n) such that $(Tx_n) \sim (x_n)$. Note that by Corollary 4.2, (x_n) can be assumed to be weakly null, and more importantly, by Lemma 4.3 every normalized block sequence (y_n) of (x_n) with positive coefficients satisfies

$$(y_n) \sim (y_n^+) \sim (y_n^-).$$

Since $\|x_n\| = 1$, we can assume that both (x_n^+) and (x_n^-) are seminormalized and weakly null sequences. Suppose next that $(Tx_n^+) \nsim (x_n^+)$. Since T is bounded, for every $n \in \mathbb{N}$ there exists some finite collection of coefficients $\{b_j^n, j \in S_n \subset \mathbb{N} \text{ finite}\}$ such that $\|\sum_{j \in S_n} b_j^n x_j^+\| = 1$, while $\|\sum_{j \in S_n} b_j^n T x_j^+\| \to 0$. Hence, there is a subsequence of the block sequence $y_n = \sum_{j \in S_n} b_j^n x_j$ such that

$$(Ty_n) \sim (Ty_n^-).$$

If the coefficients (b_j^n) in (y_n) were positive, then we would have $(y_n) \sim (y_n^-)$ by Lemma 4.3 and thus T would be invertible on the span of (y_n^-) which is a contradiction. Clearly, this can be achieved if (Tx_n^+) (or some subsequence of it) were unconditional.

At this point, let us recall that a Banach space X is said to have the *unconditional subsequence property* (*USP* in short, cf. [20]) if every weakly null sequence has an unconditional subsequence. This is the case for instance when X embeds in a space with an unconditional basis or when it has an unconditional finite-dimensional decomposition.

The following theorem, given in [7], summarizes the previous discussion.

Theorem 4.4 *Let E be a Banach lattice and X a Banach space with the USP. An operator $T : E \to X$ is LSS if and only if it is DSS.*

We make the important remark that USP here is a sufficient but not necessary condition (see Remark 4.9 below).

It is pertinent to observe that if in Theorem 4.4 the target space X is a Banach lattice, then some convexity argument yields a similar statement, as Theorem 4.6 below shows. This follows from a well-known factorization Theorem due to Maurey and Nikishin (cf. [1, Theorem 7.18]):

Theorem 4.5 (Maurey–Nikishin) *Let X be a Banach space of type $r > 1$. Suppose that $1 \leq p < r$ and that $T : X \to L_p(\mu)$ is an operator. Then T factors through $L_q(\mu)$ for any $p < q < r$. More precisely, for each $p < q < r$ there is a strictly positive density function on Ω so that $Sx = h^{-\frac{1}{p}}Tx$ defines a bounded operator from $L_p(\mu)$ into $L_q(\Omega, hd\mu)$.*

Note that if $i : L_q(hd\mu) \to L_p(hd\mu)$ is the natural inclusion and $j : L_p(hd\mu) \to L_p(\mu)$ is the isometric isomorphism defined by $j(f) = h^{-\frac{1}{p}}f$, then the above operator T satisfies $T = jiS$ (see [1, p. 167]).

Non-trivial type, which is a well-known local property in Banach space theory, turns out to be the key to avoid USP property on the range space in Theorem 4.4. The idea is simple. If $(Tx_n) \sim (x_n)$ for some disjoint sequence, then $[Tx_n]$ must be reflexive according to Proposition 4.1 and thus (Tx_n) will be weakly null. In addition, by Lemma 4.3, we have

$$(x_n) \sim (x_n^+) \sim (x_n^-) \sim (Tx_n).$$

If, additionally, $[x_n]$ has non-trivial type, then the restriction of T to $[x_n^+]$ will factor by Maurey–Nikishin's theorem through some L_q-space with $1 < q$ which will provide unconditionality for (Tx_n^+) for free. Some playing around with the Kadeč–Pełczyński's Dichotomy will do the rest. We refer the reader to [7] for the details.

Theorem 4.6 *Let E and F be Banach lattices, with F order continuous. If an LSS operator $T : E \to F$ is invertible on the span of a disjoint sequence (x_n) in E, then $[x_n]$ contains ℓ_1^n's uniformly.*

As a consequence of this result we get:

Corollary 4.7 *Let E and F be Banach lattices, and $T : E \to F$ an LSS operator. T is DSS under any of the following:*

1. *F has non-trivial type.*
2. *E has non-trivial type and F is order continuous.*

On the other hand the well-known fact that a subspace of $L_1(\mu)$ with trivial type contains a subspace isomorphic to ℓ_1 ([21], see also [22, III.C.Theorem 12]) along with Proposition 4.1 can be exploited to conclude the following:

Theorem 4.8 *Let E be a Banach lattice and $T : E \to L_1(\mu)$ an operator. Then T is LSS if and only if T is DSS.*

Remark 4.9 Theorem 4.8 shows that non-trivial type is not a necessary condition in Corollary 4.7 and, in fact, it also shows that USP condition in Theorem 4.4 is not necessary either. Indeed, notice that, for non-atomic measures μ, the space $L_1(\mu)$ is not isomorphic to a subspace of a space with an unconditional basis (cf. [16, Proposition 1.d.1]). In fact, there exist weakly null sequences in $L_1(\mu)$ without unconditional subsequences [12]. Thus, Theorem 4.8 cannot be a corollary of Theorem 4.4.

Another consequence of Theorem 4.6 that was anticipated above is the following [7, Corollary 3.6]:

Corollary 4.10 *Let E and F be Banach lattices with F order continuous. If either E or F is stable, then every LSS operator $T : E \to F$ is DSS.*

We also have the following [7, Corollary 3.5]:

Corollary 4.11 *Let F be an order continuous Banach lattice such that for every pairwise disjoint sequence (f_n) in F there is an infinite-dimensional subspace $Y \subset [f_n]$ which is complemented in F. Then every LSS operator $T : E \to F$ is DSS.*

See Remark 5.5 for examples of spaces on which these corollaries can be applied.

5 The Vector Space Structure of LSS Operators

In this section we will go back to (Q2) and show that it is in fact equivalent to (Q1). The results given summarize some of the material in [7].

It was mentioned in the introduction that DSS operators have a vector space structure. In fact it is very easy to prove the following:

Lemma 5.1 *The sum of an LSS operator with a DSS operator is an LSS operator.*

Much more involved is the proof of the following main result:

Theorem 5.2 *Let X be a Banach space. The following are equivalent:*

(a) *There is a Banach lattice E and an LSS operator $T : E \to X$ which is not DSS.*
(b) *There is a Banach lattice E and two LSS operators $T_1, T_2 : E \to X$ such that $T_1 + T_2$ is not LSS.*

The reader is referred to [7] for the precise statements and rather technical details.

There is a main difficulty underlying Theorem 5.2 which can be formulated in the following general question:

Suppose that (x_n) *is a sequence in a Banach space which can be written as* $x_n = y_n + z_n$*; what can be said about the properties of* (y_n) *and* (z_n) *in terms of those of* (x_n)*?*

In particular, in the context of question (Q1) we are faced with the problem of finding a certain kind of unconditional blocks in decompositions of an unconditional sequence (x_n). Thus, a closer inspection to the proof of Theorem 5.2 motivates the definition of Banach spaces with the *unconditional decomposition property.*

Definition 5.3 A Banach space X has the unconditional decomposition property (UDP) if for every unconditional sequence (z_n) in X which can be written as $z_n = x_n + y_n$ for some (x_n), (y_n) in X and such that

$$\sum_n a_n z_n \sim \max\left\{\sum_n a_n x_n, \sum_n a_n y_n\right\},$$

then either (x_n) or (y_n) has an unconditional positive block sequence.

The proof of Theorem 5.2 in fact reveals the following:

Corollary 5.4 *A Banach space* X *has the UDP if and only if for every Banach lattice* E*, every LSS operator* $T : E \to X$ *is DSS.*

At this junction questions (Q1) and (Q2) can be reformulated into one:

(Q3) Does every Banach space have the *UDP*?

Remarkably enough questions (Q1) and (Q2) which were originally formulated in a Banach *lattice* setting have thus been shown to be equivalent to a Banach *space* property.

Notice also that Theorem 4.4 yields that every Banach space with the *USP* has the *UDP*. Trivially, hereditarily indecomposable Banach spaces have the *UDP*. Note also that having the *UDP* is a hereditary property: every Banach space which embeds in a Banach space with the *UDP* has the *UDP*. It is also easy to check that the direct sum of a finite number of spaces with the *UDP* also has the *UDP*.

Remark 5.5 It should be noted that all the spaces arising in the previous section become immediately examples of spaces with the UDP property. Corollary 4.7 yields that every Banach lattice with non-trivial type has the UDP. Also, from Corollary 4.11 it follows that every Banach lattice with the positive Schur property (e.g., Lorentz spaces Λ_φ) or more generally, every disjointly subprojective Banach lattice (that is, a Banach lattice in which every disjoint sequence contains a complemented subsequence) must have UDP. Analogously, from Corollary 4.10 it follows that order continuous stable Banach lattices have the *UDP*. And since, by a result of [8], every Orlicz space L_φ, with an Orlicz function φ satisfying the Δ_2 condition, is stable and order continuous, in particular, these spaces also have the *UDP*.

6 Conclusion

Although unconditionality was shown to play an important role it cannot be expected to be a necessary condition for a positive answer to (Q1) as Theorem 4.8 shows.

The reader may have noticed by now that question (Q1) remains open in its more general form. No example of an LSS operator which is not DSS is known and this means that (Q2) and hence (Q3) remain equally open. The results above show where not to look for counterexamples.

Acknowledgements Support by the Spanish Government Grant No. MTM2016-76808-P is gratefully acknowledged. The third author also acknowledges the financial support from the Spanish Ministry of Economy and Competitiveness through Grant No. MTM2016-75196-P and the "Severo Ochoa Programme for Centres of Excellence in R&D" (SEV-2015-0554).

References

1. F. Albiac, N.J. Kalton, Topics in Banach space theory, in *Graduate Texts in Mathematics*, vol. 233 (Springer, New York, 2006)
2. C. Aliprantis, O. Burkinshaw, *Positive Operators* (Springer, Dordrecht, 2016)
3. J. Flores, Some remarks on disjointly strictly singular operators. Positivity **9**(3), 385–396 (2005)
4. J. Flores, F.L. Hernández, Domination by positive disjointly strictly singular operators. Proc. Am. Math. Soc. **129**(7), 1979–1986 (2001)
5. J. Flores, F.L. Hernández, N.J. Kalton, P. Tradacete, Characterizations of strictly singular operators on Banach lattices. J. Lond. Math. Soc. **79**(3), 612–630 (2009)
6. J. Flores, F.L. Hernández, Y. Raynaud, Super strictly singular and cosingular operators and related classes. J. Oper. Theory **67**(1), 121–152 (2012)
7. J. Flores, J. López-Abad, P. Tradacete, Banach lattice versions of strict singularity. J. Funct. Anal. **270**(7), 2715–2731 (2016)
8. D.J.H. Garling, Stable Banach spaces, random measures and Orlicz function spaces, in *Probability Measures on Groups*. Lecture Notes in Math, vol. 928 (Springer, Berlin, 1981), pp. 121–175
9. N. Ghoussoub, W.B. Johnson, Factoring operators through Banach lattices not containing $C(0, 1)$. Math. Z. **194**(2), 153–171 (1987)
10. F.L. Hernández, Disjointly strictly-singular operators in Banach lattices. Proc. 18 Winter School on Abstract Analysis (Srni). Acta Univ. Carolin.-Math. Phys. **31**, 35–40 (1990)
11. F.L. Hernández, B. Rodríguez-Salinas, On ℓ_p complemented copies in Orlicz spaces II. Israel J. Math. **68**(1), 27–55 (1989)
12. W.B. Johnson, B. Maurey, G. Schechtman, Weakly null sequences in L_1. J. Am. Math. Soc. **20**(1), 25–36 (2007)
13. M.I. Kadeč, A. Pełczyński, Bases, lacunary sequences and complemented subspaces in the spaces L_p. Studia Math. **21**, 161–176 (1962)
14. T. Kato, Perturbation theory for nullity deficiency and other quantities of linear operators. J. Anal. Math. **6**, 273–322 (1958)
15. J.L. Krivine, B. Maurey, Espaces de Banach stables. Israel J. Math. **39**(3), 273–281 (1981)
16. J. Lindenstrauss, L. Tzafriri, *Classical Banach Spaces I: Sequence Spaces* (Springer, Berlin, 1977)

17. J. Lindenstrauss, L. Tzafriri, *Classical Banach Spaces II: Function Spaces* (Springer, Berlin, 1979)
18. P. Meyer-Nieberg, *Banach Lattices* (Springer, Berlin, 1991)
19. B. Mityagin, A. Pełczyński, Nuclear operators and approximative dimensions, in *Proceedings of International Congress of Mathematicians Moscow* (1966), pp. 366–372
20. E. Odell, B. Zheng, On the unconditional subsequence property. J. Funct. Anal. **258**(2), 604–615 (2010)
21. H.P. Rosenthal, On subspaces of L_p. Ann. Math. **97**(2), 344–373 (1973)
22. P. Wojtaszczyk, *Banach Spaces for Analysts*. Cambridge Studies in Advanced Mathematics, vol. 25 (Cambridge University Press, Cambridge, 1991)

Asymptotics of Operator Semigroups via the Semigroup at Infinity

Jochen Glück and Markus Haase

Dedicated to Ben de Pagter on the occasion of his 65th birthday

Abstract We systematize and generalize recent results of Gerlach and Glück on the strong convergence and spectral theory of bounded (positive) operator semigroups $(T_s)_{s\in S}$ on Banach spaces (lattices). (Here, S can be an arbitrary commutative semigroup, and no topological assumptions neither on S nor on its representation are required.) To this aim, we introduce the "semigroup at infinity" and give useful criteria ensuring that the well-known Jacobs–de Leeuw–Glicksberg splitting theory can be applied to it.

Next, we confine these abstract results to positive semigroups on Banach lattices with a quasi-interior point. In that situation, the said criteria are intimately linked to the so-called AM-compact operators (which entail kernel operators and compact operators); and they imply that the original semigroup asymptotically embeds into a compact group of positive invertible operators on an atomic Banach lattice. By means of a structure theorem for such group representations (reminiscent of the Peter–Weyl theorem and its consequences for Banach space representations of compact groups) we are able to establish quite general conditions implying the strong convergence of the original semigroup.

Finally, we show how some classical results of Greiner (Sitzungsberichte der Heidelberger Akademie der Wissenschaften, Mathematisch-Naturwissenschaftliche Klasse, 55–80, 1982), Davies (J Evol Equ 5(3), 407–415, 2005), Keicher (Arch Math (Basel) 87(4), 359–367, 2006) and Arendt (Positivity 12(1), 25–44, 2008) and more recent ones by Gerlach and Glück (Convergence of positive operator semigroups, 2017) are covered and extended through our approach.

J. Glück
Ulm University, Institute of Applied Analysis, Ulm, Germany
e-mail: jochen.glueck@alumni.uni-ulm.de

M. Haase (✉)
Kiel University, Mathematisches Seminar, Kiel, Germany
e-mail: haase@math.uni-kiel.de

G. Buskes et al. (eds.), *Positivity and Noncommutative Analysis*,
Trends in Mathematics, https://doi.org/10.1007/978-3-030-10850-2_9

Keywords Convergence of operator semigroups · Jacobs–de Leeuw–Glicksberg theory · Positive semigroup representations · Positive group representations · AM-compact operators · Kernel operators · Triviality of the peripheral point spectrum

1 Introduction

In this paper we deal with the problem of finding useful criteria for the strong convergence of a bounded operator semigroup T on a Banach space E, with a special view on the asymptotics of certain semigroups of positive operators on Banach lattices. This problem is not at all new, but we refrain from even trying to give a list of relevant literature at this point. (However, cf. Sect. 7 below.) Rather, let us stress some features that distinguish our approach from most others.

Classically, asymptotics of operator semigroups focusses on *strongly continuous* one-parameter semigroups $T = (T_t)_{t\in[0,\infty)}$. However, there are important instances of operator semigroups which lack strong continuity, e.g., the heat semigroup on the space of bounded continuous functions on $\mathbb{R}$ or on the space of finite measures over $\mathbb{R}$ or, in an abstract context, dual semigroups of C_0-semigroups on non-reflexive spaces. (See the recent paper [19] for a more involved concrete example.) Hence, there is a need for results on asymptotics beyond C_0-semigroups.

Secondly, besides the "continuous time" case just mentioned, there is an even more fundamental interest in "discrete time," i.e., in the asymptotics of the powers T^n of a single operator T. From a systematic point of view, it is desirable to try to cover both cases at the same time as far as possible. This is the reason why we consider general semigroup representations $(T_s)_{s\in S}$—where $(S, +)$ is an Abelian semigroup with zero element 0—as bounded operators on a Banach space without any further topological assumptions (see Sect. 2).

It may not come as a surprise that non-trivial results about asymptotics can be obtained—even in such a general setting—by employing the so-called Jacobs–deLeeuw–Glicksberg (JdLG) theory for compact (Abelian) semitopological semigroups. In fact, the role of the JdLG-theory for asymptotics is well-established. Usually, it is applied to the semigroup

$$\mathcal{T} := \mathrm{cl}\{T_s \mid s \in S\}$$

(closure in the strong or weak operator topology) and hence rests on a "global" compactness requirement for the whole semigroup. This is appropriate for the abovementioned "classical" cases (powers of a single operator, C_0-semigroups) because there the strong compactness of $\mathcal{T}$ is necessary for the convergence of T. However, typical examples of non-continuous shift semigroups (left shift on $\mathrm{L}^\infty(0, 1)$ or $\mathrm{c}_0(\mathbb{R}_+)$) show that such semigroups may converge strongly to 0 without being relatively strongly compact. (The left shift semigroup on $\mathrm{c}_0(\mathbb{R}_0)$ is not even eventually relatively strongly compact.)

In order to cover also these more general situations, we introduce the set

$$\mathcal{T}_\infty := \bigcap_{t\in S} \mathrm{cl}\{T_{s+t} \mid s \in S\}$$

which we call the **semigroup at infinity**. It turns out that $\mathcal{T}_\infty$ is a good replacement for $\mathcal{T}$ *if* $\mathcal{T}_\infty$ is strongly compact and not empty. In particular, its minimal projection, P_∞, satisfies

$$P_\infty x = 0 \quad \Longleftrightarrow \quad \lim_{s\in S} T_s x = 0 \qquad (x \in E)$$

(see Theorem 2.2). Not surprisingly, in the mentioned "classical" cases the condition that $\mathcal{T}_\infty$ be strongly compact and not empty is actually equivalent to relative strong compactness of the original semigroup (cf. Remark 2.6), and hence the use of $\mathcal{T}_\infty$ is then—in some sense—unnecessary. In a general approach going beyond these classical cases, however, it is our means of choice.

In a next step (Sect. 3), we describe a convenient set-up that warrants the crucial property (i.e.: $\mathcal{T}_\infty$ is non-empty and compact). To this end, the new notion of *quasi-compactness relative to a subspace* is introduced. This condition generalizes the traditional notion of quasi-compactness of a semigroup and plays a central role in our first main result (Theorem 3.1).

After these completely general considerations, and from then on until the end of the paper, we confine our attention to *positive* operator semigroups on Banach *lattices*. Our second main result, Theorem 4.3, appears to be a mere instantiation of Theorem 3.1 in such a setting. However, the theorem gains its significance from the fact that the required quasi-compactness condition is intimately linked to the well-known property of *AM-compactness* of positive operators which, in turn, occurs frequently when dealing with "concrete" positive semigroups arising in evolution equations and stochastics (see Appendix A).

The main thrust of Theorem 4.3 is that it reduces the study of the asymptotic properties of certain positive semigroups to the following special case: E is an atomic Banach lattice and T embeds into a strongly compact group of positive invertible operators thereon. Hence, in the subsequent Sect. 5 we analyze this situation thoroughly and establish a structure theorem which is reminiscent of the Peter–Weyl theorem and its consequences for Banach space representations of compact groups (Theorem 5.4).

Putting the pieces together, in Sect. 6 we formulate several consequences regarding the asymptotic (and spectral-theoretic) properties of a positive semigroup $T = (T_s)_{s\in S}$ satisfying the conditions of Theorem 4.3. Particularly important here is the fact that we can identify (in Theorem 6.2) two intrinsic properties of the semigroup S that imply the convergence of T: an algebraic one (essential divisibility of S) and a topological one (under the condition that S is topological and T is continuous). Interestingly enough, both are applicable in the case $S = \mathbb{R}_+$, but none of them in the case $S = \mathbb{N}_0$.

Finally, in Sect. 7 we review the pre-history of the problem and show how the results obtained so far by other people relate to (or are covered by) our findings. Interestingly, in this pre-history the spectral–theoretic results ("triviality of the point spectrum") have taken a much more prominent role than the asymptotic results. We end the paper with a new and unifying result in this direction (Theorem 7.6).

1.1 Relevance and Relation to the Work of Others

The present work can be understood as a continuation and further development of the recent paper [14] by M. Gerlach and the first author. Many ideas in the present paper can already be found in [14], like that one can go beyond strong continuity by combining the JdLG-theory with the concept of AM-compactness; or that under AM-compactness conditions a purely algebraic property (divisibility) of the underlying semigroup suffices to guarantee the strong convergence of the representation.

However, we surpass our reference in several respects:

1) A general Banach space principle (Theorem 3.1) is established and identified as the theoretical core which underlies the results of [14]. This principle, which is based on the new notion of "quasi-compactness relative to a subspace" and on our systematic study of the "semigroup at infinity" (Theorem 2.2), has potential applications in the asymptotic theory of semigroups without any positivity assumptions.
2) As a consequence of 1), the two main results from [14], Theorem 3.7 and Theorem 3.11, are now unified. Moreover, our results hold without requiring the semigroup to have a quasi-interior fixed point.
3) A general structure theorem for representations of compact groups on atomic Banach lattices (Theorem 5.4) is established. This result is auxiliary to—but actually completely independent of—our principal enterprise, the asymptotics of operator semigroups. From a different viewpoint, it is a contribution to the theory of positive group representations as promoted by Marcel de Jeu (Leiden) and his collaborators.
4) A new spectral-theoretic result (Theorem 7.6) about the properties of unimodular eigenvalues is established.

To understand the relevance of the results obtained in this paper, one best looks into the pre-history of its predecessor [14]. We decided to place such a historical narrative *after* our systematic considerations, in Sect. 7. That gives us the possibility to then refer freely to the results proven before, and to explain in detail their relation to the results obtained earlier by other people.

1.2 Notation and Terminology

We use the letters $E, F, \ldots$ generically to denote Banach spaces or Banach lattices over the scalar field $\mathbb{K} \in \{\mathbb{R}, \mathbb{C}\}$. The space of bounded linear operators is denoted by $\mathcal{L}(E; F)$, and $\mathcal{L}(E)$ if $E = F$; the space of compact operators is $\mathcal{K}(E; F)$, and $\mathcal{K}(E)$ if $E = F$. Frequently, we shall endow $\mathcal{L}(E; F)$ with the strong operator topology (sot). To indicate this we use terms like "sot-closed," "sot-compact" or speak of "strongly closed" or "strongly compact" sets, etc. A similar convention applies when the weak operator topology (wot) is considered. Whereas for a set $A \subseteq E$ the set $\mathrm{cl}_\sigma(A)$ is the closure of A in the weak ($= \sigma(E; E')$) topology on E, the sot-closure and the wot-closure of a set $M \subseteq \mathcal{L}(E; F)$ are denoted by

$$\mathrm{cl}_{\mathrm{s}}(M) \quad \text{and} \quad \mathrm{cl}_{\mathrm{w}}(M),$$

respectively. We shall frequently use the following auxiliary result, see [10, Corollary A.5].

Lemma 1.1 *Let E be a Banach space. Then a bounded subset $M \subseteq \mathcal{L}(E)$ is relatively strongly (weakly) compact if and only if the orbit*

$$Mx := \{Tx \mid T \in M\}$$

is relatively (weakly) compact for all x from a dense subset of E.

The set $\mathcal{L}(E)$ is a semigroup with respect to operator multiplication. Operator multiplication is sot- and wot-separately continuous, and it is sot-simultaneously continuous on norm-bounded sets.

For the definition of a semigroup as well as for some elementary definitions and results from algebraic semigroup theory, see Appendix C.

We shall freely use standard results and notation from the theory of Banach lattices, with [25] and [21] being our main references. If E is a Banach lattice, the set $E_+ := \{x \in E \mid x \geq 0\}$ is its cone of positive elements. In some proofs we confine tacitly to real Banach lattices, but there should be no difficulty to extend the arguments to the complex case.

2 Representations of Abelian Semigroups

Throughout the article, S is an Abelian semigroup (written additively) containing a neutral element 0.

Observe that for each $s \in S$ the set

$$s+S := \{s + r \mid r \in S\} \subseteq S$$

is a subsemigroup of S. We turn S into a directed set by letting

$$s \le t \quad \overset{\text{def.}}{\Longleftrightarrow} \quad t \in s{+}S \quad \Longleftrightarrow \quad t{+}S \subseteq s{+}S.$$

For limits of nets $(x_s)_{s\in S}$ with respect to this direction, the notation $\lim_{s\in S} x_s$ is used. Note that $0 \le s$ for all $s \in S$.

Example Observe that in the cases $S = \mathbb{Z}_+$ and $S = \mathbb{R}_+$ the so-defined direction and the associated notion of limit coincides with the usual one.

A **representation** of S on a Banach space E is any mapping $T : S \to \mathcal{L}(E)$ satisfying

$$T(0) = \mathrm{I} \quad \text{and} \quad T(s+t) = T(s)T(t) \qquad (t, s \in S).$$

In place of $T(s)$ we also use index notation T_s, and often call $T = (T_s)_{s\in S}$ an **operator semigroup** (over S on E). The **fixed space** of the representation T is

$$\operatorname{fix}(T) := \bigcap_{s\in S} \ker(T_s - \mathrm{I}) = \{x \in E \mid T_s x = x \text{ for all } s \in S\}.$$

An operator semigroup $(T_s)_{s\in S}$ is **bounded** if

$$M_T := \sup_{s\in S} \|T_s\| < \infty.$$

Boundedness has the following useful consequence.

Lemma 2.1 *Let $T = (T_s)_{s\in S}$ be a bounded operator semigroup on the Banach space E. Then for each vector $x \in E$ the following assertions are equivalent:*

(i) $0 \in \overline{\{T_s x \mid s \in S\}}$.
(ii) $\lim_{s\in S} T_s x = 0$.

Proof Suppose that (i) holds and $\varepsilon > 0$. Then there is $s \in S$ such that $\|T_s x\| \le \varepsilon$. But then

$$\|T_t x\| \le \varepsilon\, M_T \qquad \text{for all } t \in s{+}S.$$

It follows that $\lim_{s\in S} T_s x = 0$, i.e., (ii). The converse is trivial. □

Given an operator semigroup $T = (T_s)_{s\in S}$ on a Banach space E, a subset A of E is called T-**invariant** if $T_s(A) \subseteq A$ for all $s \in S$. A closed, T-invariant subspace F of E gives rise to a **subrepresentation** by restricting the operators T_s to F. Such a subrepresentation is called **finite-dimensional** (**d-dimensional**) ($d \in \mathbb{N}$) if F is finite-dimensional (d-dimensional).

A one-dimensional subrepresentation is given by a scalar representation $\lambda : S \to \mathbb{K}$ and a non-zero vector $u \in E$ such that

$$T_s u = \lambda_s u \qquad \text{for all } s \in S.$$

The corresponding mapping λ is then called an **eigenvalue** of T, and u is called a corresponding **eigenvector**. Obviously,

$$\lambda \text{ is constant} \quad \Leftrightarrow \quad \lambda = \mathbf{1} \quad \Leftrightarrow \quad u \in \operatorname{fix}(T).$$

An eigenvalue $\lambda = (\lambda_s)_{s\in S}$ is called **unimodular** if $|\lambda_s| = 1$ for each $s \in S$. (So the constant eigenvalue $\mathbf{1}$ is unimodular.) A unimodular eigenvalue λ of T is called a **torsion eigenvalue** if there is $m \in \mathbb{N}$ such that $\lambda_s^m = 1$ for all $s \in S$.

If E is a Banach lattice, a semigroup $(T_s)_{s\in S}$ on E is called **positive** if the positive cone E_+ is T-invariant, i.e., if each operator T_s is positive. And a positive semigroup is called **irreducible** or said to **act irreducibly** on E if $\{0\}$ and E are the only T-invariant closed ideals of E. (Recall that a subspace J of E is an ideal if it satisfies: $x \in E,\ y \in J,\ |x| \le |y| \quad \Rightarrow \quad x \in J$.)

2.1 The Semigroup at Infinity

Given an operator semigroup $T = (T_s)_{s\in S}$ we write

$$T_S := \{T_s \mid s \in S\} \subseteq \mathcal{L}(E)$$

for its range, which is a subsemigroup of $\mathcal{L}(E)$. And we abbreviate

$$\mathcal{T} := \mathrm{cl}_{\mathrm{s}}\{T_s \mid s \in S\} \quad \text{and} \quad \mathcal{T}_s := \mathrm{cl}_{\mathrm{s}}\{T_t \mid t \ge s\} \quad (s \in S),$$

and call

$$\mathcal{T}_\infty := \bigcap_{s\in S} \mathcal{T}_s = \bigcap_{s\in S} \mathrm{cl}_{\mathrm{s}}\{T_t \mid t \ge s\}$$

the associated **semigroup at infinity**. In effect, $\mathcal{T}_\infty$ is the set of sot-cluster points of the net $(T_s)_{s\in S}$.

Note that $\mathcal{T}_\infty$ is multiplicative and even satisfies

$$\mathcal{T} \cdot \mathcal{T}_\infty \subseteq \mathcal{T}_\infty.$$

But it may be empty (in which case it is, according to our definition in Appendix C, not a semigroup.[1])

[1] We apologize for this little abuse of terminology.

2.2 The JdLG-Splitting Theory

One of the principal methods to prove strong convergence of a bounded semigroup is to employ the splitting theory of Jacobs, de Leeuw and Glicksberg as detailed, e.g., in [8, Chapter 16]. Usually, this theory is applied to the semigroup $\mathcal{T}$ or to its wot-counterpart $\mathrm{cl}_{\mathrm{w}}\{T_s \mid s \in S\}$. In contrast, we shall apply it to $\mathcal{T}_\infty$. If $\mathcal{T}_\infty$ is a strongly compact semigroup, the JdLG-theory tells that it contains a unique minimal idempotent, which we denote by P_∞. (Minimality means that $P_\infty \cdot \mathcal{T}_\infty$ is a minimal ideal in $\mathcal{T}_\infty$.) The range of P_∞ is denoted here by

$$E_\infty := \mathrm{ran}(P_\infty).$$

Observe that $QT_s = T_s Q$ for each $s \in S$ and each $Q \in \mathcal{T}_\infty$. In particular, E_∞ is $\mathcal{T}$-invariant.

Theorem 2.2 *Let $T = (T_s)_{s\in S}$ be a bounded operator semigroup on the Banach space E such that the associated semigroup at infinity, $\mathcal{T}_\infty$, is strongly compact and non-empty. Then the following additional assertions hold:*

a) $\mathcal{T} P_\infty = \mathcal{T}_\infty P_\infty$.
b) *T is relatively strongly compact on E_∞, i.e.,*

$$\mathcal{G} := \mathrm{cl}_{\mathrm{s}}\{T_s|_{E_\infty} \mid s \in S\} \subseteq \mathcal{L}(E_\infty)$$

is a strongly compact group of invertible operators on E_∞. Moreover,

$$\mathcal{G} = \mathcal{T}|_{E_\infty} := \{Q|_{E_\infty} \mid Q \in \mathcal{T}\}.$$

c) *For each $x \in E$ the following statements are equivalent:*
   (i) $P_\infty x = 0$.
   (ii) $0 \in \mathrm{cl}_\sigma\{T_s x \mid s \in S\}$.
   (iii) $\lim_{s\in S} T_s x = 0$.
   (iv) $Rx = 0$ *for some/all* $R \in \mathcal{T}_\infty$.
d) *If $(\lambda_s)_{s\in S}$ is a unimodular eigenvalue of T with eigenvector $0 \neq x \in E$, then $x \in E_\infty$ and there is a unique eigenvalue $\mu = (\mu_Q)_{Q\in\mathcal{G}}$ of $\mathcal{G}$ such that $\lambda_s = \mu_{T_s}$ for all $s \in S$.*
e) *If $\mu = (\mu_Q)_{Q\in\mathcal{G}}$ is an eigenvalue of $\mathcal{G}$ on E_∞, then $\lambda_s := \mu_{T_s}$ $(s \in S)$ is an unimodular eigenvalue of T.*

(We suppose $\mathbb{K} = \mathbb{C}$ for assertions d) *and* e).*)*

Proof

a) Since $\mathcal{T}\mathcal{T}_\infty \subseteq \mathcal{T}_\infty \subseteq \mathcal{T}$, we have $\mathcal{T} P_\infty = \mathcal{T} P_\infty P_\infty \subseteq \mathcal{T}_\infty P_\infty \subseteq \mathcal{T} P_\infty$.
b) By a) we have $\mathcal{T}|_{E_\infty} = \mathcal{T}_\infty|_{E_\infty}$, and the latter is a strongly compact group of invertible operators on E_∞ by the JdLG-theory. Since restriction is a sot-

continuous operator from $\mathcal{L}(E)$ to $\mathcal{L}(E_\infty; E)$, $\mathcal{T}|_{E_\infty} \subseteq \mathcal{G}$. The converse inclusion follows from $\mathcal{G}P_\infty \subseteq \mathcal{T}$, which is true because $P_\infty \in \mathcal{T}$.

c) If $Rx = 0$ for *all* $R \in \mathcal{T}_\infty$, then clearly (i) holds, and (i) implies that $Rx = 0$ for *some* $R \in \mathcal{T}_\infty$. On the other hand, this latter statement obviously implies $0 \in \overline{\{T_s x \mid s \in S\}}$, which is equivalent to $\lim_{s \in S} T_s x = 0$, i.e., (iii).

If, in turn, (iii) holds and $\varepsilon > 0$ is fixed, then there is $s \in S$ such that $\{T_t x \mid t \geq s\} \subseteq \mathrm{B}[0, \varepsilon]$. Hence, also $\mathcal{T}_\infty x \subseteq \mathrm{B}[0, \varepsilon]$. As $\varepsilon > 0$ was arbitrary, $\mathcal{T}_\infty x = \{0\}$, i.e., $Rx = 0$ for all $R \in \mathcal{T}_\infty$.

Finally, (iii) obviously implies (ii). Conversely, starting from (ii) we apply P_∞ to obtain

$$0 \in \mathrm{cl}_\sigma \{T_s P_\infty x \mid s \in S\}.$$

However, by b) the set $\{T_s P_\infty x \mid s \in S\}$ is relatively strongly compact and hence its weak and its strong closures must coincide. This yields

$$0 \in \overline{\{T_s P_\infty x \mid s \in S\}},$$

which implies $P_\infty x = P_\infty(P_\infty x) = 0$ by what we have already shown.

d) Let $0 \neq x \in E$ be an eigenvector for the unimodular eigenvalue $(\lambda_s)_{s \in S}$ of T. Define $y := x - P_\infty x$. Then $P_\infty y = 0$ and hence $T_s y \to 0$. On the other hand, since P_∞ commutes with every T_s, $T_s y = \lambda_s y$ for all $s \in S$. As $|\lambda_s| = 1$, it follows that $y = 0$ and hence $x \in E_\infty$. The remaining statement now follows easily since $\mathbb{C}x$ is T-invariant and $T_S|_{E_\infty}$ is dense in $\mathcal{G}$.

e) is obvious. □

As a corollary we obtain the following characterization of the strong convergence of a semigroup.

Corollary 2.3 *For a bounded operator semigroup $T = (T_s)_{s \in S}$ on a Banach space E the following assertions are equivalent:*

(i) *T is strongly convergent;*
(ii) *$\mathcal{T}_\infty$ is a singleton;*
(iii) *$\mathcal{T}_\infty$ is non-empty and strongly compact and acts as the identity on E_∞;*
(iv) *$\mathcal{T}_\infty$ is non-empty and strongly compact and T acts as the identity on E_∞.*

In this case: $\lim_{s \in S} T_s = P_\infty$.

Proof (i)$\Rightarrow$(ii): If T is strongly convergent with $P := \lim_{s \in S} T_s$ being its limit, then $\mathcal{T}_\infty = \{P\}$ is a singleton.

(ii)$\Rightarrow$(iii): If $\mathcal{T}_\infty = \{P\}$ is a singleton, then it is clearly non-empty and strongly compact. It follows that $P = P_\infty$, and hence $\mathcal{T}_\infty$ acts as the identity on E_∞.

(iii)$\Rightarrow$(iv): Suppose that $\mathcal{T}_\infty$ is non-empty and strongly compact and acts as the identity on E_∞. Let $Q \in \mathcal{T}_\infty$. Then, by the equivalence (i)$\Leftrightarrow$(iv) in Theorem 2.2.c), $Q(\mathrm{I} - P_\infty) = 0$ and hence $Q = QP_\infty = P_\infty$. So it follows that $\mathcal{T}_\infty = \{P_\infty\}$.

Moreover, since $T_s\mathcal{T}_\infty \subseteq \mathcal{T}_\infty$ for each $s \in S$, we obtain $T_s P_\infty = P_\infty$ and hence $T_s = \mathrm{I}$ on E_∞ for all $s \in S$.

(iv) $\Rightarrow$ (i): Suppose that (iv) holds. Then

$$\lim_{s\in S} T_s = \lim_{s\in S}(T_s P_\infty + T_s(\mathrm{I} - P_\infty)) = P_\infty + \lim_{s\in S} T_s(\mathrm{I} - P_\infty) = P_\infty$$

strongly, by the equivalence (i) $\Leftrightarrow$ (iii) of Theorem 2.2.c). □

Theorem 2.2 and its corollary yield the following strategy to prove strong convergence of an operator semigroup:

1) Show that $\mathcal{T}_\infty$ is non-empty and strongly compact.
2) Show that $\mathcal{T}_\infty$ (or, equivalently, $\mathcal{T}$) acts as the identity on $E_\infty := \mathrm{ran}(P_\infty)$. (For this one may employ the additional information that $\mathcal{T}$ acts on E_∞ as a compact group.)

Remark 2.4 Suppose that $\mathcal{T} = (T_s)_{s\in S}$ is a strongly relatively compact operator semigroup on E with minimal idempotent $P \in \mathcal{T}$. Then, of course, $\mathcal{T}_\infty$ is non-empty and strongly compact, and hence a closed ideal of $\mathcal{T}$. It follows from the minimality of P in $\mathcal{T}$ and P_∞ in $\mathcal{T}_\infty$ that

$$P\mathcal{T} \subseteq P_\infty\mathcal{T} \subseteq P_\infty\mathcal{T}_\infty \subseteq P\mathcal{T}_\infty \subseteq P\mathcal{T}.$$

Hence $P\mathcal{T} = P_\infty\mathcal{T}_\infty$, which implies that $P = P_\infty$. So, in the case that $\mathcal{T}$ is relatively strongly compact, passing to the semigroup at infinity yields the same JdLG-decomposition of E as working with $\mathcal{T}$.

We end this section with a technical, but useful characterization of the property that $\mathcal{T}_\infty$ is non-empty and compact.

Proposition 2.5 *For a bounded operator semigroup $\mathcal{T} = (T_s)_{s\in S}$ on a Banach space E the following assertions are equivalent:*

(i) *$\mathcal{T}_\infty$ is non-empty and strongly compact.*
(ii) *Every subnet of $(T_s)_{s\in S}$ has a strongly convergent subnet.*
(iii) *Every universal subnet of $(T_s)_{s\in S}$ is strongly convergent.*
(iv) *For each $x \in E$ every subnet of $(T_s x)_{s\in S}$ has a convergent subnet.*
(v) *For each $x \in E$ every universal subnet of $(T_s x)_{s\in S}$ converges.*

If S contains a cofinal sequence, then the above assertions are also equivalent to:

(vi) *For every $x \in E$ and every cofinal sequence $(s_n)_{n\in\mathbb{N}}$ in S, the sequence $(T_{s_n}x)_{n\in\mathbb{N}}$ has a convergent subsequence.*

Proof (i) $\Rightarrow$ (iv): Suppose that (i) holds and let $x \in E$. The net $(T_s(\mathrm{I} - P_\infty)x)_{s\in S}$ converges to 0 according to Theorem 2.2.c). On the other hand, the net $(T_s P_\infty x)_{s\in S}$ is contained in the compact set $\mathcal{T}_\infty P_\infty x$ due to Theorem 2.2 a), so each of its subnets has a convergent subnet. This shows (iv).

(iv) ⇒ (v): This follows since a universal net with a convergent subnet must converge.

(v) ⇒ (iii): Let $(T_{s_\alpha})_{\alpha\in I}$ be a universal subnet of $(T_s)_{s\in S}$. Then for each $x \in E$, the net $(T_{s_\alpha}x)_{\alpha\in I}$ is universal and hence, by (v), convergent. Thus, $(T_{s_\alpha})_{\alpha\in I}$ is strongly convergent.

(iii) ⇒ (ii) ⇒ (i) and (iv) ⇒ (vi) all follow from Theorem B.3.

(vi) ⇒ (i): Suppose that S admits a cofinal sequence. Then by Theorem B.3 for each $x \in E$ the set $C_x := \bigcap_{t\in S} \mathrm{cl}\{T_s x \mid s \geq t\}$ is non-empty and compact. Since $\mathcal{T}_\infty x \subseteq C_x$, it follows that $\mathcal{T}_\infty$ is strongly compact. In order to see that $\mathcal{T}_\infty$ is not empty, fix a cofinal sequence $(s_n)_n$. By (vi) and since E is metrizable, it follows that for each $x \in E$ the set $\{T_{s_n}x \mid n \in \mathbb{N}\}$ is relatively compact. Hence, $\{T_{s_n} \mid n \in \mathbb{N}\}$ is relatively strongly compact. It follows that the sequence $(T_{s_n})_n$ has a cluster point, which is a member of $\mathcal{T}_\infty$ since $(s_n)_n$ is cofinal. □

Remarks 2.6

1) Assertion (vi) in Proposition 2.5 is called **strong asymptotic compactness** in [9, p. 2636].
2) Proposition 2.5 has an interesting consequence in the "classical" cases where $S = \mathbb{N}_0$ or $S = \mathbb{R}_+$ and T is strongly continuous (cf. the Introduction). Namely, in these cases one can actually dispense with the semigroup at infinity, because $\mathcal{T}_\infty$ is strongly compact and non-empty *if and only if* T is strongly compact.

In the next section we shall present another situation in which $\mathcal{T}_\infty$ is non-empty and strongly compact.

3 The Abstract Main Result

Suppose that E and F are Banach spaces such that F is densely embedded in E:

$$F \overset{d}{\hookrightarrow} E.$$

Reference to this embedding is usually suppressed and F is simply regarded as a subspace of E. We take the freedom to consider an operator on E also as an operator from F to E. (This amounts to view $\mathcal{L}(E) \subseteq \mathcal{L}(F; E)$ via the restriction mapping.)

A semigroup $(T_s)_{s\in S}$ on E is called **F-to-E quasi-compact**, or **quasi-compact relatively to** F, if for each $\varepsilon > 0$ there is $s \in S$ and a compact operator $K : F \to E$ such that

$$\|T_s - K\|_{\mathcal{L}(F;E)} < \varepsilon.$$

Note that we do not require that K can be extended to a bounded operator on E. In effect, the condition of being F-to-E quasi-compact can be expressed as

$$\operatorname{dist}(\{T_s \mid s \in S\}, \mathcal{K}(F; E)) = 0,$$

where "dist" refers to the distance induced by the norm on $\mathcal{L}(F; E)$.

Theorem 3.1 *Let E and F be Banach spaces, with F being densely embedded into E, and let $(T_s)_{s \in S}$ be a bounded operator semigroup on E which restricts to a bounded operator semigroup on F and is F-to-E quasi-compact. Then the following assertions hold:*

a) *$\mathcal{T}_\infty$ is strongly compact and non-empty.*
b) *Each element of $\mathcal{T}_\infty$ is compact as an operator from F to E.*
c) *$\mathcal{T}$ acts on E_∞ as a sot-compact group of invertible operators.*
d) *For $x \in E$ the following assertions are equivalent:*

   (i) $\lim_{s \in S} T_s x = 0$;
   (ii) $x \in \ker P_\infty$;
   (iii) $0 \in \operatorname{cl}_\sigma\{T_s x \mid s \in S\}$.

Proof a) and b) By passing to an equivalent norm on F we may suppose that each T_s, $s \in S$, is a contraction on F. Let B_F and B_E denote the closed unit balls of E and F, respectively.

Let $\varepsilon > 0$ and choose $s \in S$ and $K \in \mathcal{K}(F; E)$ such that $\|T_s - K\|_{\mathcal{L}(F;E)} \le \varepsilon$. Then

$$T_{t+s}(\mathrm{B}_F) = T_s T_t(\mathrm{B}_F) \subseteq T_s(\mathrm{B}_F) \subseteq K(\mathrm{B}_F) + \varepsilon \mathrm{B}_E$$

for each $t \in S$, and therefore

$$\mathcal{T}_s(\mathrm{B}_F) \subseteq \overline{K(\mathrm{B}_F)} + \varepsilon \mathrm{B}_E. \tag{3.1}$$

Now, let $(T_{s_\alpha})_\alpha$ be any universal subnet of $(T_s)_{s \in S}$ (Lemma B.1) and let $x \in \mathrm{B}_F$. Then the net $(T_{s_\alpha} x)_\alpha$ is a universal net in E. Moreover, (3.1) shows that for each $\varepsilon > 0$ this net has a tail contained in the ε-neighborhood of some compact set. Hence, by Lemma B.2, it is a Cauchy net and thus convergent in E. Since F is dense in E and T is bounded, $(T_{s_\alpha} x)_\alpha$ converges for *every* $x \in E$. In other words, $(T_{s_\alpha})_\alpha$ is strongly convergent. As its limit must be a member of $\mathcal{T}_\infty$, it follows that $\mathcal{T}_\infty \neq \emptyset$.

It also follows from (3.1) that $\mathcal{T}_\infty(\mathrm{B}_F) \subseteq \overline{K(\mathrm{B}_F)} + \varepsilon \mathrm{B}_E$. As $\overline{K(\mathrm{B}_F)}$ is compact, it admits a finite ε-mesh. Hence, $\mathcal{T}_\infty(\mathrm{B}_F)$ admits a finite 2ε-mesh. Since this works for each $\varepsilon > 0$, $\mathcal{T}_\infty(\mathrm{B}_F)$ is relatively compact in E.

In particular, it follows that $\mathcal{T}_\infty \subseteq \mathcal{K}(F; E)$ and that for each $x \in F$ the orbit $\mathcal{T}_\infty x$ is relatively compact in E. Since T is bounded on E and F is dense in E, $\mathcal{T}_\infty$ is relatively strongly compact (Lemma 1.1). But $\mathcal{T}_\infty$ is strongly closed, so it is strongly compact as claimed.

c) and d) follow from a) by Theorem 2.2. □

Remark 3.2 Theorem 3.1 seems to be new even for C_0-semigroups. In that case, by Remark 2.6.b), it follows a posteriori that the C_0-semigroup is relatively compact.

In the next sections we shall see that our set-up from above has a quite natural instantiation in the context of semigroups of positive operators on Banach lattices with a quasi-interior point.

4 Positive Semigroups and AM-Compactness

From now on, we consider *positive* semigroups $T = (T_s)_{s \in S}$ on Banach lattices E. The role of F in our abstract setting from above will be taken by the *principal ideal*

$$E_y := \{x \in E \mid \text{there is } c \geq 0 \text{ such that } |x| \leq cy\}$$

for some $y \in E_+$, endowed with the natural AM-norm

$$\|x\|_y := \inf\{c \geq 0 \mid |x| \leq cy\}.$$

Since we need that $F = E_y$ is dense in E, we have to require that y is *a quasi-interior point* in E.

As we further need E_y-to-E quasi-compactness, it is natural to ask which operators on E restrict to compact operators from E_y to E. It turns out that these are precisely the *AM-compact* operators, i.e., those that map order intervals of E to relatively compact subsets of E, see Lemma A.3.

There are a couple of useful theorems that help to identify AM-compact operators. For example, operators between L^p-spaces induced by positive integral kernel functions and positive operators that "factor through L^∞-spaces" are AM-compact. (Proofs of these well-known facts are presented in Appendix A, see Theorems A.3 and A.4.)

4.1 The Range of a Positive Projection

When we apply Theorem 3.1 to a semigroup of positive operators, the resulting projection P_∞ will be positive, too. The following is a useful information about its range.

Lemma and Definition 4.1 *Let E be a Banach lattice and let P be a positive projection on E. Define*

$$\|x\|_P := \|P\,|x|\,\| \qquad (x \in \mathrm{ran}(P)).$$

Then the following assertions hold:

a) $\|\cdot\|_P$ *is an equivalent norm on* $\operatorname{ran}(P)$.
b) *The space* $\operatorname{ran}(P)$ *is a Banach lattice with respect to the order induced by* E *and the norm* $\|\cdot\|_P$. *Its modulus is given by*

$$|x|_P := P\,|x| \qquad (x \in \operatorname{ran}(P)).$$

The Banach lattice $\operatorname{ran}(P)$ endowed with the norm $\|\cdot\|_P$ as in a) and b) is denoted by $[\operatorname{ran}(P)]$.

c) *If* $y \in E_+$, *then* $P(E_y) \subseteq [\operatorname{ran}(P)]_{Py}$. *In particular, if* $y \in E_+$ *is a quasi-interior point of* E, *then* Py *is a quasi-interior point of* $[\operatorname{ran}(P)]$.

Proof This is essentially [25, Proposition III.11.5]. □

In order to obtain further insight into the relation of the closed ideals in $[\operatorname{ran}(P)]$ and in E, we define for any Banach lattice E the mapping

$$\Phi(J) := \operatorname{cl}\{x \in E \mid |x| \le y \text{ for some } y \in J_+\} \qquad (J \subseteq E).$$

If $J_+ = J \cap E_+$ is a cone, then $\Phi(J) = \Phi(J_+)$ is the smallest closed ideal in E containing J_+.

Theorem 4.2 *Let* E *be a Banach lattice and* P *a positive projection on* E, *and let* Φ *be defined as above. Then the following assertions hold:*

a) *If* I *is a closed* P*-invariant ideal in* E, *then* $P(I) = I \cap \operatorname{ran}(P)$ *is a closed ideal in* $[\operatorname{ran}(P)]$.
b) *If* J *is a closed ideal in* $[\operatorname{ran}(P)]$, *then* $\Phi(J)$ *is* P*-invariant and the smallest closed ideal in* E *containing* J. *Moreover,*

$$J = P(\Phi(J)) = \Phi(J) \cap \operatorname{ran}(P).$$

Proof

a) Let $I \subseteq E$ be a closed P-invariant ideal and let $J := I \cap \operatorname{ran}(P)$. Then J is a closed subspace of $\operatorname{ran}(P)$. And if $x \in \operatorname{ran}(P)$ and $y \in J$ with $|x|_P \le |y|_P$, it follows that

$$|x| = |Px| \le P\,|x| = |x|_P \le |y|_P = P\,|y| \in I$$

by P-invariance. Hence, $x \in J$ and therefore J is an ideal in $[\operatorname{ran}(P)]$. Moreover, again by P-invariance,

$$J = PJ = P(I \cap \operatorname{ran}(P)) \subseteq P(I) \subseteq I \cap \operatorname{ran}(P),$$

and hence $I \cap \operatorname{ran}(P) = J = P(I)$.

b) Let J be any closed ideal of $[\mathrm{ran}(P)]$. Then $\Phi(J)$ is the smallest closed ideal in E containing J. (In fact, if $x \in J$, then $|x| = |Px| \le P|x| = |x|_P \in J_+$, and hence $J \subseteq \Phi(J)$.) It is also P-invariant, for if $|x| \le y \in J_+$ then $|Px| \le P|x| \le Py = y$. This also shows that $P(\Phi(J)) \subseteq J$, and since $J \subseteq \Phi(J)$, it follows that $P(\Phi(J)) = J$. □

4.2 The Main Result for Positive Semigroups

We are now prepared for our second main theorem.

Theorem 4.3 *Let $T = (T_s)_{s\in S}$ be a bounded and positive operator semigroup on a Banach lattice E with a quasi-interior point $y \in E_+$. Suppose, in addition, that T is E_y-to-E quasi-compact and restricts to a bounded semigroup on E_y. Then the following assertions hold:*

a) *$\mathcal{T}_\infty$ is strongly compact and non-empty and consists of AM-compact operators.*
b) *$[\mathrm{ran}(P_\infty)]$ is an atomic Banach lattice with order continuous norm and quasi-interior point $P_\infty y$.*
c) *The semigroup $\mathcal{T} = \mathrm{cl_s}\{T_s \mid s \in S\}$ acts on $[\mathrm{ran}(P_\infty)]$ as a compact topological group of positive, invertible operators.*
d) *If $(T_s)_{s\in S}$ acts irreducibly on E, then $\mathcal{T}$ acts irreducibly on $[\mathrm{ran}(P_\infty)]$.*

Proof

a) follows from Theorem 3.1 and Lemma A.1.
b) Each order interval J of $E_\infty = [\mathrm{ran}(P_\infty)]$ is of the form $J = J' \cap E_\infty$, where J' is an order interval of E. Since P_∞ is AM-compact but restricts to the identity on E_∞, it follows that $J = P_\infty(J) \subseteq P_\infty(J')$ is relatively compact. By Wnuk [27, Theorem 6.1], this implies that E_∞ as a Banach lattice is atomic and has order continuous norm.
c) This follows again from Theorem 3.1.
d) Suppose that $J \ne \{0\}$ is a closed $\mathcal{T}$-invariant ideal in $[\mathrm{ran}(P_\infty)]$. Then the set

$$\{x \in E \mid |x| \le y \text{ for some } y \in J_+\}$$

is T-invariant, and hence $\Phi(J)$ is a closed T-invariant ideal in E containing J. By irreducibility, $\Phi(J) = E$, and hence $\mathrm{ran}(P_\infty) = P(E) = P(\Phi(J)) = J$ by Theorem 4.2. □

Remarks 4.4

1) The assumption that T restricts to a bounded semigroup on E_y is, for instance, satisfied if y is a **sub-fixed point** of T, i.e., if $T_t y \le y$ for all $t \in S$.
2) Certainly, if T_s is AM-compact for some $s \in S$, then T is E_y-to-E quasi-compact.

We conclude this section with the following result, essentially proved by Gerlach and Glück in [14, Lemma 3.12]. It shows that in certain situations it suffices to require merely that T_s *dominates* a non-trivial AM-compact operator for some $s \in S$.

Lemma 4.5 *Let $T = (T_s)_{s\in S}$ be a bounded, positive and irreducible semigroup on a Banach lattice E with order continuous norm and a quasi-interior sub-fixed point y of T. Suppose that there are $s \in S$ and an AM-compact operator $K \neq 0$ on E with $0 \leq K \leq T_s$. Then T is E_y-to-E quasi-compact and restricts to a bounded semigroup on E_y. In particular, Theorem 4.3 is applicable.*

5 Compact Groups of Positive Operators on Atomic Banach Lattices

In view of our general strategy, Theorem 4.3 suggests to look for criteria implying that a positive group representation on an atomic Banach lattice with order-continuous norm is trivial. To this end, we first summarize some known results about atomic Banach lattices.

5.1 Atomic Banach Lattices

Recall that an **atom** in a Banach lattice is any element $0 \neq a \in E$ such that its generated principal ideal, E_a, is one-dimensional: $E_a = \mathbb{K} \cdot a$. We denote by

$$A = A_E := \{a \in E_+ \mid a \text{ atom}, \ \|a\| = 1\}$$

the set of positive atoms of norm one. For distinct $a, \ b \in A$ one has

$$|a - b| = |a + b| = a + b \geq a$$

and hence $\|a - b\| \geq \|a\| = 1$. This shows that A is a discrete set with respect to the norm topology.

A Banach lattice E is called **atomic**, if E is the smallest band in E that contains all atoms. In other words,

$$A^d := \{x \in E \mid \ |x| \wedge a = 0 \text{ for all } a \in A\} = \{0\}.$$

For each $a \in A$ the one-dimensional subspace $E_a = \mathbb{K}a$ is a projection band, with corresponding band projection P_A given by

$$P_a x := \sup[0, x] \cap \mathbb{R}a = \sup\{t \in \mathbb{R}_+ \mid ta \leq x\} \cdot a \qquad (x \in E_+).$$

(See, e.g., [20, Thm. 26.4] and cf. [21, Prop. 1.2.11].) The next result is a consequence of [25, p.143, Ex. 7] and [21, Thm. 1.2.10]. For the convenience of the reader, we give a proof.

Theorem 5.1 *Let E be a Banach lattice and let A be its set of positive normalized atoms. Then for each finite subset $F \subseteq A$ the space*

$$\mathrm{span}(F) = \bigoplus_{a \in F} \mathbb{K}a$$

is a projection band with band projection $P_F = \sum_{a \in F} P_a$. *Suppose, in addition, that E is atomic. Then*

$$\mathrm{I}_E = \sum_{a \in A} P_a \tag{5.1}$$

as a strongly order convergent series. Each band B in E is generated (as a band) by $A \cap B$.

Remark 5.2 There are different notions of "order convergence" in the literature, see [1]. We employ the definition found in [21, Definition 1.1.9 i)]. For the case of (5.1) this simply means

$$x = \sup_F \sum_{a \in F} P_a x \qquad \text{for all } x \in E_+, \tag{5.2}$$

where the supremum is taken over all finite subsets of A.

Proof of Theorem 5.1 Fix a finite set $F \subseteq A$. If $a, b \in F$ with $a \neq b$, then $a \wedge b = 0$ and hence $P_a P_b = 0$. It follows that

$$P_F := \sum_{a \in F} P_a$$

is a projection (and F is a linearly independent set). Again by the pairwise disjointness of the elements of F,

$$\sum_{a \in F} P_a x = \bigvee_{a \in F} P_a x \le x \qquad (x \in E_+).$$

This shows that $0 \le P_F \le \mathrm{I}$, and hence P_F is a band projection [21, Lemma 1.2.8]. Since, obviously, $\mathrm{ran}(P_F) = \mathrm{span}(F)$, the first assertion is proved.

In order to prove (5.2) fix $x \in E_+$ and let $y \in E_+$ be such that $y \ge P_F x$ for all finite $F \subseteq A$. Then $y \ge P_a x$ for each $a \in A$ and hence

$$0 \le x - (x \wedge y) \le x - P_a x \perp a$$

If E is atomic, it follows that $x = x \wedge y$, i.e., $x \le y$. This yields (5.2).

Finally, let $B \subseteq E$ be any band and let $0 \le x \in B$. Then for each $a \in A$, $P_a x \in B$ (since $0 \le P_a x \le x$ and B is an ideal). Hence, either $P_a x = 0$ or $a \in B$. It follows from (5.1) that B is generated by $A \cap B$. □

With this information at hand we now turn to the representation theory.

5.2 A Structure Theorem

Let $G \subseteq \mathcal{L}(E)$ be a group of positive, invertible operators on E. (In particular, G consists of lattice homomorphisms.) Then for each $g \in G$ and $a \in A$ the element $g \cdot a \in E$ must be an atom again. In effect

$$\varphi_g(a) := \|g \cdot a\|^{-1}(g \cdot a) \in A. \tag{5.3}$$

It is easy to see that

$$\varphi : G \to \mathrm{Sym}(A), \qquad g \mapsto \varphi_g \tag{5.4}$$

is a group homomorphism from G to the group of all bijections on A. The corresponding action

$$G \times A \to A, \qquad (g, a) \mapsto \varphi_g(a)$$

is called the **induced action** of G on A. For each $a \in A$ the orbit mapping

$$G \to A, \qquad g \mapsto \varphi_g(a)$$

of the induced action is continuous (with respect to the strong operator topology on G). If, in addition, G is strongly compact, then each orbit

$$\varphi_G(a) = \{\varphi_g(a) \mid g \in G\}$$

is finite (since A is discrete). We denote by A/G the set of all these orbits. Then A/G is a partition of A into finite subsets.

Lemma 5.3 *In the described situation, suppose that G is compact. Then for* $a \in A$ *and* $g \in G$*:* $g \cdot a = a \;\Leftrightarrow\; \varphi_g(a) = a$*. Furthermore:* $g = \mathrm{I}_E \;\Leftrightarrow\; \varphi_g = \mathrm{id}_A$*.*

Proof Fix $a \in A$ and $g \in G$. If $g \cdot a = a$, then $\varphi_g(a) = a$, since $\|a\| = 1$. Conversely, suppose that $\varphi_g(a) = a$. Then $g^n \cdot a = \|g \cdot a\|^n a$ for all $n \in \mathbb{Z}$. By compactness, $\|g \cdot a\| = 1$, and hence $g \cdot a = a$ as claimed.

Suppose that $\varphi_g(a) = a$ for all $a \in A$. Then, as we have just seen, $g \cdot a = a$ for all $a \in A$. So g leaves all atoms fixed. Since g acts as a lattice isomorphism and hence is order continuous, it follows from Theorem 5.1 that $g = \mathrm{I}_E$. □

We can now prove a theorem that is reminiscent of the Peter–Weyl structure theorem and its applications to Banach space representations of compact groups.

Theorem 5.4 (Structure Theorem) *Let $E \neq \{0\}$ be an atomic Banach lattice and let A be its set of positive normalized atoms. Let $G \subseteq \mathcal{L}(E)$ be a strongly compact group of positive invertible operators on E, and let A/G be the set of orbits of elements of A under the induced action of G on A. Then the following assertions hold:*

a) *For each orbit $F \in A/G$ the band* $\mathrm{span}(F)$ *is G-invariant, the corresponding band projection P_F is G-intertwining, and G acts irreducibly on* $\mathrm{span}(F)$.
b) *If $B \neq \{0\}$ is a G-invariant band in E on which G acts irreducibly, then $B =$* $\mathrm{span}(F)$ *for some $F \in A/G$.*
c) *$I = \sum_{F\in A/G} P_F$ as a strongly order-convergent series.*
d) *In the case $\mathbb{K} = \mathbb{C}$, each eigenvalue of G on E is torsion.*
e) *If G acts irreducibly on E, then* $\dim(E) < \infty$ *and G has only finitely many eigenvalues.*

Proof

a) It is obvious that $\mathrm{span}(F)$ is G-invariant and G acts irreducibly on it. Since G consists of lattice automorphisms, also $\mathrm{span}(F)^d$ is G-invariant, and hence P_F is G-intertwining.
b) Let $B \neq \{0\}$ be any G-invariant band in E. Then B is generated (as a band) by $A \cap B$. By G-invariance, $A \cap B$ is a union of G-orbits (for the induced action), i.e., a union of elements of A/G. Hence, if G acts irreducibly on B, $A \cap B$ must coincide with precisely one G-orbit of A, i.e., $A \cap B \in A/G$.
c) follows from Theorem 5.1 since A/G is partition of A.
d) Let $\lambda : G \to \mathbb{C}$ be an eigenvalue of G on E and $0 \neq x \in E$ a corresponding eigenvector. Then λ is a continuous homomorphism, and since G is compact, λ is unimodular. By c), one must have $y := P_F x \neq 0$ for some $F \in A/G$, and by a), y is also an eigenvector corresponding to λ.
   Let $g \in G$ and $n := |F|$, the length of the (induced) G-orbit F. Then $g^{n!}$ acts (induced) on F as the identity. Hence, by Lemma 5.3, $g^{n!}$ acts (originally) as the identity on $\mathrm{span}(F)$. This yields
$$y = g^{n!}y = \lambda_g^{n!}y$$
   and hence $\lambda_g^{n!} = 1$. As $g \in G$ was arbitrary, the eigenvalue λ is torsion.
e) If G acts irreducibly on E, then b) tells that E is finite dimensional. As eigenvectors belonging to different eigenvalues have to be linearly independent, there can be only finitely many eigenvalues, as claimed. □

Remark 5.5 For the special case of Banach sequence spaces, Theorem 5.4 has been first proved by de Jeu and Wortel in [7, Theorem 5.7].

We now shall list several criteria for the group G in Theorem 5.4 to be trivial. In Sect. 6 we will translate those criteria into sufficient conditions for the strong convergence of positive operator semigroups.

A positive linear operator T on a Banach lattice E is called **strongly positive** if Tf is a quasi-interior point for every non-zero positive vector $f \in E$.

Corollary 5.6 *Let G be a strongly compact group of positive invertible operators on an atomic Banach lattice* $E \neq \{0\}$*. Then each one of the following assertions implies that* $G = \{\mathrm{I}_E\}$*:*

1) *G is divisible (cf. Appendix C).*
2) *G has no clopen subgroups different from G.*
3) *G contains a strongly positive operator.*
4) *Every finite-dimensional G-invariant band of E on which G acts irreducibly has dimension* ≤ 1.
5) $\mathbb{K} = \mathbb{C}$ *and G is Abelian and does not have any non-constant torsion eigenvalues.*

Proof By Lemma 5.3 it suffices to prove in each of the mentioned cases that the group homomorphism φ, defined in (5.4), is trivial. We fix $a \in A$ and abbreviate $F := \varphi_G(a)$.

1) G acts transitively on F, which is a finite set. By a standard result from group theory, each homomorphism from a divisible group into a finite one must be trivial (see [14, Lemma 2.3 and Proposition 2.4] for a proof). Hence $F = \{a\}$.
2) The set $H := \{g \in G \mid \varphi_g(a) = a\}$ is a clopen subgroup of G (since A is discrete), so $H = G$.
3) Suppose that $g \in G$ is strongly positive. Then $\varphi_g(a)$ is a quasi-interior point and an atom, hence $\mathrm{dom}(E) = 1$. In particular, $F = \{a\}$.
4) By Theorem 5.4, $\mathrm{span}(F)$ is a finite-dimensional G-invariant band of E on which G acts irreducibly. Hence $1 \leq |F| = \dim(\mathrm{span}(F)) \leq 1$, by assumption. If follows that $F = \{a\}$.
5) Let $m := |F|$. We may consider G as a compact Abelian group of $m \times m$-matrices acting on $\mathrm{span}(F) \cong \mathbb{C}^m$. Since G is commutative, it is simultaneously diagonalizable. Each diagonal entry in a simultaneous diagonalization is an eigenvalue of G. By Theorem 5.4 such an eigenvalue is torsion, and hence, by assumption, trivial. This means that G acts as the identity on $\mathrm{span}(F)$, which implies that $F = \{a\}$. □

6 Convergence of Positive Semigroups

We shall now combine Theorem 4.3 with the findings of the previous section to obtain general results about strong convergence of positive operator semigroups. In all results of this section we shall take the following hypotheses (those of Theorem 4.3) as a starting point:

- E is a Banach lattice;
- $\mathcal{T} = (T_s)_{s\in S}$ is a positive and bounded operator semigroup on E;
- $\mathcal{T}$ restricts to a bounded semigroup on E_y and is E_y-to-E quasi-compact for some quasi-interior point $y \in E_+$.

Let us call these our **standard assumptions** for the remainder of this paper. The standard assumptions warrant that Theorem 4.3 is applicable, and we freely make use of this fact in the following.

6.1 Spectral-Theoretic Consequences

We first draw some spectral-theoretic conclusions.

Theorem 6.1 *Suppose that an operator semigroup $(T_s)_{s\in S}$ on a complex Banach lattice E satisfies the standard assumptions. Then the following assertions hold:*

a) *Each unimodular eigenvalue of $\mathcal{T}$ is torsion.*
b) *If $\mathcal{T}$ is irreducible, then it has only finitely many unimodular eigenvalues.*
c) *$\mathcal{T}$ is strongly convergent if and only if $\mathcal{T}$ has no non-constant torsion eigenvalue.*

Proof

a) By Theorem 2.2, each unimodular eigenvalue of $\mathcal{T}$ is the restriction of an eigenvalue of

$$\mathcal{G} := \mathcal{T}|_{E_\infty},$$

where $E_\infty = [\mathrm{ran}(P_\infty)]$. Since the latter space is atomic and $\mathcal{G}$ is compact, Theorem 5.4 yields that this eigenvalue is torsion.
b) If $\mathcal{T}$ acts irreducibly on E then, by Theorem 4.3, $\mathcal{G}$ acts irreducibly on $[\mathrm{ran}(P_\infty)]$. Hence, $\mathcal{G}$ has only finitely many unimodular eigenvalues by Theorem 5.4. By Theorem 2.2, each eigenvalue of $\mathcal{T}$ is the restriction of an eigenvalue of $\mathcal{G}$. This proves the claim.
c) The "if"-part follows from Corollary 5.6. For the "only if"-part we suppose that $T_s \to P$ is strongly convergent. Then $P = P_\infty$ and $\mathcal{G} = \{\mathrm{I}_{E_\infty}\}$. Since each unimodular eigenvalue of $\mathcal{T}$ is the restriction of an eigenvalue of $\mathcal{G}$ (Theorem 2.2), $\mathcal{T}$ has no non-constant unimodular eigenvalues. □

6.2 Sufficient Conditions for Convergence

Apart from the spectral characterization of the previous theorem, Corollary 5.6 yields the following sufficient conditions for the convergence of a positive semigroup.

Theorem 6.2 *Suppose that an operator semigroup $(T_s)_{s \in S}$ on a Banach lattice E satisfies the standard assumptions. In addition, let at least one of the following conditions be satisfied:*

1) *S is essentially divisible (e.g.: S is divisible or generates a divisible group; cf. Appendix C);*
2) *S carries a topology such that T is strongly continuous and the only clopen subsemigroup of S containing 0 is S itself (e.g.: S is connected).*
3) *T_s is strongly positive for some $s \in S$.*

Then T is strongly convergent.

Proof Theorem 4.3 is applicable, so it suffices to consider the case that $E \neq \{0\}$ is atomic and has order-continuous norm and that $\mathcal{T}$ is a compact group of positive invertible operators on E. We must show that $\mathcal{T}$ acts trivially on E.

1) By Corollary 5.6 it suffices to show that $\mathcal{T}$ is divisible. But this follows from a straightforward compactness argument.
2) Let $\mathcal{H}$ be any clopen subgroup of $\mathcal{T}$. Then $H := \{s \in S \mid T_s \in \mathcal{H}\}$ is a clopen subsemigroup of S. (Note that $H \neq \emptyset$ since $\mathcal{H} \neq \emptyset$ is open and T_S is dense in $\mathcal{T}$.) By hypothesis, $H = S$, so $T_S \subseteq \mathcal{H}$. Since $\mathcal{H}$ is closed and T_S is dense in $\mathcal{T}$, it follows that $\mathcal{H} = \mathcal{T}$. Hence, $\mathcal{T} = \{I_E\}$ by Corollary 5.6.
3) If T_s is strongly positive, then $\mathcal{T}$ contains a strongly positive operator, and we conclude with the help of Corollary 5.6. □

Remarks 6.3

1) Condition 1) is satisfied, in particular, if $S = \mathbb{R}_+$ (divisible semigroup), but also if $S = \{0\} \cup [1, \infty)$ (not divisible, but generating a divisible group). Note that in the latter case, the semigroup direction is just a subordering of the natural one, but does not coincide with it. Nevertheless, the associated notions of "limit" do coincide.
2) Condition 1) is also satisfied when $S = [0, \infty)$ endowed with the semigroup operation $(a, b) \mapsto a \vee b = \max\{a, b\}$. The semigroup direction coincides with the natural ordering. This semigroup is neither divisible nor does it generate a divisible group (it is not even cancellative). However, it is essentially divisible.

   On the other hand, this example is a little artificial, as each element of S is an idempotent, and hence a representation $T = (T_s)_s$ is just a family of projections with decreasing ranges as s increases. For such semigroups, the question of convergence can often be treated by other methods.

3) The semigroup of **positive dyadic rationals** is

$$D_+ := \{0\} \cup \left\{ \tfrac{k}{2^n} \mid k, n \in \mathbb{N}_0 \right\}.$$

The semigroup direction on D_+ coincides with the usual ordering. It is easy to see that D_+ is not essentially divisible. If we endow D_+ with its natural topology, D_+ is not connected. However, D is the only clopen subsemigroup of D_+ containing 0. (Actually, apart from D_+ itself there is no other *open* subsemigroup of D_+ containing 0.)

Hence, from Theorem 6.2 it follows that each strongly continuous representation of D_+ that satisfies the standard hypotheses is strongly convergent. Without strong continuity, however, this can fail. In fact, let $D = D_+ - D_+$ denote the group generated by D_+ in the real numbers. Then D has a subgroup of index 3 (namely $3D$), so by the same construction as in the proof of [14, Theorem 2.5] we can find a positive and bounded representation of $(D, +)$ on the Banach lattice $\mathbb{R}^3$. The restriction of this representation to $(D_+, +)$ satisfies the standard conditions, but it does not converge.

4) With Theorem 6.2, Condition 3), we generalize a result of Gerlach, cf. [11, Theorem 4.3].

6.3 Lattice Subrepresentations

A closed linear subspace F of Banach lattice E is called a **lattice subspace**, if it is a Banach lattice with the order induced by E but with respect to an equivalent norm. A lattice subspace need not be a sublattice. (By Theorem 4.1, the range of a positive projection is always a lattice subspace.)

Given a representation $T = (T_s)_{s \in S}$ on a Banach lattice E, each T-invariant lattice subspace gives rise to a **lattice subrepresentation**. So the lattice subrepresentations are those subrepresentations where the underlying space is a lattice subspace.

Theorem 6.4 *Suppose that an operator semigroup $T = (T_s)_{s \in S}$ on a Banach lattice E satisfies the standard assumptions. In addition, suppose that each finite-dimensional lattice subrepresentation of T is at most one-dimensional. Then T is strongly convergent.*

Proof As in the proof of Theorem 6.2 it suffices to consider the case that $E \neq \{0\}$ is atomic and that $\mathcal{T}$ is a compact group of positive invertible operators on E. It then follows that each finite-dimensional $\mathcal{T}$-invariant band of E is at most one-dimensional. Corollary 5.6 is applicable and yields the claim. □

7 Conclusion: Some Classical Theorems Revisited

In this section we start with a little historical survey and end with demonstrating how our approach leads to far-reaching generalizations of the "classical" results.

7.1 Historical Note

In 1982, Günther Greiner in the influential paper [15] proved the following result as "Corollary 3.11":

Theorem 7.1 (Greiner [15]) *Let $T = (T_s)_{s \geq 0}$ be a positive contraction C_0-semigroup on a space $E = \mathrm{L}^p(\mathrm{X})$, $1 \leq p < \infty$, with the following properties:*

1) *There is a strictly positive T-fixed vector;*
2) *For some $s_0 > 0$ the operator T_{s_0} is a kernel operator.*

Then $\lim_{s \to \infty} T_s x$ exists for each $x \in E$.

For the proof, Greiner employed what has become known as "Greiner's 0/2-law" (see [15, Theorem 3.7] and also [16]) and a result of Axmann from [4]. Both results have involved proofs and make use of the lattice structure on the regular operators on Banach lattices with order-continuous norm. The relevance of Greiner's theorem derives from the fact that the assumptions can be frequently verified for semigroups arising in partial differential equations or in stochastics.

For a long time, Greiner's theorem stood somehow isolated within the asymptotic theory of (positive) semigroups. The "revival" of Greiner's theorem as a theoretical result began with a paper of Davies [6] from 2005. Davies showed that the peripheral point spectrum of the generator A of a C_0-semigroup T of positive contractions on a space $E = \mathrm{L}^p(\mathrm{X})$, $1 \leq p < \infty$, has to be trivial in the following cases: (1) X is countable with the counting measure and (2) X is locally compact and second countable and T has the Feller property (i.e., each T_s for $s > 0$ maps E into the space of continuous functions). Case (1) was subsequently generalized by Keicher in [18] to bounded and positive C_0-semigroups on atomic Banach lattices with order-continuous norm, and by Wolff [28] to more general atomic Banach lattices.

Shortly after, Arendt in [2] generalized Davies' results towards the following theorem.

Theorem 7.2 (Greiner [15]/Arendt [2]) *Let A be the generator of a positive contraction C_0-semigroup $T = (T_s)_{s \geq 0}$ on a space $E = \mathrm{L}^p(\mathrm{X})$, $1 \leq p < \infty$. Suppose that for some $s_0 > 0$ the operator T_{s_0} is a kernel operator. Then $\sigma_\mathrm{p}(A) \cap \mathrm{i}\mathbb{R} \subseteq \{0\}$.*

This result is "Theorem 3.1" in Arendt's paper [2]. Interestingly, as observed by Gerlach in [11], it already appears in Greiner's 1982 paper, namely in the first paragraph of his proof of Theorem 7.1 (i.e., his "Corollary 3.11"). We will thus call Theorem 7.2 the *Greiner–Arendt theorem*. Arendt points out that Theorem 7.2 implies Davies' result: in case (1) every positive operator is a kernel operator, whereas in case (2) the Feller property implies that each T_s for $s > 0$ is a kernel operator. (This follows from Bukhvalov's characterization of kernel operators, cf. [2, Corollary 2.4].)

Let us briefly sketch Arendt's proof of Theorem 7.2: If $f \in E$ is an eigenvector of A for the eigenvalue $\lambda \in \mathrm{i}\mathbb{R}$, then $T_s\,|f| \geq |f|$ for all $s \geq 0$. Since each T_s is a contraction and the norm on L^p is strictly monotone, it follows that $|f|$ is a fixed point. By restricting to the set $[\,|f| > 0\,]$ one can assume that $|f|$ is strictly positive. Next, from the weak compactness of the order interval $[0, |f|]$ it follows that the semigroup is weakly relatively compact. Then the JdLG-theory enters the scene and reduces to problem to an atomic Banach lattice with order continuous norm. Finally, Keicher's analysis from [18] shows that the dynamics there must be trivial, and hence $\lambda = 0$.

Arendt's paper is remarkable in several respects. First of all, his proof of Theorem 7.2 employs the JdLG-theory which is central also to the more recent work of Gerlach and Glück, and to the present paper. Secondly, Arendt recalls Greiner's Theorem 7.1 and gives a proof (building, as Greiner did, on Theorem 7.2) under the additional assumption that the semigroup is irreducible. (This proof appears to be the first complete one in English language, cf. [2, Remark 4.3].) Thirdly, Arendt promotes Greiner's result by illustrating its use with several concrete examples.

Most remarkable of all, however, is what is *not* written in [2]: namely that Greiner's Theorem 7.1 almost directly implies the Greiner–Arendt Theorem 7.2. Indeed, one starts exactly as in Arendt's proof until one has found the quasi-interior fixed point $|f|$; then Greiner's theorem tells that $\lim_{s\to\infty} T_s f$ exists, and hence $\lambda = 0$ follows.

In the following years the topic was taken up by M. Gerlach and J. Glück. Gerlach [11] discussed Greiner's approach in a general Banach lattice setting and extended it to semigroups that merely dominate a kernel operator; he also noted that the dominated kernel operator can be replaced by a compact operator. In their quest to find a unifying framework, and stimulated by "Corollary 3.8" in Keicher's paper [18], Gerlach and Glück in [13, 14] finally identified AM-compactness as the right property generalizing the different cases. Alongside a unification, this led also to a major simplification, since AM-compactness is much more easily shown directly than by passing through the concept of a kernel operator. (E.g., it follows directly from Theorem A.4 that a Feller operator as considered by Davies is AM-compact.)

Finally, Gerlach and Glück realized that strong continuity of the semigroup can be dispensed with, since arguments requiring time regularity can be replaced by purely algebraic ones. This led to proofs for most of the above-mentioned results for semigroups without any time regularity.

Somewhat independently from the above development, Pichór and Rudnicki proved convergence results for Markov semigroups which merely dominate a non-trivial kernel operator [22, Theorems 1 and 2]. Their results are closely related to (and earlier than) the results of Gerlach [11], but their approach is different, focusing on L^1-spaces and employing methods from stochastics. Later on, in [23, 24], these authors adapted their original results to various situations involving semigroups on L^1-spaces, with numerous applications in mathematical biology.

Remark 7.3 As noted above, the implication "Theorem 7.1 ⇒ Theorem 7.2" is almost immediate. Now, in hindsight, it becomes clear that Arendt in [2] was also very close to proving the converse implication "Theorem 7.2 ⇒ Theorem 7.1." Indeed, the weak compactness of order intervals in L^p-spaces for $1 \leq p < \infty$ implies that on such spaces a positive bounded semigroup $T = (T_s)_{s \geq 0}$ with a quasi-interior fixed point is relatively weakly compact. Hence, the "triviality of the peripheral point spectrum" asserted by Theorem 7.2 implies that T acts trivially on the "reversible" part of the corresponding JdLG-decomposition. One can then infer strong convergence of T if one knows that T is not just relatively weakly, but even relatively *strongly* compact. And the latter holds, in fact, since kernel operators are AM-compact; but in this context this was noted only later by Gerlach and Glück.

7.2 Old Theorems in a New Light

Let us now review some of the above-mentioned results in the light of our actual findings. First of all, consider the following result, which is merely an instantiation of Theorem 6.2 a), to the (divisible!) semigroup $\mathbb{R}_+$.

Theorem 7.4 *Let $T = (T_s)_{s \geq 0}$ be a positive and bounded (but not necessarily strongly continuous) semigroup on a Banach lattice E with the following properties:*

1) *There is a quasi-interior point $y \in E_+$ and $c > 0$ such that $T_s y \leq cy$ for all $s \geq 0$.*
2) *For some $s_0 > 0$ the operator T_{s_0} is AM-compact.*

Then $\lim_{s \to \infty} T_s x$ exists for each $x \in E$.

Theorem 7.4 is a slight strengthening of Theorem 4.5 from [14], where the quasi-interior point y is required to be T-fixed. It implies Greiner's Theorem 7.1 as a special case: simply note that on $E := \mathrm{L}^p(\mathrm{X})$ a strictly positive function is quasi-interior and that a kernel operator is AM-compact (Theorem A.3).

In the next result, the requirement that one of the semigroup operators is AM-compact is relaxed towards a mere domination property, however on the expenses of strengthening other hypotheses.

Theorem 7.5 *Let $T = (T_s)_{s \geq 0}$ be a positive, bounded and irreducible (but not necessarily strongly continuous) semigroup on a Banach lattice E with order continuous norm. Suppose that the conditions are satisfied:*

1) *There is a quasi-interior point $y \in E_+$ such that $T_s y \leq y$ for all $s \geq 0$.*
2) *For some $s_0 > 0$ there is an AM-compact operator $K \neq 0$ with $0 \leq K \leq T_{s_0}$.*

Then $\lim_{s \to \infty} T_s x$ exists for each $x \in E$.

Proof By Lemma 4.5, T satisfies the standard assumptions (see Sect. 6). Hence, as $\mathbb{R}_+$ is a divisible semigroup, the assertions follow from Theorem 6.2 a). □

Theorem 7.5 is a generalization of the abovementioned results [22, Theorems 1 and 2] of Pichór und Rudnicki for stochastic C_0-semigroups on L^1-spaces. For C_0-semigroups on Banach lattices, the theorem is due to Gerlach [11, Theorem 4.2].

We note that irreducibility of the semigroup can be replaced by other assumptions ensuring that the AM-compact operator K is "sufficiently large" when compared with the semigroup. A very general result of this type was proved by Gerlach and Glück in [14, Theorem 3.11].

Let us finally return to the spectral-theoretic results (by Davies, Keicher, Wolff and Greiner–Arendt) discussed above. In this direction, we establish the following general theorem.

Theorem 7.6 *Let $T = (T_s)_{s\in S}$ be a bounded and positive semigroup on a Banach lattice E. Suppose that for some $s \in S$ the operator T_s is AM-compact. Then the following assertions hold:*

a) *Each unimodular eigenvalue is torsion.*
b) *If T is irreducible, then there are only finitely many unimodular eigenvalues.*
c) *Suppose that 1) S is essentially divisible or 2) T is strongly continuous with respect to some topology on S such that the only clopen subsemigroup of S containing 0 is S itself. Then the only possible unimodular eigenvalue of T is the constant one.*

Proof We combine the classical ideas from Scheffold [26] as employed by Keicher in [18, Theorem 3.1] with the theory developed in this paper.

a) Let $\lambda = (\lambda_s)_{s\in S}$ be a unimodular eigenvalue of T, and let $0 \neq z \in E$ be a corresponding eigenvector. Abbreviate $y := |z| \in E_+$. Then

$$0 \neq y = |z| = |\lambda_s z| = |T_s z| \leq T_s\,|z| = T_s y \qquad (s \in S).$$

It follows that the net $(T_s y)_{s\in S}$ is increasing.

One can find a positive linear functional $\varphi \in E'_+$ such that $\varphi(y) > 0$. Define

$$J := \{x \in E \mid \lim_{s\in S} \varphi(T_s\,|x|) = 0\}.$$

It is routine to check that J is a closed and T-invariant ideal. Since $\varphi(T_s y) \geq \varphi(y) > 0$ for all $s \in S$, we have $y \notin J$. Moreover, $T_t y - y \in J$ since

$$\varphi(T_s\,|T_t y - y|) = \varphi T_s(T_t y - y) = \varphi(T_{t+s} y) - \varphi(T_s y) \qquad (s \in S)$$

and $(\varphi(T_s x))_{s\in S}$ is increasing and bounded.

Since J is a closed T-invariant ideal, the quotient space $E_1 := E/J$ naturally carries the structure of a Banach lattice, and the representation T on E induces a representation $\hat{T}$ on E_1 by

$$\hat{T}_s(x+J) := T_s x + J \qquad (s \in S,\ x \in E).$$

Let $\hat{z} := z + J$ and $\hat{y} := y + J$ be the equivalence classes of z and y in $E_1 = E/J$, respectively. Since the canonical surjection is a lattice homomorphism, $\hat{y} = |\hat{z}|$ in E_1. Since $y \notin J$, $\hat{z} \neq 0$. It follows that $\hat{z}$ is an eigenvector of $\hat{T}$ for the eigenvalue λ.

Since $T_s y - y \in J$ for each $s \in S$, the point $\hat{y}$ is T-fixed for the induced semigroup on E_1. Moreover, by the hypothesis and Theorem A.2, for some $s \in S$ the operator $\hat{T}_s$ is AM-compact. Hence, when we restrict to the closed ideal $E_2 := \overline{F_{\hat{y}}}$ generated by $\hat{y}$ in E_1, we find that the semigroup $\hat{T}$ restricted to E_2 satisfies the standard assumptions. Theorem 6.1 then yields that that λ must be torsion.

c) We start again as in the proof of a). By Theorem 6.2, either of the conditions 1) and 2) implies that $\hat{T}$ on E_2 is convergent. Then, by Theorem 6.1, we conclude that λ is constant. □

Theorem 7.6 generalizes the Greiner–Arendt Theorem 7.2: simply specialize $S = \mathbb{R}_+$ and note that kernel operators are AM-compact (Theorem A.3). A fortiori, it generalizes Davies' results from [6]. However, it also implies Keicher's result [18, Theorem 3.1] (but not Wolff's), as on an atomic Banach lattice with order-continuous norm all order intervals are relatively compact, and hence all bounded operators are AM-compact. Finally, Theorem 7.6 also generalizes [14, Theorem 4.19].

Acknowledgements Thank you, Ben, for having been a truly optimal boss at the Delft Institute of Applied Mathematics during my time there. You are a wonderful person, and the appreciation you showed for me and my work always made me very proud. Markus

Appendix A: AM-Compact Operators

Let E be a Banach lattice and F a Banach space. A bounded operator $T : E \to F$ is called **AM-compact** if T maps order intervals of E to relatively compact subsets of F.

The following is a useful characterization of AM-compactness in the case that E has a quasi-interior point $y \in E_+$. Recall that this means that the principal ideal

$$E_y := \{x \in E \mid \text{there is } c \geq 0 \text{ such that } |x| \leq cy\}$$

is dense in E. We endow E_y with its natural AM-norm

$$\|x\|_y := \inf\{c \geq 0 \mid |x| \leq cy\}.$$

It is well known that this turns E_y into a Banach lattice, isometrically lattice isomorphic to $C(K)$ (with y being mapped to $\mathbf{1}$) for some compact Hausdorff space K (Krein–Krein–Kakutani theorem).

Lemma A.1 *Let E be a Banach lattice with a quasi-interior point $y \in E_+$, let F be any other Banach space and $T : E \to F$ a bounded operator. Then $T : E_y \to F$ is compact if and only if $T[0, y]$ is relatively compact, if and only if T is AM-compact.*

Proof Suppose that $T : E_y \to F$ is compact. Then, $T[0, y]$ is relatively compact. If the latter is the case, then for each $c > 0$ the set

$$T[-cy, cy] = T(-cy + 2c[0, y]) = T(-cy) + 2cT[0, y]$$

is also relatively compact. Let $u \in E_+$ and $\varepsilon > 0$. Since y is a quasi-interior point, there is $c > 0$ such that $u \in [-cy, cy] + B[0, \varepsilon]$. We claim that

$$[0, u] \subseteq [-cy, cy] + B[0, \varepsilon] \tag{A.1}$$

as well. Indeed, write $u = z + r$ with $|z| \leq cy$ and $\|r\| \leq \varepsilon$ and let $0 \leq x \leq u$. Then $0 \leq x \leq |z| + |r| \leq cy + |r|$. By the decomposition property, there are $0 \leq x_1 \leq |z|$ and $0 \leq x_2 \leq |r|$ with $x = x_1 + x_2$. Hence $x \in [-cy, cy] + B[0, \varepsilon]$ as claimed.

It follows from (A.1) that

$$T[0, u] \subseteq T[-cy, cy] + B[0, \|T\|\varepsilon].$$

Since $T[-cy, cy]$ is relatively compact, it admits a finite $\|T\|\varepsilon$-mesh. Hence, $T[0, u]$ admits a finite $2\|T\|\varepsilon$-mesh. As $\varepsilon > 0$ was arbitrary, $T[0, u]$ is relatively compact.

Finally, if $[u, v]$ is any non-empty order interval of E, then $v - u \geq 0$ and $T[u, v] = Tu + T[0, v - u]$ is relatively compact, by what we have already shown. □

The following theorem shows that AM-compactness is preserved when one passes to a factor lattice with respect to an invariant closed ideal.

Theorem A.2 *Let E be a Banach lattice, let $T \in \mathcal{L}(E)$ be AM-compact, and let $I \subseteq E$ be a T-invariant closed lattice ideal in E. Then the induced operator*

$$T_/ : E/I \to E/I \qquad T_/(x + I) := Tx + I \qquad (x \in E)$$

is also AM-compact.

Proof Let $0 \leq X \in E/I$. We have to show: the order interval $[0, X]$ in E/I is mapped by $T_/$ to a relatively compact subset of E/I.

Recall from [25, Prop. II.2.6] that for $x, y \in E$ one has

$$x+I \leq y+I \iff \exists x_1, y_1 \in E : x - x_1, y - y_1 \in I, \; x_1 \leq y_1.$$

In particular, there is $x \geq 0$ such that $X = x+I$. Now, let $0 \leq Y \leq X$ be given. Then there are $y \in E$ and $z \in I$ such that $y \leq x + z$ and $Y = y+I$. Replacing z by $|z|$ we may suppose in addition that $z \geq 0$.

By the decomposition property, $y^+ = a + b$ for some elements $0 \leq a \leq x$ and $0 \leq b \leq z$. Then $b \in I$ since I is an ideal and $z \in I$. As the canonical surjection is a lattice homomorphism, $Y^+ = a+I$ and hence $T_/(Y^+) = Ta+I$. Similarly, one can show that there is $c \in [0, x]$ with $Y^- = c+I$. It follows that

$$T_/(Y) = T_/(Y^+) - T_/(Y^-) \in T([-x, x]) + I$$

and hence $T_/([0, X]) \subseteq T([-x, x])+I$. Since T is AM-compact, the set $T([-x, x])$ is relatively compact in E, and hence so is $T([-x, x]) + I$ in E/I. This implies that $T_/([0, X])$ is relatively compact, as desired. □

Examples of AM-Compact Operators

In the remainder of this appendix we consider two results that help to identify AM-compact operators. The first tells that every integral operator is AM-compact. This is a well-known consequence of abstract theory (see, e.g., the discussion at the beginning of [14, Section 4], which is based on abstract results in [21, Corollary 3.7.3] and [25, Proposition IV.9.8]). Gerlach and Glück give an elementary proof involving measure-theoretic (i.e., almost everywhere) arguments [14, Proposition A.1]. Our proof is a little less elementary, but replaces the measure theory by functional analysis.

Theorem A.3 *Let* X *and* Y *be measure spaces and let* $1 \leq p, q < \infty$. *Let, furthermore,* $k : X \times Y \to \mathbb{R}_+$ *be a measurable mapping such that by*

$$Tf(x) := \int_{\mathrm{Y}} k(x, \cdot) f \qquad (x \in X)$$

a bounded operator

$$T : \mathrm{L}^p(\mathrm{Y}) \to \mathrm{L}^q(\mathrm{X})$$

is (well) defined. Then T *is AM-compact.*

Proof Let the measure spaces be $\mathrm{X} = (X, \Sigma_{\mathrm{X}}, \mu_{\mathrm{X}})$ and $\mathrm{Y} = (Y, \Sigma_{\mathrm{Y}}, \mu_{\mathrm{Y}})$. Let $0 \leq u \in \mathrm{L}^p(\mathrm{Y})$. We have to show that $T[0, u] \subseteq \mathrm{L}^q(\mathrm{X})$ is relatively compact.

Suppose first that both measures μ_{X} and μ_{Y} are finite, $u = \mathbf{1}$ and $T\mathbf{1} \in \mathrm{L}^\infty(\mathrm{X})$. Then, in particular, $k \in \mathrm{L}^1(\mathrm{X} \times \mathrm{Y})$. Since $\mathrm{L}^1(\mathrm{X}) \otimes \mathrm{L}^1(\mathrm{Y})$ is dense in $\mathrm{L}^1(\mathrm{X} \times \mathrm{Y})$, $T : \mathrm{L}^\infty(\mathrm{Y}) \to \mathrm{L}^1(\mathrm{X})$ is compact. But T factors through $\mathrm{L}^\infty(\mathrm{X})$ (since $T\mathbf{1}$ is bounded) and since on closed L^∞-balls the L^q- and the L^1-topology coincide, $T : \mathrm{L}^\infty(\mathrm{Y}) \to \mathrm{L}^q(\mathrm{X})$ is compact. In particular, $T[0, \mathbf{1}]$ is relatively compact in $\mathrm{L}^q(\mathrm{X})$.

In the general case, consider the operator

$$S : \mathrm{L}^p(Y, \Sigma_Y, u^p \mu_Y) \to \mathrm{L}^q(\mathrm{X}), \qquad Sf := T(uf).$$

Because of $[0, u] = u \cdot [0, \mathbf{1}]$ it suffices to show that $S[0, \mathbf{1}] \subseteq \mathrm{L}^q(\mathrm{X})$ is relatively compact. Hence, one may suppose without loss of generality that μ_Y is finite and $u = \mathbf{1}$.

Under this assumption let

$$g = T\mathbf{1} := \Big(x \mapsto \int_Y k(x, y)\, \mu_Y(\mathrm{d}y)\Big) \in \mathrm{L}^q(\mathrm{X})$$

and, for $n \in \mathbb{N}$,

$$g_n := \mathbf{1}_{\left[\frac{1}{n} \le g \le n\right]} g \in \mathrm{L}^1(\mathrm{X}) \cap \mathrm{L}^\infty(\mathrm{X}).$$

Define T_n by

$$T_n f := \mathbf{1}_{\left[\frac{1}{n} \le g \le n\right]} T f = \int_Y \mathbf{1}_{\left[\frac{1}{n} \le g \le n\right]}(x)\, k(x, y) f(y)\, \mu_Y(\mathrm{d}y).$$

Observe that $T_n\mathbf{1} \in \mathrm{L}^\infty(\mathrm{X})$ and it is supported on a set of finite measure. So, by what we have shown first, $T_n : \mathrm{L}^\infty(\mathrm{Y}) \to \mathrm{L}^q(\mathrm{X})$ is compact. Now,

$$|(T - T_n) f| \le \Big(\mathbf{1}_{\left[g \le \frac{1}{n}\right]} + \mathbf{1}_{[g \ge n]}\Big)\, g\, \|f\|_{\mathrm{L}^\infty}$$

and hence

$$\|T - T_n\|_{\mathrm{L}^q \leftarrow \mathrm{L}^\infty} \le \| \ldots g\|_{\mathrm{L}^q} \to 0 \qquad (n \to \infty).$$

Consequently, also $T : \mathrm{L}^\infty(\mathrm{Y}) \to \mathrm{L}^q(\mathrm{X})$ is compact, and this concludes the proof. □

We turn to a second class of examples of AM-compact operators.

Theorem A.4 *Let* X *and* Y *be measure spaces, let* $1 \le p, q < \infty$, *and let*

$$T : \mathrm{L}^p(\mathrm{Y}) \to \mathrm{L}^q(\mathrm{X})$$

be a positive operator with the following property: There is a sequence $(A_n)_n$ *in* Σ_X *such that* $\bigcup_n A_n = X$ *and such that*

$$\mathbf{1}_{A_n} \cdot \operatorname{ran}(T) \subseteq \mathrm{L}^\infty(A_n) \qquad \textit{for all } n \in \mathbb{N}.$$

Then T *is AM-compact.*

Proof By performing the same reduction as in the proof of Theorem A.3 above, we may suppose without loss of generality that Y is a finite measure space. It remains to show that $T[0, \mathbf{1}]$ is relatively L^q-compact.

Define the projection P_n on $\mathrm{L}^q(\mathrm{X})$ by

$$P_n f := \mathbf{1}_{A_n \cap \left[T\mathbf{1} \geq \frac{1}{n} \right]} f.$$

By hypothesis and the dominated convergence theorem, $P_n T \to T$ uniformly on order intervals of $\mathrm{L}^p(\mathrm{Y})$, hence in particular with respect to the norm of $\mathcal{L}(\mathrm{L}^\infty; \mathrm{L}^q)$. Therefore, it suffices to consider the case that also X is a finite measure space, and $\mathrm{ran}(T) \subseteq \mathrm{L}^\infty(\mathrm{X})$. By the subsequent Lemma A.5 we are done. □

Lemma A.5 *Let* X *and* Y *be finite measure spaces,* $1 \leq p, q < \infty$ *and* $T : \mathrm{L}^p(\mathrm{Y}) \to \mathrm{L}^q(\mathrm{X})$ *a bounded operator with* $\mathrm{ran}(T) \subseteq \mathrm{L}^\infty(\mathrm{X})$. *Then for each* $1 < r \leq \infty$ *with* $r \geq p$ *the operator*

$$T|_{\mathrm{L}^r} : \mathrm{L}^r(\mathrm{Y}) \to \mathrm{L}^q(\mathrm{X})$$

is compact.

Proof We may suppose $1 < p = r < \infty$. By the closed graph theorem, $T : \mathrm{L}^p(\mathrm{Y}) \to \mathrm{L}^\infty(\mathrm{X})$ is bounded. If $(f_n)_n$ is a bounded sequence in $\mathrm{L}^p(\mathrm{Y})$, it has a weakly convergent subsequence (by reflexivity). Hence, $(Tf_n)_n$ has a weakly convergent subsequence in $\mathrm{L}^\infty(\mathrm{X})$. But a weakly convergent sequence in $\mathrm{L}^\infty(\mathrm{X})$ converges strongly in $\mathrm{L}^q(\mathrm{X})$ since X is a finite measure space. (Represent $\mathrm{L}^\infty(\mathrm{X}) \cong \mathrm{C}(K)$ for some compact Hausdorff space K, and observe that in $\mathrm{C}(K)$ a sequence(!) is weakly convergent if and only if it is uniformly bounded and pointwise convergent.) □

Remarks A.6

1) Actually, one can reduce Theorem A.4 to Theorem A.3 by employing Bukhvalov's characterization of integral operators [5], variants of which are, for instance, discussed in [17, Theorem 3.9], [29, Theorem 96.5], [3, Theorem 1.5] and [12, Theorem 4.2.12]
2) For $1 \leq p \leq 2$, Lemma A.5 has a much more elementary proof. Indeed, by quite elementary arguments one can show that T is a Hilbert–Schmidt operator when considered as an operator $\mathrm{L}^2(\mathrm{Y}) \to \mathrm{L}^2(\mathrm{X})$.
3) In the proof of Lemma A.5 we have used that, for a finite measure space X, a sequence which converges weakly in $\mathrm{L}^\infty(\mathrm{X})$ must converge strongly on $\mathrm{L}^q(\mathrm{X})$ for each $1 \leq q < \infty$. It would be nice to have a more elementary proof of this fact, avoiding the representation $\mathrm{L}^\infty(\mathrm{X}) \cong \mathrm{C}(K)$.

Appendix B: Universal Nets

In this appendix we treat, for the convenience of the reader, a useful but maybe not so widely known concept from general topology.

Let $(I, \leq)$ be any directed set. A subset A of I is called a **tail** of I if it is of the form $A = \{\alpha \in I \mid \alpha \geq \alpha_0\}$ for some $\alpha_0 \in I$. And it is called **cofinal** if its complement does not contain a tail. Equivalently, A is cofinal if for each $\alpha_0 \in I$ there is $\alpha \geq \alpha_0$ such that $\alpha \in A$. Clearly, each tail is cofinal.

A **tail** of a net $(x_\alpha)_{\alpha\in I}$ in a set X is a subset of the form $\{x_\alpha \mid \alpha \in A\}$, where $A \subseteq I$ is a tail of I. A net $(x_\alpha)_\alpha$ is called **universal** or an **ultranet**, if for each $Y \subseteq X$ either the set Y or its complement contains a tail of the net. If $f : X_1 \to X_2$ is a mapping and $(x_\alpha)_\alpha$ is a universal net in X_1, then $(f(x_\alpha))_\alpha$ is a universal net in X_2. Likewise, a subnet of a universal net is universal.

Lemma B.1 *Each net has a universal subnet.*

Proof Let $(x_\alpha)_\alpha$ be a net in a set X. Then the tails of the net form a filter base, hence by Zorn's lemma there is an ultrafilter $\mathcal{U}$ containing all the tails. For $\beta \in I$ and $U \in \mathcal{U}$ there is $\alpha(\beta, U) \in I$ such that $\alpha(\beta, U) \geq \beta$ and $x_{\alpha(\beta,U)} \in U$. The set $I \times \mathcal{U}$ is directed by

$$(\alpha, U) \leq (\beta, V) \quad \overset{\text{def.}}{\Longleftrightarrow} \quad \alpha \leq \beta \quad \wedge \quad V \subseteq U.$$

The net $(x_{\alpha(\beta,U)})_{\beta,U}$ is a universal subnet of $(x_\alpha)_\alpha$. □

Since a universal net in a topological space X converges if and only if it has a convergent subnet, X is compact if and only if each universal net in X converges.

Lemma B.2 *Let X be a metric space and let $(x_\alpha)_{\alpha\in I}$ be a net in X. Consider the following three assertions:*

(i) *For each $\varepsilon > 0$ there is a compact set $K \subseteq X$ such that $\mathrm{B}[K, \varepsilon]$ contains a tail of $(x_\alpha)_\alpha$.*
(ii) *For each $\varepsilon > 0$ there is $z \in X$ such that $\{\alpha \mid x_\alpha \in \mathrm{B}[z, \varepsilon]\}$ is cofinal.*
(iii) *The net $(x_\alpha)_\alpha$ is a Cauchy net.*

Then (iii) ⇒ (i) ⇒ (ii). *And if $(x_\alpha)_{\alpha\in I}$ is universal, also* (ii) ⇒ (iii). *(Hence, in the latter case, all three assertions are equivalent.)*

Proof (iii) ⇒ (i): Let $\varepsilon > 0$. Then there is $\alpha \in I$ such that $d(x_\beta, x_\gamma) \leq \varepsilon$ for all $\beta, \gamma \geq \alpha$. Hence $\mathrm{B}[\{x_\alpha\}, \varepsilon]$ contains the tail $\{x_\beta \mid \beta \geq \alpha\}$.

(i) ⇒ (ii): Let $\varepsilon > \varepsilon' > 0$. Then there is a finite subset $F \subseteq K$ such that $K \subseteq \bigcup_{z\in F} \mathrm{B}(z, \varepsilon')$. Hence, by (ii), the set $\bigcup_{z\in F} \mathrm{B}[z, 2\varepsilon]$ contains a tail of $(x_\alpha)_\alpha$. Since F is finite, there is $z \in F$ such that $\{\alpha \mid x_\alpha \in \mathrm{B}[z, 2\varepsilon]\}$ is cofinal.

(ii) ⇒ (iii): Suppose that $(x_\alpha)_\alpha$ is universal, let $\varepsilon > 0$ and pick z as in (ii). By the universality of the net either the set $A_\varepsilon := \{\alpha \mid x_\alpha \in \mathrm{B}[z, \varepsilon]\}$ or its complement contains a tail of I. But since A_ε is cofinal, the second alternative is impossible. Hence, $d(x_\alpha, x_\beta) \leq 2\varepsilon$ for all α, β from a tail of I. □

Theorem B.3 *Let $(x_\alpha)_{\alpha\in I}$ be a net in a regular topological space X and let*

$$C := \bigcap_{\beta\in I} \overline{\{x_\alpha \mid \alpha \geq \beta\}}$$

be its set of cluster points. Consider the following assertions:

(i) *Each subnet of $(x_\alpha)_{\alpha\in I}$ has a cluster point.*
(ii) *Each universal subnet of $(x_\alpha)_{\alpha\in I}$ converges.*
(iii) *For each cofinal subsequence $(\alpha_n)_{n\in\mathbb{N}}$ the sequence $(x_{\alpha_n})_{n\in\mathbb{N}}$ has a cluster point.*
(iv) *The set C is non-empty and compact.*

Then (i) $\Leftrightarrow$ (ii) $\Rightarrow$ (iii) *and* (i) $\Rightarrow$ (iv). *If, in addition, I admits a cofinal sequence and X is metrizable, then* (iii) $\Rightarrow$ (iv) *as well.*

Proof (i) $\Leftrightarrow$ (ii): This follows from Lemma B.1 and the fact that a universal net converges if and only if it has a cluster point.

(i) $\Rightarrow$ (iv): By hypothesis, the net $(x_\alpha)_{\alpha\in I}$ itself has a cluster point, so $C \neq \emptyset$. Let $(y_j)_{j\in J}$ be a universal net in C. It suffices to show that $(y_j)_j$ converges. Let

$$M := \{(\alpha, j, U) \mid \alpha \in I,\ U \text{ open in } X \text{ and } y_k \in U \text{ for all } k \geq j\}$$

be directed by

$$(\alpha, j_0, U) \leq (\beta, j_1, V) \quad \overset{\text{def.}}{\Longleftrightarrow} \quad \alpha \leq \beta \ \wedge\ j_0 \leq j_1 \ \wedge\ V \subseteq U.$$

For each $(\alpha, j, U) \in M$ one has $y_j \in U$ and hence there is

$$\alpha \leq \varphi(\alpha, j, U) \in I \quad \text{with} \quad x_{\varphi(\alpha,j,U)} \in U.$$

Then the mapping $\varphi : M \to I$ is cofinal, so $(x_{\varphi(\alpha,j,U)})_{(\alpha,j,U)\in M}$ is a subnet. By hypothesis (i), this subnet has a cluster point $y \in X$, say. We prove that $y_j \to y$.

To this end, let V be any open neighborhood of y in X. Since X is regular, there is an open set W such that $y \in W \subseteq \overline{W} \subseteq V$. It suffices to show that $\overline{W}$ contains a tail of $(y_j)_j$. Suppose that this is not the case. Then, since that net is universal, the complement $U_0 := X \setminus \overline{W}$ contains tail of $(y_j)_j$. This means that there is $j_0 \in J$ such that $y_j \in U_0$ for all $j \geq j_0$. In particular, $(\alpha, j_0, U_0) \in M$ for all $\alpha \in I$.

Fix $\alpha_0 \in I$. Then, since y is a cluster point of $(x_{\varphi(m)})_{m\in M}$ and W is an open neighborhood of y, there is $M \ni (\alpha, j, U) \geq (\alpha_0, j_0, U_0)$ with $x_{\varphi(\alpha,j,U)} \in W$. But by the construction of φ we have also $x_{\varphi(\alpha,j,U)} \in U \subseteq U_0$. Since $U_0 \cap W = \emptyset$, this yields a contradiction and the proof is complete.

(i) $\Rightarrow$ (iii): This is trivial, as the sequence $(x_{\alpha_n})_n$ is a subnet of $(x_\alpha)_\alpha$ whenever $(\alpha_n)_n$ is a cofinal sequence in I.

(iii) ⇒ (iv): Suppose that d is a metric inducing the topology of X and that I admits cofinal sequences. It then follows from (iii) that $C \neq \emptyset$. In order to see that C is compact, let $(y_n)_n$ be a sequence in C and let $(\alpha_n)_n$ be a cofinal sequence in I. Recursively, one can find a sequence $(x_{\beta_n})_n$ such that $\beta_n \geq \alpha_n$ and $d(x_{\beta_n}, y_n) \leq \frac{1}{n}$. By (iii), $(x_{\beta_n})_n$ has a cluster point. Since X is metric, this means that $(x_{\beta_n})_n$ has a convergent subsequence. But then $(y_n)_n$ also has convergent subsequence, and as C is closed, the limit of this subsequence lies in C. □

Appendix C: Some Notions from Semigroup Theory

A **semigroup** is a non-empty set S together with an associative operation $S \times S \to S$, generically called "multiplication." A subset M of a semigroup S is **multiplicative**, if $M \cdot M \subseteq M$. A non-empty multiplicative subset is a **subsemigroup**. A non-empty subset J of a semigroup S is a (two-sided) **ideal** of S if $SJ \cup JS \subseteq J$. Each ideal is a subsemigroup.

A **neutral element** in a semigroup is any element $e \in S$ such that $es = se = s$ for all $s \in S$. There is at most one neutral element; if there is none, one can adjoin one in a standard way.

A semigroup S is called **Abelian** or **commutative** if $st = ts$ for all $s, t \in S$. It is common to write Abelian semigroups additively, and denote their neutral elements by 0.

An Abelian semigroup S is called **cancellative** if it satisfies the implication

$$s + t = s + t' \quad \Rightarrow \quad t = t'$$

for all $s, t, t' \in S$. By a standard result from semigroup theory, S is faithfully embeddable into a group if and only if S is cancellative. In this case, one usually considers S as a *subset* of some group G such that $G = S - S$. The (necessarily Abelian) group G is then unique up to canonical isomorphism and is called the group **generated** by S. In each case, when we speak of a group generated by an Abelian semigroup S, we have this meaning in mind, and in particular suppose tacitly that S is cancellative.

An Abelian semigroup S with neutral element 0 is called **divisible**, if for each $s \in S$ and $n \in \mathbb{N}$ there is $t \in S$ such that $nt = s$. (This definition extends the common notion of divisibility from groups to semigroups.) And S is called **essentially divisible** if for each $s \in S$ and $n \in \mathbb{N}$ there are $t_1, t_2 \in S$ such that $nt_1 = s + nt_2$.

Each divisible semigroup and each semigroup that generates a divisible group is essentially divisible.

References

1. Y. Abramovich, G. Sirotkin, On order convergence of nets. Positivity **9**(3), 287–292 (2005)
2. W. Arendt, Positive semigroups of kernel operators. Positivity **12**(1), 25–44 (2008)
3. W. Arendt, A.V. Bukhvalov, Integral representations of resolvents and semigroups. Forum Math. **6**(1), 111–135 (1994)
4. D. Axmann, Struktur-und Ergodentheorie irreduzibler Operatoren auf Banachverbänden. PhD thesis, Eberhard-Karls-Universität zu Tübingen, 1980
5. A.V. Bukhvalov, Integral representation of linear operators. J. Math. Sci. **9**(2), 129–137 (1978)
6. E.B. Davies, Triviality of the peripheral point spectrum. J. Evol. Equ. **5**(3), 407–415 (2005)
7. M. de Jeu, M. Wortel, Compact groups of positive operators on Banach lattices. Indag. Math. (N.S.) **25**(2), 186–205 (2014)
8. T. Eisner, B. Farkas, M. Haase, R. Nagel, *Operator Theoretic Aspects of Ergodic Theory*. Graduate Texts in Mathematics, vol. 272 (Springer, Cham, 2015)
9. E.Yu. Emel'yanov, U. Kohler, F. Räbiger, M.P.H. Wolff, Stability and almost periodicity of asymptotically dominated semigroups of positive operators. Proc. Am. Math. Soc. **129**(9), 2633–2642 (2001)
10. K.-J. Engel, R. Nagel, *One-Parameter Semigroups for Linear Evolution Equations*. Graduate Texts in Mathematics, vol. 194 (Springer, New York, 2000). With contributions by S. Brendle, M. Campiti, T. Hahn, G. Metafune, G. Nickel, D. Pallara, C. Perazzoli, A. Rhandi, S. Romanelli and R. Schnaubelt
11. M. Gerlach, On the peripheral point spectrum and the asymptotic behavior of irreducible semigroups of Harris operators. Positivity **17**(3), 875–898 (2013)
12. M. Gerlach, Semigroups of Kernel operators. PhD thesis, Universität Ulm, 2014. https://doi.org/10.18725/OPARU-3263
13. M. Gerlach, J. Glück, On a convergence theorem for semigroups of positive integral operators. C. R. Math. Acad. Sci. Paris **355**(9), 973–976 (2017)
14. M. Gerlach, J. Glück, Convergence of positive operator semigroups (2017). Preprint: arxiv.org/abs/1705.01587v1
15. G. Greiner, *Spektrum und Asymptotik stark stetiger Halbgruppen positiver Operatoren*. Sitzungsberichte der Heidelberger Akademie der Wissenschaften, Mathematisch-Naturwissenschaftliche Klasse (Springer, Berlin, 1982), pp. 55–80
16. G. Greiner, R. Nagel, La loi "zéro ou deux" et ses conséquences pour le comportement asymptotique des opérateurs positifs. J. Math. Pures Appl. (9) **61**(3), 261–273 (1982)
17. J.J. Grobler, P. van Eldik, A characterization of the band of kernel operators. Quaest. Math. **4**(2), 89–107 (1980/1981)
18. V. Keicher, On the peripheral spectrum of bounded positive semigroups on atomic Banach lattices. Arch. Math. (Basel) **87**(4), 359–367 (2006)
19. M.C. Kunze, Diffusion with nonlocal Dirichlet boundary conditions on unbounded domains (2018). Preprint: arxiv.org/abs/1810.04474v1
20. W.A.J. Luxemburg, A.C. Zaanen, *Riesz Spaces. Volume I*. North-Holland Mathematical Library (North-Holland/American Elsevier, Amsterdam/New York, 1971)
21. P. Meyer-Nieberg, *Banach Lattices*. Universitext (Springer, Berlin, 1991)
22. K. Pichór, R. Rudnicki, Continuous Markov semigroups and stability of transport equations. J. Math. Anal. Appl. **249**(2), 668–685 (2000)
23. K. Pichór, R. Rudnicki, Asymptotic decomposition of substochastic operators and semigroups. J. Math. Anal. Appl. **436**(1), 305–321 (2016)
24. K. Pichór, R. Rudnicki, Asymptotic decomposition of substochastic semigroups and applications. Stoch. Dyn. **18**(1), 1850001 (2018)
25. H.H. Schaefer, *Banach Lattices and Positive Operators*. Die Grundlehren der mathematischen Wissenschaften, Band 215 (Springer, New York, 1974)
26. E. Scheffold, Das Spektrum von Verbandsoperatoren in Banachverbänden. Math. Z. **123**, 177–190 (1971)

27. W. Wnuk, *Banach Lattices with Order Continuous Norms* (Polish Scientific Publishers PWN, Warsaw, 1999)
28. M.P.H. Wolff, Triviality of the peripheral point spectrum of positive semigroups on atomic Banach lattices. Positivity **12**(1), 185–192 (2008)
29. A.C. Zaanen, *Riesz Spaces II*. North-Holland Mathematical Library, vol. 30 (North-Holland, Amsterdam, 1983)

Markov Processes, Strong Markov Processes and Brownian Motion in Riesz Spaces

Jacobus J. Grobler

To Ben de Pagter

Abstract The different definitions for Markov processes used in the classical case are studied in the abstract setting of vector lattices where they are not all equivalent. Strong Markov processes are defined and it is shown that a Brownian motion is a strong Markov process. This fact is used in a proof that a Brownian filtration is right-continuous. This holds in the classical case only for the augmentation of the Brownian filtration, but in the abstract case the augmented filtration is not larger than the original one.

Keywords Riesz space · Markov process · Strong Markov process · Brownian motion · Brownian filtration

1 Introduction

Markov processes were studied in Riesz spaces by Vardy and Watson in [15], and they mention that there are various definitions of Markov processes and the Markov property in the literature. Since they did not discuss the version that we find useful for our subsequent work, we start by proving the equivalence of the generalizations of the different definitions. We give a short discussion of conditional independence and prove a generalization of the so-called Independence Lemma (see [14, Lemma 2.3.4] and [13, Lemma A3]). It is used to give a short proof of the fact that a Brownian motion is a Markov process (something already proved in [15] in another way). Its real use is to show that a Brownian motion is a strong Markov process (see, for example, [10] for the classical case). The latter fact is used to prove that for

J. J. Grobler (✉)
Department of Finance and Investment Management, University of Johannesburg, Johannesburg, South Africa
e-mail: jacjgrobler@gmail.com

G. Buskes et al. (eds.), *Positivity and Noncommutative Analysis*,
Trends in Mathematics, https://doi.org/10.1007/978-3-030-10850-2_10

the Brownian motion (B_t), the Brownian filtration $(\mathfrak{F}_t^B, \mathbb{F}_t^B)$ is continuous. In the classical case this is true for the augmentation of the filtration, but we will remark that in our case, the Brownian filtration is equal to its augmentation.

2 Preliminaries

Let $\mathfrak{E}$ be a Dedekind complete Riesz space with weak order E and with separating order continuous dual $\mathfrak{E}_{00}^{\sim}$. We shall assume that $\mathfrak{E}$ is perfect, i.e., $(\mathfrak{E}_{00}^{\sim})_{00}^{\sim} = \mathfrak{E}$. We use the notation and definitions contained in [8, Section 2] and we refer the reader to this paper for any of the notions that are not explained in the text. We denote the *universal completion* of $\mathfrak{E}$ by $\mathfrak{E}^u$ and its *sup-completion* by $\mathfrak{E}^s$. Our notation for the band generated by $(A - B)^+$ will be $\mathfrak{B}(A > B)$ and its disjoint complement by $\mathfrak{B}(A \leq B)$. The associated band projections will be denoted in a similar manner by $\mathbb{P}(A > B)$ and $\mathbb{P}(A \leq B)$. Orth$(\mathfrak{E})$ will denote the set of all band preserving order bounded operators (called *orthomorphisms*) on $\mathfrak{E}$. The set of all order projections in $\mathfrak{E}$ will be denoted by $\mathfrak{P}_{\mathfrak{E}}$ and $\mathfrak{C}_{\mathfrak{E}} := \{\mathbb{P}E : \mathbb{P} \in \mathfrak{P}_{\mathfrak{E}}\}$.

A *conditional expectation* $\mathbb{F}$ defined on $\mathfrak{E}$ is a strictly positive order continuous linear projection that maps weak order units onto weak order units and that has as its range a Dedekind complete Riesz subspace $\mathfrak{F}$ of $\mathfrak{E}$. It will be assumed (as it may) that $\mathbb{F}(E) = E$.

The Riesz space $\mathfrak{E}$ is called $\mathbb{F}$-universally complete in $\mathfrak{E}$ (and in this case we simply say that $\mathfrak{E}$ is $\mathbb{F}$-universally complete) if, whenever $X_\alpha \uparrow$ in $\mathfrak{E}$ and if boundedness of $\mathbb{F}(X_\alpha)$ in $\mathfrak{E}$, then $X_\alpha \uparrow X$ for some $X \in \mathfrak{E}$ (see [17, Definition 2.3]). If $X_\alpha \uparrow$ in $\mathfrak{E}$ and if boundedness of $\mathbb{F}(X_\alpha)$ in $\mathfrak{E}^u$ imply that $X_\alpha \uparrow X$ for some $X \in \mathfrak{E}$, then $\mathfrak{E}$ is said to be $\mathbb{F}$-universally complete in $\mathfrak{E}^u$. It is obvious that if $\mathfrak{E}$ is $\mathbb{F}$-universally complete in $\mathfrak{E}^u$, then it is $\mathbb{F}$-universally complete in $\mathfrak{E}$. (To avoid confusion, we alert the reader to the fact that the latter notion is called $\mathbb{F}$-universally complete in [12].)

If $\mathfrak{E}$ is $\mathbb{F}$-universally complete in $\mathfrak{E}$ and if $\mathfrak{G}$ is an order closed Riesz subspace of $\mathfrak{E}$ with $\mathcal{R}(\mathbb{F}) \subset \mathfrak{G}$, then there exists a unique conditional expectation $\mathbb{F}_{\mathfrak{G}}$ on $\mathfrak{E}$ with $\mathcal{R}(\mathbb{F}_{\mathfrak{G}}) = \mathfrak{G}$ and $\mathbb{F}\mathbb{F}_{\mathfrak{G}} = \mathbb{F}_{\mathfrak{G}}\mathbb{F} = \mathbb{F}$ (see [5, 17]). In particular this result holds if $\mathfrak{E}$ is $\mathbb{F}$-universally complete in $\mathfrak{E}^u$.

The following fact will often be used. It is contained in the proof of [5, Theorem 3.4]:

Proposition 2.1 *Let $\mathfrak{E}$ be $\mathbb{F}$-universally complete in $\mathfrak{E}$ and let $\mathfrak{G}$ be an order closed Riesz subspace of $\mathfrak{E}$ such that $\mathcal{R}(\mathbb{F}) = \mathfrak{F} \subset \mathfrak{G}$. If $Y \in \mathfrak{G}$ and $X \in \mathfrak{E}$ satisfy*

$$\mathbb{F}\mathbb{P}X = \mathbb{F}\mathbb{P}Y, \text{ for all } \mathbb{P} \in \mathfrak{P}_{\mathfrak{G}}, \tag{2.1}$$

then $Y = \mathbb{F}_{\mathfrak{G}}X$. Conversely, if $Y = \mathbb{F}_{\mathfrak{G}}X$ for $X \in \mathfrak{E}$, then the condition (2.1) holds.

It follows from the remark before the proposition that for any $X \in \mathfrak{E}$ there exists a unique $Y \in \mathfrak{G}$ that satisfies this condition.

For further properties of the conditional expectation and its extension to $\mathfrak{E}^s$, we refer the reader to [8].

Let $J \subset \mathbb{F}$ be an interval. A *filtration* on $\mathfrak{E}$ is a family $(\mathbb{F}_t)_{t\in J}$ of conditional expectations satisfying $\mathbb{F}_s = \mathbb{F}_s\mathbb{F}_t$ for all $s < t$. We denote the range of $\mathbb{F}_t$ by $\mathfrak{F}_t$ and note that the ranges form a nested family of Dedekind complete subspaces of $\mathfrak{E}$ satisfying $\mathfrak{F}_s \subset \mathfrak{F}_t$ if $s < t$. We define $\mathfrak{F}_{t+} : \bigcap_{s>t} \mathfrak{F}_s$ and by $\mathfrak{F}_{t-}$ the Dedekind complete Riesz subspace generated by $\bigcup_{s<t} \mathfrak{F}_s$. There exists a unique conditional expectation $\mathbb{F}_{t+}$ mapping $\mathfrak{E}$ onto $\mathfrak{F}_{t+}$ such that $\mathbb{F}_{t+}\mathbb{F}_s = \mathbb{F}_s\mathbb{F}_{t+}$ if $s > t$ and a similar condition holds for $\mathfrak{F}_{t-}$.

A *stochastic process* in $\mathfrak{E}$ is a function $t \mapsto X_t \in \mathfrak{E}$, for $t \in J$. The process $(X_t)_{t\in J}$ is said to be *adapted to the filtration* $(\mathbb{F}_t)$ if $X_t \in \mathfrak{F}_t$ for all $t \in J$. We denote an adapted process by $(X_t, \mathfrak{F}_t, \mathbb{F}_t)_{t\in J}$, or $(X_t, \mathfrak{F}_t)_{t\in J}$. We call $(X_t, \mathfrak{F}_t)_{t\in J}$ a *supermartingale* if $\mathbb{F}_s X_t \le X_s$ for $s \le t$ and a *submartingale* if $\mathbb{F}_s X_t \ge X_s$ for $s \le t$. If equality holds, we call the process a *martingale*.

For $t \in \mathbb{R}$ we define $E_t^\ell \in \mathfrak{C}_{\mathfrak{E}}$ to be the component of the weak order unit E in the band $\mathfrak{B}(tE > X)$. The family $(E_t^\ell)_{t\in\mathbb{R}}$ is an increasing left continuous system of components of E, called the *left continuous spectral system* of X. If $\overline{E}_t^r$ is defined to be the component of E in the band $\mathfrak{B}(X > tE)$ and if we set $E_t^r := E - \overline{E}_t^r$, then the system $(E_t^r)_{t\in\mathbb{R}}$ is an increasing right-continuous system of components of E. We have $E_t^\ell \le E_t^r \le E_s^\ell$ for all $t < s$.

The identity operator $\mathbb{I}$ is a weak order unit for the space $\mathrm{Orth}(\mathfrak{E})$ and its components are order projections in $\mathfrak{E}$. If $\mathbb{S} \in \mathrm{Orth}(\mathfrak{E})$ is such that its left continuous spectral system $(\mathbb{S}_t^\ell)$ of projections satisfies $\mathbb{S}_t^\ell \in \mathfrak{P}_{\mathfrak{F}_t} = \mathfrak{P}_t$, then $\mathbb{S}$ is called an *optional time* for the filtration $(\mathbb{F}_t, \mathfrak{F}_t)_{t\in J}$. If its right-continuous spectral system $(\mathbb{S}_t^r)$ satisfies $\mathbb{S}_t^r \in \mathfrak{P}_t$, it is called a *stopping time* for the filtration. We refer the reader to [5] for properties of stopping and optional times and also for the definition of the order complete Boolean subalgebras $\mathfrak{P}_{\mathbb{S}}$ of all events prior to the stopping time $\mathbb{S}$ and $\mathfrak{P}_{\mathbb{S}+}$ of all events immediately after the optional time $\mathbb{S}$, and also for the definition of $X_{\mathbb{S}}$ where $(X_t, \mathfrak{F}_t, \mathbb{F}_t)$ is a stochastic process and $\mathbb{S}$ is a stopping time for the filtration. The order complete Riesz space generated in $\mathfrak{E}$ by the components of $\mathbb{S}E$ is denoted by $\mathfrak{F}_{\mathbb{S}}$ and we proved that if $\mathbb{S}$ is a stopping time for the filtration $(\mathfrak{F}_t, \mathbb{F}_t)$ then there exists a unique conditional expectation $\mathbb{F}_{\mathbb{S}}$ that maps $\mathfrak{E}$ onto $\mathfrak{F}_{\mathbb{S}}$ because $\mathfrak{F} \subset \mathfrak{F}_{\mathbb{S}}$.

We proved in [6, Lemma 3.7] that $\mu_X(a, b] := E_b^r - E_a^r$ defines a measure on the algebra of all left-open right-closed intervals and that it can be extended to a countably additive vector measure on the Borel σ-algebra $\mathcal{B}(\mathbb{R})$ (see also [16, Chapter XI, Section 5]). Its values are in the set $\{\mathbb{P}E \ : \ \mathbb{P} \in \mathfrak{P}(X)\}$. The measure μ_X is a Boolean measure and as such also satisfies the condition that $\mu_X(A \cap B) = \mu_X(A) \wedge \mu_X(B)$ [16, Theorem XI.5.(c)]. Since we will always be working with the right-continuous spectral system, but will have to distinguish between spectral systems of different elements, we will henceforth denote the right-continuous spectral system of the element X by $(E_t^X)_{t\in\mathbb{R}}$.

The functional calculus can be extended to bounded Borel measurable functions (see [16, p. 324]). For a bounded measurable function the integral

$$f(X) := \int_{\mathbb{R}} f(t)\, d\mu_X(t) \in \mathfrak{E}$$

exists as a limit of measurable step-functions. For an arbitrary positive measurable function the integral also exists, but its value is not necessarily in $\mathfrak{E}$, but in its supremum completion $\mathfrak{E}^s$. The Borel-measurable function f is then said to be integrable with respect to the Boolean measure μ_X whenever both $\int_{\mathbb{R}} f^+\, d\mu_X$ and $\int_{\mathbb{R}} f^-\, d\mu_X$ are elements of $\mathfrak{E}$ and in that case the integral is defined to be their difference.

3 Conditional Independence

We recall the facts about conditional independence that can be found in [7, Section 4].

Let $\mathbb{F}$ be a conditional expectation on $\mathfrak{E}$. We call two events $\mathbb{P}$ and $\mathbb{Q}$ in $\mathfrak{P}$ $\mathbb{F}$-*conditionally independent* whenever

$$\mathbb{F}(\mathbb{P}\mathbb{Q})\mathbb{F} = (\mathbb{F}\mathbb{P})(\mathbb{F}\mathbb{Q})\mathbb{F} = (\mathbb{F}\mathbb{Q})(\mathbb{F}\mathbb{P})\mathbb{F},$$

which is equivalent to $\mathbb{F}(\mathbb{P}\mathbb{Q})|_{\mathfrak{F}} = (\mathbb{F}\mathbb{P})(\mathbb{F}\mathbb{Q})|_{\mathfrak{F}} = (\mathbb{F}\mathbb{Q})(\mathbb{F}\mathbb{P})|_{\mathfrak{F}}$. By [7, Lemma 4.2] $(\mathbb{F}\mathbb{P})(\mathbb{F}\mathbb{Q})\mathbb{F} = (\mathbb{F}\mathbb{Q})(\mathbb{F}\mathbb{P})\mathbb{F}$ and so it is sufficient to define conditional independence by stating only one of the conditions.

A class $\mathfrak{C}$ of projections is $\mathbb{F}$-independent if, for every choice of a finite number of elements $\mathbb{P}_j,\ j = 1, \dots n$ of $\mathfrak{C}$ we have

$$\mathbb{F}(\prod_{j=1}^{n} \mathbb{P}_j)|_{\mathfrak{F}} = \prod_{j=1}^{n} \mathbb{F}\mathbb{P}_j|_{\mathfrak{F}}.$$

Classes $\mathfrak{C}_\alpha$ are called $\mathbb{F}$-independent if, for any choice of projections $\mathbb{P}_\alpha \in \mathfrak{C}_\alpha$, we have that this chosen class is $\mathbb{F}$-independent.

For every $t \in \mathbb{R}$ let $\mathfrak{B}_t := \mathfrak{B}(tE > X)$ with band projection $\mathbb{P}_t = \mathbb{P}(tE > X)$. Let $\mathfrak{P}(X)$ be the order complete Boolean subalgebra of $\mathfrak{P}$ generated by all $\mathbb{P}_t$. We note that $(tE - X)^+$ is an element of the Riesz space $[X, \mathfrak{F}]$ generated by the element X and $\mathfrak{F} = \mathbb{F}(\mathfrak{E})$ and so the projections $\mathbb{P}_t$ are projections in this space, i.e., $\mathfrak{P}(X) \subset \mathfrak{P}_{[X,\mathfrak{F}]}$. We say that two elements $X, Y \in \mathfrak{E}$ are $\mathbb{F}$-conditionally independent if the classes $\mathfrak{P}(X)$ and $\mathfrak{P}(Y)$ are $\mathbb{F}$-independent.

We say that the element X is independent of the algebra $\mathfrak{G}$ of projections whenever the algebras $\mathfrak{G}$ and $\mathfrak{P}(X)$ are $\mathbb{F}$-conditionally independent. This means that for any $\mathbb{Q} \in \mathfrak{G}$ and $\mathbb{P} \in \mathfrak{P}(X)$, we have that $\mathbb{P}$ and $\mathbb{Q}$ are $\mathbb{F}$-conditionally independent.

We can also define classes of elements in $\mathfrak{E}$ to be $\mathbb{F}$-conditionally independent and so it is also meaningful to refer to Riesz subspaces of $\mathfrak{E}$ to be $\mathbb{F}$-conditionally independent.

Denoting the order closed Riesz subspace generated by two subsets $\mathfrak{G}$ and $\mathfrak{H}$ in $\mathfrak{E}$ by $[\mathfrak{G}, \mathfrak{H}]$, we recall [7, Corollary 4.8] that elements X and Y in the Riesz space $\mathfrak{E}$ are $\mathbb{F}$-conditionally independent if and only if the Riesz subspaces $[X, \mathfrak{F}]$ and $[Y, \mathfrak{F}]$ of $\mathfrak{E}$ are $\mathbb{F}$-conditionally independent. Since these spaces contain $\mathfrak{F}$, we have by a result of Watson [17] that if $\mathfrak{E}$ is $\mathbb{F}$-universally complete, then there exist a unique conditional expectation $\mathbb{F}_X : \mathfrak{E} \to [X, \mathfrak{F}]$ such that $\mathbb{F}\mathbb{F}_X = \mathbb{F}_X\mathbb{F} = \mathbb{F}$.

4 Markov Processes

In the theory of Markov processes it is convenient to use some of the classical notation. The order closed $\mathbb{F}$-universally complete Riesz subspace generated in $\mathfrak{E}$ by $\mathfrak{F}$ and the elements $X_1, X_2, \ldots, X_n \in \mathfrak{E}$, that is, the space $[X_1, \ldots, X_n, \mathfrak{F}]$, will be denoted by $\mathfrak{F}_{X_1,X_2,\ldots,X_n}$. The unique conditional expectation that maps $\mathfrak{E}$ onto this subspace will be denoted by $\mathbb{F}(\cdot \mid X_1, X_2, \ldots, X_n)$. This is the conditional expectation that satisfies

$$\mathbb{F}\mathbb{F}(\cdot \mid X_1, X_2, \ldots, X_n) = \mathbb{F}(\cdot \mid X_1, X_2, \ldots, X_n)\mathbb{F} = \mathbb{F}.$$

Similarly, the order closed Riesz subspace generated by $\mathfrak{F}$ and $\{X_s : s \geq t\}$ will be denoted by $\mathfrak{F}_{X_{s,s\geq t}}$ and the conditional expectation onto this space by $\mathbb{F}(\cdot \mid X_{s,s\geq t})$. The notation $\mathfrak{F}_{X_{s,s\leq t}}$ needs no explanation. It defines a filtration on $\mathfrak{E}$ to which (X_t) is adapted and is denoted by Karatzas and Shreve [10] by $\mathfrak{F}_t^X$. We denote the conditional expectation mapping $\mathfrak{E}$ onto $\mathfrak{F}_t^X$ by $\mathbb{F}_t^X$.

We shall use the following definition of a Markov process, which is a generalization of the definition used by Ash and Gardner (see [2, Definition 4.5.1]). In the definition I_B denotes the indicator function of a set B.

Definition 4.1 Let $\mathfrak{E}$ be a Dedekind complete Riesz space with a weak order unit E. Let $(X_t, \mathfrak{F}_t, \mathbb{F}_t)_{t\in J}$ be a stochastic process adapted to the filtration $(\mathfrak{F}_t, \mathbb{F}_t)$. The process is called a *Markov process* if, for every Borel set $B \in \mathcal{B}(\mathbb{R})$ and $s, t \in J$ with $s < t$,

$$\mathbb{F}_s(I_B(X_t)) = \mathbb{F}(I_B(X_t) \mid X_s). \tag{4.1}$$

We note that the definition is equivalent to the following: *For any bounded Borel-measurable function g for which $g(X_t) \in \mathfrak{E}$, we have*

$$\mathbb{F}_s(g(X_t)) = \mathbb{F}(g(X_t) \mid X_s). \tag{4.2}$$

Also, using the definition of the measure μ_{X_t}, Eq. (4.1) can be written as

$$\mathbb{F}_s(\mu_{X_t}(B)) = \mathbb{F}(\mu_{X_t}(B) \mid X_s) \tag{4.3}$$

and the next equation as

$$\mathbb{F}_s\left(\int_{\mathbb{R}} g\, d\mu_{X_t}\right) = \mathbb{F}\left(\int_{\mathbb{R}} g\, d\mu_{X_t} \mid X_s\right). \tag{4.4}$$

If $(X_t)_{t\in J}$ is called a Markov process without mentioning a filtration, the filtration is taken to be $(\mathfrak{F}_t^X, \mathbb{F}_t^X)$. In this case Definition 4.1 becomes

$$\mathbb{F}_s^X(I_B(X_t)) = \mathbb{F}(I_B(X_t) \mid X_r, r \leq s) = \mathbb{F}(I_B(X_t) \mid X_s).$$

Proposition 4.2 *If $(X_t, \mathfrak{F}_t)$ is a Markov process, then $(X_t, \mathfrak{F}_t^X)$ is a Markov process.*

Proof We know that the conditional expectation $\mathbb{F}_s^X$ exists for each fixed s and that $\mathfrak{F}_s^X \subset \mathfrak{F}_s$. We apply Watson's theorem to conclude that there exists a unique conditional expectation operator $\overline{\mathbb{F}_s}$ mapping $\mathfrak{E}$ onto $\mathfrak{F}_s$ satisfying $\overline{\mathbb{F}_s}\mathbb{F}_s^X = \mathbb{F}_s^X\overline{\mathbb{F}_s} = \mathbb{F}_s^X$. Applying $\mathbb{F}$ to both sides of the second equality, we get $\mathbb{F}\overline{\mathbb{F}_s} = \mathbb{F}$. Also, $\mathbb{F} = \mathbb{F}_s^X\mathbb{F} = \overline{\mathbb{F}_s}\mathbb{F}_s^X\mathbb{F} = \overline{\mathbb{F}_s}\mathbb{F}$. Therefore, $\mathbb{F}\overline{\mathbb{F}_s} = \mathbb{F} = \overline{\mathbb{F}_s}\mathbb{F}$. However, $\mathbb{F}_s$ is the unique conditional expectation mapping $\mathfrak{E}$ onto $\mathfrak{F}_s$ and satisfying $\mathbb{F}_s\mathbb{F} = \mathbb{F} = \mathbb{F}\mathbb{F}_s$. This shows that $\mathbb{F}_s = \overline{\mathbb{F}_s}$ and so we have $\mathbb{F}_s\mathbb{F}_s^X = \mathbb{F}_s^X\mathbb{F}_s = \mathbb{F}_s^X$.

Apply now $\mathbb{F}_s^X$ to both sides of Definition 4.1 to get

$$\mathbb{F}_s^X\mathbb{F}_s(I_B(X_t)) = \mathbb{F}_s^X(I_B(X_t)) = \mathbb{F}_s^X\mathbb{F}(I_B(X_t)|X_s) = \mathbb{F}(I_B(X_t)|X_s)$$

since $\mathbb{F}(I_B(X_t)|X_s) \in [\mathfrak{F}, X_s] \subset [\mathfrak{F}, X_r, r \leq s]$, and again equivalently,

$$\mathbb{F}_s^X(g(X_t)) = \mathbb{F}(g(X_t)|X_s)$$

for all bounded Borel functions g. □

Since the class of intervals $(-\infty, x]$ generates the Borel σ-algebra, we can replace the Borel sets B in Definition 4.1 by these sets. Thus it becomes the condition that for all $x \in \mathbb{R}$, we have

$$\mathbb{F}_s(E_x^{X_t}) = \mathbb{F}(E_x^{X_t} \mid X_s), \tag{4.5}$$

where $E_x^{X_t} = \mathbb{P}(X_t \leq xE)E = I_{(-\infty,x]}(X_t)$.

We say that a stochastic process $(X_t)_{t\in J}$ has the Markov property with respect to a finite set $I = \{t_1 < t_2 < \cdots < t_n < t\}$ if, for all Borel sets $B \in \mathbb{R}$ we have

$$\mathbb{F}(I_B(X_t) \mid X_{t_1}, X_{t_1}, \ldots, X_{t_n}) = \mathbb{F}(I_B(X_t) \mid X_{t_n}).$$

A Markov process has the Markov property with respect to every finite subset $I \subset J$. This follows from the following more general fact.

Proposition 4.3 *If $(X_t)_{t\in J}$ is a Markov process, then it is a Markov process for every subset $I \subset J$.*

Proof Let $Y_t := X_t$ for $t \in I$. Then, for $s < t$, $s, t \in I$,

$$\mathfrak{F}_s^Y = \mathfrak{F}_{Y_r, r\in I, r\leq s} = \mathfrak{F}_{X_r, r\in I, r\leq s} \subset \mathfrak{F}_{X_u, u\leq s} = \mathfrak{F}_s^X,$$

and for the corresponding conditional expectations, we have by Watson's theorem that $\mathbb{F}_s^Y = \mathbb{F}_s^Y \mathbb{F}_s^X$. Hence, for $t > s$ $s, t \in I$, we have

$$\mathbb{F}_s^Y(I_B(X_t)) = \mathbb{F}_s^Y \mathbb{F}_s^X(I_B(X_t)) = \mathbb{F}_s^Y \mathbb{F}(I_B(X_t)|X_s) = \mathbb{F}(I_B(X_t)|X_s).$$

$(Y_t)_{t\in I}$ is a Markov process. □

We now prove a stronger inverse of the preceding result, namely, if $(X_t)_{t\in J}$ is a process such that every finite subset of J has the Markov property, then it is a Markov process. In its proof, we use the monotone class theorem applied to order complete algebras. The difference with the usual application is that we now apply it to upwards directed systems whereas in the classical case with σ-algebras, it is applied to upwards directed sequences. We use the result for order complete algebras of projections.

(a) A class of projections is called a *π-class*, if it is closed under multiplication.
(b) A class of projections is called a *d-class*, if

  (i) the identity operator $\mathbb{I}$ is in the class;
  (ii) if $\mathbb{P} \leq \mathbb{Q}$, with both $\mathbb{P}$ and $\mathbb{Q}$ in the class implies that $\mathbb{Q} - \mathbb{P} = \mathbb{Q}\mathbb{P}^d$ is in the class;
  (iii) if $\mathbb{P}_\alpha$ is in the class and if $\mathbb{P}_\alpha \uparrow \mathbb{P}$ then $\mathbb{P}$ is in the class.

(c) The monotone class theorem: If a π-class contains a d-class, then it contains the complete algebra generated by the π-class.

The only point where the proof of (c) differs from the proof in the countable case (see, for instance, [1, Theorem 1.3.9]) is in the proof that for an arbitrary set of projections $\mathbb{P}_\alpha$ their supremum is in the algebra assuming that they are elements of a d-system. Here we use the standard method of first forming the set of all finite suprema of the $\mathbb{P}_\alpha$, which is an upward directed set of projections having the same supremum as the original set.

Proposition 4.4 *Let $(X_t, \mathfrak{F}_t^X)_{t\in J}$ be a stochastic process. If $(X_t)_{t\in I}$ has the Markov property for all finite subsets $I \subset J$, then $(X_t, \mathfrak{F}_t^X)_{t\in J}$ is a Markov process.*

Proof We have to prove that for every Borel set B, and for $s < t$

$$\mathbb{F}(I_B(X_t) \mid X_r, r \leq s) = \mathbb{F}(I_B(X_t) \mid X_s).$$

In order to do that, we note that the Boolean algebra of projections generated by the Boolean algebra of projections $\mathfrak{P}_{X_r, r \leq s}$ is equal to the Boolean algebra of projections generated by all finite families of Boolean algebras $\mathfrak{P}_{X_{t_1}, X_{t_2}, \ldots, X_{tn}}$ with $t_1 < t_2 < \ldots < t_n = s$. Using Proposition 2.1 we will prove that for all projections $\mathbb{P}$ belonging to $\mathfrak{P}_{X_{s, s \leq t}}$, we have

$$\mathbb{F}(\mathbb{P} I_B(X_t)) = \mathbb{F}(\mathbb{P}(\mathbb{F}(I_B(X_t) \mid X_s)). \tag{4.6}$$

This then implies that $\mathbb{F}(I_B(X_t) \mid X_s) = \mathbb{F}(I_B(X_t) \mid X_r, r \leq s)$.

We do it using the monotone class theorem: Let $\mathfrak{P}_s$ be the set of all projections $\mathbb{P}$ satisfying Eq. (4.6). For any $\mathbb{P} \in \mathfrak{P}_{X_{t_1}, \ldots, X_{tn}}$, we have, by the definition of conditional expectation that

$$\begin{aligned} \mathbb{F}(\mathbb{P} I_B(X_t)) &= \mathbb{F}(\mathbb{P}\mathbb{F}(I_B(X_t) \mid X_{t_1}, \ldots, X_{t_n}) \\ &= \mathbb{F}(\mathbb{P}\mathbb{F}(I_B(X_t) \mid X_{t_n})) = \mathbb{F}(\mathbb{P}\mathbb{F}(I_B(X_t) \mid X_s)) \end{aligned}$$

using our assumption that (X_t) has the Markov property for finite sets. Thus, $\mathfrak{P}_{X_{t_1}, \ldots, X_{tn}} \subset \mathfrak{P}_s$. This holds for any choice of indices, and so the Boolean algebra generated by the algebras $\mathfrak{P}_{X_{t_1}, \ldots, X_{tn}}$, is a Boolean algebra of projections (and therefore a π-class) that is contained in $\mathfrak{P}_s$. It is clear that $\mathbb{I} \in \mathfrak{P}_s$, because $\mathbb{I} \in \mathfrak{P}_{X_{t_1}, \ldots, X_{tn}}$. Also, if $\mathbb{P}_\alpha \uparrow \mathbb{P}$ and every $\mathbb{P}_\alpha$ satisfies Eq. (4.6), then by the order continuity of $\mathbb{F}$, we have that $\mathbb{P}$ also satisfies Eq. (4.6). By the monotone class theorem this shows that $\mathfrak{P}_s$ contains the order complete algebra generated by all the $\mathfrak{P}_{X_{t_1}, \ldots, X_{tn}}$ for any choice of indices. As this algebra is equal to the algebra $\mathfrak{P}_{X_{s, s \leq t}}$, Eq. (4.6) holds for all $\mathbb{P}$ in this algebra and we are done. □

We note that the definition for a Markov process used by Kuo [11, Definition 10.5.4] and also generalized by Vardy and Watson [15] is that it is a process that has the Markov property for any finite subset of indices. From the two preceding results it is clear that their definition is equivalent to the definition we use.

In the derivation of the Black-Scholes formulas for the pricing of futures, the following characterization of a Markov process in the classical case is used (see [14, Definition 2.3.6 and Section 5.2.5]):

The adapted process $(X_t, \mathfrak{F}_t, \mathbb{F}_t)_{t \in J}$ is a Markov process if and only if for all $s \leq t \in J$ and for every bounded non-negative Borel-measurable function $f(t, x)$, there exists a Borel-measurable function $f(s, x)$ such that

$$\mathbb{F}_s(f(t, X_t)) = f(s, X_s).$$

A key question when considering this characterization in the general case is to determine which elements X of $\mathfrak{E}$ can be written in the form $g(X)$ with g a bounded Borel measurable function. The answer to this question is given by a result of V.I. Sobolev (see [16, Theorem XI.7.a]).

Proposition 4.5 *If $M := \{\mu_X(A) : A \in \mathcal{B}(\mathbb{R})\}$, then the bounded element $Y \in \mathfrak{E}$ can be written as $f(X)$ for some Borel-measurable function f if and only if $E_t^Y \in M$ for all $t \in \mathbb{R}$.*

Using this result, we easily show that the condition implies, in general, that the process is Markov.

Proposition 4.6 *If $(X_t, \mathfrak{F}_t, \mathbb{F}_t)_{t\in J}$ is an adapted process and if, for every positive bounded Borel measurable function $f(t, x)$, there exists a Borel-measureable function $f(s, x)$ such that for $s \leq t$, we have $\mathbb{F}_s(f(t, X_t)) = f(s, X_s)$, then the process is a Markov process.*

Proof We have, since $\mathfrak{F}_{X_s} \subset \mathfrak{F}_s$, by the properties of a conditional expectation that $\mathbb{F}(f(t, X_t) \mid X_s) = \mathbb{F}(\mathbb{F}_s(f(t, X_t)) \mid X_s)$. But, since $\mathbb{F}_s(f(t, X_t)) = f(s, X_s)$ we get, by substituting this in the equation, that

$$\mathbb{F}(f(t, X_t) \mid X_s) = \mathbb{F}(f(s, X_s) \mid X_s).$$

Since the Sobolev result implies that $f(s, X_s) \in [E, X_s] \subset [\mathfrak{F}, X_s]$, we have that $\mathbb{F}(f(s, X_s) \mid X_s) = f(s, X_s) = \mathbb{F}_s(f(t, X_t))$. So, for every positive bounded Borel-measurable function, we have

$$\mathbb{F}_s(f(t, X_t)) = \mathbb{F}(f(t, X_t) \mid X_s).$$

Hence, $(X_t, \mathfrak{F}_t, \mathbb{F}_t)_{t\in J}$ is a Markov process. □

In order to prove the converse, one has to show that for a Markov process $(X_t, \mathfrak{F}_t, \mathbb{F}_t)_{t\in J}$ and $s \leq t$, we have $\mathbb{F}(f(t, X_t) \mid X_s) = f(s, X_s)$ for every positive bounded Borel-function f. By Sobolev's result one therefore has to show that $\mathbb{F}(f(t, X_t) \mid X_s)$ belongs to the order complete Riesz space generated by components of E of the form $\mu_{X_s}(A)$, $A \in \mathcal{B}(\mathbb{R})$. However, in the general case it belongs to the order complete Riesz space $[X_s, \mathfrak{F}]$ generated by X_s and $\mathfrak{F}$ and this space need not even be contained in the first mentioned space.

Example Choose two disjoint partitions of the interval [0, 1]. Let $\mathfrak{F}$ be the space of all step-functions relative to the first partition and X a step function relative to the second partition. If Y is a step-function relative to the joint of the two partitions, then $Y \in \mathfrak{F}_X$, but the Boolean algebra of components generated by its spectral system is not necessarily contained in that of X.

The first obvious condition for the converse to hold is therefore that $\mathfrak{F}$ must be contained in the order continuous Riesz subspace of $\mathfrak{F}_t$ generated by X_t and E for every t.

Corollary 4.7 *If M denotes the values of μ_X and if $[\mathfrak{F}, X] = [X, E]$ then every bounded element $Y \in \mathfrak{E}$ can be written as $f(X)$ for some Borel-measurable function f if and only if $E_t^Y \in M$ for all $t \in \mathbb{R}$.*

Proposition 4.8 *Let $\mathfrak{E}$ be a super Dedekind complete Riesz space with weak order unit E. Let $\mathfrak{A}(X)$ be the order complete Boolean algebra of components of E generated by the components E_t^X, $t \in \mathbb{R}$. Then, for each $E_\alpha \in \mathfrak{A}(X)$, there exists a Borel measurable set $A_\alpha \in \mathcal{B}(\mathbb{R})$ such that $\mu_X(A_\alpha) = E_\alpha$, i.e., $\mathfrak{A}(X) \subset M$.*

Proof We note that for $E_{s,t} := E_t^X - E_s^X$ we have that the interval $I_{s,t} := (s,t] \in \mathcal{B}(\mathbb{R})$ satisfies $\mu_X(I_{s,t}) = E_{s,t}$. Consider the set $\mathfrak{D}$ of all $E_\alpha \in \mathfrak{X}$ that is contained in the range of μ_X. Then the set of all $E_{s,t}$ is a π-system that is contained in $\mathfrak{D}$. We claim that $\mathfrak{D}$ is a Dynkin-system:

(a) The element $E \in \mathfrak{D}$, for $\mu_X(-\infty,\infty) = E_\infty^X - E_{-\infty}^X = E - 0 = E$.
(b) Let $E_\alpha, E_\beta \in \mathfrak{D}$ with $E_\alpha \le E_\beta$. Let $\mu_X(A_\alpha) = E_\alpha$ and $\mu_X(A_\beta) = E_\beta$ with $A_\alpha, A_\beta \in \mathcal{B}(\mathbb{R})$. We first note that the disjoint complement E_α^d of E_α also belongs to $\mathfrak{D}$ because, from $E = \mu_X(\mathbb{R}) = \mu_X(A_\alpha \cup A_\alpha^c) = \mu_X(A_\alpha) + \mu_X(A_\alpha^c) = E_\alpha + \mu_X(A_\alpha^c)$ it follows that $\mu_X(A_\alpha^c) = E - E_\alpha = E_\alpha^d$. Now, $E_\beta - E_\alpha = E_\beta \wedge E_\alpha^d = \mu_X(A_\beta) \wedge \mu_X(A_\alpha^c) = \mu_X(A_\beta \cap A_\alpha^c)$ by our earlier remark that μ_X is a Boolean measure. So $E_\beta - E_\alpha \in \mathfrak{D}$.
(c) Let $E_\beta \in \mathfrak{D}$ with $E_\beta \uparrow E_0$. Since $\mathfrak{E}$ is super Dedekind complete, there exist a subsequence $E_n \uparrow E_0$. Let $\mu_X(A_n) = E_n$. We claim that $\mu_X(\bigcup_n A_n) = E_0$. Define the disjoint sequence of sets B_n in the usual way by $B_1 = A_1$, $B_{n+1} = A_{n+1} - \bigcup_{j=1}^n B_j$. Then $\bigcup_n B_n = \bigcup_n A_n = A_0$, and, since $E_n \uparrow$,

$$\mu_X(B_1) = \mu_X(A_1) = E_1, \mu_X(B_2) = E_2 - E_1, \dots, \mu_X(B_n) = E_n - E_{n-1}.$$

It follows that

$$\mu_X(A_0) = \sum_{j=1}^\infty \mu_X(B_n) = \sum_{j=1}^\infty (E_n - E_{n-1}) = E_0.$$

We therefore have that $E_0 \in \mathfrak{D}$.

It follows that $\mathfrak{D}$ contains the order complete Boolean algebra generated by the images of all left-open right-closed intervals, i.e., $\mathfrak{A}(X) \subset \mathfrak{D}$. Thus, for every $E_\alpha \in \mathfrak{A}(X)$, there exists an element $A_\alpha \in \mathcal{B}(\mathbb{R})$ such that $\mu_X(A_\alpha) = E_\alpha$. □

From the Sobolev criterion we immediately get:

Proposition 4.9 *Let $\mathfrak{E}$ be a super Dedekind complete Riesz space with weak order unit E and let X be an element of $\mathfrak{E}$. Let $\mathfrak{G}_X$ be the Dedekind complete Riesz subspace of $\mathfrak{E}$ generated by the Boolean algebra $\mathfrak{A}(X)$. Then, for every $Y \in \mathfrak{G}_X$ there exists a real valued Borel-measurable function f defined on $\mathbb{R}$ such that $Y = f(X)$.*

Applied to the case of an adapted process we therefore find an inverse of Proposition 4.6 under certain conditions.

Theorem 4.10 *Let $\mathfrak{E}$ be a super Dedekind complete Riesz space with weak order unit E. Let the adapted process $(X_t, \mathfrak{F}_t, \mathbb{F}_t)_{t\in J}$ be a Markov process such that*

$\mathfrak{F} \subset [E, X_t]$ for all t. If $s \leq t \in J$, then, for every bounded non-negative Borel-measurable function $f(t, x)$, there exists a Borel-measurable function $f(s, x)$ such that $\mathbb{F}_s(f(t, X_t)) = f(s, X_s)$.

Proof Since the process is a Markov process, we have, since $\mathfrak{F} \subset [X, E]$ by assumption, that

$$\mathbb{F}_s(f(t, X_t)) = \mathbb{F}(f(t, X_t)|X_s) = \mathbb{F}(f(t, X_t)|[X_s, \mathfrak{F}]) = \mathbb{F}(f(t, X_t)|[X_s, E]).$$

Hence, $\mathbb{F}_s(f(t, X_t)) \in \mathfrak{G}_{X_s}$, in the notation of the preceding proposition. This completes the proof. □

Let $(X_t, \mathfrak{F}_t, \mathbb{F}_t)_{t \in J}$ be a stochastic process. For any $t \in J$ consider the sets of components of the weak order unit E :

$$\mathfrak{E}_t := \{E_\alpha : E_\alpha \in \mathfrak{F}_t\}$$

$$\mathfrak{E}_{s,s \geq t} := \{\mu_{X_s}(B) : s \geq t, B \in \mathcal{B}(\mathbb{R})\} = \{I_B(X_s) : s \geq t, B \in \mathcal{B}(\mathbb{R})\}.$$

Theorem 4.11 *The adapted stochastic process $(X_t, \mathfrak{F}_t, \mathbb{F}_t)_{t \in J}$ is a Markov process if and only if $\mathfrak{E}_t$ and $\mathfrak{E}_{s,s \geq t}$ are $\mathbb{F}(\cdot \mid X_t)$ independent.*

Proof Suppose that the systems are $\mathbb{F}(\cdot \mid X_t)$ independent. Let $I_B(X_s) \in \mathfrak{E}_{s,s \geq t}$ and let $E_\alpha \in \mathfrak{E}_t$. Then

$$\begin{aligned}
\mathbb{F}(E_\alpha I_B(X_s)) &= \mathbb{F}[\mathbb{F}(E_\alpha I_B(X_s) \mid X_t)] \\
&= \mathbb{F}[\mathbb{F}(E_\alpha \mid X_t)\mathbb{F}(I_B(X_s) \mid X_t)] \\
&= \mathbb{F}[\mathbb{F}(\mathbb{F}(I_B(X_s) \mid X_t)E_\alpha \mid X_t)] \\
&= \mathbb{F}[E_\alpha \mathbb{F}(I_B(X_s) \mid X_t)].
\end{aligned}$$

Since this holds for all $E_\alpha \in \mathfrak{F}_t$, we have by Proposition 2.1 that

$$\mathbb{F}_t(I_B(X_s)) = \mathbb{F}(I_B(X_s) \mid X_t),$$

and this holds f[illegible] $\mathbb{R}$). It follows that $(X_t, \mathfrak{F}_t, \mathbb{F}_t)_{t \in J}$ is a Markov process.
[illegible] $(X_t, \mathfrak{F}_t, \mathbb{F}_t)_{t \in J}$ is a Markov process. Again let
[illegible]e then have

$$\begin{aligned}
&= \mathbb{F}[\mathbb{F}_t(E_\alpha I_B(X_s)) \mid X_t] \\
&= \mathbb{F}[E_\alpha \mathbb{F}_t(I_B(X_s)) \mid X_t] \\
&\mathbb{F}[E_\alpha \mathbb{F}(I_B(X_s) \mid X_t) \mid X_t] \\
&\mathbb{F}[E_\alpha \mid X_t]\mathbb{F}(I_B(X_s) \mid X_t).
\end{aligned}$$

[illegible] conditionally independent. □

Corollary 4.12 *If $\mathfrak{E}$ is super Dedekind complete, then every component $E_\beta \in \mathfrak{F}_{X_s, s\geq t}$ is of the form $E_\beta = \mathbb{F}_B(X_s) = \mu_{X_s}(B)$ for some $B \in \mathcal{B} \in \mathbb{R}$ and $(X_t, \mathfrak{F}_t, \mathbb{F}_t)_{t\in J}$ is a Markov process if and only if $\mathfrak{F}_t$ and $\mathfrak{F}_{X_s, s\geq t}$ are $\mathbb{F}(\cdot \mid X_t$ conditionally independent.*

The latter condition is used by Blumenthal and Getoor in [4] as the definition of a Markov process.

5 Independence Lemma

In this section, $\mathbb{F}$ will be a conditional expectation on the Dedekind complete Riesz space $\mathfrak{E}$ that has a weak order unit E. This lemma is stated for the classical case in [14] and a proof is given in [13].

Lemma 5.1 *Let $\mathfrak{F} \subset \mathfrak{G} \cap \mathfrak{H}$, with $\mathfrak{G}$ and $\mathfrak{H}$ two $\mathbb{F}$-independent Dedekind complete Riesz subspaces of $\mathfrak{E}$ and suppose that $X \in \mathfrak{G}$ and $Y \in \mathfrak{H}$. Let $\mathbb{F}_\mathfrak{G}$ and $\mathbb{F}_\mathfrak{H}$ be the conditional expectations from $\mathfrak{E}$ onto $\mathfrak{G}$ and onto $\mathfrak{H}$ satisfying $\mathbb{F}_\mathfrak{G}\mathbb{F} = \mathbb{F}\mathbb{F}_\mathfrak{G} = \mathbb{F}$ and $\mathbb{F}_\mathfrak{H}\mathbb{F} = \mathbb{F}\mathbb{F}_\mathfrak{H} = \mathbb{F}$, respectively. Then,*

$$\mathbb{F}_\mathfrak{G}[\Phi(X, Y)] = \mathbb{F}\Phi(x, Y)|_{x=X} = \mathbb{F}(\Phi(X, Y) \mid X) \tag{5.1}$$

holds for all bounded Borel measurable functions $\Phi : \mathbb{R} \times \mathbb{R} \to \mathbb{R}$.

Proof Assume the $\Phi(x, y) = \phi(x)\psi(y)$ with ϕ and ψ bounded Borel-measurable real valued functions. Then

$$\begin{aligned}\mathbb{F}_\mathfrak{G}[\Phi(X, Y)] &= \mathbb{F}_\mathfrak{G}[\phi(X)\psi(Y)] = \phi(X)\mathbb{F}_\mathfrak{G}(\psi(Y)) \\ &= \phi(X)\mathbb{F}_\mathfrak{G}\mathbb{F}_\mathfrak{H}(\psi(Y)) = \phi(X)\mathbb{F}(\psi(Y)) \text{ by [7, Corollary 4.12].}\end{aligned}$$

On the one hand, the last term can be written as

$$\phi(x)\mathbb{F}(\psi(Y))|_{x=X} = \mathbb{F}(\phi(x)\psi(Y))|_{x=X}.$$

On the other hand, since X and Y (and therefore X and $\psi(Y)$) are independent, we have that $\mathbb{F}(\psi(Y) \mid X) = \mathbb{F}(\psi(Y))$. Therefore, the right-hand side is equal to $\phi(X)\mathbb{F}(\psi(Y) \mid X) = \mathbb{F}(\phi(X)\psi(Y)) \mid X)$. This proves (5.1) for this case.

Let A and B be Borel measurable subsets in $\mathbb{R}$, and put $\phi(x) = I_A(x)$ and $\psi(y) = I_B(y)$. We then get that

$$\mathbb{F}_\mathfrak{G}[I_A(X)I_B(Y)] = \mathbb{F}I_A(x)I_B(Y)|_{x=X} = \mathbb{F}(I_A(X)I_B(Y) \mid X),$$

i.e.,

$$\mathbb{F}_{\mathfrak{G}}[I_{A\times B}(X,Y)] = \mathbb{F}I_{A\times B}(x,Y)|_{x=X} = \mathbb{F}(I_{A\times B}(X,Y)\mid X).$$

Every term in this equality can be extended to hold for indicator functions of sets in the σ-algebra generated by the elements of $\mathcal{B}(\mathbb{R})\times\mathcal{B}(\mathbb{R})$, i.e., to elements of $\mathcal{B}(\mathbb{R}\times\mathbb{R})$. Using linearity of the conditional expectation operators, the equalities extend to hold for positive step functions of two variables, positive measurable functions, and finally to bounded measurable functions. □

Remark 5.2

(1) The lemma above is often stated with the following assumptions (not mentioning the space $\mathfrak{H}$): *Let $\mathfrak{F}\subset\mathfrak{G}$, let $X\in\mathfrak{G}$, and let Y be independent of $\mathfrak{G}$. Then*

$$\mathbb{F}_{\mathfrak{G}}[\Phi(X,Y)] = \mathbb{F}\Phi(x,Y)|_{x=X} = \mathbb{F}(\Phi(X,Y)\mid X) \tag{5.2}$$

holds for all Bounded Borel measurable functions $\Phi:\mathbb{R}\times\mathbb{R}\to\mathbb{R}$. This follows from the lemma above by taking $\mathfrak{H}=[Y,\mathfrak{F}]$, i.e., the Dedekind complete Riesz space generated by Y and $\mathfrak{F}$.

(2) The independence lemma can be proved in the same manner for more than two variables.

Lemma 5.3 *Let $\mathfrak{F}\subset\mathfrak{G}$ with $\mathfrak{G}$ a Dedekind complete Riesz subspace of $\mathfrak{E}$ and let $X_1,\dots,X_m$ be elements of $\mathfrak{G}$. Let $Y_1,\dots,Y_n$ be elements of $\mathfrak{E}$ independent of $\mathfrak{G}$. If Φ is a Borel measurable function of the $n+m$ variables $x_1,\dots,x_m,y_1,\dots,y_m$, then*

$$\begin{aligned}
&\mathbb{F}_{\mathfrak{G}}[\Phi(X_1,\dots,X_m,Y_1,\dots,Y_n)]\\
&\qquad = \mathbb{F}\Phi(x_1,\dots,x_m,Y_1,\dots,Y_n)|_{x_j=X_j},\quad 1\le j\le m\\
&\qquad\qquad = \mathbb{F}[\Phi(X_1,\dots,X_m,Y_1,\dots,Y_n)\mid X_1,\dots,X_m]
\end{aligned}\tag{5.3}$$

(3) The middle term in (5.3) can be replaced by a term $\Psi(X_1,\dots,X_m)$, where $\Psi(x_1,\dots,x_m) := \mathbb{F}(\Phi(x_1,\dots,x_m,Y_1,\dots,Y_n))$.

(4) For a statement and a proof of this lemma in the classical case, see [13, Appendix A.2]. The statement for $n+m$ variables can be found in [14, Lemma 2.3.4].

We recall that an adapted process $(X_t,\mathfrak{F}_t,\mathbb{F}_t)$ is called an $\mathbb{F}$-conditional Brownian motion in $\mathfrak{E}$ if, for all $0\le s<t$ we have

1. $X_0=0$;
2. the increment X_t-X_s is conditionally independent of $\mathfrak{F}_s$;
3. $\mathbb{F}(X_t-X_s)=0$;
4. $\mathbb{F}[(X_t-X_s)^2]=(t-s)E$;
5. $\mathbb{F}[(X_t-X_s)^4]=3(t-s)^2E$.

We proved in [7] that for $0 = t_0 < t_1 < \cdots < t_n < \infty$ the increments $\{X_{t_j} - X_{t_{j-1}}\}_{j=1}^n$ are $\mathbb{F}$-independent and that the process is a martingale.

As in the classical case, a short proof of the fact that a Brownian motion is a Markov process follows from Lemma 5.1 in the form stated in Remark 5.2(1). This fact also follows from [15] where the general fact that a stochastic process with independent increments is a Markov process.

Corollary 5.4 *Every Brownian motion is a Markov process.*

Proof Let $(B_t, \mathfrak{F}_t)_{t\in[0,\infty)}$ be a Brownian motion. Let $0 < s < t$ and let f be an arbitrary bounded Borel-measurable function on $\mathbb{R}$. Write

$$f(B_t) = f(B_s + (B_t - B_s)). \tag{5.4}$$

Then $B_s \in \mathfrak{F}_s$ and $B_t - B_s$ is independent of $\mathfrak{F}_s$. Therefore, putting $\Phi(x, y) := f(x + y)$, $X = B_s$, and $Y = B_t - B_s$, we have by the independence lemma that

$$\begin{aligned}\mathbb{F}_s(f(B_t)) = \mathbb{F}_s(f(B_s + (B_t - B_s))) = \mathbb{F}_s\Phi(X, Y)\\ = \mathbb{F}(\Phi(X, Y) \mid X) = \mathbb{F}(f(B_t) \mid B_s).\end{aligned} \tag{5.5}$$

Hence $(B_t, \mathfrak{F}_t)$ has the Markov property. □

We shall prove a stronger result in the next section.

6 Strong Markov Processes

Definition 6.1 Let $(M_t, \mathfrak{F}_t)$ be a stochastic process and let $\mathbb{S}$ be a finite optional time for the filtration $(\mathfrak{F}_t, \mathbb{F}_t)$. It is called a *strong Markov process* whenever we have that for all $t > 0$, and all bounded Borel measurable functions f that

$$\mathbb{F}_{\mathbb{S}+}(f(M_{t+\mathbb{S}})) = \mathbb{F}(f(M_{t+\mathbb{S}})|M_{\mathbb{S}}).$$

We will prove that a Brownian motion is a strong Markov process.

We recall that if $\mathbb{F}$ is a conditional expectation on $\mathfrak{E}$, the $\mathbb{F}$-*conditional distribution* $(Q_x)_{x\in\mathbb{R}}$ *of an element* $X \in \mathfrak{E}$ is defined to be the image under $\mathbb{F}$ of the right-continuous spectral system of X, i.e., if $\mathbb{Q}_x := \mathbb{P}(X \le x)$, $E_x := \mathbb{Q}_x E$, then $Q_x := \mathbb{F}E_x$ (see [9, Definition 4.1] and [3]). The vector measure μ_Q defined for left-open right closed intervals $(a, b]$ by

$$\mu_Q((a, b]) := Q_b - Q_a \tag{6.1}$$

can be extended to a vector measure defined on the Borel σ-algebra $\mathcal{B}(\mathbb{R})$ and we get that $\mathbb{F}(X) = \int_{\mathbb{R}} x\, d\mu_Q(x)$ and for a measurable real function defined on an interval containing the spectral interval of X, we have $\mathbb{F}(f(X)) = \int_{\mathbb{R}} f(x)\, d\mu_Q(x)$. We have the following equivalent ways to write Eq. (6.1)

$$\mu_Q((a,b]) = Q_b - Q_a = \mathbb{F}[\mathbb{P}(a < X \leq b)E] = \mathbb{F}[\mathbb{P}(X \in (a,b])] = \mathbb{F}(I_{(a,b]}(X)).$$

Also, the $\mathbb{F}$-conditional characteristic function of X is defined as $\mathbb{F}(e^{i\lambda X}) = \int_{\mathbb{R}} e^{i\lambda x}\, d\mu_Q(x),\ \lambda \in \mathbb{R}$. From [9, Lemma 4.5] we have that if $\mathbb{F}(e^{i\lambda X}) = e^{-\frac{1}{2}\lambda^2 t}E$, then

$$\mu_Q(\Gamma) = \frac{1}{\sqrt{2\pi t}} \int_\Gamma e^{-y^2/2t}\, dyE \text{ for all } \Gamma \in \mathcal{B}(\mathbb{R}).$$

Let $0 \leq \phi \in \Phi \subset \mathfrak{E}_{00}^{\sim}$, i.e., $\phi(E) = 1$. It then follows from the above definitions that $\phi(Q_x) = \phi\mathbb{F}E_x$ is a scalar distribution function with $\phi(Q_{-\infty}) = 0$ and $\phi(Q_\infty) = 1$ for all $\phi \in \Phi^+$. Moreover,

$$\phi\mathbb{F}(X) = \int_{-\infty}^{\infty} x\, d\mu_{\phi Q}, \tag{6.2}$$

with $\mu_{\phi Q}$ the Stieltjes-Lebesgue measure on $\mathcal{B}(\mathbb{R})$ induced by the distribution function $\phi(Q_x)$. Also,

$$\phi\mathbb{F}(f(X)) = \int_{-\infty}^{\infty} f(x)\, d\mu_{\phi Q}(x) \tag{6.3}$$

and

$$\phi\mathbb{F}(e^{i\lambda X}) = \int_{-\infty}^{\infty} e^{i\lambda x}\, d\mu_{\phi Q}(x). \tag{6.4}$$

From (6.4) we get, by applying the well-known inversion formula from probability theory (see, for instance, [1, Theorem 7.3.1]) the inversion theorem for the abstract case.

Theorem 6.2 (Inversion Theorem) *Let X and Y be elements of $\mathfrak{E}$ and let $\mathbb{F}$ be a conditional expectation on $\mathfrak{E}$. If the characteristic functions of X and Y (relative to $\mathbb{F}$) are equal, then X and Y have the same conditional distribution functions relative to $\mathbb{F}$.*

If (B_t) is a Brownian motion it follows from the proof of Levy's theorem [9, Theorem 4.6] that the $\mathbb{F}$-conditional characteristic function of B_t is equal to $e^{-\frac{1}{2}\lambda^2 t}E$ and for $s < t$,

$$\mathbb{F}_s[e^{i\lambda(B_t - B_s)}] = e^{-\frac{1}{2}\lambda^2(t-s)}E. \tag{6.5}$$

Lemma 6.3 ([10, Lemma 2.6.14]) *For a Brownian motion (B_t), define the complex valued process (M_t) by*

$$M_t := \exp\left[i\lambda B_t + \frac{1}{2}\lambda^2 t\right], t \geq 0,$$

and denote its real and imaginary parts by R_t and I_t, respectively. Then $(R_t, \mathfrak{F}_t)$ and $(I_t, \mathfrak{F}_t)$ are Martingales.

Proof Let $0 \leq s < t$. Then, $\mathbb{F}_s(M_t M_s^{-1}) = E$, for

$$\begin{aligned}\mathbb{F}_s(M_t M_s^{-1}) &= \mathbb{F}_s\big[\exp(i\lambda B_t + \frac{1}{2}\lambda^2 t - i\lambda B_s - \frac{1}{2}\lambda^2 s)\big]\\ &= \big[\exp(\frac{1}{2}\lambda^2(t-s))\big]\mathbb{F}_s\big[\exp(i\lambda(B_t - B_s))\big]\\ &= \big[\exp(\frac{1}{2}\lambda^2(t-s))\big]\big[\exp(-\frac{1}{2}\lambda^2(t-s))\big]E \text{ by (6.5)}\\ &= E.\end{aligned}$$

Hence,

$$\mathbb{F}_s(M_t) = \mathbb{F}_s(M_s M_t M_s^{-1}) = M_s \mathbb{F}_s(M_t M_s^{-1}) = M_s.$$

This proves our claim. □

We will now prove the following:

Theorem 6.4 *A Brownian motion is a strong Markov process.*

Proof We prove the theorem of a Brownian motion defined on a finite interval $[0, b]$. It follows from the proof of the Doob optional sampling theorem (see [5, Theorem 6.8]) that since (M_t) is a martingale with a last element M_b, we have equality in Doob's optional sampling theorem, i.e., we have for a finite optional time $\mathbb{S}$, that

$$\mathbb{F}_{\mathbb{S}+}(M_{t+\mathbb{S}}) = M_{\mathbb{S}},$$

i.e.,

$$\mathbb{F}_{\mathbb{S}+}\big[\exp(i\lambda B_{t+\mathbb{S}} + \frac{1}{2}\lambda^2(t+\mathbb{S}))\big] = \exp(i\lambda B_{\mathbb{S}} + \frac{1}{2}\lambda^2\mathbb{S}).$$

It follows that

$$\mathbb{F}_{\mathbb{S}+}\big[\exp(i\lambda B_{t+\mathbb{S}})\big] = \exp(i\lambda B_{\mathbb{S}} - \frac{1}{2}\lambda^2 t),$$

i.e., since $\exp(i\lambda B_{\mathbb{S}}) \in \mathfrak{F}_{\mathbb{S}+}$,

$$\mathbb{F}_{\mathbb{S}+}[\exp(i\lambda(B_{t+\mathbb{S}} - B_{\mathbb{S}})] = \exp(-\frac{1}{2}\lambda^2 t)E.$$

The right-hand side of this equation is the characteristic function of a vector with normal distribution with mean $B_{\mathbb{S}}$ and variance t. Therefore, $W_t := B_{t+\mathbb{S}} - B_{\mathbb{S}}$ is a Brownian motion which is independent of $\mathfrak{F}_{\mathbb{S}+}$. Let f be a Borel measurable function and write

$$f(B_{t+\mathbb{S}}) = f(B_{\mathbb{S}} + (B_{t+\mathbb{S}} - B_{\mathbb{S}})) = \Phi(X, Y),$$

with $X = B_{\mathbb{S}}$ and $Y = B_{t+\mathbb{S}} - B_{\mathbb{S}}$ and $\mathfrak{G} = \mathfrak{F}_{\mathbb{S}+}$ in the notation of the independence lemma. This yields the required result that

$$\mathbb{F}_{\mathbb{S}+}(f(B_{t+\mathbb{S}})) = \mathbb{F}(\Phi(X, Y)|X) = \mathbb{F}(f(B_{t+\mathbb{S}})|B_{\mathbb{S}}).$$

Hence, it is a strong Markov process. □

We conclude the paper by proving that for a Brownian motion $(B_t)_{t\in[0,b]}$ the Brownian filtration $(\mathfrak{F}_t^B)$ is right-continuous. We remark that in the classical theory a somewhat larger filtration, the so-called *augmentation* of $\mathfrak{F}_t^B$, is continuous and that $\mathfrak{F}_t^B$ need not be right-continuous. This is caused by the fact that in the classical theory, filtrations can be enlarged by adding null-sets. However, in our case no new projections are created in this way since all null-sets are associated with the zero projection. Thus, in the abstract theory every filtration is equal to its augmentation and similarly, every filtration is a universal filtration (see [10, Section 2.2.7]).

Theorem 6.5 *Let (B_t) a Brownian motion. Then the filtration $(\mathfrak{F}_t^B, \mathbb{F}_t^B)$ is right-continuous.*

Proof Since a Brownian motion is a strong Markov process, we have for $s \in J = [0, b]$ and for every Borel set $A \in \mathbb{R}$, that

$$\mathbb{F}_{s+}^B(I_A(B_{s+t})) = \mathbb{F}^B(I_A(B_{s+t}) \mid B_s). \tag{6.6}$$

Let $t_0 < t_1 < \cdots < t_n < s < t_{n+1} < \cdots < t_m$ be an arbitrary choice of elements in J and let $A_0, A_1, \ldots, A_m$ be Borel measurable elements of $\mathbb{R}$. Then

$$\begin{aligned}
&\mathbb{F}_{s+}^B(I_{A_0}(B_{t_0})I_{A_1}(B_{t_1}) \cdots I_{A_m}(B_{t_m})) \\
&\quad = I_{A_0}(B_{t_0}) \cdots I_{A_n}(B_{t_n})\mathbb{F}_{s+}^B(I_{A_{n+1}}(B_{t_{n+1}}) \cdots I_{A_m}(B_{t_m})) \\
&\quad = I_{A_0}(B_{t_0}) \cdots I_{A_n}(B_{t_n})\mathbb{F}^B(I_{A_{n+1}}(B_{t_{n+1}}) \cdots I_{A_m}(B_{t_m}) \mid B_s).
\end{aligned} \tag{6.7}$$

It follows that $\mathbb{F}^B_{s+}(I_{A_0}(B_{t_0})I_{A_1}(B_{t_1})\cdots I_{A_m}(B_{t_m})) \in \mathfrak{F}^B_s$. Associating the element $I_{A_{t_j}}(B_{t_j})$ with the projection $\mathbb{P}_j$ satisfying $\mathbb{P}_j E = I_{A_{t_j}}(B_{t_j})$ we have that the elements $\prod_{j=0}^m \mathbb{P}_j$ generate the σ-algebra $\mathfrak{P}^B_b$. For any element $\mathbb{P}$ in this generating set, we have that $\mathbb{F}^B_{s+}(\mathbb{P}_j E) \in \mathfrak{F}^B_s$. The set of all elements in $\mathfrak{F}^B_b$ having this property is a Dynkin-system containing the algebra of generating elements. Hence it equals $\mathfrak{F}^B_b$. Therefore, if $\mathbb{F}(E) \in \mathfrak{F}^B_{s+}$, then

$$\mathbb{F}E = \mathbb{F}^B_{s+}(\mathbb{F}E) \in \mathfrak{F}^B_s.$$

Hence, $\mathfrak{F}^B_{s+} \subset \mathfrak{F}^B_s \subset \mathfrak{F}^B_{s+}$ and so the filtration $\mathfrak{F}^B_t$ is right-continuous. □

Acknowledgements I acknowledge with thanks the input and enthusiasm of my late friend, colleague and co-author Coenraad Labuschagne (15 May 1958–10 July 2018), who inspired me to do this research.

Financial support of the National Research Foundation of South Africa is acknowledged with thanks.

References

1. R.B. Ash, C.A. Doléans-Dade, *Probability & Measure Theory*, 2nd edn. (Harcourt/Academic, Boston/Cambridge, 2000)
2. R.B. Ash, M.F. Gardner, *Topics in Stochastic Processes* (Academic, New York, 1975)
3. Y. Azouzi, K. Ramdane, On the distribution function with respect to conditional expectation on Riesz spaces. Quest. Math. (2018). https://doi.org/10.2989/16073606.2017.1377310
4. R.M. Blumenthal, R.D. Getoor, *Markov Processes and Potential Theory* (Academic, New York, 1968)
5. J.J. Grobler, Doob's optional sampling theorem in Riesz spaces. Positivity **15**, 617–637 (2011)
6. J.J. Grobler, Jensen's and Martingale inequalities in Riesz spaces. Indag. Math. **25**, 275–295 (2014). https://doi.org/10.1016/indag.2013.02.003
7. J.J. Grobler, The Kolmogorov-Čentsov theorem and Brownian motion in vector lattices. J. Math. Anal. Appl. **410**, 891–901 (2014)
8. J.J. Grobler, C.C.A. Labuschagne, The Itô integral for martingales in vector lattices. J. Math. Anal. Appl. **450**, 1245–1274 (2017). https://doi.org/10.1016/j.jmaa.2017.01.081
9. J.J. Grobler, C.C.A. Labuschagne, Itô's rule and Lévy's theorem in vector lattices. J. Math. Anal. Appl. **455**, 979–1004 (2017). https://doi.org/10.1016/j.jmaa.2017.06.011
10. I. Karatzas, S.E. Shreve, *Brownian Motion and Stochastic Calculus*. Graduate Texts in Mathematics (Springer, New York, 1991)
11. H.-H. Kuo, *Introduction to Stochastic Integration* (Springer, New York, 2006)
12. C.C.A. Labuschagne, B.A. Watson, Discrete time stochastic integrals in Riesz spaces. Positivity **14**, 859–875 (2010)
13. R.L. Schilling, L. Partzsch, *Brownian Motion, An Introduction to Stochastic Processes* (Walter de Gruyter GmbH & Co, Berlin, 2012)
14. S.E. Shreve, *Stochastic Calculus for Finance II, Continuous-Time Models* (Springer, New York, 2004)
15. J.J. Vardy, B.A. Watson, Markov processes on Riesz spaces. Positivity **16**, 373–391 (2012)
16. B.Z. Vulikh, *Introduction to the Theory of Partially Ordered Spaces* (Wolters-Noordhoff Scientific Publishers, Groningen, 1967)
17. B.A. Watson, An Ândo-Douglas type theorem in Riesz spaces with a conditional expectation. Positivity **13**(3), 543–558 (2009)

A Solution to the Al-Salam–Chihara Moment Problem

Wolter Groenevelt

This paper is dedicated to Ben de Pagter on the occasion of his 65th birthday

Abstract We study the q-hypergeometric difference operator L on a particular Hilbert space. In this setting L can be considered as an extension of the Jacobi operator for q^{-1}-Al-Salam–Chihara polynomials. Spectral analysis leads to unitarity and an explicit inverse of a q-analog of the Jacobi function transform. As a consequence a solution of the Al-Salam–Chihara indeterminate moment problem is obtained.

Keywords q^{-1}-Al-Salam–Chihara polynomials · Little q-Jacobi function transform · q-Hypergeometric difference operator · Indeterminate moment problem

1 Introduction

For every moment problem there is a corresponding set of orthogonal polynomials. Through their three-term recurrence relation orthogonal polynomials correspond to a Jacobi operator. In particular, spectral analysis of a self-adjoint Jacobi operator leads to an orthogonality measure for the corresponding orthogonal polynomials, i.e., a solution to the moment problem. Indeterminate moment problems correspond to Jacobi operators that are not essentially self-adjoint, and in this case a self-adjoint extension of the Jacobi operator corresponds to a solution of the moment problem, see, e.g., [18, 19]. Instead of looking for self-adjoint extensions of the Jacobi operator on (weighted) $\ell^2(\mathbb{N})$, it is also useful to look for extensions of the operator to a larger Hilbert space. This method is used in, e.g., [11, 16]. A choice of

W. Groenevelt (✉)
Delft University of Technology, Delft Institute of Applied Mathematics, Delft, The Netherlands
e-mail: w.g.m.groenevelt@tudelft.nl

G. Buskes et al. (eds.), *Positivity and Noncommutative Analysis*,
Trends in Mathematics, https://doi.org/10.1007/978-3-030-10850-2_11

the larger Hilbert space often comes from the interpretation of the operator, e.g., in representation theory.

In this paper we consider an extension of the Jacobi operator for the Al-Salam–Chihara polynomials, and we obtain the spectral decomposition in a similar way as in Koelink and Stokman [16]. The Al-Salam–Chihara polynomials are a family of orthogonal polynomials introduced by Al-Salam and Chihara [1] that can be expressed as q-hypergeometric polynomials. If $q > 1$ and under particular conditions on the other parameters, see Askey and Ismail [2], they are related to an indeterminate moment problem. In case the moment problem is determinate, the polynomials are orthogonal with respect to a discrete measure. Chihara and Ismail [4] studied the indeterminate moment problem under certain conditions on the parameters. They obtained the explicit Nevanlinna parametrization, but did not derive explicit solutions. Christiansen and Ismail [5] used the Nevanlinna parametrization to obtain explicit solutions, discrete ones and absolutely continuous ones, of the indeterminate moment problem corresponding to the subfamily of symmetric Al-Salam–Chihara polynomials. Christiansen and Koelink [6] found explicit discrete solutions of the symmetric Al-Salam–Chihara moment problem exploiting the fact that the polynomials are eigenfunctions of a second-order q-difference operator acting on the variable of the polynomial (whereas the Jacobi operator acts on the degree). The solution we obtain in this paper has an absolutely continuous part and an infinite discrete part. As special cases we also obtain a solution for the symmetric Al-Salam–Chihara moment problem, and for the continuous q^{-1}-Laguerre moment problem.

The extension of the Jacobi operator we study in this paper is essentially the q-hypergeometric difference operator. The latter is a q-analog of the hypergeometric differential operator, whose spectral analysis (on the appropriate Hilbert space) leads the Jacobi function transform, see, e.g., [17]. In a similar way the q-hypergeometric difference operator corresponds to the little q-Jacobi function transform, see Kakehi, Masuda, and Ueno [12, 13] and also [15, Appendix A.2]. In this light the integral transform $\mathcal{F}$ we obtain can be considered as a second version of the little q-Jacobi function transform. The little q-Jacobi function transform has an interpretation as a spherical transform on the quantum $SU(1,1)$ group. We expect a similar interpretation for the integral transform obtained in this paper.

The organization of this paper is as follows. In Sect. 2 some notations for q-hypergeometric functions are introduced and the definition of the Al-Salam–Chihara polynomials is given. In Sect. 3 the q-difference operator L is defined on a family of Hilbert spaces. Using Casorati determinants, which are difference analogs of the Wronskian, it is shown that L with an appropriate domain is self-adjoint. In Sect. 4 eigenfunctions of L are given in terms of q-hypergeometric functions. These eigenfunctions and their Casorati determinants are used to define the Green kernel in Sect. 5, which is then used to determine the spectral decomposition of L. The discrete spectrum is only determined implicitly, except for one particular choice from the family of Hilbert spaces, where we can explicitly describe the spectrum and the spectral projections. Corresponding to this choice an integral transform $\mathcal{F}$ is defined in Sect. 6 that diagonalizes the difference operator L, and an explicit inverse transform is obtained. The inverse transform gives rise to orthogonality relations for Al-Salam–Chihara polynomials.

1.1 Notations and Conventions

Throughout the paper $q \in (0, 1)$ is fixed. $\mathbb{N}$ is the set of natural numbers including 0, and $\mathbb{T}$ is the unit circle in the complex plane. For a set $E \subseteq \mathbb{R}$ we denote by $F(E)$ the vector space of complex-valued functions on E.

2 Preliminaries

In this section we first introduce standard notations for q-hypergeometric functions from [8], state a few useful identities, and then define Al-Salam–Chihara polynomials.

2.1 q-Hypergeometric Functions

The q-shifted factorial is defined by

$$(x;q)_n = \prod_{j=0}^{n-1} 1 - xq^j, \qquad n \in \mathbb{N} \cup \{\infty\}, \quad x \in \mathbb{C},$$

where we use the convention that the empty product is equal to 1. Note that $(x;q)_n = 0$ for $x \in q^{-\mathbb{N}_{<n}}$. The following identities for q-shifted factorials will be useful later on:

$$(xq^n;q)_\infty = \frac{(x;q)_\infty}{(x;q)_n}, \qquad (xq^{-n};q)_n = (-x)^n q^{-\frac{1}{2}n(n+1)} (q/x;q)_n.$$

We also define

$$(x;q)_{-n} = \frac{(x;q)_\infty}{(xq^{-n};q)_\infty}, \qquad n \in \mathbb{N}.$$

The θ-function is defined by

$$\theta(x;q) = (x;q)_\infty (q/x;q)_\infty, \quad x \in \mathbb{C}^*.$$

Note that $\theta(x;q)=0$ for $x\in q^{\mathbb{Z}}$. We use the following notation for products of q-shifted factorials or θ-functions

$$(x_1,x_2,\ldots,x_k;q)_n=\prod_{j=1}^{k}(x_j;q)_n,$$

$$\theta(x_1,x_2,\ldots,x_k;q)=\prod_{j=1}^{k}\theta(x_j;q).$$

θ satisfies the identities

$$\begin{aligned}\theta(xq^k;q)&=(-x)^{-k}q^{-\frac{1}{2}k(k-1)}\theta(x;q),\\ \theta(-x,x;q)&=\theta(x^2;q^2),\\ \theta(x;q)&=\theta(x,qx;q^2).\end{aligned}\tag{2.1}$$

We also need the following fundamental θ-function identity:

$$\theta(xv,x/v,yw,y/w)-\theta(xw,x/w,yv,y/v)=\frac{y}{v}\theta(xy,x/y,vw,v/w).\tag{2.2}$$

The q-hypergeometric series ${}_r\varphi_s$ is defined by

$${}_r\varphi_s\left(\begin{matrix}a_1,\ldots,a_r\\ b_1,\ldots,b_s\end{matrix};q,x\right)=\sum_{n=0}^{\infty}\frac{(a_1,\ldots,a_r;q)_n}{(q,b_1,\ldots,b_s;q)_n}\left[(-1)^nq^{\frac{1}{2}n(n-1)}\right]^{s-r+1}x^n.$$

We refer to [8] for convergence properties of the series. The q-hypergeometric difference equation is given by

$$(ABx-C)\varphi(qx)+[C+q-(A+B)x]\varphi(x)+(x-q)\varphi(x/q)=0,\tag{2.3}$$

and has ${}_2\varphi_1(A,B;C;q,x)$ as a solution. For $q\uparrow 1$ this q-difference equation becomes the hypergeometric differential equation.

2.2 Al-Salam–Chihara Polynomials

The Al-Salam–Chihara polynomials in base q^{-1} are defined by

$$P_n(\lambda;b,c;q^{-1})=b^{-n}\,{}_3\varphi_2\left(\begin{matrix}q^n,b\lambda,b/\lambda\\ bc,0\end{matrix};q^{-1},q^{-1}\right),\tag{2.4}$$

which are polynomials of degree n in $\lambda+\lambda^{-1}$. They are symmetric in the parameters b and c, which follows from transformation formula [8, (III.11)]. The three-term recurrence relation is

$$(\lambda+\lambda^{-1})P_n(\lambda) = (1-bcq^{-n})P_{n+1}(\lambda) + (b+c)q^{-n}P_n(\lambda) + (1-q^{-n})P_{n-1}(\lambda), \tag{2.5}$$

where we use the convention $P_{-1} \equiv 0$. For $b+c \in \mathbb{R}$ and $bc \geq 0$ the polynomials are orthogonal with respect to a positive measure on $\mathbb{R}$. By [2, Theorem 3.2] the moment problem for these polynomials is indeterminate if and only if $b, c \in \mathbb{R}$ with $q < |b/c| < q^{-1}$, or $b = \overline{c}$.

The polynomials $P_n(\lambda; b, -b; q^{-1})$ are called symmetric Al-Salam–Chihara polynomials, and the polynomials $P_n(\lambda; q^{\frac{1}{2}(\alpha+1)}, q^{\frac{1}{2}\alpha}; q^{-1})$ with $\alpha \in \mathbb{R}$ are called continuous q^{-1}-Laguerre polynomials. The corresponding moment problems are indeterminate.

3 The Difference Operator L

In this section we define the unbounded second-order q-difference operator L, and the Hilbert space the operator L acts on. The q-difference operator is essentially the q-hypergeometric difference operator, and, on the given Hilbert space, L can be considered as an extension of the Jacobi operator for the Al-Salam–Chihara polynomials from the previous section.

Let $z \in (q, 1]$ and define

$$I^- = -q^{\mathbb{N}}, \qquad I^+ = zq^{\mathbb{Z}}, \qquad I = I^- \cup I^+.$$

We also set $I_* = I \setminus \{-1\}$. Let a and s be parameters satisfying $a \in \mathbb{R}$ and $0 < a^2 < 1$, and $s \in \mathbb{T} \setminus \{-1, 1\}$ or $s \in \mathbb{R}$ and $q < s^2 < 1$. The condition $s \notin \{-1, 1\}$ is only needed for technical purposes, and can be removed afterwards by continuity in s of the functions involved.

Definition 3.1 The second order q-difference operator $L = L_{a,s}$ on $F(I)$ is given by

$$(Lf)(x) = \frac{1}{a}\left(1+\frac{1}{x}\right)f(x/q) - \frac{s+s^{-1}}{ax}f(x) + a\left(1+\frac{1}{a^2x}\right)f(qx), \qquad x \in I_*,$$

$$(Lf)(-1) = \frac{s+s^{-1}}{a}f(-1) + a\left(1-\frac{1}{a^2}\right)f(-q).$$

Remark 3.2 Define for $x = -q^n \in I^-$

$$\phi_\lambda(x) = a^{-n}P_n(\lambda; s/a, 1/sa; q^{-1}).$$

By (2.5) $\phi_\lambda(x)$ satisfies

$$(\lambda+\lambda^{-1})\phi_\lambda(-q^n) = \frac{1}{a}\left(1-q^{-n}\right)\phi_\lambda(-q^{n-1}) - \frac{s+s^{-1}}{a}q^{-n}\phi_\lambda(-q^n) + a\left(1-\frac{q^{-n}}{a^2}\right)\phi_\lambda(-q^{n+1}),$$

so that ϕ_λ is an eigenfunction of $L|_{I^-}$. We see that L can be considered as an extension of the Jacobi operator for the Al-Salam–Chihara polynomials. Note also that the parameters $b = s/a$ and $c = 1/sa$ satisfy $b = \overline{c}$ in case $s \in \mathbb{T}$, and $q < b/c \leq 1$ in case $s \in \mathbb{R}$ with $q < s^2 < 1$, which (by symmetry in b and c) corresponds exactly to the conditions under which the moment problem is indeterminate. Note that the families of symmetric Al-Salam–Chihara polynomials and continuous q^{-1}-Laguerre polynomials are included (for $s = \pm i$, and $(a, s) = (\pm q^{-\frac{1}{2}\alpha-\frac{1}{4}}, \pm q^{\frac{1}{4}})$, respectively).

Let $\mathcal{L}^2_z = \mathcal{L}^2_z(I, w(x)d_qx)$ be the Hilbert space with inner product

$$\langle f, g\rangle_z = \int_{-1}^{\infty(z)} f(x)\overline{g(x)}w(x)\,d_qx,$$

where w is the positive weight function on I given by

$$w(x) = w(x; a; q) = \frac{(-qx; q)_\infty}{(-a^2x; q)_\infty}. \tag{3.1}$$

Here the q-integral is defined by

$$\int_{-1}^{\infty(z)} f(x)\,d_qx = (1-q)\sum_{n=0}^{\infty} f(-q^n)q^n + (1-q)\sum_{n=-\infty}^{\infty} f(zq^n)zq^n.$$

In order to show that $(L, \mathcal{D})$ is self-adjoint, with $\mathcal{D} \subseteq \mathcal{L}^2_z$ an appropriate dense domain, we need the following truncated inner product. For $k, l, m \in \mathbb{N}$ and $f, g \in F(I)$ we set

$$\begin{aligned}\langle f, g\rangle_{k,l,m} &= \left(\int_{-1}^{-q^{k+1}} + \int_{zq^{m+1}}^{zq^{-l}}\right) f(x)\overline{g(x)}w(x)d_qx \\ &= (1-q)\sum_{n=0}^{k} f(-q^n)\overline{g(-q^n)}w(-q^n)q^n \\ &\quad + (1-q)\sum_{n=-l}^{m} f(zq^n)\overline{g(zq^n)}w(zq^n)zq^n.\end{aligned} \tag{3.2}$$

Note that for $f, g \in \mathcal{L}^2_z$ we have

$$\lim_{k,l,m\to\infty} \langle f, g\rangle_{k,l,m} = \langle f, g\rangle_z. \tag{3.3}$$

The following result will be useful to establish self-adjointness of L.

Lemma 3.3 *For $k, l, m \in \mathbb{N}$ and $f, g \in F(I)$*

$$\langle Lf, g\rangle_{k,l,m} - \langle f, Lg\rangle_{k,l,m} = D(f, \overline{g})(-q^k) - D(f, \overline{g})(zq^m) + D(f, \overline{g})(zq^{-l-1}),$$

where $D(f, g) \in F(I)$ is the Casorati determinant given by

$$D(f, g)(x) = \Big(f(x)g(qx) - f(qx)g(x)\Big)a^{-1}(1-q)(1+a^2x)w(x), \tag{3.4}$$

for $x \in I$.

Proof We first consider the sum $\sum_{n=0}^{k}$ of the truncated inner product. We write $f(-q^n) = f_n$ for functions f on I^-, then

$$(Lf)_n = a^{-1}(1-q^{-n})f_{n-1} + \frac{s+s^{-1}}{a^2q^n} f_n + a(1-a^{-2}q^{-n})f_{n+1}.$$

Now we have

$$\sum_{n=0}^{k}\Big((Lf)_n\overline{g_n}w_nq^n - f_k\overline{(Lg)_n}w_nq^n\Big) =$$

$$a^{-1}\sum_{n=1}^{k} f_n\overline{g_{n-1}}(1-q^n)w_n - a^{-1}\sum_{n=1}^{k} f_{n-1}\overline{g_n}(1-q^n)w_n$$

$$+a^{-1}\sum_{n=0}^{k} f_n\overline{g_{n+1}}(1-a^2q^n)w_n - a^{-1}\sum_{n=0}^{k} f_{n+1}g_n(1-a^2q^n)w_n.$$

We use

$$(1-q^{n+1})w_{n+1} = (1-a^2q^n)w_n, \qquad n \in \mathbb{N},$$

then we see that, after shifting summation indices, the expression above is equal to

$$\Big(f_k\overline{g_{k+1}} - f_{k+1}\overline{g_k}\Big)(1-a^2q^k)a^{-1}w_k,$$

so we have obtained

$$\int_{-1}^{-q^{k+1}} \Big((Lf)(x)\overline{g(x)} - f(x)\overline{Lg(x)}\Big)w(x)d_qx = D(f, \overline{g})(-q^k).$$

The sum $\sum_{n=-l}^{m}$ is treated in the same way, using the identity

$$(1+zq^{n+1})w(zq^{n+1}) = (1+za^2q^n)w(zq^n), \qquad n \in \mathbb{Z}. \qquad \square$$

Using the identity

$$\frac{(aq^{-n};q)_n}{(bq^{-n};q)_n} = \frac{(q/a;q)_n}{(q/b;q)_n}\left(\frac{a}{b}\right)^n$$

in the definition (3.1) of the weight function w, we see that

$$w(zq^{-n}) = \frac{(-zq;q)_\infty}{(-za^2;q)_\infty}\frac{(-1/z;q)_n}{(-q/za^2;q)_n}a^{-2n}q^n,$$

so that

$$w(zq^{-n}) = \frac{\theta(-qz;q)}{\theta(-za^2;q)}a^{-2n}q^n\Big(1+\mathcal{O}(q^n)\Big), \qquad n \to \infty. \tag{3.5}$$

From this we see that if $f \in \mathcal{L}_z^2$, then

$$\lim_{n\to\infty} a^{-n} f(zq^{-n}) = 0. \tag{3.6}$$

This enables us to prove the following lemma.

Lemma 3.4 *For $f, g \in \mathcal{L}_z^2$, we have*

$$\lim_{n\to\infty} D(f,\overline{g})(zq^{-n}) = 0.$$

Proof From the definition of w we obtain

$$(1+za^2q^{-n})w(zq^{-n}) = \frac{(-zq^{1-n};q)_\infty}{(-za^2q^{1-n};q)_\infty} = \frac{(-zq;q)_\infty}{(-za^2q;q)_\infty}\frac{(-1/z;q)_n}{(-1/a^2z;q)_n}a^{-2n},$$

so that

$$(1+za^2q^{-n})w(zq^{-n}) = \frac{\theta(-zq;q)}{\theta(-za^2q;q)}a^{-2n}\Big(1+\mathcal{O}(q^n)\Big), \qquad n \to \infty.$$

Now the lemma follows from the definition of the Casorati determinant and the asymptotic behavior (3.6) of $f, g \in \mathcal{L}_z^2$. $\square$

Now we are ready to introduce an appropriate dense domain for the unbounded operator L. Let us define $\mathcal{D} \subseteq \mathcal{L}_z^2$ to be the subspace consisting of functions $f \in \mathcal{L}_z^2$

satisfying the following conditions:

- $Lf \in \mathcal{L}_z^2$
- f is bounded on $-q^{\mathbb{N}+1} \cup zq^{\mathbb{N}}$
- $\lim_{n\to\infty} f(zq^n) - f(-q^n) = 0$

Note that $\mathcal{D}$ is dense in $\mathcal{L}_z^2$, since it contains the finitely supported functions on I.

Theorem 3.5 *The operator* $(L, \mathcal{D})$ *is self-adjoint.*

Proof First we show that $(L, \mathcal{D})$ is symmetric. Let $f, g \in \mathcal{D}$, and define

$$u(x) = \frac{a}{(1-q)(1+a^2x)w(x)}, \qquad x \in I,$$

then from (3.4) we obtain

$$\begin{aligned} &u(zq^n)D(f,g)(zq^n) - u(-q^n)D(f,g)(-q^n) = \\ &\quad \left[f(zq^n) - f(-q^n)\right]g(zq^{n+1}) - f(zq^{n+1})\left[g(zq^n) - g(-q^n)\right] \\ &\quad + f(-q^n)\left[g(zq^{n+1}) - g(-q^{n+1})\right] - \left[f(zq^{n+1}) - f(-q^{n+1})\right]g(-q^n). \end{aligned} \tag{3.7}$$

From the conditions on f and g it follows that this tends to 0 as $n \to \infty$. Then using

$$\lim_{n\to\infty} u(zq^n) = \lim_{n\to\infty} u(-q^n) = \frac{a}{1-q},$$

as well as (3.3) and Lemmas 3.3, 3.4 we see that

$$\langle Lf, g\rangle_z - \langle f, Lg\rangle_z = \lim_{n\to\infty} D(f,g)(-q^n) - D(f,g)(zq^n) = 0,$$

so that $(L, \mathcal{D})$ is symmetric.

Since $(L, \mathcal{D})$ is symmetric, we have $(L, \mathcal{D}) \subseteq (L^*, \mathcal{D}^*)$, where $(L^*, \mathcal{D}^*)$ is the adjoint of the operator $(L, \mathcal{D})$. Here, by definition, $\mathcal{D}^*$ is the subspace

$$\mathcal{D}^* = \{g \in \mathcal{L}_z^2 \mid f \mapsto \langle Lf, g\rangle_z \text{ is continuous on } \mathcal{D}\}.$$

We show that $L^* = L|_{\mathcal{D}^*}$. Let f be a non-zero function with support at only one point $x \in I$ and let $g \in F(I)$, then we obtain in the same way as above that $\langle Lf, g\rangle_z = \langle f, Lg\rangle_z$. In particular, for $g \in \mathcal{D}^*$ we then have $\langle f, Lg\rangle_z = \langle f, L^*g\rangle_z$, so $(Lg)(x) = (L^*g)(x)$. This holds for all $x \in I$, hence $L^* = L|_{\mathcal{D}^*}$.

Finally we need to show that $\mathcal{D}^* \subseteq \mathcal{D}$. Let $f \in \mathcal{D}$ and let $g \in \mathcal{D}^*$. Using Lemmas 3.3, 3.4 and $L^* = L|_{\mathcal{D}^*}$ we obtain

$$\lim_{n\to\infty} D(f,\overline{g})(-q^n) - D(f,\overline{g})(zq^n) = \langle Lf, g\rangle_z - \langle f, L^*g\rangle_z = 0.$$

Since this holds for all $f \in \mathcal{D}$, we find using (3.7) that g is bounded near 0, and

$$\lim_{n\to\infty} g(zq^n) - g(-q^n) = 0,$$

hence $g \in \mathcal{D}$. □

4 Eigenfunctions of L

In this section we consider eigenspaces and eigenfunctions of L. The eigenfunctions are initially only defined on a part of I_*, and we show how to extend them to functions on I_*.

We start with some useful properties of eigenspaces of L. For $\mu \in \mathbb{C}$ we introduce the spaces

$$\begin{aligned}
V_\mu^- &= \{f \in F(I_*^-) \mid Lf = \mu f\},\\
V_\mu^+ &= \{f \in F(I^+) \mid Lf = \mu f\},\\
V_\mu &= \Big\{f \in F(I_*) \mid Lf = \mu f, \quad f(zq^n) = o(q^{-n}), \quad f(-q^n) = o(q^{-n})\\
&\qquad \text{and } f(zq^n) - f(-q^n) = o(1) \text{ for } n \to \infty\Big\}
\end{aligned} \tag{4.1}$$

Lemma 4.1 *For* $\mu \in \mathbb{C}$

(i) $\dim V_\mu^\pm = 2$.
(ii) For $f, g \in V_\mu^-$ *the Casorati determinant* $D(f, g)$ *is constant on* I_*^-.
(iii) For $f, g \in V_\mu^+$ *the Casorati determinant* $D(f, g)$ *is constant on* I^+.
(iv) For $f, g \in V_\mu$ *the Casorati determinant* $D(f, g)$ *is constant on* I_*.

Proof For (i) we write $f(tq^n) = f_n$ (with $t = -1$ or $t = z$), then we see that $Lf = \mu f$ gives a recurrence relation of the form $\alpha_n f_{n+1} + \beta_n f_n + \gamma_n f_{n-1} = \mu f_n$, with $\alpha_n, \gamma_n \neq 0$ for all n. Solutions to such a recurrence relation are uniquely determined by specifying f_n at two different points $n = l$ and $n = m$. So there are two independent solutions.

The proofs of (ii) and (iii) are the same. We prove (iii). Let $f, g \in F(I^+)$. Using the explicit expressions for the Casorati determinant and the weight function w, we find for $x \in I^+$

$$\begin{aligned}(Lf)(x)g(x) - f(x)(Lg)(x) &= a^{-1}(1+1/x)\Big(f(x/q)g(x) - f(x)g(x/q)\Big)\\ &\quad - a(1+1/a^2x)\Big(f(x)g(qx) - f(qx)g(x)\Big)\\ &= \frac{(1+1/x)D(f,g)(x/q)}{(1-q)(1+a^2x/q)w(x/q)} - \frac{a^2(1+1/a^2x)D(f,g)(x)}{(1-q)(1+a^2x)w(x)}\\ &= \frac{1}{(1-q)xw(x)}\Big(D(f,g)(x/q) - D(f,g)(x)\Big).\end{aligned}$$

Now if f and g satisfy $(Lf)(x) = \mu f(x)$ and $(Lg)(x) = \mu g(x)$, we have $D(f,g)(x/q) = D(f,g)(x)$, hence $D(f,g)$ is constant on I^+.

Statement (iv) follows from (ii) and (iii) and the fact that

$$\lim_{n\to\infty}\Big(D(f,g)(zq^n) - D(f,g)(-q^n)\Big) = 0,$$

which can be obtained by rewriting the difference of the Casorati determinants similar as in the proof of Theorem 3.5. □

We can now introduce explicit eigenfunctions of L (in the algebraic sense). Comparing L given in Definition 3.1 to the q-hypergeometric difference equation (2.3) we find that

$$\psi_\lambda(x;s) = \psi_\lambda(x;a,s\,|\,q) = s^{-n}\,{}_2\varphi_1\left(\begin{matrix}a\lambda/s,\, a/\lambda s\\ q/s^2\end{matrix};q,-qx\right), \qquad |x| < q^{-1}, \tag{4.2}$$

with $x = tq^n$, $t \in \{-1, z\}$, is a solution of the eigenvalue equation

$$Lf = \mu(\lambda)f, \qquad \mu(\lambda) = \lambda + \lambda^{-1}, \tag{4.3}$$

on $I_{*,\leq 1}$. From $L_{a,s} = L_{a,s^{-1}}$ it follows immediately that $\psi_\lambda(x;a,s^{-1}\,|\,q)$ is also a solution of (4.3) on $I_{*,\leq 1}$. Note that it does not follow from the q-hypergeometric difference equation that ψ_λ is a solution to (4.3) for $x = -1$. In fact, we show later on that for generic λ it is not a solution for $x = -1$. Using (4.3) the solutions $\psi_\lambda(x;s^{\pm 1})$ can be extended to functions on I_* still satisfying (4.3). We denote these extensions still by $\psi_\lambda(x;s^{\pm 1})$. Later on we give explicit expressions for the extensions. Note that both functions $\psi_\lambda(x;s^{\pm 1})$ can be considered as little q-Jacobi functions [13].

Lemma 4.2 *For $\lambda \in \mathbb{C}^*$, $\psi_\lambda(\cdot;s)$ and $\psi_\lambda(\cdot;s^{-1})$ are in $V_{\mu(\lambda)}$.*

Proof We already know that ψ_λ is an eigenfunction of L, so we only need to check that ψ_λ has the behavior near 0 as stated in the definition of $V_{\mu(\lambda)}$. From (4.2) we find for $t \in \{-1, z\}$

$$\psi_\lambda(tq^n; s) = s^{-n}\Big(1 + \mathcal{O}(q^n)\Big), \qquad n \to \infty,$$

so that $\psi_\lambda(tq^n; s^{\pm 1}) = o(q^{-n})$ by the conditions on s, and

$$\psi_\lambda(zq^n; s^{\pm 1}) - \psi_\lambda(-q^n; s^{\pm 1}) = o(1).$$

□

Another solution of (4.3) is

$$\Psi_\lambda(x) = \Psi_\lambda(x; a, s \,|\, q) = (a\lambda)^{-n} \,{}_2\varphi_1\left(\begin{matrix} a\lambda/s, as\lambda \\ q\lambda^2 \end{matrix}; q, -\frac{q}{a^2 x}\right), \qquad x = zq^n > q/a^2,$$

and clearly $\Psi_{1/\lambda}(x)$ also satisfies (4.3). Both are solutions on $zq^{-\mathbb{N}_{\geq k}}$ for k large enough. For $n \to \infty$ we have

$$\Psi_\lambda(zq^{-n}) = (a\lambda)^n\Big(1 + \mathcal{O}(q^n)\Big), \tag{4.4}$$

so by the asymptotic behavior (3.5) of the weight function w, for $|\lambda| < 1$ the function Ψ_λ is an L^2-function at ∞. The solutions $\Psi_{\lambda^{\pm 1}}$ can be used to give the explicit extension of ψ_λ on I^+;

$$\psi_\lambda(x; s) = b_z(\lambda; s)\Psi_\lambda(x) + b_z(1/\lambda; s)\Psi_{1/\lambda}(x), \tag{4.5}$$

with

$$b_z(\lambda; s) = b_z(\lambda; a, s \,|\, q) = \frac{(q/as\lambda, a/s\lambda; q)_\infty}{(1/\lambda^2, q/s^2; q)_\infty} \frac{\theta(-qaz\lambda/s; q)}{\theta(-qz; q)},$$

which follows from the three-term transformation formula [8, (III.32)].

Next we give a solution ϕ_λ of the eigenvalue equation (4.3) for all $x \in I$.

Definition 4.3 Define

$$\phi_\lambda(x) = \phi_\lambda(x; a, s \,|\, q) = d(\lambda; s)\psi_\lambda(x; a, s \,|\, q) + d(\lambda; 1/s)\psi_\lambda(x; a, 1/s \,|\, q),$$

with

$$d(\lambda; s) = d(\lambda; a, s \,|\, q) = \frac{(as\lambda, as/\lambda; q)_\infty}{(a^2, s^2; q)_\infty}.$$

Note that ϕ_λ is invariant under $\lambda \leftrightarrow 1/\lambda$ and $s \leftrightarrow 1/s$. Since ψ_λ is a solution of (4.3) for $x \in I_*$, it is clear that ϕ_λ is also a solution of (4.3) for $x \in I_*$. We will show that ϕ_λ is an eigenfunction of L on the whole of I, i.e. including the point -1, by identifying $\phi_\lambda(-q^n)$ with an Al-Salam–Chihara polynomial.

Lemma 4.4 *For $n \in \mathbb{N}$ we have*

$$\phi_\lambda(-q^n) = a^{-n} P_n(\lambda; s/a, 1/sa; q^{-1}).$$

As a consequence, $L\phi_\lambda(x) = \mu(\lambda)\phi_\lambda(x)$ for all $x \subset I$.

Proof We use the three-term transformation formula for ${}_3\varphi_2$-functions [8, (III.34)] with $A = q^{-n}$, $n \in \mathbb{N}$. Letting $D \to \infty$ gives

$$
\begin{aligned}
{}_3\varphi_1\left(\begin{matrix} q^{-n}, B, C \\ E \end{matrix}; q, \frac{Eq^n}{BC}\right) &= \frac{(E/B, E/C; q)_\infty}{(E, E/BC; q)_\infty}\, {}_2\varphi_1\left(\begin{matrix} B, C \\ BCq/E \end{matrix}; q, q^{n+1}\right) \\
&\quad + \frac{(B, C; q)_\infty}{(E, BC/E; q)_\infty}\left(\frac{E}{BC}\right)^n {}_2\varphi_1\left(\begin{matrix} E/B, E/C \\ Eq/BC \end{matrix}; q, q^{n+1}\right).
\end{aligned}
$$

Using the identity $(A; q^{-1})_n = (1/A; q)_n(-A)^n q^{-\frac{1}{2}n(n-1)}$, we find the transformation

$$
{}_3\varphi_1\left(\begin{matrix} q^{-n}, B, C \\ E \end{matrix}; q, \frac{Eq^n}{BC}\right) = {}_3\varphi_2\left(\begin{matrix} q^n, 1/B, 1/C \\ 1/E, 0 \end{matrix}; q^{-1}, q^{-1}\right).
$$

From these formulas with $B = a\lambda/s$, $C = a/\lambda s$, $E = 1/a^2$, we find

$$
\phi_\lambda(-q^n; a, s\,|\,q) = s^n\, {}_3\varphi_2\left(\begin{matrix} q^n, s\lambda/a, s/\lambda a \\ a^2, 0 \end{matrix}; q^{-1}, q^{-1}\right),
$$

which we recognize as the Al-Salam–Chihara polynomial $a^{-n}P_n(\lambda; s/a, 1/sa; q^{-1})$, see (2.4). From the recurrence relation for these polynomials it follows that ϕ_λ satisfies $(L\phi_\lambda)(x) = \mu(\lambda)\phi_\lambda(x)$, for $x \in I^-$ (including -1), so ϕ_λ is a solution for (4.3) on I. □

Note that the eigenspace

$$W_\mu^- = \{f \in F(I^-) \mid Lf = \mu f\}$$

is one-dimensional, since the value of $f \in W_\mu^-$ at the point -1 completely determines f on I^-. So ϕ_λ is the only solution (up to a multiplicative constant) to the eigenvalue equation (4.3) on I^-, hence also the only function in $V_{\mu(\lambda)}$ which is a solution to (4.3) on I.

We also need the expansion of ϕ_λ into $\Psi_{\lambda^{\pm 1}}$ and we need to determine an explicit expression for the Casorati determinant $D(\phi_\lambda, \Psi_\lambda)$, which we need later on to describe the resolvent for L. Determining the expansion is a straightforward calculation.

Lemma 4.5 *For* $x \in I^+$,

$$\phi_\lambda(x) = c_z(\lambda)\Psi_\lambda(x) + c_z(\lambda^{-1})\Psi_{\lambda^{-1}}(x),$$

with

$$\begin{aligned} c_z(\lambda) &= c_z(\lambda; a, s \,|\, q) \\ &= \frac{(as/\lambda, a/s\lambda; q)_\infty}{(a^2, \lambda^{-2}; q)_\infty \theta(-qz; q)} \left(\frac{\theta(as\lambda, -qaz\lambda/s; q)}{\theta(s^2; q)} + \frac{\theta(a\lambda/s, -qasz\lambda; q)}{\theta(s^{-2}; q)} \right). \end{aligned}$$

Proof From (4.5) and Definition 4.3 we find

$$\phi_\lambda = c_z(\lambda)\Psi_\lambda + c_z(\lambda^{-1})\Psi_{\lambda^{-1}}, \qquad \text{on } I^+,$$

with

$$c_z(\lambda) = d(\lambda; a, s \,|\, q) b_z(\lambda; a, s \,|\, q) + d(\lambda; a, 1/s \,|\, q) b_z(\lambda; a, 1/s \,|\, q).$$

From the explicit expression for $d(\lambda)$ and $b_z(\lambda)$ we obtain the expression for $c_z(\lambda)$. □

Lemma 4.6 *For* $x \in I^+$,

$$D\big(\psi_\lambda(\cdot; s), \psi_\lambda(\cdot; 1/s)\big)(x) = a(1-q)(s^{-1} - s)$$

and

$$D(\phi_\lambda, \Psi_\lambda)(x) = K_z\, c_z(\lambda^{-1})(\lambda^{-1} - \lambda),$$

with

$$K_z = (1-q)\frac{\theta(-z; q)}{\theta(-za^2; q)}.$$

Proof The Casorati determinant $D(\Psi_{\lambda^{-1}}, \Psi_\lambda)$ is constant on I^+, so we may determine it by letting $x \to \infty$. Now from the asymptotic behavior (4.4) of $\Psi_\lambda(zq^{-k})$ when $k \to \infty$, the definition (3.4) of the Casorati determinant, and the asymptotic behavior of $(1 + za^2q^{-k})w(zq^{-k})$ for $k \to \infty$, see Lemma 3.4, we obtain

$$D(\Psi_{\lambda^{-1}}, \Psi_\lambda)(x) = K_z(\lambda^{-1} - \lambda).$$

With this result we can compute $D\big(\psi_\lambda(\cdot;s),\Psi_\lambda\big)$. First expand $\psi_\lambda(x;s)$ in terms of $\Psi_{\lambda^{\pm1}}(x)$ using (4.5), then we find

$$D(\psi_\lambda(\cdot;s),\Psi_\lambda)(x) = b_z(1/\lambda;s)D(\Psi_{1/\lambda},\Psi_\lambda)(x) = K_z b_z(1/\lambda;s)(\lambda-\lambda^{-1}).$$

This leads to

$$\begin{aligned} D\big(\psi_\lambda(\cdot;s),&\psi_\lambda(\cdot;1/s)\big)(x) \\ &= b_z(\lambda;1/s)D(\psi_\lambda(\cdot;s),\Psi_\lambda)(x) + b_z(1/\lambda;1/s)D(\psi_\lambda(\cdot;s),\Psi_{1/\lambda})(x) \\ &= K_z\Big(b_z(1/\lambda;s)b_z(\lambda;1/s) - b_z(\lambda;s)b_z(1/\lambda;1/s)\Big)(\lambda-\lambda^{-1}). \end{aligned}$$

Writing this out and using the fundamental θ-function identity (2.2) with

$$(x,y,v,w) = (a,-az,\lambda/s,\lambda s)$$

we find

$$D\big(\psi_\lambda(\cdot;s),\psi_\lambda(\cdot;1/s)\big)(x) = \frac{sK_z(\lambda-\lambda^{-1})}{az\lambda}\,\frac{\theta(-a^2z,-1/z,\lambda^2,1/s^2;q)}{(\lambda^{\pm2},qs^{\pm2};q)_\infty\theta(-qz;q)^2}.$$

This expression simplifies using identities for θ-functions.

Finally, using the c-function expansion from Lemma 4.5 we have

$$D(\phi_\lambda,\Psi_\lambda) = c_z(\lambda^{-1})D(\Psi_{\lambda^{-1}},\Psi_\lambda) = K_z\,c_z(\lambda^{-1})(\lambda^{-1}-\lambda). \qquad \square$$

Note that so far we only have expressions for the Casorati determinants on I^+. These expressions are actually valid on I_*, which follows from the following result.

Proposition 4.7 *The set* $\{\psi_\lambda(\cdot;s),\psi_\lambda(\cdot;1/s)\}$ *is a basis for* $V_{\mu(\lambda)}$.

Proof From Lemma 4.2 we know that $\psi_\lambda(\cdot;s^{\pm1})\in V_{\mu(\lambda)}$. Then by Lemma 4.1 the Casorati determinant $D\big(\psi_\lambda(\cdot;s),\psi_\lambda(\cdot;1/s)\big)$ is constant on I_*, so the value from Lemma 4.6 is valid on I_*. It is nonzero since we assumed $s\neq\pm1$, so $\psi_\lambda(\cdot;s)$ and $\psi_\lambda(\cdot;1/s)$ are linearly independent, and since $\dim V_{\mu(\lambda)}=2$ they form a basis. $\square$

Now Ψ_λ extends to a function on I_* by expanding it in terms of $\psi_\lambda(x;s^{\pm1})$. This can be done explicitly, but we do not need the explicit expansion. Then $\Psi_\lambda,\phi_\lambda\in V_{\mu(\lambda)}$, so as a consequence we obtain the following result.

Corollary 4.8 *On* I_*,

$$D(\phi_\lambda,\Psi_\lambda) = K_z\,c_z(\lambda^{-1})(\lambda^{-1}-\lambda), \qquad \lambda\in\mathbb{C}^*.$$

5 The Spectral Decomposition of L

In this section we determine the spectral decomposition for L as a self-adjoint operator on $\mathcal{L}_z^2$. For $z = 1$ the spectral projections are completely explicit.

To determine the spectral measure E for the self-adjoint operator $(L, \mathcal{D})$ we use the following formula, see [7, Theorem XII.2.10],

$$\langle E(a,b)f, g\rangle_z = \lim_{\delta\downarrow 0}\lim_{\varepsilon\downarrow 0} \frac{1}{2\pi i}\int_{a+\delta}^{b-\delta} \Big(\langle R_{\gamma+i\varepsilon} f, g\rangle_z - \langle R_{\gamma-i\varepsilon} f, g\rangle_z\Big) d\gamma, \tag{5.1}$$

for $a < b$ and $f, g \in \mathcal{L}_z^2$. Here $R_\gamma = (L-\gamma)^{-1}$ is the resolvent for L. $R_{\mu(\lambda)}$ can be expressed using the Green kernel $G_\lambda \in F(I \times I)$ given by

$$G_\lambda(x,y) = \begin{cases} \dfrac{\phi_\lambda(x)\Psi_\lambda(y)}{D_\lambda}, & x \le y, \\[2ex] \dfrac{\phi_\lambda(y)\Psi_\lambda(x)}{D_\lambda}, & x > y, \end{cases}$$

where $D_\lambda = D(\phi_\lambda, \Psi_\lambda)$ (see Corollary 4.8) and we assume $\lambda \in \mathbb{D} \setminus ($-
$\mathbb{D}$ is the open unit disc. Note that $G_\lambda(\cdot, y) \in \mathcal{L}_z^2$ for $y \in I$. Now t'
given by

$$(R_{\mu(\lambda)} f)(y) = \langle f, G_\lambda(\cdot, y)\rangle_z,$$

so that for $f, g \in \mathcal{L}_z^2$

$$\begin{aligned} &\langle R_{\mu(\lambda)} f, g\rangle_z \\ &= \iint_{I\times I} f(x)\overline{g(y)} G_\lambda(x,y)\, w(x)w(y)\, d_qx d_qy \\ &= \iint_{\substack{(x,y)\in I\times I\\ x\le y}} \frac{\phi_\lambda(x)\Psi_\lambda(y)}{D_\lambda}\Big(f(x)\overline{g(y)} + f(y)\overline{g(} \end{aligned}$$

The proof that $R_{\mu(\lambda)}$ is indeed the resolv
Proposition 6.1].

Let $\mu^{-1} : \mathbb{C} \setminus \mathbb{R} \to \mathbb{D} \setminus (-1, 1)$ '
then $\gamma = \mu(e^{i\psi})$ for a unique $\psi \in$

lir
ε'

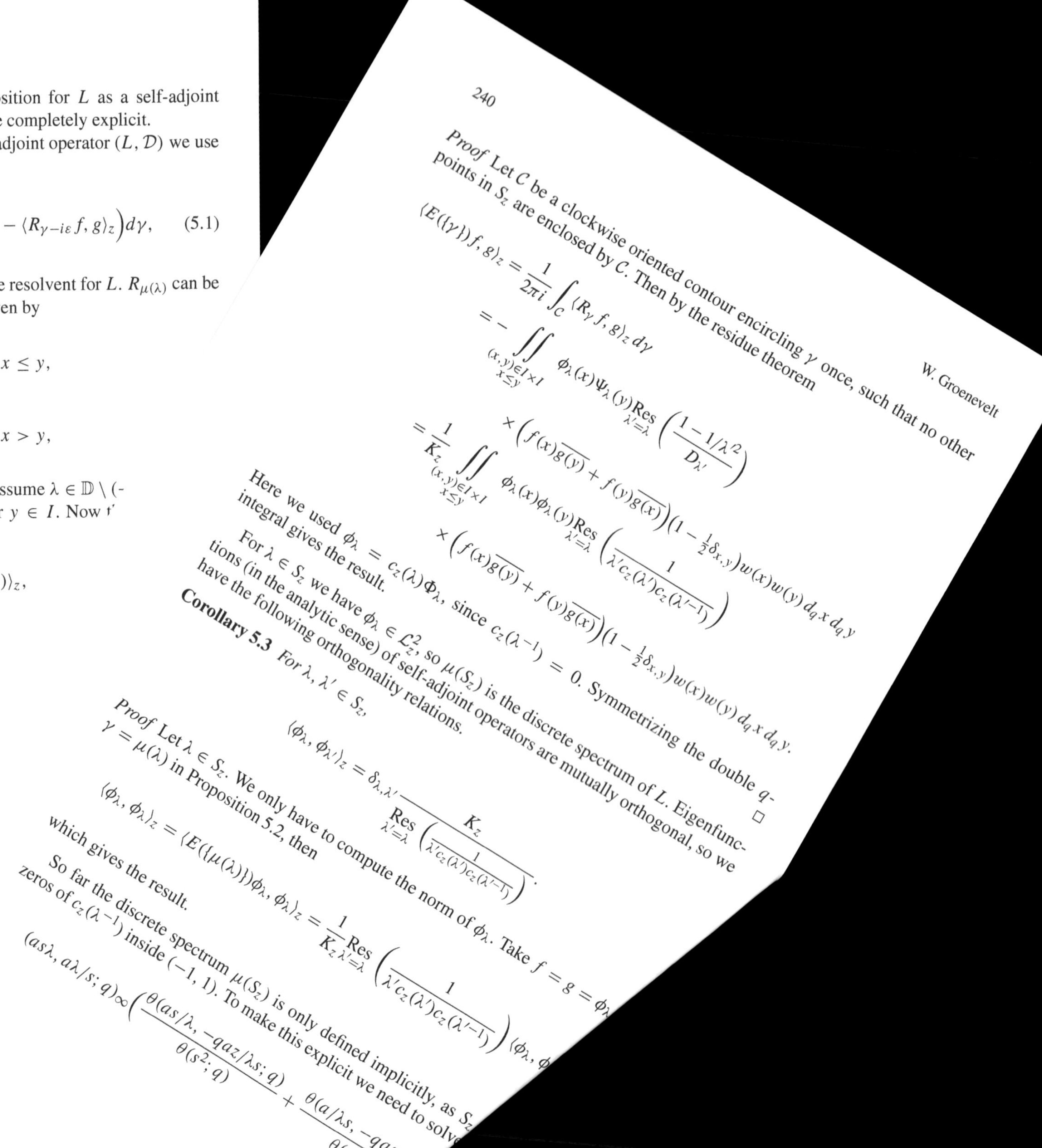

Proof Let C be a clockwise oriented contour encircling γ once, such that no other points in S_z are enclosed by C. Then by the residue theorem

$$\begin{aligned}\langle E(\{\gamma\})f, g\rangle_z &= \frac{1}{2\pi i}\int_C \langle R_\gamma f, g\rangle_z\, d\gamma \\ &= -\iint_{\substack{(x,y)\in I\times I\\ x\le y}} \phi_\lambda(x)\Psi_\lambda(y)\operatorname*{Res}_{\lambda'=\lambda}\left(\frac{1-1/\lambda'^2}{D_{\lambda'}}\right) \\ &\quad\times\Big(f(x)\overline{g(y)} + f(y)\overline{g(x)}\Big)\big(1-\tfrac12\delta_{x,y}\big)w(x)w(y)\, d_qx\, d_qy \\ &= \frac{1}{K_z}\iint_{\substack{(x,y)\in I\times I\\ x\le y}} \phi_\lambda(x)\phi_\lambda(y)\operatorname*{Res}_{\lambda'=\lambda}\left(\frac{1}{\lambda' c_z(\lambda')c_z(\lambda'^{-1})}\right) \\ &\quad\times\Big(f(x)\overline{g(y)} + f(y)\overline{g(x)}\Big)\big(1-\tfrac12\delta_{x,y}\big)w(x)w(y)\, d_qx\, d_qy.\end{aligned}$$

Here we used $\phi_\lambda = c_z(\lambda)\Phi_\lambda$, since $c_z(\lambda^{-1}) = 0$. Symmetrizing the double q-integral gives the result. □

For $\lambda \in S_z$ we have $\phi_\lambda \in \mathcal{L}_z^2$, so $\mu(S_z)$ is the discrete spectrum of L. Eigenfunctions (in the analytic sense) of self-adjoint operators are mutually orthogonal, so we have the following orthogonality relations.

Corollary 5.3 *For $\lambda, \lambda' \in S_z$,*

$$\langle \phi_\lambda, \phi_{\lambda'}\rangle_z = \delta_{\lambda,\lambda'}\frac{K_z}{\operatorname*{Res}_{\lambda'=\lambda}\left(\frac{1}{\lambda' c_z(\lambda')c_z(\lambda'^{-1})}\right)}.$$

Proof Let $\lambda \in S_z$. We only have to compute the norm of ϕ_λ. Take $f = g = \phi_\lambda$
$\gamma = \mu(\lambda)$ in Proposition 5.2, then

$$\langle \phi_\lambda, \phi_\lambda\rangle_z = \langle E(\{\mu(\lambda)\})\phi_\lambda, \phi_\lambda\rangle_z = \frac{1}{K_z}\operatorname*{Res}_{\lambda'=\lambda}\left(\frac{1}{\lambda' c_z(\lambda')c_z(\lambda'^{-1})}\right)\langle \phi_\lambda, \phi$$

which gives the result.

So far the discrete spectrum $\mu(S_z)$ is only defined implicitly, as S_z
zeros of $c_z(\lambda^{-1})$ inside $(-1, 1)$. To make this explicit we need to solv

$$(as\lambda, a\lambda/s; q)_\infty\Big(\frac{\theta(as/\lambda, -qaz/\lambda s; q)}{\theta(s^2; q)} + \frac{\theta(a/\lambda s, -qasz/\lambda}{\theta(s^{-2}; q)}$$

see Lemma 4.5. The factor $(as\lambda, a\lambda/s; q)_\infty$ has $\frac{1}{as}q^{-\mathbb{N}} \cup \frac{s}{a}q^{-\mathbb{N}}$ as the set of zeros. But it does not seem possible to give explicit expressions for the zeros of the second factor, the sum of θ-functions, except in case $z = 1$.

Lemma 5.4 *For $z = 1$,*

$$c_1(\lambda) = \frac{(as/\lambda, a/s\lambda; q)_\infty \theta(a^2\lambda^2 q; q^2)}{(a^2, \lambda^{-2}; q)_\infty \theta(qs^2; q^2)}.$$

As a consequence, $c_1(\lambda^{-1}) = 0$ if and only if $\lambda \in \Gamma$ with

$$\Gamma = \frac{1}{as}q^{-\mathbb{N}} \cup \frac{s}{a}q^{-\mathbb{N}} \cup aq^{\frac{1}{2}+\mathbb{Z}} \cup (-a)q^{\frac{1}{2}+\mathbb{Z}}.$$

Proof For $z = 1$ we can simplify the expression for $c_z(\lambda)$ using identities for θ-functions from Sect. 2. We have

$$\theta(-qaz\lambda s^{\pm 1}; q) = \theta(-qa\lambda s^{\pm 1}; q) = \frac{1}{a\lambda s^{\pm 1}}\theta(-a\lambda s^{\pm 1}; q),$$

so that

$$c_1(\lambda) = \frac{s}{a\lambda}\frac{(as/\lambda, a/s\lambda; q)_\infty}{(a^2, \lambda^{-2}; q)_\infty \theta(-q, s^2; q)}\Big(\theta(as\lambda, -a\lambda/s; q) - \theta(a\lambda/s, -as\lambda; q)\Big).$$

Use the fundamental θ-function identity (2.2) with

$$(x, y, v, w) = (a\lambda s^{\frac{1}{2}}, a\lambda s^{-\frac{1}{2}}, s^{\frac{1}{2}}, -s^{-\frac{1}{2}})$$

for a fixed root $s^{\frac{1}{2}}$ of s, to obtain

$$\theta(as\lambda, -a\lambda/s; q) - \theta(a\lambda/s, -as\lambda; q) = \frac{a\lambda}{s}\frac{\theta(a^2\lambda^2, s, -1, -s; q)}{\theta(a\lambda, -a\lambda; q)}.$$

Then simplifying the remaining expressions using (2.1) gives the result. □

Recall that we assumed $0 < a^2 < 1$, and $s \in \mathbb{T}$ or $s \in \mathbb{R}$ with $q < s^2 < 1$. Using Lemma 5.4 we can read off the zeros of $c_1(\lambda^{-1})$ that are inside $(-1, 1)$. This gives the following result for the spectrum.

Proposition 5.5 *The spectrum of the self-adjoint q-difference operator L on $\mathcal{L}^2_1$ is*

$$[-2, 2] \cup \mu(S_1),$$

where for $s \in \mathbb{T} \setminus \{-1, 1\}$ or $s \in \mathbb{R}$ such that $|s/a| \geq 1$

$$S_1 = \left\{\pm aq^{m+\frac{1}{2}} \mid m \in \mathbb{Z} \text{ such that } -1 < aq^{m+\frac{1}{2}} < 1\right\},$$

and for $s \in \mathbb{R}$ *such that* $|s/a| < 1$

$$S_1 = \left\{ \frac{s}{a} \right\} \cup \left\{ \pm a q^{m+\frac{1}{2}} \mid m \in \mathbb{Z} \text{ such that } -1 < a q^{m+\frac{1}{2}} < 1 \right\}.$$

6 The Integral Transform

In this section we arrive at the main result. We define an integral transform $\mathcal{F}$ that can be considered as a q-analog of the Jacobi function transform. We show that $\mathcal{F}$ is unitary and we determine the inverse of $\mathcal{F}$. As a consequence we find orthogonality relations for q^{-1}-Al-Salam–Chihara polynomials, and hence we have a solution to the corresponding moment problem. Throughout this section we assume $z = 1$ and omit all the subscripts z; in particular, $\mathcal{L}^2 = \mathcal{L}^2_1$ and $c(z) = c_1(z)$.

We define the integral transform $\mathcal{F}$ as follows.

Definition 6.1 Let $\mathcal{D}_0 \subseteq \mathcal{L}^2$ be the subset consisting of finitely supported functions. $\mathcal{F} : \mathcal{D}_0 \to F(\mathbb{T} \cup S)$ is given by

$$(\mathcal{F}f)(\lambda) = \int_{-1}^{\infty(1)} f(x)\phi_\lambda(x)w(x)\, d_qx, \qquad \lambda \in \mathbb{T} \cup S.$$

Let ν be the measure defined by

$$\int f(\lambda)\, d\nu(\lambda) = \frac{1}{4K\pi i} \int_{\mathbb{T}} f(\lambda) W(\lambda) \frac{d\lambda}{\lambda} + \frac{1}{K} \sum_{\lambda \in S} f(\lambda) \hat{W}(\lambda),$$

where $\mathbb{T}$ is oriented in the counter-clockwise direction,

$$K = (1-q)\frac{\theta(-1;q)}{\theta(-a^2;q)},$$

$$W(\lambda) = \frac{1}{|c(\lambda)|^2} = \left| \frac{(as/\lambda, a/s\lambda; q)_\infty \theta(a^2\lambda^2 q; q^2)}{(a^2, \lambda^{-2}; q)_\infty \theta(qs^2; q^2)} \right|^2, \qquad \lambda \in \mathbb{T},$$

and

$$\hat{W}(\lambda) = \operatorname*{Res}_{\lambda'=\lambda} \left(\frac{1}{\lambda' c(\lambda') c(\lambda'^{-1})} \right), \qquad \lambda \in S,$$

with $S = S_1$ is given in Proposition 5.5. The residues can be computed explicitly, which gives the following expressions: for $\pm aq^{m+\frac{1}{2}} \in S$,

$$\hat{W}(\pm aq^{m+\frac{1}{2}}) = \frac{(a^2, a^2, a^2q, a^{-2}q^{-1}; q)_\infty \theta(qs^2, q/s^2; q^2)}{2(q^2; q^2)_\infty^2 (\pm a^2 q^{\frac{1}{2}} s, \pm a^2 q^{\frac{1}{2}}/s, \pm q^{-\frac{1}{2}} s, \pm q^{-\frac{1}{2}}/s; q)_\infty \theta(a^4q^2; q^2)}$$

$$\times \frac{1 - a^2q^{2m+1}}{1 - a^2q} \frac{(\pm a^2 q^{\frac{1}{2}} s, \pm a^2 q^{\frac{1}{2}}/s; q)_m}{(\pm q^{\frac{3}{2}} s, \pm q^{\frac{3}{2}}/s; q)_m} q^{m(m+1)},$$

and if $s/a \in S$,

$$\hat{W}(s/a) = \frac{(a^2, s^2/a^2; q)_\infty \theta(s^2q; q^2)}{(q, s^2; q)_\infty \theta(a^4q/s^2; q^2)}.$$

Let $\mathcal{H}$ be the Hilbert space consisting of functions f satisfying $f(\lambda) = f(\lambda^{-1})$ ν-almost everywhere, with inner product

$$\langle f, g \rangle_{\mathcal{H}} = \int f(\lambda) \overline{g(\lambda)}\, d\nu(\lambda).$$

Define $\mathcal{G} : \mathcal{H} \to F(I)$ by

$$(\mathcal{G}g)(x) = \langle g, \phi_\bullet(x) \rangle_{\mathcal{H}}, \qquad x \in I.$$

We can now formulate the main result.

Theorem 6.2 *$\mathcal{F}$ extends uniquely to a unitary operator $\mathcal{F} : \mathcal{L}^2 \to \mathcal{H}$, with inverse $\mathcal{G}$.*

As a consequence we have orthogonality relations for the functions $\phi_\bullet(x)$ with respect to the measure ν. Since $\phi_\lambda(-q^n)$ is an Al-Salam–Chihara polynomial by Lemma 4.4, ν is a solution of the corresponding indeterminate moment problem. Note that the polynomials are not dense in $\mathcal{H}$, so the measure ν is not an N-extremal solution.

Corollary 6.3 *The set $\{\phi_\bullet(x) \mid x \in I\}$ is an orthogonal basis for $\mathcal{H}$, with*

$$\langle \phi_\bullet(x), \phi_\bullet(y) \rangle_{\mathcal{H}} = \delta_{x,y} \frac{1}{|x| w(x)}.$$

In particular, the q^{-1}-Al-Salam–Chihara polynomials $P_n(\lambda) = P_n(\lambda; s/a, 1/sa; q^{-1})$ satisfy

$$\int P_n(\lambda) P_{n'}(\lambda)\, d\nu(\lambda) = \delta_{n,n'} \frac{a^{2n}}{q^n w(-q^n)}.$$

Remark 6.4 The kernel ϕ_λ defined in Definition 4.3 is a linear combination of the functions $\psi_\lambda(\cdot; s^{\pm 1})$, which are essentially little q-Jacobi functions. Furthermore, $\mathcal{F}$ diagonalizes the q-hypergeometric difference operator L considered as an unbounded operator on $\mathcal{L}^2$, i.e.

$$\mathcal{F} L \mathcal{F}^{-1} = M_\mu,$$

where M_μ is multiplication by $\mu(\,\cdot\,)$ on $\mathcal{H}$. So we may consider $\mathcal{F}$ as another little q-Jacobi function transform.

Remark 6.5 As already mentioned in Sect. 3 the symmetric Al-Salam–Chihara polynomials and the q-Laguerre polynomials are special cases of the Al-Salam–Chihara polynomials we consider, so we also obtained a solution for the corresponding moment problems. These solutions seem to be new. We can also (formally) obtain a solution of the Al-Salam–Carlitz II moment problem. Assume $|s/a| \geq 1$, then from Proposition 5.5 we see that the operator $aL_{a,s}$ has spectrum

$$[-2a, 2a] \cup \left\{ \pm (a^2 q^{m+\frac{1}{2}} + q^{-m-\frac{1}{2}}) \mid m \in \mathbb{Z} \text{ such that } -1 < aq^{m+\frac{1}{2}} < 1 \right\}.$$

For $a \to 0$ the continuous spectrum shrinks to $\{0\}$ and the discrete spectrum becomes $\{\pm q^{-m-\frac{1}{2}} \mid m \in \mathbb{Z}\}$. From the recurrence relation for $a^{-n} P_n(\pm a q^{m+\frac{1}{2}})$ we find that the polynomials $p_n(\pm q^{m+\frac{1}{2}}) = \lim_{a \to 0} a^{-n} P_n(\pm a q^{m+\frac{1}{2}}; s/a, 1/as; q^{-1})$ satisfy

$$\pm q^{-m-\frac{1}{2}} p_n = -q^{-n} p_{n+1} + (s + s^{-1}) q^{-n} p_n + (1 - q^{-n}) p_{n-1}.$$

Comparing this to the recurrence relation for the Al-Salam–Carlitz II polynomials $V_n^{(a)}(x; q)$ [14, §14.25], we find that $p_n(\lambda) = (-s)^n q^{\frac{1}{2}n(n-1)} V_n^{(s^{-2})}(\lambda; q)$. Letting $a \to 0$ in the orthogonality relations for $a^{-n} P_n(\pm a q^{m+\frac{1}{2}}; s/a, 1/as; q^{-1})$ we find that the polynomials p_n satisfy

$$\sum_{m \in \mathbb{Z}} p_n(q^{-m-\frac{1}{2}}) p_{n'}(q^{-m-\frac{1}{2}}) \frac{q^{m(m+1)}}{(q^{\frac{3}{2}} s, q^{\frac{3}{2}}/s; q)_m}$$
$$+ p_n(-q^{-m-\frac{1}{2}}) p_{n'}(-q^{-m-\frac{1}{2}}) \frac{q^{m(m+1)}}{(-q^{\frac{3}{2}} s, -q^{\frac{3}{2}}/s; q)_m} = \delta_{n,n'} N_n,$$

where N_n can be determined explicitly. This is a special case of the solutions found in [10].

Let us now turn to the proof of Theorem 6.2, which takes several steps. From the results of the previous section we obtain the following result.

Proposition 6.6 *The map $\mathcal{F}$ extends uniquely to an isometry $\mathcal{F} : \mathcal{L}^2 \to \mathcal{H}$.*

Proof Let $f_1, f_2 \in \mathcal{D}_0$, then by Propositions 5.1, 5.2 and the definition of the inner product $\langle \cdot, \cdot \rangle_{\mathcal{H}}$,

$$\langle f_1, f_2 \rangle = \langle E(\mathbb{R}) f_1, f_2 \rangle = \langle \mathcal{F} f_1, \mathcal{F} f_2 \rangle_{\mathcal{H}},$$

from which it follows that $\mathcal{F}$ extends to an isometry $\mathcal{L}^2 \to \mathcal{H}$. □

We show that $\mathcal{F}$ is unitary by showing that $\mathcal{G}$ is indeed the inverse of $\mathcal{F}$. Note that $\phi_\lambda(x) = (\mathcal{F} d_x)(\lambda)$, where $d_x \in \mathcal{L}^2$ is given by $d_x(y) = \frac{\delta_{x,y}}{w(x)|x|}$. So $\phi_\bullet(x) \in \mathcal{H}$ by Proposition 6.6, and we see that $\mathcal{G} g$ exists for all $g \in \mathcal{H}$. Using the functions d_x it is easy to show that $\mathcal{G}$ is a left-inverse of $\mathcal{F}$.

Lemma 6.7 $\mathcal{G}\mathcal{F} = \mathrm{id}_{\mathcal{L}^2}$.

Proof Let $f \in \mathcal{L}^2$ and $x \in I$. Using $(\mathcal{F} d_x)(\lambda) = \phi_\lambda(x)$ and Proposition 6.6 we have

$$(\mathcal{G}\mathcal{F} f)(x) = \langle \mathcal{F} f, \mathcal{F} d_x \rangle_{\mathcal{H}} = \langle f, d_x \rangle = f(x).$$ □

It requires more work to show that $\mathcal{G}$ is also a right inverse of $\mathcal{F}$. We use a classical method [3, 9], which is essentially approximating with the Fourier transform. We need the truncated inner product, see (3.2), of ϕ_λ and $\phi_{\lambda'}$ with $\lambda \neq \lambda'$. We first derive results about these inner products that will be useful later on.

Lemma 6.8 *For $l \in \mathbb{N}$ and $\lambda, \lambda' \in \mathbb{C}^*$ with $\mu(\lambda) \neq \mu(\lambda')$, the limit*

$$\langle \phi_\lambda, \phi_{\lambda'} \rangle_l = \lim_{k \to \infty} \langle \phi_\lambda, \phi_{\lambda'} \rangle_{k,l,k}$$

exists and for $l \to \infty$,

$$\langle \phi_\lambda, \phi_{\lambda'} \rangle_l = K \sum_{\varepsilon, \eta \in \{-1,1\}} \frac{(\lambda^\varepsilon - \lambda'^\eta)(\lambda^\varepsilon \lambda'^\eta)^l c(\lambda^\varepsilon) c(\lambda'^\eta)}{\mu(\lambda) - \mu(\lambda')} \Big(1 + \mathcal{O}(q^l)\Big).$$

Proof From Lemma 3.3 we obtain

$$\begin{aligned}\big(\mu(\lambda) - \mu(\lambda')\big) \langle \phi_\lambda, \phi_{\lambda'} \rangle_{k,l,m} &= \\ &D(\phi_\lambda, \phi_{\lambda'})(-q^k) - D(\phi_\lambda, \phi_{\lambda'})(q^m) + D(\phi_\lambda, \phi_{\lambda'})(q^{-l-1}).\end{aligned}$$

Using $\phi_\lambda \in V_{\mu(\lambda)}$, see (4.1), we obtain from the expression (3.4) for the Casorati determinant that

$$\lim_{k \to \infty} D(\phi_\lambda, \phi_{\lambda'})(-q^k) - D(\phi_\lambda, \phi_{\lambda'})(q^k) = 0,$$

which shows that

$$\langle \phi_\lambda, \phi_{\lambda'} \rangle_l = \frac{D(\phi_\lambda, \phi_{\lambda'})(q^{-l-1})}{\mu(\lambda) - \mu(\lambda')}.$$

The asymptotic behavior of this Casorati determinant can be obtained, using the c-function expansion in Lemma 4.5, from the asymptotic behavior of the Casorati determinant $D(\Psi_\lambda, \Psi_{\lambda'})(q^{-l-1})$. From (3.4) and the asymptotic behavior (4.4) and (3.5) of Ψ_λ and w, we obtain

$$D(\Psi_\lambda, \Psi_{\lambda'})(q^{-l-1}) = K(\lambda - \lambda')(\lambda\lambda')^l\big(1 + \mathcal{O}(q^l)\big), \qquad l \to \infty,$$

from which the result follows. □

Next we show that $\langle \phi_\lambda, \phi_{\lambda'} \rangle_l$ has a reproducing property similar to the Dirichlet kernel. Let

$$C_0(\mathbb{T}) = \{g \in C(\mathbb{T}) \mid g(-1) = g(1) = 0\}.$$

Proposition 6.9 *For* $g \in C_0(\mathbb{T})$

$$\lim_{l\to\infty} \frac{1}{4\pi i} \int_{\mathbb{T}} g(\lambda) \langle \phi_\lambda, \phi_{\lambda'} \rangle_l \frac{d\lambda}{\lambda} = \begin{cases} K\, g(\lambda')|c(\lambda')|^2, & \lambda' \in \mathbb{T} \setminus \{-1, 1\}, \\ 0, & \lambda' \in S. \end{cases}$$

Proof First consider $\lambda' = e^{i\theta'}$ with $\theta' \in (0, \pi)$. Using Lemma 6.8 and substitution we have, for $l \to \infty$,

$$\begin{aligned} I_l[g](\lambda') &= \frac{1}{4\pi i} \int_{\mathbb{T}} g(\lambda) \langle \phi_\lambda, \phi_{\lambda'} \rangle_l \frac{d\lambda}{\lambda} \\ &= \frac{K}{2\pi} \sum_{\varepsilon,\eta \in \{-1,1\}} \int_0^\pi g(e^{i\theta}) \Big(F_l^{\varepsilon,\eta}(\theta, \theta') + \mathcal{O}(q^l) \Big)\, d\theta, \end{aligned}$$

where

$$F_l^{\varepsilon,\eta}(\theta, \theta') = \frac{(e^{i\varepsilon\theta} - e^{i\eta\theta'}) e^{il(\varepsilon+\eta)} c(e^{i\varepsilon\theta}) c(e^{i\eta\theta'})}{2\cos(\theta) - 2\cos(\theta')}.$$

Since $\lambda \mapsto c(\lambda)$ is continuous on $\mathbb{T} \setminus \{-1, 1\}$, the function $\theta \mapsto F_l^{\varepsilon,\eta}(\theta, \theta')$ is continuous on $(0, \pi) \setminus \{\theta'\}$. If $\mathrm{sgn}(\varepsilon) = \mathrm{sgn}(\eta)$ the singularity at θ' is removable, so that these terms vanish in the limit by the Riemann-Lebesgue lemma. This gives

$$\lim_{l\to\infty} I_l[g](e^{i\theta'}) = \lim_{l\to\infty} \frac{K}{2\pi} \int_0^\pi g(e^{i\theta}) \Big(F_l^{1,-1}(\theta, \theta') + F_l^{-1,1}(\theta, \theta') \Big)\, d\theta,$$

where dominated convergence is used to get rid of the $\mathcal{O}(q^l)$-terms. Furthermore, using trigonometric identities we obtain

$$F_l^{1,-1}(\theta,\theta') + F_l^{-1,1}(\theta,\theta') = \frac{[c(e^{-i\theta})c(e^{i\theta'}) - c(e^{i\theta})c(e^{-i\theta'})](e^{-i\theta} - e^{i\theta'})}{4\sin(\frac{1}{2}(\theta+\theta'))\sin(\frac{1}{2}(\theta-\theta'))} e^{il(\theta'-\theta)} + c(e^{i\theta})c(e^{i\theta'})D_l(\theta-\theta'), \tag{6.1}$$

where

$$D_l(t) = \frac{\sin((2l+1)t)}{\sin(\frac{1}{2}t)}$$

is the Dirichlet kernel. Note that the first term in (6.1) has a removable singularity at $\theta = \theta'$, so using the Riemann-Lebesgue lemma again this term vanishes in the limit $l \to \infty$. Then from the well-known limit property of the Dirichlet kernel we obtain

$$\lim_{l\to\infty} I_l[g](e^{i\theta'}) = Kg(e^{i\theta'})c(e^{i\theta'})c(e^{-i\theta'}),$$

which gives the result for $\lambda' \in \mathbb{T} \setminus \{-1, 1\}$.

Next let $\lambda' \in S$. In this case $c(1/\lambda') = 0$ and then since $|\lambda'| < 1$, we obtain from Lemma 6.8

$$\lim_{l\to\infty} \langle \phi_\lambda, \phi_{\lambda'} \rangle_l = K \lim_{l\to\infty} \sum_{\varepsilon\in\{-1,1\}} \frac{(\lambda^\varepsilon - \lambda')(\lambda^\varepsilon\lambda')^l c(\lambda^\varepsilon)c(\lambda')}{\mu(\lambda) - \mu(\lambda')} \left(1 + \mathcal{O}(q^l)\right) = 0,$$

from which the result follows. □

Now we are ready to prove Theorem 6.2.

Proof of Theorem 6.2 From Proposition 6.6 and Lemma 6.7 we already know that $\mathcal{F}$ is an isometry with left-inverse $\mathcal{G}$, so we have to show that $\mathcal{G}$ is a right-inverse of $\mathcal{F}$. Let $\mathcal{H}_0 \subseteq \mathcal{H}$ be the dense subspace consisting of functions $g \in \mathcal{H}$ such that $g|_{\mathbb{T}} \in C_0(\mathbb{T})$ and $g|_S$ has finite support. Then for $\lambda' \in \mathbb{T} \cup S$ and $g \in \mathcal{H}_0$,

$$\begin{aligned}(\mathcal{F}\mathcal{G}g)(\lambda') &= \int_{-1}^{\infty(1)} \phi_{\lambda'}(x) \int g(\lambda)\phi_\lambda(x)\,d\nu(\lambda)\,w(x)\,d_qx \\ &= \lim_{l\to\infty} \frac{1}{4K\pi i} \int_{\mathbb{T}} g(\lambda)\langle \phi_\lambda, \phi_{\lambda'} \rangle_l \frac{d\lambda}{\lambda|c(\lambda)|^2} \\ &\quad + \frac{1}{K} \sum_{\lambda\in S} g(\lambda)\langle \phi_\lambda, \phi_{\lambda'} \rangle \operatorname*{Res}_{\hat\lambda=\lambda} \left(\frac{1}{\hat\lambda c(\hat\lambda)c(\hat\lambda^{-1})} \right).\end{aligned}$$

Using Proposition 6.9 for the integral part and Corollary 5.3 for the sum part, we find that this equals $g(\lambda')$. Since $\mathcal{H}_0$ is dense in $\mathcal{H}$ we find $\mathcal{F}\mathcal{G} = \mathrm{Id}_{\mathcal{H}}$. □

References

1. W.A. Al-Salam, T.S. Chihara, Convolutions of orthonormal polynomials. SIAM J. Math. Anal. **7**(1), 16–28 (1976)
2. R. Askey, M.E.H. Ismail, Recurrence relations, continued fractions, and orthogonal polynomials. Mem. Am. Math. Soc. **49**(300), iv+108 (1984)
3. B.L.J. Braaksma, B. Meulenbeld, Integral transforms with generalized Legendre functions as kernels. Compos. Math. **18**, 235–287 (1967)
4. T.S. Chihara, M.E.H. Ismail, Extremal measures for a system of orthogonal polynomials. Constr. Approx. **9**(1), 111–119 (1993)
5. J.S. Christiansen, M.E.H. Ismail, A moment problem and a family of integral evaluations. Trans. Am. Math. Soc. **358**(9), 4071–4097 (2006)
6. J.S. Christiansen, E. Koelink, Self-adjoint difference operators and symmetric Al-Salam-Chihara polynomials. Constr. Approx. **28**(2), 199–218 (2008)
7. N. Dunford, J.T. Schwartz, *Linear Operators Part II* (Interscience, New York, 1963)
8. G. Gasper, M. Rahman, *Basic Hypergeometric Series*, 2nd edn. (Cambridge University Press, Cambridge, 2004)
9. F. Götze, Verallgemeinerung einer Integraltransformation von Mehler-Fock durch den von Kuipers und Meulenbeld eingefürten Kern $P_k^{m,n}(z)$. Indag. Math. **27**, 396–404 (1965)
10. W. Groenevelt, Orthogonality relations for Al-Salam-Carlitz polynomials of type II. J. Approx. Theory **195**, 89–108 (2015)
11. W. Groenevelt, E. Koelink, The indeterminate moment problem for the q-Meixner polynomials. J. Approx. Theory **163**(7), 838–863 (2011)
12. T. Kakehi, Eigenfunction expansion associated with the Casimir operator on the quantum group $\mathrm{SU}_q(1,1)$. Duke Math. J. **80**, 535–573 (1995)
13. T. Kakehi, T. Masuda, K. Ueno, Spectral analysis of a q-difference operator which arises from the quantum $\mathrm{SU}(1,1)$ group. J. Operator Theory **33**, 159–196 (1995)
14. R. Koekoek, P.A. Lesky, R. Swarttouw, *Hypergeometric Orthogonal Polynomials and Their q-Analogues*. Springer Monographs in Mathematics (Springer, Berlin, 2010)
15. E. Koelink, J.V. Stokman, Fourier transforms on the quantum SU(1,1) group, With an appendix by Mizan Rahman. Publ. Res. Inst. Math. Sci. **37**(4), 621–715 (2001)
16. E. Koelink, J.V. Stokman, The big q-Jacobi function transform. Constr. Approx. **19**, 191–235 (2003)
17. T.H. Koornwinder, Jacobi functions and analysis on noncompact semisimple Lie groups, in *Special Functions: Group Theoretical Aspects and Applications*, ed. by R.A. Askey, T.H. Koornwinder, W. Schempp (D. Reidel Publ. Comp., Dordrecht, 1984), pp. 1–85
18. K. Schmüdgen, *The Moment Problem*. Graduate Texts in Mathematics, vol. 277 (Springer, Cham, 2017)
19. B. Simon, The classical moment problem as a self-adjoint finite difference operator. Adv. Math. **137**(1), 82–203 (1998)

The Katowice Problem for Analysts

Klaas Pieter Hart

For my long term neighbour

Abstract The Katowice Problem is well known among topologists and set theorists. The aim of this paper is to make it known among analysts and to give Ben de Pagter something to think about in his retirement.

Keywords Banach algebra · ℓ_∞ · ℓ_∞/c_0 · Boolean algebra · $\mathcal{P}(\mathbb{N})$ · $\mathcal{P}(\mathbb{N})/\mathit{fin}$ · Čech-Stone compactification · $\beta\mathbb{N}$ · $\beta\mathbb{N} \setminus \mathbb{N}$

1 Introduction

The Katowice Problem, as posed by Marian Turzański, is about Čech-Stone remainders of discrete spaces. For the purposes of this paper it suffices to know that for a discrete space X its Čech-Stone compactification, βX, is a compact Hausdorff space that contains X as a dense subset and that has the property that disjoint subsets of X have disjoint closures in βX. The *remainder* (or *growth*) is $\beta X \setminus X$ and it is generally denoted X^*.

Let X and Y be two infinite sets, endowed with the discrete topology. The problem under consideration asks

The Katowice Problem If the remainders X^* and Y^* are homeomorphic must there be a bijection between X and Y?

This problem has its origins in Parovichenko's paper [11] where a topological characterization of $\mathbb{N}^*$ is given, under the assumption of the Continuum Hypothesis. The obvious question then is whether such characterizations are possible for the

K. P. Hart (✉)
Faculty EEMCS, TU Delft, Delft, The Netherlands
e-mail: k.p.hart@tudelft.nl; http://fa.its.tudelft.nl/~hart

G. Buskes et al. (eds.), *Positivity and Noncommutative Analysis*,
Trends in Mathematics, https://doi.org/10.1007/978-3-030-10850-2_12

remainders of other discrete spaces, and a natural side question is what can be said if the remainders are homeomorphic.

For more information on the Čech-Stone compactification of discrete spaces, and in particular on $\beta\mathbb{N}$, we refer to Van Mill's survey [13].

In this paper we discuss various equivalent versions of the Katowice Problem, algebraic and analytic. We also summarise what is known about the problem: the answer is positive except for the case of the first two infinite cardinal numbers. That last remaining case has withstood many attacks thus far and it is hoped that an analytic approach may shed new light on the problem.

2 Another Road to the Problem

One can arrive at the Katowice Problem by a purely algebraic road. This road starts with an elementary exercise: given a bijection f between two sets X and Y, construct a bijection between their power sets $\mathcal{P}(X)$ and $\mathcal{P}(Y)$. The solution is easy: define F by $F(A) = f[A]$.

Now turn this exercise around: given a bijection F between the power sets $\mathcal{P}(X)$ and $\mathcal{P}(Y)$ of the sets X and Y, construct a bijection between X and Y.

This second exercise is way more difficult than the first one. The case of finite sets is easily dispensed with, it seems: the function $n \mapsto 2^n$ is injective on the set of natural numbers, so if the finite sets X and Y have the same numbers of subsets then X and Y will have the same number of elements. However, this does not solve the problem as required: from the given bijection F construct another bijection. And that last thing cannot be done.

To see this we consider Cohen's original proof that the Continuum Hypothesis is not provable from the axioms of ZFC. In the resulting model there is a bijection between the power sets of the first two infinite cardinal numbers, ω_0 and ω_1, yet there is, of course, no bijection between these sets themselves. This implies that without using additional properties about the sets in question it is not possible to turn a bijection between the power sets into a bijection between the sets themselves.

I recommend Kunen's book [10, Chapter VII] for an exposition of Cohen's method.

The situation changes if one considers additional structure. The power set of a set is partially ordered by inclusion and it is even a Boolean algebra, with $\cap$ and $\cup$ as its operations.

Now, from an isomorphism $F : \mathcal{P}(X) \to \mathcal{P}(Y)$ it is quite easy to extract a bijection between X and Y. Indeed, the singleton subsets of X are the atoms of the algebra $\mathcal{P}(X)$; where $a \in \mathcal{P}(X)$ is an atom if $a > \emptyset$ and whenever $a = b \cup c$ one must have $b = a$ or $c = a$. The isomorphism F must then contain a bijection between the respective sets of atoms, which then is a bijection between X and Y of course.

To get to the Katowice Problem we ask what happen when we hide the atoms, that is, when we set the atoms equal to zero. In algebraic terms this amounts to taking the ideal, *fin*, of finite sets and considering the quotient algebra $\mathcal{P}(X)/\mathit{fin}$.

The Katowice Problem If the Boolean algebras $\mathcal{P}(X)/\mathit{fin}$ and $\mathcal{P}(Y)/\mathit{fin}$ are isomorphic is there then a bijection between X and Y?

That this is indeed a reformulation of the Katowice Problem follows readily using M. H. Stone's duality theorem for Boolean algebras. The space βX is the Stone space of the Boolean algebra $\mathcal{P}(X)$ and the Čech-Stone remainder X^* is the Stone space of $\mathcal{P}(X)/\mathit{fin}$.

More information about Stone's duality theorem can be found in Koppelberg's book [8, Chapter 3].

3 Rings and Banach Algebras and Lattices

Yet another way of looking at the Katowice Problem is via the function space ℓ_∞. That is, for every set X we consider $\ell_\infty(X)$, the set of all bounded real- (or complex-)valued functions on the set X.

One can consider $\ell_\infty(X)$ as a ring, by defining addition and multiplication pointwise, and as a Banach-algebra, by endowing it with the supremum norm $\|\cdot\|_\infty$, and, most importantly, as a Banach lattice under the pointwise order.

In all three cases, if $\ell_\infty(X)$ and $\ell_\infty(Y)$ are isomorphic, then there is a bijection between X and Y. This follows by applying the theorem of Gel'fand and Kolmogoroff [5] in the case of rings, or that of Gel'fand and Neumark [6] in the case of Banach-algebras, or that of Kakutani [7] and Krein and Krein [9] in the case of Banach lattices.

These theorems represent $\ell_\infty(X)$ as the ring, or Banach-algebra, or Banach lattice of continuous functions, respectively, on a certain compact Hausdorff space. In this case that space is just βX, the Čech-Stone compactification of the discrete space X.

Just as in the case of Boolean algebras one can hide the finite sets by taking the quotient $\ell_\infty(X)/c_0$ by the ideal or subalgebra or ideal, respectively, of functions that vanish at infinity, where $f : X \to \mathbb{R}$ *vanishes at infinity* if for every $\varepsilon > 0$ the set $\{x : |f(x)| \geqslant \varepsilon\}$ is finite.

The quotient $\ell_\infty(X)/c_0$ corresponds to the ring, or Banach-algebra, or Banach lattice of continuous functions on the Čech-Stone remainder X^* and thus we come to a version reformulation of the Katowice Problem that, I hope, is of interest to analysts.

The Katowice Problem If the Banach lattices $\ell_\infty(X)/c_0$ and $\ell_\infty(Y)/c_0$ are isomorphic, is there then a bijection between X and Y?

4 What Is Known?

To begin: the Generalised Continuum Hypothesis (GCH) implies that the Katowice Problem has a positive answer. The Boolean algebraic version makes this clear: the Boolean algebra $\mathcal{P}(X)/fin$ has cardinality $2^{|X|}$, and the GCH implies that the function $\kappa \mapsto 2^\kappa$ is injective on the class of cardinal numbers.

In fact, much more is known. In joint work Balcar and Frankiewicz established that the answer is actually positive without any additional set-theoretic assumptions, *when the two sets are both uncountable*. More precisely

Theorem **([1, 4])** *If the remainders X^* and Y^* are homeomorphic and the sets X and Y are uncountable, then there is a bijection between X and Y.*

In fact, this theorem leaves just one pair of cardinal numbers for which the problem is still open: the first two infinite cardinals numbers ω_0 and ω_1. I use 'cardinal numbers' rather than 'sets' because it would become increasingly cumbersome to formulate everything in terms of arbitrary sets.

The cardinal numbers form a class of well-ordered sets against which all other sets are measured: for every set X there is one cardinal number κ such that there is a bijection between X and κ. We write $\kappa = |X|$ and call κ the cardinal number of X. By using the word 'one' we implicitly specify that there are no bijections between distinct cardinal numbers.

All this reduces the Katowice Problem to one final case.

Main Problem Prove that $\ell_\infty(\omega_0)/c_0$ and $\ell_\infty(\omega_1)/c_0$ are not isomorphic.

This formulation reflects this author's preferred solution of this problem. I am fully aware of the possibility that it is relatively consistent with ZFC that $\ell_\infty(\omega_0)/c_0$ and $\ell_\infty(\omega_1)/c_0$ are isomorphic. However, to me that would be too shocking to be true.

In the next section I will list some consequences derived from the assumption that the two lattices are isomorphic. The reader will see that this has almost developed into a game where someone derives a consequence and someone else shows that that consequence does not lead to a contradiction, not even when combined with earlier consequences.

5 Some Consequences

Most of the consequences have been obtained in the Boolean algebraic setting so we adopt that language from now on. Thus, our standing assumption is that there is an isomorphism $\gamma : \mathcal{P}(\omega_0)/fin \to \mathcal{P}(\omega_1)/fin$.

The first consequence of this is straightforward: the cardinalities of the Boolean algebras are the same, so we obtain

Consequence 1 $2^{\aleph_0} = 2^{\aleph_1}$.

We have already seen that this consequences does not lead to a contradiction.

To describe the other consequences we introduce some notation. First we change the underlying sets to $\mathbb{Z} \times \omega_0$ and $\mathbb{Z} \times \omega_1$, where $\mathbb{Z}$ is the set of integers.

In the product $\mathbb{Z} \times \omega_1$ we distinguish a few special sets:

1. $V_n = \{n\} \times \omega_1$ ($n \in \mathbb{Z}$), the vertical lines
2. $H_\alpha = \mathbb{Z} \times \{\alpha\}$ ($\alpha \in \omega_1$), the horizontal lines
3. $E_\alpha = \mathbb{Z} \times [\alpha, \omega_1)$ ($\alpha \in \omega_1$), the end segments, here $[\alpha, \omega_1)$ is a convenient short hand for the set $\{\beta \in \omega_1 : \beta \geqslant \alpha\}$

If A is a subset of $\mathbb{Z} \times \omega_0$ or $\mathbb{Z} \times \omega_1$, then A^* denotes its equivalence class modulo *fin*.

Back in $\mathcal{P}(\mathbb{Z} \times \omega_0)$ we choose sets v_n, h_α and e_α such that $\gamma(v_n^*) = V_n^*$, $\gamma(h_\alpha^*) = H_\alpha^*$, and $\gamma(e_\alpha^*) = E_\alpha^*$. The relations between the sets V_n, H_α and E_α are mirrored by those between the sets v_n, h_α and e_α. For example, $H_\alpha \cap H_\beta = \emptyset$ if $\alpha < \beta$, so in $\mathcal{P}(\mathbb{Z} \times \omega_0)/fin$ we have $h_\alpha^* \wedge h_\beta^* = 0$; for the sets themselves this means that $h_\alpha \cap h_\beta$ belongs to *fin*. We write this as $h_\alpha \cap h_\beta =^* \emptyset$ and say that h_α and h_β are *almost disjoint*.

Likewise when $\alpha < \beta$ we have $E_\beta \subseteq E_\alpha$ and hence $e_\beta^* \leqslant e_\alpha^*$, which means that $e_\beta \setminus e_\alpha$ is finite; we abbreviate the latter by $e_\beta \subseteq^* e_\alpha$. In fact $E_\alpha \setminus E_\beta$ is infinite, hence so is $e_\alpha \setminus e_\beta$; we should therefore actually write $e_\beta \subset^* e_\alpha$.

The sets v_n are also almost disjoint but we may alter each of them by a finite set (and apply a bijection from $\mathbb{Z} \times \omega_0$ to itself) to achieve that $v_n = \{n\} \times \omega_0$ for all n.

One of the (admittedly feeble) reasons for believing that $\mathcal{P}(\omega_0)/fin$ and $\mathcal{P}(\omega_1)/fin$ are *not* isomorphic is the shape of the two products: $\mathbb{Z} \times \omega_0$ looks like a squat 2×1-rectangle, while $\mathbb{Z} \times \omega_1$ is a rectangle with the same base that is much taller than the first product. It seems inconceivable we can squeeze the uncountable stack of H_α's into the flat rectangle that is $\mathbb{Z} \times \omega_0$.

We turn to a less trivial consequence of having our supposed isomorphism γ.

Consider the end segments E_α and their companion sets e_α. Since $V_n \cap E_\alpha$ is always uncountable, the intersection $v_n \cap e_\alpha$ is always infinite. This means that we can define $f_\alpha : \mathbb{Z} \to \omega_0$ by

$$f_\alpha(n) = \min\{m : \langle n, m \rangle \in e_\alpha\}$$

The resulting sequence $\langle f_\alpha : \alpha \in \omega_1 \rangle$ of functions has two interesting properties.

We just saw that $e_\beta \setminus e_\alpha$ is finite and $e_\alpha \setminus e_\beta$ is infinite when $\alpha < \beta$. From this we get the first property: $f_\alpha(n) \leqslant f_\beta(n)$ for all but finitely many n. This is generally abbreviated as $f_\alpha \leqslant^* f_\beta$.

For the second let $f : \mathbb{Z} \to \omega_0$ be arbitrary and consider the set

$$l_f = \{\langle n, m\rangle : m \leqslant f(n)\}.$$

In $\mathcal{P}(\mathbb{Z} \times \omega_1)$ choose a set L_f such that $\gamma(l_f^*) = L_f^*$.

For every n the intersection $v_n \cap l_f$ is finite, hence so is $V_n \cap L_f$. This implies that there is an α such that $E_\alpha \cap L_f = \emptyset$. Back at the ω_0-side we find that $e_\alpha \cap l_f$ is finite. But this then implies that $f(n) < f_\alpha(n)$ for all but finitely many n.

We see that $\langle f_\alpha : \alpha \in \omega_1\rangle$ is both increasing and cofinal with respect to the (quasi-)order $\leqslant^*$. Such a sequence is called an ω_1-scale.

Consequence 2 There is an ω_1-scale.

The existence of an ω_1-scale follows from the Continuum Hypothesis (CH) but it is also consistent with the latter's negation and, specifically, also with $2^{\aleph_0} = 2^{\aleph_1}$.

The next consequence involves the horizontal lines H_α and their counterparts, the h_α. We have already seen that the h_α form an *almost disjoint family*; we now show that it is a very special such family.

Suppose that for every α we choose a subset x_α of h_α. At the side of ω_1 we choose $X_\alpha \subseteq H_\alpha$, such that $\gamma(x_\alpha^*) = X_\alpha^*$ and we take the union $X = \bigcup_{\alpha\in\omega_1} X_\alpha$. Then we know that $X \cap H_\alpha = X_\alpha$ for all α. If we then choose x with $\gamma(x^*) = X^*$, then this x will satisfy $x \cap h_\alpha =^* x_\alpha$ for all α, where $=^*$ means 'almost equal' (again: but for finitely many points).

We say that $\{h_\alpha : \alpha < \omega_1\}$ is a *uniformizable* almost disjoint family.

Consequence 3 There is a uniformizable almost disjoint family of cardinality $\aleph_1$ (also called a strong Q-sequence).

The existence of a uniformizable almost disjoint family implies the equality $2^{\aleph_0} = 2^{\aleph_1}$: to code a subset Y of ω_1 apply the previous paragraph with $x_\alpha = h_\alpha$ if $\alpha \in Y$ and $x_\alpha = \emptyset$ if $\alpha \notin Y$ to obtain x_Y. The map $Y \mapsto x_Y$ is injective from $\mathcal{P}(\mathbb{Z} \times \omega_1)$ into $\mathcal{P}(\mathbb{Z} \times \omega_0)$.

That the assumptions $2^{\aleph_0} = 2^{\aleph_1}$ and 'there is an ω_1-scale' do not together lead to $0 = 1$ is an easy exercise for anyone who has learned the rudiments of forcing (use the random real model). To show the same thing for the combination of 'there is a strong Q-sequence' and 'there is an ω_1-scale' is not such an easy exercise. But it can be done, see [2] for a proof.

We treat one more consequence, which involves automorphisms of the Boolean algebras. An algebra like $\mathcal{P}(X)/\mathit{fin}$ has many automorphisms: every permutation of X and, more generally, every bijection between co-finite subsets of X determines an automorphism of $\mathcal{P}(X)/\mathit{fin}$. Such automorphisms are called trivial. It is a remarkable result of Shelah's [12, Chapter IV] that it is consistent that all automorphisms of $\mathcal{P}(\omega_0)/\mathit{fin}$ are trivial. This was extended to *all* infinite sets X by Veličković in [14].

Using our (postulated) isomorphism γ we can show that both $\mathcal{P}(\omega_0)$ and $\mathcal{P}(\omega_1)$ must have *non-trivial* automorphisms. We describe these automorphisms and refer to [3] for proofs.

For the first take the shift on ω_0 (which is set theory's set of natural numbers): $\sigma(n) = n + 1$. The automorphism that we get by transplanting the automorphism $A^* \mapsto \sigma[A]^*$ to $\mathcal{P}(\omega_1)/\mathit{fin}$ is non-trivial.

For the second take the map $\tau : \mathbb{Z} \times \omega_1 \to \mathbb{Z} \times \omega_1$ given by $\tau(n, \alpha) = \langle n+1, \alpha \rangle$. Transplanting $A^* \mapsto \tau[A]^*$ to $\mathcal{P}(\mathbb{Z} \times \omega_0)/\mathit{fin}$ results in a non-trivial automorphism.

Consequence 4 Both $\mathcal{P}(\omega_0)/\mathit{fin}$ and $\mathcal{P}(\omega_0)/\mathit{fin}$ have non-trivial automorphisms.

The paper [3] contains some more consequences and references to other sources of information about the Katowice Problem.

References

1. B. Balcar, R. Frankiewicz, To distinguish topologically the spaces m^*. II. Acad. Pol. Sci. Bull. Sér. Sci. Math. Astron. Phys. **26**(6), 521–523 (1978). English, with Russian summary
2. D. Chodounský, Strong-Q-sequences and small $\mathfrak{d}$. Topology Appl. **159**(13), 2942–2946 (2012). https://doi.org/10.1016/j.topol.2012.05.012
3. D. Chodounský, A. Dow, K.P. Hart, H. de Vries, The Katowice problem and autohomeomorphisms of ω_0^*. Topol. Appl. **213**, 230–237 (2016). https://doi.org/10.1016/j.topol.2016.08.006
4. R. Frankiewicz, To distinguish topologically the space m^*. Acad. Pol. Sci. Bull. Sér. Sci. Math. Astron. Phys. **25**(9), 891–893 (1977). English, with Russian summary
5. I.M. Gel'fand, A.N. Kolmogoroff, On rings of continuous functions on topological spaces. C.R. Dokl. Acad. Sci. URSS **22**, 11–15 (1939)
6. I.M. Gel'fand, M. Neumark, On the imbedding of normed rings into the ring of operators in Hilbert space. Mat. Sb. Novaya Seriya **12**, 197–213 (1943). English
7. S. Kakutani, Concrete representation of abstract (M)-spaces (A characterization of the space of continuous functions). Ann. Math. (2) **42**, 994–1024 (1941). https://doi.org/10.2307/1968778
8. S. Koppelberg, in *Handbook of Boolean Algebras*, vol. 1, ed. by J.D. Monk, R. Bonnet (North-Holland Publishing Co., Amsterdam, 1989)
9. M. Krein, S. Krein, On an inner characteristic of the set of all continuous functions defined on a bicompact Hausdorff space. C.R. Dokl. Acad. Sci. URSS (N.S.) **27**, 427–430 (1940)
10. K. Kunen, *Set Theory. An Introduction to Independence Proofs*. Studies in Logic and the Foundations of Mathematics, vol. 102 (North-Holland Publishing Co., Amsterdam, 1980)
11. I.I. Parovičenko, A universal bicompact of weight $\aleph$. Sov. Math. Dokl. **4**, 592–595 (1963). Russian original: Ob odnom universal′nom bikompakte vesa $\aleph$, Dokl. Akad. Nauk SSSR **150**, 36–39 (1963)
12. S. Shelah, *Proper Forcing*. Lecture Notes in Mathematics, vol. 940 (Springer, Berlin, 1982)
13. J. van Mill, An introduction to $\beta\omega$, in *Handbook of Set-Theoretic Topology*, ed. by K. Kunen, J.E. Vaughan (North-Holland, Amsterdam, 1984), pp. 503–567
14. B. Veličković, OCA and automorphisms of $\mathcal{P}(\omega)$/fin. Topol. Appl. **49**(1), 1–13 (1993). https://doi.org/10.1016/0166-8641(93)90127-Y

Onefold and Twofold Ellis–Gohberg Inverse Problems for Scalar Wiener Class Functions

M. A. Kaashoek and F. van Schagen

Dedicated to Ben de Pagter on the occasion of his 65th birthday

Abstract The theory of onefold and twofold Ellis-Gohberg inverse problems for Wiener functions on the real line is specified further for the scalar case. Assuming the left onefold problem or the right onefold problem to be solvable, a necessary and sufficient condition is given for the twofold problem to be solvable. In that case the left and the right onefold problem are both solvable, and the solutions are equal. An example shows that the latter is not always true, i.e., the left and the right onefold problem are solvable does not imply that the solutions are equal.

Keywords Inverse problem for orthogonal functions · Wiener algebra · Toeplitz operator · Hankel operator · Rational functions · Inner-outer factorization · Realization

1 Introduction

The problems we shall be dealing with concern generalizations of the inverse theorem for Szegő orthogonal polynomials to Wiener class functions. The first of these generalizations has been presented by R.L. Ellis and I. Gohberg in the seminal paper [2] and extended further in their book [3]. For that reason we shall use the term Ellis-Gohberg inverse problem or simply EG inverse problem for the type of problems considered in the present paper.

M. A. Kaashoek (✉) · F. van Schagen
Department of Mathematics, Vrije Universiteit Amsterdam, Amsterdam, The Netherlands
e-mail: m.a.kaashoek@vu.nl; f.van.schagen@vu.nl

G. Buskes et al. (eds.), *Positivity and Noncommutative Analysis*,
Trends in Mathematics, https://doi.org/10.1007/978-3-030-10850-2_13

To formulate the EG inverse problem for scalar functions on the real line requires some preliminaries. The starting point is given by four functions:

$$\alpha(\lambda) = 1 + \int_0^\infty e^{i\lambda t} a(t)\, dt\,, \quad a \in L^1(\mathbb{R}_+), \tag{1.1}$$

$$\beta(\lambda) = \int_0^\infty e^{i\lambda t} b(t)\, dt\,, \quad b \in L^1(\mathbb{R}_+), \tag{1.2}$$

$$\gamma(\lambda) = \int_{-\infty}^0 e^{i\lambda t} c(t)\, dt\,, \quad c \in L^1(\mathbb{R}_-), \tag{1.3}$$

$$\delta(\lambda) = 1 + \int_{-\infty}^0 e^{i\lambda t} d(t)\, dt\,, \quad d \in L^1(\mathbb{R}_-). \tag{1.4}$$

We shall refer to $\{\alpha, \beta, \gamma, \delta\}$ as the *data set*. Note that these functions are Fourier transforms of absolutely integrable functions on the real line and hence they are Wiener functions (cf., [5, p. 834] or [1, Section 9.1]).

To formulate the corresponding inverse problems some further terminology is needed. If ρ is a function on $\mathbb{R}$ given by

$$\rho(\lambda) = \int_{-\infty}^\infty e^{i\lambda t} r(t)\, dt \quad \text{with } r \in L^1(\mathbb{R}), \tag{1.5}$$

then ρ_+ and ρ_- are the functions defined by

$$\rho_+(\lambda) = \int_0^\infty e^{i\lambda t} r(t)\, dt\,, \quad \rho_-(\lambda) = \int_{-\infty}^0 e^{i\lambda t} r(t)\, dt\,.$$

The function ρ_+ extends to an analytic function on the open upper half plane (which will be denoted $\mathbb{C}_+$) and ρ_- to an analytic function on the open lower half plane (which will be denoted $\mathbb{C}_-$). Moreover the value at infinity of these functions is zero. Finally, the adjoint ρ^* of a function ρ given by (1.5) is defined by

$$\rho^*(\lambda) = \int_{-\infty}^\infty e^{-i\lambda t}\, \overline{r(t)}\, dt = \int_{-\infty}^\infty e^{i\lambda t}\, \overline{r(-t)}\, dt\,.$$

We are now ready to state the EG inverse problems we shall be dealing with. The *left onefold EG inverse problem associated with the data* $\{\alpha, \gamma\}$ is to find a function φ such that φ is of the form

$$\varphi(\lambda) = \int_0^\infty e^{i\lambda t} h(t)\, dt\,, \quad \text{for some} \quad h \in L^1(\mathbb{R}_+), \tag{1.6}$$

and φ satisfies

$$\big(\varphi^*\alpha + \gamma\big)_- = 0 \quad \text{and} \quad (\alpha - 1 + \varphi\gamma)_+ = 0. \tag{1.7}$$

(Here 1 stands for the constant function with value 1.) The *right onefold EG inverse problem associated with* $\{\beta, \delta\}$ is to find a function φ such that φ is of the form (1.6) and φ satisfies

$$(\varphi\delta + \beta)_{+} = 0 \quad \text{and} \quad \left(\delta - 1 + \varphi^{*}\beta\right)_{-} = 0. \tag{1.8}$$

The *twofold EG inverse problem associated with* $\{\alpha, \beta, \gamma, \delta\}$ is to find a function φ of the form (1.6) such that both (1.7) and (1.8) are satisfied. It is known (see [7, Theorem 4.1]) that in this scalar case the solution of the onefold inverse problem, left or right, if it exists, is unique. The same is true for the twofold inverse problem, even in the matrix-valued case (see [9, Theorem 1.2]).

Clearly, a solution for the twofold EG inverse problem is a solution for both the left and right onefold EG inverse problem. Conversely, if the solutions of the left and right onefold EG inverse problems exist and are equal, then, obviously, the converse statement is also true. So the question arises, can it happen that the left and right onefold EG inverse problems are solvable but the solutions are not equal? The answer is positive and an example will be given in Sect. 3. In the next section we shall present a necessary and sufficient condition for the left and right onefold EG inverse problems to have the same solution.

The above function theory problems can also be formulated as problems involving structured Banach space operators. The latter is essential for solving the problems. To see this connection, let $h \in L^1(\mathbb{R}_+)$, put $h^*(t) = \overline{h(-t)}$, and let H and H_* be the operators defined by

$$H : L^1(\mathbb{R}_-) \to L^1(\mathbb{R}_+), \quad (Hf)(t) = \int_{-\infty}^{0} h(t-s)f(s)\,ds, \quad t \geq 0;$$

$$H_* : L^1(\mathbb{R}_+) \to L^1(\mathbb{R}_-), \quad (H_*f)(t) = \int_{0}^{\infty} h^*(t-s)f(s)\,ds, \quad t \leq 0.$$

These operators are compact Hankel operators. Next put

$$\varphi(\lambda) = \int_0^\infty e^{i\lambda t} h(t)\,dt\,, \quad h \in L^1(\mathbb{R}_+).$$

Then φ is a solution of the twofold EG inverse problem associated with $\{\alpha, \beta, \gamma, \delta\}$ if and only if

$$\begin{bmatrix} I & H \\ H_* & I \end{bmatrix}\begin{bmatrix} a \\ c \end{bmatrix} = \begin{bmatrix} 0 \\ -h^* \end{bmatrix} \quad \text{and} \quad \begin{bmatrix} I & H \\ H_* & I \end{bmatrix}\begin{bmatrix} b \\ d \end{bmatrix} = \begin{bmatrix} -h \\ 0 \end{bmatrix}.$$

These operator theory connections play an essential role in solving EG inverse problems; see, e.g., [3, Chapter 12] or [7] and references therein.

2 A Necessary and Sufficient Condition

In this section α, β, γ, δ are the functions given by (1.1)–(1.4). If the twofold EG inverse problem associated with these data is solvable, then necessarily the following identities are satisfied:

$$\alpha^*\alpha - \gamma^*\gamma = 1, \quad \delta^*\delta - \beta^*\beta = 1, \quad \alpha^*\beta = \gamma^*\delta. \tag{2.1}$$

If the left (or right) onefold inverse problem is solvable, then necessarily the first (or second) identity in (2.1) holds true. For the left onefold case, this result is due to R.L. Ellis and I. Gohberg; see formula (2.5) in [3, Section 12.2]. The right onefold result is proved in a similar way. See also [6, Theorem 4.1].

We are now ready to state our main theorem.

Theorem 2.1 *Let $\{\alpha, \beta, \gamma, \delta\}$ be a data set, and assume the identities in* (2.1) *are satisfied. If the left onefold EG inverse problem associated with the pair α, γ has a solution or the right onefold EG inverse problem associated with the pair β, δ has a solution, then the twofold EG inverse problem associated with the data set $\{\alpha, \beta, \gamma, \delta\}$ is solvable if and only if α and δ^* have the same zeros on $\mathbb{C}_+$, with the multiplicities being taken into account. Moreover, in that case, both the left onefold EG inverse problem and right onefold EG inverse problem have a solution, and these two solutions are the same.*

It will be convenient to first prove the following lemma.

Lemma 2.2 *Let $\{\alpha, \beta, \gamma, \delta\}$ be a data set, and assume the identities in* (2.1) *are satisfied. Then $\alpha\alpha^* = \delta\delta^*$ and $\beta\beta^* = \gamma\gamma^*$. Furthermore, $\alpha = \delta^*$ if and only if α and δ^* have the same zeros on $\mathbb{C}_+$, multiplicities taken into account, and in that case $\beta = \gamma^*$ too.*

Proof From, e.g., [9, Remark 2.3] we know that together the three identities in (2.1) are equivalent to the following three identities:

$$\alpha\alpha^* - \beta\beta^* = 1, \quad \delta\delta^* - \gamma\gamma^* = 1, \quad \alpha\gamma^* = \beta\delta^*. \tag{2.2}$$

For the sake of completeness we add the proof. First note that the identities in (2.1) can be rewritten as a product of 2×2 matrices:

$$\begin{bmatrix} \alpha(\lambda)^* & -\gamma(\lambda)^* \\ \beta(\lambda)^* & -\delta(\lambda)^* \end{bmatrix} \begin{bmatrix} \alpha(\lambda) & -\beta(\lambda) \\ \gamma(\lambda) & -\delta(\lambda) \end{bmatrix} = \begin{bmatrix} 1 & 0 \\ 0 & 1 \end{bmatrix}, \quad \lambda \in \mathbb{R}. \tag{2.3}$$

Since the entries are scalar, this implies that also

$$\begin{bmatrix} \alpha(\lambda) & -\beta(\lambda) \\ \gamma(\lambda) & -\delta(\lambda) \end{bmatrix} \begin{bmatrix} \alpha(\lambda)^* & -\gamma(\lambda)^* \\ \beta(\lambda)^* & -\delta(\lambda)^* \end{bmatrix} = \begin{bmatrix} 1 & 0 \\ 0 & 1 \end{bmatrix}, \quad \lambda \in \mathbb{R}. \tag{2.4}$$

The latter identity shows that the three identities in (2.2) are satisfied.

Given (2.1) and (2.2), and using the fact that the functions involved are scalar we see that

$$\alpha\alpha^* = 1 + \beta\beta^* = \delta^*\delta \quad \text{and} \quad \beta\beta^* = \alpha\alpha^* - 1 = \delta^*\delta - 1 = \gamma\gamma^*.$$

We proceed with the two final statements. Since $\alpha\alpha^* = 1+\beta\beta^* = \delta^*\delta$, we know (see Section 2 in [7]) that α and δ^* admit factorizations

$$\alpha = w_{\text{sp}}\theta_\alpha \quad \text{and} \quad \delta^* = w_{\text{sp}}\theta_\delta,$$

where θ_α and θ_δ are scalar proper rational bi-inner functions which have the value one at infinity, and w_{sp} is the spectral factor of $\alpha\alpha^* = \delta^*\delta$. Since the spectral factor w_{sp} has no zeros on $\mathbb{C}_+$, it follows that the zeros of the function α on $\mathbb{C}_+$ are equal to the zeros of θ_α, and similarly, the zeros of the functions δ^* on $\mathbb{C}_+$ are equal to the zeros of θ_δ on $\mathbb{C}_+$. In all cases the multiplicities of the zeros are taken into account. But two proper rational bi-inner functions which have the value one at infinity have the same zeros on $\mathbb{C}_+$, again taking into account the multiplicities, if and only if the two functions coincide. The latter follows from the fact that a proper scalar rational bi-inner function which has the value one at infinity is given by

$$\theta(\lambda) = \Pi_{j=1}^{n} \left(\frac{\lambda - a_j}{\lambda - \bar{a}_j} \right), \tag{2.5}$$

where $a_1, \ldots, a_n$ are the zeros of θ on $\mathbb{C}_+$. In the case when θ has no zeros on $\mathbb{C}_+$, then $n = 0$ and the right-hand side of (2.5) reduces to one, i.e., θ is identically equal to one. Thus α and δ^* have the same zeros on $\mathbb{C}_+$, multiplicities being taken into account, if and only if $\alpha = \delta^*$. Finally, if $\alpha = \delta^*$, then the third identity in (2.1) implies that $\beta = \gamma^*$. □

The proof of Theorem 2.1 requires some additional preliminaries, taken from Sections 2 and 3 in [7].

The first equality in (2.1) implies that $\alpha(\lambda) \neq 0$ for all $\lambda \in \mathbb{R}$ and $\alpha(\infty) = 1$. Moreover, using the Riemann-Lebesgue lemma, we know that

$$\alpha^{-1}(\lambda) = \frac{1}{\alpha(\lambda)} = 1 + \int_{-\infty}^{\infty} e^{i\lambda t} a^\times(t)\, dt \quad \text{for some} \quad a^\times \in L^1(\mathbb{R}).$$

Therefore, in what follows $(\alpha^{-1} - 1)_-$ is the function given by

$$(\alpha^{-1} - 1)_-(\lambda) := \int_{-\infty}^{0} e^{i\lambda t} a^\times(t)\, dt\,, \quad \lambda \in \mathbb{C}_-. \tag{2.6}$$

In a similar way, using δ^* in place of α and the second equality in (2.1) in place of the first, we have

$$\delta^{-*}(\lambda) = \frac{1}{\delta^*(\lambda)} = 1 + \int_{-\infty}^{\infty} e^{i\lambda t} d^\times(t)\, dt \quad \text{for some} \quad d^\times \in L^1(\mathbb{R}),$$

and

$$(\delta^{-*} - 1)_-(\lambda) := \int_{-\infty}^{0} e^{i\lambda t} d^\times(t)\, dt\,, \quad \lambda \in \mathbb{C}_-. \tag{2.7}$$

From Proposition 2.6 in [7] it follows that $(\alpha^{-1} - 1)_-$ and $(\delta^{-*} - 1)_-$ are rational functions which are analytic on the open lower half plane $\mathbb{C}_-$. Moreover, both functions are proper and the value at infinity is equal to zero. But then we can use the classical realization theory of rational functions to conclude that there exist matrices A_1, B_1, C_1 and A_2, B_2, C_2 such that

$$\left(\alpha^{-1} - 1\right)_-(\lambda) = C_1(\lambda I_{n_1} - A_1)^{-1} B_1, \quad \text{and} \tag{2.8}$$

$$\left(\delta^{-*} - 1\right)_-(\lambda) = C_2(\lambda I_{n_2} - A_2)^{-1} B_2. \tag{2.9}$$

Moreover, we may assume that the above realizations are minimal, that is, we have

$$\operatorname{Im}\begin{bmatrix} B_j & A_j B_j & \cdots & A_j^{n_j-1} B_j \end{bmatrix} = \mathbb{C}^{n_j}, \ \operatorname{Ker}\begin{bmatrix} C_j \\ C_j A_j \\ \vdots \\ C_j A_j^{n_j-1} \end{bmatrix} = \{0\}\ (j = 1, 2).$$

Using the minimality of the above realizations, it follows that the poles of α^{-1} in $\mathbb{C}_+$ are equal to the eigenvalues of A_1, with the multiplicities being taken into account. Similarly, the eigenvalues of A_2 coincide with the poles of $(\delta^*)^{-1}$, again multiplicities included. So we have, using Lemma 2.2 above, that

$$A_1 \text{ is similar to } A_2 \quad \Longleftrightarrow \quad \alpha = \delta^*. \tag{2.10}$$

It also follows that the eigenvalues of both A_1 and A_2 are in the upper half plane, and therefore the eigenvalues of A_1^* and A_2^* are in the lower half plane. Since γ and β^* are analytic on the open lower half plane, the functional calculus implies that $\gamma(A_1^*)$ and $\beta^*(A_2^*)$ are well-defined matrices of sizes $n_1 \times n_1$ and $n_2 \times n_2$, respectively. In fact, these matrices are given by

$$\gamma(A_1^*) := \frac{1}{2\pi i} \int_\Lambda \gamma(\lambda)(\lambda I_{n_1} - A_1^*)^{-1}\, d\lambda, \tag{2.11}$$

$$\beta(A_2^*) := \frac{1}{2\pi i} \int_\Lambda \beta^*(\lambda)(\lambda I_{n_2} - A_2^*)^{-1}\, d\lambda, \tag{2.12}$$

where Λ is a Cauchy contour in the open lower half plane with the eigenvalues of A_1^* and A_2^* in its inner domain.

Lemma 2.3 *If the left onefold EG inverse problem associated with the pair* $\{\alpha, \gamma\}$ *has a solution, then the matrix* $\gamma(A_1^*)$ *defined by* (2.11) *is invertible, and if the right onefold EG inverse problem associated with the pair* $\{\beta, \delta\}$ *has a solution, then the matrix* $\beta^*(A_2^*)$ *defined by* (2.12) *is invertible*

Proof The fact that the matrices $\gamma(A_1^*)$ and $\beta^*(A_2^*)$ are invertible follows from [7, Theorem 4.1], using the one but last paragraph of [7, Section 2]. For the sake of completeness we repeat some of the main arguments.

Assume the left onefold EG inverse problem associated with the pair $\{\alpha, \gamma\}$ has a solution. Then [7, Theorem 4.1] tells us that the functions α^* and γ have no common zero in the open lower half plane $\mathbb{C}_-$. Because of the minimality of the realization (2.8), the eigenvalues of A_1^* coincide with the zeros of α^* in $\mathbb{C}_-$. But then the functional calculus (see [4, Section I.3]) implies that $\gamma(A_1^*)$ is invertible. In a similar way one shows that the eigenvalues of A_2^* coincide with the zeros of δ in $\mathbb{C}_-$, and the functional calculus shows that $\beta^*(A_2^*)$ is invertible. □

Proof of Theorem 2.1 We split the proof into two parts.

Part 1 In this part we assume that the twofold EG inverse problem associated with the data $\{\alpha, \beta, \gamma, \delta\}$ has a solution, φ say. Then φ is a solution of the left onefold EG inverse problem associated with $\{\alpha, \gamma\}$, and φ is a solution of the right onefold EG inverse problem associated with $\{\beta, \delta\}$. Let A_1 and A_2 be given by (2.8) and (2.9). Then [8, formula (1)], the corrected version of formula (4.1) in [7, Theorem 4.1] implies that

$$\varphi(\lambda) = -\left(\alpha^{-*}\gamma^*\right)_+(\lambda) - B_1^*(\lambda I_{n_1} - A_1^*)^{-1}\gamma(A_1^*)^{-1}C_1^*, \tag{2.13}$$

$$\varphi(\lambda) = -\left(\delta^{-1}\beta\right)_+(\lambda) - B_2^*(\lambda I_{n_2} - A_2^*)^{-1}\left(\beta^*(A_2^*)\right)^{-1}C_2^*. \tag{2.14}$$

Since (2.1) is satisfied, we know that $\alpha^{-*}\gamma^* = \beta\delta^{-1} = \delta^{-1}\beta$ (the second identity holds because we deal with the scalar case). It follows that the first terms in the right-hand side of (2.13) and (2.14) are equal. So we have

$$B_1^*(\lambda I_{n_1} - A_1^*)^{-1}\gamma(A_1^*)^{-1}C_1^* = B_2^*(\lambda I_{n_2} - A_2^*)^{-1}\left(\beta^*(A_2^*)\right)^{-1}C_2^*. \tag{2.15}$$

Since the realizations (2.8) and (2.9) are minimal, the realizations in the left and right-hand side of (2.15) are minimal too. Indeed, since $\gamma(A_1^*)^{-1}$ commutes with A_1^* and $\beta^*(A_2^*)^{-1}$ with A_2^* we have

$$\operatorname{rank}\begin{bmatrix} \gamma(A_1^*)^{-1}C_1^* \\ A_1^*\,\gamma(A_1^*)^{-1}C_1^* \\ \vdots \\ (A_1^*)^{n_1-1}\,\gamma(A_1^*)^{-1}C_1^* \end{bmatrix} = \operatorname{rank}\left(\gamma(A_1^*)^{-1}\begin{bmatrix} C_1^* \\ A_1^*C_1^* \\ \vdots \\ (A_1^*)^{n_1-1}\,C_1^* \end{bmatrix}\right) = n_1,$$

and

$$\operatorname{rank}\begin{bmatrix} \beta^*(A_2^*)^{-1}C_2^* \\ A_2^*\,\beta^*(A_2^*)^{-1}C_2^* \\ \vdots \\ (A_2^*)^{n_2-1}\,\beta^*(A_2^*)^{-1}C_2^* \end{bmatrix} = \operatorname{rank}\left(\beta^*(A_2^*)^{-1}\begin{bmatrix} C_2^* \\ A_2^*C_2^* \\ \vdots \\ (A_2^*)^{n_2-1}\,C_2^* \end{bmatrix}\right) = n_2,$$

which proves that the realizations in the left and right-hand side of (2.15) are minimal. It follows that A_1 is similar to A_2, and Lemma 2.2 tells us that α and δ^* have the same zeros on $\mathbb{C}_+$, as desired.

Part 2 In this part we assume that the left onefold EG inverse problem associated with the pair α, γ has a solution or the right onefold EG inverse problem associated with the pair β, δ has a solution. Furthermore, we assume that α and δ^* have the same zeros on $\mathbb{C}_+$. Then Lemma 2.2 tells us that $\alpha = \delta^*$ and $\beta = \gamma^*$. Thus, φ is a solution of the left onefold EG inverse problem associated with $\{\alpha, \gamma\}$ implies that φ is a solution of the right onefold EG inverse problem associated with $\{\beta, \delta\}$, and conversely. Thus our assumption implies that φ is the solution of the twofold EG inverse problem, which completes the proof. □

3 A Few Examples

Example 1 In this example we assume that the functions a, b, c, d appearing in (1.1)–(1.4) are zero. Thus

$$\alpha = 1, \quad \beta = 0, \quad \gamma = 0, \quad \delta = 1. \tag{3.1}$$

The related twofold EG inverse problem is solvable and its unique solution is the function $\varphi = 0$. Indeed, for this example with $\varphi = 0$ we have

$$\varphi^*\alpha + \gamma = 0, \quad \alpha + \varphi\gamma = 1, \quad \varphi\delta + \beta = 0, \quad \delta + \varphi^*\beta = 1.$$

In particular, for this choice of the data the identities (1.7) and (1.8) are satisfied. Hence $\varphi = 0$ is the solution.

If one replaces $\delta \equiv 1$ by $\delta(\lambda) = (\lambda + i)(\lambda - i)^{-1}$, keeping α, β, γ as in (3.1), then the three identities in (2.1) are satisfied. Furthermore, as above, the left onefold EG inverse problem is solvable and its solution is $\varphi = 0$. But the right onefold EG inverse problem and the twofold EG inverse problem are not solvable. Note that in this case

$$\delta(\lambda) = 1 + \int_{-\infty}^{0} e^{i\lambda t} d(t)\, dt\,, \quad d(t) = -2e^t, \quad t \le 0.$$

Example 2 Let ℓ and m be positive real numbers such that $m^2 = 1 + \ell^2$. Given these numbers we define

$$\alpha(\lambda) = \frac{\lambda - im}{\lambda + i}, \qquad \beta(\lambda) = \frac{\ell}{\lambda + i}, \tag{3.2}$$

$$\gamma(\lambda) = \frac{\ell}{\lambda - i}, \qquad \delta(\lambda) = \frac{\lambda + im}{\lambda - i}. \tag{3.3}$$

Note that functions $\alpha, \beta, \gamma, \delta$ are of the form (1.1)–(1.4). More precisely, the functions $\alpha, \beta, \gamma, \delta$ are also given by

$$\alpha(\lambda) = 1 + \int_0^\infty e^{i\lambda t} a(t)\, dt\,, \quad a(t) = -(m+1)e^{-t}, \quad t \geq 0, \tag{3.4}$$

$$\beta(\lambda) = \int_0^\infty e^{i\lambda t} b(t)\, dt\,, \quad b(t) = -i\ell e^{-t}, \quad t \geq 0, \tag{3.5}$$

$$\gamma(\lambda) = \int_{-\infty}^0 e^{i\lambda t} c(t)\, dt\,, \quad c(t) = i\ell e^{t}, \quad t \leq 0, \tag{3.6}$$

$$\delta(\lambda) = 1 + \int_{-\infty}^0 e^{i\lambda t} d(t)\, dt\,, \quad d(t) = -(m+1)e^{t}, \quad t \leq 0. \tag{3.7}$$

The fact that ℓ and m are real numbers implies that $\alpha^* = \delta$ and $\gamma^* = \beta$. Hence the conditions appearing in Lemma 2.2 are satisfied.

Furthermore, for the data set $\{\alpha, \beta, \gamma, \delta\}$ given by (3.2) and (3.3) the three identities in (2.1) are satisfied. It is straightforward to check that the first identity in (2.1) is satisfied. Indeed, we have

$$\alpha^*(\lambda)\alpha(\lambda) = \alpha(\bar{\lambda})^*\alpha(\lambda) = \frac{(\lambda + im)(\lambda - im)}{(\lambda - i)(\lambda + i)} = \frac{\lambda^2 + m^2}{\lambda^2 + 1};$$

$$\gamma^*(\lambda)\gamma(\lambda) + 1 = \gamma(\bar{\lambda})^*\gamma(\lambda) + 1 = \frac{\ell^2}{(\lambda + i)(\lambda - i)} + 1 = \frac{\lambda^2 + \ell^2 + 1}{\lambda^2 + 1}.$$

Since $m^2 = 1 + \ell^2$ by assumption, it follows that $\alpha^*(\lambda)\alpha(\lambda) = \gamma^*(\lambda)\gamma(\lambda) + 1$. The second identity in (2.1) follows from the first because $\alpha^* = \delta$ and $\gamma^* = \beta$. The latter two identities also imply that the third identity in (2.1) is fulfilled. Thus all three identities in (2.1) are fulfilled.

For the present example the left onefold EG inverse problem, the right onefold EG inverse problem, and the twofold EG inverse problem have the same solution. Indeed, the fact that α and γ^* do not have a common zero on $\mathbb{C}_+$ allows us to use [7, Theorem 4.1] to show that the left onefold EG inverse problem associated with the pair $\{\alpha, \gamma\}$ has a unique solution, φ say. Furthermore, since $\alpha = \delta^*$, Theorem 2.1 tells us that φ is also the (unique) solution of the twofold EG inverse problem associated with the data set $\{\alpha, \beta, \gamma, \delta\}$ given by (3.2) and (3.3). But then φ is also the (unique) solution of the right onefold EG inverse problem associated with the pair $\{\beta, \delta\}$.

Using formula (2.13) the unique solution φ can be computed explicitly. To do this, note that in this case $\alpha^{-1}(\lambda) - 1 = C_1(\lambda - A_1)^{-1}B_1$, with $A_1 = im$, $B_1 = 1$, and $C_1 = i(m+1)$. So we get

$$\left(\alpha^{-*}\gamma^*\right)_+(\lambda) = \frac{\ell(\lambda - i)}{(\lambda + im)(\lambda + i)}, \quad \gamma(A_1^*)^{-1} = \frac{-i(m+1)}{\ell},$$

and

$$B_1^*(\lambda - A_1^*)^{-1}\gamma(A_1^*)^{-1}C_1^* = \frac{-(m+1)^2}{\ell(\lambda + im)}.$$

But then (2.13) tells us that

$$\varphi(\lambda) = \frac{2(m+1)}{\ell(\lambda + i)}. \tag{3.8}$$

Given formula (3.8) one can show by direct checking that indeed φ satisfies the four conditions in (1.7) and (1.8).

Example 3 In this example the left and right onefold EG inverse problems are solvable but the solutions are not equal, and hence the twofold problem is not solvable.

Let ℓ, m, and n be positive real numbers such that $m^2 = 1 + \ell^2$ and $n \neq m$. Given these numbers we define

$$a(t) = \frac{\lambda - im}{\lambda + i}, \quad \beta(\lambda) = \frac{\ell}{\lambda + i}, \tag{3.9}$$

$$\gamma(\lambda) = \frac{\ell(\lambda + in)}{(\lambda - i)(\lambda - in)}, \quad \delta(\lambda) = \frac{(\lambda + im)(\lambda + in)}{(\lambda - i)(\lambda - in)}. \tag{3.10}$$

Note that functions $\alpha, \beta, \gamma, \delta$ are of the form (1.1)–(1.4). In fact, α, β are as in (3.4) and (3.5), and γ, δ are given by

$$\gamma(\lambda) = \int_{-\infty}^{0} e^{i\lambda t}c(t)\,dt, \quad c(t) = i\ell\frac{1+n}{1-n}e^t - i\ell\frac{2n}{1-n}e^{nt}, \quad t \leq 0, \tag{3.11}$$

$$\delta(\lambda) = 1 + \int_{-\infty}^{0} e^{i\lambda t}d(t)\,dt, \quad d(t) = -(m+1)\frac{(1+n)}{1-n}e^t + + \frac{2n(n+m)}{1-n}e^{nt}, \quad t \leq 0. \tag{3.12}$$

To see better the difference with (3.2) and (3.3), let $\alpha_1, \beta_1, \gamma_1, \delta_1$ be the functions defined by (3.2) and (3.3), and put $\theta(\lambda) = (\lambda + in)(\lambda - in)^{-1}$. Then

$$\alpha = \alpha_1, \quad \beta = \beta_1, \quad \gamma = \gamma_1\theta, \quad \delta = \delta_1\theta, \quad \text{and} \quad \theta\theta^* = \theta^*\theta = 1.$$

Using the results of Example 2 it is then straightforward to show that for the data set $\{\alpha, \beta, \gamma, \delta\}$ defined by (3.9) and (3.10) the three identities in (2.1) hold true. Furthermore, since $m \neq n$ by assumption, the functions α and γ^* have no common zero in $\mathbb{C}_+$, and we know from [7, Theorem 4.1] that the left onefold EG inverse problem associated with the pair $\{\alpha, \gamma\}$ is solvable. Similarly, using that β and δ^* have no common zero in $\mathbb{C}_+$, Theorem 4.1 in [7] also shows that the right onefold EG inverse problem associated with the pair $\{\beta, \delta^*\}$ is solvable too.

However the two solutions are not equal. Indeed, if these two solutions would be equal, then the twofold EG inverse problem associated with the data $\alpha, \beta, \gamma, \delta$ would be solvable, and hence, by Theorem 2.1, the functions α and δ^* would have the same zeros on $\mathbb{C}_+$, taking multiplicities into account. But α has only one zero in $\mathbb{C}_+$, namely $\lambda = im$, and δ^* has two different zeros in $\mathbb{C}_+$, namely $\lambda = im$ and $\lambda = in$. Recall that $m \neq n$ by assumption. Thus the twofold EG inverse problem is not solvable while the left and right onefold problems are solvable.

References

1. A. Böttcher, B. Silbermann, *Analysis of Toeplitz operators* (Springer, Berlin, 2006)
2. R.L. Ellis, I. Gohberg, Orthogonal systems related to infinite Hankel matrices. J. Funct. Anal. **109**, 155–198 (1992)
3. R.L. Ellis, I. Gohberg, *Orthogonal Systems and Convolution Operators*. Oper. Theory Adv. Appl., vol. 140 (Birkhäuser Verlag, Basel, 2003)
4. I. Gohberg, S. Goldberg, M.A. Kaashoek, *Classes of Linear Operators*, Volume I. Oper. Theory Adv. Appl., vol. 49 (Birkhäuser Verlag, Basel, 1990)
5. I. Gohberg, S. Goldberg, M.A. Kaashoek, *Classes of Linear Operators*, Volume II. Oper. Theory Adv. Appl., vol. 63 (Birkhäuser Verlag, Basel, 1993)
6. M.A. Kaashoek, F. van Schagen, Ellis-Gohberg identities for certain orthogonal functions II: Algebraic setting and asymmetric versions, West Memorial Issue. Proc. Math. Royal Irish Acad. **113A**(2), 107–130 (2013)
7. M.A. Kaashoek, F. van Schagen, The Ellis-Gohberg inverse problem for matrix-valued Wiener functions on the line. Oper. Matrices **10**(4), 1009–1042 (2016)
8. M.A. Kaashoek, F. van Schagen, Errata for The Ellis-Gohberg inverse problem for matrix-valued Wiener functions on the line. Oper. Matrices **12**(4), 1199–1200 (2018)
9. S. ter Horst, M.A. Kaashoek, F. van Schagen, The twofold Ellis-Gohberg inverse problem in an abstract setting and applications. Oper. Theory Adv. Appl., vol. 272, 189–246 (2019)

Relatively Uniform Convergence in Partially Ordered Vector Spaces Revisited

Anke Kalauch and Onno van Gaans

To Ben de Pagter, on the occasion of his 65th birthday, with admiration and gratitude

Abstract We consider relatively uniform convergence of nets in a partially ordered vector space. We give an example of a set V in a space X where adding the limits of nets in V that converge in X does not produce the closure of V in X. The closure of a set can be constructed by adding limits if that process is repeated by transfinite induction. We also consider closed sets and complete spaces and show that they coincide with sequentially closed sets and sequentially complete spaces, respectively.

Keywords Net · Partially ordered vector space · Relatively uniform closure · Relatively uniform convergence · Riesz space · Sequence

1 Introduction

In the analysis of an Archimedean partially ordered vector space with an order unit the norm induced by the unit plays an important role. If an order unit does not exist, the notion of relatively uniform convergence is often used instead. Although many features of relatively uniform convergence are common knowledge, written accounts of them seem to be scarce and scattered; see, e.g., [5, 6, 11, 12], [13, Chapter 7], and [3, Section 2.7]. Relatively uniform convergence is considered in several books on Riesz spaces as well, e.g., [1, 2, 10]. In the present note we intend

A. Kalauch
Technische Universität Dresden, Fakultät Mathematik, Institut für Analysis, Dresden, Germany
e-mail: Anke.Kalauch@tu-dresden.de

O. van Gaans (✉)
Leiden University, Mathematical Institute, Leiden, The Netherlands
e-mail: vangaans@math.leidenuniv.nl

G. Buskes et al. (eds.), *Positivity and Noncommutative Analysis*,
Trends in Mathematics, https://doi.org/10.1007/978-3-030-10850-2_14

to collect a few of the fundamental properties of relatively uniform convergence. The results are presented in the general setting of partially ordered vector spaces. We do not claim any of the results to be new in principle, but we believe that combining them and reproducing them in general partially ordered vector spaces may be convenient. We consider relatively uniform convergence of nets. Usually, the scope in the literature is restricted to sequences. We deal with closed sets and completeness. It turns out that these notions for sequences and nets coincide.

For the standard notions and theory of partially ordered vector spaces and Riesz spaces, we refer to [1, 2, 9, 10, 14].

For a net $(\lambda_\alpha)_{\alpha\in A}$ in $\mathbb{R}_+ := \{t \in \mathbb{R} \colon t \geq 0\}$ we denote $\lambda_\alpha \downarrow 0$ if $(\lambda_\alpha)_{\alpha\in A}$ is decreasing (i.e., $\alpha \geq \beta$ implies $\lambda_\alpha \leq \lambda_\beta$) and $\inf\{\lambda_\alpha \colon \alpha \in A\} = 0$. If A is a partially ordered set and $\alpha_0 \in A$, then we denote $A_{\geq\alpha_0} := \{\alpha \in A \colon \alpha \geq \alpha_0\}$. The concepts of the next definition seem to be due to E.H. Moore [11, p. 344], who introduced them for sequences in a context of function spaces.

Definition 1.1 Let (X, K) be a partially ordered vector space, let $(x_\alpha)_{\alpha\in A}$ be a net in X and $x \in X$.

(i) For a $u \in K$ we say that $(x_\alpha)_{\alpha\in A}$ *u-converges* to x if there exists a net $(\lambda_\beta)_{\beta\in B}$ in $\mathbb{R}_+$ with $\lambda_\beta \downarrow 0$ such that for every $\beta \in B$ there exists $\alpha_\beta \in A$ with

$$-\lambda_\beta u \leq x_\alpha - x \leq \lambda_\beta u \quad \text{for all } \alpha \in A_{\geq\alpha_\beta}.$$

(ii) We say that $(x_\alpha)_{\alpha\in A}$ *converges relatively uniformly* to x if there exists $u \in K$ such that $(x_\alpha)_{\alpha\in A}$ *u-converges* to x. In this case, u is called a *regulator* for the convergence.

Recall that a partially ordered vector space (X, K) is *Archimedean* if for every $x, y \in X$ with $nx \leq y$ for all $n \in \mathbb{N}$ one has that $x \leq 0$. If the space (X, K) is Archimedean, then the limit x in Definition 1.1(i) is unique; see, e.g., [9, Proposition 1.5.36(ii)] or, for sequences, [3, below Definition 2.56].

Various other definitions of relatively uniform convergence of nets are in use. The next proposition considers three of them and states that in Archimedean spaces they are equivalent to our definition.

Proposition 1.2 *Let (X, K) be an Archimedean partially ordered vector space, let $(x_\alpha)_{\alpha\in A}$ be a net in X, let $x \in X$, and let $u \in K$. The following statements are equivalent:*

(a) *The net $(x_\alpha)_{\alpha\in A}$ u-converges to x.*
(b) *There exist $\alpha_0 \in A$ and a net $(\lambda_\alpha)_{\alpha\in A_{\geq\alpha_0}}$ with $\lambda_\alpha \downarrow 0$ such that for every $\alpha \in A_{\geq\alpha_0}$ we have $-\lambda_\alpha u \leq x_\alpha - x \leq \lambda_\alpha u$.*
(c) *For every $\varepsilon > 0$ there exists $\alpha_\varepsilon \in A$ such that for every $\alpha \in A_{\geq\alpha_\varepsilon}$ we have $-\varepsilon u \leq x_\alpha - x \leq \varepsilon u$.*
(d) *For every $n \in \mathbb{N}$ there exists $\alpha_n \in A$ such that for every $\alpha \in A_{\geq\alpha_n}$ we have $-\frac{1}{n}u \leq x_\alpha - x \leq \frac{1}{n}u$.*

Proof First we show that (a) implies (d). Indeed, according to (a) there is a net $(\lambda_\beta)_{\beta\in B}$ in $\mathbb{R}_+$ with $\lambda_\beta \downarrow 0$ such that for every $\beta \in B$ there exists $\alpha_\beta \in A$ with $-\lambda_\beta u \leq x_\alpha - x \leq \lambda_\beta u$ for every $\alpha \in A_{\geq\alpha_\beta}$. Let $n \in \mathbb{N}$. There exists $\beta_n \in B$ such that $\lambda_{\beta_n} < \frac{1}{n}$. Take $\alpha_n := \alpha_{\beta_n}$. For every $\alpha \in A_{\geq\alpha_n}$ we have $x_\alpha - x \leq \lambda_{\beta_n} u \leq \frac{1}{n}u$ and, similarly, $x_\alpha - x \geq -\frac{1}{n}u$.

To see that (d) implies (c), let $\varepsilon > 0$. Choose $n \in \mathbb{N}$ with $\frac{1}{n} < \varepsilon$ and choose α_ε to be α_n as in (d). Then the desired statement holds.

The implication (c)⇒(b) is contained in [9, Proposition 1.5.38(i)]. Here we use that X is Archimedean.

It is clear that (b) implies (a). □

The notions of convergence in each of the statements (b), (c), and (d) of Proposition 1.2 yield corresponding notions of relatively uniform convergence, just as in Definition 1.1(ii). It follows that these notions of relatively uniform convergence are equivalent, provided X is Archimedean.

Remark 1.3

(i) Note that in Proposition 1.2 (d) trivially implies (a) and that the proof of (a)⇒(d) does not need that X is Archimedean. Hence the equivalence of (a) and (d) is true in every partially ordered vector space X.
(ii) Statement (d) of Proposition 1.2 yields that in the definition of u-convergence of a net $(x_\alpha)_{\alpha\in A}$ we can always choose $B = \mathbb{N}$ and $\lambda_n = \frac{1}{n}$. In particular, if we consider u-convergence of a sequence $(x_n)_{n\in\mathbb{N}}$, then $(\lambda_\beta)_{\beta\in B}$ may be chosen to be a sequence. Thus, Definition 1.1 is equivalent to the definition of relatively uniform convergence for sequences given in [3] or, for Riesz spaces, [1, 2]. The formulation in Proposition 1.2(c) corresponds to the definition of relatively uniformly convergent sequences in [10, Theorem 16.2].

Let (X, K) be a partially ordered vector space. An element $u \in K$ is called an *order unit* of X if for every $x \in X$ there exists $\lambda \in \mathbb{R}_+$ such that $x \leq \lambda u$. Then there also exists $\mu \in \mathbb{R}_+$ such that $-\mu u \leq x \leq \mu u$. Note that if X has an order unit, then X is directed, i.e., $X = K - K$. For an order unit u of X,

$$\|x\|_u := \inf\{\lambda \in \mathbb{R}_+ \colon \; -\lambda u \leq x \leq \lambda u\}, \quad x \in X,$$

defines a seminorm $\|\cdot\|_u$ on X. If X is Archimedean, then $\|\cdot\|_u$ is a norm on X, called the *order unit norm*.

For an element $u \in K$, let us denote

$$X_u := \{x \in X \colon \exists\lambda \in \mathbb{R} \text{ such that } -\lambda u \leq x \leq \lambda u\}.$$

Then X_u is a linear subspace of X and u is an order unit of X_u.

Lemma 1.4 *Let (X, K) be a partially ordered vector space, let $(x_\alpha)_{\alpha\in A}$ be a net in X, and let $x \in X$. If $u \in K$ is such that $x_\alpha, x \in X_u$ for all $\alpha \in A$, then*

$$(x_\alpha)_{\alpha\in A} \text{ } u\text{-converges to } x \text{ if and only if } \|x_\alpha - x\|_u \to 0.$$

Proof Suppose $(x_\alpha)_{\alpha\in A}$ u-converges to x. For every $n \in \mathbb{N}$ there exists $\alpha_n \in A$ such that $-\frac{1}{n}u \leq x_\alpha - x \leq \frac{1}{n}u$ for all $\alpha \in A_{\geq \alpha_n}$. Then $\|x_\alpha - x\|_u \leq \frac{1}{n}$ for all $\alpha \in A_{\geq\alpha_n}$, hence $\|x_\alpha - x\|_u \to 0$.

Conversely, suppose $\|x_\alpha - x\|_u \to 0$. For every $n \in \mathbb{N}$ there exists $\alpha_n \in A$ such that $\|x_\alpha - x\|_u < \frac{1}{n}$ for every $\alpha \in A_{\geq\alpha_n}$. Then $-\frac{1}{n}u \leq x_\alpha - x \leq \frac{1}{n}u$ for all $\alpha \in A_{\geq\alpha_n}$, so that $(x_\alpha)_{\alpha\in A}$ u-converges to x. □

Addition and scalar multiplication are continuous for relatively uniform convergence in the following sense.

Proposition 1.5 *Let (X, K) be a partially ordered vector space, let $(x_\alpha)_{\alpha\in A}$ and $(y_\beta)_{\beta\in B}$ be nets in X and let $x, y \in X$ be such that $(x_\alpha)_{\alpha\in A}$ converges relatively uniformly to x and $(y_\beta)_{\beta\in B}$ converges relatively uniformly to y. Then the net $(x_\alpha + y_\beta)_{(\alpha,\beta)\in A\times B}$ converges relatively uniformly to $x + y$ and, for every $t \in \mathbb{R}$, the net $(tx_\alpha)_{\alpha\in A}$ converges relatively uniformly to tx. Here the index set $A \times B$ is endowed with coordinatewise order.*

Proof There are $u, v \in K$ and for every $n \in \mathbb{N}$ there are $\alpha_n \in A$ and $\beta_n \in B$ with

$$-\tfrac{1}{n}u \leq x_\alpha - x \leq \tfrac{1}{n}u \quad \text{and} \quad -\tfrac{1}{n}v \leq y_\beta - y \leq \tfrac{1}{n}v$$

for all $\alpha \in A_{\geq\alpha_n}$ and $\beta \in B_{\geq\beta_n}$. Then for every $(\alpha, \beta) \in A \times B_{\geq(\alpha_n,\beta_n)}$ we have

$$-\tfrac{1}{n}(u + v) \leq x_\alpha + y_\beta - (x + y) \leq \tfrac{1}{n}(u + v).$$

Hence $(x_\alpha + y_\beta)_{(\alpha,\beta)\in A\times B}$ $(u + v)$-converges to $x + y$.

Also, for $\alpha \in A_{\geq\alpha_n}$ we have $-\frac{1}{n}|t|u \leq tx_\alpha - tx \leq \frac{1}{n}|t|u$, so that $(tx_\alpha)_{\alpha\in A}$ $|t|u$-converges to tx. □

We proceed by showing that order bounded operators are continuous for relatively uniform convergence. Let (X, K) be a partially ordered vector space. For $a, b \in X$ with $a \leq b$ we denote $[a, b] := \{x \in X : a \leq x \leq b\}$. Recall that a subset V of X is called *order bounded* if there are $a, b \in X$ with $a \leq b$ such that $V \subseteq [a, b]$. Now let (X, K_X) and (Y, K_Y) be partially ordered vector spaces. A linear map $T : X \to Y$ is called an *order bounded operator* if for every order bounded set $U \subseteq X$ the set $T[U]$ is order bounded in Y. Note that if Y is directed and $T : X \to Y$ is an order bounded operator, then for every $u \in K_X$ there exists $v \in K_Y$ such that T maps $[-u, u]$ into $[-v, v]$.

Proposition 1.6 *Let (X, K_X) and (Y, K_Y) be partially ordered vector spaces with Y directed and let $T: X \to Y$ be an order bounded operator.*

(i) *T maps relatively uniformly convergent nets to relatively uniformly convergent nets.*
(ii) *For every $u \in K_X$ there exists $v \in K_Y$ such that $T[X_u] \subseteq Y_v$ and $T|_{X_u}$ is continuous with respect to $\|\cdot\|_u$ on X_u and $\|\cdot\|_v$ on Y_v.*

Proof

(i) Let $(x_\alpha)_{\alpha\in A}$ be a net in X and let $x \in X$ be such that $(x_\alpha)_{\alpha\in A}$ converges relatively uniformly to x. Then there is $u \in K_X$ and for every $n \in \mathbb{N}$ there is $\alpha_n \in A$ such that $-\frac{1}{n}u \le x_\alpha - x \le \frac{1}{n}u$ for all $\alpha \in A_{\ge\alpha_n}$. Take $v \in K_Y$ such that T maps $[-u, u]$ into $[-v, v]$. For every $n \in \mathbb{N}$ and $\alpha \in A_{\ge\alpha_n}$ we have $-u \le nx_\alpha - nx \le u$, so $-v \le nTx_\alpha - nTx \le v$, hence $-\frac{1}{n}v \le Tx_\alpha - Tx \le \frac{1}{n}v$. Thus, the net $(Tx_\alpha)_{\alpha\in A}$ v-converges to Tx.
(ii) Let $u \in K_X$ and take $v \in K_Y$ such that T maps $[-u, u]$ into $[-v, v]$. Then $T[X_u] \subseteq Y_v$. If $x \in X$ and $\lambda \in \mathbb{R}$ with $\lambda > 0$ are such that $-\lambda u \le x \le \lambda u$, then $-\lambda v \le Tx \le \lambda v$, hence $\|Tx\|_v \le \|x\|_u$.

□

2 Closed Sets and Closures

Let (X, K) be a partially ordered vector space. For a set $V \subseteq X$ we denote the set of relatively uniform limits of nets in V by

$$\mathrm{L}(V) := \{x \in X\colon \exists (v_\alpha)_{\alpha\in A} \text{ in } V \text{ such that } v_\alpha \to x \text{ relatively uniformly}\}.$$

We say that V is *relatively uniformly closed* in X if $\mathrm{L}(V) = V$. The *relatively uniform closure* of V is defined by

$$\overline{V} := \bigcap\{C \subseteq X\colon C \text{ is relatively uniformly closed and } V \subseteq C\}.$$

The next example presents a set V such that $\mathrm{L}(V)$ is not relatively uniformly closed. In particular, $\mathrm{L}(V)$ is then strictly smaller than the relatively uniform closure of V.

Example 2.1 Consider

$$X = \{x\colon \mathbb{Z} \to \mathbb{R}\colon \exists c \in \mathbb{R}_+,\ p \in \mathbb{N} \text{ such that } |x(n)| \le c|n|^p \text{ for all } n \in \mathbb{Z}\}.$$

Then X is an Archimedean Riesz space. For $k, m \in \mathbb{N}$ define

$$v_k^m = k^m \mathbb{1}_{\{-k\}} + \left(1 + \tfrac{1}{m}\right) \mathbb{1}_{\{1,\ldots,m\}} \in X.$$

Let

$$V := \{v_k^m \colon\ k, m \in \mathbb{N}\},$$
$$W := \{x \in X \colon\ \exists (v_\alpha)_{\alpha \in A} \text{ in } V \text{ such that } v_\alpha \to x \text{ relatively uniformly in } X\},$$

that is, $W = \mathrm{L}(V)$. We show that W is not relatively uniformly closed in X.

We first show that for every $m \in \mathbb{N}$ we have $x^m := \left(1 + \frac{1}{m}\right) \mathbb{1}_{\{1,\ldots,m\}} \in W$. Indeed, let $u^m(i) := |i|^{m+1}$, $i \in \mathbb{Z}$. Then $u^m \in X$ and for every $k \in \mathbb{N}$ we have

$$0 \le v_k^m - x^m = k^m \mathbb{1}_{\{-k\}} = \tfrac{1}{k} k^{m+1} \mathbb{1}_{\{-k\}} \le \tfrac{1}{k} u^m,$$

so $(v_k^m)_k$ converges relatively uniformly to x^m. Hence $x^m \in W$.

Next we show that $x^m \to \mathbb{1}_{\mathbb{N}}$ relatively uniformly in X. Indeed, let $u(i) := |i|$, $i \in \mathbb{Z}$. Then $u \in X$ and

$$|x^m - \mathbb{1}_{\mathbb{N}}| = \tfrac{1}{m} \mathbb{1}_{\{1,\ldots,m\}} + \mathbb{1}_{\{m+1,m+2,\ldots\}} \le \tfrac{1}{m} u + \tfrac{1}{m+1} u \le \tfrac{2}{m} u,$$

so $x^m \to \mathbb{1}_{\mathbb{N}}$ as $m \to \infty$.

To establish that W is not relatively uniformly closed, we show that $\mathbb{1}_{\mathbb{N}} \notin W$. For a proof, let $u \in X$ with $u \ge 0$. Take $c \in \mathbb{R}$ with $c > 0$ and $p \in \mathbb{N}$ such that

$$|u(n)| \le c|n|^p \quad \text{for all } n \in \mathbb{Z}.$$

Choose $\varepsilon > 0$ with

$$\varepsilon < \min\left\{\tfrac{1}{p(u(1)+1)}, \tfrac{1}{c}\right\}.$$

Suppose there are $k, m \in \mathbb{N}$ such that

$$|v_k^m - \mathbb{1}_{\mathbb{N}}| \le \varepsilon u,$$

then $|v_k^m(1) - \mathbb{1}_{\mathbb{N}}(1)| \le \varepsilon u(1)$, so

$$\tfrac{1}{m} \le \varepsilon u(1) \le \tfrac{1}{p(u(1)+1)} u(1) < \tfrac{1}{p},$$

so $m > p$. Also, $|v_k^m(-k) - \mathbb{1}_{\mathbb{N}}(-k)| \le \varepsilon u(-k)$, so

$$k^m \le \varepsilon u(-k) \le \varepsilon c k^p < \tfrac{1}{c} c k^p = k^p,$$

so $m < p$, which is a contradiction. Hence $|v_k^m - \mathbb{1}_{\mathbb{N}}| > \varepsilon u$ for every $k, m \in \mathbb{N}$. Therefore, there can be no net in V converging relatively uniformly to $\mathbb{1}_{\mathbb{N}}$. Thus, $\mathbb{1}_{\mathbb{N}} \notin W$ and W is not relatively uniformly closed. In conclusion, $\mathbb{1}_{\mathbb{N}} \in \overline{V} \setminus \mathrm{L}(V)$.

If we replace W by

$$\tilde{W} := \{x \in X\colon \exists (v_n)_{n\in\mathbb{N}} \text{ in } V \text{ such that } v_n \to x \text{ relatively uniformly in } X\},$$

then it follows similarly that $\tilde{W}$ is not sequentially closed.

In view of Lemma 1.4, one could wonder if one could obtain the relatively uniform closure of a set V in a partially ordered vector space (X, K) by taking the closure of $V \cap X_u$ in X_u relative to $\|\cdot\|_u$ and then take the union over all $u \in K$. Example 2.1 shows that such a procedure fails. Indeed, in Example 2.1, for every $u \in X$ with $u \geq 0$ we have that $\mathbb{1}_{\mathbb{N}}$ is not contained in the closure of $V \cap X_u$ in X_u with respect to $\|\cdot\|_u$, but $\mathbb{1}_{\mathbb{N}}$ is in the relatively uniform closure of V.

The closure of V can be constructed by means of transfinite recursion and the axiom of choice.

Theorem 2.2 *Let (X, K) be a partially ordered vector space and let $V \subseteq X$. For each ordinal α we define a set V_α by means of transfinite recursion:*

$$\begin{aligned} V_\alpha &:= V \quad \text{if } \alpha = 0, \\ V_{\alpha+1} &:= \mathrm{L}(V_\alpha), \\ V_\alpha &:= \bigcup_{\beta<\alpha} V_\beta \quad \text{if } \alpha \text{ is a limit ordinal.} \end{aligned}$$

There exists an ordinal α such that $V_\alpha = \overline{V}$.

Proof Observe that, by construction, $V_\alpha \subseteq V_\beta$ for ordinals α and β with $\alpha \leq \beta$ and that $V \subseteq V_\alpha \subseteq \overline{V}$ for each ordinal α. We show that there exists an ordinal α such that $V_{\alpha+1} = V_\alpha$. Then V_α is relatively uniformly closed and as $V \subseteq V_\alpha \subseteq \overline{V}$ it then follows that $\overline{V} \subseteq \overline{V_\alpha} = V_\alpha \subseteq \overline{V}$, so $V_\alpha = \overline{V}$.

The axiom of choice provides a well-order on X. Let ξ be the ordinal that is order isomorphic to this well-ordered set. Then the cardinality of ξ equals the cardinality of X. For ordinals α and β with $\alpha < \beta$ we have $V_\alpha \subseteq V_\beta$, so

$$\mathcal{V} := \{V_{\alpha+1} \setminus V_\alpha\colon \alpha \leq \xi + 1\}$$

consists of mutually disjoint subsets of X. Therefore, the cardinality of $\mathcal{V}$ is at most the cardinality of X. Thus at least one element of $\mathcal{V}$ must be empty, which means that there exists an ordinal α with $\alpha \leq \xi + 1$ such that $V_{\alpha+1} \setminus V_\alpha$ is empty, and therefore $V_{\alpha+1} = V_\alpha$. □

Corollary 2.3 *Let (X, K) be a partially ordered vector space and $V \subseteq X$. If V is convex, then $\overline{V}$ is convex.*

Proof We use the construction of $\overline{V}$ given in Theorem 2.2. As addition and scalar multiplication are continuous for relatively uniform convergence of nets, we have that if $W \subseteq X$ is convex, then $\mathrm{L}(W)$ is convex. Hence if α is an ordinal for which

V_α is convex, then $V_{\alpha+1}$ is convex as well. Further, if β is a limit ordinal and V_α is convex for every $\alpha < \beta$, then V_β is convex. Indeed, let $x_1, x_2 \in V_\alpha$ and $t \in [0, 1]$. Then there are ordinals $\alpha_1 < \beta$ and $\alpha_2 < \beta$ such that $x_1 \in V_{\alpha_1}$ and $x_2 \in V_{\alpha_2}$. Without loss of generality we may assume that $\alpha_2 \leq \alpha_1$. Then $x_1, x_2 \in V_{\alpha_1}$ and as V_{α_1} is convex it follows that $tx_1 + (1-t)x_2 \in V_{\alpha_1} \subseteq V_\beta$. Hence V_β is convex. It now follows by transfinite induction that V_α is convex for every ordinal α. □

Remark 2.4 Relatively uniform closures play a role in the construction of the tensor product of directed Archimedean partially ordered vector spaces. The projective cone, which is the cone generated by the elementary tensors of positive elements, is too small. The appropriate cone in the tensor product is called the *Archimedean tensor cone* in [7, Section 2]. It can be characterized and constructed as the intersection of all relatively uniformly closed wedges that contain the projective cone; see [9, Theorem 2.7.10]. It follows from Corollary 2.3 that the Archimedean tensor cone is exactly the relatively uniform closure of the projective cone.

Proposition 2.5 *Let (X, K) be a partially ordered vector space and let $V \subseteq X$. Then V is relatively uniformly closed if and only if V is relatively uniformly sequentially closed.*

Proof We only need to show that sequential closedness implies closedness. So, suppose that V is relatively uniformly sequentially closed, let $(x_\alpha)_{\alpha\in A}$ be a net in V, and let $x \in X$ be such that $(x_\alpha)_{\alpha\in A}$ converges relatively uniformly to x. There exists $u \in K$ such that for every $n \in \mathbb{N}$ there is $\alpha_n \in A$ with $-\frac{1}{n}u \leq x_\alpha - x \leq \frac{1}{n}u$ for all $\alpha \in A_{\geq\alpha_n}$. Then the sequence $(x_{\alpha_n})_{n\in\mathbb{N}}$ converges relatively uniformly to x. As V is relatively uniformly sequentially closed, it follows that $x \in V$. Hence V is relatively uniformly closed. □

3 Completeness

Let (X, K) be a partially ordered vector space. For an element $u \in K$ a net $(x_\alpha)_{\alpha\in A}$ is said to be a *u-Cauchy net* if for every $\varepsilon > 0$ there exists $\alpha_\varepsilon \in A$ such that for every $\alpha, \beta \in A_{\geq\alpha_\varepsilon}$ we have

$$-\varepsilon u \leq x_\alpha - x_\beta \leq \varepsilon u.$$

The net $(x_\alpha)_{\alpha\in A}$ is a u-Cauchy net if and only if the net $(x_\alpha - x_\beta)_{(\alpha,\beta)\in A\times A}$ u-converges to 0. We say that $(x_\alpha)_{\alpha\in A}$ is a *relatively uniformly Cauchy net* if there exists $u \in K$ such that $(x_\alpha)_{\alpha\in A}$ is a u-Cauchy net.

The partially ordered vector space (X, K) is said to be *relatively uniformly complete* if every relatively uniformly Cauchy net in X is relatively uniformly convergent in X. There is an alternative definition that is used more commonly (see, e.g., [10]): X is relatively uniformly complete if for every $u \in K$ every u-Cauchy net is u-convergent. The next lemma yields that these two definitions are equivalent in Archimedean spaces; see also [4, Lemma 1.12], where the result is given for sequences in Archimedean lattice ordered groups.

Lemma 3.1 *Let (X, K) be an Archimedean partially ordered vector space and let $u \in K$. If $(x_\alpha)_{\alpha\in A}$ is a u-Cauchy net in X and $x \in X$ is such that $x_\alpha \to x$ relatively uniformly in X, then $(x_\alpha)_{\alpha\in A}$ is u-convergent to x.*

Proof Let $v \in K$ be such that for every $n \in \mathbb{N}$ there is $\alpha_n \in A$ with

$$-\tfrac{1}{n}v \le x_\alpha - x \le \tfrac{1}{n}v \quad \text{for all } \alpha \in A_{\ge\alpha_n}.$$

Let $u \in K$ be such that for every $n \in \mathbb{N}$ there is $\beta_n \in A$ with

$$-\tfrac{1}{n}u \le x_\alpha - x_{\alpha'} \le \tfrac{1}{n}u \quad \text{for all } \alpha, \alpha' \in A_{\ge\beta_n}.$$

For every $n \in \mathbb{N}$ choose $\gamma_n \in A$ with $\gamma_n \ge \alpha_n$ and $\gamma_n \ge \beta_k$ for every $k \in \{1, \dots, n\}$. For every $n \in \mathbb{N}$ and $\alpha \in A_{\ge\gamma_n}$ we have for every $m \in \mathbb{N}_{\ge n}$ that

$$x_\alpha - x = (x_\alpha - x_{\gamma_m}) + (x_{\gamma_m} - x) \le \tfrac{1}{n}u + \tfrac{1}{m}v.$$

As X is Archimedean, we obtain $x_\alpha - x \le \frac{1}{n}u$. Similarly, $x_\alpha - x \ge -\frac{1}{n}u$. Thus, $(x_\alpha)_{\alpha\in A}$ is u-convergent to x. □

The following relations between relatively uniform closedness and relatively uniform completeness hold.

Proposition 3.2 *Let (X, K) be an Archimedean partially ordered vector space and let D be a linear subspace of X.*

(i) *If X is relatively uniformly complete and D is relatively uniformly closed in X, then D is relatively uniformly complete.*
(ii) *If D is majorizing in X and D is relatively uniformly complete, then D is relatively uniformly closed in X.*

Proof

(i) Let $(x_\alpha)_{\alpha\in A}$ be a relatively uniformly Cauchy net in D. Choose $u \in D \cap K$ such that $(x_\alpha)_{\alpha\in A}$ is a u-Cauchy net in D. Then $(x_\alpha)_{\alpha\in A}$ is a u-Cauchy net in X, hence, by Lemma 3.1, $(x_\alpha)_{\alpha\in A}$ is u-convergent in X to some $x \in X$. As D is relatively uniformly closed, we have $x \in D$. Since $u \in D$, it follows that $(x_\alpha)_{\alpha\in A}$ converges relatively uniformly to x in D.
(ii) Let $(x_\alpha)_{\alpha\in A}$ be a net in D and let $x \in X$ be such that $(x_\alpha)_{\alpha\in A}$ converges relatively uniformly to x in X. Choose $u \in K$ such that $(x_\alpha)_{\alpha\in A}$ is u-convergent to x. As D is majorizing in X, there is $v \in D$ with $v \ge u$ and then $(x_\alpha)_{\alpha\in A}$ is also v-convergent to x. It follows that $(x_\alpha)_{\alpha\in A}$ is a v-Cauchy net in D, hence

there is a $y \in D$ such that $(x_\alpha)_{\alpha\in A}$ v-converges to y in D. Then $(x_\alpha)_{\alpha\in A}$ also v-converges to y in X and as X is Archimedean we obtain that $y = x$. Thus, $x \in D$ and D is relatively uniformly closed in X.

□

The next example shows that in Proposition 3.2(ii) the condition that D be majorizing in X is not superfluous.

Example 3.3 Let $X := l^\infty(\mathbb{N})$ be the vector space of all bounded functions on $\mathbb{N}$ endowed with the pointwise order and let

$$D := \{x \in X\colon\ \exists N \in \mathbb{N} \text{ such that } x(n) = 0 \text{ for all } n \in \mathbb{N}_{\geq N+1}\}.$$

Then X is an Archimedean partially ordered vector space and D is a linear subspace of X. Actually, X is even a Riesz space and D is a Riesz subspace of X. Observe that D is not majorizing in X. We show that D is relatively uniformly complete, but not relatively uniformly closed in X. For a proof of the completeness, let $(x_\alpha)_{\alpha\in A}$ be a relatively uniformly Cauchy net in D. Take $u \in D$ with $u \geq 0$ such that $(x_\alpha)_{\alpha\in A}$ is a u-Cauchy net in D. For every $\varepsilon > 0$ there is $\alpha_\varepsilon \in A$ such that

$$-\varepsilon u \leq x_\alpha - x_{\alpha'} \leq \varepsilon u \text{ for all } \alpha, \alpha' \in A_{\geq \alpha_\varepsilon}.$$

There is $N \in \mathbb{N}$ such that $u(n) = 0$ and $x_{\alpha_1}(n) = 0$ for all $n \in \mathbb{N}_{\geq N+1}$. Then, for every $\alpha \in A_{\geq \alpha_1}$ and $n \geq N + 1$ we have

$$|x_\alpha(n)| = |x_\alpha(n) - x_{\alpha_1}(n)| \leq u(n) = 0,$$

so $x_\alpha(n) = 0$. For each $n \in \{1, \ldots, N\}$, we have that $(x_\alpha(n))_{\alpha\in A}$ is a Cauchy net in $\mathbb{R}$ and therefore convergent to a $\xi_n \in \mathbb{R}$. Define $x\colon \mathbb{N} \to \mathbb{R}$ by $x(n) := \xi_n$ for $n \in \{1, \ldots, N\}$ and $x(n) := 0$ for $n \in \mathbb{N}_{\geq N+1}$. Then $x \in D$ and $(x_\alpha)_{\alpha\in A}$ u-converges to x. Thus, D is relatively uniformly complete.

To see that D is not relatively uniformly closed in X, let $x \in X$ be given by $x(n) := \frac{1}{n}$, $n \in \mathbb{N}$. For $m \in \mathbb{N}$, define $x_m(n) := \frac{1}{n}$ for $n \in \{1, \ldots, m\}$ and $x_m(n) := 0$ for $n \in \mathbb{N}_{\geq m+1}$. Let $u(n) := 1$ for all $n \in \mathbb{N}$. Then $(x_m)_{m\in\mathbb{N}}$ is a sequence in D, $u \in X$, $u \geq 0$, and for every $m \in \mathbb{N}$ we have

$$|x_m(n) - x(n)| \leq \tfrac{1}{m+1} = \tfrac{1}{m+1}u(n) \text{ for all } n \in \mathbb{N},$$

so $(x_m)_{m\in\mathbb{N}}$ u-converges to x in X. As $x \notin D$, it follows that D is not relatively uniformly closed in X.

In [3] relatively uniform completeness is defined by sequences rather than nets. Indeed, in [3, Definition 2.57] a partially ordered vector space is called uniformly complete if for every $u \geq 0$ every u-Cauchy sequence is u-convergent. To distinguish from our definition with nets, we call a partially ordered vector space (X, K) *relatively uniformly sequentially complete* if every relatively uniformly

Cauchy sequence in X is relatively uniformly convergent in X. The next proposition shows that in Archimedean spaces sequential completeness is equivalent to the completeness defined by nets.

Proposition 3.4 *Let (X, K) be an Archimedean partially ordered vector space. Then X is relatively uniformly complete if and only if X is relatively uniformly sequentially complete.*

Proof We only have to show that sequential completeness implies completeness. So, suppose that X is relatively uniformly sequentially complete and let $(x_\alpha)_{\alpha\in A}$ be a relatively uniformly Cauchy net in X. Then there is $u \in K$ and for every $n \in \mathbb{N}$ there is $\alpha_n \in A$ with $-\frac{1}{n}u \leq x_\alpha - x_{\alpha'} \leq \frac{1}{n}u$ for all $\alpha, \alpha' \in A_{\geq\alpha_n}$. Inductively, we can choose α_n such that $\alpha_{n+1} \geq \alpha_n$ for all $n \in \mathbb{N}$. Then $(x_{\alpha_n})_{n\in\mathbb{N}}$ is a u-Cauchy sequence in X. As X is relatively uniformly sequentially complete, there exists $x \in X$ such that $(x_{\alpha_n})_{n\in\mathbb{N}}$ converges relatively uniformly to x. According to Lemma 3.1, the sequence $(x_{\alpha_n})_{n\in\mathbb{N}}$ actually u-converges to x.

We next show that $(x_\alpha)_{\alpha\in A}$ u-converges to x. Let $n \in \mathbb{N}$ and $\alpha \in A_{\geq\alpha_n}$. For every $\varepsilon > 0$ there is $m \in \mathbb{N}_{\geq n}$ with $-\varepsilon u \leq x_{\alpha_m} - x \leq \varepsilon u$ and then

$$x_\alpha - x \leq (x_\alpha - x_{\alpha_m}) + (x_{\alpha_m} - x) \leq \tfrac{1}{n}u + \varepsilon u \leq \tfrac{1}{n}u + \varepsilon u.$$

As X is Archimedean, we obtain that $x_\alpha - x \leq \frac{1}{n}u$. Similarly, $x_\alpha - x \geq -\frac{1}{n}u$. Hence, $(x_\alpha)_{\alpha\in A}$ u-converges to x, which means that $(x_\alpha)_{\alpha\in A}$ is relatively uniformly convergent. Thus, X is relatively uniformly complete. □

In [1, Lemma 1.56] it is stated that every σ-Dedekind complete Riesz space is relatively uniformly sequentially complete. With the aid to Proposition 3.4, we conclude the following.

Proposition 3.5 *Every σ-Dedekind complete Riesz space is relatively uniformly complete.*

The importance of relatively uniform closedness and relatively uniform completeness becomes, for instance, clear from the prominent role that they play in the article [8] by Huijsmans and de Pagter.

References

1. C.D. Aliprantis, O. Burkinshaw, *Locally Solid Riesz Spaces with Applications to Economics*, 2nd edn. (AMS, Providence, 2003)
2. C.D. Aliprantis, O. Burkinshaw, *Positive Operators* (Springer, Dordrecht, 2006)
3. C.D. Aliprantis, R. Tourky, *Cones and Duality* (AMS, Providence, 2007)
4. Š. Černák, J. Lihová, On a relative uniform completion of an Archimedean lattice ordered group. Math. Slovaca **59**(2), 231–250 (2009)
5. V.W. Fedorov, Representation of relatively uniform and order convergence topologies by an inductive limit. Moscow Univ. Math. Bull. **68**(5), 221–230 (2013)
6. H. Gordon, Relative uniform convergence. Math. Annalen **153**, 418–427 (1964)

7. J.J. Grobler, C.C.A. Labuschagne, The tensor product of Archimedean ordered vector spaces. Math. Proc. Camb. Philos. Soc. **104**, 331–345 (1988)
8. C.B. Huijsmans, B. de Pagter, Subalgebras and Riesz subspaces of an f-algebra. Proc. Lond. Math. Soc. **48**(1), 161–174 (1984)
9. A. Kalauch, O. van Gaans, *Pre-Riesz Spaces* (De Gruyter, Berlin, 2018)
10. W.A.J. Luxemburg, A.C. Zaanen, *Riesz Spaces I* (North-Holland Publishing, Amsterdam, 1971)
11. E.H. Moore, On the foundations of the theory of linear integral equations. Bull. Am. Math. Soc. **18**, 334–362 (1912)
12. M. Schroder, Relatively uniform convergence and Stone-Weierstrass approximation. Math. Chron. **14**, 55–73 (1985)
13. Y.-C. Wong, K.-F. Ng, *Partially Ordered Topological Vector Spaces* (Clarendon Press, Oxford, 1973)
14. B.Z. Wulich, in *Geometrie der Kegel in normierten Räumen*, ed. by M.R. Weber (De Gruyter, Berlin, 2017)

Dedekind Complete and Order Continuous Banach $C(K)$-Modules

Arkady Kitover and Mehmet Orhon

This paper is dedicated to our dear friend, Professor Ben de Pagter, on the occasion of his 65th birthday

Abstract We extend the notions of Dedekind complete and σ-Dedekind complete Banach lattices to Banach $C(K)$-modules. As our main result we prove for these modules an analogue of Lozanovsky's well-known characterization of Banach lattices with order continuous norm.

Keywords Banach lattice · Dedekind complete · Order continuous · Banach $C(K)$-module

1 Introduction

This paper continues the investigation of the properties of finitely generated Banach $C(K)$-modules (see the definition preceding Theorem 4.7) undertaken by the authors in the papers [7–9]. There is good reason to consider such modules as the nearest relatives of Banach lattices. Indeed, in the above-mentioned papers the authors proved that the well-known criteria of reflexivity, weak sequential completeness, and dual Radon–Nikodym property established for Banach lattices remain valid for finitely generated Banach $C(K)$-modules, while, as it is also well known, for arbitrary Banach spaces this is not the case.

A. Kitover (✉)
Department of Mathematics, Community College of Philadelphia, Philadelphia, PA, USA
e-mail: akitover@ccp.edu

M. Orhon
Department of Mathematics and Statistics, University of New Hampshire, Durham, NH, USA
e-mail: mo@unh.edu

G. Buskes et al. (eds.), *Positivity and Noncommutative Analysis*,
Trends in Mathematics, https://doi.org/10.1007/978-3-030-10850-2_15

It provides some hope that it is possible to develop a meaningful theory of finitely generated Banach $C(K)$-modules parallel to the one of Banach lattices, and the current paper can be considered as another small step in this direction.

It is a well-known result of Lozanovsky [10] that a Dedekind complete Banach lattice has order continuous norm (in short: is order continuous) if and only if it does not contain a copy of l^∞. It is sufficient actually to require that the Banach lattice is σ-Dedekind complete (see, e.g., [15]), but some condition of this type is, of course, necessary, because, e.g., $C[0, 1]$, being separable, does not contain a copy of l^∞, but the standard norm on it is not order continuous.

Now, the following questions arise.

1. What is a natural analogue of Dedekind completeness for Banach $C(K)$-modules?
2. The same question for σ-Dedekind completeness.
3. What should be considered as an analog of order continuity for Banach $C(K)$-modules?
4. Does an analog of Lozanovsky's result remain valid for **finitely generated** Banach $C(K)$-modules?

As the reader will see, the answer to the first question is provided by the well-known notion of Kaplansky module (see Definition 3.9).

The second question is more involved, and our answer to it makes use of some deep ideas of the late A.I. Veksler [14]

An answer to the third question, as well as a positive answer to question four (see Theorems 4.6 and 4.7) represents the main results of the current paper.

After this short introduction we will proceed with some basic definitions, concepts, and notations.

2 Preliminaries

Let K be a compact Hausdorff space and $C(K)$ be the Banach algebra of all complex (or real)-valued continuous functions on K. Let X be a Banach space over the field $\mathbb{C}$ of complex numbers or over the field $\mathbb{R}$ of real numbers. Let m be a unital bounded algebra homomorphism of $C(K)$ into the algebra $L(X)$ of all bounded linear operators on X. The triple $(C(K), m, X)$ is called a Banach $C(K)$-module.

Given a Banach $C(K)$-module we can define an equivalent norm on X,

$$\|x\|' = sup\{\|ax\|, a \in C(K), \|a\| \leq 1\}.$$

With respect to this norm the homomorphism m is a contraction. Since m is an algebra homomorphism its kernel, $\ker m$ is a closed ideal in the algebra $C(K)$. By replacing $C(K)$ with $C(K') \cong C(K)/\ker m$ where $K' = \{k \in K : a(k) = 0\ \forall a \in \ker m\}$ we can always assume that m is one-to-one. The following lemma was proved in [5, Lemma 2].

Lemma 2.1 *Let $m : C(K) \to L(X)$ be a contractive homomorphism. Then*

1. $a, b \in C(K)$, $|a| \leq |b| \Rightarrow \|m(a)x\| \leq \|m(b)x\|$ *for any* $x \in X$.
2. *If m is one-to-one, then it is an isometry.*

Remark 2.2 We would like to emphasize that while in the statement of our main result we will suppose only that m is a bounded homomorphism, in its proof, in virtue of Lemma 2.1 we will assume that it is an isometry.

Definition 2.3 Let $x \in X$. The cyclic subspace $X(x)$ in X is defined as

$$X(x) = cl\{m(a)x,\ a \in C(K)\},$$

where cl denotes the closure of a set.

A Banach $C(K)$-module is called a **cyclic Banach space** if there is an $x \in X$ such that $X = X(x)$.

The following important and well-known lemma is a consequence of more general results proved in [4] and [1]. Special cases were considered earlier by Veksler [14] and Schaefer [13]. A direct proof is in [11, Lemma 2]

Lemma 2.4 *Let $m : C(K) \to L(X)$ be a contractive unital homomorphism. Let $X = X(x_0)$ for some $x_0 \in X$, i.e. X is a cyclic Banach space with a cyclic vector x_0. Then*

1. *X can be represented as a Banach lattice with quasi-interior point x_0.*
2. *The cone of the positive elements in X is the closure in X of the set $\{m(a)x_0 : a \in C(K), a \geq 0\}$.*
3. *The center Z(X) of the Banach lattice X can be identified with the closure of $m(C(K))$ in the weak operator topology on $L(X)$.*
4. *The unit ball of $Z(X)$ is the closure of the unit ball of $m(C(K))$ in the weak operator topology.*
5. *If x is a quasi-interior point in the Banach lattice X, then for the order ideal I_x generated by x we have $I_x = Z(X)x$.*

Another important tool needed for our results is the following lemma [5, Lemma 4(3), equivalence $(b) \Leftrightarrow (c)$].

Lemma 2.5 *Let $m : C(K) \to L(X)$ be a bounded unital algebra homomorphism. Then the following statements are equivalent:*

1. *For each $x \in X$ the map $a \to m(a)x, a \in C(K)$, is compact, provided that $C(K)$ is endowed with its norm topology and X with its weak topology.*
2. *The homomorphism m can be extended uniquely to an algebra homomorphism $\hat{m} : C(K)'' \to L(X)$ which is continuous provided that $C(K)''$ is endowed with the weak-star topology and $L(X)$ with the weak operator topology.*

Remark 2.6

(a) Note that $C(K)'' \cong C(S)$ where S is a hyperstonian compact space. The proof of Lemma 2.5 given in [5] requires the use of Arens extension of the $C(K)$-module multiplication on X to a $C(K)''$-module multiplication on X' and X''. Since we will not need the use of this result other than in the above form, we will not introduce Arens extensions here. However, we encourage the interested reader to look up the straightforward exposition and the proof in [5, pp. 75–76].
(b) The property (1) in Lemma 2.5 is called the **weakly compact action** of $m(C(K))$ on X. We will use this term in the sequel.

3 Dedekind Complete and σ-Dedekind Complete Cyclic Banach Spaces

In this section we will characterize cyclic Banach spaces which, when represented as Banach lattices in accordance with Lemma 2.4 are Dedekind complete or σ-Dedekind complete. A characterization of σ-Dedekind complete cyclic Banach spaces was obtained by Veksler in [14]. Here we will simplify Veksler's proof by using Lemmas 2.1 and 2.4. But we need to emphasize that the main idea is still the one used by Veksler.

For the remainder of this section we will assume that the compact space K is **totally disconnected**. We will denote by $\mathcal{B}$ the Boolean algebra of all the idempotents in $C(K)$. Clearly, $\mathcal{B}$ consists of characteristic functions of the clopen subsets of K. The identically one function $\mathbf{1}$ and the identically zero function $\mathbf{0}$ correspond to the identity and the zero of the Boolean algebra $\mathcal{B}$. Let $m : C(K) \to L(X)$ be a bounded unital algebra homomorphism, notice that the closed subset K' of K defined in Section 2 as the set of common zeros of functions from $\ker m$, is also a totally disconnected compact Hausdorff space with the relative topology inherited from K.

Definition 3.1 Let $x \in X$. An idempotent $e_x \in \mathcal{B}$ is called the **carrier projection** of x if $m(e_x)x = x$ and

$$e \in \mathcal{B},\ m(e)x = x \Rightarrow e_x \le e \text{ in } C(K).$$

Remark 3.2 Such a projection is called by Veksler the support of x (see [14]). It is worth noticing that when $m(\mathcal{B})$ is a Bade complete Boolean algebra of projections on X (see Definition 4.2 below) the above definition coincides with the Bade's definition of carrier projection in [3].

Lemma 3.3 *Let $x \in X$ and e_x be the carrier projection of x. Then for any $a \in C(K)$, $m(a)x = 0$ if and only if $ae_x = 0$.*

Proof Suppose $ae_x = 0$. Then $m(a)x = m(a)m(e_x)x = m(ae_x)x = 0$.

Conversely, initially suppose that for some $e \in \mathcal{B}$, $m(e)x = 0$. Then $m((1 - e))x = x$ and therefore $e_x \leq 1 - e$. It follows that $ee_x = 0$. Now assume that for some non-negative $a \in C(K)$ we have $m(a)x = 0$ and that for some $t \in K$, $a(t) > 0$. Then, since K is totally disconnected, there are some $\varepsilon > 0$ and $e \in \mathcal{B}$ such that $e(t) = 1$ and $0 \leq \varepsilon e \leq a$. Then by Lemma 2.1 (1), $\varepsilon\|m(e)x\| \leq \|m(a)x\| = 0$. Therefore $ee_x = 0$ and $e_x(t) = 0$, hence $ae = 0$.

Finally, if $m(a)x = 0$ for some $a \in C(K)$, then applying again Lemma 2.1 (1) we see that $m(|a|)x = 0$ hence $|a|e_x = 0$, and therefore $ae_x = 0$. □

Definition 3.4 We will say that a Banach $C(K)$-module X is a **Veksler module** if any $x \in X \setminus \{0\}$ has a carrier projection $e_x \in C(K)$.

Remark 3.5 Cyclic Banach spaces with the property stated in Definition 3.4 were introduced by Veksler in [14]. They were also considered by Rall [12] who called the corresponding Boolean algebra $\mathcal{B}$ τ-complete.

Assume that X is a Veksler module. For every $x \in X \setminus \{0\}$ we will denote the clopen support of e_x in K by K_x. Clearly, K_x is the Stone representation space of the Boolean algebra $e_x\mathcal{B}$.

Let us recall that a compact Hausdorff space K is called quasi-Stonian if it is basically disconnected, i.e. the closure of every open F_σ set in K is open. It is also worth recalling that the following properties are equivalent:

1. K is quasi-Stonian.
2. $C(K)$ is σ-Dedekind complete.
3. Every non-negative sequence bounded from above in $C(K)$ has a supremum in $C(K)$.
4. Every principal band in $C(K)$ is a projection band.

Lemma 3.6 *Let X be a Veksler module with respect to the Boolean algebra $\mathcal{B}$. Then for every $x \in X \setminus \{0\}$, the Banach algebra $C(K_x)$ is a Veksler module with respect to the Boolean algebra $e_x\mathcal{B}$. Moreover, the compact space K_x is quasi-Stonian.*

Proof Let $x \in X$ and let $e_x \in \mathcal{B}$ be the corresponding carrier projection. Let $a \in C(K_x)$. We extend a as a continuous function on K by letting $a \equiv 0$ on $K \setminus K_x$. We will identify a and its extension on K. We claim that $e_{m(a)x}$ is the carrier projection for a in $e_x\mathcal{B}$. Note that $m(1 - e_x)m(a)x = 0$. Therefore $(1 - e_x)e_{m(a)x}=0$ and $e_{m(a)x} \in e_x\mathcal{B}$. Now suppose that for some $b \in C(K_x)$, $ba = 0$. Then $m(b)m(a)x = 0$, and consequently $be_{m(a)x} = 0$. Conversely, suppose $be_{m(a)x} = 0$. Then $m(b)m(a)x = 0$ and therefore $bae_x = 0$. Hence $e_{m(a)x}$ is the carrier projection of $a \in C(K_x)$. This also means that $\{a\}^{\perp} = (1 - e_{m(a)x})C(K_x)$. Therefore the principal band $\{a\}^{\perp\perp} = e_{m(a)x}C(K_x)$ is a projection band and the space K_x is quasi-Stonian. □

The critical property of cyclic subspaces of a Veksler module is stated in the next lemma due to Veksler [14]. For the sake of completeness we will give a simplified proof of this result.

Lemma 3.7 *Let X be a cyclic Veksler module with a cyclic vector x_0. If $x_n \to x$, $y_n \to y$ and for each n, $e_{x_n} e_{y_n} = 0$ then $e_x e_y = 0$.*

Proof Because m is an isometry throughout this proof we will identify $a \in C(K)$ and $m(a) \in L(X)$. We assume that by Lemma 2.4, X has been represented as a Banach lattice with quasi interior point x_0. Since $C(K) \subseteq Z(X)$, we have for some $a \in C(K)$ and $x \in X$ that $ax = 0$ if and only if $a|x| = 0$. Therefore the carrier projections of x and $|x|$ are the same. So it is sufficient to consider non-negative sequences $\{x_n\}$ and $\{y_n\}$. Also, since $x_n \to x$ implies $e_x x_n \to e_x x = x$, we can assume that $e_{x_n} \le e_x$ and $e_{y_n} \le e_y$. Next we notice that $e_{x_n} x \to x$ and $e_{y_n} y \to y$. Indeed,

$$\|e_{x_n} x - x\| \le \|e_{x_n}(x - x_n)\| + \|x_n - x\| \le 2\|x_n - x\|.$$

We may assume that $\|x - e_{x_n} x\| < \frac{1}{2^n}$ and $\|y - e_{y_n} y\| < \frac{1}{2^n}$ for $n \in \mathbb{N}$. For any $m \in \mathbb{N}$ consider the decreasing sequence $x_n^m = (e_{x_m} e_{x_{m+1}} \dots e_{x_n})x$ where $n \ge m$. Let $m \le n$ and $p \in \mathbb{N}$. Then

$$x_n^m - x_{n+p}^m = (e_{x_m} e_{x_{m+1}} \dots e_{x_n})[(1 - e_{x_{n+1}}) + e_{x_{n+1}}(1 - e_{x_{n+2}}) + \dots$$

$$+ e_{x_{n+1}} e_{x_{n+2}} \dots e_{x_{n+p-1}}(1 - e_{x_{n+p}})]x.$$

Therefore $\|x_n^m - x_n^{m+p}\| \le \frac{1}{2^n}$ and we can assume that $x_n^m \underset{n\to\infty}{\to} x^m \in X$. Similarly $y_n^m \to y^m \in X$. Since both sequences are decreasing, for all $n \ge m$ we have $e_{x^m} \le e_{x_n}$ and $e_{y^m} \le e_{y_n}$.

It is clear that $\{x^m\}$ and $\{y^m\}$ are increasing sequences and that x and y are upper bounds for each sequence, respectively. Next consider for any fixed m the increasing sequence $\{x - x_n^m\}_{n \ge m}$. We have

$$x - x_n^m = [(1 - e_{x_m}) + e_{x_m}(1 - e_{x_{m+1}}) + e_{x_m} e_{x_{m+1}}(1 - e_{x_{m+2}}) + \dots$$

$$+ e_{x_m} e_{x_{m+1}} \dots e_{x_{n-1}}(1 - e_{x_n})]x.$$

It follows easily that $\|x - x_n^m\| < \frac{1}{2^{m-1}}$. Hence, passing to the limit, we have $\|x - x^m\| \le \frac{1}{2^{m-1}}$. Therefore $x^m \to x$ and $y^m \to y$. Since $e_{y^m} e_{x^n} = 0$ for all $n \ge m$, we have $e_{y^m} x^n = 0$, $n \ge m$. Hence, $e_{y^m} x = 0$ and therefore $e_{y^m} e_x = 0$ for all m. Then $e_x y^m = 0$ for all m, hence $e_x y = 0$. It follows that $e_x e_y = 0$. □

Lemma 3.8 (Veksler) *A cyclic Veksler module represented as a Banach lattice is σ-Dedekind complete.*

Proof It follows from Lemmas 3.7 and 3.6 that $m(C(K))$ is weak-operator closed. To see this, note that by Lemma 2.4, the weak operator closure of $C(K)$ is $Z(X) = C(\hat{K})$. Since $\mathcal{B} \subset C(K) \subseteq C(\hat{K})$ it is sufficient to prove that $\mathcal{B}$ separates the points of $\hat{K}$. Let s, t be two distinct points in $\hat{K}$. Then there is $\Phi \in C(\hat{K})$ such that

$-1 \leq \Phi \leq 1$, $\Phi(s) = 1$, and $\Phi(t) = -1$. Let x_0 be a cyclic vector in X. Without loss of generality we will assume that the carrier projection of x_0 is $\mathbf{1} \in \mathcal{B}$. There is a sequence $\{a_n\}$ in $C(K)$ such that $\|a_n x_0 - \Phi x_0\| \to 0$. Since $|a_n^+ - \Phi^+| \leq |a_n - \Phi|$ and $|a_n^- - \Phi^-| \leq |a_n - \Phi|$, where the lattice operations are considered in $C(\hat{K}) = Z(X)$, we have $\|(a_n^+ - \Phi^+)x_0\| \leq \|(a_n - \Phi)x_0\|$ and $\|(a_n^- - \Phi^-)x_0\| \leq \|(a_n - \Phi)x_0\|$. Thus $a_n^+ x_0 \to \Phi^+ x_0$ and $a_n^- x_0 \to \Phi^- x_0$ in norm in X. Let $e_n, e_n' \in \mathcal{B}$ be the carrier projections of a_n^+ and a_n^- in $C(K)$, respectively (they exist because K is quasi-Stonian). Then clearly $e_n' \leq 1 - e_n$. By Lemma 3.6, $e_{a_n^+ x_0} = e_n$ and $e_{a_n^- x_0} = e_n'$. Let e and e' be carrier projections in $\mathcal{B}$ of $\Phi^+ x_0$ and $\Phi^- x_0$, respectively. By Lemma 3.7, $ee' = 0$. So $e\Phi^- x_0 = 0$ and, because x_0 is a $C(K)$-cyclic vector, $e\Phi^- = 0$, as an operator in $Z(X)$, hence $e(t) = 0$. Moreover, $(1 - e)\Phi^+ x_0 = 0$ hence $(1 - e)\Phi^+ = 0$ and $e(s) = 1$. Thus, $K = \hat{K}$ and $Z(X) = C(K)$.

Assume that $x_n \in X_+$ and $x_n \leq y \in X$, $n \in \mathbb{N}$. Let $z = x_0 \vee y$, then z is a quasi-interior element in X and $x_n \in I_z$ where I_z is the order ideal in X generated by the quasi-interior element z. Then, by Lemma 2.4(5) (actually, by the Krein-Kakutani theorem), $I_z = Z(X)z$. Hence there are $a_n \in C(K) = Z(X)$ such that $0 \leq a_n \leq 1$ and $x_n = a_n z$. Recall that K is quasi-Stonian, hence $C(K)$ is σ-Dedekind complete. Let $a = \sup_{n \in \mathbb{N}} a_n \in C(K)$. It is immediate to see that $az = \sup_{n \in \mathbb{N}} a_n z \in X$. □

It is well known that if E is a Dedekind complete vector lattice its center $Z(E) = C(K)$ is also Dedekind complete, hence K is a Stonian (extremally disconnected) compact space. An analog of this property for Banach $C(K)$-modules was first considered by Kaplansky [6].

Definition 3.9 A Banach $C(K)$-module X is called a **Kaplansky module** if it satisfies the following two conditions:

1. The compact space K is Stonian.
2. For any $x \in X$ and for any non-negative set $\{a_\alpha\}$ bounded above in $C(K)$ the following implication holds:

$$a_\alpha x = 0, \text{ for all } \alpha \Rightarrow ax = 0, \text{ where } a = \sup a_\alpha.$$

It is easy to see that any Kaplansky module is also a Veksler module and that any Dedekind complete Banach lattice X is a Kaplansky module over its center $Z(X)$. The lemma below shows that the converse of the last statement is true for cyclic Kaplansky modules.

Lemma 3.10 *Let X be a cyclic Kaplansky module over $C(K)$ and let $x_0 \in X$ be a cyclic element of X. Then when represented as a Banach lattice with the quasi-interior point x_0, X is Dedekind complete.*

Proof Because, as a Kaplansky module, X is also a Veksler module we can apply Lemma 3.8 to conclude that $Z(X) = C(K)$. Note that $C(K)$ is Dedekind complete. Then repeat the argument from the proof of Lemma 3.8 to see that any subset of X_+ that is bounded above has a supremum in X. □

4 Order Continuous Banach $C(K)$-Modules

In this section we will extend to Banach $C(K)$-modules the following classic result of Lozanovsky [10].

Theorem 4.1 *Let E be a σ-Dedekind complete Banach lattice. The following conditions are equivalent:*

1. *The original lattice norm on E is order continuous.*
2. *E does not contain ℓ^∞ as a closed subspace.*
3. *E does not contain ℓ^∞ as a closed sublattice.*

Throughout this section we assume that X is a Banach space either over $\mathbb{R}$ or over $\mathbb{C}$, K is a compact Hausdorff space, and $m : C(K) \to L(X)$ is a bounded injective unital algebra homomorphism (and therefore an isometry).

Definition 4.2 Let K be a Stonian compact Hausdorff space and $\mathcal{B}$ is the Boolean algebra of all idempotents in $C(K)$. We say that $\mathcal{B}$ is a **Bade-complete Boolean algebra of projections** on X if for any increasing net $\{e_\alpha\}$ in $\mathcal{B}$ and for any $x \in X$ we have $\|(e - e_\alpha)x\| \to 0$, where $e = \sup e_\alpha$.

Remark 4.3 The above definition is equivalent to the definition of a complete Boolean algebra of projections on a Banach space as given by Bade [3].

We start with our main result concerning cyclic Banach spaces.

Theorem 4.4 *Let X be a cyclic Banach $C(K)$-module, x_0 be a cyclic vector in X, and let $\mathcal{B}$ be the Boolean algebra of all idempotents in $C(K)$. The following conditions are equivalent:*

1. *The compact space K is totally disconnected, X is a Veksler module, and X does not contain a copy of ℓ^∞.*
2. *The compact space K is Stonian, X is a Kaplansky module, and X does not contain a copy of ℓ^∞.*
3. *$m(C(K))$ is weak operator closed, and X, when represented as a Banach lattice, has order continuous norm.*
4. *The compact space K is Stonian and $\mathcal{B}$ is a Bade complete Boolean algebra of projections on X.*
5. *The compact space K is hyperstonian and m is ($w^\star$, weak-operator)-continuous.*

Proof (2) $\Rightarrow$ (1). This implication follows from the fact that a Kaplansky module is a Veksler module.

(1) $\Rightarrow$ (3). Note that when X is represented as a Banach lattice with a quasi-interior point x_0, it is σ-Dedekind complete and $m(C(K)) = Z(X)$ (see Lemma 3.8 and its proof). Therefore $m(C(K))$ is weak operator closed. Next, since X does not contain any copy of ℓ^∞, X has order continuous norm by Theorem 4.1.

(3) $\Rightarrow$ (2). It is well known that X has order continuous norm implies that X is Dedekind complete and by Theorem 4.1 it does not contain a copy of ℓ^∞.

(3) $\Rightarrow$ (4). Since $m(C(K))$ is weak operator closed, by Lemma 2.4 (3) we have $m(C(K)) = Z(X)$. As we have already noticed, X is Dedekind complete. Therefore $Z(X)$ is Dedekind complete and K is Stonian. Moreover, if $\{e_\alpha\}$ is an increasing net in $\mathcal{B}$ and $e = \sup e_\alpha$ then for any non-negative $x \in X$ we have $\sup e_\alpha x = ex$. Therefore, $\|e_\alpha x - ex\| \to 0$. It follows that $\mathcal{B}$ is a Bade complete Boolean algebra of projections on X.

(5) $\Rightarrow$ (3). The fact that K is hyperstonian implies that $C(K)$ is a dual Banach lattice. Let us denote its predual by $C(K)_\star$. The definition of order in $C(K)$ by means of its predual implies that whenever $0 \leq a_\alpha$ is an increasing net and $\sup a_\alpha = a$ in $C(K)$ then $a_\alpha \to a$ in the $w^\star$ topology in $C(K)$. Since m is ($w^\star$, weak-operator)-continuous , we have $a_\alpha x \to ax$ in the weak topology in X. In particular if for every α, $a_\alpha x = 0$ then $ax = 0$ and therefore X is a Kaplansky module. Hence, by Lemma 3.10, when represented as a Banach lattice, X is Dedekind complete and $m(C(K)) = Z(X)$. Next assume that $\{x_\alpha\}$ is an increasing net bounded above with non-negative elements in X and that $x = \sup x_\alpha$. Then there is an increasing net $\{a_\alpha\}$ of non-negative elements in $C(K)$ such that $x_\alpha = a_\alpha x$. It is immediate to see that $\sup a_\alpha e_x = e_x$, the carrier projection of x. Then the net $\{a_\alpha e_x\}$ converges to e_x in the $w^\star$ topology in $C(K)$ and therefore, $\{x_\alpha\}$ converges to x in the weak topology in X. Duality implies that there is a sequence $\{y_n\}$ in X such that every y_n is a convex combination of elements from the net $\{x_\alpha\}$ and $\|y_n - x\| \to 0$. Since $\{x_\alpha\}$ is an increasing net of non-negative elements, it follows that the net $\{x_\alpha\}$ converges to x in norm, hence the norm on X is order continuous.

(4) $\Rightarrow$ (5). Suppose $x \in X$, $x' \in X'$ and define $\mu_{x,x'} \in C(K)'$ as

$$\mu_{x,x'}(a) = x'(ax), \ a \in C(K).$$

We want to show that each such functional is order continuous on $C(K)$. Because the set of these functionals is clearly total on $C(K)$, it would follow that K is hyperstonian. It is sufficient to show that for any closed nowhere dense subset D of K we have $\mu_{x,x'}(D) = 0$ where we identify $\mu_{x,x'}$ with the corresponding finite regular Borel measure on K. Initially assume that $\mu_{x,x'} \geq 0$. Let us fix D and let us consider the collection $\{\tau\}$ of clopen subsets of K disjoint from D and ordered by inclusion. Let $e_\tau \in \mathcal{B}$ be the characteristic function of the subset τ. Then $\{e_\tau\}$ is an increasing net in $\mathcal{B}$. Because the union of all the sets in the collection $\{\tau\}$ is $K \setminus D$, an open dense subset of K, we have $\sup e_\tau = 1$. Recalling that $\mathcal{B}$ is Bade complete we see that $\|(1 - e_\tau)x\| \to 0$. From the inequalities

$$0 \leq \mu_{x,x'}(D) \leq \mu_{x,x'}(K \setminus \tau) = \mu_{x,x'}(1 - e_\tau) = x'((1 - e_\tau)x)$$

we conclude that $\mu_{x,x'}(D) = 0$.

Now denote by X_R the Banach space X considered as a real Banach space and by X'_R the space of all real valued bounded linear functionals on X_R. We claim that for any $x' \in X'_R$ and for any $x \in X$ there exists $e_+ \in \mathcal{B}$ such that $\mu_{e_+x,x'} = \mu^+_{x,x'}$

and $-\mu_{(1-e_+)x,x'} = \mu^-_{x,x'}$. Given $\mu_{x,x'}$ let us, without loss of generality, assume that both $\mu^+_{x,x'}$ and $\mu^-_{x,x'}$ are not equal to zero. We can assume, without loss of generality, that for some $e \in \mathcal{B}$ we have $\mu_{x,x'}(e) > 0$ and that the set $\mathcal{N} = \{e' \in e\mathcal{B} : \mu_{x,x'}(e') < 0\}$ is not empty. By Zorn's lemma there is a maximal collection Ω of pairwise disjoint elements of $\mathcal{N}$. Let $\{e_\alpha\}$ be an increasing net in $\mathcal{B}$ formed by finite sums of elements of Ω. Let $e' \in \mathcal{B}$ be the supremum of $\{e_\alpha\}$. Since $\mathcal{B}$ is Bade complete on X, we have $\|(e' - e_\alpha)x\| \to 0$. Also, since for every α, $\mu_{x,x'}(e_\alpha) < 0$ and the net $\{\mu_{x,x'}(e_\alpha)\}$ is decreasing, we have

$$\mu_{x,x'}(e') = \lim \mu_{x,x'}(e_\alpha) < 0.$$

Then for any projection $p \in (e - e')\mathcal{B}$ we have $\mu_{x,x'}(p) \geq 0$. Indeed, otherwise we arrive at a contradiction with the maximality of Ω. Thus we have proved that the set

$$\mathcal{P} = \{e \in \mathcal{B} : e' \in e\mathcal{B} \Rightarrow \mu_{x,x'}(e') \geq 0\}$$

is not empty. Applying Zorn's lemma again, we can find a maximal collection $\mathcal{E}$ of pairwise disjoint elements in $\mathcal{P}$. Let $e_+ = \sup_{e \in \mathcal{E}} e \in \mathcal{B}$. We claim that $e_+ \in \mathcal{P}$. Indeed, let $e' \in \mathcal{B}$. Then for any $e \in \mathcal{E}$ we have $\mu_{x,x'}(e'e) \geq 0$. Bade completeness of $\mathcal{B}$ on X implies that $\mu_{x,x'}(e'e_+) \geq 0$, hence $e_+ \in \mathcal{P}$. It is easy to conclude from the maximality of $\mathcal{E}$ that for any idempotent $e \in (1 - e_+)\mathcal{B}$, we have $\mu_{x,x'}(e) \leq 0$. It follows that both $\mu_{e_+x,x'}$ and $-\mu_{(1-e_+)x,x'}$ are positive linear functionals on $C_R(K)$ (the space of all real-valued continuous functions on K). Hence, as we have already proved, both of them, and therefore their difference $\mu_{x,x'}$ are order continuous on $C_R(K)$. Since the map m is one-to-one, it is clear the set of functionals $\{\mu_{x,x'} : x \in X_R, x' \in X'_R\}$ is total on $C_R(K)$. Therefore K is hyperstonian and $C(K)$ is a dual Banach lattice. Since each linear functional in X' is a linear combination over $\mathbb{C}$ of functionals from X'_R every functional $\mu_{x,x'}, x \in X, x' \in X'$ is order continuous on $C(K)$ and therefore belongs to its predual $C(K)_\star$. Therefore if $a_\alpha \to a$ in $w^\star$-topology in $C(K)$ then $m(a_\alpha) \to m(a)$ in the weak operator topology in $L(X)$. □

We are now ready to state and prove our main result for general Banach $C(K)$-modules. In connection with this we would like to make the following remark:

Remark 4.5

(a) In condition (3) of Theorem 4.4 we cannot dispense with the requirement that $m(C(K))$ is weak-operator closed. Indeed, c_0 is a cyclic Banach c-module and has order continuous norm. But $Z(c_0) = w - cl(m(c_0)) = \ell^\infty \neq m(c_0)$.
(b) As the reader will see, in our next result, Theorem 4.6, where we consider general Banach $C(K)$-modules, we assume from the very beginning that $m(C(K))$ is weak-operator closed. The following example illustrates that we have to be careful with regard to this condition. We consider c as a Banach c-module; then $m(c) = Z(c)$ is closed in the weak operator topology in $L(c)$ but when we restrict m to the cyclic subspace c_0 as we have just seen $m(c)$ is not weak operator closed in $L(c_0)$.

Theorem 4.6 *Let $m : C(K) \to L(X)$ be an injective bounded unital algebra homomorphism such that $m(C(K))$ is weak-operator closed. The following conditions are equivalent:*

1. *(a) K is totally disconnected and*
   (b) X is a Veksler module, and
   (c) no cyclic subspace of X contains a copy of ℓ^∞.
2. *(a) K is Stonian and*
   (b) X is a Kaplansky module, and
   (c) no cyclic subspace of X contains a copy of ℓ^∞.
3. *Each cyclic subspace of X when represented as a Banach lattice has order continuous norm.*
4. *(a) K is Stonian and*
   (b) the set $\mathcal{B}$ of all idempotents in $C(K)$ is a Bade complete Boolean algebra of projections on X.
5. *(a) K is hyperstonian and*
   (b) the map m is ($w^\star$, weak operator) continuous.

Proof (2) $\Rightarrow$ (1). This implication is trivial.

(1) $\Rightarrow$ (3). It follows from (1) and Lemma 3.8 that for any $x \in X$ the cyclic subspace $X(x)$ when represented as a Banach lattice is σ-Dedekind complete and $Z(X(x)) = C(K_x)$ where K_x is the support of e_x in K. Because $X(x)$ does not contain a copy of ℓ^∞ by Theorem 4.1 ((2) $\Rightarrow$ (1)), the implication is proved.

(3) $\Rightarrow$ (5). First we claim that (3) implies that $m(C(K))$ has weakly compact action on X. Let $x \in X$. Then $X(x)$ has order continuous norm when represented as a Banach lattice. Notice that for every $a \in C(K)$ the cyclic space $X(x)$ is invariant for the operator $m(a)$. Then the correspondence $a \to m(a)|X(x)$ defines the algebraic homomorphism $m_x : C(K) \to L(X(x))$. Let $C(K_x) \cong C(K)/\ker m_x$ and let $\dot{m}_x : C(K_x) \to L(X(x))$ be the unital injective algebra homomorphism induced by m_x. By Lemma 2.4 the closure of $\dot{m}_x(C(K_x))$ in the weak operator topology in $L(X(x))$ can be identified with $Z(X(x))$. Let L be the Stone representation space of $Z(X(x))$. Clearly $C(K_x)$ can be identified with a closed subalgebra of $C(L)$, the map $\dot{m}_x$ can be extended to an algebraic unital injective homomorphism $\tilde{m}_x : C(L) \to L(X(x))$, and $\tilde{m}_x(C(L)) = Z(X)$ hence $\tilde{m}_x(C(L))$ is weak operator closed in $L(X(x))$. Applying the implication (3) $\Rightarrow$ (5) from Theorem 4.4 we obtain that the Stone representation space L of $Z(X(x))$ is hyperstonian and the embedding $\tilde{m}_x : C(L) \to L(X(x))$ is ($w^\star$, weak-operator)-continuous. Because $\tilde{m}_x(a)x = m(a)x$, when $a \in C(K)$, we have that for any $x \in X$ the map $a \to m(a)x$ is weakly compact, and thus the claim is proved.

Then, by Lemma 2.5 there is a unique extension of m, $\hat{m} : C(K)'' \to L(X)$ that is ($w^\star$, weak-operator)-continuous. Since $C(K)''$ is a dual AM-space its Stone representation space is hyperstonian. Because $\ker \hat{m}$ is a $w^\star$-closed ideal in $C(K)''$ we have that $C(K)''/\ker \hat{m} = C(S)$ where S is a hyperstonian compact space. This means that as well as being an isometric unital algebra homomorphism the map $\hat{m} : C(S) \to L(X)$ is ($w^\star$, weak-operator) continuous. Therefore X is a Kaplansky $C(S)$-module. We claim that this implies that $\hat{m}(C(S))$ is weak-operator closed in $L(X)$. For each $x \in X$, let e_x denote the carrier projection of x and S_x the clopen subset of S that is the support of e_x. Note that $C(S_x) = e_x C(S) = C(S)/(\ker \hat{m}|X(x))$. Since $\ker \hat{m}|X(x)$ is $w^\star$-closed, $C(S_x)$ is a dual Banach space and S_x is hyperstonian. By Lemma 3.10, $X(x)$ when represented as a Banach lattice is Dedekind complete and $Z(X(x)) = C(S_x)$ is weak-operator closed in $L(X(x))$. Consider $T \in L(X)$ which is in the weak-operator closure of $\hat{m}(C(S))$. Then for any $x \in X$ the cyclic subspace $X(x)$ is T-invariant and $T|X(x) \in Z(X(x))$. Therefore for each $x \in X$ there is $a_x \in C(S_x) = e_x C(S)$ such that $T(x) = \hat{m}(a_x)x$. Then as proved (by a standard argument) in [4, Lemma 2] there is $a \in C(S)$ such that $e_x a = a_x, x \in X$ and therefore $T(x) = \hat{m}(a)x, x \in X$. Thus $T \in \hat{m}(C(S))$ and as claimed $\hat{m}(C(S))$ is weak-operator closed. But $m(C(K))$ is weak-operator closed, $C(K)$ is $w^\star$-dense in C(S), and $\hat{m}$ is ($w^\star$, weak operator)- continuous. Hence $m(C(K)) = \hat{m}(C(S))$ and K is homeomorphic to S.

(5) $\Rightarrow$ (4). Suppose that $\{e_\alpha\}$ is an increasing net in $\mathcal{B}$ such that $\sup e_\alpha = e \in \mathcal{B}$. Then $e_\alpha \to e$ in the $w^\star$ topology in $C(K)$. Let $x \in X$. Because m is ($w^\star$, weak-operator)-continuous we have $e_\alpha x \to ex$ in the weak topology in X. Duality implies that for a sequence $\{a_n\}$ of convex combinations of e_α we have $\|ex - a_n x\| \to 0$. But the net $\{e_\alpha\}$ is increasing and therefore for a fixed n and for all sufficiently large α we have $\|ex - e_\alpha x\| \leq \|ex - a_n x\|$. Hence $\|ex - e_\alpha x\| \to 0$ and $\mathcal{B}$ is a Bade complete Boolean algebra of projections on X.

(4) $\Rightarrow$ (3). It follows immediately from (4) that X is a Kaplansky Banach $C(K)$-module. It remains to notice that for each $x \in X$ the Boolean algebra $e_x\mathcal{B}$ is Bade complete on $X(x)$ and apply the implication (4) $\Rightarrow$ (3) from Theorem 4.4.

(4) $\Rightarrow$ (2) We have already proved that (4) implies that X is a Kaplansky module and that each cyclic subspace $X(x)$ when represented as a Banach lattice has order continuous norm. By Theorem 4.1, $X(x)$ does not contain any copy of ℓ^∞. □

Our next theorem relates to the case of finitely generated Banach $C(K)$-modules. It would be just a corollary of Theorem 4.6 except for the fact that it contains two new equivalences that are not true in the general case.

Recall (see [7]) that a Banach $C(K)$ module is called **finitely generated** if there are $x_1, \dots, x_n \in X$ such that the linear span of cyclic subspaces $X(x_i), i = 1, \dots, n$ is dense in X. The elements $x_1, \dots, x_n$ are called **generators** of X.

Theorem 4.7 *Let X be a finitely generated Banach $C(K)$-module such that m is injective and $m(C(K))$ is weak operator closed in $L(X)$. Then the conditions (1)–(5) of Theorem 4.6 are equivalent to the following two conditions:*

- $(1A)$ *K is totally disconnected, X is a Veksler module, and X does not contain a copy of ℓ^∞.*
- $(2A)$ *K is Stonian, X is a Kaplansky module, and X does not contain a copy of ℓ^∞.*

Proof Equivalence of (1)–(5) follows from Theorem 4.6. It is obvious that $(2A) \Rightarrow (1A) \Rightarrow (1)$. Therefore the theorem will be proved to show that (2A) follows from conditions (1)–(5).

Hence we can assume that K is hyperstonian, $m : C(K) \to L(X)$ is ($w^\star$, weak-operator)-continuous, and the Boolean algebra $\mathcal{B}$ of all idempotents in $C(K)$ is Bade complete on X. Furthermore, no cyclic subspace of X contains a copy of ℓ^∞. Under these conditions we have to show that X does not contain a copy of ℓ^∞. We will prove it by induction on the number of generators of X. When $n = 1$, X is cyclic and therefore does not contain a copy of ℓ^∞. Suppose that whenever X has n generators the conditions (1)–(5) imply (2A). Let X have a minimum of $n + 1$ generators $\{x_0, x_1, \dots, x_n\}$. Consider the submodule Y of X with the generators $\{x_1, \dots, x_n\}$. The map m generates in an obvious way the map $m_Y : C(K) \to L(Y)$. Clearly $\ker m_Y$ is $w^\star$-closed and therefore $C(K)/\ker m_Y \cong C(K_Y)$ where K_Y is a clopen subset of K. Hence $C(K_Y)$ is a dual Banach lattice. Let e_Y be the characteristic function of K_Y. Then $e_Y\mathcal{B}$ is Bade complete on Y. Therefore $m_Y(C(K_Y))$ is weak-operator closed in $L(Y)$ and Y satisfies conditions (1)–(5). Then by the induction hypothesis no closed subspace of Y is isomorphic to ℓ^∞. Next we consider the factor X/Y with elements $[x] = x + Y, x \in X$. Because Y is a $C(K)$-submodule of X the map $m_{X/Y} : C(K) \to L(X/Y)$ is well defined. It is clear that $\ker m_{X/Y}$ is a $w^\star$-closed ideal in $C(K)$ and therefore $C(K)/\ker m_{X/Y} \cong C(K_0)$ where K_0 is a clopen subset of K. Moreover $C(K_0)$ is a dual Banach lattice. Notice that $X/Y = X/Y([x_0])$ is a cyclic Banach space and because the map $m_{X/Y} : C(K_0) \to L(X/Y)$ is ($w^\star$, weak-operator)-continuous we see that the Boolean algebra $e_0\mathcal{B}$, where e_0 is the characteristic function of K_0, is Bade complete on X/Y. Hence by Theorem 4.4 X/Y has order continuous norm when represented as a Banach lattice. Then by Theorem 4.1 X/Y does not contain any copy of ℓ^∞. Since not containing ℓ^∞ is a three-space property [2], X does not contain a copy of ℓ^∞, and the proof is complete. □

Remark 4.8 In Theorem 4.7 we cannot dispense with the condition that X is finitely generated. Indeed, ℓ^∞ considered as a $\mathbb{C}$-module satisfies conditions (1)–(5) of Theorem 4.6 but it does contain a copy of itself.

Our final result relates to the dual Radon–Nikodym property and is a corollary of Theorem 4.7 and [9, Theorem 3.4].

Theorem 4.9 *Let K be a totally disconnected compact space and X be a finitely generated Veksler $C(K)$-module. Assume also that $m(C(K))$ is weak-operator closed in $L(X)$. Then the following conditions are equivalent:*

1. *X' has the Radon–Nikodym property.*
2. *X does not contain a copy of ℓ^1.*
3. *No cyclic subspace of X contains a copy of ℓ^1.*

Proof Without loss of generality we may assume that m is injective. The implications (1) $\Rightarrow$ (2) $\Rightarrow$ (3) are trivial. Assume (3). Since ℓ^∞ contains a copy of ℓ^1 no cyclic subspace of X contains a copy of ℓ^∞. By Theorem 4.7 the Boolean algebra $\mathcal{B}$ of all idempotents in $C(K)$ is Bade complete. Now Theorem 3.4 ((3) $\Rightarrow$ (1)) in [9] implies that X' has the Radon–Nikodym property. □

References

1. Yu.A. Abramovich, E.L. Arenson, A.K. Kitover, in *Banach $C(K)$-Modules and Operators Preserving Disjointness*. Pitman Research Notes in Mathematical Series, vol. 277 (Longman Scientific and Technical, 1992)
2. J.M.F. Castillo, M. Gonzalez, in *Three-Space Problems in Banach Space Theory*. Lecture Notes in Mathematics, vol. 1667 (Springer, 1997)
3. N. Dunford, J.T. Schwartz, *Linear Operators, Part III: Spectral Operators* (Wiley, New York, 1971)
4. D. Hadwin, M. Orhon, Reflexivity and approximate reflexivity for Boolean algebras of projections. J. Funct. Anal. **87**, 348–358 (1989)
5. D. Hadwin, M. Orhon, A noncommutative theory of Bade functionals. Glasgow Math. J. **33**, 73–81 (1991)
6. I. Kaplansky, Modules over operator algebras. Am. J. Math. **75**(4), 839–853 (1953)
7. A.K. Kitover, M. Orhon, Reflexivity of Banach $C(K)$-modules via reflexivity of Banach lattices. Positivity **18**(3), 475–488 (2014)
8. A.K. Kitover, M. Orhon, Weak sequential completeness in Banach $C(K)$-modules of finite multiplicity. Positivity **21**(2), 739–753 (2017)
9. A.K. Kitover, M. Orhon, The dual Radon-Nikodym property for finitely generated Banach $C(K)$-modules. Positivity **22**(2), 587–596 (2018)
10. G.Ya. Lozanovskii, On isomorphous Banach structures. Sib. Math. J. **10**(1), 64–68 (1969)
11. H. Önder, M. Orhon, Transitive operator algebras on the n-fold direct sum of a cyclic Banach space. J. Oper. Theory **22**, 99–107 (1989)
12. C. Rall, Uber Boolesche Algebren von Projektionen. Math. Z. **153**(3), 199–217 (1977)
13. H.H. Schaefer, *Banach Lattices and Positive Operators* (Springer, Berlin, 1974)
14. A.I. Veksler, Cyclic Banach spaces and Banach lattices. Sov. Math. Dokl. **14**, 1773–1779 (1973)
15. W. Wnuk, *Banach Lattices with Order Continuous Norm* (PWN, Warszawa, 1999)

Matrix Valued Laguerre Polynomials

Erik Koelink and Pablo Román

Dedicated to Ben de Pagter on the occasion of his retirement. In admiration of his mathematics, personal leadership and calmness

Abstract Matrix valued Laguerre polynomials are introduced via a matrix weight function involving several degrees of freedom using the matrix nature. Under suitable conditions on the parameters the matrix weight function satisfies matrix Pearson equations, which allow to introduce shift operators for these polynomials. The shift operators lead to explicit expressions for the structures of these matrix valued Laguerre polynomials, such as a Rodrigues formula, the coefficients in the three-term recurrence, differential operators, and expansion formulas.

Keywords Matrix polynomial · Orthogonal polynomials · Matrix differential equation · Pearson equation · Shift operator · Rodrigues formula

1 Introduction

Matrix valued polynomials have a long history, introduced by M.G. Krein in the 1940s in a study of the matrix moment problem as well as in the study of operators with higher deficiency indices [33, 34]. Since its introduction several other applications have been linked to matrix valued orthogonal polynomials, such as relations and applications to, e.g., spectral theory of higher order recurrences, scattering theory and differential operators, see, e.g., [3, 12, 14–16]. As in the study

E. Koelink (✉)
IMAPP, Radboud Universiteit, Nijmegen, The Netherlands
e-mail: e.koelink@math.ru.nl

P. Román
CIEM FaMAF, Universidad Nacional de Córdoba, Córdoba, Argentina
e-mail: roman@famaf.unc.edu.ar

G. Buskes et al. (eds.), *Positivity and Noncommutative Analysis*,
Trends in Mathematics, https://doi.org/10.1007/978-3-030-10850-2_16

of classical, i.e. scalar-valued, polynomials there are several possible approaches, and this can be roughly divided in general methods and studying special classes. More general methods can be methods such as studying the general moment problem, general approximation properties, etc. The special classes are typically introduced as arising in or related to the study of other topics, such as solutions to differential operators, in terms of hypergeometric functions, numerical analysis, representation theory, applications in mathematical physics, etc. This is particularly true for the matrix valued case of orthogonal polynomials, but the theory is not as fully developed as that of orthogonal polynomials. We refer to Damanik et al. [7] for an overview of the theory and for references.

Important examples of the special cases are the ones that arise in representation theory via matrix valued spherical functions. Even though the general spherical functions go back at least to the work of Godement in the 1950s, see references in [13], the relation to matrix valued orthogonal polynomials is of a much later time. A first step was made by Koornwinder [32] realizing vector-valued orthogonal polynomials, and next Grünbaum et al. [18] studied matrix valued orthogonal polynomials related to the symmetric pair $(\mathrm{SU}(3), \mathrm{U}(2))$ using differential operators. These papers have been motivational to understand the situation of matrix valued orthogonal polynomials in connection to representation theory, and in particular we refer to [1, 21, 28, 29, 31, 37, 38].

It is well-known that in the scalar-case many polynomials in the Askey scheme, see [4, 25, 26], have an interpretation in representation theory, but typically only for a very limited set of parameters involved. It is then an important question how to extend to a general set of parameters, and this is a non-trivial question since usually the limited set is finite. An important example is the extension of multivariable spherical functions on Riemannian symmetric spaces to the general solutions of integrable systems in terms of Heckman-Opdam functions, see Heckman's lectures in [20]. For the case of the extension of [28, 29] from matrix valued Chebyshev polynomials to matrix valued Gegenbauer polynomials, this is done in [30].

In [30] it turns out that one of the essential features that make the analytic extension work is the appropriate LDU-decomposition of the matrix weight. Here L is a unipotent lower triangular matrix consisting of Gegenbauer polynomials, and the diagonal matrix D is intimately related to the weight of the classical Gegenbauer polynomials. These features make it possible to introduce matrix valued analogues of shift operators, which make it possible to derive properties in an explicit fashion, cf. [20] for the importance of shift operators in the multivariable setting. Moreover, it turns out that the inverse matrix polynomial L has a simple expression in terms of Gegenbauer polynomials, due to Cagliero and Koornwinder [5]. It is then a natural question to find the appropriate analogues for the simplest classical polynomials in the Askey scheme, i.e. the Hermite and Laguerre polynomials. For the Hermite polynomials it is possible to use a limit from Gegenbauer polynomials, and this suggests how to deal with the appropriate matrix valued analogue of the Hermite polynomials, see [23]. It turns out that then the matrix polynomial L is, up to a constant matrix, a matrix exponential, and it is shown that, for certain values of the parameters, it is related to examples by Grünbaum and Durán in [8, 10, 11].

Again, as shown in [23], the shift operators allow us to be explicit in describing the properties of these matrix polynomials.

The purpose of this paper is to describe explicitly the matrix valued analogue of the Laguerre polynomials following the approach sketched above. So we introduce in Sect. 3 the matrix weight function in terms of an LDU-decomposition where the matrix polynomial L is a unipotent lower triangular matrix with entries given in terms of Laguerre polynomials in analogy with the LDU-decomposition of the matrix weight in [30]. The corresponding monic matrix valued Laguerre polynomials are eigenfunctions of a second order matrix valued differential operator which is symmetric with respect to the weight, see Sect. 4. In Sect. 5 we use the degrees of freedom introduced in the matrix weight to impose two non-linear conditions, which allow us to derive Pearson equations for the weight. For specific values of the parameters, the matrix-valued Laguerre polynomials are closely related to the family studied in [9]. In turn, in Sect. 6 this allows us to calculate the squared norm explicitly, and we give an Rodrigues formula to generate the polynomials. We naturally obtain from the Pearson equations a matrix valued differential operator which is factored as lowering and raising operator. We calculate its Darboux transform in terms of the matrix differential operator obtained before. As an application of the shift operators, we give expressions for the matrix coefficients in the three-term recurrence relation and we apply it to find an expansion formula for the matrix valued Laguerre polynomials. We briefly describe three families of solutions to the conditions. In particular, this family of matrix valued orthogonal polynomials fits into the setting of Cantero et al. [6], since the derivative of the matrix valued Laguerre polynomials forms a family of matrix valued orthogonal polynomials as well. However, [6] does not contain any explicit family of arbitrary size of comparable complexity as in this paper. For convenience, Sect. 2 describes some basic results on matrix valued orthogonal polynomials.

As mentioned above, the matrix valued analogue of the Laguerre polynomials in this paper is not motivated by a limit transition of the matrix valued Gegenbauer polynomials [30], which find their origin in group theory and the use of shift operators. This is the case for the Hermite polynomials of [23], and it turns out that the matrix valued Gegenbauer polynomials and matrix valued Hermite polynomials share some features, such as the simple structure of the matrices in the three-term recurrence relation. Indeed, the B_n and C_n of (2.2) are tridiagonal and diagonal, respectively. This allows, with the results of [27, Lemma 3.1], to give a simple proof of the triviality of $\mathcal{A}_W$ and A_W, see Sect. 2. In the case of the matrix valued Laguerre polynomials the expressions for B_n and C_n are explicit, but more complex, see Proposition 6.6 and a proof of the irreducibility along these lines does not seem to work. We conjecture that $\mathcal{A}_W$ and A_W are trivial as well, but we have no full proof. Another important difference is that in the case of the matrix valued Hermite, respectively Gegenbauer, polynomials, the entries of these polynomials can be explicitly calculated in terms of the scalar Hermite polynomials and Hahn polynomials, respectively Gegenbauer and Racah polynomials, see [23, Thm. 3.13], [30, Thm. 3.4]. In the Laguerre setting of this paper, this is not possible because the eigenvalue matrix of Proposition 4.3 is not diagonal.

2 Preliminaries

In this section we give some background on matrix valued orthogonal polynomials and we fix the notation, for more information we refer to, e.g., [7, 10, 17].

We consider a complex $N \times N$ matrix valued integrable function $W: (a, b) \to M_N(\mathbb{C})$, and we assume that $W(x) > 0$ almost everywhere and such that the weight has finite moments of all orders. Here $M_N(\mathbb{C})$ is the algebra of complex $N \times N$-matrices. Integration of a matrix valued function is separate integration of each matrix entry, so that the integral is a matrix. Here we allow the endpoints a and b to be infinite. By $M_N(\mathbb{C})[x]$ we denote the algebra over $\mathbb{C}$ of all polynomials in x with coefficients in $M_N(\mathbb{C})$. The weight matrix W induces a Hermitian sesquilinear form

$$\langle P, Q\rangle = \int_a^b P(x)W(x)Q(x)^* dx \in M_N(\mathbb{C}), \tag{2.1}$$

such that for all $P, Q, R \in M_N(\mathbb{C})[x]$, $T \in M_N(\mathbb{C})$ and $a, b \in \mathbb{C}$ the following properties are satisfied

$$\langle aP+bQ, R\rangle = a\langle P, R\rangle + b\langle Q, R\rangle, \qquad \langle TP, Q\rangle = T\langle P, Q\rangle, \langle P, Q\rangle^* = \langle Q, P\rangle.$$

Moreover

$$\langle P, P\rangle = 0 \iff P = 0,$$

see, for instance, [17]. Given a weight matrix W there exists a unique sequence of monic orthogonal polynomials $\{P_n\}_{n\geq 0}$ in $M_N(\mathbb{C})[x]$. Another sequence of $\{R_n\}_{n\geq 0}$ of orthogonal polynomials in $M_N(\mathbb{C})[x]$ is of the form $R_n(x) = A_n P_n(x)$ for some $A_n \in \mathrm{GL}_N(\mathbb{C})$.

Orthogonal polynomials satisfy a three-term recurrence relation and for monic orthogonal polynomials $\{P_n\}_{n\geq 0}$ we have

$$xP_n(x) = P_{n+1}(x) + B_n(x)P_n(x) + C_n P_{n-1}(x), \quad n \geq 0, \tag{2.2}$$

where $P_{-1} = 0$ and B_n, C_n are matrices depending on n and not on x.

We say that a matrix weight W is reducible to weights of smaller size if there exists a constant matrix M such that $MW(x)M^* = \mathrm{diag}(W_1(x), \ldots, W_k(x))$ for all $x \in [a, b]$, where $W_1, \ldots, W_k$, $k > 1$, are weights of size less than N. In such a case, the real vector space

$$\mathcal{A}_W = \{Y \in \mathrm{Mat}_N(\mathbb{C}) \mid YW(x) = W(x)Y^*, \quad \text{for all } x \in [a, b]\},$$

is non-trivial. On the other hand, if the commutant algebra

$$A_W = \{Y \in \mathrm{Mat}_N(\mathbb{C}) \mid YW(x) = W(x)Y, \quad \text{for all } x \in [a, b]\},$$

is nontrivial, then the weight W is reducible via a unitary matrix M. The relation between $\mathcal{A}_W$ and A_W is the following; if $\mathcal{A}_W$ is $*$-invariant, the weight W reduces to weights of smaller size if and only if the commutant algebra A_W is not trivial, see [27].

3 Matrix Valued Laguerre-Type Polynomials

We introduce a family of matrix valued weight functions by giving an explicit LDU-decomposition, and in Sect. 5 we give conditions on the parameters involved. The matrix polynomial L in this decomposition is a unipotent lower triangular matrix polynomial, whose entries are multiples of Laguerre polynomials independent of the dimension of the matrix. This structure is inspired by a family of matrix valued orthogonal polynomials related to the symmetric pair $(G, K) = (\mathrm{SU}(2) \times \mathrm{SU}(2), \mathrm{SU}(2))$ (K diagonally embedded), which involves a lower triangular matrix whose entries are Gegenbauer polynomials, see [28–30], as well as by the extension to matrix valued q-orthogonal polynomials related to the quantum analogue of (G, K), see [2].

In the paper the integer $N \geq 1$ is fixed. Let $\mu = (\mu_1, \ldots, \mu_N)$ be a sequence of non-zero coefficients and $\alpha > 0$. Then $L_\mu^{(\alpha)}$ is the $N \times N$ unipotent lower triangular matrix defined by

$$L_\mu^{(\alpha)}(x)_{m,n} = \begin{cases} \frac{\mu_m}{\mu_n} L_{m-n}^{(\alpha+n)}(x), & m \geq n, \\ 0 & n < m. \end{cases} \tag{3.3}$$

The Laguerre polynomials $L_n^{(\alpha)}(x)$ can be defined by the generating function

$$(1-t)^{-1-\alpha} \exp\left(\frac{xt}{t-1}\right) = \sum_{n=0}^{\infty} L_n^{(\alpha)}(x)\, t^n, \tag{3.4}$$

see, e.g., [35, 18.12.13], [25, (1.11.10)]. Note that the μ dependence in (3.3) is by conjugation with the matrix $S_\mu = \mathrm{diag}(\mu_1, \ldots, \mu_N)$, so that $L_\mu^{(\alpha)}(x) = S_\mu L^{(\alpha)}(x) S_\mu^{-1}$, where $L^{(\alpha)}(x)$ is (3.3) with $\mu_i = 1$ for all i. It is a direct consequence of the formula $\frac{d}{dx} L_n^{(\alpha)} = -L_{n-1}^{(\alpha+1)}$, see, e.g., [25, (1.11.6)], that the matrix $L^{(\alpha)}$ satisfies

$$\frac{dL^{(\alpha)}}{dx}(x) = L^{(\alpha)}(x)\, A, \quad A = -\sum_{k=2}^{N} E_{k,k-1} \implies L^{(\alpha)}(x) = L^{(\alpha)}(0) e^{xA}. \tag{3.5}$$

By conjugation we have $A_\mu = S_\mu A S_\mu^{-1}$, $\frac{dL_\mu^{(\alpha)}}{dx}(x) = L_\mu^{(\alpha)}(x)\,A_\mu$ and $L_\mu^{(\alpha)}(x) = L_\mu^{(\alpha)}(0)e^{xA_\mu}$. Because of the dependence of the parameter of the Laguerre polynomial on n in (3.3), $L^{(\alpha)}(x)$ does not commute with A, see Lemma 4.2.

Remark 3.1 Comparing the expression $L^{(\alpha)}(x)$ of (3.3), (3.5) with the corresponding L operator in the matrix valued Hermite case, see [23, (3.1), Prop. 3.1], we see that the exponential part of $L^{(\alpha)}(x)$ and of the L in [23] is the same up to a factor -2. So, denoting L of [23] for the moment by H, we see that $L^{(\alpha)}(-2x) = L^{(\alpha)}(0)H(0)^{-1}H(x)$. Taking the (m,n)-entry we get connection coefficients relating Laguerre polynomials and Hermite polynomials, explicitly

$$L_{m-n}^{(\alpha+n)}(-2x) = \sum_{p=n}^{m} \big(L^{(\alpha)}(0)H(0)^{-1}\big)_{m,p} \frac{1}{(p-n)!} H_{p-n}(x)$$

where H_n are the standard Hermite polynomials [25, §1.13]. Since $H(0)^{-1}$ is known by Ismail et al. [23, Prop. 3.1], we can rewrite the $\big(L^{(\alpha)}(0)H(0)^{-1}\big)_{m,p}$ explicitly as a sum, and we obtain

$$L_{m-n}^{(\alpha+n)}(-2x) = \\ \sum_{p=n}^{m} \frac{(\alpha+p+1)_{m-p}}{(m-p)!\,(p-n)!}\, {}_2F_2\left(\begin{matrix} \frac{1}{2}(p-m), \frac{1}{2}(p-m+1) \\ \frac{1}{2}(\alpha+p+1), \frac{1}{2}(\alpha+p+2) \end{matrix}; 1\right) H_{p-n}(x). \tag{3.6}$$

Observe that the inverse of $L_\mu^{(\alpha)}$ is a unipotent lower triangular polynomial matrix function. Hence, its inverse is a unipotent lower triangular polynomial matrix function as well. In order to obtain an explicit expression for $(L_\mu^{(\alpha)}(x))^{-1}$, we need the following result of [24, §1], which can be obtained from the generating function (3.4) for the Laguerre polynomials.

Lemma 3.2 *For $i, j \in \mathbb{N}$, $i \geq j$, the following inversion formula for the Laguerre polynomials holds true.*

$$\sum_{k=j}^{i} L_{i-k}^{(-\alpha-i-1)}(-x) L_{k-j}^{(\alpha+j)}(x) = \delta_{i,j}.$$

Corollary 3.3 *The inverse of $L_\mu^{(\alpha)}$ is given explicitly by*

$$(L_\mu^{(\alpha)}(x)^{-1})_{m,n} = \begin{cases} \frac{\mu_m}{\mu_n} L_{m-n}^{(-\alpha-m-1)}(-x), & m \geq n, \\ 0 & n < m. \end{cases}$$

Proof For $r \geq s$, we have

$$\left(L^{(\alpha)}(x)^{-1}L^{(\alpha)}(x)\right)_{r,s} = \sum_{k=1}^{N}(L^{(\alpha)}(x))^{-1}_{r,k}L^{(\alpha)}(x)_{k,s}$$

$$= \sum_{k=s}^{r} L^{(-\alpha-r-1)}_{r-k}(-x)L^{(\alpha+s)}_{k-s}(x) = \delta_{r,s}.$$

The case $r < s$ follows with an analogous calculation, and conjugate with S_μ. □

Note that Lemma 3.2 follows directly from a generating function, and the same holds for the Hermite case [23]. For the other cases [29, 30] this is more involved, and the necessary result has been obtained by Cagliero and Koornwinder [5] and for the Gegenbauer case a q-analogue is given by Aldenhoven [1].

Writing $L^{(\alpha)}_\mu(x)$ as the product of a constant matrix times an exponential function as in (3.5) allows us to give an explicit relation between the matrices with different parameters.

Lemma 3.4 *Let $L^{(\alpha)}_\mu(x)$ be the matrix polynomial as in* (3.3) *and let $\lambda > 0$. Then*

$$L^{(\alpha)}_\mu(x) = M^{(\alpha,\lambda)}_\mu L^{(\lambda)}_\mu(x), \qquad M^{(\alpha,\lambda)}_\mu = L^{(\alpha)}_\mu(x)L^{(\lambda)}_\mu(x)^{-1} = L^{(\alpha)}_\mu(0)L^{(\lambda)}_\mu(0)^{-1}.$$

Moreover, $M^{(\alpha,\lambda)}_\mu$ is explicitly given by

$$M^{(\alpha,\lambda)}_\mu = \sum_{k=0}^{N-1}\frac{(\alpha-\lambda)_k}{k!}(-1)^k A^k_\mu.$$

Note that the series for $M^{(\alpha,\lambda)}_\mu$ terminates, since A_μ is nilpotent. Using the binomial theorem it can also be expressed as

$$M^{(\alpha,\lambda)}_\mu = (1+A_\mu)^{\lambda-\alpha}.$$

Proof Note that it suffices to deal with the case $\mu_i = 1$ for all i, and next conjugate by S_μ. Since $L^{(\alpha)}(x)$ and $(L^{(\lambda)}(x))^{-1}$ are unipotent lower triangular matrices, $M^{(\alpha,\lambda)}$ is unipotent lower triangular. Moreover, $M^{(\alpha,\lambda)}$ is constant, since the exponentials cancel. Now the (r,s)-entry of $M^{(\alpha,\lambda)}L^{(\lambda)}(x) = L^{(\alpha)}(x)$, for $r > s$, is

$$\sum_{k=s}^{r} M^{(\alpha,\lambda)}_{r,k}L^{(\lambda+s)}_{k-s}(x) = L^{(\alpha+s)}_{r-s} = \sum_{k=s}^{r}\frac{(\alpha-\lambda)_{r-k}}{(r-k)!}L^{(\lambda+s)}_{k-s},$$

where we use [35, 18.18.18] in order to write the Laguerre polynomials with parameter $\lambda + s$ in terms of Laguerre polynomials with parameter $\alpha + s$. Note that the required identity [35, 18.18.18] is a direct consequence of the same generating function (3.4) and the binomial theorem. We obtain

$$M_{r,k}^{(\alpha,\lambda)} = \frac{(\alpha-\lambda)_{r-k}}{(r-k)!}.$$

Conjugating by S_μ, the lemma follows. □

We are now ready to introduce the weight matrix. For $\nu > 0$ we define the matrix

$$W_\mu^{(\alpha,\nu)}(x) = L_\mu^{(\alpha)}(x)\, T^{(\nu)}(x)\, L_\mu^{(\alpha)}(x)^*, \qquad T^{(\nu)}(x) = e^{-x}\sum_{k=1}^{N} x^{\nu+k}\delta_k^{(\nu)}\, E_{k,k}. \tag{3.7}$$

Here the parameters $\delta_k^{(\nu)}$, $1 \le k \le N$, are to be determined later, see conditions (5.20) and (5.22) in Sect. 5. For now we assume the condition $\delta_k^{(\nu)} > 0$, $1 \le k \le N$, so that the weight is positive definite. Moreover, since $\nu > 0$ the factor $e^{-x}x^{\nu+k}$ in the entries of the diagonal matrix $T^{(\nu)}$ guarantees that all the moments exist. Note that even $\nu > -2$ suffices for the existence of moments as well, but we require $\nu > 0$ in all examples in Sect. 6.1. By $\{P_n^{(\alpha,\nu)}\}_n$ we denote the sequence of monic orthogonal polynomials with respect to $W_\mu^{(\alpha,\nu)}$. We suppress the μ-dependence in the notation for the polynomials and the related quantities, such as the squared norms, coefficients in the three-term recurrence relation, etc. Note that the structure in $T^{(\nu)}$ is motivated by the results of [29, 30].

Observe that for real μ_i's we have

$$W_\mu^{(\alpha,\nu)}(x) = L_\mu^{(\alpha)}(x)T^{(\nu)}(x)L_\mu^{(\alpha)}(x)^* = L_\mu^{(\alpha)}(0)e^{xA_\mu}T^{(\nu)}(x)e^{xA_\mu^*}L_\mu^{(\alpha)}(0)^*. \tag{3.8}$$

From now on we assume that the coefficients μ_i are real and non-zero for all i.

Using Lemma 3.4 we obtain

$$W_\mu^{(\alpha,\nu)} = M_\mu^{(\alpha,\lambda)}\, W_\mu^{(\lambda,\nu)}(M_\mu^{(\alpha,\lambda)})^*, \tag{3.9}$$

for $\alpha, \lambda > 0$. Denote by $H_n^{(\alpha,\nu)}$ the n-th squared norms for the monic orthogonal polynomials:

$$H_n^{(\alpha,\nu)} = \int_0^\infty P_n^{(\alpha,\nu)}(x)W_\mu^{(\alpha,\nu)}(x)P_n^{(\alpha,\nu)}(x)^*dx,$$

so that by (3.9) we have the following relation for the squared norms with different parameters

$$H_n^{(\alpha,\nu)} = M_\mu^{(\alpha,\lambda)}H_n^{(\lambda,\nu)}(M_\mu^{(\alpha,\lambda)})^*. \tag{3.10}$$

Recall that the squared norms satisfy $H_n^{(\alpha,\nu)} > 0$.

Proposition 3.5 *If we let $\alpha = \nu$, then the 0-th moment $H_0^{(\nu,\nu)}$ is the diagonal matrix*

$$(H_0^{(\nu,\nu)})_{j,j} = \int_0^\infty (W_\mu^{(\nu)}(x))_{j,j}\, dx = \frac{\mu_j^2 \Gamma(\nu+j+1)}{(j-1)!} \sum_{k=1}^{j} \frac{\delta_k^{(\nu)}}{\mu_k^2} (-1)^{k+1}(-j+1)_{k-1}.$$

Proof It follows from the definition of the weight (3.7) and the orthogonality relations for the Laguerre polynomials that

$$\begin{aligned}(H_0^{(\nu,\nu)})_{i,j} &= \int_0^\infty (W_\mu^{(\nu,\nu)}(x))_{i,j}\, dx \\ &= \sum_{k=1}^{\min(i,j)} \frac{\mu_i \mu_j}{\mu_k^2} \delta_k^{(\nu)} \int_0^\infty L_{i-k}^{(\nu+k)}(x) L_{j-k}^{(\nu+k)}(x) x^{\nu+k} e^{-x}\, dx \\ &= \delta_{i,j} \mu_j^2 \Gamma(\nu+j+1) \sum_{k=1}^{N} \frac{\delta_k^{(\nu)}}{\mu_k^2 (j-k)!}.\end{aligned}$$

Now the proposition follows by rewriting the factor $(j-k)!$. □

4 A Symmetric Second Order Differential Operator

A standard technique in order to deal with matrix valued polynomials is to obtain a matrix valued differential operator having the matrix valued orthogonal polynomials as eigenfunctions, see, e.g., [8, 10, 19, 23, 30]. We obtain a second-order matrix valued differential operator which is symmetric with respect to $W_\mu^{(\nu)}$ and which preserves polynomials and its degree, by establishing a conjugation to a diagonal matrix differential operator using the approach of [30].

Let F_2, F_1, F_0 be matrix valued polynomials of degrees two, one and zero, respectively. Assume that we have a matrix valued second-order differential operator D which acts on a matrix valued $C^2([0,\infty))$-function Q by

$$QD = \left(\frac{d^2 Q}{dx^2}\right)(x)\, F_2(x) + \left(\frac{dQ}{dx}\right)(x)\, F_1(x) + Q(x) F_0(x). \tag{4.11}$$

For a positive definite matrix valued weight W with finite moments of all orders, we say that D is symmetric with respect to W if for all matrix valued $C^2([0,\infty))$-functions G, H we have

$$\int_0^\infty (GD)(x) W(x) (H(x))^*\, dx = \int_0^\infty G(x) W(x) ((HD)(x))^* dx.$$

By [10, Thm 3.1], D is symmetric with respect to W if and only if the boundary conditions

$$\lim_{x\to a} F_2(x)W(x) = 0 = \lim_{x\to b} F_2(x)W(x), \tag{4.12}$$

$$\lim_{x\to a} F_1(x)W(x) - \frac{d(F_2W)}{dx}(x) = 0 = \lim_{x\to b} F_1(x)W(x) - \frac{d(F_2W)}{dx}(x) \tag{4.13}$$

and the symmetry conditions

$$F_2(x)W(x) = W(x)\big(F_2(x)\big)^*, \qquad 2\frac{d(F_2W)}{dx}(x) - F_1(x)W(x) = W(x)\big(F_1(x)\big)^*, \tag{4.14}$$

$$\frac{d^2(F_2W)}{dx^2}(x) - \frac{d(F_1W)}{dx}(x) + F_0(x)W(x) = W(x)\big(F_0(x)\big)^* \tag{4.15}$$

for almost all $x \in (a, b)$ hold.

Remark 4.1 Suppose that the differential operator D is symmetric with respect to a weight matrix of the form $W_\mu^{(\nu)}(x) = L(x)\,T(x)\,L(x)^*$, and let $\widetilde{D} = \frac{d^2}{dx^2}\widetilde{F}_2 + \frac{d}{dx}\widetilde{F}_1 + \widetilde{F}_0$ be the second-order differential operator obtained by conjugation of D by L. Then for all C^2-matrix valued functions Q we have

$$\frac{d^2(QL)}{dx^2}(x)\widetilde{F}_2(x) + \frac{d(QL)}{dx}(x)\widetilde{F}_1(x) + (QL)(x)\widetilde{F}_0(x) = \big(QD\big)(x)L(x), \tag{4.16}$$

where the coefficients F_i and $\widetilde{F}_i$ are related by

$$F_2L = L\widetilde{F}_2, \quad F_1L = 2\frac{dL}{dx}\widetilde{F}_2 + L\widetilde{F}_1, \quad F_0L = \frac{d^2L}{dx^2}\widetilde{F}_2 + \frac{dL}{dx}\widetilde{F}_1 + L\widetilde{F}_0, \tag{4.17}$$

see the discussion in [30, §4]. Moreover, it follows from [30, Prop. 4.2] that D is symmetric with respect to W if and only if $\widetilde{D}$ is symmetric with respect to T.

In order to obtain the explicit expression for a symmetric differential operator having the matrix valued orthogonal polynomials $P_n^{(\nu)}$ as eigenfunctions, we need to control commutation relations with some explicit matrices. In particular, J is the diagonal matrix $J_{k,k} = k$. Note that $JS_\mu = S_\mu J$.

Lemma 4.2 *The following commutation relations for $L_\mu^{(\alpha)}$, A_μ and J hold.*

$$\begin{aligned} L_\mu^{(\alpha)}(x)J(L_\mu^{(\alpha)}(x))^{-1} &= x(A_\mu+1)^{-1} - x + (\alpha+J)A_\mu + J, \\ (L_\mu^{(\alpha)}(x))^{-1}JL_\mu^{(\alpha)}(x) &= J - (\alpha+J-x)A_\mu - xA_\mu^2, \\ L_\mu^{(\alpha)}(x)(1-A_\mu)(L_\mu^{(\alpha)}(x))^{-1} &= (1+A_\mu)^{-1}. \end{aligned}$$

Note that in particular, the last equality gives $[A_\mu, L_\mu^{(\alpha)}(x)] = A_\mu L_\mu^{(\alpha)}(x) A_\mu$.

Proof It suffices to prove the lemma for $\mu_i = 1$ for all i. We first prove the last identity. If we multiply the third equation by $L^{(\alpha)}(x)$ on the right and by $A+1$ on the left, we see that it suffices to show $(A+1)L^{(\alpha)}(x)A = AL^{(\alpha)}(x)$. The (r,s)-entry of this equation is given by

$$L_{r-s-2}^{(\alpha+s+1)}(x) - L_{r-s-1}^{(\alpha+s+1)}(x) = -L_{r-s-1}^{(\alpha+s)}(x),$$

which is, e.g., [35, 18.9.14].

Similarly, the first equation is equivalent to

$$(A+1)L^{(\alpha)}(x)J = ((\alpha + J - x)A + AJ + J + \alpha A^2 + AJA)L^{(\alpha)}(x).$$

The (r,s)-entry of this equation, after simplifying and regrouping terms, gives

$$\begin{aligned} - sL_{r-s-1}^{(\alpha+s)}(x) + L_{r-s}^{(\alpha+s)}(x) &= rL_{r-s}^{(\alpha+s)}(x) \\ &\quad + (x - \alpha - 2r + 1)L_{r-s-1}^{(\alpha+s)}(x) + (\alpha + r - 1)L_{r-s-2}^{(\alpha+s)}(x). \end{aligned}$$

Observe that all the Laguerre polynomials have the same parameter and this follows from the three-term recurrence relation of the Laguerre polynomials, see, e.g., [25, (1.11.3)].

The second equation follows from the other two. Use the third equation in the first one twice to rewrite the first equation as

$$\begin{aligned} L^{(\alpha)}(x)J(L^{(\alpha)}(x))^{-1} &= -xL^{(\alpha)}(x)A(L^{(\alpha)}(x))^{-1} + \alpha A \\ &\quad + JL^{(\alpha)}(x)(1-A)^{-1}(L^{(\alpha)}(x))^{-1} \end{aligned}$$

and isolating J from the last term gives

$$J = L^{(\alpha)}(x)\big(J(1-A) + xA(1-A) - \alpha(L^{(\alpha)}(x))^{-1}AL^{(\alpha)}(x)(1-A)\big)(L^{(\alpha)}(x))^{-1}$$

and use the last equation once more to see $(L^{(\alpha)}(x))^{-1}AL^{(\alpha)}(x)(1-A) = A$. Rewriting gives the second equation. □

Now we are ready to write explicitly a symmetric second order differential operator.

Proposition 4.3 *Let $D^{(\alpha,\nu)}$ be given by*

$$D^{(\alpha,\nu)} = \left(\frac{d^2}{dx^2}\right) F_2^{(\alpha,\nu)}(x) + \left(\frac{d}{dx}\right) F_1^{(\alpha,\nu)}(x) + F_0^{(\alpha,\nu)}(x),$$

with $F_2^{(\alpha,\nu)}(x) = x$ *and*

$$F_1^{(\alpha,\nu)}(x) = -x(A_\mu+1)^{-1}+\nu+J+1+(\alpha+J)A_\mu,\ F_0^{(\alpha,\nu)}(x) = (\alpha-\nu)(A_\mu+1)^{-1}-J.$$

Then $D^{(\alpha,\nu)}$ *is symmetric with respect to* $W_\mu^{(\alpha,\nu)}$. *Moreover,*

$$P_n^{(\alpha,\nu)} D^{(\alpha,\nu)} = \Gamma_n^{(\alpha,\nu)} P_n^{(\alpha,\nu)}, \qquad \Gamma_n^{(\alpha,\nu)} = (-n+\alpha-\nu)(A_\mu+1)^{-1}-J, \qquad n \in \mathbb{N}.$$

Note that the eigenvalue matrix $\Gamma_n^{(\alpha,\nu)}$ is a lower triangular matrix, which also depends on the choice of the sequence μ.

Proof Let us consider the differential operator $\widetilde{D}^{(\alpha,\nu)} = \frac{d^2}{dx^2}\widetilde{F}_2^{(\alpha,\nu)} + \frac{d}{dx}\widetilde{F}_1^{(\alpha,\nu)} + \widetilde{F}_0^{(\alpha,\nu)}$ obtained by conjugation of $D^{(\alpha,\nu)}$ by the matrix $L_\mu^{(\alpha)}$. Then it follows from (4.17) that the coefficients $\widetilde{F}_i^{(\alpha,\nu)}$ are given by $\widetilde{F}_2^{(\alpha,\nu)}(x) = x$ and

$$\begin{aligned} \widetilde{F}_1^{(\alpha,\nu)} &= (L_\mu^{(\alpha)})^{-1}\left(F_1^{(\nu)} L_\mu^{(\alpha)} - 2x\frac{dL_\mu^{(\alpha)}}{dx}\right), \\ \widetilde{F}_0^{(\alpha,\nu)} &= (L_\mu^{(\alpha)})^{-1}\left(F_0^{(\alpha,\nu)} L_\mu^{(\alpha)} - x\frac{d^2L_\mu^{(\alpha)}}{dx^2} - \frac{dL_\mu^{(\alpha)}}{dx}\widetilde{F}_1^{(\alpha,\nu)}\right). \end{aligned} \tag{4.18}$$

The derivatives in (4.18) can be evaluated by (3.5). It follows from the definition of $F_1^{(\alpha,\nu)}$ that

$$\begin{aligned} \widetilde{F}_1^{(\alpha,\nu)}(x) = \nu + 1 - xL_\mu^{(\alpha)}(x)^{-1}(A_\mu+1)^{-1}L_\mu^{(\alpha)}(x) + L_\mu^{(\alpha)}(x)^{-1}JL_\mu^{(\alpha)}(x) \\ + L_\mu^{(\alpha)}(x)^{-1}(\alpha+J)A_\mu L_\mu^{(\alpha)}(x) - 2xA_\mu. \end{aligned}$$

Use the first equation of Lemma 4.2 to rewrite the term $L_\mu^{(\alpha)}(x)^{-1}(\alpha+J)A_\mu L_\mu^{(\alpha)}(x)$. Similarly the last equation of Lemma 4.2 is used, and we obtain $\widetilde{F}_1^{(\alpha,\nu)}(x) = \nu + J + 1 - x$.

Similarly, using (3.5) and Lemma 4.2 and the expression for $\widetilde{F}_1^{(\alpha,\nu)}$ we obtain

$$\begin{aligned} \widetilde{F}_0^{(\alpha,\nu)}(x) = (\alpha-\nu)(1-A_\mu) - L_\mu^{(\alpha)}(x)^{-1}JL_\mu^{(\alpha)}(x) - xA_\mu^2 - A_\mu(\nu+J+1-x) \\ = (\alpha-\nu) - J. \end{aligned}$$

By Remark 4.1, in order to prove that $D^{(\alpha,\nu)}$ is symmetric with respect to $W_\mu^{(\alpha,\nu)}$ it is enough to prove that $\widetilde{D}^{(\alpha,\nu)}$ is symmetric with respect to $T^{(\nu)}$, i.e. we need to show that the boundary conditions (4.12), (4.13) and the symmetry conditions (4.14), (4.15) hold true with W replaced by $T^{(\nu)}$ and the F's replaced by the corresponding $\widetilde{F}^{(\alpha,\nu)}$'s. Since $\widetilde{F}^{(\alpha,\nu)}$'s are polynomials, the weight involves the exponential e^{-x} we see that for $\nu > -1$, so in particular for $\nu > 0$, the boundary conditions (4.12), (4.13) are satisfied.

The symmetry equations (4.14) and (4.15) are diagonal conditions, and can be verified by a simple calculation.

Hence $D^{(\alpha,\nu)}$ is symmetric with respect to the weight matrix $W_\mu^{(\alpha,\nu)}$. Since $D^{(\alpha,\nu)}$ preserves polynomials and the degree of the polynomials, $P_n^{(\alpha,\nu)} D^{(\alpha,\nu)}$ are also orthogonal with respect to $W_\mu^{(\alpha,\nu)}$, so that $P_n^{(\alpha,\nu)} D^{(\alpha,\nu)} = \Gamma_n^{(\alpha,\nu)} P_n^{(\alpha,\nu)}$ for some matrix $\Gamma_n^{(\alpha,\nu)}$, which is obtained by considering the leading coefficient. □

Remark 4.4 It follows from the proof Proposition (4.3) that matrix valued polynomials $P_n^{(\alpha,\nu)} L_\mu^{(\alpha)}$ are polynomial eigenfunctions of the diagonal second-order differential operator $\widetilde{D}^{(\alpha,\nu)}$ with eigenvalue $\Gamma_n^{(\alpha,\nu)} = -n(A_\mu+1)^{-1}+(\alpha-\nu)(A_\mu+1)^{-1} - J$. More precisely, we have

$$x\frac{d^2(P_n^{(\alpha,\nu)} L_\mu^{(\alpha)})}{dx^2}(x) + \frac{d(P_n^{(\alpha,\nu)} L_\mu^{(\alpha)})}{dx}(x)(\nu+J+1-x) - (P_n^{(\alpha,\nu)} L_\mu^{(\alpha)})(x)J$$
$$= \Gamma_n^{(\alpha,\nu)} (P_n^{(\alpha,\nu)} L_\mu^{(\alpha)})(x).$$

Observe that although the differential operator $\widetilde{D}^{(\alpha,\nu)}$ is diagonal, the eigenvalue $\Gamma_n^{(\alpha,\nu)}$ is a full lower triangular matrix so that the previous equation gives a coupled system of differential equations for the entries of $P_n^{(\alpha,\nu)} L_\mu^{(\alpha)}$. This is in contrast with the case of matrix valued Gegenbauer polynomials studied in [30]. An analogous result for a differential operator and involving a diagonal eigenvalue allows to determine the entries of the analogue of $P_n^{(\alpha,\nu)} L_\mu^{(\alpha)}$ as a single Gegenbauer polynomial. This allowed to find explicit expressions for the polynomials, see [30, §5.2]. The situation of the Gegenbauer setting of [30] is repeated for the matrix valued Hermite polynomials [23, §3].

5 The Matrix Valued Pearson Equation

In order to establish the existence of shift operators, the Pearson equations are essential. We derive the matrix valued Pearson equations for the family of weights $W_\mu^{(\alpha,\nu)}$ under explicit non-linear conditions relating the coefficients of the sequence μ and the coefficients in $T^{(\nu)}$.

First we need certain relations involving the function e^{xA_μ}. Recall the diagonal matrix J; $J_{k,l} = \delta_{k,l}k$. Then $[J, A_\mu] = A_\mu$ and $[J, A_\mu^*] = -A_\mu^*$, so that

$$e^{-xA_\mu} J e^{xA_\mu} = xA_\mu + J, \qquad e^{-xA_\mu^*} J e^{xA_\mu^*} = -xA_\mu^* + J, \tag{5.19}$$

For this we use that the left-hand side of the first expression is a matrix valued polynomial in x, since A_μ is nilpotent. Its derivative $e^{-xA_\mu}[J, A_\mu]e^{xA_\mu} = A_\mu$ is constant, and the first formula follows. For the second equation of (5.19) we take adjoints and replace x by $-x$ in the first formula.

Now we need to impose conditions on the sequence $\{\mu_i\}_i$ and the coefficients $\delta_k^{(\nu)}$. We consider the diagonal matrix $\Delta^{(\nu)} = \mathrm{diag}(\delta_1^{(\nu)}, \ldots, \delta_N^{(\nu)})$, so that $(T^{(\nu)})_{k,k} = e^{-x} x^{\nu+k} (\Delta^{(\nu)})_{k,k}$. We assume that there exist coefficients $c^{(\nu)}$ and $d^{(\nu)}$ such that $\delta_k^{(\nu+1)} = (k\, d^{(\nu)} + c^{(\nu)})\, \delta_k^{(\nu)}$ for all $k = 1, \ldots, N$. In other words, we assume that

$$\Delta^{(\nu+1)} = (d^{(\nu)} J + c^{(\nu)})\, \Delta^{(\nu)}. \tag{5.20}$$

Note that $d^{(\nu)}, c^{(\nu)} \geq 0$, since $\delta_k^{(\nu)} > 0$. In view of (5.19), the condition (5.20) implies that

$$e^{-xA_\mu^*} (\Delta^{(\nu)})^{-1} \Delta^{(\nu+1)} e^{xA_\mu^*} = e^{-xA_\mu^*} (d^{(\nu)} J + c^{(\nu)}) e^{xA_\mu^*} = d^{(\nu)} (-xA_\mu^* + J) + c^{(\nu)}. \tag{5.21}$$

This is the main ingredient in Proposition 5.1.

Proposition 5.1 *Let $W_\mu^{(\alpha,\nu)}$ be the weight matrix given in* (3.7) *and assume that* (5.20) *holds true. Then*

$$\Phi^{(\alpha,\nu)}(x) = (W_\mu^{(\alpha,\nu)}(x))^{-1} W_\mu^{(\alpha,\nu+1)}(x),$$

is a matrix valued polynomial of degree two.

Proof Using (3.8), the fact that $T^{(\nu)}(x)^{-1} T^{(\nu+1)}(x) = x(\Delta^{(\nu)})^{-1} \Delta^{(\nu+1)}$ and (5.21), we obtain

$$\begin{aligned}
(W_\mu^{(\alpha,\nu)}(x))^{-1} W_\mu^{(\alpha,\nu+1)}(x) &= x(L_\mu^{(\alpha)}(x)^*)^{-1} (\Delta^{(\nu)})^{-1} \Delta^{(\nu+1)} L_\mu^{(\alpha)}(x)^* \\
&= x(L_\mu^{(\alpha)}(0)^*)^{-1} e^{-xA_\mu^*} (d^{(\nu)} J + c^{(\nu)}) e^{xA_\mu^*} L_\mu^{(\alpha)}(0)^* \\
&= -d^{(\nu)} x^2 (L_\mu^{(\alpha)}(0)^*)^{-1} A_\mu^* L_\mu^{(\alpha)}(0)^* + \\
&\qquad + x\left(d^{(\nu)} (L_\mu^{(\alpha)}(0)^*)^{-1} J L_\mu^{(\alpha)}(0)^* + c^{(\nu)}\right).
\end{aligned}$$

□

Now we assume that the coefficients μ_k and $\delta_k^{(\nu)}$ satisfy the relation

$$\frac{\mu_{k+1}^2}{\mu_k^2} = d^{(\nu)} k(N-k) \frac{\delta_{k+1}^{(\nu)}}{\delta_k^{(\nu+1)}}, \qquad k = 1, \ldots, N-1. \tag{5.22}$$

Note that, since the coefficients μ_k are independent of ν, we require the right-hand side of (5.22) to be independent of ν.

Proposition 5.2 *Let $W_\mu^{(\alpha,\nu)}$ be the weight matrix given in* (3.7) *and assume that the conditions* (5.20) *and* (5.22) *hold true. Then*

$$\Psi^{(\alpha,\nu)}(x) = (W_\mu^{(\alpha,\nu)}(x))^{-1}\frac{dW_\mu^{(\alpha,\nu+1)}}{dx}(x),$$

is a matrix valued polynomial of degree one.

Proof Using (3.8) we obtain

$$\begin{aligned} L_\mu^{(\alpha)}(0)^*\Psi^{(\alpha,\nu)}(x)(L_\mu^{(\alpha)}(0)^*)^{-1} =& e^{-xA_\mu^*}(T^{(\nu)}(x))^{-1}A_\mu T^{(\nu+1)}(x)e^{xA_\mu^*} \\ &+ e^{-xA_\mu^*}(T^{(\nu)})^{-1}\frac{dT^{(\nu+1)}}{dx}(x)e^{xA_\mu^*} + e^{-xA_\mu^*}(T^{(\nu)})^{-1}T^{(\nu+1)}e^{xA_\mu^*}A_\mu^*. \end{aligned}$$

It follows from (5.20) and (5.19) that

$$\begin{aligned} e^{-xA_\mu^*}T^{(\nu)}(x)^{-1}T^{(\nu+1)}(x)e^{xA_\mu^*}A_\mu^* &= xe^{-xA_\mu^*}(\Delta^{(\nu)})^{-1}\Delta^{(\nu+1)}e^{xA_\mu^*}A_\mu^* \\ &= -x^2d^{(\nu)}(A_\mu^*)^2 + xd^{(\nu)}JA_\mu^* + xc^{(\nu)}A_\mu^*, \end{aligned} \tag{5.23}$$

and

$$\begin{aligned} &e^{-xA_\mu^*}(T^{(\nu)})^{-1}\frac{dT^{(\nu+1)}}{dx}(x)e^{xA_\mu^*} = x^2d^{(\nu)}(A_\mu^* + (A_\mu^*)^2) \\ &- x((A_\mu^*+1)(d^{(\nu)}J + c^{(\nu)}) + d^{(\nu)}(\nu+J+1)A_\mu^*) + (\nu+J+1)(d^{(\nu)}J + c^{(\nu)}). \end{aligned} \tag{5.24}$$

We observe that the term $x^2d^{(\nu)}(A_\mu^*)^2$ of the right-hand side of (5.24) cancels with the term of degree two in (5.23).

Now we note that $(T^{(\nu)}(x))^{-1}A_\mu T^{(\nu+1)}(x) = x(\Delta^{(\nu)})^{-1}A_\mu\Delta^{(\nu+1)}$. On the other hand, the matrix $[(\Delta^{(\nu)})^{-1}A_\mu\Delta^{(\nu+1)}, A_\mu^*]$ is a diagonal matrix whose k-th diagonal entry is given by

$$[(\Delta^{(\nu)})^{-1}A_\mu\Delta^{(\nu+1)}, A^*]_{k,k} = \frac{\mu_{k+1}^2\delta_k^{(\nu+1)}}{\mu_k^2\delta_{k+1}^{(\nu)}} - \frac{\mu_k^2\delta_{k-1}^{(\nu+1)}}{\mu_{k-1}^2\delta_k^{(\nu)}}.$$

By (5.22) we verify that $[(\Delta^{(\nu)})^{-1} A_\mu \Delta^{(\nu+1)}, A_\mu^*] = 2d^{(\nu)}J - d^{(\nu)}(N+1)$. This leads to

$$\begin{aligned}\frac{d}{dx}\left(e^{-xA_\mu^*}(\Delta^{(\nu)})^{-1} A_\mu \Delta^{(\nu+1)} e^{xA_\mu^*}\right) &= e^{-xA_\mu^*}\left[(\Delta^{(\nu)})^{-1} A_\mu \Delta^{(\nu+1)}, A_\mu^*\right] e^{xA_\mu^*} \\ &= e^{-xA_\mu^*}(2d^{(\nu)}J - d^{(\nu)}(N+1))e^{xA_\mu^*} \\ &= -2xd^{(\nu)}A_\mu^* + 2d^{(\nu)}J - d^{(\nu)}(N+1).\end{aligned}$$

Therefore we have that

$$\begin{aligned}e^{-xA_\mu^*}(\Delta^{(\nu)})^{-1} A_\mu \Delta^{(\nu+1)} e^{xA_\mu^*} &= \\ -d^{(\nu)}x^2 A_\mu^* + xd^{(\nu)}(2J - N - 1) &+ (\Delta^{(\nu)})^{-1} A_\mu \Delta^{(\nu+1)}\end{aligned} \quad (5.25)$$

Adding (5.23)–(5.25) shows that $\Psi^{(\alpha,\nu)}$ is a polynomial of degree one. □

For future reference we state Corollary 5.3 as an immediate consequence of the proofs of Propositions 5.1 and 5.2.

Corollary 5.3 *Assuming the conditions* (5.20) *and* (5.22), *the matrix valued polynomials*

$$\begin{aligned}L_\mu^{(\alpha)}(0)^*\Phi^{(\alpha,\nu)}(x)(L_\mu^{(\alpha)}(0)^*)^{-1} &= -d^{(\nu)}x^2 A_\mu^* + x\left(d^{(\nu)}J + c^{(\nu)}\right), \\ L_\mu^{(\alpha)}(0)^*\Psi^{(\alpha,\nu)}(x)(L_\mu^{(\alpha)}(0)^*)^{-1} &= x\left(d^{(\nu)}(J - A_\mu^*(J+\nu+1) - N - 1) - c^{(\nu)}\right) \\ &\quad + \left((\nu + J + 1)(d^{(\nu)}J + c^{(\nu)}) + (\Delta^{(\nu)})^{-1} A_\mu \Delta^{(\nu+1)}\right).\end{aligned}$$

satisfy the Pearson equations

$$\Phi^{(\alpha,\nu)}(x) = (W_\mu^{(\alpha,\nu)}(x))^{-1} W_\mu^{(\alpha,\nu+1)}(x), \quad \Psi^{(\alpha,\nu)}(x) = (W_\mu^{(\alpha,\nu)}(x))^{-1}\frac{dW_\mu^{(\alpha,\nu+1)}}{dx}(x).$$

Remark 5.4 Upon replacing $\delta_k^{(\nu+1)} = (kd^{(\nu)} + c^{(\nu)})\delta_k^{(\nu)}$ in (5.22), we can iterate the resulting identity to obtain

$$\frac{\delta_k^{(\nu)}}{\mu_k^2} = \frac{(1 + \frac{c^{(\nu)}}{d^{(\nu)}})_{k-1}}{(k-1)!(N-k+1)_{k-1}} \frac{\delta_1^{(\nu)}}{\mu_1^2}. \quad (5.26)$$

This relation can now be used to evaluate explicitly the 0-th moment $H_0^{(\nu,\nu)}$ given in Proposition (3.5). Indeed,

$$
\begin{aligned}
(H_0^{(\nu,\nu)})_{j,j} &= \frac{\mu_j^2\,\delta_1^{(\nu)}\,\Gamma(\nu+j+1)}{\mu_1^2\,(j-1)!}\sum_{k=1}^{N}\frac{(1+\frac{c^{(\nu)}}{d^{(\nu)}})_{k-1}(-j+1)_{k-1}}{(N-k+1)_{k-1}(k-1)!} \\
&= \frac{\mu_j^2\,\delta_1^{(\nu)}\,\Gamma(\nu+j+1)}{\mu_1^2\,(j-1)!}\,{}_2F_1\left(\begin{matrix}1+\frac{c^{(\nu)}}{d^{(\nu)}},\,-(j-1)\\ -N+1\end{matrix};1\right) \\
&= \frac{\mu_j^2\,\delta_1^{(\nu)}\,\Gamma(\nu+j+1)(-N-\frac{c^{(\nu)}}{d^{(\nu)}})_{j-1}}{\mu_1^2\,(j-1)!(-N-1)_{j-1}},
\end{aligned}
$$

where the ${}_2F_1$ is summed by the Chu-Vandermonde identity.

6 Shift Operators

In this section we use the Pearson equations to give explicit lowering and rising operators for the polynomials $P_n^{(\alpha,\nu)}$. Next we exploit the existence of the shift operators to give an explicit Rodrigues formula, to calculate the squared norms as well as the coefficients in the three-term recurrence relation. Moreover, we find another matrix valued differential operator to which the matrix polynomials are eigenfunctions. For this explicit matrix valued differential operator it is possible to perform a Darboux transform, and we give an explicit expression for the Darboux transformation. We end by obtaining a Burchnall type identity, see [23], for the matrix valued Laguerre polynomials, and by showing that there are at least three families of solutions to the non-linear conditions (5.20) and (5.22). In this section we assume that these conditions are satisfied, and hence that the Pearson equations of Corollary 5.3 hold.

For matrix valued functions P and Q, we denote by

$$
\langle P, Q\rangle^{(\alpha,\nu)} = \int_0^\infty P(x)W_\mu^{(\alpha,\nu)}(x)Q(x)^*\,dx,
$$

whenever the integral converges. Moreover, we have

$$
\begin{aligned}
\langle \frac{dP}{dx}, Q\rangle^{(\alpha,\nu+1)} &= \int_0^\infty \frac{dP}{dx}(x)W_\mu^{(\alpha,\nu+1)}(x)\big(Q(x)\big)^*\,dx \\
&= -\int_0^\infty P(x)W_\mu^{(\alpha,\nu)}(x)\Psi^{(\alpha,\nu)}(x)Q(x)^*\,dx - \int_0^\infty P(x)W_\mu^{(\nu)}(x)\Phi^{(\alpha,\nu)}(x)\frac{dQ^*}{dx}(x)\,dx \\
&= -\,\langle P, QS^{(\alpha,\nu)}\rangle^{(\alpha,\nu)},
\end{aligned}
$$

where $S^{(\alpha,\nu)}$ is the first order matrix valued differential operator

$$(QS^{(\alpha,\nu)})(x) = \frac{dQ}{dx}(x)(\Phi^{(\alpha,\nu)}(x))^* + Q(x)(\Psi^{(\alpha,\nu)}(x))^*. \tag{6.27}$$

Note that we have to assume that the decay at 0 and at ∞ is sufficiently large, which is the case for, e.g., polynomials P and Q.

In particular, if we set $P(x) = P_n^{(\alpha,\nu)}(x)$ and $Q(x) = x^k$, considering the degrees of $\Phi^{(\alpha,\nu)}$ and $\Psi^{(\alpha,\nu)}$ we obtain $\langle \frac{dP_n^{(\alpha,\nu)}}{dx}, Q\rangle^{(\alpha,\nu+1)} = 0$ for all $k \in \mathbb{N}$, $k < n$. Since the leading coefficient of $\frac{dP_n^{(\alpha,\nu)}}{dx}$ is non-singular, we conclude that $\{\frac{dP_n^{(\alpha,\nu)}}{dx}\}_n$ is a sequence of matrix valued orthogonal polynomials with respect to $W_\mu^{(\alpha,\nu+1)}$. Similarly, the sequence $\{P_n^{(\alpha,\nu+1)}S^{(\alpha,\nu)}\}_n$ is a sequence of matrix valued orthogonal polynomials with respect to $W_\mu^{(\alpha,\nu)}$.

Proposition 6.1 *Assume that the conditions* (5.20) *and* (5.22) *are satisfied for all* ν *of the form* $\nu_0 + k$, $k \in \mathbb{N}$, *for some fixed* ν_0. *The first order differential operator* $S^{(\alpha,\nu)}$ *given in* (6.27) *satisfies*

$$\langle \frac{dP}{dx}, Q\rangle^{(\alpha,\nu+1)} = -\langle P, QS^{(\alpha,\nu)}\rangle^{(\alpha,\nu)},$$

for matrix valued polynomials P *and* Q. *Moreover*

$$\frac{dP_n^{(\alpha,\nu)}}{dx}(x) = n\, P_{n-1}^{(\alpha,\nu+1)}(x), \qquad (P_{n-1}^{(\alpha,\nu+1)}S^{(\alpha,\nu)})(x) = K_n^{(\alpha,\nu)}P_n^{(\alpha,\nu)}(x),$$

where the matrices $K_n^{(\alpha,\nu)}$ *are invertible and are explicitly given by*

$$L_\mu^{(\alpha)}(0)^{-1}K_n^{(\alpha,\nu)}L_\mu^{(\alpha)}(0) = d^{(\nu)}(J - (J+\nu+n)A_\mu - N - 1) - c^{(\nu)}.$$

Proof Taking into account the preceding discussion, we have that $\frac{dP_n^{(\alpha,\nu)}}{dx}$ is a multiple of $P_{n-1}^{(\alpha,\nu+1)}$ and $P_{n-1}^{(\alpha,\nu+1)}S^{(\alpha,\nu)}$ is a multiple of $P_n^{(\alpha,\nu)}(x)$, these multiples follow from the leading coefficients. Now we only need to show that $K_n^{(\alpha,\nu)}$ is invertible. Observe that $L_\mu^{(\alpha)}(0)^{-1}K_n^{(\alpha,\nu)}L_\mu^{(\alpha)}(0)$ is a lower triangular matrix whose j-th diagonal entry is given by

$$(L_\mu^{(\alpha)}(0)^{-1}K_n^{(\alpha,\nu)}L_\mu^{(\alpha)}(0))_{j,j} = d^{(\nu)}(j - (N+1)) - c^{(\nu)}.$$

These entries are strictly negative, since $c^{(\nu)}$ and $d^{(\nu)}$ are positive. So invertibility follows. □

We note that the matrix $\widetilde{K}_n^{(\alpha,\nu)} = L_\mu^{(\alpha)}(0)^{-1}K_n^{(\alpha,\nu)}L_\mu^{(\alpha)}(0)$ is actually a function of $\nu + n$ so that $\widetilde{K}_{n-j}^{(\alpha,\nu+j)} = \widetilde{K}_n^{(\alpha,\nu)}$ for all $j \leq n$. Now conjugating with $L_\mu^{(\alpha)}(0)^{-1}$

we obtain that

$$K_{n-j}^{(\alpha,\nu+j)} = K_n^{(\alpha,\nu)}, \qquad j \leq n. \tag{6.28}$$

Theorem 6.2 *Assume that the conditions of Proposition 6.1. The polynomials $P_n^{(\alpha,\nu)}$ satisfy the following Rodrigues formula:*

$$P_n^{(\alpha,\nu)}(x) = G_n^{(\alpha,\nu)} \left(\frac{d^n W_\mu^{(\alpha,\nu+n)}}{dx^n}(x) \right) W_\mu^{(\alpha,\nu)}(x)^{-1},$$

where $G_n^{(\alpha,\nu)} = (K_n^{(\alpha,\nu)})^{-1} \cdots (K_1^{(\alpha,\nu+n-1)})^{-1} = (K_n^{(\alpha,\nu)})^{-n}$. The squared norm $H_n^{(\alpha,\nu)}$ is given by

$$\begin{aligned} H_n^{(\alpha,\nu)} &= (-1)^n\, n!\, (K_n^{(\alpha,\nu)})^{-n} H_0^{(\alpha,\nu+n)} \\ &= (-1)^n\, n!\, M_\mu^{(\alpha,\nu+n)} (K_n^{(\nu+n,\nu)})^{-n} H_0^{(\nu+n,\nu+n)} (M_\mu^{(\alpha,\nu+n)})^*, \end{aligned}$$

where $M_\mu^{(\nu+n,\nu)}$ is the matrix given in (3.9).

Proof Observe that $(QS^{(\alpha,\nu)})(x) = \frac{d(QW_\mu^{(\alpha,\nu+1)})}{dx}(x)\big(W_\mu^{(\alpha,\nu)}(x)\big)^{-1}$ by Corollary 5.3. Iterating gives

$$\big(QS^{(\alpha,\nu+n-1)} \cdots S^{(\alpha,\nu)}\big)(x) = \frac{d^n(QW_\mu^{(\alpha,\nu+n)})}{dx^n}(x)\big(W_\mu^{(\alpha,\nu)}(x)\big)^{-1}.$$

Now taking $Q(x) = P_0^{(\alpha,\nu+n)}(x) = 1$ and using Proposition 6.1 repeatedly we obtain the Rodrigues formula.

Finally, for the squared norm we observe that

$$nH_{n-1}^{(\alpha,\nu+1)} = n\langle P_{n-1}^{(\alpha,\nu+1)}, P_{n-1}^{(\alpha,\nu+1)}\rangle^{(\alpha,\nu+1)} = \langle \frac{dP_n^{(\alpha,\nu)}}{dx}, P_{n-1}^{(\alpha,\nu+1)}\rangle^{(\alpha,\nu+1)} =$$
$$\langle P_n^{(\alpha,\nu)}, P_{n-1}^{(\alpha,\nu+1)} S^{(\alpha,\nu)}\rangle^{(\alpha,\nu)} = \langle P_n^{(\alpha,\nu)}, P_n^{(\alpha,\nu)}\rangle^{(\alpha,\nu)} (K_n^{(\alpha,\nu)})^* = H_n^{(\alpha,\nu)} (K_n^{(\alpha,\nu)})^*,$$

so that $H_n^{(\alpha,\nu)} = n\,(K_n^{(\alpha,\nu)})^{-1} H_{n-1}^{(\alpha,\nu+1)}$, where we have used that $H_n^{(\alpha,\nu)}$ is self-adjoint for all $n \in \mathbb{N}$. Iterating and using (6.28) and (3.10) gives the expressions for the squared norm. □

Corollary 6.3 *Assume the conditions of Proposition 6.1. The second-order differential operator*

$$\mathcal{D}^{(\alpha,\nu)} = S^{(\alpha,\nu)} \circ \frac{d}{dx} = \left(\frac{d^2}{dx^2}\right) \Phi^{(\alpha,\nu)}(x)^* + \left(\frac{d}{dx}\right) \Psi^{(\alpha,\nu)}(x)^*,$$

is symmetric with respect to the weight $W_\mu^{(\alpha,\nu)}$. *Moreover, for all* $n \in \mathbb{N}$ *we have*

$$P_n^{(\alpha,\nu)}\mathcal{D}^{(\alpha,\nu)} = \Lambda_n^{(\alpha,\nu)} P_n^{(\alpha,\nu)}, \qquad \Lambda_n^{(\alpha,\nu)} = nK_n^{(\alpha,\nu)}.$$

Moreover, the operators $\mathcal{D}^{(\alpha,\nu)}$ *and* $D^{(\alpha,\nu)}$ *commute.*

Proof The fact that $\mathcal{D}^{(\alpha,\nu)}$ is symmetric with respect to $W_\mu^{(\alpha,\nu)}$ follows directly from the factorization $\mathcal{D}^{(\alpha,\nu)} = S^{(\alpha,\nu)} \circ \frac{d}{dx}$ and Proposition 6.1. Then the orthogonal polynomials $P_n^{(\alpha,\nu)}$ are eigenfunctions of $\mathcal{D}^{(\alpha,\nu)}$ and the eigenvalue is obtained by looking at the leading coefficients.

In order to prove that $\mathcal{D}^{(\alpha,\nu)}$ and $D^{(\alpha,\nu)}$ commute, we will show that the corresponding eigenvalues $\Gamma_n^{(\alpha,\nu)}$ and $\Lambda_n^{(\alpha,\nu)}$ commute, see [36, Cor. 4.4]. It is then enough to show that the following matrices commute:

$$\widetilde{\Lambda}_n^{(\alpha,\nu)} = (L_\mu^{(\alpha)}(0))^{-1}\Lambda_n^{(\alpha,\nu)}L_\mu^{(\alpha)}(0), \qquad \widetilde{\Gamma}_n^{(\alpha,\nu)} = (L_\mu^{(\alpha)}(0))^{-1}\Gamma_n^{(\alpha,\nu)}L_\mu^{(\alpha)}(0).$$

Using the explicit expressions of $\Gamma_n^{(\alpha,\nu)}$, Propositions 4.2 and 6.1 we obtain

$$\widetilde{\Gamma}_n^{(\alpha,\nu)} - (\alpha-\nu-n) = (J+\nu+n)A_\mu - J = -\frac{\widetilde{\Lambda}_n^{(\alpha,\nu)} - c^{(\nu)}}{d^{(\nu)}} - N - 1.$$

So $\Gamma_n^{(\alpha,\nu)}$ and $\Lambda_n^{(\alpha,\nu)}$ commute for all n. □

Remark 6.4 Corollary 6.3 states that the differential operator $\mathcal{D}^{(\alpha,\nu)}$ has a factorization, where first the lowering operator $\frac{d}{dx}$ is applied and next the raising operator $S^{(\alpha,\nu)}$. The Darboux transform of such a differential operator is obtained by interchanging the order of the lowering and raising operator. This gives a differential operator $\widetilde{\mathcal{D}}^{(\alpha,\nu)} = \frac{d}{dx} \circ S^{(\nu)}$ which has the orthogonal polynomials $P_n^{(\alpha,\nu+1)}$ as eigenfunctions. Explicitly, we have

$$\left(P_n^{(\alpha,\nu+1)}\tilde{\mathcal{D}}^{(\alpha,\nu)}\right)(x) = \frac{d^2P_n^{(\alpha,\nu+1)}}{dx^2}(x)\Phi^{(\alpha,\nu)}(x)^* +$$

$$\frac{dP_n^{(\alpha,\nu+1)}}{dx}(x)\left(\frac{d\Phi^{(\alpha,\nu)}}{dx}(x)^* + \Psi^{(\alpha,\nu)}(x)^*\right) + P_n^{(\alpha,\nu+1)}(x)\frac{d\Psi^{(\alpha,\nu)}}{dx}(x)^* =$$

$$\Xi_n^{(\alpha,\nu+1)}P_n^{(\alpha,\nu+1)}(x),$$

where the eigenvalue $\Xi_n^{(\alpha,\nu+1)}$ is given by

$$\Xi_n^{(\alpha,\nu+1)} = n^2\mathrm{lc}(\Phi^{(\alpha,\nu)})^* + (n+1)\,\mathrm{lc}(\Psi^{(\alpha,\nu)})^*.$$

Proposition 6.5 *With the notations as in Proposition 4.3, Corollary 6.3 and Remark 6.4 and assuming* $\frac{c^{(\nu)}}{d^{(\nu)}} = \nu + \rho$ *for some constant* ρ *we have*

$$\frac{1}{d^{(\nu)}} P\tilde{\mathcal{D}}^{(\alpha,\nu)} = \frac{1}{d^{(\nu+1)}} P\mathcal{D}^{(\alpha,\nu+1)} - PD^{(\alpha,\nu+1)} + \alpha - N - 2\nu - 2 - \rho.$$

Note that the assumption on the quotient is satisfied in all the examples of Sect. 6.1.

Proof The proof is a bit involved, but essentially straightforward. First, to use the explicit expression of Remark 6.4 we need the explicit expressions for $\Phi^{(\alpha,\nu)}(x)^*$, $\Psi^{(\alpha,\nu)}(x)^*$ and compare the difference for ν and $\nu+1$. From Corollary 5.3 we have

$$\frac{1}{d^{(\nu)}}\Phi^{(\alpha,\nu)}(x)^* - \frac{1}{d^{(\nu+1)}}\Phi^{(\alpha,\nu+1)}(x)^* = x\left(\frac{c^{(\nu)}}{d^{(\nu)}} - \frac{c^{(\nu+1)}}{d^{(\nu+1)}}\right) I = -xI,$$

$$\begin{aligned}\frac{1}{d^{(\nu)}}\left(\frac{d\Phi^{(\alpha,\nu)}}{dx}(x)\right)^* &= L_\mu^{(\alpha)}(0)\Big(-2xA_\mu + J + \nu + \rho\Big)L_\mu^{(\alpha)}(0)^{-1} \\ &= -2x\big(1-(1+A_\mu)^{-1}\big) + (\alpha+J)A_\mu + J + \nu + \rho\end{aligned}$$

using Lemma 4.2. To consider $\Psi^{(\alpha,\nu)}(x)^*$ we observe that $\frac{1}{d^{(\nu)}}(\Delta^{(\nu)})^{-1}A_\mu\Delta^{(\nu+1)}$ is actually independent of ν, because of (5.22). So we find from Corollary 5.3 and Lemma 4.2

$$\begin{aligned}&\frac{1}{d^{(\nu)}}\Psi^{(\alpha,\nu)}(x)^* - \frac{1}{d^{(\nu+1)}}\Psi^{(\alpha,\nu+1)}(x)^* \\ &= L_\mu^{(\alpha)}(0)\Big(x(1+A_\mu) - 2J - 2 - 2\nu - \rho\Big)L_\mu^{(\alpha)}(0)^{-1} \\ &= x + x(1-(1+A_\mu)^{-1}) - 2J - 2(\alpha+J)A_\mu - 2 - 2\nu - \rho.\end{aligned}$$

Finally, by Corollary 5.3 and Lemma 4.2

$$\begin{aligned}\frac{1}{d^{(\nu)}}\left(\frac{d\Psi^{(\alpha,\nu)}}{dx}(x)\right)^* &= L_\mu^{(\alpha)}(0)\Big(J - (J+\nu+1)A_\mu - N - 1 - \frac{c^{(\nu)}}{d^{(\nu)}}\Big)L_\mu^{(\alpha)}(0)^{-1} \\ &= (\alpha+J)A_\mu + J - ((\alpha+J)A_\mu + J + \nu + 1)(1-(1+A_\mu)^{-1}) - N - 1 - \nu - \rho \\ &= ((\alpha+J)A_\mu + J + \nu + 1)(1+A_\mu)^{-1} - N - 2\nu - 2 - \rho \\ &= J + (\alpha A_\mu + \nu + 1)(1+A_\mu)^{-1} - N - 2\nu - 2 - \rho.\end{aligned}$$

Collecting these expressions in the differential operator in Remark 6.4 gives

$$\frac{1}{d^{(\nu)}} P\tilde{\mathcal{D}}^{(\alpha,\nu)}(x) - \frac{1}{d^{(\nu+1)}} P\mathcal{D}^{(\alpha,\nu+1)}(x) =$$

$$\frac{d^2P}{dx^2}(x)(-xI) + \frac{dP}{dx}(x)\Big(x(1+A_\mu)^{-1} - J - (\alpha+J)A - 2 - \nu\Big) +$$

$$P(x)\Big(J + (\alpha A_\mu + \nu + 1)(1+A_\mu)^{-1} - N - 2\nu - 2 - \rho\Big) =$$

$$-(PD^{(\alpha,\nu+1)})(x) + \alpha - N - 2\nu - 2 - \rho. \qquad \square$$

As a next application of the shift operators, we calculate the matrix coefficients in the three-term recurrence for the monic polynomials explicitly. As in (2.2), the monic matrix valued Laguerre-type polynomials $P_n^{(\alpha,\nu)}$ satisfy a three-term recurrence relation of the form

$$xP_n^{(\alpha,\nu)}(x) = P^{(\alpha,\nu)}(x) + B_n^{(\alpha,\nu)}P_n^{(\alpha,\nu)}(x) + C_n^{(\alpha,\nu)}P_{n-1}^{(\alpha,\nu)}(x).$$

Proceeding as in [30, §5.3], the coefficients $B_n^{(\alpha,\nu)}$ and $C_n^{(\alpha,\nu)}$, are given by

$$B_n^{(\alpha,\nu)} = X_n^{(\alpha,\nu)} - X_{n+1}^{(\alpha,\nu)}, \qquad C_n^{(\alpha,\nu)} = H_n^{(\alpha,\nu)}(H_{n-1}^{(\alpha,\nu)})^{-1}, \tag{6.29}$$

where $X_n^{(\alpha,\nu)}$ is the one-but-leading coefficient of $P_n^{(\alpha,\nu)}$.

Proposition 6.6 *Assume the conditions of Proposition 6.1. The coefficients of the three-term recurrence relation for the monic Laguerre-type orthogonal polynomials are given by*

$$B_n^{(\alpha,\nu)} = n\,(K_1^{(\alpha,\nu+n-1)})^{-1}(\Psi^{(\alpha,\nu+n-1)}(0))^* - (n+1)\,(K_1^{(\alpha,\nu+n)})^{-1}(\Psi^{(\alpha,\nu+n)}(0))^*,$$

$$C_n^{(\alpha,\nu)} = -nM_\mu^{(\alpha,\nu+n)}(K_n^{(\nu+n,\nu)})^{-n}H_0^{(\nu+n,\nu+n)}(1-A_\mu^*)(H_0^{(\nu+n-1,\nu+n-1)})^{-1} \times (K_{n-1}^{(\nu+n-1,\nu)})^{n-1}(M_\mu^{(\alpha,\nu+n-1)})^{-1}$$

Proof By taking the derivative of $P_n^{(\alpha,\nu)}$ with respect to x and using Proposition 6.1, we find that $(n-1)X_n^{(\alpha,\nu)} = nX_{n+1}^{(\alpha,\nu)}$ which gives $X_n^{(\alpha,\nu)} = nX_1^{(\alpha,\nu+n-1)}$. Using the Rodrigues formula we obtain $P_1^{(\alpha,\nu)}(x) = G_1^{(\alpha,\nu)}(\Psi^{(\alpha,\nu)}(x))^*$. Evaluating at $x=0$ gives

$$X_n^{(\alpha,\nu)} = n\,(K_1^{(\alpha,\nu+n-1)})^{-1}(\Psi^{(\alpha,\nu+n-1)}(0))^*.$$

Replacing $X_n^{(\alpha,\nu)}$ in (6.29) we obtain the expression for $B_n^{(\alpha,\nu)}$.

On the other hand, using the expression for the norm in Theorem 6.2, we find that

$$H_n^{(\alpha,\nu)}(H_{n-1}^{(\alpha,\nu)})^{-1} = -nM_\mu^{(\alpha,\nu+n)}(K_n^{(\nu+n,\nu)})^{-n}H_0^{(\nu+n,\nu+n)}(M_\mu^{(\alpha,\nu+n)})^* \times ((M_\mu^{(\alpha,\nu+n-1)})^*)^{-1}(H_0^{(\nu+n-1,\nu+n-1)})^{-1}(K_{n-1}^{(\nu+n-1,\nu)})^{n-1}(M_\mu^{(\alpha,\nu+n-1)})^{-1}.$$

Now we use Lemma 3.4 to write $L_\mu^{(\nu+n)}(0)^*(L_\mu^{(\nu+n-1)}(0))^*)^{-1} = (1-A_\mu^*)$ and this completes the proof of the proposition. □

As a final application of the shift operators we discuss briefly an expansion formula for the matrix valued Laguerre polynomials arising from the Burchnall formula for matrix valued polynomials satisfying a Rodrigues formula. Note that the matrix valued Laguerre polynomials form an example of the general conditions in [23, §4]. In particular, Burchnall's formula [23, Thm. 4.1] applies. Assuming $\frac{c^{(\nu)}}{d^{(\nu)}} = \nu + \rho$ as in Proposition 6.5, we see from Corollary 5.3 that

$$\left(\Phi^{(\alpha,\nu)}(x)\cdots\Phi^{(\alpha,\nu+k-1)}(x)\right)^* = (-1)^k x^k \left(\prod_{p=0}^{k-1} d^{(\nu+p)}\right)(J - xA_\mu + \nu + \rho)_k.$$

Moreover, since the matrices $G_n^{(\alpha,\nu)}$ in Theorem 6.2 are powers, the result of [23, Cor 4.2] simplifies and we obtain Corollary 6.7.

Corollary 6.7 *Under the conditions of Proposition 6.5 we have the following expansion formula for the matrix valued Laguerre polynomials.*

$$(K_{n+m}^{(\alpha,\nu)})^{-n}P_{n+m}^{(\alpha,\nu)}(x) = \sum_{k=0}^{m\wedge n}(-1)^k\binom{n}{k}\binom{m}{k}k!\Big(\prod_{p=0}^{k-1}d^{(\nu+p)}\Big)x^k \times P_{m-k}^{(\alpha,\nu+n-k)}(x)(K_n^{(\alpha,\nu)})^{n-k}P_{n-k}^{(\alpha,\nu+k)}(x)L_\mu^{(\alpha)}(0)(J - xA_\mu + \nu + \rho)_k L_\mu^{(\alpha)}(0)^{-1}$$

Remark 6.8 In the scalar case the Burchnall identities for some subclasses of the Askey scheme, notably Hermite, Laguerre, Meixner-Pollaczek, Krawtchouk, Meixner and Charlier polynomials, can be used to find expressions for the orthogonal polynomials for the corresponding Toda modification of the weight, i.e. multiplication by e^{-xt}, see [22, Prop. 7.1]. These are precisely the cases where it is easy to "glue" on the exponent e^{-xt} to the classical weight function. In the matrix case, the Toda modification of the matrix weight leads to solutions of the non-abelian Toda lattice, see [23, §2.1] and references given there, and this is worked out in detail for the matrix valued Hermite polynomials in [23, §5]. In the case of the matrix valued Laguerre polynomials, however, we are not led to a corresponding solution of the non-abelian Toda lattice. Essentially, the Burchnall approach fails due to the fact that $\Phi^{(\alpha,\nu)}\ldots\Phi^{(\alpha,\nu+k-1)}$ is a polynomial of degree $2k$ (instead of k), see the discussion in [22]. Even though in the matrix valued case it is straightforward to glue on the exponential e^{-xt} to the matrix weight $W_{\mu}^{(\alpha,\nu)}(x)$, it leads to a linear change in the parameters μ and $\delta_k^{(\nu)}$, and since, the conditions (5.20) and (5.22) form non-linear conditions, the Pearson equations do not hold for the Toda modification. So in particular, Proposition 6.6 is no longer valid for the Toda modified matrix weight.

6.1 Examples

We cannot give all the solutions to the (5.20) and (5.22) in general, due to the non-linearity of these relations. However, if we choose coefficients μ_k such that the quotient μ_{k+1}^2/μ_k^2 coincides with some of the factors in the product $k(N-k)$ of the right-hand side of (5.22), we can give explicit examples of Laguerre-type matrix valued orthogonal polynomials. In the case of Hermite-type matrix valued orthogonal polynomials given in [23] there are two nonlinear conditions. The first one is the same as (5.20) and the second one differs from (5.22) by a factor of $\frac{1}{2}$ on the right-hand side. The examples in this paper are obtained in the same way as in [23].

Example If we assume that $\mu_{k+1} = \sqrt{N-k}\,\mu_k$, then the left-hand side of (5.22) coincides with the factor $N-k$ on the right-hand side. Thus $\mu_k = \sqrt{(N-k+1)_k}$ for $k = 1, \ldots, N$. Then (5.20) and (5.22) give the following recurrence relations

$$\delta_k^{(\nu+1)} = (d^{(\nu)}k + c^{(\nu)})\delta_k^{(\nu)}, \qquad \delta_{k+1}^{(\nu)} = \left(k + \frac{c^{(\nu)}}{d^{(\nu)}}\right)\delta_k^{(\nu)}. \tag{6.30}$$

One solution to (6.31) is given by

$$\delta_k^{(\nu)} = \Gamma(\nu)(\nu)_k, \qquad c^{(\nu)} = \nu, \qquad d^{(\nu)} = 1.$$

Example If we assume that $\mu_{k+1} = \sqrt{k(N-k)}\,\mu_k$, then the left-hand side of (5.22) coincides with the factor $k(N-k)$ on the right-hand side. Then $\mu_k = \sqrt{(k-1)!(N-k+1)_{k-1}}$, $k = 1, \ldots, N$. The relations (5.20) and (5.22) give the following recurrence relations

$$\delta_k^{(\nu+1)} = (d^{(\nu)}k + c^{(\nu)})\delta_k^{(\nu)}, \qquad \delta_{k+1}^{(\nu)} = \left(k + \frac{c^{(\nu)}}{d^{(\nu)}}\right)\delta_k^{(\nu)}. \tag{6.31}$$

A solution to (6.31) is given by

$$\delta_k^{(\nu)} = \lambda^{\nu}\,\Gamma(\nu+k) = (\nu)_k\lambda^{\nu}\Gamma(\nu), \qquad c^{(\nu)} = \nu\lambda, \qquad d^{(\nu)} = \lambda.$$

for some fixed $\lambda > 0$.

Example Now we take $\mu_k = 1$ for all $k = 1, \ldots, N$. Therefore the relations (5.20) and (5.22) are given by

$$1 = \frac{d^{(\nu)}k(N-k)}{(d^{(\nu)}k + c^{(\nu)})} \frac{\delta_{k+1}^{(\nu)}}{\delta_k^{(\nu)}}, \qquad \delta_k^{(\nu+1)} = (d^{(\nu)}k + c^{(\nu)})\delta_{k+1}^{(\nu)}$$

for which

$$d^{(\nu)} = \rho, \qquad c^{(\nu)} = C + \nu\rho, \qquad \delta_k^{(\nu)} = \frac{(1+\nu+C/\rho)_{k-1}}{(k-1)!\,(N-k+1)_{k-1}} \rho^\nu\, \Gamma(1+\nu+C/\rho)$$

with $\rho > 0$, $\nu > 0$ and $C \geq 0$ gives a solution meeting all the conditions.

Acknowledgements E.K. gratefully acknowledges the support and hospitality of FaMAF at Universidad Nacional de Córdoba and the support of an Erasmus+ travel grant. The work of P.R. was supported by Radboud Excellence Fellowship, CONICET grant PIP 112-200801-01533, FONCyT grant PICT 2014-3452 and by SeCyT-UNC.

References

1. N. Aldenhoven, Explicit matrix inverses for lower triangular matrices with entries involving continuous q-ultraspherical polynomials. J. Approx. Theory **199**, 1–12 (2015)
2. N. Aldenhoven, E. Koelink, P. Román, Matrix valued orthogonal polynomials related to the quantum analogue of $(\mathrm{SU}(2) \times \mathrm{SU}(2), diag)$. Ramanujan J. **43**, 243–311 (2017)
3. A.I. Aptekarev, E.M. Nikishin, The scattering problem for a discrete Sturm-Liouville operator. Mat. USSR Sbornik **49**, 325–355 (1984)
4. R. Askey, J. Wilson, Some basic hypergeometric orthogonal polynomials that generalize Jacobi polynomials. Mem. Am. Math. Soc. **54**(319), 1–55 (1985)
5. L. Cagliero, T.H. Koornwinder, Explicit matrix inverses for lower triangular matrices with entries involving Jacobi polynomials. J. Approx. Theory **193**, 20–38 (2015)
6. M.J. Cantero, L. Moral, L. Velázquez, Matrix orthogonal polynomials whose derivatives are also orthogonal. J. Approx. Theory **146**, 174–211 (2007)
7. D. Damanik, A. Pushnitski, B. Simon, The analytic theory of matrix orthogonal polynomials. Surv. Approx. Theory **4**, 1–85 (2008)
8. A.J. Durán, A method to find weight matrices having symmetric second-order differential operators with matrix leading coefficient. Constr. Approx. **29**(2), 181–205 (2009)
9. A.J. Durán, M. de la Iglesia, Some examples of orthogonal matrix polynomials satisfying odd order differential equations. J. Approx. Theory **150**(2), 153–174 (2008)
10. A.J. Durán, F.A. Grünbaum, Orthogonal matrix polynomials satisfying second-order differential equations. Int. Math. Res. Not. **2004**, 461–484 (2004)
11. A.J. Durán, F.A. Grünbaum, Structural formulas for orthogonal matrix polynomials satisfying second-order differential equations. I. Constr. Approx. **22**(2), 255–271 (2005)
12. A.J. Durán, W. Van Assche, Orthogonal matrix polynomials and higher-order recurrence relations. Linear Algebra Appl. **219**, 261–280 (1995)
13. R. Gangolli, V.S. Varadarajan, *Harmonic Analysis of Spherical Functions on Real Reductive Groups*. Ergebnisse der Mathematik und ihrer Grenzgebiete, vol. 101 (Springer, Berlin, 1988)
14. J.S. Geronimo, Scattering theory and matrix orthogonal polynomials on the real line. Circuits Syst. Signal Process. **1**, 471–495 (1982)

15. W. Groenevelt, E. Koelink, A hypergeometric function transform and matrix valued orthogonal polynomials. Constr. Approx. **38**, 277–309 (2013)
16. W. Groenevelt, M.E.H. Ismail, E. Koelink, Spectral theory and matrix valued orthogonal polynomials. Adv. Math. **244**, 91–105 (2013)
17. F.A. Grünbaum, J. Tirao, The algebra of differential operators associated to a weight matrix. Integral Equ. Oper. Theory **58**, 449–475 (2007)
18. F.A. Grünbaum, I. Pacharoni, J. Tirao, Matrix valued spherical functions associated to the complex projective plane. J. Funct. Anal. **188**, 350–441 (2002)
19. F.A. Grünbaum, M.D. de la Iglesia, A. Martínez-Finkelshtein, Properties of matrix orthogonal polynomials via their Riemann-Hilbert characterization. Symmetry Integrability Geom. Methods Appl. **7** (2011), paper 098, 31 pp.
20. G. Heckman, H. Schlichtkrull, *Harmonic Analysis and Special Functions on Symmetric Spaces*. Perspectives in Mathematics, vol. 16 (Academic, San Diego, 1994)
21. G. Heckman, M. van Pruijssen, Matrix valued orthogonal polynomials for Gelfand pairs of rank one. Tohoku Math. J. (2) **68**, 407–436 (2016)
22. M.E.H. Ismail, E. Koelink, P. Román, Generalized Burchnall-type identities for orthogonal polynomials and expansions. Symmetry Integrability Geom. Methods Appl. **14**, 072, 24 pp. (2018)
23. M.E.H. Ismail, E. Koelink, P. Román, Matrix valued Hermite polynomials, Burchnall formulas and non-abelian Toda lattice. arXiv:1811.07219
24. R. Koekoek, Inversion formulas involving orthogonal polynomials and some of their applications, in *Special Functions*, ed. by M.E.H. Ismail, C. Dunkl, R. Wong (World Scientific, River Edge, 2000), pp. 166–180
25. R. Koekoek, R.F. Swarttouw, The Askey-scheme of hypergeometric orthogonal polynomials and its *q-analogue*, online at http://aw.twi.tudelft.nl/~koekoek/askey.html, Report 98-17, Technical University Delft, 1998
26. R. Koekoek, P.A. Lesky, R.F. Swarttouw, *Hypergeometric Orthogonal Polynomials and Their q-Analogues* (Springer, Berlin, 2010)
27. E. Koelink, P. Román, Orthogonal vs. non-orthogonal reducibility of matrix valued measures. Symmetry Integrability Geom. Methods Appl. **12**, 008, 9 pp. (2016)
28. E. Koelink, M. van Pruijssen, P. Román, Matrix-valued orthogonal polynomials related to $(\mathrm{SU}(2) \times \mathrm{SU}(2), \mathrm{diag})$. Int. Math. Res. Not. **2012**, 5673–5730 (2012)
29. E. Koelink, M. van Pruijssen, P. Román, Matrix-valued orthogonal polynomials related to $(\mathrm{SU}(2) \times \mathrm{SU}(2), \mathrm{diag})$*, II*. Publ. Res. Inst. Math. Sci. **49**, 271–312 (2013)
30. E. Koelink, A.M. de los Ríos, P. Román, Matrix-valued Gegenbauer-type type polynomials. Constr. Approx. **46**, 459–487 (2017)
31. E. Koelink, M. van Pruijssen, P. Román, Matrix elements of irreducible representations of $\mathrm{SU}(n+1) \times \mathrm{SU}(n+1)$ and multivariable matrix valued orthogonal polynomials. arXiv:1706.01927
32. T.H. Koornwinder, Matrix elements of irreducible representations of $\mathrm{SU}(2)\times\mathrm{SU}(2)$ and vector-valued orthogonal polynomials. SIAM J. Math. Anal. **16**, 602–613 (1985)
33. M.G. Krein, Infinite J-matrices and a matrix moment problem. Dokl. Akad. Nauk SSSR **69**, 125–128 (1949)
34. M.G. Krein, Fundamental aspects of the representation theory of Hermitian operators with deficiency index (m, m). AMS Transl. Ser. 2 **97**, 75–143 (1971)
35. F.W.J. Olver, A.B. Olde Daalhuis, D.W. Lozier, B.I. Schneider, R.F. Boisvert, C.W. Clark, B.R. Miller, B.V. Saunders (eds.), Digital Library of Mathematical Functions (2010). http://dlmf.nist.gov
36. I. Pacharoni, P. Román, A sequence of matrix valued orthogonal polynomials associated to spherical functions. Constr. Approx. **28**, 127–147 (2008)
37. M. van Pruijssen, Multiplicity free induced representations and orthogonal polynomials. Int. Math. Res. Not. **2018**, 2208–2239 (2018)
38. M. van Pruijssen, P. Román, Matrix-valued classical pairs related to compact Gelfand pairs of rank one. Symmetry Integrability Geom. Methods Appl. **10**, 113, 28 pp. (2014)

Weighted Noncommutative Banach Function Spaces

L. E. Labuschagne and C. Steyn

Dedicated to Prof Ben de Pagter on the occasion of his 65th birthday

Abstract We review the concept of a weighted noncommutative Banach function space. This concept constitutes a generalisation of the by now well-known theory of noncommutative Banach function spaces associated with a semifinite von Neumann algebra. In this review we remind the reader of the quantum statistical problem which gave birth to this concept, we investigate the extent to which a weighted theory of measurable operators agrees with the standard theory, we explore competing methods for defining such spaces, before finally describing the monotone interpolation theory of such spaces.

Keywords Banach function space · Noncommutative · Weighted · Decreasing rearrangement · K-method · Monotone interpolation space

The contribution of L. E. Labuschagne is based on research partially supported by the National Research Foundation (IPRR Grant 96128). Any opinion, findings and conclusions or recommendations expressed in this material are those of the author, and therefore the NRF does not accept any liability in regard thereto.

L. E. Labuschagne (✉) · C. Steyn
DST-NRF CoE in Mathematical and Statistical Sciences (NWU-Node), Unit for BMI, School of Mathematical and Statistical Sciences, NWU, Potchefstroom, South Africa

School of Computer, Statistical and Mathematical Sciences, NWU, Potchefstroom, South Africa
e-mail: louis.labuschagne@nwu.ac.za; steyn.claud@gmail.com

G. Buskes et al. (eds.), *Positivity and Noncommutative Analysis*,
Trends in Mathematics, https://doi.org/10.1007/978-3-030-10850-2_17

1 Introduction

In some sense the embryonic concept of a noncommutative Banach function space dates all the way back to von Neumann [16]. However, it was the 1989 paper of Dodds et al. [2] that proved to be the crucial catalyst for popularising this theory. Buoyed by the parallel development of the theory of noncommutative decreasing rearrangements (see, for example, [6]), this paper sparked a wide-ranging research effort which saw the theory of noncommutative Banach function spaces blossoming into the rich burgeoning theory that it is today; a theory which includes a highly refined theory of Köthe duality [4], real interpolation [4], weak compactness [5], to name just a few topics.

Aspects of this theory were subsequently extended to the context of quasi-Banach spaces by Xu [17]. Spurred on by the objective of applications to quantum statistical mechanics, Labuschagne and Majewski [8] recently pushed the theory in yet another direction, by introducing the notion of weighted noncommutative Banach function spaces. It is the analysis of this class of spaces that is the topic of [13] and of the present paper. In the present paper we will firstly revise the key aspects of [13] which was to investigate competing ways of defining weighted non-commutative Banach function spaces and show their equivalence. Then in the next section revise the ideas from statistical mechanics that lead to the introduction of these spaces, before going on to show that for $\widetilde{\mathcal{M}}$, the weighted topology of convergence in measure agrees with the standard topology of convergence in measure. Then finally we investigate the interpolation theory of these spaces, and ultimately use that theory to show that all quantum Markov dynamical maps canonically induce a contractive action on each weighted noncommutative Banach function space.

Throughout $\mathcal{M}$ will be a semi-finite von Neumann algebra, equipped with a semi-finite, faithful, normal trace τ. Such von Neumann algebras may be extended to the space of τ-measurable operators $\widetilde{\mathcal{M}}$ consisting of all the operators affiliated with $\mathcal{M}$ for which there exists a projection $p \in \mathcal{M}$ with $\tau_x(1-p) \leq \delta$ for some $\delta > 0$, such that $p\mathcal{H} \subset D(a)$ (in which case $\|ap\|_\infty < \infty$). It is well-known that $\widetilde{\mathcal{M}}$ is a complete metrisable *-algebra (with respect to the strong sum and strong product) under the topology of convergence in measure, which is generated by a neighbourhood base at 0 made up of sets of the form $\mathcal{N}(\epsilon, \delta) = \{a \in \widetilde{\mathcal{M}} : \exists p \in \mathbb{P}(M), p\mathcal{H} \subset D(a), \|ap\| \leq \epsilon, \tau(1-p) \leq \delta\}$ where $\epsilon, \delta > 0$ (see [11, 14, 15]). The space $\widetilde{\mathcal{M}}$ is the noncommutative analogue of the completion of $L^\infty(X, \Sigma, \nu)$ under the topology of convergence in measure in the classical setting. By substituting $\widetilde{\mathcal{M}}$ for the measurable functions and the trace τ for the integral, the classical theory of Banach function spaces may be extended to the noncommutative setting with a remarkable degree of faithfulness to the original.

A key ingredient in this theory is the concept of noncommutative decreasing rearrangements of elements of $\widetilde{\mathcal{M}}$. This construct is provided by the so-called generalised singular value function, which for a given element $a \in \widetilde{\mathcal{M}}$ and $t \in [0, \infty)$ is defined by $\mu_t(a) = \inf\{s \geq 0 : \tau(\mathbb{1} - e_s(|a|)) \leq t\}$ where $e_s(|a|)$, $s \in \mathbb{R}$ is the spectral resolution of $|a|$. The function $t \to \mu_t(a)$ will generally be denoted by $\mu(a)$. We proceed to briefly review the concept of a Banach function space of measurable functions on $(0, \infty)$ (see [2]). Though there are subtly different ways in which one can approach the theory, at its most basic level, one starts by defining a Banach function norm ρ on $L^0(X, \Sigma, m)$ to be a mapping $\rho : L^0_+ \to [0, \infty]$ satisfying

[F1] $\rho(f) = 0$ if and only if $f = 0$ a.e.
[F2] $\rho(\lambda f) = \lambda\rho(f)$ for all $f \in L^0_+, \lambda > 0$.
[F3] $\rho(f + g) \leq \rho(f) + \rho(g)$ for all $f, g \in L^0_+$.
[F4] $f \leq g$ implies $\rho(f) \leq \rho(g)$ for all $f, g \in L^0_+$.

Such a ρ may be extended to all of L^0 by setting $\rho(f) = \rho(|f|)$, in which case we may then define $L^\rho(X, \Sigma, m) = \{f \in L^0(X, \Sigma, m) : \rho(f) < \infty\}$. If indeed $L^\rho(X, \Sigma, mu)$ turns out to be a Banach space when equipped with the norm $\rho(\cdot)$, we refer to it as a Banach function space.

A subclass of Banach function spaces exhibiting a high degree of regularity are the so-called Orlicz spaces. We briefly revise this notion. By the term *Orlicz function* we understand a convex function $\Psi : [0, \infty) \to [0, \infty]$ satisfying $\Psi(0) = 0$ and $\lim_{u\to\infty} \Psi(u) = \infty$, which is neither identically zero nor infinite valued on all of $(0, \infty)$, and which is left continuous at $b_\Psi = \sup\{u > 0 : \Psi(u) < \infty\}$. Each such function induces a complementary Orlicz function Ψ^* which may be defined by $\Psi^*(u) = \sup_{v>0}(uv - \Psi(v))$.

The Orlicz space $L^\Psi(X, \Sigma, m)$ associated with Ψ on some σ-finite measure space (X, Σ, m) is defined to be the set

$$L^\Psi = \{a \in L^0 : \Psi(\lambda|a|) \in L^1 \quad \text{for some} \quad \lambda = \lambda(a) > 0\}.$$

This space turns out to be a linear subspace of $L^0(X, \Sigma, m)$, which becomes a Banach space when equipped with either the Luxemburg-Nakano norm

$$\|a\|_\Psi = \inf\{\lambda > 0 : \|\Psi(|a|/\lambda)\|_1 \leq 1\}.$$

or the equivalent Orlicz norm given by

$$\|a\|^0_\Psi = \sup\{\textstyle\int |ab| \, \mathrm{dm} : b \in L^{\Psi^*}, \|b\|_{\Psi^*}\}.$$

Using the above context Dodds et al. [2] formally defined the noncommutative space $L^\rho(\widetilde{\mathcal{M}})$ to be $L^\rho(\widetilde{\mathcal{M}}) = \{f \in \widetilde{\mathcal{M}} : \mu(f) \in L^\rho(0, \infty)\}$ and showed that if

- for any $(f_n) \cup \{f\} \subset L^\rho$ we have that $\rho(f_n) \uparrow \rho(f)$ whenever $0 \leq f_n \uparrow f$ a.e.
- and $L^\rho(0, \infty)$ is rearrangement-invariant in the sense that the situation $f \in L^\rho$, $g \in L^0$ and $\mu(f) = \mu(g)$ forces $g \in L^\rho$ and $\rho(g) = \rho(f)$,

then $L^\rho(\widetilde{\mathcal{M}})$ is a Banach space when equipped with the norm $\|f\|_\rho = \rho(\mu(f))$. The space $L^\rho(\widetilde{\mathcal{M}})$ is said to be *fully symmetric* if for any $f \in L^\rho(\widetilde{\mathcal{M}})$ and $g \in \widetilde{\mathcal{M}}$ the property

$$\int_0^\alpha \mu_t(|g|)\,\mathrm{d}t \leq \int_0^\alpha \mu_t(|f|)\,\mathrm{d}t \quad \text{for all} \quad \alpha > 0$$

ensures that $g \in L^\rho(\widetilde{\mathcal{M}})$ with $\rho(g) \leq \rho(f)$. The property of having $\int_0^\alpha \mu_t(|g|)\,\mathrm{d}t \leq \int_0^\alpha \mu_t(|f|)\,\mathrm{d}t$ for all $\alpha > 0$ is referred to as subordination, and commonly denoted by $g \prec\prec f$.

By restricting attention to function norms satisfying some natural additional restrictions (see, for example, the approach of Bennett and Sharpley [1]), one may obtain a theory where the resultant class of Banach function spaces is all fully symmetric. *In this paper we will follow the doctrine of Bennett and Sharpley, and hence when we speak of a Banach function norm ρ on $L^0(X, \Sigma, m)$, we will hereafter additionally demand that ρ also satisfy the requirements below:*

[F5] *If $f_n \uparrow f$ m-ae, then $\rho(f_n) \uparrow \rho(f)$.*
[F6] *For any measurable subset E of X, $m(E) < \infty \Rightarrow \rho(\chi_E) < \infty$.*
[F7] *For any measurable subset E of X there exists a constant C_E depending only on E and ρ, such that $m(E) < \infty \Rightarrow \int_E f\,dm \leq C_E\rho(f)$ for any $f \in L^0_+$.*

To develop a weighted theory, one starts with an a priori given $x \in (L^1 + L^\infty)(\widetilde{\mathcal{M}})$, and instead of the trace τ, use the quantity

$$\tau_x : \widetilde{\mathcal{M}} \to \mathbb{R} : a \mapsto \int_0^\infty \mu_t(a)\mu_t(x)dt. \tag{1.1}$$

It was shown in [8, Proposition 3.10] that the above functional is "trace-like" in the sense that it satisfies all the requirements for being a finite normal faithful trace, except for the fact that for any $a, b \in \widetilde{\mathcal{M}}$ we at best have that $\tau_x(a + b) \leq \tau(a) + \tau(b)$ instead of additivity. However τ_x is "close enough" to being a trace, for us to be able to use this quantity to construct spaces which closely mimic the behaviour of "tracial" Banach function spaces. (We observe that the theory in [8] was developed under the assumption that $x \in L^1(\widetilde{\mathcal{M}})$. However an examination of the relevant proofs in [8] reveals that all we need for that theory to go through is that $\int_0^t \mu_s(x)\,\mathrm{d}s < \infty$ for any $t > 0$. As can be seen from [4, Proposition 2.6], it is exactly the elements of $(L^1 + L^\infty)(\widetilde{\mathcal{M}})$ that satisfy this requirement. Hence we may freely assume that all the relevant results in [8] hold in this more general setting.) The weighted theory is developed by substituting τ_x for τ, and function norms on $L^0((0, \infty), \nu)$ for function norms on $L^0((0, \infty), \lambda)$, where ν is the measure given by $d\nu(t) = \mu_t(x)\mathrm{d}t$.

2 Descriptions of Weighted Noncommutative Banach Function Spaces

We start our investigation by introducing the notion of τ_x-measurable operators, and of a weighted non-commutative decreasing rearrangement corresponding to some fixed $x \in (L^1 + L^\infty)(\widetilde{\mathcal{M}})$.

Definition 2.1 A closed operator a affiliated with $\mathcal{M}$ is τ_x-measurable if and only if for all $\delta > 0$ there exists a projection $p \in \mathcal{M}$ such that $p\mathcal{H} \subset D(a)$, $\|ap\| < \infty$ and $\tau_x(1-p) \leq \delta$. We denote the set of all τ_x-measurable operators $\widetilde{\mathcal{M}}_x$.

For $a \in \widetilde{\mathcal{M}}_x$, we define the *weighted noncommutative decreasing rearrangement* to be the function $t \mapsto \mu_t(a, x) = \inf\{\|ap\| : p \in \mathbb{P}(\mathcal{M}), \tau_x(1-p) \leq t\}$. As in the tracial case, it can be shown that $\mu_t(a, x) = \inf\{s \geq 0 : \tau_x(e_{(s,\infty)}(|a|)) \leq t\}$, where $e_{(s,\infty)}(|a|)$ is a spectral projection of a.

For the proof of the following two theorems, we refer to Theorem 2.5 and Theorem 3.7 of [13].

Theorem 2.2 *An operator a is τ-measurable if and only if a is τ_x-measurable.*

Theorem 2.3 *Let $a \in \widetilde{\mathcal{M}}$ and consider $\mu_t(a) \in L^0([0,\infty), \nu)$. Then the decreasing rearrangement of $\mu(a)$ with respect to ν is $\mu(a, x)$.*

We recall the definition of a weighted noncommutative Banach function space.

Definition 2.4 ([8]) Let $x \in (L^1+L^\infty)(\widetilde{\mathcal{M}})$, and let ρ be a rearrangement-invariant Banach function norm on $L^0((0,\infty), \mu_t(x)dt)$. The weighted noncommutative Banach function space associated with ρ is then defined to be $L^\rho_x(\widetilde{\mathcal{M}}, \tau) = \{a \in \widetilde{\mathcal{M}} : \mu(a) \in L^\rho((0,\infty), \mu_t(x)dt)\}$.

The above definition was presented in [8] by Labuschagne and Majewski. They then went on to prove that the associated Banach function norm induces a canonical norm on the space $L^\rho_x(\widetilde{\mathcal{M}}, \tau)$. As can be seen, this theory very closely parallels the theory of noncommutative Banach function spaces as introduced by Dodds et al. [2]. However, the proof of existence presented by Dodds, Dodds and de Pagter requires some significant modification before one can obtain the conclusion stated below.

Theorem 2.5 ([8, Theorem 3.7]) *For any Banach function norm ρ on $L^0((0,\infty), \mu_t(x)dt)$, the quantity $\|a\|_\rho = \rho(\mu(a))$ defines a norm on $L^\rho_x(\widetilde{\mathcal{M}}, \tau)$ under which this space is a Banach space.*

An alternative definition, presented in [13], is possible.

Definition 2.6 Let $x \in L^1(\widetilde{\mathcal{M}})$ and ρ a Banach function norm on $L^0(\mathbb{R}^+)$. The space $L^\rho(\widetilde{\mathcal{M}}, \tau_x)$ is the space of all $a \in \widetilde{\mathcal{M}}$ such that $\mu(a, x) \in L^\rho(\mathbb{R}^+)$.

These two definitions can be shown to define the same spaces. We refer to [13] for the proof.

Theorem 2.7 *Let $L^{\rho}_x(\widetilde{\mathcal{M}})$ be a weighted noncommutative Banach function space. Then there exists a rearrangement invariant Banach function norm $\bar{\rho}$ such that $L^{\rho}_x(\widetilde{\mathcal{M}}) = L^{\bar{\rho}}(\widetilde{\mathcal{M}}, \tau_x)$.*

Conversely, for the space $L^{\rho}(\widetilde{\mathcal{M}}, \tau_x)$, there exists a weighted noncommutative Banach function space $L^{\bar{\rho}}_x(\widetilde{\mathcal{M}})$ such that $L^{\rho}(\widetilde{\mathcal{M}}, \tau_x) = L^{\bar{\rho}}_x(\widetilde{\mathcal{M}})$.

In the case of Orlicz spaces, the above correspondence can be made more exact. Specifically given an Orlicz function Ψ, we have that $L^{\Psi}(\widetilde{\mathcal{M}}, \tau_x) = L^{\Psi}_x(\widetilde{\mathcal{M}})$ with full agreement of the (Luxemburg) norms.

3 Application: The Space of Regular Observables

We start by reminding the reader of the definition of the classical regular model (cf [12]).

Definition 3.1 The classical regular statistical model consists of the measure space $\{\Omega, \Sigma, \nu\}$ (where ν is called the reference measure), the state space

$$\mathcal{S}_\nu = \{f \in L^1(\nu) : f > 0 \quad \nu - a.s.,\ E_1(f) = 1\},$$

and the set of measurable functions $L^0(\Omega, \Sigma, \nu)$.

Within this framework, regular random variables are those having all moments finite. It turns out that the random variables having this property can be selected by means of the Laplace transform.

Definition 3.2 The set of all random variables such that

1. $\hat{u}_f(t) = \int \exp(tu) f d\nu, \qquad (t \in \mathbb{R})$ is well defined in a neighbourhood of the origin 0,
2. and $E_f(u) = \int u.f\, d\nu = 0$,

will be denoted by L_f and called the regular random variables.

As is noted by Pistone and Sempi [12] *all the moments of every $u \in L_f$ exist and are the values at* 0 *of the derivatives of $\hat{u}_f$*. So the above definition provides a scheme for selecting random variables such that for each u, any moment $E_f(u^n)$ is well defined, with the random variables with 0 expectation being identified as the "regular" random variables. The following result of Pistone and Sempi provides an interesting connection to weighted Banach function spaces:

Theorem 3.3 (Pistone-Sempi, [12]) *L_f is the closed subspace of the Orlicz space $L^{\cosh -1}(f \cdot \nu)$ of zero expectation random variables.*

Definition 3.4 (Quantum Statistical Model) For semifinite algebras the quantum statistical model consists of the quantum "measure space" $(\mathcal{M}, \tau)$, quantum "densities" with respect to τ in the form of the set $(\mathcal{M}^{+,1}_{*,0})$ of all norm one functionals of the form $\omega_x = \tau(x\cdot)$ where $x \in (\mathcal{M} \cap L^1(\mathcal{M}, \tau))^+$, and the set of τ-measurable operators $\widetilde{\mathcal{M}}$.

Using Pistone and Sempi's work as a template, one may select some $x \in (M_{*,0}^{+,1})$ to play the role of a reference measure. The selection of regular quantum observables then in some sense amounts to the selection of states that behave well with respect to the state described by x. Using the rearrangement of x as a weight, we follow Pistone and Sempi in describing regularity in terms of the Laplace transform.

Definition 3.5 The space L_x^{quant} of regular quantum observables consists of all $g \in \widetilde{\mathcal{M}}$ for which the transform

$$\widehat{\mu_x^g(t)} = \int \exp(t\mu_s(g))\mu_s(x)ds, \qquad t \in \mathbb{R} \tag{3.2}$$

is well-defined in a neighbourhood of the origin.

We then obtain the following noncommutative analogue of Pistone and Sempi's result. The second assertion in this theorem is new.

Theorem 3.6 (Labuschagne and Majewski [8]) *Given some $x \in (\mathcal{M}_{*,0}^{+,1})$, $f \in \widetilde{\mathcal{M}}$ is a regular quantum observable in the sense that $t \to \int_0^\infty \exp(t\mu_s(f))\mu_s(x)\,ds$ exists in a neighbourhood of 0 if and only if $f \in L_x^{\cosh -1}(\mathcal{M}, \tau)$. Moreover $\{f \in L_x^{\cosh -1} : \tau(fx) = 0\}$ is a closed subspace of $L_x^{\cosh -1}(\mathcal{M}, \tau)$.*

Proof The first claim is essentially what was proven in Theorem 3.8 of [8]. To see the second first note that we may assume that $\tau(x) = 1$, which would make $\mu_t(x)dt$ a probability measure on $[0, \infty)$. Let $f \in L_x^{\cosh -1}$ be given. By [6, Theorem 4.2(iii)],

$$|\tau(fx)| \leq \tau(|fx|) = \int_0^\infty \mu_t(fx)\,dt \leq \int_0^\infty \mu_t(f)\mu_t(x)\,dt.$$

Let $L_\Psi((0, \infty), \mu_t(x)dt)$ be the Köthe dual of $L^{\cosh -1}([0, \infty), \mu_t(x)dt)$. Since $\mu_t(x)dt$ is a probability measure, L^∞ injects continuously into $L_\Psi((0, \infty), \mu_t(x)\,dt)$. If we combine this fact with Hölder's inequality, we get that for some $C > 0$,

$$\begin{aligned}|\tau(fx)| \leq \tau(|fx|) &\leq \int_0^\infty \mu_t(f)\mu_t(x)\,dt \\ &\leq \|\mu(f)\|_{\cosh -1}.\|1\|_\Psi^O \\ &\leq C\|\mu(f)\|_{\cosh -1}.\|1\|_\infty \\ &= C\|f\|_{\cosh -1}.\end{aligned}$$

□

Remark 3.7 It is of interest to note that the Köthe dual of the Orlicz space $L^{\cosh -1}$ is isomorphic to the space $L\log(L+1)$. The space $L\log(L+1)$ too has some significance for statistical mechanics in that the space $L^1 \cap L\log(L+1)$ is the natural home for states with well-defined entropy. Thus the pair $\langle L^{\cosh -1}, L\log(L+1)\rangle$ provides a very elegant formalism for studying statistical mechanics. This aspect is explored in more detail in the papers [9] and [10].

4 Topologies of Convergence in Measure for Weighted Spaces

Recall that the topology of convergence in measure, which we will denote by γ_{cm}, has a neighbourhood basis of zero consisting of sets of the form

$$\mathcal{N}(\epsilon,\delta) = \{a \in \widetilde{\mathcal{M}} : \exists p \in \mathbb{P}(M), p\mathcal{H} \subset D(a), \|ap\| \le \epsilon, \tau(1-p) \le \delta\}$$

for $\epsilon, \delta > 0$. By replacing τ with τ_x, we can similarly define a family of sets $\{\mathcal{N}_x(\epsilon,\delta) : \epsilon > 0, \delta > 0\}$ where

$$\mathcal{N}_x(\epsilon,\delta) = \{a \in \widetilde{\mathcal{M}} : \exists p \in \mathbb{P}(M), p\mathcal{H} \subset D(a), \|ap\| \le \epsilon, \tau_x(1-p) \le \delta\}.$$

We will need the following lemma. This result is well known for traces and the proof in our context is nearly identical. The subadditivity does play a small role in the proof, but luckily does not have an effect on the desired conclusion.

Lemma 4.1 *For* $p_1, p_2, \ldots, p_n \in \mathcal{M}_P$,

$$\tau_x(\vee_{i=1}^n p_i) \le \sum_{i=1}^n \tau_x(p_i).$$

Proof First note that for all $i \le n$ we have that $\tau_x(p_i) \le \tau_x(1) < \infty$. For $n = 1$, the result is trivial. Suppose then that the lemma holds for the case $n = k$, i.e.

$$\tau_x(\vee_{i=1}^k p_i) \le \sum_{i=1}^k \tau_x(p_i).$$

Then by the Kaplansky formula $\vee_{i=1}^{k+1} p_i - p_{k+1} \sim \vee_{i=1}^k p_i - (p_{k+1} \wedge \vee_{i=1}^k p_i)$, whence

$$\begin{aligned}
\tau_x(\vee_{i=1}^{k+1} p_i) - \tau_x(p_{k+1}) &\le \tau_x(\vee_{i=1}^{k+1} p_i - p_{k+1}) \\
&= \tau_x(\vee_{i=1}^k p_i - (p_{k+1} \wedge \vee_{i=1}^k p_i) \\
&\le \tau_x(\vee_{i=1}^k p_i)
\end{aligned}$$

Therefore $\tau_x(\vee_{i=1}^{k+1} p_i) \leq \tau_x(\vee_{i=1}^{k} p_i)+\tau_x(p_{k+1})$, and so by the induction hypothesis we have that $\tau_x(\vee_{i=1}^{n} p_i) \leq \sum_{i=1}^{n} \tau_x(p_i)$ for all $n \in \mathbb{N}$. □

Lemma 4.2 *For all $\epsilon, \epsilon_1, \epsilon_2, \delta, \delta_1, \delta_2 > 0$ and $\lambda \in \mathbb{C}$, we have*

1. $\mathcal{N}_x(|\lambda|\epsilon, \delta) = \lambda\mathcal{N}_x(\epsilon, \delta)$
2. $\epsilon_1 < \epsilon_2$ *and* $\delta_1 < \delta_2$ *implies that* $\mathcal{N}_x(\epsilon_1, \delta_1) \subset \mathcal{N}_x(\epsilon_2, \delta_2)$
3. $\mathcal{N}_x(\epsilon_1 \wedge \epsilon_2, \delta_1 \wedge \delta_2) \subset \mathcal{N}_x(\epsilon_1, \delta_1) \cap \mathcal{N}_x(\epsilon_2, \delta_2)$
4. $\mathcal{N}_x(\epsilon_1, \delta_1) + \mathcal{N}_x(\epsilon_2, \delta_2) \subset \mathcal{N}_x(\epsilon_1 + \epsilon_2, \delta_1 + \delta_2)$.

Proof All the above claims follow fairly directly from the definition of the neighbourhoods $\mathcal{N}_x(\epsilon, \delta)$.We prove the final claim, leaving the proofs of the others as an exercise. Let $a \in \mathcal{N}_x(\epsilon_1, \delta_1)$ and $b \in \mathcal{N}_x(\epsilon_2, \delta_2)$ be given. Then there exists a projection $p_1 \in \mathcal{M}$ such that $p_1\mathcal{H} \subset D(a)$, $\|ap_1\| \leq \epsilon_1$, and $\tau_x(1 - p_1) \leq \delta_1$. Similarly there exists a projection $p_2 \in \mathcal{M}$ such that $p_2\mathcal{H} \subset D(b)$, $\|bp_2\| \leq \epsilon_2$, and $\tau_x(1 - p_2) \leq \delta_2$. Let $p = p_1 \wedge p_2$. Then

$$p\mathcal{H} \subset p_1\mathcal{H} \cap p_2\mathcal{H} \subset D(a) \cap D(b) = D(a + b),$$

$$\|(a + b)p\| \leq \|ap\| + \|bp\| \leq \epsilon_1 + \epsilon_2,$$

and

$$\tau_x(1 - p) = \tau_x((p_1 \wedge p_2)^{\perp}) = \tau_x(p_1^{\perp} \vee p_2^{\perp}) \leq \tau_x(1 - p_1) + \tau_x(1 - p_2) \leq \delta_1 + \delta_2.$$

It follows that $a+b \in \mathcal{N}_x(\epsilon_1+\epsilon_2, \delta_1+\delta_2)$ from which the desired result follows. □

As in the tracial case showed in [15], it follows from the above that the family of sets $\{\mathcal{N}_x(\epsilon, \delta) : \epsilon, \delta\}$ form a neighbourhood bases of zero for a vector topology on $\mathcal{M}$. We will denote this vector topology by $\gamma_{cm,x}$.

Theorem 4.3 *The family of sets $\{\mathcal{N}_x(\epsilon, \delta) : \epsilon > 0, \delta > 0\}$ is a neighbourhood basis at zero for γ_{cm}, i.e. $\gamma_{cm,x} = \gamma_{cm}$.*

Proof Let $\mathcal{N}(\epsilon, \delta)$ be given. Let λ_0 be given with $\lambda_0 < \int_0^\delta \mu_t(x)\mathrm{dt}$. We will show that then $\mathcal{N}_x(\epsilon, \lambda_0) \subset \mathcal{N}(\epsilon, \delta)$. Given $a \in \mathcal{N}_x(\epsilon, \lambda_0)$, there exists a projection $q \in \mathbb{P}(\mathcal{M})$ such that $q\mathcal{H} \subset D(a)$, $\|aq\| \leq \epsilon$, and $\tau_x(1 - q) \leq \lambda_0$. Now since

$$\begin{aligned}
\int_0^{\tau(1-q)} \mu_t(x)\mathrm{dt} &= \int_0^\infty \chi_{(0,\tau(1-q))}(t)\mu_t(x)\mathrm{dt} \\
&= \int_0^\infty \mu_t(1 - q)\mu_t(x)\mathrm{dt} \\
&= \tau_x(1 - q) \leq \lambda_0 < \int_0^\delta \mu_t(x)\mathrm{dt},
\end{aligned}$$

it is clear that $\tau(1 - q) < \delta$. Therefore $a \in \mathcal{N}(\epsilon, \delta)$, whence $\mathcal{N}_x(\epsilon, \lambda_0) \subset \mathcal{N}(\epsilon, \delta)$ as claimed.

Let $\mathcal{N}_x(\epsilon, \delta)$ be given. Select $\lambda > 0$ such that $\int_0^{\lambda} \mu_t(x)\mathrm{dt} < \delta$. We show that then $\mathcal{N}(\epsilon, \lambda) \subset \mathcal{N}_x(\epsilon, \delta)$. Let $a \in \mathcal{N}(\epsilon, \lambda)$ be given. Then there exists a projection $q \in \mathbb{P}(\mathcal{M})$ such that $q\mathcal{H} \subset D(a)$, $\|aq\| \leq \epsilon$, $\tau(1-q) \leq \lambda$ and therefore $\tau_x(1-q) = \int_0^{\tau(1-q)} \mu_t(x)\mathrm{dt} \leq \int_0^{\lambda} \mu_t(x)\mathrm{dt} < \delta$. Therefore $a \in \mathcal{N}_x(\epsilon, \delta)$, or equivalently $\mathcal{N}(\epsilon, \lambda) \subset \mathcal{N}_x(\epsilon, \delta)$ as claimed. □

We will present the proofs of Lemmas 4.4, 4.5 and 4.7 for the convenience of the reader, even though the proofs are nearly identical to the proofs of the equivalent statement in the tracial case.

Lemma 4.4 *$a \in \mathcal{N}_x(\epsilon, \delta)$ if and only if $\tau_x(e_{(\epsilon,\infty)}(|a|)) \leq \delta$.*

Proof Let $a \in \widetilde{\mathcal{M}}$ be given with $\tau_x(e_{(\epsilon,\infty)}(|a|)) \leq \delta$ for some $\epsilon, \delta > 0$. Take $p = e_{(\epsilon,\infty)}(|a|)$. Then $p\mathcal{H} \subset D(a)$, $\||a|p\| \leq \epsilon$, and $\tau_x(e_{(\epsilon,\infty)}(|a|)) \leq \delta$. Hence $a \in \mathcal{N}_x(\epsilon, \delta)$.

Conversely suppose $a \in \mathcal{N}_x(\epsilon, \delta)$. Then of course $|a| \in \mathcal{N}_x(\epsilon, \delta)$. So there exists a projection $p \in \mathbb{P}(\mathcal{M})$ such that $p\mathcal{H} \subset D(|a|)$, $\||a|p\| \leq \epsilon$, and $\tau_x(1-p) \leq \delta$. Now for all $\xi \in p\mathcal{H}$, $\||a|\xi\| = \||a|p\xi\| \leq \epsilon\|\xi\|$. But for all $\xi \in e_{(\epsilon,\infty)}(|a|)\mathcal{H}$ we have that $\||a|\xi\| = \||a|e_{(\epsilon,\infty)}(|a|)\xi\| \geq \epsilon\|\xi\|$. It follows that $p \wedge e_{(\epsilon,\infty)}(|a|) = 0$ and therefore that $\tau_x(e_{(\epsilon,\infty)}(|a|)) \leq \tau_x(1-p) \leq \delta$. □

Lemma 4.5 *For $a \in \widetilde{\mathcal{M}}$, $\tau_x(e_{(\epsilon,\infty)}(|a|)) \leq t \Leftrightarrow \mu_t(a, x) \leq \epsilon$.*

Proof If $\tau_x(e_{(\epsilon,\infty)}(|a|)) \leq t$ then $\mu_t(a, x) = \inf\{s > 0 : \tau_x(e_{(s,\infty)}(|a|) \leq t\} \leq \epsilon$. Conversely let $\mu_t(a, x) = \inf\{s > 0 : \tau_x(e_{(s,\infty)}(|a|) \leq t\} \leq \epsilon$, then by Steyn [13, Lemma 3.2 and Proposition 3.4], we have that $\tau_x(e_{(\epsilon,\infty)}(|a|)) \leq \tau_x(e_{(\mu_t(a,x),\infty)}(|a|)) \leq t$. □

Lemma 4.6 *$a \in \mathcal{N}_x(\epsilon, t)$ if and only if $\mu_t(a, x) \leq \epsilon$.*

Proof $a \in \widetilde{\mathcal{M}}_x(\epsilon, t) \Leftrightarrow \tau_x(e_{(\epsilon,\infty)}(|a|)) \leq t \Leftrightarrow \mu_t(a, x) \leq \epsilon$. □

Lemma 4.7 *Let (a_i) be a sequence in $\widetilde{\mathcal{M}}$ and $a \in \widetilde{\mathcal{M}}$. Then the following are equivalent:*

1. $a_i \to a$ *in* $\widetilde{\mathcal{M}}$
2. $\mu_t(a_i - a, x) \to 0$ *for all* $t > 0$
3. $\mu_t(a_i - a) \to 0$ *for all* $t > 0$.

Proof That (1) is equivalent to (3) is well known and proved in [6, Lemma 3.1]. The proof of the equivalence of (1) and (2) is identical to that of [6, Lemma 3.1], though for the sake of the reader we reiterate it here in somewhat more detail. The proof essentially follows by observing that for any net $(a_i) \subset \widetilde{\mathcal{M}}$, we have that

$$\begin{aligned} a_i \to a \text{ in } \widetilde{\mathcal{M}} &\Leftrightarrow \text{for all } \epsilon, t > 0 \text{ there exists an } i_0 \in I \text{ such that } a_i - a \in \mathcal{N}_x(\epsilon, t) \\ &\qquad \text{whenever } i \geq i_0 \\ &\Leftrightarrow \text{for all } \epsilon, t > 0 \text{ there exists an } i \in I \text{ such that } \mu_t(a_i - a, x) \leq \epsilon \\ &\qquad \text{whenever } i \geq i_0 \\ &\Leftrightarrow \mu_t(a_i - a, x) \to 0. \end{aligned}$$

□

Lemma 4.8 *Let $a \in \widetilde{\mathcal{M}}$. Then $\mu_t(a, x) = 0$ when $t > \tau(x)$.*

Proof Let $|a| = \int_0^\infty s e_{(s,\infty)}(|a|)$ be the spectral decomposition of $|a|$ and $t > \tau(x) = \int_0^\infty \mu_s(x)\mathrm{d}s$. Then $\mu_t(a, x) = \inf\{s \geq 0 : \tau_x(e_{(s,\infty)}(|a|)) \leq t\}$. But we have that $\tau_x(e_{(s,\infty)}(|a|)) \leq \tau_x(\mathbf{1}) = \tau(x) \leq t$ for all $s \geq 0$. Therefore $\mu_t(a, x) = 0$. □

For the sake of context, we remind the reader that the measure ν appearing in the following Lemma is given by $d\nu(t) = \mu_t(x)dt$, and that $\mu(f, \nu)$ is the decreasing rearrangement of f with respect to ν.

Lemma 4.9 *Let $0 \leq f \in L^0((0, \infty), dt)$ be non-increasing and $t > 0$. Then $\mu(\chi_{(0,t)} f, \nu) = \chi_{(0,t')}\mu(f, \nu)$ where $t' = \int_0^t \mu_s(x)ds$.*

Proof If $\theta < f(t)$, then since f is non-increasing, we have that $d_\theta(\chi_{(0,t)} f, \nu) = \nu\{s \geq 0 : (\chi_{(0,t)} f)(s) > \theta\} = \nu(0, t) = \int_0^t \mu_s(x)\mathrm{d}s = t'$. If however $\theta > f(t)$, then $d_\theta(\chi_{(0,t)} f, \nu) = d_\theta(f, \nu)$. So

$$d_\theta(\chi_{(0,t)} f, \nu) = \begin{cases} t' & 0 < \theta < f(t) \\ d_\theta(f, \nu) & \theta > f(t). \end{cases} \tag{4.3}$$

Recall that $\mu_s(\chi_{(0,t)} f, \nu) = \inf\{\theta \geq 0 : d_\theta(\chi_{(0,t)} f, \nu) \leq s\}$. First let $0 < s < t'$. Now $\theta \mapsto d_\theta(\chi_{(0,t)} f, \nu)$ is non-increasing. From Eq. (4.3) it is clear that if $\psi \in \{\theta \geq 0 : d_\theta(\chi_{(0,t)} f, \nu) \leq s < t'\}$, then $d_\psi(f, \nu) = d_\psi(\chi_{(0,t)} f, \nu) \leq s$, and so $\psi \in \{\theta \geq 0 : d_\theta(f, \nu) \leq s\}$. From this it follows that $\mu_s(\chi_{(0,t)} f, \nu) \geq \mu_s(f, \nu)$, since $\mu_s(f, \nu) = \inf\{\theta > 0 : d_\theta(f, \nu) \leq s\}$. If, on the other hand, $\psi \in \{\theta \geq 0 : d_\theta(f, \nu) \leq s < t'\}$, then of course $d_\psi(f, \nu) \leq s < t'$, which by Eq. (4.3) and the fact that $r \to d_r(\chi_{(0,t)} f, \nu)$ is decreasing, can only be true if $\psi > f(t)$, in which case it will then follow that $d_\psi(\chi_{(0,t)} f, \nu) = d_\psi(f, \nu) \leq s$. Therefore $\psi \in \{\theta \geq 0 : d_\theta(\chi_{(0,t)} f, \nu) \leq s < t'\}$. From this we have that $\mu_s(f, \nu) \geq \mu_s(\chi_{(0,t)} f, \nu)$, and hence that $\mu_s(\chi_{(0,t)} f, \nu) = \mu_s(f, \nu)$.

Suppose $s \geq t'$, then $d_\theta(\chi_{(0,t)} f, \nu) \leq t' \leq s$ for all $\theta \geq 0$, and so $\mu_s(\chi_{(0,t)} f, \nu) = 0$. Therefore $\mu(\chi_{(0,t)} f, \nu) = \chi_{(0,t')}\mu(f, \nu)$. □

Proposition 4.10 *There exists a non-decreasing, continuous map $(0, \infty) \mapsto (0, \tau(x)] : t \mapsto \int_0^t \mu_s(x)ds = t'$ such that for all non-increasing $0 \leq f \in L^0((0, \infty), ds)$ and $t > 0$, we have that*

$$\int_0^t f(s)\mu_s(x)ds = \int_0^{t'} \mu_s(f, \nu)ds.$$

Proof We have that

$$\begin{aligned}\int_0^t f(s)\mu_s(x)\mathrm{d}s &= \int_0^\infty \chi_{(0,t)}(s) f(s)\mu_s(x)\mathrm{d}s \\ &= \int_0^\infty \mu_s(\chi_{(0,t)} f, \nu)\mathrm{d}s\end{aligned}$$

$$= \int_0^\infty \chi_{(0,t')}(s)\mu_s(f, \nu)\mathrm{ds}$$

$$= \int_0^{t'} \mu(f, \nu)\mathrm{ds}.$$

□

Corollary 4.11 *If $a, b \in \widetilde{\mathcal{M}}$ are given such that $a \prec\prec b$, then $a \prec\prec_x b$ in the sense that for all $t > 0$, $\int_0^t \mu_s(a, x)ds \leq \int_0^t \mu_s(b, x)ds$.*

Proof We have that for all $t > 0$, $\int_0^t \mu_s(a)\mathrm{ds} \leq \int_0^t \mu_s(b)\mathrm{ds}$. By Hardy's lemma [1, Proposition 2.3.6] we then have that for all $t > 0$, $\int_0^t \mu_s(a)\mu_s(x)\mathrm{ds} \leq \int_0^t \mu_s(b)\mu_s(x)\mathrm{ds}$.

Now let $t' > 0$. Suppose that $t' > \tau(x)$. From Lemma 4.8 we have that $\mu_s(a, x) = 0 = \mu_s(b, x)$ for all $s > \tau(x)$. Therefore $\int_0^{t'} \mu_s(a, x)\mathrm{ds} = \int_0^\infty \mu_s(a, x)\mathrm{ds} = \tau_x(a) \leq \tau_x(b) = \int_0^\infty \mu_s(b, x)\mathrm{ds} = \int_0^{t'} \mu_s(b, x)\mathrm{ds}$.

Now suppose that $0 < t' < \tau(x)$. The map $t \mapsto \int_0^t \mu_s(x)\mathrm{ds}$ is non-decreasing and continuous, with 0 mapping onto 0, and $\lim_{t\to\infty} \int_0^t \mu_s(x)\mathrm{ds} = \int_0^\infty \mu_s(x)\mathrm{ds} = \tau(x)$. So by the intermediate value theorem, there exists $t > 0$ such that $t \mapsto t'$. On considering Theorem 2.3 alongside Proposition 4.10, it now follows that $\int_0^t \mu_s(a)\mu_s(x)\mathrm{ds} = \int_0^{t'} \mu_s(\mu(a), \nu)\mathrm{ds} = \int_0^{t'} \mu_s(a, x)\mathrm{ds}$, and similarly that $\int_0^t \mu_s(b)\mu_s(x)\mathrm{ds} = \int_0^{t'} \mu_s(b, x)\mathrm{ds}$. The known inequality $\int_0^t \mu_s(a)\mu_s(x)\mathrm{ds} \leq \int_0^t \mu_s(b)\mu_s(x)\mathrm{ds}$ therefore corresponds to the statement that

$$\int_0^{t'} \mu_s(a, x)\mathrm{ds} \leq \int_0^{t'} \mu_s(b, x)\mathrm{ds}.$$

□

5 Weighted Spaces as Interpolation Spaces

In unfolding the interpolation theory of weighted noncommutative Banach function spaces, we will rely on the theory of monotone interpolation spaces of an abstract Banach couple (X_0, X_1). We will in particular describe the exact monotone interpolation spaces of the Banach couple $(L^1_x(\widetilde{\mathcal{M}}), \mathcal{M})$ in this section. This theory is based on the K-method of interpolation, and an excellent introduction thereto may be found in [1, Chapter 5].

We remind the reader that a pair of Banach spaces (X_0, X_1) is called a Banach couple if indeed both may be embedded in a common Hausdorff topological vector space in which it is possible to make sense of X_0+X_1 and $X_0 \cap X_1$. It is well-known that in this setting the spaces $X_0 + X_1$ and $X_0 \cap X_1$ turn out to be Banach spaces in their own right [1]. A bounded linear map T on $X_0 + X_1$ is called admissible if on both of X_0 and X_1 it restricts to a bounded linear map. The admissible norm of such a map is then given by $\|T\|_{\mathcal{A}} = \max(\|T\|_0, \|T\|_1)$. Any Banach space X which allows for the continuity of the embeddings $X_0 \cap X_1 \hookrightarrow X \hookrightarrow X_0 + X_1$ is called an intermediate space of the couple (X_0, X_1). If an intermediate space X has the property that every admissible map induces a bounded action on X, it is called an interpolation space of the couple (X_0, X_1). If it has the additional property that the norm on X of any admissible map T is majorised by $\|T\|_{\mathcal{A}}$, we say that the space X is an exact interpolation space.

We start our investigation by reviewing some essential concepts from the theory of monotone interpolation spaces. We first prove that the pair $(L^1_x(\mathcal{M}), \mathcal{M})$ is indeed a Banach couple. This can be seen by noting that by Definition 2.6, the space $(L^1 + L^\infty)_x(\widetilde{\mathcal{M}})$ consists of all $a \in \widetilde{\mathcal{M}}$ for which $\int_0^1 \mu_s(a, x)\, ds < \infty$. From the technology outlined in Sect. 2, it is clear that both $L^1_x(\mathcal{M})$ and $\mathcal{M}$ contractively map into this space.

The K-functional for the Banach couple $(L^1_x(\widetilde{\mathcal{M}}), \mathcal{M})$ is for each $t > 0$ defined to be

$$K(a, t) = \inf\{\|a_1\|_{1,x} + t\|a_2\|_\infty\}.$$

Since $t \mapsto K(a, t)$ is a concave function, we can write it in the form

$$K(a, t) = K(a, 0+) + \int_0^t k(a, s)\mathrm{ds}$$

for a unique non-negative, right-continuous, non-increasing function $t \mapsto k(a, t)$ – the so-called k-functional.

Definition 5.1 Let (X_0, X_1) be a compatible couple of Banach spaces. An intermediate space X of (X_0, X_1) is said to be *monotone* if and only if the condition

$$K(g, t) \leq K(f, t) \text{ for each } t > 0 \text{ where } f \in X, g \in X_0 + X_1,$$

ensures that $g \in X$ with $\|g\|_X \leq \|f\|_X$.

The very first step in achieving our stated outcome is to gain a deeper understanding of the structure of the K-functional for this pair. In the tracial case, we have that $K(a, t) = \int_0^t \mu_s(a)\mathrm{ds}$ for all $a \in \widetilde{\mathcal{M}}$ [6, p. 289] and therefore that $k(a, t) = \mu_t(a)$.

To show that the equivalent facts hold for the weighted case, we need the following facts, the proofs of which are identical to those for [6, Lemma 2.5].

Lemma 5.2 *For $a, b \in \widetilde{\mathcal{M}}$ and $x \in (L^1 + L^\infty)(\widetilde{\mathcal{M}})$,*

1. $\lim_{t\downarrow 0} \mu_t(a, x) = \|x\|$.
2. $\mu_t(a, x) = \mu_t(|a|, x) = \mu_t(a^*, x)$ *and* $\mu_t(\lambda a, x) = |\lambda|\mu_t(a, x)$.
3. $\mu_{t+s}(a + b, x) \leq \mu_t(a, x) + \mu_s(b, x)$.
4. $\mu_{t+s}(ab, x) \leq \mu_t(a, x)\mu_s(b, x)$.

Although the proof of the following result is essentially just a copy of the corresponding proof in [6], the strategic significance of this result leads us to state the proof in full.

Theorem 5.3 *Let $a \in L^1_x(\widetilde{\mathcal{M}}) + \mathcal{M}$. Then*

$$\int_0^t \mu_s(a, x)ds = \inf\{\|a_1\|_{1,x} + t\|a_2\|_\infty\}$$

where the infimum is taken over all decompositions $a = a_1 + a_2$ for where $a_1 \in L^1_x(\widetilde{\mathcal{M}})$ and $a_2 \in \mathcal{M}$.

Proof Fix $t > 0$. Let $a = a_1 + a_2$ be a decomposition of a where $a_1 \in L^1_x(\widetilde{\mathcal{M}})$ and $a_2 \in \mathcal{M}$, and let $0 < \alpha < 1$ be given. Then $\mu_s(a, x) \leq \mu_{\alpha s}(a_1, x) + \mu_{(1-\alpha)s}(a_2, x) \leq \mu_{\alpha s}(a, x) + \|a_2\|$, whence

$$\begin{aligned} \int_0^t \mu_s(a, x)ds &\leq \int_0^t \mu_{\alpha s}(a_1, x)ds + t\|a_2\|_\infty \\ &\leq \int_0^\infty \mu_{\alpha s}(a_1, x)ds + t\|a_2\|_\infty \\ &= \alpha^{-1} \int_0^\infty \mu_s(a_1, x)ds + t\|a_2\|_\infty. \end{aligned}$$

On letting $\alpha^{-1} \uparrow 1$, we then get that $\int_0^t \mu_s(a, x)\mathrm{d}t \leq \|a_1\|_{1,x} + t\|a_2\|_\infty$.

It remains to prove the reverse inequality. Again we fix $t > 0$. Let $a = u|a|$ be the polar decomposition of a and $|a| = \int_0^\infty \lambda\, de_t$ be the spectral decomposition of $|a|$. Set $\alpha = \mu_t(a, x)$, $a_1 = u \int_\alpha^\infty (\lambda - \alpha)de_t$ and $a_2 = a - a_1$. For the function

$$f(\lambda) = \begin{cases} 0 & 0 \leq \lambda \leq \alpha \\ \lambda - \alpha & \lambda \geq \alpha \end{cases}$$

we then have that $|a_1| = f(|a|)$. It then follows that

$$\mu_s(a_1, x) = f(\mu_s(a, x)) = \begin{cases} \mu_s(a, x) - \alpha & 0 < s < t \\ 0 & s \geq t \end{cases}.$$

We also have that $\|a_2\| \leq \alpha$. Therefore

$$\begin{aligned} \|a_1\|_{1,x} + t\|a_2\| &\leq \int_0^\infty \mu_s(a_1, x)ds + t\alpha \\ &= \int_0^t (\mu_s(a, x) - \alpha)ds + t\alpha \\ &= \int_0^t \mu_s(a, x)ds, \end{aligned}$$

from which the result then follows. □

From the above it is clear that for the pair $(L^1_x(\widetilde{\mathcal{M}}), \mathcal{M})$, the k-functional for a given $a \in L^1_x(\widetilde{\mathcal{M}})$ is just $t \to \mu_t(a, x)$. As we shall see, the exact monotone interpolation spaces of the couple $(L^1_x(\mathcal{M}), \mathcal{M})$ are described in terms of the k-functional, and the concept of Riesz-Fischer norms, the latter of which we now introduce.

Definition 5.4 A Banach function norm ρ on $L^{0+}([0,\infty), m)^+$, where m is Lebesgue measure, is called a *Riesz-Fischer* norm on $([0,\infty), m)$ if ρ satisfies the following conditions for all $f, g \in L^{0+}([0,\infty), m)^+$, $\alpha > 0$ and E a Borel set:

1. If f and g are equimeasurable, then $\rho(f) = \rho(g)$.
2. $\rho(\chi_E) < \infty$ whenever $m(E) < \infty$.
3. Whenever $m(E) < \infty$, there exists a constant C_E dependent only on ρ and E such that $\int_E f \mathrm{dm} \leq C_E \rho(f)$ for all $f \in L^0([0,\infty), dt)^+$.
4. $\rho(\sum_n f_n) \leq \sum_n \rho(f_n)$ for each sequence (f_n) in $L^{0+}([0,\infty), dt)$.

If ρ also satisfies the condition that $\rho(f) \leq \rho(g)$ whenever $\int_0^t f^*(s)\, dm(s) \leq \int_0^t g^*(s)\, dm(s)$ for any $t > 0$ (where f^* and g^* denote the rearrangements of f and g), then ρ is said to be a *monotone* Riesz-Fischer norm.

Definition 5.5 Let (X_0, X_1) be a compatible couple of Banach spaces, and let ρ be a monotone Riesz-Fischer norm on $([0,\infty), \mathscr{B}, m)$. The space $(X_0, X_1)_\rho$ is defined to be the space of all $f \in \overline{(X_0 \cap X_1)}^{X_0} + X_1$ for which $\rho(k(f, \cdot; X_0, X_1)) < \infty$.

For the Banach couple $(L^1(\widetilde{\mathcal{M}}), \mathcal{M})$, the density of $\mathcal{M} \cap L^1(\widetilde{\mathcal{M}})$ in $L^1(\widetilde{\mathcal{M}})$ ensures that in this context the spaces defined above are simply the spaces $L^\rho(\widetilde{\mathcal{M}})$.

Since for all $a \in L^1_x(\widetilde{\mathcal{M}}) + \mathcal{M}$ we have that $K(a, t) = \int_0^t \mu_s(a, x)\mathrm{ds}$, it follows from the general theory of the K-method of interpolation that $L^1_x(\widetilde{\mathcal{M}}) \cap \mathcal{M}$ is similarly dense in $L^1_x(\widetilde{\mathcal{M}})$ (see [1, Theorem V.1.15]). So for the Banach couple $(L^1_x(\widetilde{\mathcal{M}}), \mathcal{M})$, we similarly have that the spaces defined using the above formalism are just the spaces $L^\rho_x(\widetilde{\mathcal{M}})$. (In this regard observe that it easily follows from condition 3 of Definition 5.4 that $\int_0^1 f^*(s)\, dm(s) < \infty$ for all $f \in L^\rho([0,\infty), m)$ or equivalently that $f \in (L^1 + L^\infty)([0,\infty), m)$ if $f \in L^\rho([0,\infty), m)$. In the context of weighted Banach function spaces, this condition ensures that for any Riesz-Fischer norm ρ, the requirement that $\mu(a, x) \in L^\rho([0,\infty), m)$ forces $a \in L^1_x(\widetilde{\mathcal{M}}) + \mathcal{M}$. So for a Banach function norm ρ, the prescription in the preceding definition produces all of $L^\rho_x(\widetilde{\mathcal{M}})$ as defined in Definition 2.4.)

To achieve the promised description of monotone interpolation spaces, we need the concept of a Gagliardo couple.

Definition 5.6 A Banach couple (X_0, X_1) is a *Gagliardo* couple if:

- for all $f \in X_0 + X_1$ the quantity $\lim_{t\to\infty} K(f, t)$ is finite if and only if $f \in X_0$. If so, then $\lim_{t\to\infty} K(f, t) = \|f\|_{X_0}$.
- for all $f \in X_0 + X_1$ the quantity $\lim_{t\to 0} t^{-1} K(f, t)$ is finite if and only if $f \in X_1$. If so, then $\lim_{t\to 0} t^{-1} K(f, t) = \|f\|_{X_1}$.

The significance of the above concept is that the Banach couple $(L^1_x(\widetilde{\mathcal{M}}), \mathcal{M})$ proves to be a Gagliardo couple. To see this observe that for any $a \in L^1_x(\widetilde{\mathcal{M}})$, we have by Theorem 5.3 that $\lim_{t\to\infty} K(f,t) = \int_0^\infty \mu_s(a,x)\mathrm{d}s = \tau_x(a)$. The first requirement of Definition 5.6 is therefore clearly satisfied. For the second, recall that for any $a \in \widetilde{\mathcal{M}}$ we have that $\|a\|_\infty = \lim_{t\downarrow 0}\mu_t(a,x)$. So given $a \in \mathcal{M}$, this right-continuity of $t \to \mu_t(a,x)$ at 0 with $\mu_0(a,x) = \|a\|_\infty$ ensures that then $\lim_{t\downarrow 0} t^{-1}K(f,t) = \lim_{t\downarrow 0} t^{-1}\int_0^t \mu_s(a,x)\mathrm{d}s = \|a\|_\infty$. Conversely suppose that for some $a \in (L^1_x(\widetilde{\mathcal{M}}) + \mathcal{M})$ we have that $\lim_{t\downarrow 0} t^{-1}K(f,t) = \lim_{t\downarrow 0} t^{-1}\int_0^t \mu_s(a,x)\mathrm{d}s < \infty$. For each $t > 0$, the fact that $s \to \mu_s(a,x)$ is non-decreasing ensures that $t^{-1}\int_0^t \mu_s(a,x)\mathrm{d}s \geq t^{-1}\int_0^t \mu_t(a,x)\mathrm{d}s = \mu_t(a,x)$. Since then $\|a\|_\infty = \sup_{t>0}\mu_t(a,x) = \lim_{t\downarrow 0}\mu_t(a,x) \leq \lim_{t\downarrow 0} t^{-1}\int_0^t \mu_s(a,x)\mathrm{d}s < \infty$, we have that $a \in \mathcal{M}$ as required.

For a general Banach couple the space $(X_0, X_1)_\rho$ (where ρ is a monotone Riesz-Fischer norm on $L^{0+}([0,\infty), m)^+$) is known to be an exact monotone interpolation space [1, Theorem V.3.1]. The importance of the above observation lies in the fact that if the pair (X_0, X_1) is a Gagliardo couple, the converse is also true (see part (i) of [1, Theorem V.3.7])! So if we apply these observations to the pair $(L^1_x(\widetilde{\mathcal{M}}), \mathcal{M})$, we obtain the following conclusion:

Theorem 5.7 *For the Banach couple $(L^1_x(\widetilde{\mathcal{M}}), \mathcal{M})$ every space of the form $L^\rho_x(\widetilde{\mathcal{M}})$, where ρ is a monotone Riesz-Fischer norm on $L^{0+}([0,\infty), m)^+$ is an exact interpolation space. Conversely if X is an exact monotone interpolation space, then there exists a monotone Riesz-Fischer norm ρ on $L^{0+}([0,\infty), m)^+$, such that $X = L^\rho_x(\widetilde{\mathcal{M}})$.*

In conclusion we use the above technology to show that quantum dynamical maps associated with the system described by $\mathcal{M}$ have a canonical action on each of the weighted Banach function spaces. For us such quantum dynamical maps will consist of the maps described by the definition below.

Definition 5.8 We define a positive map $T : \mathcal{M} \to \mathcal{M}$ to be a (sub-)Markov map, if T is normal, $T(\mathbb{1}) \leq \mathbb{1}$, and $\tau \circ T \leq \tau$.

Remark 5.9 It is known that all maps of the form described above extend canonically to maps with a contractive action on $L^1(\mathcal{M}, \tau)$ [7, Theorem 5.1]. As can be seen from [3, Proposition 2.1] and the discussion preceding it, this in turn ensures that T extends canonical to a contractive map on the space $(L^1 + L^\infty)(\widetilde{\mathcal{M}})$ which is "admissible" for the Banach couple $(L^1(\widetilde{\mathcal{M}}), \mathcal{M})$, and that for this canonical extension we have that $T(a) \prec\prec a$ for every $a \in (L^1 + L^\infty)(\widetilde{\mathcal{M}})$.

Theorem 5.10 *For any Riesz-Fischer norm ρ on $L^{0+}([0,\infty), m)^+$, each Markov map $T : \mathcal{M} \to \mathcal{M}$ will canonically induce a contractive map on the space $L^\rho_x(\widetilde{\mathcal{M}})$.*

Proof Let ρ and T be as in the hypothesis. We have already noted in the preceding remark that T canonically extends to a map on $(L^1 + L^\infty)(\widetilde{\mathcal{M}})$ for which we have that $T(a) \prec\prec a$ for each $a \in (L^1 + L^\infty)(\widetilde{\mathcal{M}})$. But by Corollary 4.11 we will then have that $T(a) \prec\prec_x a$ for each $a \in \mathcal{M}$. Since by Theorem 5.7, the space

$L^1_x(\widetilde{\mathcal{M}})$ is a monotone intermediate space of the pair $(L^1_x(\mathcal{M}), \mathcal{M})$ in the sense of Definition 5.1, this in turn ensures that $\|T(a)\|_{1,x} \leq \|a\|_{1,x}$ for each $a \in \mathcal{M}$. Thus T canonically extends to a contractive map on $L^1_x(\widetilde{\mathcal{M}})$. But by Theorem 5.3, contractivity of T on both $\mathcal{M}$ and $L^1_x(\widetilde{\mathcal{M}})$ ensures that $T(a) \prec\prec_x a$ for each $a \in L^1_x(\widetilde{\mathcal{M}})$. The fact that $\|T(a)\|_{\rho,x} \leq \|a\|_{\rho,x}$ for each $a \in L^\rho_x(\widetilde{\mathcal{M}})$ is then a direct consequence of Theorem 5.7. □

References

1. C. Bennett, R. Sharpley, *Interpolation of Operators* (Academic, Boston, 1988)
2. P.G. Dodds, T.K.-Y. Dodds, B. de Pagter, Non-commutative Banach function spaces. Math. Z. **201**, 583–597 (1989)
3. P.G. Dodds, T. K.-Y. Dodds, B. de Pagter, Fully symmetric operator spaces. Integral Equ. Oper. Theory **15**, 942–972 (1992)
4. P.G. Dodds, T.K.-Y. Dodds, B. de Pagter, Noncommutative Köthe duality. Trans. Am. Math. Soc. **339**, 717–750 (1993)
5. P.G. Dodds, F.A. Sukochev, G. Schlüchtermann, Weak compactness criteria in symmetric spaces of measurable operators. Math. Proc. Camb. Philos. Soc. **131**, 363–384 (2001)
6. T. Fack, H. Kosaki, Generalized s-numbers of τ-measurable operators. Pac. J. Math. **123**, 269–300 (1986)
7. U. Haagerup, M. Junge, Q. Xu, A reduction method for noncommutative L^p-spaces and applications. Trans. Am. Math. Soc. **362**, 2125–2165 (2010)
8. L.E. Labuschagne, W.A. Majewski, Maps on non-commutative Orlicz spaces. Ill. J. Math. **55**, 1053–1081 (2011)
9. L.E. Labuschagne, W.A. Majewski, Quantum dynamics on Orlicz spaces. arXiv:1605.01210 [math-ph]
10. W.A. Majewski, L.E. Labuschagne, On applications of Orlicz Spaces to Statistical Physics. Ann. Henri Poincaré **15**, 1197–1221 (2014)
11. E. Nelson, Notes on non-commutative integration. J. Funct. Anal. **15**, 103–116 (1974)
12. G. Pistone, C. Sempi, An infinite-dimensional geometric structure on the space of all the probability measures equivalent to a given one. Ann. Stat. **23**, 1543–1561 (1995)
13. C. Steyn, An alternative approach to weighted non-commutative Banach function spaces. arXiv:1810.12753 [math.OA]
14. M. Takesaki, *Theory of Operator Algebras, Vol I,II,III* (Springer, New York, 2003)
15. M. Terp, L^p spaces associated with von Neumann algebras. Københavs Universitet, Mathematisk Institut, Rapport No 3a (1981)
16. J. von Neumann, Some matrix inequalities and metrization of matrix space. Tomsk Univ. Rev. **1**, 286–300 (1937)
17. Q. Xu, Analytic functions with values in lattices and symmetric spaces of measurable operators. Math. Proc. Camb. Philos. Soc. **109**, 541–563 (1991)

Majorization for Compact and Weakly Compact Polynomials on Banach Lattices

Yongjin Li and Qingying Bu

Dedicated to Professor Ben de Pagter on the occasion of his 65th birthday

Abstract For Banach lattices E and F, let $P,\ Q : E \to F$ be n-homogeneous polynomials such that $0 \leq Q \leq P$. We prove that if F has an order continuous norm and P is weakly compact then Q is also weakly compact. We also prove that if F is an atomic Banach lattice with an order continuous norm and P is compact then Q is also compact.

Keywords Banach lattice · Regular homogeneous polynomial · Fremlin tensor product

1 Introduction

Let E and F be Banach lattices. The classical Wickstead theorem states that if either E^* or F has an order continuous norm then every positive linear operator from E to F dominated by another positive weakly compact linear operator is weakly compact (see, e.g., [1, p. 296, Theorem 5.31]). And the classical Dodds-Fremlin theorem states that if both E^* and F have order continuous norms then every positive linear operator from E to F dominated by another positive compact linear operator is compact (see, e.g., [12, p. 222, Theorem 3.7.13]). Wickstead [18] obtained necessary and sufficient conditions for which the Dodds-Fremlin theorem holds.

Y. Li
Department of Mathematics, Sun Yat-sen University, Guangzhou, People's Republic of China
e-mail: stslyj@mail.sysu.edu.cn

Q. Bu (✉)
Department of Mathematics, University of Mississippi, University, MS, USA
e-mail: qbu@olemiss.edu

G. Buskes et al. (eds.), *Positivity and Noncommutative Analysis*,
Trends in Mathematics, https://doi.org/10.1007/978-3-030-10850-2_18

For a positive integer n, it is known that each regular n-homogeneous polynomial $P : E \to F$ induces a unique regular linear operator $\widetilde{P} : \hat{\otimes}_{n,s,|\pi|}E \to F$, where $\hat{\otimes}_{n,s,|\pi|}E$ is the Fremlin projective symmetric tensor product of E (see, e.g. [3, 8, 9, 16]). In this paper, we first show that if $P : E \to F$ is (weakly) compact then its linearization $\widetilde{P} : \hat{\otimes}_{n,s,|\pi|}E \to F$ is also (weakly) compact under some conditions. Then through the relationship between the (weak) compactness of P and the (weak) compactness of $\widetilde{P}$, we obtain polynomial versions for the Wickstead theorem and the Dodds-Fremlin theorem.

2 Preliminaries

For a vector space X and $n \in \mathbb{N}$, let $\otimes_n X$ denote the n-fold algebraic tensor product of X. For $x_1 \otimes \cdots \otimes x_n \in \otimes_n X$, let $x_1 \otimes_s \cdots \otimes_s x_n$ denote its symmetrization, that is,

$$x_1 \otimes_s \cdots \otimes_s x_n = \frac{1}{n!} \sum_{\sigma \in \pi(n)} x_{\sigma(1)} \otimes \cdots \otimes x_{\sigma(n)},$$

where $\pi(n)$ is the group of permutations of $\{1, \ldots, n\}$. Let $\otimes_{n,s} X$ denote the n-fold algebraic symmetric tensor product of X, that is, the linear span of $\{x_1 \otimes_s \cdots \otimes_s x_n : x_1, \ldots, x_n \in X\}$ in $\otimes_n X$. Define $\theta : X \to \otimes_{n,s} X$ by $\theta(x) = x \otimes \overbrace{\cdots}^{n} \otimes x$ for each $x \in X$. Then θ is an n-homogeneous polynomial. It is known that each $u \in \otimes_{n,s} X$ admits a representation $u = \sum_{i=1}^{m} \lambda_i \theta(x_i)$, where $\lambda_1, \ldots, \lambda_m$ are scalars and $x_1, \ldots, x_m$ are vectors in X.

In addition, if X is a Banach space, the *projective symmetric tensor norm* on $\otimes_{n,s} X$ is defined by

$$\|u\|_{s,\pi} = \inf\left\{ \sum_{i=1}^{m} |\lambda_i| \, \|x_i\|^n : u = \sum_{i=1}^{m} \lambda_i \theta(x_i), \lambda_i \in \mathbb{R}, x_i \in X \right\},$$

and the *n-fold projective symmetric tensor product* of X, denoted by $\hat{\otimes}_{n,s,\pi} X$, is the completion of $\otimes_{n,s} X$ under the norm $\|\cdot\|_{s,\pi}$.

For Banach spaces X and Y, let $\mathcal{L}(X; Y)$ denote the space of all continuous linear operators from X to Y and $\mathcal{P}(^n X; Y)$ denote the space of all continuous n-homogeneous polynomials from X to Y. Each $P \in \mathcal{P}(^n X; Y)$ induces a unique linear operator $\widetilde{P} : \hat{\otimes}_{n,s,\pi} X \to Y$, called the *linearization* of P, such that $P = \widetilde{P} \circ \theta$. Moreover, $\mathcal{P}(^n X; Y)$ is isometrically isomorphic to $\mathcal{L}(\hat{\otimes}_{n,s,\pi} X; Y)$ under the mapping $P \mapsto \widetilde{P}$.

For an n-homogeneous polynomial $P : X \to Y$, let $T_P : X^n \to Y$ denote its associated symmetric n-linear operator. We have the following *polarization formula*

$$T_P(x_1, \ldots, x_n) = \frac{1}{2^n n!} \sum_{\delta_i = \pm 1} \delta_1 \cdots \delta_n P\Big(\sum_{i=1}^{n} \delta_i x_i\Big), \quad x_1, \ldots, x_n \in X.$$

Recall that an n-homogeneous polynomial $P : X \to Y$ is called (*weakly*) *compact* if P takes the closed unit ball of X to a relatively (weakly) compact subset of Y. It is known (see, e.g., [14, Proposition 3.4] and [15]) that $P : X \to Y$ is (weakly) compact if and only if its linearization $\widetilde{P} : \hat{\otimes}_{n,s,\pi} X \to Y$ is (weakly) compact.

For basic knowledge about homogeneous polynomials and symmetric tensor products, we refer to [6, 7, 13, 15].

For a vector lattice E, let E^+ denote its positive cone. The algebraic symmetric tensor product $\otimes_{n,s} E$ with the positive cone generated by $\{\theta(x) : x \in E^+\}$ is an ordered vector space. The *n-fold Fremlin vector lattice symmetric tensor product* of E is a pair $(\bar{\otimes}_{n,s} E, \theta)$ such that

(a) $\bar{\otimes}_{n,s} E$ is a vector lattice in which $\otimes_{n,s} E$ is embedded as a linear subspace, and the n-homogeneous polynomial $\theta : E \to \bar{\otimes}_{n,s} E$ is a lattice homomorphism, and
(b) for any vector lattice F and any n-homogeneous polynomial $P : E \to F$ that is a lattice homomorphism, there exists a unique linear operator $\widetilde{P} : \bar{\otimes}_{n,s} E \to F$, called the *linearization* of P, such that $P = \widetilde{P} \circ \theta$ and $\widetilde{P}$ is also a lattice homomorphism.

If, in addition, E is a Banach lattice, the *positive projective symmetric tensor norm* on $\bar{\otimes}_{n,s} E$ is defined by

$$\|u\|_{s,|\pi|} = \inf\Big\{\sum_{i=1}^{m} \lambda_i \|x_i\|^n : \lambda_i > 0, x_i \in E^+, |u| \le \sum_{i=1}^{m} \lambda_i \theta(x_i)\Big\}$$

for each $u \in \bar{\otimes}_{n,s} E$. Then $\|\cdot\|_{s,|\pi|}$ is a lattice norm on $\bar{\otimes}_{n,s} E$ and
(c) $\otimes_{n,s} E$ is norm dense in $\bar{\otimes}_{n,s} E$, and
(d) the cone generated by $\{\theta(x) : x \in E^+\}$ is norm dense in $(\bar{\otimes}_{n,s} E)^+$.

Let $\hat{\otimes}_{n,s,|\pi|} E$ denote the completion of $\bar{\otimes}_{n,s} E$ under the lattice norm $\|\cdot\|_{s,|\pi|}$. Then $\hat{\otimes}_{n,s,|\pi|} E$ is a Banach lattice, called the *n-fold Fremlin projective symmetric tensor product*, or the *n-fold positive projective symmetric tensor product* of E.

For vector lattices E and F, an n-linear operator $T : E^n \to F$ is called (i) *positive* if $T(x_1, \ldots, x_n) \in F^+$ whenever $x_1, \ldots, x_n \in E^+$, and (ii) *regular* if it is a difference of two positive n-linear operators. An n-homogeneous polynomial $P : E \to F$ is called (i) *positive* if its associated symmetric n-linear operator T_P is positive, and (ii) *regular* if it is a difference of two positive n-homogeneous polynomials. If, in addition, E and F are Banach lattices with F Dedekind complete, let $\mathcal{L}^r(E; F)$ denote the space of all regular linear operators from E to F and $\mathcal{P}^r({}^nE; F)$ denote the space of all regular n-homogeneous polynomials from E to F (with their own regular norm $\|\cdot\|_r$). It is known that each $P \in \mathcal{P}^r({}^nE; F)$ has its linearization $\widetilde{P} \in \mathcal{L}^r(\hat{\otimes}_{n,s,|\pi|}E; F)$ such that $P = \widetilde{P} \circ \theta$. Moreover, $\mathcal{P}^r({}^nE; F)$ is lattice isometric to $\mathcal{L}^r(\hat{\otimes}_{n,s,|\pi|}E; F)$ under the mapping $P \mapsto \widetilde{P}$.

For basic knowledge about the Fremlin vector lattice tensor product, the Fremlin projective tensor product, and regular homogeneous polynomials, we refer to [3, 8, 9, 16].

3 A Polynomial Version of the Wickstead Theorem

For a Banach lattice E, E^* denotes its topological dual and B_E denotes its closed unit ball. Let $B_E^+ = B_E \cap E^+$. For a subset A of E, let $co(A)$ denote the convex hull of A and $\overline{sco}(A)$ denote the closed solid convex hull of A.

For Banach lattices E and F, let $P : E \to F$ be a regular n-homogeneous polynomial and $\widetilde{P} : \hat{\otimes}_{n,s,|\pi|}E \to F$ be its linearization. Since $P = \widetilde{P} \circ \theta$, it follows that the weak compactness of $\widetilde{P}$ implies the weak compactness of P. In the following theorem we will give a condition under which the weak compactness of P implies the weak compactness of $\widetilde{P}$. First we need the following lemma.

Lemma 3.1 *Let E, F be Banach lattices, $P : E \to F$ be a positive n-homogeneous polynomial, and $\widetilde{P} : \hat{\otimes}_{n,s,|\pi|}E \to F$ be its linearization. Then*

$$\widetilde{P}[B_{\hat{\otimes}_{n,s,|\pi|}E}] \subseteq \overline{sco}(P[B_E^+]).$$

Proof First we show that $B_{\hat{\otimes}_{n,s,|\pi|}E} = \overline{sco}(\otimes_{n,s}B_E^+)$, where $\otimes_{n,s}B_E^+ := \{\theta(x) : x \in B_E^+\}$. To end this, we only need to show that $B_{\hat{\otimes}_{n,s,|\pi|}E} \subseteq \overline{sco}(\otimes_{n,s}B_E^+)$. Note that $\bar{\otimes}_{n,s}E$ is dense in $\hat{\otimes}_{n,s,|\pi|}E$. We only need to show that if $u \in \bar{\otimes}_{n,s}E$ with $\|u\|_{s,|\pi|} \leq 1$ then $u \in \overline{sco}(\otimes_{n,s}B_E^+)$. In the case that $\|u\|_{s,|\pi|} < 1$, there exist $\lambda_i > 0$ and $x_i \in E^+$ for $i = 1, \ldots, m$ such that $\sum_{i=1}^m \lambda_i \|x_i\|^n < 1$ and $|u| \leq v := \sum_{i=1}^m \lambda_i \theta(x_i)$. Let $t_i = \lambda_i \|x_i\|^n$ and $y_i = x_i/\|x_i\|$ for $i = 1, \ldots, m$. Then $v = \sum_{i=1}^m t_i \theta(y_i) \in co(\otimes_{n,s}B_E^+)$ and hence, $u \in sol(co(\otimes_{n,s}B_E^+)) \subseteq \overline{sco}(\otimes_{n,s}B_E^+)$. In the case that $\|u\|_{s,|\pi|} = 1$, let $u_k = ku/(k+1)$. Then $\|u_k\|_{s,|\pi|} < 1$ and $u_k \to u$ as $k \to \infty$. It follows from the first part that $u \in \overline{sco}(\otimes_{n,s}B_E^+)$. Therefore, we have proven that $B_{\hat{\otimes}_{n,s,|\pi|}E} = \overline{sco}(\otimes_{n,s}B_E^+)$. Since P is positive, it follows that $\widetilde{P}$ is positive as well. Thus

$$\widetilde{P}[B_{\hat{\otimes}_{n,s,|\pi|}E}] = \widetilde{P}[\overline{sco}(\otimes_{n,s}B_E^+)] \subseteq \overline{sco}(\widetilde{P}[\otimes_{n,s}B_E^+]) = \overline{sco}(P[B_E^+]).$$

The proof is complete. □

If a positive n-homogeneous polynomial $P : E \to F$ is weakly compact, then $P[B_E^+]$ is a positive and relatively weakly compact subset of F. By the Krein-

Smulian theorem, $co(P[B_E^+])$ is also a positive and relatively weakly compact subset of F. Moreover, if the norm on F is order continuous, then it follows from [12, Proposition 2.5.12] that $\overline{sco}(P[B_E^+])$ is a weakly compact subset of F. The following theorem comes straightforwardly from Lemma 3.1.

Theorem 3.2 *Let E be a Banach lattice and F be a Banach lattice with an order continuous norm. Then a positive n-homogeneous polynomial $P : E \to F$ is weakly compact if and only if its linearization $\widetilde{P} : \hat{\otimes}_{n,s,|\pi|}E \to F$ is weakly compact.*

From the Wickstead theorem (see, e.g., [1, p. 296, Theorem 5.31]) and from Theorem 3.2 it is straightforward to obtain a polynomial version of the Wickstead theorem as follows.

Corollary 3.3 *Let E be a Banach lattice and F be a Banach lattice with an order continuous norm. Let $P,\ Q : E \to F$ be n-homogeneous polynomials such that $0 \leq Q \leq P$. If P is weakly compact, then Q is also weakly compact.*

Remark 3.4 For the linear case of Corollary 3.3, that is, both P and Q are linear operators, the Wickstead theorem says that if the norm on E^* is order continuous then Corollary 3.3 holds for any Banach lattice F (the norm on F is not necessary to be order continuous). However, it is no longer true for the polynomial case by the following example.

Example For $1 \leq p \leq n$, define n-homogeneous polynomials $P, Q : \ell_p \to \ell_\infty$ by

$$Q\big((a_k)_k\big) = \Big(\sum_{i=1}^{k} a_i^n\Big)_k \quad \text{and} \quad P\big((a_k)_k\big) = \Big(\sum_{i=1}^{\infty} a_i^n\Big)_k$$

for every $(a_k)_k \in \ell_p$. Then $0 \leq Q \leq P$ and P is compact (which is a rank one operator). Let $e_k = (0, \ldots, 0, 1, 0, 0, \ldots)$ only the k-th coordinate is one and others are zeros. Then

$$Q(e_k) = (0, \ldots, 0, \overset{(k)}{1}, 1, \ldots).$$

Thus $Q(e_k) \downarrow 0$ and $\|Q(e_k)\| = 1$ for every $k \in \mathbb{N}$. By the Dini theorem (see, e.g., [12, p. 33]), the sequence $Q(e_k)$ has no weakly convergent subsequences and hence, Q is not weakly compact.

4 Polynomial Versions of the Dodds-Fremlin Theorem

For Banach lattices E and F, let $P : E \to F$ be a regular n-homogeneous polynomial and $\widetilde{P} : \hat{\otimes}_{n,s,|\pi|}E \to F$ be its linearization. Since $P = \widetilde{P} \circ \theta$, it follows that the compactness of $\widetilde{P}$ always implies the compactness of P. In this section we consider three cases under which the compactness of P implies the compactness of $\widetilde{P}$.

Moreover, if P is positive and compact, then $P[B_E^+]$ is a positive and relatively compact subset of F. By the Mazur theorem, $co(P[B_E^+])$ is also a positive and relatively compact subset of F. Note that if F is an atomic space with an order continuous norm then the solid convex hull of a compact set in F is also compact (see, e.g., [17, Theorem 5]). The following theorem follows straightforwardly from Lemma 3.1.

Theorem 4.1 *Let E be a Banach lattice and F be an atomic Banach lattice with an order continuous norm. Then a positive n-homogeneous polynomial $P : E \to F$ is compact if and only if its linearization $\widetilde{P} : \hat{\otimes}_{n,s,|\pi|} E \to F$ is compact.*

From the Dodds-Fremlin theorem (see, e.g., [18]) and from Theorem 4.1 it is straightforward to obtain a polynomial version of the Dodds-Fremlin theorem as follows.

Corollary 4.2 *Let E be a Banach lattice and F be an atomic Banach lattice with an order continuous norm. Let $P,\ Q : E \to F$ be n-homogeneous polynomials such that $0 \le Q \le P$. If P is compact then Q is also compact.*

In the next theorem we give another condition on F under which the compactness of P implies the compactness of $\widetilde{P}$.

Theorem 4.3 *Let E be a Banach lattice and F be an AL-space. Then a positive n-homogeneous polynomial $P : E \to F$ is compact if and only if its linearization $\widetilde{P} : \hat{\otimes}_{n,s,|\pi|} E \to F$ is compact.*

Proof To show the linear operator $\widetilde{P} : \hat{\otimes}_{n,s,|\pi|} E \to F$ is compact, it suffices to show that its adjoint operator $\widetilde{P}^* : F^* \to (\hat{\otimes}_{n,s,|\pi|} E)^* = \mathcal{P}^r({}^n E; \mathbb{R})$ is compact. Since F^* does not contain a subspace isomorphic to ℓ_1, it suffices to show that $\widetilde{P}^*$ is Dunford-Pettis operator.

Take a sequence $y_k^* \in F^*$ such that $y_k^* \to 0$ weakly in F^*. Since F is an AL-space, F^* is an AM-space with an order unit and hence, the lattice operations on F^* are weakly sequentially continuous. Thus $|y_k^*| \to 0$ weakly in F^*. Since P is compact, $P[B_E]$ is a totally bounded subset of F. For every $\varepsilon > 0$ there exist $x_1, \ldots, x_m \in B_E$ such that

$$P[B_E] \subseteq \bigcup_{i=1}^{m} U(P(x_i), \varepsilon),$$

where

$$U(P(x_i), \varepsilon) = \{y \in F : \|y - P(x_i)\| < \varepsilon\}.$$

Note that $|y_k^*| \to 0$ weak* in F^*. There exists $N \in \mathbb{N}$ such that for any $k > N$,

$$|\langle P(x_i), |y_k^*|\rangle| < \varepsilon, \quad i = 1, \ldots, m.$$

For any $x \in B_E$, there exists $i \in \{1, \ldots, m\}$ such that

$$\|P(x) - P(x_i)\| < \varepsilon.$$

Thus for any $k > N$,

$$\begin{aligned} |\langle P(x), |y_k^*|\rangle| &\leq |\langle P(x) - P(x_i), |y_k^*|\rangle| + |\langle P(x_i), |y_k^*|\rangle| \\ &\leq c \cdot \|P(x) - P(x_i)\| + \varepsilon < (1 + c)\varepsilon, \end{aligned}$$

where $c = \sup_k \|y_k^*\|$. Since $\widetilde{P}^*$ is positive, it follows that for any $k > N$,

$$\begin{aligned} \|\widetilde{P}^*(y_k^*)\|_r \leq \|\widetilde{P}^*(|y_k^*|)\|_r &= \|\widetilde{P}^*(|y_k^*|)\| \\ &= \sup\{|\widetilde{P}^*(|y_k^*|)(x)| : x \in B_E\} \\ &= \sup\{|\langle P(x), |y_k^*|\rangle| : x \in B_E\} < (1 + c)\varepsilon, \end{aligned}$$

which shows that $\widetilde{P}^*(y_k^*) \to 0$ in $\mathcal{P}^r({}^nE; \mathbb{R})$ and hence, $\widetilde{P}^*$ is Dunford-Pettis. □

From Theorem 4.3 we have another polynomial version of the Dodds-Fremlin theorem as follows.

Corollary 4.4 *Let E be a Banach lattice and F be an AL-space. Let $P,\ Q : E \to F$ be n-homogeneous polynomials such that $0 \leq Q \leq P$. If P is compact, then Q is also compact.*

Proof From the end of proof of Theorem 4.3, we have

$$\|\widetilde{Q}^*(y_k^*)\|_r \leq \|\widetilde{Q}^*(|y_k^*|)\|_r \leq \|\widetilde{P}^*(|y_k^*|)\|_r < (1 + c)\varepsilon.$$

Thus $\widetilde{Q}^*$ is a Dunford-Pettis operator and hence, a compact operator, which implies that $\widetilde{Q}$ and hence, Q is a compact operator. □

In the next theorem we show that for a special class of homogeneous polynomials, their compactness implies the compactness of their linearization. Recall that an n-homogeneous polynomial $P : E \to F$ is called *orthogonally additive* if $P(x + y) = P(x) + P(y)$ for any disjoint elements $x, y \in E$.

Theorem 4.5 *Let E be a σ-Dedekind complete Banach lattice and F be a Dedekind complete Banach lattice. If $P : E \to F$ is compact and orthogonally additive, then its linearization $\widetilde{P} : \hat{\otimes}_{n,s,|\pi|}E \to F$ is compact.*

Proof To show that the linear operator $\widetilde{P} : \hat{\otimes}_{n,s,|\pi|}E \to F$ is compact, it suffices to show that its adjoint operator $\widetilde{P}^* : F^* \to (\hat{\otimes}_{n,s,|\pi|}E)^* = \mathcal{P}^r({}^nE; \mathbb{R})$ is compact. Since $P[B_E]$ is totally bounded, for each $\varepsilon > 0$, there exist $z_1, \ldots, z_k \in B_E$ such that

$$P[B_E] \subseteq \bigcup_{i=1}^{k} U(P(z_i), \varepsilon/3). \tag{4.1}$$

Note that $D := \{(\langle P(z_i), y^*\rangle)_{i=1}^k : y^* \in B_{F^*}\}$ is a bounded subset of ℓ_∞^k and hence, totally bounded. Thus there exist $y_1^*, \ldots, y_m^* \in B_{F^*}$ such that

$$D \subseteq \bigcup_{j=1}^{m} U((\langle P(z_i), y_j^*\rangle)_{i=1}^k, \varepsilon/3). \tag{4.2}$$

Now pick any $y^* \in B_{F^*}$. By (4.2) there exists y_j^* for some $1 \le j \le m$ such that

$$|\langle P(z_i), y^*\rangle - \langle P(z_i), y_j^*\rangle| \le \varepsilon/3, \qquad i = 1, \ldots, k. \tag{4.3}$$

For every $x \in B_E$, by (4.1) there exists z_i for some $1 \le i \le k$ such that

$$\|P(x) - P(z_i)\| \le \varepsilon/3. \tag{4.4}$$

Thus by (4.3) and (4.4),

$$\begin{aligned} |(\widetilde{P}^*(y^*) - \widetilde{P}^*(y_j^*))(x)| &= |\langle P(x), y^* - y_j^*\rangle| \\ &\le |\langle P(x) - P(z_i), y^* - y_j^*\rangle| + |\langle P(z_i), y^* - y_j^*\rangle| \\ &\le \|P(x) - P(z_i)\| \cdot (\|y^*\| + \|y_j^*\|) + \varepsilon/3 \le \varepsilon. \end{aligned}$$

It follows that

$$\|\widetilde{P}^*(y^*) - \widetilde{P}^*(y_j^*)\| \le \varepsilon.$$

Since P is orthogonally additive, it is easy to see that $\widetilde{P}^*(y^*) - \widetilde{P}^*(y_j^*) : E \to \mathbb{R}$ is orthogonally additive. It follows from [5, Corollary 4.10] that

$$\left\|\widetilde{P}^*(y^*) - \widetilde{P}^*(y_j^*)\right\|_r \le \frac{n^n}{n!} \cdot \left\|\widetilde{P}^*(y^*) - \widetilde{P}^*(y_j^*)\right\| \le \frac{n^n}{n!}\varepsilon,$$

which shows that

$$\widetilde{P}^*[B_{F^*}] \subseteq \bigcup_{j=1}^{m} U(\widetilde{P}^*(y_j^*), \frac{n^n}{n!}\varepsilon),$$

and hence, $\widetilde{P}^*$ is compact. □

Let E be a reflexive atomic Banach lattice. By Bu and Wong [4, Theorem 6.8] we have that $E\hat{\otimes}_{|\pi|}E$ contains no sublattice isomorphic to ℓ_1 if and only if every positive linear operator from E to E^* is compact, which, by Ji et al. [10, Lemma 5.1], is equivalent to that every positive bilinear operator from $E \times E$ to $\mathbb{R}$ is weakly sequentially continuous. Similarly, we have that $E\hat{\otimes}_{s,|\pi|}E$ contains no sublattice isomorphic to ℓ_1 if and only if every positive symmetric bilinear operator from

$E \times E$ to $\mathbb{R}$ is weakly sequentially continuous, which, by the polarization formula, is equivalent to that every positive 2-homogeneous polynomial from E to $\mathbb{R}$ is weakly sequentially continuous. Using mathematical induction, we conclude that $\hat{\otimes}_{n,s,|\pi|}E$ contains no sublattice isomorphic to ℓ_1 if and only if every positive n-homogeneous polynomial from E to $\mathbb{R}$ is weakly sequentially continuous. Note that a Banach lattice contains no sublattice isomorphic to ℓ_1 if and only if the norm on its dual space is order continuous (see, e.g., [12, Theorem 2.4.14]). It follows from Theorem 4.5 and the Dodds-Fremlin theorem (see, e.g., [18]) that we obtain the third polynomial version of the Dodds-Fremlin theorem as follows.

Corollary 4.6 *Let F be a Banach lattice and E be a reflexive atomic Banach lattice such that every positive n-homogeneous polynomial from E to $\mathbb{R}$ is weakly sequentially continuous. Let $P,\ Q : E \to F$ be n-homogeneous polynomials such that $0 \le Q \le P$. If P is orthogonally additive and compact then Q is also compact.*

Remark 4.7 It is known that if $p > n$ then every continuous n-homogeneous scalar-valued polynomial on ℓ_p is weakly sequentially continuous (see, e.g., [2]). Thus Corollary 4.6 holds for $E = \ell_p\ (p > n)$ and any Banach lattice F. From the example in Sect. 3 it is seen that Corollary 4.6 is no longer true for $E = \ell_p\ (1 < p \le n)$ and $F = \ell_\infty$.

It is worthwhile to mention here that Kusraeva [11] also obtained other polynomial versions of the Dodds-Fremlin theorem for orthogonally additive and compact polynomials.

Acknowledgement Both authors "Yongjin Li and Qingying Bu" are supported by the NNSF (No. 11571378) of China.

References

1. C.D. Aliprantis, O. Burkinshaw, *Positive Operators* (Springer, Dordrecht, 2006)
2. F. Bombal, On polynomial properties in Banach spaces. Atti Sem. Dis. Univ. Modena **41**, 135–146 (1996)
3. Q. Bu, G. Buskes, Polynomials on Banach lattices and positive tensor products. J. Math. Anal. Appl. **388**, 845–862 (2012)
4. Q. Bu, W.C. Wong, Some geometric properties inherited by the positive tensor products of atomic Banach lattices. Indag. Math. **23**, 199–213 (2012)
5. Q. Bu, G. Buskes, Y. Li, Abstract M- and abstract L-spaces of polynomials on Banach lattices. Proc. Edinb. Math. Soc. (2) **58**, 617–629 (2015)
6. S. Dineen, *Complex Analysis on Infinite Dimensional Spaces* (Springer, London, 1999)
7. K. Floret, Natural norms on symmetric tensor products of normed spaces. Note Mat. **17**, 153–188 (1997)
8. D.H. Fremlin, Tensor products of Archimedean vector lattices. Am. J. Math. **94**, 778–798 (1972)
9. D.H. Fremlin, Tensor products of Banach lattices. Math. Ann. **211**, 87–106 (1974)
10. D. Ji, K. Navoyan, Q. Bu, Bases in the space of regular multilinear operators on Banach lattices. J. Math. Anal. Appl. **457**, 803–821 (2018)

11. Z.A. Kusraeva, On compact domination of homogeneous orthogonally additive polynomials. Sib. Math. J. **57**, 519–524 (2016)
12. P. Meyer-Nieberg, *Banach Lattices* (Springer, New York, 1991)
13. J. Mujica, *Complex Analysis in Banach Spaces*. North-Holland Mathematics Studies, vol. 120 (North-Holland, Amsterdam 1986)
14. J. Mujica, Linearization of bounded holomorphic mappings on Banach spaces. Trans. Am. Math. Soc. **324**, 867–887 (1991)
15. R.A. Ryan, Applications of Topological Tensor Products to Infinite Dimensional Holomorphy, Doctoral thesis, Trinity College Dublin, 1980
16. A.R. Schep, Factorization of positive multilinear maps. Ill. J. Math. **28**, 579–591 (1984)
17. A.W. Wickstead, Compact subsets of partially ordered Banach spaces. Math. Ann. **212**, 271–284 (1975)
18. A.W. Wickstead, Converses for the Dodds-Fremlin and Kalton-Saab theorems. Math. Proc. Camb. Philos. Soc. **120**, 175–179 (1996)

The UMD Property for Musielak–Orlicz Spaces

Nick Lindemulder, Mark Veraar, and Ivan Yaroslavtsev

Dedicated to Ben de Pagter on the occasion of his 65th birthday

Abstract In this paper we show that Musielak–Orlicz spaces are UMD spaces under the so-called Δ_2 condition on the generalized Young function and its complemented function. We also prove that if the measure space is divisible, then a Musielak–Orlicz space has the UMD property if and only if it is reflexive. As a consequence we show that reflexive variable Lebesgue spaces $L^{p(\cdot)}$ are UMD spaces.

Keywords UMD · Musielak–Orlicz spaces · Variable L^p-spaces · Young functions · Vector-valued martingales · Variable Lebesgue spaces

1 Introduction

Theclass of Banach spaces X with the UMD (unconditional martingale differences) property is probably the most important one for vector-valued analysis. Harmonic and stochastic analysis in UMD spaces can be found in [6, 8, 23, 34] and references therein. Among other things the UMD property of X implies the following results in X-valued harmonic analysis:

- Marcinkiewicz/Mihlin Fourier multiplier theorems (see [8], [23, Theorem 5.5.10] and [24, Theorem 8.3.9]);
- the T_b-theorem (see [21]);
- the L^p-boundedness of the lattice maximal function (see [34, Theorem 3]);

N. Lindemulder · M. Veraar (✉) · I. Yaroslavtsev
Delft Institute of Applied Mathematics, Delft University of Technology, Delft, The Netherlands
e-mail: N.Lindemulder@tudelft.nl; M.C.Veraar@tudelft.nl; I.S.Yaroslavtsev@tudelft.nl

G. Buskes et al. (eds.), *Positivity and Noncommutative Analysis*,
Trends in Mathematics, https://doi.org/10.1007/978-3-030-10850-2_19

and in X-valued stochastic analysis:

- the L^p-boundedness of martingale transforms (see [6, 23]);
- the continuous time Burkholder–Davis–Gundy inequalities (see [36–38]);
- the lattice Doob maximal L^p-inequality (see [34, 37]).

Most of the classical reflexive spaces are UMD spaces. A list of known spaces with UMD can be found in [23, p. 356]. On the other hand, a relatively simple to state space without UMD is given by $X = L^p(L^q(L^p(L^q(\ldots)\ldots)))$ with $1 < p \neq q < \infty$ (see [33]). The latter space is not only reflexive, but also uniformly convex.

In [17] and [28] it has been shown that an Orlicz space L^Φ has the UMD property if and only if it is reflexive. In [17] the proof is based on an interpolation argument and in [28] a more direct argument is given which uses known Φ-analogues of the defining estimates in UMD. The Musielak-Orlicz spaces of course include all Orlicz spaces but also the important class of variable Lebesgue spaces $L^{p(\cdot)}$.

It seems that a study of the UMD property of Musielak-Orlicz spaces L^Φ and even $L^{p(\cdot)}$ is not available in the literature yet. In the present paper we show that under natural conditions on Φ, the Musielak-Orlicz space L^Φ has UMD. We did not see how to prove this by interpolation arguments and instead we use an idea from [28]. Even in the Orlicz case our proof is simpler, and at the same time it provides more information on the UMD constant.

Theorem 1.1 *Assume that $\Phi, \Psi : T \times [0,\infty) \to [0,\infty]$ are complementary Young functions which both satisfy the Δ_2 condition. Then the Musielak-Orlicz space $L^\Phi(T)$ is a UMD space.*

This theorem is a special case of Theorem 3.1 below, in which we also have an estimate for the UMD constant in terms of the constants appearing in the Δ_2 condition for Φ, Ψ. The result implies the following new result for the variable Lebesgue spaces.

Corollary 1.2 *Assume $1 < p_0 < p_1 < \infty$ and $p : T \to [p_0, p_1]$ is measurable. Then $L^{p(\cdot)}(T)$ is a UMD space.*

In the case the measure space is divisible, one can actually characterize the UMD property in terms of Δ_2 and even in terms of reflexivity (see Corollary 3.3 below).

In the Orlicz setting (i.e. Φ does not dependent on T) the noncommutative analogue of [17, 28] was obtained in [15, Corollary 1.8]. It would be interesting to obtain the noncommutative analogues of our results as well. Details on non-commutative analysis and interpolation theory can be found in the forthcoming book [16].

Notation For a number $p \in [1,\infty]$ we write $p' \in [1,\infty]$ for its Hölder conjugate which satisfies $\frac{1}{p} + \frac{1}{p'} = 1$. For a random variable f, $\mathbb{E}(f)$ denotes the expectation of f.

2 Preliminaries

2.1 *Musielak–Orlicz Spaces*

For details on Orlicz spaces we refer to [27, 35] and references therein. Details on Musielak–Orlicz spaces can be found in [14, 25, 26, 30, 40].

Let X be a Banach space and let (T, Σ, μ) be a σ-finite measure space. We say that a measurable function $\Phi : T \times [0, \infty) \to [0, \infty]$ is a *Young function* if for each $t \in T$,

1. $\Phi(t, 0) = 0$, $\exists x_1, x_2 > 0$ s.t. $\Phi(t, x_1) > 0$ and $\Phi(t, x_2) < \infty$;
2. $\Phi(t, \cdot)$ is increasing, convex and left-continuous.

As a consequence of the above $\lim_{x\to\infty} \Phi(t, x) = \infty$.

A function Φ with the above properties is a.e. differentiable, the right-derivative $\varphi := \partial_x \Phi$ is increasing and

$$\Phi(t, x) = \int_0^x \varphi(t, \lambda) d\lambda, \quad t \in T, \ x \in \mathbb{R}_+.$$

Note that the function $\varphi(t, \cdot)$ has a right-continuous version since any increasing function has at most countably many discontinuities, so $\varphi(t, \lambda) = \lim_{\varepsilon \to 0} \varphi(t, \lambda + \varepsilon)$ for each $t \in T$ for a.e. $\lambda \in [0, \infty)$.

For a strongly measurable function $f : T \to X$ we say that $f \in L^\Phi(T; X)$ if there exists a $\lambda > 0$ such that

$$\int_T \Phi(t, \|f(t)\|_X/\lambda) \, d\mu(t) < \infty.$$

The space $L^\Phi(T; X)$ equipped with the norm

$$\|f\|_{L^\Phi(T;X)} := \inf\Big\{\lambda > 0 : \int_T \Phi(t, \|f(t)\|_X/\lambda) \, d\mu(t) \le 1\Big\} \tag{2.1}$$

is a Banach space. Here as usual we identify functions which are almost everywhere identical. The space $L^\Phi(T; X)$ is called the X-valued *Musielak-Orlicz space* associated with Φ.

The following norm will also be useful in the sequel.

$$\|f\|_{X,\Phi} := \inf_{\lambda>0} \frac{1}{\lambda}\Big[1 + \int_T \Phi(t, \lambda\|f(t)\|_X) \, d\mu(t)\Big]. \tag{2.2}$$

It is simple to check that this gives an equivalent norm (see [22, Lemma 2.1])

$$\|f\|_{L^\Phi(T;X)} \le \|f\|_{X,\Phi} \le 2\|f\|_{L^\Phi(T;X)}. \tag{2.3}$$

In case $X = \mathbb{R}$ or $X = \mathbb{C}$, we write $L^\Phi(T)$ for the above space.

Example 2.1 Let $p : T \to [1, \infty)$ be a measurable function and let $\Phi(t, \lambda) = |\lambda|^{p(t)}$. Then $L^{\Phi}(T)$ coincides with the variable Lebesgue space $L^{p(\cdot)}$.

Next we recall condition Δ_2 from [30, Theorem 8.13]. There it was used to study the dual space of the Musielak–Orlicz space and to prove uniform convexity and in particular reflexivity (see [30, Section 11]). Let $L^1_+(T) \subset L^1(T)$ be the set of all nonnegative integrable functions.

Definition 2.2 A Young function $\Phi : T \times [0, \infty) \to [0, \infty]$ is said to be in Δ_2 if there exists a $K > 1$ and an $h \in L^1_+(T)$ such that for a.a. $t \in T$

$$\Phi(t, 2\lambda) \leq K\Phi(t, \lambda) + h(t), \quad \lambda \in [0, \infty).$$

Note that $\Phi \in \Delta_2$ implies that $\Phi(t, \lambda) < \infty$ for almost all $t \in T$ and all $\lambda \in [0, \infty)$. Unlike is standard for Young's function independent of T, the condition Δ_2 depends on the measure space; namely, if one has that $\mu(T) = \infty$ and h does not depend on $t \in T$, then $h = 0$.

If $\Phi : T \times [0, \infty) \to [0, \infty]$ is a Young function, we define its *complemented function* $\Psi : T \times [0, \infty) \to [0, \infty]$ by the Legendre transform

$$\Psi(t, x) = \sup_{y \geq 0}(xy - \Phi(t, y)).$$

Then Ψ is a Young function as well. Moreover, one can check that the complemented function of $\Psi(t, \cdot)$ equals $\Phi(t, \cdot)$.

Example 2.3 Let the notations be as in Example 2.1. Then the following statements hold.

(i) Φ is in Δ_2 if and only if $p \in L^{\infty}(T)$, in which case Φ satisfies the Δ_2-condition with $K = 2^{\|p\|_{\infty}}$ and $h = 0$.
(ii) The complemented function Ψ to Φ is given by

$$\Psi(t, x) = x^{p'(t)}\mathbf{1}_{\{p>1\}\times[0,\infty)}(t, x) + \infty \cdot \mathbf{1}_{\{p=1\}\times(1,\infty)}(t, x),$$

where $p'(t) = p(t)'$ is the Hölder conjugate.

In particular, Φ and Ψ are both in Δ_2 if and only if $1 < \operatorname{ess\,inf} p \leq \operatorname{ess\,sup} p < \infty$, in which case $\Psi(t, x) = x^{p'(t)}$ for a.a. $t \in T$ and all $x \in [0, \infty)$.

Proof Let us only give the proof of Example (2.3)(i). If $p \in L^{\infty}(T)$, then, for a.a. $t \in T$,

$$\Phi(t, 2\lambda) = 2^{p(t)}\Phi(t, \lambda) \leq 2^{\|p\|_{\infty}}\Phi(t, \lambda), \quad \lambda \in [0, \infty).$$

Conversely assume that Φ is in Δ_2. Let K and h be as in the Δ_2 condition for Φ. Then, for a.a. $t \in T$ and all $\lambda \in [0, \infty)$,

$$2^{p(t)}\Phi(t, \lambda) = \Phi(t, 2\lambda) \leq K\Phi(t, \lambda) + h(t)$$

and thus

$$(2^{p(t)} - K)\Phi(t, \lambda) \le h(t).$$

As $\lim_{\lambda\to\infty} \Phi(t, \lambda) = \infty$, this implies that $2^{p(t)} \le K$ for a.a. $t \in T$. Hence, $p \in L^\infty(T)$. □

By the properties of the functions Φ and Ψ one can check that for $\varphi = \partial_x \Phi$ and $\psi = \partial_x \Psi$ (where φ and ψ are taken right-continuous in x), we have $\varphi^{-1}(t, \cdot) = \psi(t, \cdot)$, where

$$\varphi^{-1}(t, y) = \sup\{x : \varphi(t, x) \le y\} \quad y \ge 0. \tag{2.4}$$

Note that $\psi(t, \varphi(t, x)) \ge x$ and $\varphi(t, \psi(t, x)) \ge x$ because of the above choices.

Recall Young's inequality (see [27, Section I.2] or [35, Proposition 15.1.2]) for a.a. $t \in T$,

$$xy \le \Phi(t, x) + \Psi(t, y), \quad x, y \ge 0 \tag{2.5}$$

with equality if and only if $y = \varphi(t, x)$ or $x = \psi(t, y)$.

Lemma 2.4 *Let $\Phi : T \times [0, \infty) \to [0, \infty)$ be a Young function and let Ψ be its complemented function. If $\Phi \in \Delta_2$ with constant $K > 1$ and $h \in L^1_+(T)$, then for almost all $t \in T$,*

$$\Psi(t, \lambda) \le \frac{K}{K-1}\lambda\psi(t, \lambda) + \frac{1}{K}h(t), \quad \lambda \ge 0.$$

Proof We use a variation of the argument in [27, Section 1.4]. By the Δ_2 condition there exist $K > 1$ and $h \in L^1_+(T)$ such that for almost all $t \in T$ and all $\lambda \ge 0$

$$K\Phi(t, \lambda) + h(t) \ge \Phi(t, 2\lambda) = \int_0^{2\lambda} \varphi(t, x)dx \ge \int_\lambda^{2\lambda} \varphi(t, x)dx \ge \lambda\varphi(t, \lambda),$$

where we used the fact that $\varphi(t, \cdot)$ is increasing. Using the identity case of (2.5) we obtain

$$K\lambda\varphi(t, \lambda) - K\Psi(t, \varphi(t, \lambda)) + h(t) \ge \lambda\varphi(t, \lambda)$$

Therefore,

$$\frac{K\Psi(t, \varphi(t, \lambda))}{\lambda\varphi(t, \lambda)} \le K - 1 + \frac{h(t)}{\lambda\varphi(t, \lambda)}.$$

Taking $\lambda = \psi(t,x)$ and using the estimates below (2.4) and the fact that $y \mapsto \frac{\Psi(t,y)}{y}$ is increasing (see [27, (1.18)]) we obtain

$$\begin{aligned}\frac{K\Psi(t,x)}{x\psi(t,x)} &\le \frac{K\Psi(t,\varphi(t,\psi(t,x)))}{\psi(t,x)\varphi(t,\psi(t,x))}\\ &\le K-1+\frac{h(t)}{\psi(t,x)\varphi(t,\psi(t,x))} \le K-1+\frac{h(t)}{x\psi(t,x)}.\end{aligned}$$

We may conclude that

$$\Psi(t,x) \le \frac{K-1}{K}x\psi(t,x) + \frac{1}{K}h(t).$$

□

2.2 *UMD Spaces*

For details on UMD spaces the reader is referred to [6, 34] and the monographs [23, 24]. Let $(\Omega, \mathcal{A}, (\mathscr{F}_n)_{n\ge 0}, \mathbb{P})$ denote a filtered probability space which is rich enough in the sense that it supports an i.i.d. sequence $(\varepsilon_n)_{n\ge 0}$ such that $\mathbb{P}(\varepsilon_n = 1) = \mathbb{P}(\varepsilon_n = -1) = \frac{1}{2}$ for each $n \ge 0$. Such a sequence is called a Rademacher sequence.

For a sequence of random variables $f = (f_n)_{n\ge 0}$ with values in X, we write $f_n^* = \sup_{k\le n}\|f_k\|_X$ and $f^* = \sup_{k\ge 0}\|f_k\|_X$. Moreover, if $\epsilon = (\epsilon_n)_{n\ge 0}$ is a sequence of signs, we write $(\epsilon * f)_n = \sum_{k=0}^n \epsilon_k(f_k - f_{k-1})$, where $f_{-1} = 0$.

We say that X is a UMD space if there exists a $p \in (1,\infty)$ and $\beta \in [1,\infty)$ such that for all L^p-martingales $f = (f_n)_{n\ge 0}$ and all sequences of signs $\epsilon = (\epsilon_n)_{n\ge 0}$ we have that

$$\|(\epsilon * f)_n\|_{L^p(\Omega;X)} \le \beta\|f_n\|_{L^p(\Omega;X)}, \quad n \ge 0,$$

where the least admissible constant β is denoted by $\beta_{p,X}$ and is called *the UMD constant*. If the above holds for some $p \in (1,\infty)$, then it holds for all $p \in (1,\infty)$. Examples and counterexamples of UMD spaces have been mentioned in the introduction. Every UMD space is (super-)reflexive (see [23, Theorem 4.3.8]).

We say that $f = (f)_{n\ge 0}$ is a *Paley–Walsh martingale* if f is a martingale with respect to the filtration $(\mathscr{F}_n)_{n\ge 0}$ with $\mathscr{F}_0 = \{\varnothing, \Omega\}$ and $\mathscr{F}_n = \sigma\{\varepsilon_k : 1 \le k \le n\}$ for some Rademacher sequence $(\varepsilon_k)_{k\ge 0}$ and if $f_0 = 0$.

The following result follows from [4, Theorems 1.1 and 3.2].

Proposition 2.5 *Let X be a Banach space. Then X is a UMD space if and only if for all Paley–Walsh martingales f and all sequences of signs ϵ we have*

$$\sup_{n\geq 0} \|f_n\|_{L^\infty(\Omega;X)} < \infty \implies \mathbb{P}(\sup_{n\geq 0} \|(\epsilon * f)_n\|_X < \infty) > 0.$$

We will also need the following lemma which allows to estimate the ϵ-transform for different functions than $\Phi(x) = |x|^p$. This lemma is a straightforward extension [3, p. 1001] where the case $b = 0$ was considered. Moreover, since we only state it for Paley–Walsh martingales it follows from [6, Proof of (10)].

Lemma 2.6 *Assume X is a UMD space. Let $\Phi : [0,\infty) \to [0,\infty)$ be a Young function and assume that there exist constants $K > 1$ and $b \geq 0$ such that*

$$\Phi(2\lambda) \leq K\Phi(\lambda) + b, \quad \lambda \geq 0. \tag{2.6}$$

Let $f = (f_n)_{n\geq 0}$ be a Paley–Walsh martingale, $\epsilon = (\epsilon_n)_{n\geq 0}$ be a sequence of signs, and set $g := \epsilon * f$. Then there exists a constant $C_{K,X} \geq 0$ only depending on K and (the UMD constant of) X such that

$$\mathbb{E}\Phi(g^*) \leq C_{K,X}(\mathbb{E}\Phi(f^*) + b).$$

Remark 2.7 To obtain Lemma 2.6 in the case of general martingales (as it is done in [3, p. 1001]), one can use the Davis decomposition to reduce to a bad part and a good part of f. To estimate the bad part of the Davis decomposition one can use [7, Theorem 3.2 and the proof of Theorem 2.1] (see [31, Proposition A-3-5] and [29, Theorem 53] for a simpler proof).

Recall that X is a UMD space if and only if it is ζ-*convex*, i.e. there exists a biconvex function $\zeta : X \times X \to \mathbb{R}$ such that $\zeta(0,0) > 0$ and $\zeta(x,y) \leq \|x+y\|$ for all $x, y \in X$ with $\|x\| = \|y\| = 1$ (see [4, 5, 23]). By the ζ-function we will usually mean the *optimal ζ-function* which can be defined as the supremum over all admissible ζ's, and this obviously satisfies the required conditions.

The following theorem can be found in [6, equation (20)].

Theorem 2.8 (Burkholder) *Let X be a UMD Banach space and let $\zeta : X \times X \to \mathbb{R}$ be an optimal ζ-function (i.e. $\zeta(0,0)$ is maximal). For any $1 < p < \infty$ one then has that*

$$\frac{1}{\zeta(0,0)} \leq \beta_{p,X} \leq \frac{72}{\zeta(0,0)}\frac{(p+1)^2}{p-1}. \tag{2.7}$$

The following lemma follows from [4, p. 49].

Lemma 2.9 *Let X be a UMD Banach space and let $\zeta : X \times X \to \mathbb{R}$ be an optimal ζ-function (i.e. $\zeta(0,0)$ is maximal). Then for any $\varepsilon > 0$ there exists an X-valued Paley–Walsh martingale $f = (f_n)_{n\geq 1}$ which starts in zero and a sequence of signs*

$\epsilon = (\epsilon_n)_{n\geq 1}$ *such that* $\mathbb{P}(g^* > 1) = 1$ *and* $\sup_{n\geq 1} \mathbb{E}\|f_n\| \leq \frac{\zeta(0,0)}{2} + \varepsilon$, *where* $g := \epsilon * f$.

Remark 2.10 Let us compute an upper bound for $C_{K,X}$ in Lemma 2.6. Let $M \geq 1$ be the least integer such that $2^{-M} \leq \frac{\zeta(0,0)}{48K}$. Fix $\beta := 2$ and $\delta := 2^{-M}$. Then by formula [3, (1.8)] one has that for f and g from Lemma 2.6

$$\mathbb{P}(g^* > 2\lambda, f^* \leq 2^{-M}\lambda) \leq \varepsilon \mathbb{P}(g^* > \lambda), \quad \lambda > 0,$$

where $\varepsilon = 3c\delta/(\beta - \delta - 1) \leq 1/(2K)$, and where we used the fact that the constant c from [3, (1.2)] can be bounded from above by $4/\zeta(0,0)$ by [32, Theorem 3.26 and Lemma 3.23] (see also [39]). Note that by (2.6)

$$\Phi(\beta\lambda) \leq K\Phi(\lambda) + b, \quad \Phi(\delta^{-1}\lambda) \leq K^M\Phi(\lambda) + bMK^M, \quad \lambda > 0,$$

where one needs to iterate (2.6) M times in order to get the latter inequality. Therefore by exploiting [2, proof of Lemma 7.1] one has the following analogue of the formula [2, (7.6)]

$$\mathbb{E}\Phi(2^{-1}g^*) \leq \varepsilon\mathbb{E}\Phi(g^*) + K^M\mathbb{E}\Phi(f^*) + bMK^M,$$

and by using the fact that $\mathbb{E}\Phi(g^*) \leq K\mathbb{E}\Phi(2^{-1}g^*) + b$ and the fact that $\varepsilon K \leq 1/2$ one has that

$$\mathbb{E}\Phi(g^*) \leq 2K^{M+1}\mathbb{E}\Phi(f^*) + 2b(1 + MK^{M+1}) \leq 2(MK^{M+1} + 1)(\mathbb{E}\Phi(f^*) + b),$$

so $C_{K,X} \leq 2(MK^{M+1} + 1)$, where M can be taken $[\log_2 \frac{48K}{\zeta(0,0)}] + 1$. Of course this bound is not optimal.

3 Musielak-Orlicz Spaces Are UMD Spaces

The main result of this paper is the following.

Theorem 3.1 *Assume X is a UMD space. Let $\Phi, \Psi : T \times [0,\infty) \to [0,\infty)$ be complemented Young functions which both satisfy Δ_2. Then the Musielak-Orlicz space $L^\Phi(T; X)$ is a UMD space.*

Moreover, if $\Phi \in \Delta_2$ with constant K_Φ and $h_\Phi \in L^1_+(T)$ and $\Psi \in \Delta_2$ with constant K_Ψ and $h_\Psi \in L^1_+(T)$, then for the optimal ζ-function $\zeta : L^\Phi(T; X) \times L^\Phi(T; X) \to \mathbb{R}$ (see the discussion preceding Theorem 2.8) one has that

$$\zeta(0,0) \geq \frac{1}{6K_\Psi C_{K_\Phi,X} C_h}, \tag{3.8}$$

and

$$\beta_{p,L^{\Phi}(T;X)} \leq 432 K_{\Psi} C_{K_{\Phi},X} C_h \frac{(p+1)^2}{p-1}, \tag{3.9}$$

where $C_{K_{\Phi},X}$ *is as in Lemma 2.6 and* $C_h := 2 + \|h_{\Phi}\|_{L^1(T)} + \frac{1}{K_{\Psi}}\|h_{\Psi}\|_{L^1(T)}$.

This result is well-known in the case of $\Phi(x) = |x|^p$, and then it is a simple consequence of Fubini's theorem which allows to write $L^p(\Omega; L^p(T; X)) = L^p(T; L^p(\Omega; X))$ and to apply the UMD property of X pointwise a.e. in T (see [23, Proposition 4.2.15]). Such a Fubini argument is necessarily limited to L^p-spaces. Indeed, the Kolmogorov–Nagumo theorem says that for Banach function spaces E and F one has $E(F) = F(E)$ isomorphically, if and only if E and F are weighted L^p-spaces (see [1, Theorem 3.1]).

To prove Theorem 3.1 we will use several results from the preliminaries. Moreover, we will need the following scalar-valued result which is a well-known version of Doob's maximal inequality for a certain class of Young functions.

Proposition 3.2 *Suppose that* $\Phi : [0,\infty) \to [0,\infty]$ *is a Young function with a right-continuous derivative* $\varphi : [0,\infty) \to [0,\infty)$ *and that there exists a* $q \in (1,\infty)$ *and* $c \in [0,\infty)$ *such that*

$$\Phi(\lambda) \leq \frac{1}{q}\lambda\varphi(\lambda) + c, \quad \lambda \geq 0.$$

Then for all nonnegative submartingales $(f_n)_{n\geq 0}$

$$\mathbb{E}\Phi(f_n^*) \leq \mathbb{E}\Phi(q' f_n) + c, \quad n \geq 0.$$

In particular, $\|f_n^*\|_{\Phi} \leq q'(1+c)\|f_n\|_{\Phi}$.

Proof The result for $c = 0$ is proved in [13, estimate (104.5)], and the case $c > 0$ follows by a simple modification of that argument. The final assertion follows from the obtained estimate since for any $\lambda > 0$ we have

$$\begin{aligned}\|f_n^*\|_{\Phi} &\leq \lambda^{-1}(1 + \mathbb{E}\Phi(\lambda f_n^*)) \\ &\leq \lambda^{-1}(c + 1 + \mathbb{E}\Phi(\lambda q' f_n)) \\ &\leq (c+1)\lambda^{-1}(1 + \mathbb{E}\Phi(\lambda q' f_n)).\end{aligned}$$

Taking the infimum over all $\lambda > 0$ yields the required conclusion. □

Proof of Theorem 3.1 Let $Y := L^{\Phi}(T; X)$. In order to prove the theorem we will use Proposition 2.5. Let $f = (f_n)_{n\geq 0}$ be a Paley–Walsh martingale with values in Y. Let $\epsilon = (\epsilon_n)_{n\geq 0}$ be a sequence of signs and $g := \epsilon * f$. We will show that $g^* < \infty$ a.s. For this it is enough to show that

$$\mathbb{E} \sup_{n\geq 0} \|g_n\|_Y \leq K_\Psi C_{K_\Phi,X} C_h \sup_{n\geq 0} \|f_n\|_{L^\infty(\Omega;Y)}, \tag{3.10}$$

where $C_{K_\Phi,X}$ is as in Lemma 2.6 and $C_h = 2 + \|h_\Phi\|_{L^1(T)} + \frac{1}{K_\Psi}\|h_\Psi\|_{L^1(T)}$. By homogeneity we can assume $\sup_{n\geq 0} \|f_n\|_{L^\infty(\Omega;Y)} = 1$.

We know that $K_\Phi > 1$ and a function $h_\Phi \in L^1_+(T)$ satisfy the following inequality

$$\Phi(t, 2\lambda) \leq K_\Phi \Phi(t, \lambda) + h_\Phi(t), \quad \lambda \in [0, \infty),\ t \in T. \tag{3.11}$$

Since Ψ satisfies Δ_2 with constant $K_\Psi > 1$ and $h_\Psi \in L^1_+(T)$ it follows from Lemma 2.4 that

$$\Phi(t, \lambda) \leq \frac{K_\Psi - 1}{K_\Psi} \lambda\varphi(t, \lambda) + \frac{1}{K_\Psi} h_\Psi(t), \quad \lambda \in [0, \infty), \quad t \in T. \tag{3.12}$$

One can check that for a.e. $t \in T$, $f(t)$ is an X-valued martingale and $g_n(t) = (\epsilon * (f(t)))_n$ (use that f is a Paley–Walsh martingale). Therefore, first applying (3.11) and Lemma 2.6 and then (3.12) and Proposition 3.2 to the submartingale $(\|f_k(t)\|_X)_{k\geq 0}$ gives that for almost all $t \in T$,

$$\begin{aligned}\mathbb{E}\Phi(t, \sup_{k\leq n} \|g_k(t)\|_X) &\leq C_{K_\Phi,X} \left[\mathbb{E}\Phi(t, \sup_{k\leq n} \|f_k(t)\|_X) + h_\Phi(t)\right]\\ &\leq C_{K_\Phi,X} \left[\mathbb{E}\Phi(t, K_\Psi \|f_n(t)\|_X) + h_\Phi(t) + \tfrac{1}{K_\Psi} h_\Psi(t)\right].\end{aligned}$$

The same holds with (f, g) replaced by $(\lambda f, \lambda g)$ for any $\lambda > 0$. Integrating over $t \in T$ (and using (2.2)) we find that

$$\begin{aligned}\mathbb{E} \sup_{k\leq n} \|g_k\|_{X,\Phi} &\leq \mathbb{E} \sup_{k\leq n} \frac{1}{\lambda}\Big(1 + \int_T \Phi(t, \lambda\|g_k(t)\|_X) d\mu(t)\Big)\\ &\overset{(*)}{\leq} \mathbb{E}\frac{1}{\lambda}\Big(1 + \int_T \Phi(t, \sup_{k\leq n} \lambda\|g_k(t)\|_X) d\mu(t)\Big)\\ &\leq C_{K_\Phi,X} \mathbb{E}\frac{1}{\lambda}\Big(1 + \int_T \Phi(t, K_\Psi\lambda\|f_n(t)\|_X) d\mu(t) + \|h_\Phi\|_{L^1(T)} + \|\tfrac{1}{K_\Psi} h_\Psi\|_{L^1(T)}\Big),\end{aligned}$$

where $(*)$ follows form the fact that $\sup \int \leq \int \sup$ and the fact that the map $\lambda \mapsto \Phi(t, \lambda)$ is increasing in $\lambda \geq 0$. Since $\int_T \Phi(t, \|f_n(t)\|_X) d\mu(t) \leq 1$ a.s. by (2.1) and the assumption $\|f_n\|_{L^\infty(\Omega;Y)} \leq 1$, it follows by setting $\lambda = 1/K_\Psi$ that

$$\mathbb{E} \sup_{k \leq n} \|g_k\|_{X,\Phi} \leq K_\Psi C_{K_\Phi,X} C_h.$$

Now the required estimate (3.10) follows from (2.3).

For proving (3.8) and (3.9) we will use Lemma 2.9. By the first part of the proof, $Y = L^\Phi(T;X)$ is UMD. Fix $\varepsilon > 0$. Then by Lemma 2.9 there exists a Y-valued Paley–Walsh martingale $f = (f_n)_{n\geq 0}$ which starts in zero and a sequence of signs $\epsilon = (\epsilon_n)_{n\geq 0}$ such that $\mathbb{P}(g^* > 1) = 1$ and $\sup_{n\geq 0} \mathbb{E}\|f_n\|_Y \leq \frac{\zeta(0,0)}{2} + \varepsilon$, where $g := \epsilon * f$. By Burkholder [4, Lemma 3.1] there exist discrete Y-valued Paley–Walsh martingales $F = (F_n)_{n\geq 0}$ and $G = (G_n)_{n\geq 0}$ such that $G = \epsilon * F$, $\mathbb{P}(G^* > 1) \leq 1/2$, and

$$\sup_{n\geq 0} \|F_n\|_{L^\infty(\Omega;Y)} \leq 6 \sup_{n\geq 0} \mathbb{E}\|f_n\|_Y.$$

Therefore, by (3.10),

$$\begin{aligned}\frac{1}{2} \leq \mathbb{E}G^* &\leq K_\Psi C_{K_\Phi,X} C_h \sup_{n\geq 0} \|F_n\|_{L^\infty(\Omega;Y)} \\ &\leq 6K_\Psi C_{K_\Phi,X} C_h \sup_{n\geq 0} \mathbb{E}\|f_n\|_Y \leq 3K_\Psi C_{K_\Phi,X} C_h(\zeta(0,0) + 2\varepsilon),\end{aligned}$$

so letting $\varepsilon \to 0$ gives (3.8). Equation (3.9) follows from (3.8) and (2.7). □

We recover the following result of [17] and [28]. Recall that a measure space (T, Σ, μ) is divisible if for every $A \in \Sigma$ and $t \in (0,1)$ there exist sets $B, C \in \Sigma$ such that $B, C \subseteq A$, $\mu(B) = t\mu(A)$ and $\mu(C) = (1-t)\mu(A)$. The divisibility condition is only needed in the implication (ii)⇒(iii).

Corollary 3.3 *Let $X \neq \{0\}$ be a Banach space and assume that T is divisible and σ-finite. Suppose $\Phi, \Psi : T \times [0,\infty) \to [0,\infty]$ are complementary Young functions. Then the following are equivalent:*

(i) $L^\Phi(T;X)$ is a UMD space;
(ii) $L^\Phi(T)$ is reflexive and X is a UMD space;
(iii) Φ and Ψ both satisfy Δ_2 and X is a UMD space.

For the proof we will need the following lemma which follows from [26, Theorem 2.2] and [25, Theorem 4.7].

Lemma 3.4 *Let $\Phi, \Psi : T \times [0,\infty) \to [0,\infty]$ be complementary Young functions. Then there exists a decomposition $L^\Phi(T)^* = L^\Psi(T) \oplus \Lambda$ of the dual of $L^\Phi(T)$ into a direct sum of two Banach spaces, where $g \in L^\Psi(T)$ acts on $L^\Phi(T)$ in the* *following way:*

$$\langle f, g\rangle = \int_T fg \, \mathrm{d}\mu, \quad f \in L^{\Phi}(T).$$

Proof of Corollary 3.3 (i)⇒(ii): Fix $h \in L^{\Phi}(T)$ and $x \in X$ of norm one. Then $L^{\Phi}(T)$ and X can be identified with the closed subspaces $L^{\Phi}(T) \otimes x$ and $h \otimes X$ of the UMD space $L^{\Phi}(T; X)$, respectively, and therefore have UMD themselves. In particular, $L^{\Phi}(T)$ is reflexive.

(ii)⇒(iii): We show that Φ satisfies Δ_2. The proof for Ψ is similar. By Lemma 3.4, $L^{\Phi}(T)^* = L^{\Psi}(T) \oplus \Lambda$, so

$$L^{\Phi}(T)^{**} = L^{\Psi}(T)^* \oplus \Lambda^* \supseteq L^{\Phi}(T) \oplus \Lambda^*,$$

where the latter inclusion follows from Lemma 3.4 and means that $L^{\Phi}(T) \oplus \Lambda^*$ is a closed subspace of $L^{\Psi}(T)^* \oplus \Lambda^*$, and hence of $L^{\Phi}(T)^{**}$. Thus $\Lambda = 0$ due to the reflexivity of $L^{\Phi}(T)$, and since T is divisible the desired statement follows from [26, Corollary 1.7.4].

(iii)⇒(i): This follows from Theorem 3.1. □

As a consequence of the above results many other spaces are UMD as well. Indeed, it suffices to be isomorphic to a closed subspace (or quotient space) of an $L^{\Phi}(T; X)$ space with UMD. This applies to the Musielak–Orlicz variants of Sobolev, Besov, and Triebel–Lizorkin spaces.

Remark 3.5 A result of Rubio de Francia (see [34, p. 214]) states that for a Banach function space E and a Banach space X one has that $E(X)$ is a UMD space if and only if E and X are both UMD spaces. Therefore, it actually suffices to consider $X = \mathbb{R}$ in the proof of Theorem 3.1. Since our argument works in the vector-valued case without difficulty, we consider that setting from the start.

For the variable Lebesgue spaces we obtain the following consequence. For a measurable mapping $p : T \to [1, \infty]$ we will write $p_+ = \|p\|_{L^{\infty}(T)}$ and $p_- = \|1/p\|^{-1}_{L^{\infty}(T)}$.

Corollary 3.6 *Let $X \neq \{0\}$ be a Banach space and assume T is divisible and σ-finite. Assume $p : T \to [1, \infty]$ is measurable. Then the following assertions are equivalent.*

(i) $L^{p(\cdot)}(T; X)$ is a UMD space;
(ii) $L^{p(\cdot)}(T)$ is reflexive and X is a UMD space;
(iii) $p_- > 1$ and $p_+ < \infty$ and X is a UMD space.

The result that $L^{p(\cdot)}(T)$ is reflexive if and only if $p_- > 1$ and $p_+ < \infty$ can also be found in [12, Proposition 2.79&Corollary 2.81] and [14, Remark 3.4.8].

Proof This is an immediate consequence of Corollary 3.3 and Example 2.3. □

Remark 3.7 Let $Y := L^{p(\cdot)}(T; X)$. Let us bound $\zeta(0, 0)$ from below using (3.8). Note that by Example 2.3 one has that $K_\Phi = p_+$, $K_\Psi = p'_-$, and $h_\Phi = h_\Psi = 0$, so

$$\zeta(0, 0) \geq \frac{1}{6K_\Psi C_{K_\Phi, X} C_h} = \frac{1}{3p'_- C_{p_+, X}},$$

where an upper bound for $C_{p_+, X}$ can be found using Remark 2.10. From this one can obtain an upper bound for the UMD constant using (2.7).

In [20] the analytic Radon–Nikodym (ARNP) and analytic UMD (AUMD) properties are shown to hold for Musielak-Orlicz spaces $L^\Phi(T)$ where Φ satisfies a condition which is slightly more restrictive than Δ_2. To end the paper we want to state a related conjecture about spaces satisfying a randomized version of UMD. In order to introduce it let $(\Omega', \mathscr{A}', \mathbb{P}')$ be a second probability space with a Rademacher sequence $\varepsilon' = (\varepsilon'_n)_{n \geq 1}$. A Banach space X is said to be a UMD^-_{PW} space if there is a $p \in [1, \infty)$ and a constant $C \geq 0$ such that for all Paley–Walsh martingales f,

$$\|f\|_{L^p(\Omega; X)} \leq \|\varepsilon' * f\|_{L^p(\Omega \times \Omega'; X)}.$$

This property turns out to be p-independent, and it gives a more general class of Banach spaces than the UMD spaces (see [9–11, 18]). For instance, L^1 is a UMD^-_{PW} space.

Conjecture 3.8 Assume $\Phi : T \times [0, \infty) \to [0, \infty)$ is a Young function such that $\Phi \in \Delta_2$. Then $L^\Phi(T)$ is UMD^-_{PW}.

The conjecture is open also in the case Φ is not dependent on T. If $\Phi : [0, \infty) \to [0, \infty)$ is merely continuous, increasing to infinity and $\Phi(0) = 0$ and satisfies Δ_2, then the same question can be asked. However, in this case $L^\Phi(T)$ is not a Banach space, but only a quasi-Banach space. Some evidence for the conjecture can be found in [11, Theorem 4.1] and [19, Theorem 1.1] where analogues of Lemma 2.6 can be found (only $\Phi \in \Delta_2$ is needed in the proof). Doob's inequality plays a less prominent role for UMD^- because of [11, Lemma 2.2]. Similar questions can be asked for the possibly more restrictive "decoupling property" of a quasi-Banach space X introduced in [10, 11].

Acknowledgements The authors would like to thank Emiel Lorist and Jan van Neerven for helpful comments. The authors "Nick Lindemulder" and "Mark Veraar" were supported by the Vidi subsidy 639.032.427 of the Netherlands Organisation for Scientific Research (NWO).

References

1. A. Boccuto, A.V. Bukhvalov, A.R. Sambucini, Some inequalities in classical spaces with mixed norms. Positivity **6**(4), 393–411 (2002)
2. D.L. Burkholder, Distribution function inequalities for martingales. Ann. Probab. **1**, 19–42 (1973)
3. D.L. Burkholder, A geometrical characterization of Banach spaces in which martingale difference sequences are unconditional. Ann. Probab. **9**(6), 997–1011 (1981)
4. D.L. Burkholder, Martingale transforms and the geometry of Banach spaces, in *Probability in Banach Spaces, III (Medford, Mass., 1980)*. Lecture Notes in Mathematics, vol. 860 (Springer, Berlin, 1981), pp. 35–50
5. D.L. Burkholder, A geometric condition that implies the existence of certain singular integrals of Banach-space-valued functions, in *Conference on Harmonic Analysis in Honor of Antoni Zygmund, Vol. I, II (Chicago, Ill., 1981)*. Wadsworth Mathematics Series (Wadsworth, Belmont, 1983), pp. 270–286
6. D.L. Burkholder, Martingales and singular integrals in Banach spaces, in *Handbook of the Geometry of Banach Spaces, Vol. I* (North-Holland, Amsterdam, 2001), pp. 233–269
7. D.L. Burkholder, B.J. Davis, R.F. Gundy, Integral inequalities for convex functions of operators on martingales, in *Berkeley Symposium on Mathematical Statistics and Probability* (University of California Press, Berkeley, 1972), pp. 223–240
8. Ph. Clément, B. de Pagter, F. A. Sukochev, H. Witvliet, Schauder decompositions and multiplier theorems. Studia Math. **138**(2), 135–163 (2000)
9. S.G. Cox, S. Geiss, On decoupling in Banach spaces. arXiv:1805.12377 (2018)
10. S.G. Cox, M.C. Veraar, Some remarks on tangent martingale difference sequences in L^1-spaces. Electron. Commun. Probab. **12**, 421–433 (2007)
11. S.G. Cox, M.C. Veraar, Vector-valued decoupling and the Burkholder-Davis-Gundy inequality. Illinois J. Math. **55**(1), 343–375 (2011)
12. D.V. Cruz-Uribe, A. Fiorenza, *Variable Lebesgue Spaces: Foundations and Harmonic Analysis*. Applied and Numerical Harmonic Analysis (Birkhäuser/Springer, Heidelberg, 2013)
13. C. Dellacherie, P.-A. Meyer, *Probabilities and Potential*. North-Holland Mathematics Studies, vol. 29 (North-Holland, Amsterdam, 1978)
14. L. Diening, P. Harjulehto, P. Hästö, M. Ružička, *Lebesgue and Sobolev Spaces with Variable Exponents*. Lecture Notes in Mathematics, vol. 2017 (Springer, Heidelberg, 2011)
15. P.G. Dodds, F.A. Sukochev, Contractibility of the linear group in Banach spaces of measurable operators. Integr. Equ. Oper. Theory **26**(3), 305–337 (1996)
16. P.G. Dodds, B. de Pagter, F.A. Sukochev, Theory of noncommutative integration. unpublished monograph, to appear
17. D.L. Fernandez, J.B. Garcia, Interpolation of Orlicz-valued function spaces and U.M.D. property. Studia Math. **99**(1), 23–40 (1991)
18. D.J.H. Garling, Random martingale transform inequalities, in *Probability in Banach Spaces 6 (Sandbjerg, 1986)*. Progress in Probability, vol. 20 (Birkhäuser, Boston, 1990), pp. 101–119
19. P. Hitczenko, S.J. Montgomery-Smith, Tangent sequences in Orlicz and rearrangement invariant spaces. Math. Proc. Cambridge Philos. Soc. **119**(1), 91–101 (1996)
20. Y.-L. Hou, P.D. Liu, Two geometrical properties of vector-valued Musielak-Orlicz spaces. Acta Anal. Funct. Appl. **1**(1), 11–15 (1999)
21. T.P. Hytönen, The vector-valued nonhomogeneous Tb theorem. Int. Math. Res. Not. IMRN **2014**(2), 451–511 (2014)
22. T.P. Hytönen, M.C. Veraar, On Besov regularity of Brownian motions in infinite dimensions. Probab. Math. Stat. **28**(1), 143–162 (2008)
23. T.P. Hytönen, J.M.A.M. van Neerven, M.C. Veraar, L. Weis, *Analysis in Banach Spaces. Vol. I. Martingales and Littlewood-Paley Theory*. Ergebnisse der Mathematik und ihrer Grenzgebiete. 3. Folge., vol. 63 (Springer, Berlin, 2016)

24. T.P. Hytönen, J.M.A.M. van Neerven, M.C. Veraar, L. Weis, *Analysis in Banach Spaces. Vol. II. Probabilistic Methods and Operator Theory*. Ergebnisse der Mathematik und ihrer Grenzgebiete. 3. Folge., vol. 67 (Springer, Berlin, 2017)
25. A. Kozek, Orlicz spaces of functions with values in Banach spaces. Comment. Math. Prace Mat. **19**(2), 259–288 (1976/1977)
26. A. Kozek, Convex integral functionals on Orlicz spaces. Comment. Math. Prace Mat. **21**(1), 109–135 (1980)
27. M.A. Krasnosel'skiĭ, J.B. Rutickiĭ, *Convex Functions and Orlicz Spaces*. Translated from the first Russian edition by Leo F. Boron (P. Noordhoff, Groningen, 1961)
28. P.D. Liu, Spaces in which martingale difference sequences are unconditional. J. Syst. Sci. Math. Sci. **9**(3), 251–259 (1989)
29. P.-A. Meyer, *Martingales and Stochastic Integrals. I*. Lecture Notes in Mathematics, vol. 284 (Springer, Berlin, 1972)
30. J. Musielak, *Orlicz Spaces and Modular Spaces*. Lecture Notes in Mathematics, vol. 1034 (Springer, Berlin, 1983)
31. J. Neveu, *Discrete-Parameter Martingales*. Revised edition (North-Holland, Amsterdam, 1975)
32. A. Osękowski, *Sharp Martingale and Semimartingale Inequalities*. Instytut Matematyczny Polskiej Akademii Nauk. Monografie Matematyczne (New Series) [Mathematics Institute of the Polish Academy of Sciences. Mathematical Monographs (New Series)], vol. 72 (Birkhäuser/Springer Basel AG, Basel, 2012)
33. Y. Qiu, On the UMD constants for a class of iterated $L_p(L_q)$ spaces. J. Funct. Anal. **263**(8), 2409–2429 (2012)
34. J.L. Rubio de Francia, Martingale and integral transforms of Banach space valued functions, in *Probability and Banach Spaces (Zaragoza, 1985)*. Lecture Notes in Mathematics, vol. 1221 (Springer, Berlin, 1986), pp. 195–222
35. B.-Z.A. Rubshtein, G.Y. Grabarnik, M.A. Muratov, Y.S. Pashkova, *Foundations of Symmetric Spaces of Measurable Functions. Lorentz, Marcinkiewicz and Orlicz Spaces*. Developments in Mathematics, vol. 45 (Springer, Cham, 2016)
36. J.M.A.M. van Neerven, M.C. Veraar, L.W. Weis, Stochastic integration in UMD Banach spaces. Ann. Probab. **35**(4), 1438–1478 (2007)
37. M.C. Veraar, I.S. Yaroslavtsev, Pointwise properties of martingales with values in Banach function spaces. arXiv:1803.11063 (2018)
38. I.S. Yaroslavtsev, Burkholder–Davis–Gundy inequalities in UMD Banach spaces. arXiv:1807.05573 (2018)
39. I.S. Yaroslavtsev, Weak L^1-estimates for weakly differentially subordinated martingales. In preparation
40. A.C. Zaanen, *Riesz Spaces. II*. North-Holland Mathematical Library, vol. 30 (North-Holland, Amsterdam, 1983)

The ℓ^s-Boundedness of a Family of Integral Operators on UMD Banach Function Spaces

Emiel Lorist

Dedicated to Ben de Pagter on the occasion of his 65th birthday

Abstract We prove the ℓ^s-boundedness of a family of integral operators with an operator-valued kernel on UMD Banach function spaces. This generalizes and simplifies earlier work by Gallarati, Veraar and the author, where the ℓ^s-boundedness of this family of integral operators was shown on Lebesgue spaces. The proof is based on a characterization of ℓ^s-boundedness as weighted boundedness by Rubio de Francia.

Keywords ℓ^s-boundedness · Integral operator · Banach function space · Muckenhoupt weights · Hardy–Littlewood maximal operator · UMD

1 Introduction

Over the past decades there has been a lot of interest in the L^p-maximal regularity of PDEs. Maximal L^p-regularity of the abstract Cauchy problem

$$\begin{cases} u'(t) + Au(t) = f(t), & t \in (0, T] \\ u(0) = x, \end{cases} \tag{1.1}$$

where A is a closed operator on a Banach space X, means that for all $f \in L^p((0, T]; X)$ the solution u has "maximal regularity", i.e. both u' and Au are in $L^p((0, T]; X)$. Maximal L^p-regularity can, for example, be used to solve quasi-linear and fully nonlinear PDEs by linearization techniques combined with the contraction mapping principle, see, e.g., [1, 6, 29, 34].

E. Lorist (✉)
Delft Institute of Applied Mathematics, Delft University of Technology, Delft, The Netherlands
e-mail: e.lorist@tudelft.nl

G. Buskes et al. (eds.), *Positivity and Noncommutative Analysis*,
Trends in Mathematics, https://doi.org/10.1007/978-3-030-10850-2_20

In the breakthrough work of Weis [40, 41], an operator theoretic characterization of maximal L^p-regularity on UMD Banach spaces was found in terms of the $\mathcal{R}$-boundedness of the resolvents of A on a sector. $\mathcal{R}$-boundedness is a random boundedness condition on a family of operators which is a strengthening of uniform boundedness. We refer to [7, 20] for more information on $\mathcal{R}$-boundedness.

In [11, 12] Gallarati and Veraar developed a new approach to maximal L^p-regularity for the case where the operator A in (1.1) is time-dependent and $t \mapsto A(t)$ is merely assumed to be measurable. In this new approach $\mathcal{R}$-boundedness is once again one of the main tools. For their approach the $\mathcal{R}$-boundedness of the family of integral operators $\{I_k : k \in \mathcal{K}\}$ on $L^p(\mathbb{R}; X)$ is required. Here I_k is defined for $f \in L^p(\mathbb{R}; X)$ as

$$I_k f(t) := \int_{-\infty}^{t} k(t-s)T(t,r)f(r)\,\mathrm{d}r, \qquad t \in \mathbb{R},$$

where $T(t,r)$ is the two-parameter evolution family associated to $A(t)$ and $\mathcal{K}$ contains all kernels $k \in L^1(\mathbb{R})$ such that $|k| * |g| \leq Mg$ for all simple $g : \mathbb{R} \to \mathbb{C}$.

In the literature there are many $\mathcal{R}$-boundedness results for integral operators, see [20, Chapter 8] for an overview. However none of these are applicable to the operator family of $\{I_k : k \in \mathcal{K}\}$. Therefore in [13] Gallarati, Veraar and the author show a sufficient condition for the $\mathcal{R}$-boundedness of $\{I_k : k \in \mathcal{K}\}$ on $L^p(\mathbb{R}; X)$ in the special case where $X = L^q$. This is done through the notion of ℓ^s-boundedness, which states that for all finite sequences $(I_{k_j})_{j=1}^n$ in $\{I_k : k \in \mathcal{K}\}$ and $(x_j)_{j=1}^n$ in X we have

$$\Big\|\Big(\sum_{j=1}^{n}|I_{k_j}x_j|^s\Big)^{1/s}\Big\|_X \lesssim \Big\|\Big(\sum_{j=1}^{n}|x_j|^s\Big)^{1/s}\Big\|_X.$$

For $s = 2$ this notion coincides with $\mathcal{R}$-boundedness as a consequence of the Kahane-Khintchine inequalities.

Our main contribution is the generalization of the main result in [13] to the setting of UMD Banach function spaces X. For the proof we will follow the general scheme of [13] with some simplifications. As in case $X = L^q$, for any UMD Banach function space the notions of ℓ^2-boundedness and $\mathcal{R}$-boundedness coincide, so the following theorem in particular implies the $\mathcal{R}$-boundedness of $\{I_k : k \in \mathcal{K}\}$.

Theorem 1.1 *Let X be a* UMD *Banach function space and $p \in (1, \infty)$. Let $T : \mathbb{R} \times \mathbb{R} \to \mathcal{L}(X)$ be such that the family of operators*

$$\big\{T(t,r) : t, r \in \mathbb{R}\big\}$$

is ℓ^s-bounded for all $s \in (1, \infty)$. Then $\{I_k : k \in \mathcal{K}\}$ is ℓ^s-bounded on $L^p(\mathbb{R}; X)$ for all $s \in (1, \infty)$.

We will prove Theorem 1.1 in a more general setting in Sect. 3. In particular we allow weights in time, which in applications, for example, allow rather rough initial values (see, e.g., [22, 25, 30, 35]).

For certain UMD Banach function spaces the ℓ^s-boundedness assumption in Theorem 1.1 can be checked by weighted extrapolation techniques, see Corollary 3.5 and Remark 3.6.

Notation For a measure space (S, μ) we denote the space of all measurable functions by $L^0(S)$. We denote the Lebesgue measure of a Borel set $E \in \mathcal{B}(\mathbb{R}^d)$ by $|E|$. For Banach spaces X and Y we denote the vector space of bounded linear operators from X to Y by $\mathcal{L}(X, Y)$ and we set $\mathcal{L}(X) := \mathcal{L}(X, X)$. For an operator family $\Gamma \subset \mathcal{L}(X, Y)$ we set $\Gamma^* := \{T^* : T \in \Gamma\}$. For $p \in [1, \infty]$ we let $p' \in [1, \infty]$ be such that $\frac{1}{p} + \frac{1}{p'} = 1$.

Throughout the paper we write $C_{a,b,\cdots}$ and $\phi_{a,b,\cdots}$ to denote a constant and a nondecreasing function on $[1, \infty)$, respectively, which only depend on the parameters $a, b, \cdots$ and the dimension d and which may change from line to line.

2 Preliminaries

2.1 Banach Function Spaces

Let (S, μ) be a σ-finite measure space. An order ideal X of $L^0(S)$ equipped with a norm $\|\cdot\|_X$ is called a *Banach function space* if it has the following properties:

1. *Compatibility:* if $\xi, \eta \in X$ with $|\xi| \leq |\eta|$, then $\|\xi\|_X \leq \|\eta\|_X$.
2. *Fatou property:* if $0 \leq \xi_n \uparrow \xi$ for $(\xi_n)_{n=1}^\infty$ in X, $\xi \in L^0(S)$ and $\sup_{n\in\mathbb{N}} \|\xi_n\|_X < \infty$, then $\xi \in X$ and $\|\xi\|_X = \sup_{n\in\mathbb{N}} \|\xi_n\|_X$.

Note that the compatibility is actually a special case of the Fatou property. Moreover a Banach function space is complete, see [42, Section 30]. Without loss of generality we may always assume that X has a weak order unit, i.e. there is a $\xi \in X$ with $\xi > 0$ pointwise a.e., see [42, Section 67].

A Banach function space is called *order continuous* if for any sequence $0 \leq \xi_n \uparrow \xi \in X$ we have $\|\xi_n - \xi\|_X \to 0$. Every reflexive Banach function space is order continuous. Order continuity ensures that the dual of X is also a Banach function space. For a thorough introduction to Banach function spaces, we refer to [3] or [42].

A Banach function space X is said to be *p-convex* for $p \in [1, \infty]$ if

$$\Big\| \Big(\sum_{j=1}^{n} |\xi_k|^p \Big)^{1/p} \Big\|_X \leq \Big(\sum_{j=1}^{n} \|\xi_j\|_X^p \Big)^{1/p}$$

for all $\xi_1, \cdots, \xi_n \in X$, with the sums replaced by suprema if $p = \infty$. The defining inequality for p-convexity often includes a constant, but X can always be renormed such that this constant equals 1. If a Banach function space is p-convex for some $p \in [1, \infty]$, then X is also q-convex for all $q \in [1, q]$.

For a p-convex Banach function space X we can define another Banach function space by

$$X^p := \big\{|\xi|^p \operatorname{sgn} \xi : \xi \in X\big\} = \big\{\xi \in L^0(S) : |\xi|^{1/p} \in X\big\}$$

equipped with the norm $\|\xi\|_{X^p} := \big\||\xi|^{1/p}\big\|_X^p$. We refer the interested reader to [27, section 1.d] for an introduction to p-convexity.

2.2 ℓ^s-Boundedness

Let X and Y be Banach functions spaces and let $\Gamma \subseteq \mathcal{L}(X, Y)$ be a family of operators. We say that Γ is *ℓ^s-bounded* if for all finite sequences $(T_j)_{j=1}^n$ in Γ and $(x_j)_{j=1}^n$ in X we have

$$\Big\|\Big(\sum_{j=1}^n |T_j x_j|^s\Big)^{1/s}\Big\|_Y \leq C \Big\|\Big(\sum_{j=1}^n |x_j|^s\Big)^{1/s}\Big\|_X.$$

with the sums replaced by suprema if $s = \infty$. The least admissible constant C will be denoted by $[\Gamma]_{\ell^s}$.

Implicitly ℓ^s-boundedness is a classical tool in harmonic analysis for operators on L^p-spaces (see, e.g., [14, Chapter V] and [16, 17]). For Banach function spaces the notion was introduced in [40] under the name $\mathcal{R}_s$-boundedness, underlining its connection to the more well-known notion of $\mathcal{R}$-boundedness. An extensive study of ℓ^s-boundedness can be found in [23] and for a comparison between ℓ^2-boundedness and $\mathcal{R}$-boundedness we refer to [24].

Lemma 2.1 *Let X and Y be Banach function spaces and let $\Gamma \subseteq \mathcal{L}(X, Y)$.*

(i) Let $1 \leq s_0 < s_1 \leq \infty$ and assume that X and Y are order continuous. If Γ is ℓ^{s_0}- and ℓ^{s_1}-bounded, then Γ is ℓ^s-bounded for all $s \in [s_0, s_1]$ with $[\Gamma]_{\ell^s} \leq \max\big\{[\Gamma]_{\ell^{s_0}}, [\Gamma]_{\ell^{s_1}}\big\}$.

(ii) Let $s \in [1, \infty]$ and assume that Γ is ℓ^s-bounded. Then the adjoint family Γ^* is $\ell^{s'}$-bounded with $[\Gamma^*]_{\ell^{s'}} = [\Gamma]_{\ell^s}$.

Proof Lemma 2.1(i) follows from Calderón's theory of complex interpolation of vector-valued function spaces, see [5] or [23, Proposition 2.14]. Lemma 2.1(ii) is direct from the identification $X(\ell^s_n)^* = X^*(\ell^{s'}_n)$, see [27, Section 1.d] or [23, Proposition 2.17]. □

The following characterization of ℓ^s-boundedness for $s \in [1, \infty)$ will be one of the key ingredients of our main result. This characterization relating ℓ^s-boundedness to a certain weighted boundedness comes from the work of Rubio de Francia [14, 36, 37].

Proposition 2.2 *Let $s \in [1, \infty)$ and let X and Y be s-convex order continuous Banach function spaces over (S_X, μ_X) and (S_Y, μ_Y), respectively. Let $\Gamma \subseteq \mathcal{L}(X)$ and take $C > 0$. Then the following are equivalent:*

(i) Γ is ℓ^s-bounded with $[\Gamma]_{\ell^s} \leq C$.
(ii) For all nonnegative $u \in (Y^s)^*$, there exists a nonnegative $v \in (X^s)^*$ with $\|v\|_{(Y^s)^*} \leq \|u\|_{(X^s)^*}$ and

$$\Big(\int_{S_Y} |T(\xi)|^s u \, d\mu_Y\Big)^{1/s} \leq C \Big(\int_{S_X} |\xi|^s v \, d\mu_X\Big)^{1/s}$$

for all $\xi \in X$ and $T \in \Gamma$.

Proof The statement is a combination of [37, Lemma 1, p. 217] and [14, Theorem VI.5.3], which for $X = Y$ is proven [2, Lemma 3.4]. The statement for $X \neq Y$ can be extracted from the proof of [2, Lemma 3.4] and can in full detail be found in [28, Proposition 6.1.3]. □

2.3 Muckenhoupt Weights

A locally integrable function $w : \mathbb{R}^d \to (0, \infty)$ is called a *weight*. For $p \in (1, \infty)$ and a weight w we let $L^p(w)$ be the space of all $f \in L^0(\mathbb{R}^d)$ such that

$$\|f\|_{L^p(w)} := \Big(\int_{\mathbb{R}^d} |f|^p w\Big)^{1/p} < \infty.$$

We will say that a weight w lies in the *Muckenhoupt class* A_p and write $w \in A_p$ if it satisfies

$$[w]_{A_p} := \sup_Q \frac{1}{|Q|} \int_Q w \cdot \Big(\frac{1}{|Q|} \int_Q w^{1-p'}\Big)^{p-1} < \infty,$$

where the supremum is taken over all cubes $Q \subseteq \mathbb{R}^d$ with sides parallel to the coordinate axes.

Lemma 2.3 *Let $p \in (1, \infty)$ and $w \in A_p$.*

(i) $w \in A_q$ for all $q \in (p, \infty)$ with $[w]_{A_q} \leq [w]_{A_p}$.
(ii) $w^{1-p'} \in A_{p'}$ with $[w]_{A_p}^{1/p} = [w^{1-p'}]_{A_p'}^{1/p'}$.
(iii) $w \in A_{p-\varepsilon}$ for $\varepsilon = \frac{1}{\phi_p([w]_{A_p})}$ with $[w]_{A_{p-\varepsilon}} \leq \phi_p([w]_{A_p})$.

The first two properties of Lemma 2.3 follow directly from the definition. The third is, for example, proven in [17, Exercise 9.2.4]. For a more thorough introduction to Muckenhoupt weights, we refer to [17, Chapter 9].

2.4 *The UMD Property*

A Banach space X is said to have the UMD property if the martingale difference sequence of any finite martingale in $L^p(\Omega; X)$ is unconditional for some (equivalently all) $p \in (1, \infty)$. We will work with UMD Banach function spaces, of which standard examples include reflexive Lebesgue, Lorentz and Orlicz spaces. In this Festschrift it is shown that reflexive Musielak-Orlicz spaces, so in particular reflexive variable Lebesgue spaces, have the UMD property, see [26]. The UMD property implies reflexivity, so in particular L^1 and L^∞ do not have the UMD property. For a thorough introduction to the theory of UMD Banach spaces, we refer to [19, 33].

For an order continuous Banach function space X over (S, μ) there is also a characterization of the UMD property in terms of the *lattice Hardy–Littlewood maximal operator*, which for simple functions $f : \mathbb{R}^d \to X$ is given by

$$\widetilde{M} f(x) := \sup_{Q \ni x} \frac{1}{|Q|} \int_Q |f(y)| \, \mathrm{d}y, \qquad x \in \mathbb{R}^d$$

where the supremum is taken pointwise in S and over all cubes $Q \subseteq \mathbb{R}^d$ with sides parallel to the coordinate axes (see [15] or [18, Lemma 5.1] for a detailed definition of $\widetilde{M}$). It is a deep result by Bourgain [4] and Rubio de Francia [37] that X has the UMD property if and only if $\widetilde{M}$ is bounded on $L^p(\mathbb{R}^d; X)$ and $L^p(\mathbb{R}^d; X^*)$ for some (equivalently all) $p \in (1, \infty)$. For weighted L^p-spaces we have the following proposition, which was proven in [15]. The increasing dependence on $[w]_{A_p}$ is shown in [18, Corollary 5.3]. Recall that $\phi_{a,b,\dots}$ denotes a nondecreasing function on $[1, \infty)$.

Proposition 2.4 *Let X be a* UMD *Banach function space, $p \in (1, \infty)$ and $w \in A_p$. Then for all $f \in L^p(w; X)$ we have*

$$\|\widetilde{M} f\|_{L^p(w;X)} \leq \phi_{X,p}\big([w]_{A_p}\big) \|f\|_{L^p(w;X)}.$$

The UMD property of a Banach function space X also implies that X^q has the UMD property for a $q > 1$, which is a deep result by Rubio de Francia [37, Theorem 4].

Proposition 2.5 *Let X be a* UMD *Banach function space. Then there is a $p > 1$ such that X is p-convex and X^q is a* UMD *Banach function space for all $q \in [1, p]$.*

3 Integral Operators with an Operator-Valued Kernel

Before turning to our main result on the ℓ^s-boundedness of a family of integral operators on $L^p(w; X)$ with operator-valued kernels, we will first study the ℓ^s-boundedness of a family of convolution operators on $L^p(w; X)$ with scalar-valued kernels. For this define

$$\mathcal{K} := \{k \in L^1(\mathbb{R}^d) : |k| * |f| \leq Mf \text{ a.e. for all simple } f : \mathbb{R}^d \to \mathbb{C}\}.$$

As an example any radially decreasing $k \in L^1(\mathbb{R}^d)$ with $\|k\|_{L^1(\mathbb{R}^d)} \leq 1$ is an element of $\mathcal{K}$. For more examples see [16, Chapter 2] and [39, Proposition 4.6].

Let X be a Banach function space. For a kernel $k \in \mathcal{K}$ and a simple function $f : \mathbb{R}^d \to X$ we define

$$T_k f(x) := k * f(x) = \int_{\mathbb{R}^d} k(x-y) f(y) \, \mathrm{d}y, \qquad x \in \mathbb{R}^d.$$

As

$$\|T_k f\|_X \leq |k| * \|f\|_X \leq M\big(\|f\|_X\big),$$

and since the Hardy–Littlewood maximal operator M is bounded on $L^p(w)$ for all $p \in (1, \infty)$ and $w \in A_p$, T_k extends to a bounded linear operator on $L^p(w; X)$ by density. This argument also shows that the family of convolution operators given by $\Gamma := \{T_k : k \in \mathcal{K}\}$ is uniformly bounded on $L^p(w; X)$.

If X is a UMD Banach function space we can say more. The following lemma was first developed by van Neerven, Veraar and Weis in [38, 39] in connection to stochastic maximal regularity. As in [38, 39], the endpoint case $s = 1$ will play a major role in the proof of our main theorem in the next section.

Proposition 3.1 *Let X be a* UMD *Banach function space, $s \in [1, \infty]$, $p \in (1, \infty)$ and $w \in A_p$. Then $\Gamma = \{T_k : k \in \mathcal{K}\}$ is ℓ^s-bounded on $L^p(w; X)$ with*

$$[\Gamma]_{\ell^s} \leq \phi_{X,p}\big([w]_{A_p}\big).$$

The proof is a weighted variant of [39, Theorem 4.7], which for the special case where X is an iterated Lebesgue space is presented in [13, Proposition 3.6]. For convenience of the reader, we sketch the proof in the general case.

Proof As X is reflexive and therefore order continuous, $\widetilde{M}$ is well-defined on $L^p(w; X)$ and we have $T_k f \leq \widetilde{M} f$ pointwise a.e. for all simple $f : \mathbb{R}^d \to X$.

If $s = \infty$ take simple functions $f_1, \cdots, f_n \in L^p(w; X)$ and $k_1, \cdots, k_n \in \mathcal{K}$. Using Proposition 2.4 we have

$$\begin{aligned}\Big\| \sup_{1\leq j\leq n} |T_{k_j} f_j| \Big\|_{L^p(w;X)} &\leq \Big\| \sup_{1\leq j\leq n} \widetilde{M} f_j(x) \Big\|_{L^p(w;X)} \\ &\leq \Big\| \widetilde{M}\Big(\sup_{1\leq j\leq n} |f_j|\Big)(x) \Big\|_{L^p(w;X)} \\ &\leq \phi_{X,p}\big([w]_{A_p}\big) \Big\| \sup_{1\leq j\leq n} |f_j| \Big\|_{L^p(w;X)}.\end{aligned}$$

The result now follows by the density of simple functions in $L^p(w; X)$.

If $s = 1$ we use duality. Note that since X is reflexive we have $L^p(w; X)^* = L^{p'}(w'; X^*)^*$ with $w' = w^{1-p'}$ under the duality pairing

$$\langle f, g\rangle_{L^p(w;X), L^{p'}(w';X^*)} = \int_{\mathbb{R}^d} \langle f(x), g(x)\rangle_{X,X^*} \, \mathrm{d}x \tag{3.2}$$

by Lemma 2.3(ii) and [19, Corollary 1.3.22]. One can routinely check that $T_k^* = T_{\tilde{k}}$ with $\tilde{k}(x) = k(-x)$ and that $k \in \mathcal{K}$ if and only if $\tilde{k} \in \mathcal{K}$. Since X^* is also a UMD Banach function space (see [19, Proposition 4.2.17]) we know from the case $s = \infty$ that the adjoint family Γ^* is ℓ^∞-bounded on $L^{p'}(\mathbb{R}^d, w'; X^*)$, so the result follows by Lemma 2.1(ii). Finally if $s \in (1, \infty)$ the result follows by Lemma 2.1(i). □

With these preparations done we can now introduce the family of integral operators with operator-valued kernel that we will consider. Let X and Y be a Banach function space and let $\mathcal{T}$ be a family of operators $\mathbb{R}^d \times \mathbb{R}^d \to \mathcal{L}(X, Y)$ such that $(x, y) \mapsto T(x, y)\xi$ is measurable for all $T \in \mathcal{T}$ and $\xi \in X$. The integral operators that we will consider are for simple $f : \mathbb{R}^d \to X$ given by

$$I_{k,T} f(x) = \int_{\mathbb{R}^d} k(x-y) T(x, y) f(y) \, \mathrm{d}y$$

with $k \in \mathcal{K}$ and $T \in \mathcal{T}$. If $\|T(x, y)\|_{\mathcal{L}(X,Y)} \leq C$ for all $T \in \mathcal{T}$ and $x, y \in \mathbb{R}^d$, we have

$$\|I_{k,T} f\|_X \leq C\, |k| * \|f\|_X \leq C\, M\big(\|f\|_X\big).$$

So as before $I_{k,T}$ extends to a bounded linear operator from $L^p(w; X)$ to $L^p(w; Y)$ for all $p \in (1, \infty)$ and $w \in A_p$, and

$$\mathcal{I}_{\mathcal{T}} := \big\{ I_{k,T} : k \in \mathcal{K}, T \in \mathcal{T} \big\}$$

is uniformly bounded. For the details, see [13, Lemma 3.9].

If X and Y are Hilbert spaces, this implies that $\mathcal{I}_{\mathcal{T}}$ is also ℓ^2-bounded from $L^2(\mathbb{R}^d; X)$ to $L^2(\mathbb{R}^d; Y)$, as these notions coincide on Hilbert spaces. However if X and Y are not Hilbert spaces, but a UMD Banach function space or if we move to weighted L^p-spaces, the ℓ^2-boundedness of $\mathcal{I}_{\mathcal{T}}$ is much more delicate.

Our main theorem is a quantitative and more general version of Theorem 1.1 in the introduction:

Theorem 3.2 *Let X and Y be a* UMD *Banach function spaces and let $p, s \in (1, \infty)$. Let $\mathcal{T}$ be a family of operators $\mathbb{R}^d \times \mathbb{R}^d \to \mathcal{L}(X, Y)$ such that*

(i) $(x, y) \mapsto T(x, y)\xi$ is strongly measurable for all $T \in \mathcal{T}$ and $\xi \in X$,
(ii) the family of operators $\widetilde{\mathcal{T}} := \{T(x, y) : T \in \mathcal{T},\ x, y \in \mathbb{R}^d\}$ is ℓ^σ-bounded for all $\sigma \in (1, \infty)$.

Then $\mathcal{I}_{\mathcal{T}}$ is ℓ^s-bounded from $L^p(w; X)$ to $L^p(w; Y)$ for all $w \in A_p$ with

$$\begin{aligned}
[\mathcal{I}_{\mathcal{T}}]_{\ell^s} &\le \phi_{X,Y,p}\big([w]_{A_p}\big) \max\big\{[\widetilde{\mathcal{T}}]_{\ell^\sigma}, [\widetilde{\mathcal{T}}]_{\ell^{\sigma'}}\big\}, \qquad \sigma = 1 + \frac{1}{\phi_{p,s}\,[w]_{A_p}} \\
&\le \phi_{X,Y,\mathcal{T},p,s}\big([w]_{A_p}\big).
\end{aligned}$$

We will first prove a result assuming the ℓ^s-boundedness of $\widetilde{\mathcal{T}}$ for a fixed $s \in [1, \infty)$.

Proposition 3.3 *Fix $1 \le s \le r < p < \infty$ and let X and Y be s-convex Banach function spaces such that X^s has the* UMD *property. Let $\mathcal{T}$ be a family of operators $\mathbb{R}^d \times \mathbb{R}^d \to \mathcal{L}(X, Y)$ such that*

(i) $(x, y) \mapsto T(x, y)\xi$ is strongly measurable for all $T \in \mathcal{T}$ and $\xi \in X$,
(ii) the family of operators $\widetilde{\mathcal{T}} := \{T(x, y) : T \in \mathcal{T},\ x, y \in \mathbb{R}^d\}$ is ℓ^s-bounded.

Then $\mathcal{I}_{\mathcal{T}}$ is ℓ^s-bounded from $L^p(w; X)$ to $L^p(w; Y)$ for all $w \in A_{p/s}$ with

$$[\mathcal{I}_{\mathcal{T}}]_{\ell^s} \le \phi_{X,p,r}\big([w]_{A_{p/s}}\big)[\widetilde{\mathcal{T}}]_{\ell^s}.$$

Proof Let (S_X, μ_X) and (S_Y, μ_Y) be the measure spaces associated to X and Y, respectively. For $j = 1, \cdots, n$ take $I_j \in \mathcal{I}_{\mathcal{T}}$ and let $k_j \in \mathcal{K}$ and $T_j \in \mathcal{T}$ be such that $I_j = I_{k_j,T_j}$. Fix simple functions $f_1, \cdots, f_n \in L^p(w; X)$ and note that

$$\Big\|\Big(\sum_{j=1}^n |I_j f_j|^s\Big)^{1/s}\Big\|_{L^p(w;Y)} = \Big\|\sum_{j=1}^n |I_j f_j|^s\Big\|_{L^{p/s}(w;Y^s)}^{1/s}. \tag{3.3}$$

Fix $x \in \mathbb{R}^d$, then by Hahn–Banach we can find a nonnegative $u_x \in (Y^s)^*$ with $\|u_x\|_{(X^s)^*} = 1$ such that

$$\Big\|\sum_{j=1}^n |I_j f_j(x)|^s\Big\|_{Y^s} = \sum_{j=1}^n \int_{S_Y} |I_j f_j(x)|^s u_x \,\mathrm{d}\mu_Y. \tag{3.4}$$

With Proposition 2.2 we can then find a nonnegative $v_x \in (X^s)^*$ with $\|v_x\|_{(X^s)^*} \leq 1$ such that

$$\int_{S_Y} |T_j(x,y)\xi|^s v_x \, \mathrm{d}\mu_Y \leq [\widetilde{\mathcal{T}}]_{\ell^s} \int_{S_X} |\xi|^s v_x \, \mathrm{d}\mu_X \tag{3.5}$$

for $j = 1, \cdots, n$, $y \in \mathbb{R}^d$ and $\xi \in X$. Since $\|k_j\|_{L^1(\mathbb{R}^d)} \leq 1$ by [39, Lemma 4.3], Holder's inequality yields

$$|I_j f_j(x)|^s \leq \int_{\mathbb{R}^d} |k_j(x-y)||T_j(x,y) f_j(y)|^s \, \mathrm{d}y. \tag{3.6}$$

Applying (3.6) and (3.5) successively we get

$$\begin{aligned}
\sum_{j=1}^n \int_{S_Y} \left|I_j f_j(x)\right|^s u_x \, \mathrm{d}\mu_Y &\leq \sum_{j=1}^n \int_{S_Y} \int_{\mathbb{R}^d} |k_j(x-y)||T_j(x,y) f_j(y)|^s \, \mathrm{d}y \, u_x \, \mathrm{d}\mu_Y \\
&= \sum_{j=1}^n \int_{\mathbb{R}^d} |k_j(x-y)| \int_{S_Y} |T_j(x,y) f_j(y)|^s \, u_x \, \mathrm{d}\mu_Y \, \mathrm{d}y \\
&\leq [\widetilde{\mathcal{T}}]_{\ell^s} \sum_{j=1}^n \int_{S_X} \int_{\mathbb{R}^d} |k_j(x-y)||f_j(y)|^s \, \mathrm{d}y \, v_x \, \mathrm{d}\mu_X \\
&\leq [\widetilde{\mathcal{T}}]_{\ell^s} \Big\| \sum_{j=1}^n (|k_j| * |f_j|^s)(x) \Big\|_{X^s},
\end{aligned}$$

using duality and $\|v_x\|_{(X^s)^*} \leq 1$ in the last step. We can now use the ℓ^1-boundedness result of Proposition 3.1, since $(X^s)^*$ has the UMD property by [20, Proposition 4.2.17]. Combined with (3.3) and (3.4) we obtain

$$\begin{aligned}
\Big\| \Big(\sum_{j=1}^n |I_j f_j|^s \Big)^{1/s} \Big\|_{L^p(w;Y)} &\leq [\widetilde{\mathcal{T}}]_{\ell^s} \Big\| \sum_{j=1}^n |k_j| * |f_j|^s \Big\|_{L^{p/s}(w;X^s)}^{\frac{1}{s}} \\
&\leq \phi_{X,p/s}\big([w]_{A_{p/s}}\big) \, [\widetilde{\mathcal{T}}]_{\ell^s} \Big\| \sum_{j=1}^n |f_j|^s \Big\|_{L^{p/s}(w;X^s)}^{1/s} \\
&\leq \phi_{X,p,r}\big([w]_{A_{p/s}}\big) [\widetilde{\mathcal{T}}]_{\ell^s} \Big\| \Big(\sum_{j=1}^n |f_j|^s \Big)^{\frac{1}{s}} \Big\|_{L^p(w;X)},
\end{aligned}$$

where we can pick the increasing function ϕ in the last step independent of s, since the increasing function in Proposition 3.1 depends continuously on p. This can, for example, be seen by writing out the exact dependence on p in Theorem 2.4 using [18, Theorem 1.3] and [31, Theorem 3.1]. □

Using this preparatory proposition, we will now prove Theorem 3.2. Recall that $\phi_{a,b,\dots}$ denotes a nondecreasing function on $[1, \infty)$.

Proof of Theorem 3.2 Let $w \in A_p$. We shall prove the theorem in three steps.

Step 1 First we shall prove the theorem very small $s > 1$. By Proposition 2.5 we know that there exists a $\sigma_{X,Y} \in (1, p)$ such that X and Y are s-convex and X^s has the UMD property for all $s \in [1, \sigma_X]$. By Lemma 2.3(iii) we can then find a $\sigma_{p,w} \in (1, \sigma_{X,Y}]$ such that for all $s \in [1, \sigma_{p,w}]$

$$[w]_{A_{p/s}} \leq [w]_{A_{p/\sigma_{p,w}}} \leq \phi_p([w]_{A_p})$$

Let $\sigma_1 = \min\{\sigma_{X,Y}, \sigma_{p,w}\}$, then by Proposition 3.3 we know that $\mathcal{I}_{\mathcal{T}}$ is ℓ^s-bounded from $L^p(w; X)$ to $L^p(w; Y)$ for $s \in (1, \sigma_1]$ with

$$[\mathcal{I}_{\mathcal{T}}]_{\ell^s} \leq \phi_{X,p,\sigma_{X,Y}}([w]_{A_{p/s}})[\widetilde{\mathcal{T}}]_{\ell^s} \leq \phi_{X,Y,p}([w]_{A_p})[\widetilde{\mathcal{T}}]_{\ell^s}. \tag{3.7}$$

Step 2 Now we use a duality argument to prove the theorem for large $s < \infty$. As noted in the proof of Proposition 3.1, we have $L^p(w; X)^* = L^{p'}(w'; X^*)$ with $w' = w^{1-p'}$ under the duality pairing as in (3.2) and similarly for Y. Furthermore X^* and Y^* have the UMD property.

It is routine to check that under this duality $I_{k,T}^* = I_{\tilde{k},\tilde{T}}$ with $\tilde{k}(x) = k(-x)$ and $\tilde{T}(x, y) = T^*(y, x)$ for any $I_{k,T} \in \mathcal{I}_{\mathcal{T}}$. Trivially $\tilde{k} \in \mathcal{K}$ if and only if $k \in \mathcal{K}$ and by Proposition 3.1(ii) the adjoint family $\widetilde{\mathcal{T}}^*$ is $\ell^{\sigma'}$-bounded with

$$[\widetilde{\mathcal{T}}^*]_{\ell^{\sigma'}} = [\widetilde{\mathcal{T}}]_{\ell^\sigma}$$

for all $\sigma \in (1, \infty)$. Therefore, it follows from step 1 that there is a $\sigma_2 > 1$ such that $\mathcal{I}_{\mathcal{T}}^*$ is ℓ^s-bounded from $L^{p'}(w'; Y^*)$ to $L^{p'}(w'; X^*)$ for all $s \in (1, \sigma_2]$. Using Proposition 3.1(ii) again, we deduce that $\mathcal{I}_{\mathcal{T}}$ is ℓ^s-bounded from $L^p(w; X)$ to $L^p(w; Y)$ for all $s \in [\sigma_2', \infty)$ with

$$[\mathcal{I}_{\mathcal{T}}]_{\ell^s} = [\mathcal{I}_{\mathcal{T}}^*]_{\ell^{s'}} \leq \phi_{X,Y,p}([w]_{A_p})[\widetilde{\mathcal{T}}]_{\ell^s}. \tag{3.8}$$

Step 3 We can finish the proof by an interpolation argument for $s \in (\sigma_1, \sigma_2')$. By Proposition 2.2(i) we get for $s \in (\sigma_1, \sigma_2')$ that $\mathcal{I}_{\mathcal{T}}$ is ℓ^s-bounded from $L^p(w; X)$ to $L^p(w; Y)$ with

$$[\mathcal{I}_{\mathcal{T}}]_{\ell^s} \leq \phi_{X,Y,p}([w]_{A_p}) \max\Big\{[\widetilde{\mathcal{T}}]_{\ell^{\sigma_1}}, [\widetilde{\mathcal{T}}]_{\ell^{\sigma_2'}}\Big\}. \tag{3.9}$$

Now note that by Lemma 2.3 there is a $\sigma \in (1, \infty)$ such that $\sigma < \sigma_1, \sigma_2$ and $\sigma < s < \sigma'$ and

$$\sigma = 1 + \frac{1}{\phi_{p,s}([w]_{A_p})}.$$

Thus combining (3.7), (3.8) and (3.9) we obtain

$$[\mathcal{I}_{\mathcal{T}}]_{\ell^s} \leq \phi_{X,Y,p}\big([w]_{A_p}\big) \max\big\{[\widetilde{\mathcal{T}}]_{\ell^\sigma}, [\widetilde{\mathcal{T}}]_{\ell^{\sigma'}}\big\} \leq \phi_{X,Y,\mathcal{T},p,s}\big([w]_{A_p}\big),$$

using the fact that $t \mapsto \max\big\{[\widetilde{\mathcal{T}}]_{\ell^t}, [\widetilde{\mathcal{T}}]_{\ell^{t'}}\big\}$ is increasing for $t \to 1$ by Proposition 2.2(i). This proves the theorem.

□

Remark 3.4

- From Theorem 3.2 one can also conclude that $\mathcal{I}_{\mathcal{T}}$ is $\mathcal{R}$-bounded, since $\mathcal{R}$- and ℓ^2-boundedness coincide if X and Y have the UMD property, see, e.g., [20, Theorem 8.1.3].
- The UMD assumptions in Theorem 3.2 are necessary. Indeed already if $X = Y$, $w = 1$ and if $\widetilde{\mathcal{T}}$ only contains the identity operator, it is shown in [21] that the ℓ^2-boundedness of $\mathcal{I}_{\mathcal{T}}$ implies the UMD property of X.
- The main result of [13] is Theorem 3.2 for the special case $X = Y = L^q(S)$. In applications to systems of PDEs one needs Theorem 3.2 on $L^q(S; \mathbb{C}^n)$ with $s = 2$, see, e.g., [11]. This could be deduced from the proof of [13, Theorem 3.10], by replacing absolute values by norms in $\mathbb{C}^n$. In our more general statement the case $L^q(S; \mathbb{C}^n)$ is included, since $L^q(S; \mathbb{C}^n)$ is a UMD Banach function space over $S \times \{1, \cdots, n\}$.

If $X = Y$ is a rearrangement invariant Banach function space on $\mathbb{R}^e$, we can check the ℓ^σ-boundedness of $\widetilde{\mathcal{T}}$ for all $\sigma \in (1, \infty)$ by weighted extrapolation. Examples of such Banach function spaces are Lebesgue, Lorentz and Orlicz spaces. See [27, Section 2.a] for an introduction to rearrangement invariant Banach function spaces.

Corollary 3.5 *Let X be a rearrangement invariant* UMD *Banach function space on $\mathbb{R}^e$ and let $p, s \in (1, \infty)$. Let $\mathcal{T}$ be a family of operators $\mathbb{R}^d \times \mathbb{R}^d \to \mathcal{L}(X)$ such that*

(i) $(x, y) \mapsto T(x, y)\xi$ is strongly measurable for all $T \in \mathcal{T}$ and $\xi \in X$,
(ii) for some $q \in (1, \infty)$ and all $v \in A_q$ we have

$$\sup_{T \in \mathcal{T},\, x,y \in \mathbb{R}^d} \|T(x, y)\|_{\mathcal{L}(L^q(v))} \leq \phi_{\mathcal{T},q}\big([v]_{A_q}\big).$$

Then $\mathcal{I}_{\mathcal{T}}$ is ℓ^s-bounded on $L^p(w; X)$ for all $w \in A_p$ with

$$[\mathcal{I}_{\mathcal{T}}]_{\ell^s} \leq \phi_{X,Y,\mathcal{T},p,q,s}\big([w]_{A_p}\big).$$

Note that in Corollary 3.5 we need that $T(x, y)$ is well-defined on $L^q(v)$ for all $T \in \mathcal{T}$ and $x, y \in \mathbb{R}^d$. This is indeed the case, since $X \cap L^q(v)$ is dense in $L^q(v)$.

Proof Let Y be the linear span of

$$\{\mathbf{1}_K \xi : K \subseteq \mathbb{R}^e \text{ compact}, \ \xi \in X \cap L^\infty(\mathbb{R}^e)\}.$$

Then $Y \subseteq L^q(v)$ for all $v \in A_p$ and Y is dense in X by order continuity. Define

$$\mathcal{F} := \big\{\big(|T(x,y)\xi|, |\xi|\big) : T \in \mathcal{T},\ x, y \in \mathbb{R}^d,\ \xi \in Y\big\}.$$

Note that X has upper Boyd index $q_X < \infty$ by the UMD property (see [20, Proposition 7.4.12] and [27, Section 2.a]). So we can use the extrapolation result for Banach function spaces in [10, Theorem 2.1] to conclude that for $\sigma \in (1, \infty)$

$$\Big\|\Big(\sum_{j=1}^{n} |T_j(x_j, y_j)\xi_j|^\sigma\Big)^{1/\sigma}\Big\|_X \leq C_{\mathcal{T},q} \Big\|\Big(\sum_{j=1}^{n} |\xi_j|^\sigma\Big)^{1/\sigma}\Big\|_X$$

for any $T_j \in \mathcal{T}$, $x_j, y_j \in \mathbb{R}^d$ and $\xi_j \in Y$ for $j = 1, \cdots, n$. By the density this extends to $\xi_j \in X$, so

$$\{T(x, y) : x, y \in \mathbb{R}^d, T \in \mathcal{T}\}$$

is ℓ^σ-bounded for all $\sigma \in (1, \infty)$. Therefore the corollary follows from Theorem 3.2. □

Remark 3.6

- A sufficient condition for the weighted boundedness assumption in Corollary 3.5 is that $T(x, y)\xi \leq C M\xi$ for all $T \in \mathcal{T}$, $x, y \in \mathbb{R}^d$ and $\xi \in L^q(\mathbb{R}^e)$, which follows directly from [17, Theorem 9.1.9].
- Corollary 3.5 holds more generally for UMD Banach function spaces X such that the Hardy–Littlewood maximal operator is bounded on both X and X^* (see [9, Theorem 4.6]). For example, the variable Lebesgue spaces $L^{p(\cdot)}$ satisfy this assumption if $p_+, p_- \in (1, \infty)$ and $p(\cdot)$ satisfies a certain continuity condition, see [8, 32].
- The conclusion of Corollary 3.5 also holds for $X(v)$ for all $v \in A_{p_X}$ where p_X is the lower Boyd index of X and $X(v)$ is a weighted version of X, see [10, Theorem 2.1].

Acknowledgements The author would like to thank Mark Veraar and Jan van Neerven for carefully reading the draft version of this paper. Author Emiel Lorist is supported by the VIDI subsidy 639.032.427 of the Netherlands Organisation for Scientific Research (NWO).

References

1. H. Amann, Maximal regularity and quasilinear parabolic boundary value problems, in *Recent Advances in Elliptic and Parabolic Problems* (World Scientific Publishing, Hackensack, 2005), pp. 1–17
2. A. Amenta, E. Lorist, and M.C. Veraar, Rescaled extrapolation for vector-valued functions. Publ. Mat. **63**(1), 155–182 (2019)
3. C. Bennett, R. Sharpley, *Interpolation of Operators*. Pure and Applied Mathematics, vol. 129 (Academic Press, Boston, 1988)
4. J. Bourgain, Extension of a result of Benedek, Calderón and Panzone. Ark. Mat. **22**(1), 91–95 (1984)
5. A.P. Calderón, Intermediate spaces and interpolation, the complex method. Studia Math. **24**, 113–190 (1964)
6. P. Clément, J. Prüss, An operator-valued transference principle and maximal regularity on vector-valued L_p-spaces, in *Evolution Equations and Their Applications in Physical and Life Sciences (Bad Herrenalb, 1998)*. Lecture Notes in Pure and Applied Mathematics, vol. 215 (Dekker, New York, 2001), pp. 67–87
7. P. Clément, B. de Pagter, F.A. Sukochev, H. Witvliet, Schauder decompositions and multiplier theorems. Studia Math. **138**(2), 135–163 (2000)
8. D. Cruz-Uribe, A. Fiorenza, C.J. Neugebauer, The maximal function on variable L^p spaces. Ann. Acad. Sci. Fenn. Math. **28**(1), 223–238 (2003)
9. D.V. Cruz-Uribe, J.M. Martell, C. Pérez, *Weights, Extrapolation and the Theory of Rubio de Francia*. Operator Theory: Advances and Applications, vol. 215 (Birkhäuser/Springer Basel AG, Basel, 2011)
10. G.P. Curbera, J. García-Cuerva, J.M. Martell, C. Pérez, Extrapolation with weights, rearrangement-invariant function spaces, modular inequalities and applications to singular integrals. Adv. Math. **203**(1), 256–318 (2006)
11. C. Gallarati, M.C. Veraar, Evolution families and maximal regularity for systems of parabolic equations. Adv. Differ. Equ. **22**(3–4), 169–190 (2017)
12. C. Gallarati, M.C. Veraar, Maximal regularity for non-autonomous equations with measurable dependence on time. Potential Anal. **46**(3), 527–567 (2017)
13. C. Gallarati, E. Lorist, M.C. Veraar, On the ℓ^s-boundedness of a family of integral operators. Rev. Mat. Iberoam. **32**(4), 1277–1294 (2016)
14. J. García-Cuerva, J.L. Rubio de Francia, *Weighted Norm Inequalities and Related Topics*. North-Holland Mathematics Studies, vol. 116 (North-Holland, Amsterdam, 1985). Notas de Matemática, 104
15. J. García-Cuerva, R. Macías, J.L. Torrea, The Hardy-Littlewood property of Banach lattices. Israel J. Math. **83**(1–2), 177–201 (1993)
16. L. Grafakos, *Classical Fourier Analysis*. Graduate Texts in Mathematics, vol. 249, 2nd edn. (Springer, New York, 2008)
17. L. Grafakos, *Modern Fourier Analysis*. Graduate Texts in Mathematics, vol. 250, 2nd edn. (Springer, New York, 2009)
18. T.S. Hänninen, E. Lorist, Sparse domination for the lattice Hardy–Littlewood maximal operator. Proc. Am. Math. Soc. **147**(1), 271–284 (2019)
19. T.P. Hytönen, J.M.A.M. van Neerven, M.C. Veraar, L. Weis, *Analysis in Banach Spaces. Volume I: Martingales and Littlewood-Paley Theory*. Ergebnisse der Mathematik und ihrer Grenzgebiete, vol. 63 (Springer, Berlin, 2016)
20. T.P. Hytönen, J.M.A.M. van Neerven, M.C. Veraar, L. Weis, *Analysis in Banach Spaces. Volume II: Probabilistic Methods and Operator Theory*. Ergebnisse der Mathematik und ihrer Grenzgebiete, vol. 67 (Springer, Berlin, 2017)
21. N.J. Kalton, E. Lorist, L. Weis, Euclidean structures (In preparation)
22. M. Köhne, J. Prüss, M. Wilke, On quasilinear parabolic evolution equations in weighted L_p-spaces. J. Evol. Equ. **10**(2), 443–463 (2010)

23. P. Kunstmann, A. Ullmann, $\mathcal{R}_s$-sectorial operators and generalized Triebel-Lizorkin spaces. J. Fourier Anal. Appl. **20**(1), 135–185 (2014)
24. S. Kwapień, M.C. Veraar, L. Weis, R-boundedness versus γ-boundedness. Ark. Mat. **54**(1), 125–145 (2016)
25. N. Lindemulder, Maximal regularity with weights for parabolic problems with inhomogeneous boundary data. arXiv:1702.02803 (2017)
26. N. Lindemulder, M.C. Veraar, I.S. Yaroslavtsev, The UMD property for Musielak–Orlicz spaces, in *Positivity and Noncommutative Analysis – Festschrift in Honour of Ben de Pagter on the Occasion of His 65th Birthday*. Trends in Mathematics (Birkhäuser Verlag, Basel, 2019)
27. J. Lindenstrauss, L. Tzafriri, *Classical Banach Spaces. II*. Ergebnisse der Mathematik und ihrer Grenzgebiete, vol. 97 (Springer, Berlin, 1979)
28. E. Lorist, Maximal functions, factorization, and the $\mathcal{R}$-boundedness of integral operators. Master's Thesis, Delft University of Technology, Delft, the Netherlands (2016)
29. A. Lunardi, *Analytic Semigroups and Optimal Regularity in Parabolic Problems*. Progress in Nonlinear Differential Equations and their Applications, vol. 16 (Birkhäuser Verlag, Basel, 1995)
30. M. Meyries, R. Schnaubelt, Maximal regularity with temporal weights for parabolic problems with inhomogeneous boundary conditions. Math. Nachr. **285**(8–9), 1032–1051 (2012)
31. K. Moen, Sharp weighted bounds without testing or extrapolation. Arch. Math. **99**(5), 457–466 (2012)
32. A. Nekvinda, Hardy-Littlewood maximal operator on $L^{p(x)}(\mathbb{R})$. Math. Inequal. Appl. **7**(2), 255–265 (2004)
33. G. Pisier, *Martingales in Banach Spaces*. Cambridge Studies in Advanced Mathematics, vol. 155 (Cambridge University Press, Cambridge, 2016)
34. J. Prüss, Maximal regularity for evolution equations in L_p-spaces. Conf. Semin. Mat. Univ. Bari **285**, 1–39 (2003)
35. J. Prüss, G. Simonett, Maximal regularity for evolution equations in weighted L_p-spaces. Arch. Math. **82**(5), 415–431 (2004)
36. J.L. Rubio de Francia, Factorization theory and A_p weights. Am. J. Math. **106**(3), 533–547 (1984)
37. J.L. Rubio de Francia, Martingale and integral transforms of Banach space valued functions, in *Probability and Banach Spaces (Zaragoza, 1985)*. Lecture Notes in Mathematics, vol. 1221 (Springer, Berlin, 1986), pp. 195–222
38. J.M.A.M. van Neerven, M.C. Veraar, L. Weis, Stochastic maximal L^p-regularity. Ann. Probab. **40**(2), 788–812 (2012)
39. J.M.A.M. van Neerven, M.C. Veraar, L. Weis, On the R-boundedness of stochastic convolution operators. Positivity **19**(2), 355–384 (2015)
40. L. Weis, A new approach to maximal L_p-regularity, in *Evolution Equations and Their Applications in Physical and Life Sciences (Bad Herrenalb, 1998)*. Lecture Notes in Pure and Applied Mathematics, vol. 215 (Dekker, New York, 2001), pp. 195–214
41. L. Weis, Operator-valued Fourier multiplier theorems and maximal L_p-regularity. Math. Ann. **319**(4), 735–758 (2001)
42. A.C. Zaanen, *Integration* (North-Holland/Interscience Publishers Wiley, Amsterdam/New York, 1967). Completely revised edition of An introduction to the theory of integration

Backward Stochastic Evolution Equations in UMD Banach Spaces

Qi Lü and Jan van Neerven

Dedicated to Ben de Pagter on the occasion of his 65th birthday

Abstract Extending results of Pardoux–Peng and Hu–Peng, we prove well-posedness results for backward stochastic evolution equations in UMD Banach spaces.

Keywords Backward stochastic evolution equations · Brownian filtration · Stochastic integration in UMD Banach spaces · γ-radonifying operators · γ-boundedness

1 Introduction

In this paper we extend the classical results of Pardoux and Peng [25] and Hu and Peng [14] on backward stochastic differential equations to the UMD-valued setting.

We consider backward stochastic evolution equations (BSEEs) of the form

$$\begin{cases} \mathrm{d}U(t) + AU(t)\,\mathrm{d}t = f(t, U(t), V(t))\,\mathrm{d}t + V(t)\,\mathrm{d}W(t), & t \in [0, T], \\ U(T) = u_T, \end{cases} \tag{BSEE}$$

Q. Lü
School of Mathematics, Sichuan University, Chengdu, China
e-mail: lu@scu.edu.cn

J. van Neerven (✉)
Delft Institute of Applied Mathematics, Delft University of Technology, Delft, The Netherlands
e-mail: J.M.A.M.vanNeerven@TUDelft.nl

G. Buskes et al. (eds.), *Positivity and Noncommutative Analysis*,
Trends in Mathematics, https://doi.org/10.1007/978-3-030-10850-2_21

where $-A$ is the generator of a C_0-semigroup $S = (S(t))_{t \geqslant 0}$ on a UMD Banach space X and $W = (W(t))_{t \in [0,T]}$ is a standard Brownian motion. Our results extend to finite-dimensional Brownian motions and, more generally, to cylindrical Brownian motions without difficulty, but we do not pursue this here in order to keep the presentation as simple as possible. Denoting by $\mathbb{F} = \{\mathscr{F}_t\}_{t \in [0,T]}$ the augmented filtration generated by the Brownian motion W, the final value u_T is taken from $L^p(\Omega, \mathscr{F}_T; X)$, the closed subspace $L^p(\Omega; X)$ of all functions having a strongly $\mathscr{F}_T$-measurable pointwise defined representative. The mapping f is assumed to be $\mathbb{F}$-adapted and to satisfy suitable integrability and Lipschitz continuity requirements with respect to the natural norm arising from the L^p-stochastic integral in X. We will be interested in L^p-solutions (U, V) with values in X.

BSEEs, as infinite dimensional extensions of backward stochastic differential equations, arise in many applications related to stochastic control. For instance, the Duncan–Mortensen–Zakai filtration equation for the optimal control problem of partially observed stochastic differential equations is a linear BSEE (see, e.g., [4]); in order to establish the maximum principle for the optimal control problem of stochastic evolution equations one needs to introduce a linear BSEE as the adjoint equation (see, e.g., [22, 37]); in the study of controlled non-Markovian SDEs the stochastic Hamilton–Jacobi–Bellman equation is a class of fully nonlinear BSEEs (see, e.g., [11, 26]); and when the coefficients of the stochastic differential equation describing the stock price are random processes, the stochastic version of the Black-Scholes formula for option pricing is a BSEE (see, e.g., [23]).

In a Hilbert space setting, BSEEs have already been studied in [14]; see also [1, 2, 12, 20–22] and the references cited therein. In [9, 23, 24] the existence of a solution in the Sobolev space $W^{m,2}$ is obtained, in [3, 10] the existence of a solution in L^q, and in [29] the existence of a solution in Hölder spaces.

In the present paper, we study BSEEs in the abstract framework of evolution equations on UMD Banach spaces. The main results in [9, 10, 23, 24] are covered by our results. Furthermore, our results can be used to show the well-posedness of many other backward stochastic partial differential equations, such as $2m$-order backward stochastic parabolic equations.

The second-named author would like to use this opportunity to express warm-felt gratitude to Ben for invaluable mentorship and support throughout an entire mathematical career. Thanks for all, Ben!

2 Preliminaries

In this section we recall some useful concepts and results which will be used in the course of the paper. Proofs and more details, as well as references to the literature, can be found in the papers [5, 18, 30, 34], the lecture notes [7, 19], and the monographs [15, 16, 27].

Unless stated otherwise, all vector spaces are assumed to be real. We will always identify Hilbert spaces with their duals by means of the Riesz representation theorem.

2.1 γ-Boundedness

Let X and Y be Banach spaces and let $\{\gamma_n\}_{n\geqslant 1}$ be Gaussian sequence (i.e., a sequence of independent real-valued standard Gaussian random variables).

Definition 2.1 A family $\mathscr{T}$ of bounded linear operators from X to Y is called γ-*bounded* if there exists a constant $C \geqslant 0$ such that for all finite sequences $\{x_n\}_{n=1}^N$ in X and $\{T_n\}_{n=1}^N$ in $\mathscr{T}$ we have

$$\mathbb{E}\Big\|\sum_{n=1}^N \gamma_n T_n x_n\Big\|^2 \leqslant C^2\mathbb{E}\Big\|\sum_{n=1}^N \gamma_n x_n\Big\|^2.$$

Clearly, every γ-bounded family of bounded linear operators from X to Y is uniformly bounded and $\sup_{t\in\mathscr{T}}\|T\|_{\mathscr{L}(X;Y)} \leqslant C$, the constant appearing in the above definition. In the setting of Hilbert spaces both notions are equivalent and the above inequality holds with $C = \sup_{t\in\mathscr{T}}\|T\|_{\mathscr{L}(X;Y)}$.

γ-Boundedness is the Gaussian analogue of R-boundedness, obtained by replacing Gaussian variables by Rademacher variables. This notion was introduced and thoroughly studied in the seminal paper [6].

2.2 γ-Radonifying Operators

Let H be a Hilbert space with inner product $(\cdot|\cdot)$ and X a Banach space. Let $H\otimes X$ denote the linear space of all finite rank operators from H to X. Every element in $H \otimes X$ can be represented in the form $\sum_{n=1}^N h_n \otimes x_n$, where $h_n \otimes x_n$ is the rank one operator mapping the vector $h \in H$ to $(h|h_n)x_n \in X$. By a Gram-Schmidt orthogonalisation argument we may always assume that the sequence $\{h_n\}_{n=1}^N$ is orthonormal in H.

Definition 2.2 The Banach space $\gamma(H, X)$ is the completion of $H\otimes X$ with respect to the norm

$$\Big\|\sum_{n=1}^N h_n \otimes x_n\Big\|_{\gamma(H,X)} := \Big(\mathbb{E}\Big\|\sum_{n=1}^N \gamma_n x_n\Big\|^2\Big)^{1/2}, \tag{2.1}$$

where $\{h_n\}_{n=1}^N$ is orthonormal in H and $\{\gamma_n\}_{n=1}^N$ is a Gaussian sequence.

Since the distribution of a Gaussian vector in $\mathbb{R}^N$ is invariant under orthogonal transformations, the quantity on the right-hand side of (2.1) is independent of the representation of the operator as a finite sum of the form $\sum_{n=1}^N h_n \otimes x_n$ as long as $\{h_n\}_{n=1}^N$ is orthonormal in H. Therefore, the norm $\|\cdot\|_{\gamma(H,X)}$ is well defined.

Remark 2.3 By the Kahane-Khintchine inequalities [16, Theorem 6.2.6], for all $0 < p < \infty$ there exists a universal constant κ_p, depending only on p, such that for all Banach spaces X and all finite sequences $\{x_n\}_{n=1}^N$ in X we have

$$\frac{1}{\kappa_p}\Big(\mathbb{E}\Big\|\sum_{n=1}^N \gamma_n x_n\Big\|^p\Big)^{1/p} \leqslant \Big(\mathbb{E}\Big\|\sum_{n=1}^N \gamma_n x_n\Big\|^2\Big)^{1/2} \leqslant \kappa_p\Big(\mathbb{E}\Big\|\sum_{n=1}^N \gamma_n x_n\Big\|^p\Big)^{1/p}.$$

As a consequence, for $1 \leqslant p < \infty$ the norm

$$\Big\|\sum_{n=1}^N h_n \otimes x_n\Big\|_{\gamma^p(H,X)} := \Big(\mathbb{E}\Big\|\sum_{n=1}^N \gamma_n x_n\Big\|^p\Big)^{1/p},$$

with $\{h_n\}_{n=1}^N$ orthonormal in H, is an equivalent norm on $\gamma(H, X)$. Endowed with this equivalent norm, the space is denoted by $\gamma^p(H, X)$.

For any Hilbert space H we have a natural isometric isomorphism

$$\gamma(H, X) = \mathscr{L}_2(H, X),$$

where $\mathscr{L}_2(H, X)$ is the space of all Hilbert-Schmidt operators from H to X. Furthermore, for $1 \leqslant p < \infty$ and σ-finite measures μ we have an isometric isomorphism of Banach spaces

$$\gamma^p(H, L^p(\mu; X)) \simeq L^p(\mu; \gamma^p(H; X)) \tag{2.2}$$

which is obtained by associating with $f \in L^p(\mu; \gamma(H; X))$ the mapping $h' \mapsto f(\cdot)h'$ from H to $L^p(\mu; X)$ [16, Theorem 9.4.8]. In particular, upon identifying $\gamma(H, \mathbb{R})$ with H, we obtain an isomorphism of Banach spaces

$$\gamma(H, L^p(\mu)) \simeq L^p(\mu; H).$$

When I is an interval in the real line, for brevity we write

$$\gamma(I; X) := \gamma(L^2(I), X).$$

Definition 2.4 A strongly measurable function $f : I \to X$ is said to *define an element of* $\gamma(I; X)$ if $\langle f, x^*\rangle \in L^2(I)$ for all $x^* \in X^*$ and the Pettis integral operator

$$g \mapsto \int_I f(t)g(t)\,\mathrm{d}t$$

belongs to $\gamma(I; X)$.

Observe that the condition $\langle f, x^* \rangle \in L^2(I)$ for all $x^* \in X^*$ ensures that fg is Pettis integrable for all $g \in L^2(I)$; see [16, Definition 9.2.3] and the discussion following it.

Throughout the paper we fix a final time $0 < T < \infty$. For any $f \in \gamma(0, T; X)$ it is possible to define a $\frac{1}{2}$-Hölder continuous function $[0, T] \ni t \mapsto \int_0^t f(s)\,\mathrm{d}s \in X$ as follows. We begin by observing that integration operator $I_{s,t} : \phi \mapsto \int_s^t f(r)\,\mathrm{d}r$ is bounded from $L^2(0, T)$ to $\mathbb{R}$ and has norm $(t-s)^{1/2}$. Therefore, by the Kalton–Weis extension theorem [16, Theorem 9.6.1] the mapping $\widetilde{I}_{s,t} : \phi \otimes x \mapsto (I_{s,t}\phi) \otimes x$ has a unique extension to a bounded linear operator from $\gamma(0, T; X)$ to X of the same norm: $\|\widetilde{I}_{s,t}\|_{\mathscr{L}(\gamma(0,T;X),X)} = \|I_{s,t}\|_{\mathscr{L}(L^2(0,T),\mathbb{R})} = (t-s)^{1/2}$. We now define, for $g \in \gamma(0, T; X)$,

$$\int_s^t f(s)\,\mathrm{d}s := \widetilde{I}_{s,t} f.$$

Noting that $\widetilde{I}_{0,t} f - \widetilde{I}_{0,s} f = \widetilde{I}_{s,t} f$, we see that $t \mapsto \int_0^t f(s)\,\mathrm{d}s$ is Hölder continuous of order $\frac{1}{2}$ and

$$\left\| \int_s^t f(s)\,\mathrm{d}s \right\| \leqslant (t-s)^{1/2} \|f\|_{\gamma(0,T;X)}. \tag{2.3}$$

Remark 2.5 We are abusing notation slightly here, as the above integral notation is only formal since elements in $\gamma(0, T; X)$ cannot in general be represented as functions. For the sake of readability this notation will be used throughout the paper.

Treating t as a variable, we may also use the Kalton–Weis extension theorem to extend $f \mapsto \int_0^{\cdot} f(s)\,\mathrm{d}s$ (viewed as a bounded operator on $L^2(0, T)$ of norm $T/\sqrt{2}$) to a bounded operator on $\gamma(0, T; X)$ of the same norm. With the same slight abuse of notation this may be expressed as

$$\left\| t \mapsto \int_0^t f(s)\,\mathrm{d}s \right\|_{\gamma(0,T;X)} \leqslant \frac{T}{\sqrt{2}} \|f\|_{\gamma(0,T;X)}.$$

We will need the following elaboration on this theme, which is of some independent interest. Put

$$\Delta := \{(s, t) \in (0, T) \times (0, T) : 0 < s \leqslant t < T\}.$$

Lemma 2.6 *Let X and Y be Banach spaces and assume that Y does not contain a closed subspaces isomorphic to c_0.*

(1) *Let $M : (0, T) \to \mathscr{L}(X, Y)$ be a function with the property that $t \mapsto M(t)x$ is strongly measurable for all $x \in X$ and assume that M has γ-bounded range, with γ-bound $\gamma(M)$. Then the function*

$$\Phi f : t \mapsto \int_0^t M(t-s) f(s) \,\mathrm{d}s, \quad f \in L^2(0, T) \otimes X,$$

defines an element of $\gamma(0, T; Y)$ of norm

$$\|\Phi f\|_{\gamma(0,T;Y)} \leqslant T\gamma(M)\|f\|_{\gamma(0,T;X)}.$$

(2) *Let $M : \Delta \to \mathscr{L}(X, Y)$ be a function with the property that $(s, t) \mapsto M(s, t)x$ is strongly measurable for all $x \in X$ and assume that M has γ-bounded range, with γ-bound $\gamma(M)$. The function*

$$\Phi f : t \mapsto \int_0^t M(s, t) f(s, t) \,\mathrm{d}s, \quad f \in L^2(\Delta) \otimes X,$$

defines an element of $\gamma(0, T; Y)$ of norm

$$\|\Phi f\|_{\gamma(0,T;Y)} \leqslant T^{1/2}\gamma(M)\|f\|_{\gamma(\Delta;X)}.$$

As a consequence, the mappings $f \mapsto \Phi f$ extend uniquely to bounded operators from $\gamma(0, T; X)$ to $\gamma(0, T; Y)$ and from $\gamma(\Delta; X)$ to $\gamma(0, T; Y)$, respectively, of norms at most $T\gamma(M)$ and $T^{1/2}\gamma(M)$, respectively.

Proof We begin with the proof of (1). The estimate

$$\int_0^T \int_0^t |g(t-s)|^2 \,\mathrm{d}s \,\mathrm{d}t \leqslant T\|g\|_2^2$$

shows that the mapping $J_1 : g \mapsto [(s, t) \mapsto g(t-s)]$ is bounded from $L^2(0, T)$ to $L^2(\Delta_T)$ of norm at most $T^{1/2}$. By the Kalton–Weis extension theorem, it extends to a bounded operator from $\gamma(0, T; X)$ to $\gamma(\Delta; X)$ of the same norm. By the Kalton–Weis multiplier theorem [16, Theorem 9.5.1], the pointwise multiplier M (acting in the variable s, so that $[(s, t) \mapsto g(t-s)]$ is mapped to $[(s, t) \mapsto M(s)g(t-s)]$) extends to a bounded operator from $\gamma(\Delta; X)$ to $\gamma(\Delta; Y)$ of norm at most $\gamma(M)$. Next, the estimate

$$\int_0^T \left|\int_0^t f(s, t) \,\mathrm{d}s\right|^2 \mathrm{d}t \leqslant T \int_0^T \int_0^t |h(s, t)|^2 \,\mathrm{d}s \,\mathrm{d}t$$

shows that the mapping $J_2 : h \mapsto [t \mapsto \int_0^t h(s,t)\,ds]$ is bounded from $L^2(\Delta_T)$ to $L^2(0,T)$ of norm at most $T^{1/2}$. By the Kalton–Weis extension theorem, it extends to a bounded operator from $\gamma(\Delta; Y)$ to $\gamma(0,T;Y)$ of the same norm. The mapping $f \mapsto \Phi f$ in the statement of the lemma factorises as $\Phi = J_2 \circ M \circ J_1$ and therefore extends to a bounded operator from $\gamma(0,T;X)$ to $\gamma(0,T;Y)$ of norm at most $T\gamma(M)$.

(2): This is proved similarly, except that the first step of the proof can now be skipped. □

2.3 UMD Spaces and the Upper Contraction Property

We next introduce the class of Banach spaces in which we will be working.

Definition 2.7 A Banach space X is called a *UMD space* if for some (equivalently, for all) $1 < p < \infty$ there is a constant $C_{p,X} \geqslant 0$ such that for all finite X-valued L^p-martingale difference sequences $\{d_n\}_{n=1}^N$ on a probability space Ω and sequences of signs $\{\epsilon_n\}_{n=1}^N$ one has

$$\mathbb{E}\Big\|\sum_{n=1}^N \varepsilon_n d_n\Big\|^p \leqslant C_{p,X}^p \mathbb{E}\Big\|\sum_{n=1}^N d_n\Big\|^p, \quad \forall N \geqslant 1.$$

Every Hilbert space and every space $L^p(\mu)$ with $1 < p < \infty$ is a UMD space. If X is a UMD space, then the spaces $L^p(\mu; X)$ are UMD for all $1 < p < \infty$. Moreover, X is a UMD space if and only X^* is a UMD space. Every UMD space is reflexive (and in fact super-reflexive); it follows that spaces such as c_0, $C(K)$, ℓ^∞, $L^\infty(\mu)$, ℓ^1, $L^1(\mu)$, and all Banach spaces containing isomorphic copies of one of these spaces fail the UMD property (apart from the trivial cases giving rise to finite-dimensional spaces, i.e., when K is finite or μ is supported on finitely many atoms).

Definition 2.8 A Banach space X has the *upper contraction property* if for some (equivalently, for all) $1 \leqslant p < \infty$ there is a constant $C_{p,X} \geqslant 0$ such that for all finite sequences $\{x_{mn}\}_{m,n=1}^{M,N}$ in X and all Gaussian sequences $\{\gamma'_m\}_{m=1}^M$ and $\{\gamma''_n\}_{n=1}^N$ on independent probability spaces Ω' and Ω'' and $\{\gamma_{m,n}\}_{m,n=1}^{M,N}$ on a probability space Ω, we have

$$\mathbb{E}\Big\|\sum_{m=1}^M\sum_{n=1}^N \gamma_{mn}x_{mn}\Big\|^p \leqslant C_{p,X}^p \mathbb{E}'\mathbb{E}''\Big\|\sum_{m=1}^M\sum_{n=1}^N \gamma'_m\gamma''_n x_{mn}\Big\|^p.$$

By interchanging the two double sums one obtains the related *lower contraction property*, and a Banach space is said to have the *Pisier contraction property* if it has both the upper and lower contraction property. In the present paper we only need the upper contraction property.

Every Hilbert space and every Banach lattice with finite cotype (in particular, every space $L^p(\mu)$ with $1 \leqslant p < \infty$) has the Pisier contraction property. If X has the upper (resp. lower, Pisier) contraction property, then the spaces $L^p(\mu; X)$ have the upper (resp. lower, Pisier) contraction property for all $1 \leqslant p < \infty$. Moreover, if X is K-convex, then X has the upper (resp. lower, Pisier) contraction property if and only X^* has the lower (resp. upper, Pisier) contraction property. Every Banach space with type 2 has the upper contraction property. The reader is referred to [16, Section 7.6] for proofs and more details.

The following lemma translates the above definition into the language of γ-radonification. A proof is obtained by noting that for functions in $L^2(0,T) \otimes L^2(0,T) \otimes X$ the lemma follows from the estimate of the definition, and the general case follows from it by approximation.

Lemma 2.9 *If X is a Banach space with the upper contraction property, then for all $f \in L^2(0,T) \otimes L^2(0,T) \otimes X$ we have*

$$\|f\|_{\gamma((0,T)\times(0,T);X)} \leqslant C_{p,X} \|f\|_{\gamma(0,T;\gamma(0,T;X))}.$$

2.4 Stochastic Integration

Let $\mathbb{F} = (\mathscr{F}_t)_{t\in[0,T]}$ be a filtration in Ω. An X-valued *$\mathscr{F}$-adapted step process* is a finite linear combination of indicator processes of the form $\mathbf{1}_{(s,t)\times F} \otimes x$ with $F \in \mathscr{F}_s$ and $x \in X$. The space

$$L^p_{\mathbb{F}}(\Omega; \gamma(0,T;X))$$

is defined as the closure in $L^p(\Omega; \gamma(0,T;X))$ of the X-valued $\mathscr{F}$-adapted step processes. The following result is from [32].

Lemma 2.10 *If the process $\phi : [0,T] \times \Omega \to X$ is $\mathbb{F}$-adapted and defines an element of $L^p(\Omega; \gamma(0,T;X))$, then it defines an element of $L^p_{\mathbb{F}}(\Omega; \gamma(0,T;X))$.*

From the point of view of stochastic integration, the raison d'être for UMD spaces is the following result of [32].

Theorem 2.11 (Itô Isomorphism) *Let X be a UMD space and let $1 < p < \infty$. For all $\mathbb{F}$-adapted elementary processes $\phi \in L^p(\Omega; \gamma(0,T;X))$ we have*

$$\mathbb{E}\Big\| \int_0^T \phi \, \mathrm{d}W \Big\|^p \eqsim_p \mathbb{E} \sup_{t\in[0,T]} \Big\| \int_0^t \phi \, \mathrm{d}W \Big\|^p \eqsim_{p,X} \|\phi\|^p_{L^p(\Omega;\gamma(0,T;X))}$$

with implied constants depending only on p and X.

As an immediate consequence, the stochastic integral can be extended to arbitrary integrands in $L^p_{\mathbb{F}}(\Omega;\gamma(0,T;X))$, with the same two-sided bound on their L^p-moments. It can furthermore be shown (see [13]) that the UMD property is necessary in Theorem 2.11 in the sense that it is implied by the validity of the statement in the theorem.

Remark 2.12 For $\phi \in L^p_{\mathbb{F}}(\Omega;\gamma(0,T;X))$ we denote by $\int_0^T \phi\,\mathrm{d}W$ the unique extension of the stochastic integral as guaranteed by the theorem. For $t \in [0,T]$ we write $\int_0^t \phi\,\mathrm{d}W := \int_0^T \mathbf{1}_{(0,t)}\phi\,\mathrm{d}W$.

3 Backward Stochastic Evolution Equations: Well-Posedness

Let us now take up our main topic, the study of the backward stochastic evolution equation (BSEE)

$$\begin{cases} \mathrm{d}U(t) + AU(t)\,\mathrm{d}t = f(t,U(t),V(t))\,\mathrm{d}t + V(t)\,\mathrm{d}W(t), & t \in [0,T], \\ U(T) = u_T. \end{cases} \tag{BSEE}$$

The function f also depends on the underlying probability space, but following common practice we suppress this from the notation. The following standing assumptions, or, when this is explicitly indicated, a selection of them, will be in force throughout the remainder of the paper:

(H1) X is a UMD Banach space and $1 < p < \infty$;
(H2) $\mathbb{F} = \{\mathscr{F}_t\}_{t\in[0,T]}$ is the augmented filtration generated by the Brownian motion $W = (W(t))_{t\in[0,T]}$;
(H3) u_T belongs to $L^p(\Omega,\mathscr{F}_T;X)$;
(H4) A generates a C_0-semigroup $S = \{S(t)\}_{t\geqslant 0}$ on X;
(H5) the set $\{S(t)\}_{t\in[0,T]}$ is γ-bounded.

If X is isomorphic to a Hilbert space, (H5) follows from (H4). If X is a UMD space, (H4) and (H5) are fulfilled when A has maximal L^p-regularity on $[0,T]$. Recall that a densely defined, closed operator A acting in a Banach space X has *maximal L^p-regularity on* $[0,T]$ if there exists a constant $C \geqslant 0$ such that for every $f \in C_c(0,T)\otimes \mathsf{D}(A)$ there exists a strongly measurable function $u : [0,T] \to X$ with the following properties:

1. u takes values in $\mathsf{D}(A)$ almost everywhere and Au belongs to $L^p(0,T;X)$;
2. for almost all $t \in (0,T)$ we have

$$u(t) + \int_0^t Au(s)\,\mathrm{d}s = \int_0^t f(s)\,\mathrm{d}s;$$

3. we have the estimate

$$\|Au\|_{L^p(I;X)} \leqslant C\|f\|_{L^p(0,T;X)},$$

with a constant $C \geqslant 0$ independent of f.

A systematic discussion of maximal L^p-regularity is given in [8], where among other things it is shown that if A has maximal L^p-regularity, then A generates an (analytic) C_0-semigroup. In particular, maximal L^p-regularity implies that (H4) holds. A celebrated result of Weis [36] states that a densely defined closed operator A in a UMD space X has maximal L^p-regularity and only if $-A$ generates an analytic C_0-semigroup on X which is γ-bounded on some sector in the complex plane containing the positive real axis. In particular this implies that (H5) holds.

Examples of operators with maximal L^p-regularity include most second-order elliptic operators on $\mathbb{R}^d$ or on sufficiently smooth bounded domains in $\mathbb{R}^d$ with various boundary conditions, provided the coefficients satisfy appropriate smoothness assumptions. For more details, the reader is referred to [7, 8, 17, 19, 28].

Below we will consider the three special cases where (a) $A = 0$ and the process $f : [0, T]\times\Omega\times X\times X \to X$ only depends on the first two variables, (b) the process $f : [0, T] \times \Omega \times X \times X \to X$ only depends on the first two variables, and (c) no additional restrictions are imposed. The precise assumptions on f will depend on the case under consideration, but in each of the three cases they coincide with, or are special cases of, the following condition:

(H6) The function $f : [0, T] \times \Omega \times X \times X \to X$ has the following properties:

1. f is jointly measurable in the first two variables and continuous in the third and fourth;
2. for all $U, V \in L^p_{\mathbb{F}}(\Omega; \gamma(0, T; X))$ the process

$$f(\cdot, U, V) : (t, \omega) \mapsto f(t, \omega, U(t, \omega), V(t, \omega))$$

defines an element of $L^p_{\mathbb{F}}(\Omega; \gamma(0, T; X))$;

3. there is a constant $C \geqslant 0$ such that for all $U, V \in L^p_{\mathbb{F}}(\Omega; \gamma(0, T; X))$ we have

$$\begin{aligned}&\|f(\cdot, U, V)\|_{L^p(\Omega;\gamma(0,T;X))}\\ &\qquad \leqslant C(1 + \|U\|_{L^p(\Omega;\gamma(0,T;X))} + \|V\|_{L^p(\Omega;\gamma(0,T;X))});\end{aligned}$$

4. there is a constant $L \geqslant 0$ such that for all $U, U', V, V' \in L^p_{\mathbb{F}}(\Omega; \gamma(0, T; X))$ we have

$$\begin{aligned}&\|f(\cdot, U, V) - f(\cdot, U', V')\|_{L^p(\Omega;\gamma(0,T;X))}\\ &\qquad \leqslant L(\|U - U'\|_{L^p(\Omega;\gamma(0,T;X))} + \|V - V'\|_{L^p(\Omega;\gamma(0,T;X))}).\end{aligned}$$

A closely related notion of γ-Lipschitz continuity has been introduced and studied in [33]. In the same way as in this reference one shows that if X has type 2 (e.g., if X is a Hilbert space or a space $L^p(\mu)$ with $2 \leqslant p < \infty$), then the usual linear growth and Lipschitz conditions

$$\|f(t, \omega, x, y)\| \leqslant C_f(1 + \|x\| + \|y\|),$$
$$\|f(t, \omega, x, y) - f(t, x', y')\| \leqslant L_f(\|x - x'\| + \|y - y'\|),$$

imply that f satisfies (H6).

Definition 3.1 Assume (H1)–(H6). A *mild L^p-solution* to the problem (BSEE) is a pair (U, V), where U and V are continuous $\mathbb{F}$-adapted processes defining elements in $L^p_{\mathbb{F}}(\Omega; \gamma(0, T; X))$ such that

$$U(t) + \int_t^T S(s-t) f(s, U(s), V(s))\, \mathrm{d}s + \int_t^T S(s-t) V(s)\, \mathrm{d}W(s) = S(T-t) u_T,$$

where the identity is to be interpreted in the sense explained in Sect. 2.2.

Assumptions (H5) and (H6) imply, via the Kalton–Weis multiplier theorem, that if $U, V \in L^p_{\mathbb{F}}(\Omega; \gamma(0, T; X))$, then for each $t \in [0, T]$ the mappings $s \mapsto S(s-t) f(s, U(s), V(s))$ and $s \mapsto S(s-t)V(s)$ define elements in $L^p_{\mathbb{F}}(\Omega; \gamma(t, T; X))$. Therefore by (2.3) the integral

$$\int_t^T S(s-t) f(s, U(s), V(s))\, \mathrm{d}s$$

is well defined as an element of $L^p(\Omega; X)$, and by Theorem 2.11 the same is true for the stochastic integral

$$\int_t^T S(s-t) V(s)\, \mathrm{d}W(s).$$

Thus, in hindsight, the identity in Definition 3.1 admits an interpretation in $L^p(\Omega; X)$ pointwise in $t \in [0, T]$, and it is of interest to ask about time regularity of U.

Proposition 3.2 *Assume (H1)–(H6). If (U, V) is a mild L^p-solution to the problem* (BSEE)*, then U belongs to $C([0, T]; L^p(\Omega; X))$.*

Proof It is not hard to see that $t \mapsto \int_t^T S(s-t) f(s, U(s), V(s))\, \mathrm{d}s$ belongs to $L^p(\Omega; C([0, T]; X))$ (and hence to $C([0, T]; L^p(\Omega; X))$). Indeed, arguing pathwise, it suffices to note that for all g in the dense subspace $L^2(0, T) \otimes X$ of

$\gamma(0,T;X)$ the mapping $t \mapsto \int_t^T S(s-t)g(s)\,ds$ is continuous and satisfies

$$\sup_{t\in[0,T]} \Big\| \int_t^T S(s-t)g(s)\,ds \Big\| \leqslant \sup_{t\in[0,T]} (T-t)^{1/2} \|s \mapsto S(t-s)g(s)\|_{\gamma(T-t,T;X)} \leqslant T^{1/2}\gamma(S)\|g\|_{\gamma(0,T;X)}$$

using (2.3), where $\gamma(S)$ is the γ-bound of $\{S(t) : t \in [0,T]\}$. Similarly the mapping $t \mapsto \int_t^T S(s-t)V(s)\,dW(s)$ is seen to belong to $C([0,T]; L^p(\Omega;X))$. Indeed for adapted X-valued step processes V, which are dense in $L^p_{\mathbb{F}}(\Omega, \gamma(0,T;X))$, the mapping $t \mapsto \int_t^T S(s-t)V(s)\,dW(s)$ is continuous and satisfies

$$\sup_{t\in[0,T]} \Big\| t \mapsto \int_t^T S(s-t)V(s)\,dW(s) \Big\|_{L^p(\Omega;X)} \lesssim_{p,X} \sup_{t\in[0,T]} \|s \mapsto S(s-t)V(s)\|_{L^p(\Omega;\gamma(T-t,T;X))} \leqslant \gamma(S)\|V\|_{L^p(\Omega;\gamma(0,T;X))}$$

using Theorem 2.11. □

From the proof we see that U is in $L^p(\Omega; C([0,T];X))$ if and only if $t \mapsto \int_t^T S(s-t)V(s)\,dW(s)$ is in $L^p(\Omega; C([0,T];X))$, but the latter is not to be expected unless we make additional conditions implying maximal estimates for stochastic convolutions (such as in [35, Section 4]).

3.1 The Case $A = 0$, $f(t,\omega,x,y) = f(t,\omega)$

We begin by considering the problem

$$\begin{cases} dU(t) = f(t)\,dt + V(t)\,dW(t), & t \in [0,T], \\ U(T) = u_T, \end{cases} \tag{3.1}$$

assuming (H1)–(H3) as well as
(H6)′ f defines an element of $L^p_{\mathbb{F}}(\Omega; \gamma(0,T;X))$.
We comment on this assumption in Remark 3.4 below. Even though (3.1) is a special case of the problem (3.5) considered in the next subsection, it is instructive to treat it separately.

Following the ideas of [25] we define the X-valued process M by

$$M(t) := \mathbb{E}\Big(u_T - \int_0^T f(s)\,ds \Big| \mathscr{F}_t\Big).$$

By [32, Theorems 4.7, 5.13] this is a continuous L^p-martingale with respect to $\mathbb{F}$ in X and there exists a unique $V \in L^p_{\mathbb{F}}(\Omega; \gamma(0, T; X))$ such that

$$M(t) = M(0) + \int_0^t V \, \mathrm{d}W. \tag{3.2}$$

By [32, Theorems 4.5, 5.12] and the observations in Sect. 2.2 combined with Lemma 2.10, both M and the $\mathbb{F}$-adapted process

$$U(t) := M(t) + \int_0^t f(s) \, \mathrm{d}s \tag{3.3}$$

belong to $L^p_{\mathbb{F}}(\Omega; \gamma(0, T; X))$.

Proposition 3.3 *Let (H1)–(H3) and (H6)′ be satisfied. Then the problem* (3.1) *admits a unique mild L^p-solution (U, V). It is given by the pair constructed in* (3.2) *and* (3.3).

Proof Let U and V be defined by (3.2) and (3.3). We have already checked that U and V belong to $L^p_{\mathbb{F}}(\Omega; \gamma(0, T; X))$. To show that (U, V) is an L^p-solution, note that

$$\begin{aligned}
U(t) &+ \int_t^T f(s) \, \mathrm{d}s + \int_t^T V \, \mathrm{d}W \\
&= \Big(M(t) + \int_0^t f(s) \, \mathrm{d}s\Big) + \int_t^T f(s) \, \mathrm{d}s + (M(T) - M(t)) \\
&= \int_0^T f(s) \, \mathrm{d}s + M(T) \\
&= \int_0^T f(s) \, \mathrm{d}s + \Big(u_T - \int_0^T f(s) \, \mathrm{d}s\Big) \\
&= u_T.
\end{aligned}$$

Concerning uniqueness, suppose $(\widetilde{U}, \widetilde{V})$ is another L^p-solution. Then

$$\widetilde{U}(t) - U(t) + \int_t^T (\widetilde{V} - V) \, \mathrm{d}W = 0 \quad \forall t \in [0, T]. \tag{3.4}$$

Taking conditional expectations with respect to $\mathscr{F}_t$ it follows that $\widetilde{U}(t) - U(t) = 0$, where we used [32, Proposition 4.3] to see that the conditional expectation of the stochastic integral vanishes. Uniqueness of V is already implicit in the uniqueness part of (3.2). It also follows from (3.4), where $\widetilde{U} = U$ gives $\int_t^T (\widetilde{V} - V) \, \mathrm{d}W = 0$

for all $t \in [0, T]$. Taking $t = 0$ and taking L^p-means, using [32, Theorem 3.5] it follows that

$$\|\widetilde{V} - V\|_{L^p(\Omega;\gamma(0,T;X))} \eqsim_{p,X} \mathbb{E}\Big\| \int_0^T (\widetilde{V} - V)\,\mathrm{d}W\Big\|^p = 0,$$

and therefore $\widetilde{V} = V$ in $L^p(\Omega; \gamma(0, T; X))$. □

Remark 3.4 The reader may check that, *mutatis mutandis*, Proposition 3.3 admits a version when (H6)′ is replaced by the simpler condition $f \in L^p_{\mathbb{F}}(\Omega; L^1(0, T; X))$. That the integral in (3.3) defines an element of $L^p_{\mathbb{F}}(\Omega; \gamma(0, T; X))$ then follows from [16, Proposition 9.7.1] . The motivation for the present formulation of (H6)′ is that it is a special case of the assumption (H6) needed in the final section where mixed L^p-L^1 conditions do not seem to work.

3.2 The Case $f(t, \omega, x, y) = f(t, \omega)$

We now consider the problem

$$\begin{cases} \mathrm{d}U(t) + AU(t)\,\mathrm{d}t = f(t)\,\mathrm{d}t + V(t)\,\mathrm{d}W(t), & t \in [0, T], \\ U(T) = u_T, \end{cases} \tag{3.5}$$

assuming (H1)–(H4) and (H6)′. Our proof of the well-posedness of the problem (3.5) relies on the following lemma, where s and σ denote two time variables; the dependence on ω is suppressed. To give a meaning to the expression in the second condition below we recall from (2.2) the isomorphism of Banach spaces

$$\gamma(0, T; L^p(\Omega; Y)) \eqsim_p L^p(\Omega; \gamma(0, T; Y)).$$

This isomorphism allows us to interpret, in condition (2) below, k as an element of $\gamma(0, T; L^p_{\mathbb{F}}(\Omega; \gamma(0, T; X)))$.

Lemma 3.5 *Let (H1), (H2), and (H6)′ be satisfied. There exists a unique* $k \in L^p_{\mathbb{F}}(\Omega; \gamma(0, T; \gamma(0, T; X)))$ *satisfying the following conditions:*

(1) *almost surely, k is supported on the set $\{(s, \sigma) \in [0, T] \times [0, T] : \sigma \leqslant s\}$;*
(2) *for almost all $s \in [0, T]$ we have*

$$f(s) = \mathbb{E}f(s) + \int_0^s k(s, \sigma)\,\mathrm{d}W(\sigma) \;\text{ in } L^p(\Omega; X);$$

(3) *we have the estimate*

$$\|k\|_{L^p(\Omega;\gamma(0,T;\gamma(0,T;X)))} \lesssim_{p,X} \|f\|_{L^p(\Omega;\gamma(0,T;X)))}.$$

The precise meaning of condition (1) is that for almost all $\omega \in \Omega$, the operator $k(\omega) \in \gamma(0, T; \gamma(0, T; X))$ vanishes on all $f \in L^2(0, T) \otimes L^2(0, T)$, which, as functions on $(0, T) \times (0, T)$, are supported on the set $\{(s, \sigma) \in (0, T) \times [0, T] : \sigma > s\}$.

Proof Since by assumption $f \in L^p_{\mathbb{F}}(\Omega; \gamma(0, T; X))$, we may pick a sequence of adapted step processes $\{f_n\}_{n=1}^{\infty}$ such that $f_n \to f$ in $L^p(\Omega; \gamma(0, T; X)))$ as $n \to \infty$. For each $n \geqslant 1$ we then may write

$$f_n(s, \omega) = \sum_{i=0}^{N_n-1} \mathbf{1}_{[t_{n,i}, t_{n,i+1})}(s)\xi_{n,i}(\omega)$$

where $\{t_{n,0}, t_{n,1}, \cdots, t_{n,N_n}\}$ is a partition of $[0, T]$ and the random variables $\xi_{n,i} \in L^p(\Omega; X)$ are strongly $\mathscr{F}_{t_{n,i}}$-measurable. By [32, Theorem 3.5] there exist $k_{n,i} \in L^p_{\mathbb{F}}(\Omega; \gamma(0, t_{n,i}; X))$ such that

$$\xi_{n,i} = \mathbb{E}\xi_{n,i} + \int_0^{t_{n,i}} k_{n,i}\, \mathrm{d}W.$$

In what follows we will identify $k_{n,i}$ with elements of $L^p_{\mathbb{F}}(\Omega; \gamma(0, T; X))$ in the natural way. Put

$$k_n(s, \sigma) := \sum_{i=0}^{N_n-1} \mathbf{1}_{[t_{n,i}, t_{n,i+1})}(s)\mathbf{1}_{[0,t_{n,i})}(\sigma)k_{n,i}(\sigma).$$

Each k_n satisfies the support condition of (1) and

$$f_n(s) = \mathbb{E}f_n(s) + \int_0^s k_n(s, \sigma)\, \mathrm{d}W(\sigma). \tag{3.6}$$

Choose an orthonormal basis $\{h_j\}_{j\geqslant 1}$ for $L^2(0, T)$ and let $\{\gamma'_j\}_{j\geqslant 1}$ be a Gaussian sequence on an independent probability space $(\Omega', \mathbb{P}')$. Then, by [16, Theorem 9.1.17], the Itô isomorphism of Theorem 2.11, and the stochastic Fubini theorem (see, e.g., [31]) and keeping in mind the support properties, we have

$$
\begin{aligned}
&\big\|s \mapsto k_n(s,\cdot) - k_m(s,\cdot)\big\|^p_{\gamma(0,T;L^p(\Omega;\gamma(0,T;X)))} \\
&\eqsim_p \mathbb{E}'\Big\| \sum_{j\geqslant 1} \gamma_j' \int_0^T h_j(s)(k_n(s,\cdot) - k_m(s,\cdot))\,\mathrm{d}s\Big\|^p_{L^p(\Omega;\gamma(0,T;X))} \\
&\eqsim_{p,X} \mathbb{E}'\mathbb{E}\Big\| \sum_{j\geqslant 1} \gamma_j' \int_0^T \int_0^T h_j(s)(k_n(s,\sigma) - k_m(s,\sigma))\,\mathrm{d}s\,\mathrm{d}W(\sigma)\Big\|^p \\
&= \mathbb{E}\mathbb{E}'\Big\| \sum_{j\geqslant 1} \gamma_j' \int_0^T h_j(s) \int_0^s (k_n(s,\sigma) - k_m(s,\sigma))\,\mathrm{d}W(\sigma)\,\mathrm{d}s\Big\|^p \\
&\eqsim_p \mathbb{E}\Big\| s \mapsto \int_0^s (k_n(s,\sigma) - k_m(s,\sigma))\,\mathrm{d}W(\sigma)\Big\|^p_{\gamma(0,T;X)} \\
&= \mathbb{E}\big\|s \mapsto [f_n(s) - f_m(s) - (\mathbb{E}f_n(s) - \mathbb{E}f_m(s))]\big\|^p_{\gamma(0,T;X)},
\end{aligned}
\tag{3.7}
$$

and therefore

$$
\big\|s \mapsto k_n(s,\cdot) - k_m(s,\cdot)\big\|_{\gamma(0,T;L^p(\Omega;\gamma(0,T;X)))} \lesssim_{p,X} \|f_n - f_m\|_{L^p(\Omega;\gamma(0,T;X))}.
$$

Since $\{f_n\}_{n=1}^{\infty}$ is a Cauchy sequence in $\gamma(0,T;L^p(\Omega;X))$, the estimate (3.7) implies that $\{k_n\}_{n=1}^{\infty}$ is a Cauchy sequence in $\gamma(0,T;L^p(\Omega;\gamma(0,T;X)))$. Let $k \in \gamma(0,T;L^p(\Omega;\gamma(0,T;X))) \eqsim L^p(\Omega;\gamma(0,T;\gamma(0,T;X)))$ be its limit. By adaptedness of the k_n we have $L^p_{\mathbb{F}}(\Omega;\gamma(0,T;\gamma(0,T;X)))$, and by passing to the limit $n \to \infty$ in (3.6), assertions (1) and (2) are obtained.

Similar to (3.7) we have

$$
\|s \mapsto k_n(s,\cdot)\|_{\gamma(0,T;L^p(\Omega;\gamma(0,T;X)))} \lesssim_{p,X} \|f_n\|_{\gamma(0,T;L^p(\Omega;X))}. \tag{3.8}
$$

Letting $n \to \infty$ in (3.8) we obtain assertion (3). □

Proposition 3.6 *Let (H1)–(H5) and (H6)′ be satisfied and assume in addition that X has the upper contraction property. Then the problem* (3.5) *admits a unique mild L^p-solution (U,V).*

Proof We extend the argument of [14] to the UMD setting. As in Sect. 3.1, by martingale representation in UMD spaces there is a unique element $\phi \in L^p_{\mathbb{F}}(\Omega;\gamma(0,T;X))$ such that for all $t \in [0,T]$,

$$
\mathbb{E}(u_T|\mathscr{F}_t) = \mathbb{E}u_T + \int_0^t \phi\,\mathrm{d}W \quad \text{in } L^p(\Omega;X). \tag{3.9}
$$

Put

$$U(t) := \mathbb{E}\Big(S(T-t)u_T - \int_t^T S(s-t)f(s)\,\mathrm{d}s \,\Big|\, \mathscr{F}_t\Big).$$

Let $k \in L^p_{\mathbb{F}}(\Omega; \gamma(0,T;\gamma(0,T;X)))$ be the kernel obtained from Lemma 3.5. Then for almost all $s \in [0,T]$ we have

$$f(s) = \mathbb{E}f(s) + \int_0^s k(s,\sigma)\,\mathrm{d}W(\sigma). \tag{3.10}$$

By (3.9) (applied to t and T and subtracting the results),

$$u_T - \mathbb{E}(u_T|\mathscr{F}_t) = \int_t^T \phi\,\mathrm{d}W. \tag{3.11}$$

The definition of U, together with (3.10) and (3.11), implies that

$$\begin{aligned} U(t) &= \mathbb{E}(S(T-t)u_T|\mathscr{F}_t) - \Big(\int_t^T S(s-t)\Big(\mathbb{E}f(s) + \int_0^s k(s,\sigma)\,\mathrm{d}W(\sigma)\Big)\Big|\mathscr{F}_t\Big)\,\mathrm{d}s \\ &= S(T-t)\mathbb{E}(u_T|\mathscr{F}_t) - \int_t^T S(s-t)\Big(\mathbb{E}f(s) + \int_0^t k(s,\sigma)\,\mathrm{d}W(\sigma)\Big)\,\mathrm{d}s \\ &= S(T-t)\Big(u_T - \int_t^T \phi\,\mathrm{d}W\Big) - \int_t^T S(s-t)\Big(f(s) - \int_t^s k(s,\sigma)\,\mathrm{d}W(\sigma)\Big)\,\mathrm{d}s. \end{aligned} \tag{3.12}$$

We will analyse the two terms on the right-hand side separately.

Since by assumption $\{S(t) : t \in [0,T]\}$ is γ-bounded, we may apply the Kalton-Weis multiplier theorem [16, Theorem 9.5.1] to see that $t \mapsto S(T-t)\mathbb{E}u_T$ defines an element of $L^p(\Omega, \gamma(0,T;X))$. By Lemma 2.10 it then defines an element of $L^p_{\mathbb{F}}(\Omega, \gamma(0,T;X))$. Also, by [32, Theorem 4.5], $t \mapsto \int_t^T \phi\,\mathrm{d}W$ defines an element of $L^p(\Omega, \gamma(0,T;X))$, and by another appeal to γ-boundedness, the same is true for

$$t \mapsto S(T-t)\int_t^T \phi\,\mathrm{d}W.$$

By Lemma 2.10 this mapping defines an element of $L^p_{\mathbb{F}}(\Omega, \gamma(0,T;X))$.

We now turn to the second term in the right-hand side of (3.12) and consider the two terms in the integral separately. For the first term we observe that

$$t \mapsto \int_t^T S(s-t)f(s)\,\mathrm{d}s$$

belongs to $L^p(\Omega; \gamma(0,T;X))$ by Lemma 2.6(1). Turning to the second term in the integral, to see that the mapping

$$t \mapsto \int_t^T S(s-t) \int_t^s k(s,\sigma)\,\mathrm{d}W(\sigma)\,\mathrm{d}s$$

defines an element of $L^p(\Omega; \gamma(0,T;X))$ we apply the stochastic Fubini theorem, the isomorphism $L^p(\Omega; \gamma(0,T;X)) \eqsim \gamma(0,T; L^p(\Omega;X))$, Theorem 2.11, the isomorphism once more, Lemma 2.6(2), the Kalton–Weis multiplier theorem, and the upper contraction property. This leads to the estimate

$$\begin{aligned}
&\Big\| t \mapsto \int_t^T S(s-t) \int_t^s k(s,\sigma)\,\mathrm{d}W(\sigma)\,\mathrm{d}s \Big\|_{L^p(\Omega;\gamma(0,T;X))} \\
&\quad = \Big\| t \mapsto \int_t^T \int_\sigma^T S(s-t)k(s,\sigma)\,\mathrm{d}s\,\mathrm{d}W(\sigma) \Big\|_{L^p(\Omega;\gamma(0,T;X))} \\
&\quad \eqsim_{p,X} \Big\| t \mapsto \int_t^T \int_\sigma^T S(s-t)k(s,\sigma)\,\mathrm{d}s\,\mathrm{d}W(\sigma) \Big\|_{\gamma(0,T;L^p(\Omega;X))} \\
&\quad \eqsim_{p,X} \Big\| t \mapsto \Big[\sigma \mapsto \int_\sigma^T S(s-t)k(s,\sigma)\,\mathrm{d}s\Big] \Big\|_{\gamma(0,T;L^p(\Omega;\gamma(0,T;X)))} \\
&\quad \eqsim_{p,X} \Big\| t \mapsto \Big[\sigma \mapsto \mathbf{1}_{\{t \leqslant \sigma\}} S(\sigma - t) \int_\sigma^T S(s-\sigma)k(s,\sigma)\,\mathrm{d}s\Big] \Big\|_{L^p(\Omega;\gamma(0,T;\gamma(0,T;X)))} \\
&\quad \leqslant \gamma(S) \Big\| t \mapsto \Big[\sigma \mapsto \int_\sigma^T S(s-\sigma)k(s,\sigma)\,\mathrm{d}s\Big] \Big\|_{L^p(\Omega;\gamma(0,T;\gamma(0,T;X)))} \\
&\quad = T^{1/2}\gamma(S) \Big\| \sigma \mapsto \int_\sigma^T S(s-\sigma)k(s,\sigma)\,\mathrm{d}s \Big\|_{L^p(\Omega;\gamma(0,T;X)))} \\
&\quad \lesssim_{p,X} T\gamma(S)^2 \|k\|_{L^p(\Omega;\gamma(\Delta;X))}, \\
&\quad \eqsim_{p,X} T\gamma(S)^2 \|k\|_{L^p(\Omega;\gamma(0,T;\gamma(0,T;X)))}.
\end{aligned} \tag{3.13}$$

Collecting what has been proved, it follows that $U \in L^p_{\mathbb{F}}(\Omega; \gamma(0,T;X))$, the adaptedness of U being a consequence of Lemma 2.10 and the representation given by the first identity in (3.12).

By the stochastic Fubini theorem,

$$\begin{aligned}
U(t) &= S(T-t)u_T - \int_t^T S(s-t)f(s)\,\mathrm{d}s - \int_t^T S(T-t)\phi(\sigma)\,\mathrm{d}W(\sigma) \\
&\quad + \int_t^T \int_\sigma^T S(s-t)k(s,\sigma)\,\mathrm{d}s\,\mathrm{d}W(\sigma) \\
&= S(T-t)u_T - \int_t^T S(s-t)f(s)\,\mathrm{d}s - \int_t^T S(\sigma-t)V(\sigma)\,\mathrm{d}W(\sigma),
\end{aligned}$$

where

$$\sigma \mapsto V(\sigma) := S(T-\sigma)\phi(\sigma) + \int_\sigma^T S(s-\sigma)k(s,\sigma)\,ds \tag{3.14}$$

is $\mathbb{F}$-adapted. It remains to be checked that the process V defines an element of $L^p_{\mathbb{F}}(\Omega; \gamma(0,T;X))$. This can be done by repeating the arguments which showed the corresponding result for U.

Next we prove the uniqueness of the solution. The proof is very similar to the one for $A = 0$. Suppose $(\widetilde{U}, \widetilde{V})$ is another L^p-solution to (3.5). Then from the definition of the mild solution to (3.5), we find that

$$\widetilde{U}(t) - U(t) + \int_t^T S(s-t)(\widetilde{V}(s) - V(s))\,dW(s) = 0 \tag{3.15}$$

for all $t \in [0,T]$ By taking conditional expectations with respect to $\mathscr{F}_t$ for (3.15), we see that $\widetilde{U}(t) - U(t) = 0$. Thus $\int_t^T S(s-t)(\widetilde{V}(s) - V(s))\,dW(s) = 0$ for all $t \in [0,T]$. Taking L^p-means, using [32, Theorem 3.5] it follows that

$$\|S(\cdot - t)(\widetilde{V}(\cdot) - V(\cdot))\|^p_{L^p(\Omega;\gamma(0,T;X))} \eqsim_{p,X} \mathbb{E}\Big\| \int_t^T S(s-t)(\widetilde{V}(s) - V(s))\,dW(s)\Big\|^p = 0.$$

Hence, for any $t \in [0,T]$, in $L^p(\Omega; \gamma(t,T;X))$ we obtain the equality

$$S(\cdot - t)\widetilde{V}(\cdot) = S(\cdot - t)V(\cdot).$$

To deduce from this that $\widetilde{V} = V$ in $L^p(\Omega; \gamma(0,T;X))$ we argue pathwise and prove that if $v \in \gamma(0,T)$ satisfies $S(\cdot - t)v(\cdot) = 0$ in $\gamma(t,T)$ for all $t \in [0,T]$, then $v = 0$. Fix an integer $N \geqslant 1$ and set $t_j = jT/N$ for $j = 0, 1, \ldots, N$. Multiplying the identity $S(\cdot - t_j)v(\cdot) = 0$ by $S(t_{j+1} - (\cdot - t_j))$ on $I_j := [t_j, t_{j+1}]$ it follows that $S(T/N)v(\cdot) = 0$ as an element of $\gamma(t_j, t_{j+1}; X)$, $j = 0, 1, \ldots, N-1$, and therefore $S(T/N)v(\cdot) = 0$ as an element of $\gamma(0,T;X)$. Now we can apply [16, Proposition 9.4.6] to deduce that $v = 0$ as an element of $\gamma(0,T;X)$. □

3.3 The General Case

In the final section we consider the problem

$$\begin{cases} dU(t) + AU(t)\,dt = f(t, U(t), V(t))\,dt + V(t)\,dW(t), & t \in [0,T], \\ U(T) = u_T, \end{cases} \tag{3.16}$$

under the assumptions (H1)–(H6).

Theorem 3.7 *Let (H1)–(H6) be satisfied and assume in addition that X has the upper contraction property. Then the problem* (3.16) *admits a unique mild L^p-solution (U, V).*

Proof Following the ideas of [25] the existence proof proceeds by a Picard iteration argument, where the existence and uniqueness in each iteration follows from the well-posedness of the problem (3.5) considered in the previous subsection.

Step 1 In this step we prove the existence of an L^p-solution on the interval $I_\delta := [T-\delta, T]$ for $\delta \in (0, T)$ small enough.

Set $U_0 = 0$ and $V_0 = 0$ and define the pair $(U_{n+1}, V_{n+1}) \in L^p_{\mathbb{F}}(\Omega; \gamma(I_\delta; X)) \times L^p_{\mathbb{F}}(\Omega; \gamma(I_\delta; X))$ inductively as the unique mild L^p-solution of the problem

$$\begin{cases} \mathrm{d}U(t) = -AU(t)\,\mathrm{d}t + f(t, U_n(t), V_n(t))\,\mathrm{d}t + V_n(t)\,\mathrm{d}W(t), & t \in I_\delta, \\ U(T) = u_T. \end{cases}$$

Note that at each iteration the function $t \mapsto g_n(t) := f(t, U_n(t), V_n(t))$ defines an element of $L^p_{\mathbb{F}}(\Omega; \gamma(I_\delta; X))$ by (H6) with norm

$$\|g_n\|_{L^p_{\mathbb{F}}(\Omega;\gamma(I_\delta;X))} \leqslant C(1 + \|U_n\|_{L^p_{\mathbb{F}}(\Omega;\gamma(I_\delta;X))} + \|V_n\|_{L^p_{\mathbb{F}}(\Omega;\gamma(I_\delta;X))})$$

with a constant $C \geqslant 0$ independent of U_n and V_n. By Proposition 3.6,

$$\|U_1 - U_0\|_{L^p_{\mathbb{F}}(\Omega;\gamma(I_\delta;X))} = \|U_1\|_{L^p_{\mathbb{F}}(\Omega;\gamma(I_\delta;X))} \leqslant C(\|g_0\|_{L^p_{\mathbb{F}}(\Omega;\gamma(I_\delta;X))} + \|u_T\|_{L^p(\Omega;X)}),$$

$$\|V_1 - V_0\|_{L^p_{\mathbb{F}}(\Omega;\gamma(I_\delta;X))} = \|V_1\|_{L^p_{\mathbb{F}}(\Omega;\gamma(I_\delta;X))} \leqslant C(\|g_0\|_{L^p_{\mathbb{F}}(\Omega;\gamma(I_\delta;X))} + \|u_T\|_{L^p(\Omega;X)}),$$

where $C \geqslant 0$ is a constant independent of f and u_T.

For $n \geqslant 1$, by (3.12) we can estimate

$$\begin{aligned}
&\|U_{n+1} - U_n\|_{L^p_{\mathbb{F}}(\Omega;\gamma(I_\delta;X))} \\
&\leqslant \Big\| t \mapsto \int_t^T S(s-t)(g_n(s) - g_{n-1}(s))\,\mathrm{d}s \Big\|_{L^p_{\mathbb{F}}(\Omega;\gamma(I_\delta;X))} \\
&\quad + \Big\| t \mapsto \int_t^T S(s-t) \int_t^s (k_n(s,\sigma) - k_{n-1}(s,\sigma))\,\mathrm{d}W(\sigma)\,\mathrm{d}s \Big\|_{L^p_{\mathbb{F}}(\Omega;\gamma(I_\delta;X))} \\
&= (I) + (II).
\end{aligned}$$

We estimate these terms separately. To estimate (I) we use Lemma 2.6(1) with $[0, T]$ replaced by I_δ:

$$\begin{aligned}(I) &= \Big\| t \mapsto \int_t^T S(s-t)(g_n(s) - g_{n-1}(s))\,\mathrm{d}s \Big\|_{L^p(\Omega;\gamma(I_\delta;X))} \\ &\leqslant \delta\gamma(S)\|g_n - g_{n-1}\|_{L^p(\Omega;\gamma(I_\delta;X))} \\ &\leqslant L\delta\gamma(S)(\|U_n - U_{n-1}\|_{L^p_{\mathbb{F}}(\Omega;\gamma(I_\delta;X))} + \|V_n - V_{n-1}\|_{L^p_{\mathbb{F}}(\Omega;\gamma(I_\delta;X))}),\end{aligned}$$

where $\gamma(S)$ is the γ bound of $\{S(t) : t \in [0, T]\}$ and L the Lipschitz constant in (H6). To estimate (II) we proceed as in (3.13), again with $[0, T]$ replaced by I_δ:

$$\begin{aligned}(II) &= \Big\| t \mapsto \int_t^T S(s-t) \int_t^s (k_n(s,\sigma) - k_{n-1}(s,\sigma))\,\mathrm{d}W(\sigma)\,\mathrm{d}s \Big\|_{L^p(\Omega;\gamma(I_\delta;X))} \\ &\leqslant \delta^{1/2}\gamma(S)\Big\| \sigma \mapsto \int_\sigma^T S(s-\sigma)(k_n(s,\sigma) - k_{n-1}(s,\sigma))\,\mathrm{d}s \Big\|_{L^p(\Omega;\gamma(I_\delta;X)))} \\ &= \delta^{1/2}\gamma(S)\|V_{n+1} - V_n\|_{L^p_{\mathbb{F}}(\Omega;\gamma(I_\delta;X))},\end{aligned}$$

using (3.10) and (3.14) in the last step. Moreover, by Lemmas 2.6(2) and 2.9, and 3.5,

$$\begin{aligned}&\|V_{n+1} - V_n\|_{L^p_{\mathbb{F}}(\Omega;\gamma(I_\delta;X))} \\ &\quad\leqslant \delta^{1/2}\gamma(S)\|k_n - k_{n-1}\|_{L^p(\Omega;\gamma(\Delta_\delta;X))} \\ &\quad\eqsim_{p,X} \delta^{1/2}\gamma(S)\|k_n - k_{n-1}\|_{L^p(\Omega;\gamma(I_\delta;\gamma(I_\delta;X)))} \\ &\quad\lesssim_{p,X} \delta^{1/2}\gamma(S)\|g_n - g_{n-1}\|_{L^p_{\mathbb{F}}(\Omega;\gamma(I_\delta;X))} \\ &\quad= \delta^{1/2}\gamma(S)\|f(\cdot, U_n(\cdot), V_n(\cdot)) - f(\cdot, U_{n-1}(\cdot), V_{n-1}(\cdot))\|_{L^p_{\mathbb{F}}(\Omega;\gamma(I_\delta;X))} \\ &\quad\leqslant L\delta^{1/2}\gamma(S)(\|U_n - U_{n-1}\|_{L^p_{\mathbb{F}}(\Omega;\gamma(I_\delta;X))} + \|V_n - V_{n-1}\|_{L^p_{\mathbb{F}}(\Omega;\gamma(I_\delta;X))}).\end{aligned}$$

Combining all estimates, we see that, if δ is small enough, the sequences $\{U_n\}_{n\geqslant 1}$ and $\{V_n\}_{n\geqslant 1}$ converge in $L^p_{\mathbb{F}}(\Omega; \gamma(I_\delta; X))$ to limits U and V. It is clear that the pair (U, V) is an L^p-solution on the interval I_δ.

Step 2 The arguments in Step 1 show that we always obtain a unique mild L^p-solution if δ is small enough. Since the estimates involve constants that are independent of T, δ, and u_T, the proof may be repeated with I_δ replaced by any interval $[T - 2\delta, T - \delta]$. In this way we can obtain a global existence result by partitioning $[0, T]$ into finitely many such intervals, and successively solving the backwards equation proceeding 'from the right to the left'. This gives us solutions for the backward equation on each sub-interval, and it is easy to check that a global solution is obtained by patching together these local solutions.

Step 3 Finally we prove the uniqueness of the solution. The proof is very similar to the one for $A = 0$. Suppose $(\widetilde{U}, \widetilde{V})$ is another L^p-solution to (3.16). Then from the definition of the mild solution to (3.16), we find that

$$\widetilde{U}(t) - U(t) + \int_t^T S(s-t)(\widetilde{V}(s) - V(s))\,\mathrm{d}W(s) = 0 \tag{3.17}$$

for all $t \in [0, T]$. By taking conditional expectations with respect to $\mathscr{F}_t$ for (3.17), we see that $\widetilde{U}(t) - U(t) = 0$. Thus $\int_t^T S(s-t)(\widetilde{V} - V)\,\mathrm{d}W(s) = 0$ for all $t \in [0, T]$. Taking L^p-means, using [32, Theorem 3.5] it follows that

$$\|S(\cdot - t)(\widetilde{V} - V)\|^p_{L^p(\Omega;\gamma(0,T;X))} \eqsim_{p,X} \mathbb{E}\Big\|\int_t^T S(s-t)(\widetilde{V}(s) - V(s))\,\mathrm{d}W(s)\Big\|^p = 0.$$

Hence, for any $t \in [0, T]$, in $\gamma(t, T)$ we obtain the equality

$$S(\cdot - t)\widetilde{V}(\cdot) = S(\cdot - t)V(\cdot).$$

As before this proves that $\widetilde{V} = V$. □

Acknowledgements The authors thank Mark Veraar for helpful comments. The Qi Lü author is supported by the NSF of China under grant 11471231 and Grant MTM2014-52347 of the MICINN, Spain. This paper was started while the author Jan van Neerven visited Sichuan University. He would like to thank the School of Mathematics for its kind hospitality.

References

1. V. Anh, J. Yong, *Backward Stochastic Volterra Integral Equations in Hilbert Spaces*. Differential and Difference Equations and Applications (Hindawi, New York, 2006), pp. 57–66
2. V. Anh, W. Grecksch, J. Yong, Regularity of backward stochastic Volterra integral Equations in Hilbert Spaces. Stoch. Anal. Appl. **29**(1), 146–168 (2011)
3. M. Azimi, Banach Space Valued Stochastic Integral Equations and Their Optimal Control. PhD Thesis, University of Halle-Wittenberg (2018)
4. A. Bensoussan, *Stochastic Control of Partially Observed Systems* (Cambridge University Press, Cambridge, 1992)
5. D.L. Burkholder, Martingales and singular integrals in banach spaces, in *Handbook of the Geometry of Banach Spaces*, vol. I (North-Holland, Amsterdam, 2001), pp. 233–269
6. Ph. Clément, B. de Pagter, F. Sukochev, H. Witvliet, Schauder decompositions and multiplier theorems. Stud. Math. **138**(2), 135–163 (2000)
7. R. Denk, M. Hieber, J. Prüss, R-boundedness, Fourier multipliers and problems of elliptic and parabolic type. Mem. Am. Math. Soc. **166**(788), 114 pp. (2003)
8. G. Dore, Maximal regularity in L^p spaces for an abstract Cauchy problem. Adv. Differ. Equ. **5**(1–3), 293–322 (2000)

9. K. Du, S. Tang, Strong solution of backward stochastic partial differential equations in C^2 domains. Probab. Theory Relat. Fields **154**, 255–285 (2012)
10. K. Du, J. Qiu, S. Tang, L^p theory for super-parabolic backward stochastic partial differential equations in the whole space. Appl. Math. Optim. **65**, 175–219 (2012)
11. N. Englezos, I. Karatzas, Utility maximization with habit formation: dynamic programming and stochastic PDEs. SIAM J. Control Optim. **48**, 481–520 (2009)
12. M. Fuhrman, G. Tessitore, Infinite horizon backward stochastic differential equations and elliptic equations in Hilbert spaces. Ann. Probab. **32**(1B), 607–660 (2004)
13. D.H.J. Garling, Brownian motion and UMD-spaces, in *Probability and Banach Spaces* (Springer, Berlin, 1986), pp. 36–49
14. Y. Hu, S. Peng, Adapted solution of a backward semilinear stochastic evolution equation. Stoch. Anal. Appl. **9**, 445–459 (1991)
15. T.P. Hytönen, J.M.A.M. van Neerven, M.C. Veraar, L.W. Weis, *Analysis in Banach Spaces, Vol. I: Martingales and Littlewood–Paley Theory*. Ergebnisse der Mathematik und ihrer Grenzgebiete (3. Folge), vol. 63 (Springer, Cham, 2016)
16. T.P. Hytönen, J.M.A.M. van Neerven, M.C. Veraar, L.W. Weis, *Analysis in Banach Spaces, Vol. II: Probabilistic Methods and Operator Theory*. Ergebnisse der Mathematik und ihrer Grenzgebiete (3. Folge), vol. 67 (Springer, Cham, 2017)
17. T.P. Hytönen, J.M.A.M. van Neerven, M.C. Veraar, L.W. Weis, *Analysis in Banach Spaces, Vol. III: Harmonic and Stochastic Analysis* (In preparation)
18. N.J. Kalton, L.W. Weis, The H^∞-calculus and square function estimates, in *Nigel J. Kalton Selecta* ed. by F. Gesztesy, G. Godefroy, L. Grafakos, I. Verbitsky. Contemporary Mathematicians (Birkhäuser-Springer, Basel, 2016)
19. P.C. Kunstmann, L.W. Weis, Maximal L^p-regularity for parabolic equations, Fourier multiplier theorems and H^∞-functional calculus, in *Functional Analytic Methods for Evolution Equations*. Lecture Notes in Mathematics, vol. 1855 (Springer, Berlin, 2004), pp. 65–311
20. Q. Lü, X. Zhang, Well-posedness of backward stochastic differential equations with general filtration. J. Differ. Equ. **254**(8), 3200–3227 (2013)
21. Q. Lü, X. Zhang, *General Pontryagin-Type Stochastic Maximum Principle and Backward Stochastic Evolution Equations in Infinite Dimensions*. SpringerBriefs in Mathematics (Springer, Cham, 2014)
22. Q. Lü, X. Zhang, Transposition method for backward stochastic evolution equations revisited, and its application. Math. Control Relat. Fields **5**(3), 529–555 (2015)
23. J. Ma, J. Yong, Adapted solution of a degenerate backward SPDE, with applications. Stochastic Process. Appl. **70**, 59–84 (1997)
24. J. Ma, J. Yong, On linear, degenerate backward stochastic partial differential equations. Probab. Theory Relat. Fields **113**, 135–170 (1999)
25. E. Pardoux, S. Peng, Adapted solution of backward stochastic equation. Systems Control Lett. **14**, 55–61 (1990)
26. S. Peng, Stochastic Hamilton–Jacobi–Bellman equations. SIAM J. Control Optim. **30**, 284–304 (1992)
27. G. Pisier, *Martingales in Banach Spaces*. Cambridge Studies in Advanced Mathematics, vol. 155 (Cambridge University Press, Cambridge, 2016)
28. J. Prüss, G. Simonett, *Moving Interfaces and Quasilinear Parabolic Evolution Equations*. Monographs in Mathematics, vol. 105 (Birkhäuser, Basel, 2016)
29. S. Tang, W. Wei, On the Cauchy problem for backward stochastic partial differential equations in Hölder spaces. Ann. Probab. **44**, 360–398 (2016)
30. J.M.A.M. van Neerven, γ-Radonifying operators–a survey, in *Spectral Theory and Harmonic Analysis (Canberra, 2009)*. Proceedings of Centre for Mathematical Analysis, Australian National University, vol. 44 (Australian National University, Canberra, 2010), pp. 1–62
31. J.M.A.M. van Neerven, M.C. Veraar, On the stochastic Fubini theorem in infinite dimensions, in *Stochastic Partial Differential Equations and Applications - VII* (Levico Terme, 2004). Lecture Notes in Pure and Applied Mathematics, vol. 245 (CRC Press, Boca Raton, 2005)

32. J.M.A.M. van Neerven, M.C. Veraar, L.W. Weis, Stochastic integration in UMD Banach spaces. Ann. Probab. **35**, 1438–1478 (2007)
33. J.M.A.M. van Neerven, M.C. Veraar, L.W. Weis, Stochastic evolution equations in UMD Banach spaces. J. Funct. Anal. **255**, 940–993 (2008)
34. J.M.A.M. van Neerven, M.C. Veraar, L.W. Weis, Stochastic integration in Banach spaces – a survey, in *Stochastic Analysis: A Series of Lectures* (Lausanne, 2012). Progress in Probability, vol. 68 (Birkhäuser Verlag, Basel, 2015)
35. M.C. Veraar, L.W. Weis, A note on maximal estimates for stochastic convolutions. Czechoslovak Math. J. **61**(136), 743–758 (2011)
36. L.W. Weis, Operator-valued Fourier multiplier theorems and maximal L_p-regularity. Math. Ann. **319**(4), 735–758 (2001)
37. X. Zhou, A duality analysis on stochastic partial differential equations. J. Funct. Anal. **103**(2), 275–293 (1992)

On the Lipschitz Decomposition Problem in Ordered Banach Spaces and Its Connections to Other Branches of Mathematics

Miek Messerschmidt

Dedicated to the occasion of Ben de Pagter's 65th birthday

Abstract Consider the following still-open problem: for any Banach space X, ordered by a closed generating cone $C \subseteq X$, do there always exist Lipschitz functions $\cdot^+ : X \to C$ and $\cdot^- : X \to C$ satisfying $x = x^+ - x^-$ for every $x \in X$?

We discuss the connections of this problem to a large number of other branches of mathematics: set-valued analysis, selection theorems, the non-linear geometry of Banach spaces, Ramsey theory, Lipschitz function spaces, duality theory, and tensor products of Banach spaces. We give numerous equivalent reformulations of the problem, and, through known examples, provide circumstantial evidence that the above question could be answered in the negative.

Keywords Ordered Banach spaces · Lipschitz geometry · Lipschitz function spaces · Duality theory · Tensor products of Banach spaces

1 Introduction

This paper is a brief survey on what we will term the *Lipschitz decomposition problem* (Problem 1.1 below) in general ordered Banach spaces. To the author's knowledge, this problem is unsolved and has remained open for number of years. The problem came to the author's attention in 2013, but may well be older. While the terminology employed in the statement of the problem is fairly standard, explicit definitions are provided in Sect. 2.

M. Messerschmidt (✉)
Department of Mathematics and Applied Mathematics, University of Pretoria, Hatfield, Pretoria, South Africa
e-mail: mmesserschmidt@gmail.com; miek.messerschmidt@up.ac.za

G. Buskes et al. (eds.), *Positivity and Noncommutative Analysis*,
Trends in Mathematics, https://doi.org/10.1007/978-3-030-10850-2_22

Problem 1.1 Which one of the following mutually exclusive statements is true?

1. For every Banach space X, ordered by a closed generating cone $C \subseteq X$, there exist Lipschitz functions $\cdot^+ : X \to C$ and $\cdot^- : X \to C$ satisfying $x = x^+ - x^-$ for every $x \in X$.
2. There exists a Banach space X, ordered by a closed generating cone $C \subseteq X$, for which there exist no Lipschitz functions $\cdot^+ : X \to C$ and $\cdot^- : X \to C$ satisfying $x = x^+ - x^-$ for every $x \in X$.

We will say the Lipschitz decomposition problem is solved positively if the statement (1) in Problem 1.1 is true. It is known that certain ordered Banach spaces do admit such Lipschitz functions (cf. Sect. 3), hence we introduce the following terminology.

Definition 1.2 (Lipschitz Decomposition Property) We will say that a Banach space X, ordered by closed generating cone $C \subseteq X$, has the *Lipschitz decomposition property,* if there exist Lipschitz functions $\cdot^+ : X \to C$ and $\cdot^- : X \to C$ satisfying $x = x^+ - x^-$ for every $x \in X$.

The Lipschitz decomposition problem has connections to a large number of other branches of mathematics: set-valued analysis, selection theorems, the Lipschitz- and uniform geometry of Banach spaces, Ramsey theory, Lipschitz function spaces, duality theory, and tensor products of Banach spaces. Discussing the relevance of each of these subjects to the problem at hand, including some very recent developments, is the main aim of this paper.

After some brief preliminaries in Sect. 2, we first establish the trivial cases of the Lipschitz decomposition problem in Sect. 3: Banach lattices and order unit spaces all have the Lipschitz decomposition property.

In Sect. 4, we show, similarly to modern proofs of the Bartle-Graves Theorem (cf. [3, Corollary 17.67]) as an application of Michael's Selection Theorem (Theorem 4.2), how the general problem may be translated into the language of set-valued analysis and selection theorems. A positive solution of the problem is equivalent to the existence of Lipschitz right inverses of a specific quotient map from a complete metric cone onto a Banach space (cf. Proposition 4.3).

This question, of the existence of Lipschitz right inverses of quotient maps, is discussed in Sect. 5 and is intimately related to the uniform- and Lipschitz geometry of Banach spaces and in particular the Lipschitz isomorphism problem: "Are Lipschitz isomorphic Banach spaces necessarily linearly isomorphic?" The existence of Lipschitz right inverses of quotient maps is a crucial part in constructing examples of Banach spaces that are Lipschitz isomorphic but not linearly isomorphic (cf. [1] and [15]). On the other hand, of particular interest are two previously known examples, one due to Lindenstrauss and Aharoni [1], and one due to Kalton [22] (in this paper: Examples 5.3 and 5.4), of quotient maps of Banach spaces that do not admit uniformly continuous or Lipschitz right inverses. These two examples are the only such examples known to the author and provide some circumstantial evidence that the Lipschitz decomposition problem might be resolved negatively, perhaps by employing similar constructions. As such, these examples are presented in some detail. Apart from the mentioned relevance to the Lipschitz decomposition problem, the techniques employed in these examples are of independent interest. Both (arguably) employ some form of Ramsey theory.

In Sect. 6 we introduce Lipschitz function spaces and Banach-space-valued Lipschitz function spaces. Through an easy observation, we show that the Lipschitz decomposition property can be transferred to equivalent statements on such function spaces. It has long been known that scalar-valued Lipschitz function spaces are dual Banach spaces, with a Free Lipschitz space (also called the Arens-Eels space) as a predual [35]. Very recently in [17], dual-Banach-space-valued Lipschitz function spaces were also shown to be dual Banach spaces, and furthermore, having a projective tensor product (with a Free Lipschitz space as tensor factor) as predual.

The observation that dual-Banach-space-valued Lipschitz function spaces have projective tensor products as preduals connects the Lipschitz decomposition problem to tensor products and the geometric duality theory of ordered Banach spaces. In Sect. 7 we show that the Lipschitz decomposition property for dual Banach spaces transfers in general to equivalent statements regarding the geometry of the projective tensor cone in a projective tensor product (having a Free Lipschitz space as tensor factor). This raises further questions about the structure of projective tensor cones in projective tensor products having a Free Lipschitz space as tensor factor, and for general projective tensor products of ordered Banach spaces, which are of relevance to the Lipschitz decomposition problem.

2 Preliminaries

We will assume that all vector spaces are over $\mathbb{R}$. Let V be a vector space. A set $C \subseteq V$ will be called a *cone* if both $C + C \subseteq C$ and $\lambda C \subseteq C$ hold for all $\lambda \geq 0$. A standard exercise establishes a bijection between cones and translation—and positive homogeneous pre-orders on V, cf. [3, Section 1.1]. We will say that C is *generating in* V, if $V = C - C$. For a subset $S \subseteq V$ we define the *conical span* of S, denoted $\operatorname{cspan} S$, as the set of all elements of the form $\sum_{j=1}^{n} \lambda_j s_j$ with $n \in \mathbb{N}$, and, for all $j \in \{1, \ldots, n\}$, having $\lambda_j \geq 0$ and $s_j \in S$. For a topological vector space W with Hausdorff topology τ, we denote the topological dual of W by W^*, or $(W, \tau)^*$ if confusion may arise. The closure of a set $S \subseteq W$ will be denoted by $\overline{S}$, or $\overline{S}^{\tau}$ if confusion may arise. If W is a normed space with norm $\|\cdot\|$, we will use the symbol $\|\cdot\|$ as stand-in for the norm-topology. For a cone $C \subseteq W$, we define the dual cone $C^* := \{\phi \in W^* \mid \forall c \in C,\ \phi(c) \geq 0\}$. Let X and Y be Banach spaces. Unless indicated otherwise, Banach spaces are assumed to be endowed with their norm topology. We denote the open unit ball, closed unit ball and unit sphere of X, respectively, by $\mathbb{B}_X$, $\mathbf{B}_X$, and $\mathbf{S}_X$. The space of bounded linear operators from X to Y will be denoted $B(X, Y)$ and endowed with the usual operator norm.

For a metric space (M, d) and a Banach space X, a function $f : M \to X$ will be said to be *Lipschitz* if there exists some constant $K > 0$, so that $\|f(a) - f(b)\| \leq K d(a, b)$ for all $a, b \in M$. For a Lipschitz function $f : M \to X$, we define the

Lipschitz constant of f by

$$L(f) := \inf\{K \in \mathbb{R} \mid \forall a, b \in M, \ \|f(a) - f(b)\| \leq Kd(a, b)\}.$$

3 The Trivial Solutions

We point out the trivial solutions to the Lipschitz decomposition problem.

Proposition 3.1 *Every Banach lattice has the Lipschitz decomposition property.*

Proof An easy exercise shows that the canonical maps $\cdot^+ : X \to X$ and $\cdot^- : X \to X$ defined by $x^+ := x \vee 0$ and $x^- := (-x) \vee 0$ are Lipschitz [34, Theorem II.5.2]. □

Proposition 3.2 *Let X be a Banach space and $C \subseteq X$ a closed generating cone. If there exists some $u \in C$ so that for every $x \in X$ there exists some $\lambda \geq 0$ with $x \in (-\lambda u + C) \cap (\lambda u - C)$, then X has the Lipschitz decomposition property.*

Proof This follows from the fact that order units are interior points of C [4, Theorem 2.8]. Since $X = \bigcup_{n\in\mathbb{N}}(-nu + C) \cap (nu - C)$, by the Baire Category Theorem, the order interval $(-u+C)\cap(u-C)$ has a non-empty interior. Let $\alpha > 0$ and $v \in X$ be such that

$$2^{-1}\alpha\mathbf{S}_X \subseteq \alpha\mathbb{B}_X \subseteq (v - u + C) \cap (v + u - C) \subseteq (v + u - C).$$

Set $w := 2\alpha^{-1}(v + u)$, and since $0 \in \mathbb{B}_X$ we have $w \in C$. For every $x \in \mathbf{S}_X$, we have $x \in w - C$ and hence define $x^+ := w \in C$ and $x^- := w - x \in C$. Clearly $x = x^+ - x^-$. Furthermore, it is easily seen that the maps $x \mapsto x^\pm$ are Lipschitz on $\mathbf{S}_X$. By applying the reverse triangle inequality, the positive homogeneous extensions defined as $X \ni x \mapsto \|x\| (x/\|x\|)^\pm$ can be seen to be Lipschitz on all of X. □

In particular, the previous result shows that all finite dimensional spaces ordered by closed generating cones have the Lipschitz decomposition property.

4 Set-Valued Analysis and Selection Theorems

A naive approach to solving the Lipschitz decomposition problem positively in general would be to mimic the proof of the following theorem.

Theorem 4.1 *Let X be a Banach space and $C \subseteq X$ a closed cone. The following are equivalent:*

1. *The cone C is generating in X.*
2. *There exists a constant $\alpha > 0$ so that, for every $x \in X$, there exist $a, b \in C$ so that $x = a - b$ and $\|a\| + \|b\| \leq \alpha \|x\|$.*

3. *There exists a constant $\alpha > 0$ and continuous positively homogeneous functions $\cdot^+ : X \to C$ and $\cdot^- : X \to C$ so that $x = x^+ - x^-$ and $\|x^+\| + \|x^-\| \leq \alpha \|x\|$.*
4. *For every topological space T, the closed cone of all bounded continuous functions on T taking values in C is generating in the Banach space $C(T, X)$ of all bounded continuous X-valued functions on T with the uniform norm.*

Proof Outline It is clear that (4) $\Leftrightarrow$ (3) $\Rightarrow$ (2) $\Rightarrow$ (1). That (1) implies (2) is the Klee-Andô Theorem [23, (3.2)], [5, Lemma 3], which is in essence a generalization of the usual open mapping theorem for Banach spaces, cf. [10] and [27]. That (2) implies (3) follows from Michael's Selection Theorem (stated below) [10, Corollary 4.2, Theorem 4.5], [27, Corollary 5.6]. □

Theorem 4.2 (Michael's Selection Theorem [3, Theorem 17.66]**)** *Let P be a paracompact space and X a Banach space. If the set-valued map $\Phi : P \to 2^X$ is closed-, convex-, and non-empty-valued and is lower hemi-continuous* (cf. [3, Definition 17.2]*), then Φ has a continuous selection, i.e., there exists a continuous function $f : P \to X$ so that $f(a) \in \Phi(a)$ for all $a \in P$.*

The approach to proving (2) implies (3) in Theorem 4.1 is to consider $C \times C$ as a subset of the ℓ^1-direct sum $X \oplus X$ and the continuous additive and positively homogeneous surjection $\Sigma : C \times C \to X$ defined as $\Sigma(c, d) := c - d$ for $c, d \in C$. The set-valued map $x \mapsto \Sigma^{-1}\{x\}$ can then be shown to satisfy the hypothesis of Michael's Selection Theorem. This idea is directly related to modern proofs of the Bartle-Graves Theorem, stating that continuous linear surjections (read quotient maps) between Banach spaces always have continuous (not necessarily linear) right inverses [3, Corollary 17.67].

The Lipschitz decomposition problem can then easily be rephrased as "Does the map Σ always have a Lipschitz right inverse?":

Proposition 4.3 *Let X be a Banach space, ordered by a closed generating cone $C \subseteq X$. The following are equivalent:*

1. *The space X has the Lipschitz decomposition property.*
2. *The continuous additive positively homogeneous surjection $\Sigma : C \times C \to X$, as defined above, has a Lipschitz right inverse.*
3. *The set-valued map $X \ni x \mapsto \Sigma^{-1}\{x\}$ has a Lipschitz selection.*

We note here that the surjection $\Sigma : C \times C \to X$ is in essence a quotient map from the complete metric cone[1] $C \times C$ onto X, prompting the question "When do quotients of complete metric cones onto Banach spaces have Lipschitz right inverses?" (Not always, cf. Remark 5.5 below). Since every Banach space is a complete metric cone in its own right, this is a more general question than "When

[1]For our current purpose a closed cone inside a Banach space is sufficient. See the more general definitions: [10, Definitions 2.2 and 2.3].

do quotient maps from Banach spaces onto Banach spaces have Lipschitz right inverses?" (Also, not always, cf. Examples 5.3 and 5.4 below).

One seemingly reasonable (but doomed) approach to solving the Lipschitz decomposition problem positively would be to attempt to prove a Lipschitz version of Michael's Selection Theorem by replacing the word "paracompact" with "metric" and making suitable adjustments. An observation that lends support to this idea is the appealing Lipschitz-like properties possessed by set-valued maps like $x \mapsto \Sigma^{-1}\{x\}$ (cf. [26]). It is however known that no general Lipschitz version of Michael's Selection can exist [39]. It is possible to do better than mere continuity, but only negligibly: one can obtain a selection that is continuous and pointwise Lipschitz on a dense set (cf. [26, Corollary 4.5]). The existence of a general Lipschitz version of Michael's Selection would contradict Examples 5.3 and 5.4 presented in the next section. Hence, if this approach is to be successful in resolving the Lipschitz decomposition problem positively, then it *must* leverage specific properties of the map Σ as defined above.

5 Non-Linear Geometry of Banach Spaces and Ramsey Theory

In this section we indicate some of the connections of the Lipschitz decomposition problem to the nonlinear geometry of Banach spaces. This is a vast and highly active research subject and we will restrict ourselves to material that we deem to be relevant to the Lipschitz decomposition problem. We refer the reader to Lindenstrauss and Benyamini's book [8], and Chapter 14 of the recent second edition of Albiac and Kalton's book [2] for a more thorough treatment of the subject.

The previous section reduced the Lipschitz decomposition problem to the following question "When do quotient maps from complete metric cones/Banach spaces onto Banach spaces have Lipschitz right inverses?" This question is closely related to the Lipschitz/uniform isomorphism problem for Banach spaces and its partial resolution (cf. the survey [24]):

Problem 5.1 Let X and Y be Banach spaces and $A : X \to Y$ a non-linear Lipschitz (uniformly continuous) bijection with Lipschitz (uniformly continuous) inverse. Are X and Y necessarily linearly isomorphic?

There exist Banach spaces (like ℓ^p-spaces [20]) for which the above question is answered affirmatively. However, examples exist of non-separable Banach spaces that are Lipschitz isomorphic, but not linearly isomorphic. Currently, the only known method for constructing such spaces revolves around constructing a space X with closed non-complemented subspace $E \subseteq X$ so that X/E is non-separable and the quotient map $q : X \to X/E$ has a Lipschitz right inverse. The first such example was constructed by Lindenstrauss and Aharoni in [1] and the argument is closely related to that of showing the non-complementability of c_0 in ℓ^∞ (see [37]).

In a more systematic fashion, Kalton and Godefroy used this method to show that every weakly compactly generated non-separable Banach space will give rise to a pair of Lipschitz isomorphic Banach spaces that are not linearly isomorphic [15, Corollary 4.4].

There exist examples of separable uniformly isomorphic Banach spaces that are not linearly isomorphic (cf. [32]). At present it is not known whether there exist separable Banach spaces that are Lipschitz isomorphic but not linearly isomorphic. By proving the following theorem, Kalton and Godefroy also eliminated the method described above in attacking in the problem in the separable case:

Theorem 5.2 ([15, Corollary 3.2]) *Let X be a Banach space with $E \subseteq X$ a closed subspace with X/E separable. Then the quotient map $q : X \to X/E$ has a Lipschitz right inverse if and only if E is complemented in X.*

Keeping in mind the relevance of the existence/non-existence of Lipschitz right inverses to quotient maps, as discussed in the previous section, below we present two examples of linear quotient maps that do not admit Lipschitz right inverses (or even uniformly continuous right inverses). Although interesting in their own right, these examples provide some circumstantial evidence toward a negative solution to the Lipschitz decomposition problem. We therefore present these examples in quite some detail, but, in the interest of flow, will favor rather informal language to describe some of the most technical details.

Example 5.4 below, due to Kalton, relies on a Ramsey-type graph coloring theorem. Arguably Example 5.3 below, due to Lindenstrauss and Aharoni, also relies on Ramsey theoretic ideas in a broader sense of the term: for some fixed notion of regularity and some large collection of objects, for any choice of object made from the large collection of objects, there necessarily exists a related object (often sub-object) with the mentioned notion of regularity.

We begin with Aharoni and Lindenstrauss' example. Let $D_{\mathbb{Q}}[0, 1]$ be the càdlàg space of all bounded $\mathbb{R}$-valued functions on $[0, 1]$ with the uniform norm that only admit discontinuities at rational points in $[0, 1]$; are right continuous everywhere; with left limits existing everywhere on $[0, 1]$. We note that for every $\varepsilon > 0$, an element in $f \in D_{\mathbb{Q}}[0, 1]$ can have only finitely many discontinuities larger than ε, else, by the Bolzano-Weierstrass Theorem, there would exist a point where f is not right continuous. Therefore, with $C[0, 1]$ denoting the closed subspace of all continuous $\mathbb{R}$-valued functions on $[0, 1]$, the quotient $D_{\mathbb{Q}}[0, 1]/C[0, 1]$ is isomorphic to $c_0([0, 1] \cap \mathbb{Q})$, with the coordinates of $q(f) \in c_0([0, 1] \cap \mathbb{Q})$ measuring (half) the size of the discontinuities of $f \in D_{\mathbb{Q}}[0, 1]$.

Example 5.3 (Aharoni-Lindenstrauss [1], [8, Example 1.2]*)* The quotient map $q : D_{\mathbb{Q}}[0, 1] \to c_0([0, 1] \cap \mathbb{Q})$ does not admit a Lipschitz right inverse.

Sketch of Proof Two Ramsey theoretic ideas are employed; both are straightforward:

The first Ramsey Theoretic idea is: for every $f \in D_{\mathbb{Q}}[0, 1]$, around every irrational number in $[0, 1]$ there exists an open set U on which f "varies very little" and necessarily has only "small" discontinuities on U.

The second Ramsey Theoretic idea is: for distinct elements $x, y \in c_0([0,1]\cap\mathbb{Q})$ and every non-empty open set $U \subseteq [0,1]$ and $\alpha \in (0,1)$, there exists some $r \in \mathbb{Q}\cap U$ so that $\|x-y\pm\alpha\|x-y\|e_r\| = \|x-y\|$. I.e., no matter the open set U, there exists a rational number in U where a rather "large" perturbation of $x-y$ does not affect the norm.

Assuming the existence of a Lipschitz right inverse $\tau : c_0([0,1]\cap\mathbb{Q}) \to D_{\mathbb{Q}}[0,1]$ of the quotient q, for every $\varepsilon > 0$, there exist distinct elements $x, y \in c_0([0,1]\cap\mathbb{Q})$ so that $\|\tau(x)-\tau(y)\|_\infty > (L(\tau)-\varepsilon)\,\|x-y\|$. Hence there exists a point in $[0,1]$ (which, by right continuity, we may assume to be irrational) at which $\tau(x)$ and $\tau(y)$ take on values that are "far apart" (at least $(L(\tau)-\varepsilon)\,\|x-y\|$). By the first Ramsey Theoretic idea above, there exists an open set U, on which both $\tau(x)$ and $\tau(y) \in D_{\mathbb{Q}}[0,1]$ "vary very little" on U, and by construction take on values that are "far apart" on U. In particular their average $2^{-1}(\tau(x)+\tau(y))$ also "varies very little" on U.

By the second Ramsey theoretic idea, there exists a rational number $r \in U\cap\mathbb{Q}$, which allows the construction of a metric midpoint $z \in c_0([0,1]\cap\mathbb{Q})$ through a "large" perturbation at r, so that $\|x-z\| = \|y-z\| = 2^{-1}\|x-y\|$. This necessarily implies that $\tau(z) \in D_{\mathbb{Q}}[0,1]$ has a "large" discontinuity at r.

At the same time, using the Lipschitz-ness of τ, we obtain both the following inequalities:

$$\|\tau(y)-\tau(z)\|_\infty \le 2^{-1}L(\tau)\,\|x-y\|_\infty \quad\text{and}\quad \|\tau(x)-\tau(z)\|_\infty \le 2^{-1}L(\tau)\,\|x-y\|_\infty .$$

However, on U, the functions $\tau(y)$ and $\tau(x)$ take on values that are "far apart," (at least a distance of $(L(\tau)-\varepsilon)\,\|x-y\|$). So the only way that the above two inequalities can hold is with $\tau(z)$ taking on values that are "very close"[2] (within $2^{-1}\varepsilon\,\|x-y\|$) to the average $2^{-1}(\tau(x)+\tau(y))$ on U. But since $2^{-1}(\tau(x)+\tau(y))$ "varies very little" on U, so does $\tau(z)$, which contradicts $\tau(z)$ having a "large" discontinuity at $r \in U$. □

The next example, due to Kalton, proceeds through showing that every Banach lattice X whose unit ball embeds uniform continuously into ℓ^∞ with uniform continuous inverse necessarily has the property that every increasing transfinite sequence (meaning: indexed by the first uncountable ordinal) in X must eventually be constant (cf. Theorem 5.7). Kalton called this property *The Monotone Transfinite Sequence Property*. This is proven through a Lipschitz-Ramsey Theorem (stated as Theorem 5.6 below), its proof proceeding through an argument in ordinal combinatorics. With this result, by showing that ℓ^∞/c_0 does not have The Monotone Transfinite Sequence Property [22, Theorem 4.2] (again using similar methods to usual techniques employed in showing the non-complementability of c_0 in ℓ^∞), one arrives at:

[2]Although very seldomly used in analysis, this is purely an observation about real numbers: let $a, b \in \mathbb{R}$ be distinct real numbers and let $K, \varepsilon > 0$ satisfy $0 < (K-\varepsilon) < |a-b| \le K$. If $c \in \mathbb{R}$ is such that $|a-c| \le 2^{-1}K$ and $|b-c| \le 2^{-1}K$, then $\left|c - 2^{-1}(a+b)\right| \le 2^{-1}\varepsilon$.

Example 5.4 (Kalton [22, Theorem 4.2]*)* The quotient map $q : \ell^\infty \to \ell^\infty/c_0$ has no uniformly continuous right inverse.

Remark 5.5 This also provides an example of a continuous additive positively homogenous surjection from a complete metric cone onto a Banach space that has no uniform continuous right inverse. Since Σ as defined in the previous section is Lipschitz, the surjection $q \circ \Sigma : \ell^\infty_+ \times \ell^\infty_+ \to \ell^\infty/c_0$ cannot have a uniformly continuous right inverse, as that would contradict Example 5.4. We note however that the quotient ℓ^∞/c_0 is actually a Banach lattice by [34, Proposition II.5.4], and so does have the Lipschitz decomposition property.

We sketch the essentials of the proof of Kalton's Monotone Transfinite Sequence Theorem (Theorem 5.7). Let I be any set. For $n \in \mathbb{N}$, by $I^{[n]}$ we denote the set of all n-element subsets of I. Let Ω be the first uncountable ordinal. The set $\Omega^{[n]}$ is made into a graph, by defining the set of edges as all pairs of distinct elements $a, b \in \Omega^{[n]}$ which *interlace*: for which either $a_1 \leq b_1 \leq a_2 \leq b_2 \leq \dots a_n \leq b_n$ or $b_1 \leq a_1 \leq b_2 \leq \dots b_n \leq a_n$ (where we write $a = \{a_1, a_2, \dots, a_n\}$ with $a_1 < a_2 < \dots < a_n$). With the least path length metric, the graph $\Omega^{[n]}$ is a bounded metric space with diameter n.

Theorem 5.6 (Kalton's Lipschitz-Ramsey Theorem [22, Theorem 3.6]**)** *For any $n \in \mathbb{N}$, let $c : \Omega^{[n]} \to \ell^\infty$ be Lipschitz, with Lipschitz constant L. Then there exists some $\xi \in \ell^\infty$ and uncountable set $\Theta \subseteq \Omega$, so that, for all $a \in \Theta^{[n]}$, $\|c(a) - \xi\|_\infty \leq L/2$.*

Translating into coloring language, one may interpret the previous result as follows: coloring the vertices of the graph $\Omega^{[n]}$ by elements in ℓ^∞ through a Lipschitz map $c : \Omega^{[n]} \to \ell^\infty$ with Lipschitz constant L, in general any subset of vertices $A \subseteq \Omega^{[n]}$ with diameter n has a *color diameter,* $\operatorname{diam} c(A)$, of at most nL (read: "can have rather large variations in color: of the order of the diameter of the graph $\Omega^{[n]}$"). However, by the previous theorem, one is assured of an uncountable set of ordinals $\Theta \subseteq \Omega$, so that the color diameter, $\operatorname{diam} c(\Theta^{[n]})$, of $\Theta^{[n]} \subseteq \Omega^{[n]}$ is at most the Lipschitz constant L of the coloring c (read: "has not so large variations in color, and is independent of the diameter of $\Theta^{[n]}$"). It is important to notice here that $\Theta^{[n]}$ has diameter n, since $\Theta \subseteq \Omega$ is an uncountable set of countable ordinals.

Theorem 5.7 (Kalton's Monotone Transfinite Sequence Theorem [22, Theorem 4.1]**)** *If the closed unit ball $\mathbf{B}_X$ of a Banach lattice X embeds uniform continuously (with uniformly continuous inverse) into ℓ^∞, then X has the Monotone Transfinite Sequence Property.*

Sketch of Proof Let $f : \mathbf{B}_X \to \ell^\infty$ be a uniformly continuous embedding with uniformly continuous inverse $g : f(\mathbf{B}_X) \to \mathbf{B}_X$. Let $(x_\mu)_{\mu\in\Omega} \subseteq X$ be an increasing transfinite sequence. Since transfinite increasing sequences are always bounded, we may assume $(x_\mu)_{\mu\in\Omega} \subseteq \mathbf{B}_X$ without loss.

Let $n \in \mathbb{N}$ be arbitrary and define the averaging map $A_n : \Omega^{[n]} \to \mathbf{B}_X$ by $A_n(\{a_1, \dots a_n\}) := \frac{1}{n}\sum_{j=1}^n x_{a_j}$. The coloring map $c_n := f \circ A_n : \Omega^{[n]} \to \ell^\infty$ can be verified to be Lipschitz (we omit the details), and we denote its Lipschitz constant by L_n. Furthermore it can be verified that the sequence (L_n) converges to zero as $n \to \infty$. By Kalton's Lipschitz-Ramsey Theorem (Theorem 5.6), there exists some uncountable set of ordinals $\Theta_n \subseteq \Omega$, so that, for all $a, b \in \Theta_n^{[n]}$ we have $\|c_n(a) - c_n(b)\|_\infty \leq L_n$.

Now, for any $n \in \mathbb{N}$, fix any $a \in \Theta_n^{[n]}$. There exists $\mu_n \in \Omega$ with $\max a < \mu_n$. For all $\nu \in \Omega$ satisfying $\max a < \mu_n < \nu$, since Θ_n is an uncountable set, there exists $b \in \Theta_n^{[n]}$ so that $\max a < \mu_n < \nu < \min b$. Using the monotonicity of the norm of the Banach lattice X, we obtain

$$\left\|x_\nu - x_{\mu_n}\right\| \leq \left\| \frac{1}{n} \sum_{j=1}^{n} x_{b_j} - \frac{1}{n} \sum_{j=1}^{n} x_{a_j} \right\| = \|g \circ c_n(b) - g \circ c_n(a)\| .$$

However $\sup_{a,b \in \Theta_n^{[n]}} \|c_n(a) - c_n(b)\|_\infty \leq L_n \to 0$ as $n \to \infty$, and, since g is uniformly continuous, we have $\sup_{a,b \in \Theta_n^{[n]}} \|g \circ c_n(b) - g \circ c_n(a)\| \to 0$ as $n \to \infty$. Taking $\mu_0 := \sup_{n \in \mathbb{N}} \mu_n \in \Omega$, we see, for all $\nu \in \Omega$ with $\mu_0 < \nu$, again by monotonicity of the norm of X, that $\left\|x_\nu - x_{\mu_0}\right\| \leq \left\|x_\nu - x_{\mu_n}\right\| \to 0$ as $n \to \infty$. Therefore $\left\|x_\nu - x_{\mu_0}\right\| = 0$ for all $\mu_0 < \nu$, and hence the increasing transfinite sequence $(x_\mu)_{\mu \in \Omega}$ is eventually constant. □

Remark 5.8 We point out that only the monotonicity of the norm of the Banach lattice was used in the above result. By a straightforward adaptation of the proof, the same is true if we replace the Banach lattice X by an ordered Banach space X for which there exists a constant $\alpha > 0$, so that for all $a, b \in X$, the inequality $0 \leq a \leq b$ implies $\|a\| \leq \alpha \|b\|$.

6 Lipschitz Function Spaces

We begin with some basic definitions.

Let X be a Banach space and (M, d) be a metric space. We will assume M to be pointed with some point in M fixed and labeled as $0_M \in M$. Viewing X as a metric space, we will always take $0_X := 0 \in X$. The *X-valued Lipschitz space* $\text{Lip}_0(M, X)$ is defined as the set of all Lipschitz functions $f : M \to X$ satisfying $f(0_M) = 0$. The map $L(\cdot)$, as defined in Sect. 2, is a norm on $\text{Lip}_0(M, X)$. A standard exercise will show $\text{Lip}_0(M, X)$ endowed with $L(\cdot)$ is again Banach space. As usual, if $X = \mathbb{R}$, then we will write $\text{Lip}_0(M)$ for $\text{Lip}_0(M, \mathbb{R})$. For a closed cone $C \subseteq X$ we will define $\text{Lip}_0(M, C) := \text{Lip}_0(M, X) \cap C^M$. This cone is certainly non-empty as for every $c \in C$, the map $M \ni a \mapsto d(0_M, a)c$ is in $\text{Lip}_0(M, C)$.

We refer the reader to Weaver's book [35] for a treatment of scalar-valued Lipschitz function spaces. Of note is the order structure of $\text{Lip}_0(M)$ when ordered by the *standard cone* $\text{Lip}_0(M)_+ := \left\{ f \in \text{Lip}_0(M) \,\middle|\, \forall a \in M,\ f(a) \geq 0 \right\}$. The space $\text{Lip}_0(M)$ is then lattice ordered [35, Proposition 1.5.5]. However, depending on the structure of the metric space M, the space $\text{Lip}_0(M)$ need not be a Banach lattice in general, as order intervals need not be norm-bounded. E.g., For any $\alpha > 0$, one can easily construct a (say piecewise affine) function $g \in \mathrm{Lip}_0([0, 1])$ so that, with $f(x) := x$, the function g satisfies $0 \leq g \leq f$, while $L(f) < \alpha L(g)$.

The following easy observation connects the Lipschitz decomposition property to the structure of Lipschitz function spaces:

Theorem 6.1 *Let X be a Banach space, ordered by a closed generating cone $C \subseteq X$. The following are equivalent:*

1. *The space X has the Lipschitz decomposition property.*
2. *The cone $\mathrm{Lip}_0(X, C)$ is generating in $\mathrm{Lip}_0(X, X)$.*
3. *The cone $\mathrm{Lip}_0(\mathbf{S}_X \cup \{0\}, C)$ is generating in $\mathrm{Lip}_0(\mathbf{S}_X \cup \{0\}, X)$.*
4. *For every metric space M, the cone $\mathrm{Lip}_0(M, C)$ is generating in $\mathrm{Lip}_0(M, X)$.*

Proof It is clear that (4) implies (2) and (3), and (2) implies (1).

We prove (1) implies (4). Let $\cdot^+ : X \to C$ and $\cdot^- : X \to C$ be Lipschitz maps such that $x = x^+ - x^-$ for all $x \in X$. By considering the positive homogeneous extension of the restrictions $\cdot^{\pm}|_{\mathbf{S}_X}$, we may assume that $0^{\pm} = 0$. Hence, for any $f \in \mathrm{Lip}_0(M, X)$, by defining $f^{\pm} \in \mathrm{Lip}_0(M, C)$ as $f^{\pm}(a) := f(a)^{\pm}$ for all $a \in M$, we obtain $f^{\pm} \in \mathrm{Lip}_0(M, C)$ and $f = f^+ - f^-$. That (3) implies (4) is proven similarly. □

A classical result in the scalar-valued case, due to de Leeuw, is that $\mathrm{Lip}_0(M)$ is always a dual Banach space (first proven in [11] for $M = \mathbb{R}$. See [35, Chapter 2] for a general treatment). Very recently, Weaver showed in [36] that, under mild conditions on the metric space M, preduals of $\mathrm{Lip}_0(M)$ are unique up to isometric isomorphism.

One construction of a predual of $\mathrm{Lip}_0(M)$ by Kalton and Godefroy [15] is through the Dixmier-Ng Theorem [29, Theorem 1] as the closed linear span of the evaluation maps $\{\delta_a : a \in M\}$ in $\mathrm{Lip}_0(M)^*$, with norm inherited from $\mathrm{Lip}_0(M)^*$. Denoted by $F(M)$, this space is called the *Free Lipschitz space* (or often *Lipschitz free-space*). We define the *standard cone* in $F(M)$ as the norm-closure of cspan $\{\delta_a \mid a \in M\}$ and denote it by $F(M)_+$. It is straightforward to verify that $F(M)^*_+ = \mathrm{Lip}_0(M)_+$.

The name Free Lipschitz space is apt. The Free Lipschitz space mimics the universal property of the free group over a set of symbols, of inducing a homomorphism on the free group from any function mapping the set of symbols into another group G. Here, from any Lipschitz map from M into a Banach space X, a linear map is induced, mapping the Free Lipschitz space $F(M)$ into the Banach space X. Explicitly:

Theorem 6.2 *Let M be a metric space. For every Banach space X and for every $f \in \mathrm{Lip}_0(M, X)$, there exists a unique bounded linear operator $T_f : F(M) \to X$ with $\|T_f\| = L(f)$ making the following diagram commute:*

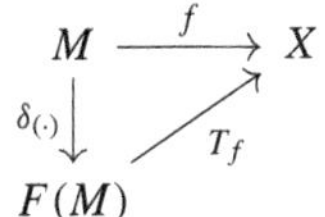

Furthermore, the map $f \mapsto T_f$ *is an isometric isomorphism from* $\mathrm{Lip}_0(M, X)$ *onto* $B(F(M), X)$.

The previous result is straightforward, and has been proven independently a number of times, cf. [31, Theorem 1], [35, Theorem 2.2.4], [21, Lemma 3.2].

An elementary verification will also show, if $C \subseteq X$ is a norm-closed cone, then the isometric isomorphism from $\mathrm{Lip}_0(M, X)$ onto $B(F(M), X)$ maps the cone $\mathrm{Lip}_0(M, C)$ onto the cone $\{T \in B(F(M), X) \mid TF(M)_+ \subseteq C\}$.

Vector-valued Lipschitz spaces has seen much recent developments. Of interest here is the observation by Guerrero, Lopez-Perez, and Ruedo Zoca that, as in the scalar case, with X a Banach space, the space $\mathrm{Lip}_0(M, X^*)$ is always a dual Banach space, having a vector-valued version of the Free Lipschitz space as predual [17]. Explicitly, with $\delta_a \otimes x(f) := f(a)(x)$ for all $a \in M, x \in X$, and $f \in \mathrm{Lip}_0(M, X^*)$, the *X-valued Free Lipschitz space* $F(M, X)$ is defined as the closed linear span of $\{\delta_a \otimes x \mid a \in M,\ x \in X\}$ in $\mathrm{Lip}_0(M, X^*)^*$. Moreover, viewing $B(F(M), X^*)$ as the dual of the projective tensor product $F(M)\overline{\otimes}_\pi X$ [33, Section 2.2], the isometric isomorphism between the spaces $\mathrm{Lip}_0(M, X^*)$ and $B(F(M), X^*)$ from Theorem 6.2 is wk*-to-wk* continuous, so that by [14, Exercise 3.60], the space $F(M, X)$ is isometrically isomorphic to the projective tensor product $F(M)\overline{\otimes}_\pi X$, cf. [17, Proposition 1.1].

Through the geometric duality theory of cones in Banach spaces, this last observation allows for the Lipschitz decomposition property to be characterized by the geometry of a polar cone in projective tensor products with a Free Lipschitz space as tensor factor. This is discussed in the next section.

7 Duality of Ordered Banach Spaces and Tensor Products

We introduce the following preliminaries to describe the geometric duality theory of ordered Banach spaces. Let V and W be vector spaces and $\langle \cdot \mid \cdot \rangle : V \times W \to \mathbb{R}$ a bilinear map. If $\{\, \langle \cdot \mid w \rangle \mid w \in W\}$ and $\{\, \langle v \mid \cdot \rangle \mid v \in V\}$ separate the points of V and W, respectively, we call (V, W) a *dual pair* and $\langle \cdot \mid \cdot \rangle$ a *duality*. By $\sigma(V, W)$ and $\sigma(W, V)$ we denote the smallest topologies for which all elements of $\{\, \langle \cdot \mid w \rangle \mid w \in W\}$ and $\{\, \langle v \mid \cdot \rangle \mid v \in V\}$ are continuous functionals, respectively. For $A \subseteq V$ and $B \subseteq W$ we define the *one-sided polars* of A *and* B *with respect to the dual pair (V,W)* by $A^{\odot} := \{w \in W \mid \forall a \in A,\ \langle a \mid w \rangle \leq 1\}$ and $B^{\odot} := \{v \in V \mid \forall b \in B,\ \langle v \mid b \rangle \leq 1\}$, respectively.

The following lemma forms the basis of geometric duality theory of ordered Banach spaces. Its proof is an elementary exercise in applications of the Hahn-Banach Separation Theorem.

Lemma 7.1 ([25, Lemmas 2.1 and 2.4]) *Let* (V, W) *be a dual pair. Let* $A, B \subseteq V$ *and* $C \subseteq V$ *a cone and* $\{A_i\}_{i \in I}$ *a collection of nonempty subsets of* V. *Then*

1. *The set* $A^{\odot}$ *contains zero, is convex and is* $\sigma(W, V)$*-closed in* W.

2. *If* $A \subseteq B$*, then* $B^{\odot} \subseteq A^{\odot}$.
3. *For* $\lambda > 0$*, we have* $(\lambda A)^{\odot} = \lambda^{-1}(A^{\odot})$.
4. *The set* $(\bigcup_{i\in I} A_i)^{\odot}$ *equals* $\bigcap_{i\in I} A_i^{\odot}$.
5. *The set* $A^{\odot\odot}$ *equals the* $\sigma(V, W)$*-closed convex hull of* $\{0\} \cup A$.
6. *If, for every* $i \in I$*, the set* A_i *is* $\sigma(V, W)$*-closed, convex and contains zero, then* $\left(\bigcap_{i\in I} A_i\right)^{\odot}$ *equals the* $\sigma(W, V)$*-closed convex hull of* $\bigcup_{i\in I} A_i^{\odot}$.
7. *The set* $C^{\odot} \subseteq W$ *is a* $\sigma(W, V)$*-closed cone and* $C^{\odot} = -C^*$.
8. *If* A *is* $\sigma(V, W)$*-closed, convex and contains zero and* C *is* $\sigma(V, W)$*-closed, then* $(A \cap C)^{\odot} = \overline{(A^{\odot} + C^{\odot})}^{\sigma(W,V)}$.
9. *If* A *is convex and contains zero, then* $(A + C)^{\odot} = A^{\odot} \cap C^{\odot}$.
10. *If* A *is* $\sigma(V, W)$*-closed, convex and contains zero, then* $A = \bigcap_{\lambda>1} \lambda A$.

Typical of the geometric duality theory of ordered Banach spaces are theorems like Theorem 7.2, below, which relate the geometry of a cone with that of its dual cone. This result, and results that are closely related to it, are well-known and have long been studied by many authors. See, for example, the following list of references, which is not claimed to be exhaustive: [6, 7, 9, 13, 16, 19, 28, 30].

The oldest references that the author is aware of are that of Grosberg and Krein from 1939 [16], and of Ellis from 1964 [13, Theorem 8] which assert the following:

Theorem 7.2 *Let* $\alpha > 0$*,* X *be a Banach space and* $C \subseteq X$ *a norm closed cone.*

1. *The following are equivalent:*
   (a) *For every* $v, w \in X$*, if* $x \in (v+C)\cap(w-C)$*, then* $\|x\| \leq \alpha \max\{\|v\|, \|w\|\}$.
   (b) *For every* $\eta \in X^*$*, there exist* $\psi, \phi \in C^*$ *so that* $\eta = \psi - \phi$ *and* $\|\phi\| + \|\psi\| \leq \alpha \|\eta\|$.
2. *The following are equivalent:*
   (a) *For every* $\varepsilon > 0$ *and* $x \in X$*, there exist* $a, b \in C$ *so that* $x = a - b$ *and* $\|a\| + \|b\| \leq (\alpha + \varepsilon) \|x\|$.
   (b) *For every* $\phi, \psi \in X^*$*, if* $\eta \in (\phi + C^*) \cap (\psi - C^*)$*, then* $\|\eta\| \leq \alpha \max\{\|\phi\|, \|\psi\|\}$.

As indicated by the list of references above, Theorem 7.2 and similar results have been rediscovered quite a few times, and the terminology employed by different authors tend to be quite fragmented. Also proofs of analogues of Theorem 7.2(2) tend to become somewhat involved, relying on technical results (see, for example, the proof of [13, Theorem 8]; proof of [28, Proposition 6]; Lemmas 1.1.3 and 1.1.5 and Theorem 1.1.4 from [7], and Proposition 1.3.1, Lemma 1.3.2 and Theorem 2.1.5 from [6]). These results do require some investment to verify and understand.

We will prove Theorem 7.2, as a special case of a result from [25]. Our reason for providing this proof is to abstract the main idea of the proof as purely an application of the one-sided polar calculus, and to demonstrate that the reason for the previously mentioned technicalities, when boiled down, is the "wanting to erase a weak-closure" in a certain set inclusion (see the Proof of Theorem 7.2 below).

Crucial in performing this mentioned "closure erasure" successfully are Lemmas 7.3 and 7.4 below.

Before stating these lemmas, we introduce some further terminology from Jameson [19, Appendix]: for a topological vector space (V, τ) we say a set $A \subseteq V$ is *τ-cs-compact* if, for any sequences $(x_n) \subseteq A$ and $(\lambda_n) \subseteq [0, 1]$ satisfying $\sum_{n=1}^{\infty} \lambda_n = 1$, the series $\sum_{j=1}^{\infty} \lambda_n x_n$ converges with respect to the τ-topology and its limit is an element of A. We say a set $A \subseteq V$ is *τ-cs-closed,* if for any sequences $(x_n) \subseteq A$ and $(\lambda_n) \subseteq [0, 1]$ satisfying $\sum_{n=1}^{\infty} \lambda_n = 1$, if the series $\sum_{j=1}^{\infty} \lambda_n x_n$ converges with respect to the τ-topology, then its limit is an element of A.

The first lemma is a characterization of Banach spaces as having exactly norm-cs-compact closed unit balls. Its proof is an easy exercise in using the absolute convergence characterization of Banach spaces, cf. [14, Lemma 1.22].

Lemma 7.3 *A normed space is a Banach space if and only if its closed unit ball is norm-cs-compact.*

The second lemma is the main tool in "closure erasure". We note explicitly that the reader should notice in Lemma 7.4 the erasure of the weak-closure by paying an arbitrarily small price in scaling up the set A.

Lemma 7.4 *Let X be a Banach space, $D \subseteq X$ and $A \subseteq X$ a $\sigma(X, X^*)$-cs-closed set. Let $G \subseteq X$ be a $\sigma(X, X^*)$-cs-compact set. If $G \subseteq D \subseteq \overline{A}^{\sigma(X,X^*)}$ and, for all $r > 0$ and $d \in D$, we have $(d + rG) \cap A \neq \emptyset$, then $D \subseteq \lambda A$ for all $\lambda > 1$.*

To the author's knowledge, this lemma is originally due to Batty and Robinson [7, Lemma 1.1.3], and proven in a slightly more general form in [25, Lemma 2.3].

We now turn to a proof Theorem 7.2, here adapted as a special case from [25, Theorem 3.4]. The main idea is to reformulate the statements into equivalent statements involving specific set inclusions. The proof then becomes an exercise in the one-sided polar calculus (cf. Lemma 7.1) and "closure erasure" by applying Lemma 7.4.

Proof of Theorem 7.2 With "$\oplus_p$" denoting the usual ℓ^p-direct sum for $1 \leq p \leq \infty$, we consider the canonical duality of the spaces $X \oplus_\infty X$ and $X^* \oplus_1 X^*$. We define the following sets

$$\Xi_\infty := \{(a, b) \in X \oplus_\infty X \mid a = b\},$$
$$\Xi_1 := \Xi_\infty \cap \mathbf{B}_{X \oplus_\infty X},$$
$$C \oplus_\infty (-C) := \{(a, b) \in X \oplus_\infty X \mid -b, a \in C\},$$

and

$$\Sigma_1 := \left\{(\phi, \psi) \in X^* \oplus_1 X^* \mid \|\phi + \psi\| \leq 1\right\},$$
$$\Sigma_0 := \left\{(\phi, \psi) \in X^* \oplus_1 X^* \mid \phi + \psi = 0\right\},$$
$$(-C^*) \oplus_1 C^* := \left\{(\phi, \psi) \in X^* \oplus_1 X^* \mid -\phi, \psi \in C^*\right\}.$$

We prove (1). Assume (1)(a). This is equivalent to

$$(\mathbf{B}_{X\oplus_\infty X} + C \oplus_\infty (-C)) \cap \Xi_\infty \subseteq \alpha\Xi_1.$$

With wk $:= \sigma(X \oplus_\infty X, X^* \oplus_1 X^*)$, the above inclusion implies[3]

$$\overline{(\mathbf{B}_{X\oplus_\infty X} + C \oplus_\infty (-C))}^{\text{wk}} \cap \Xi_\infty \subseteq \alpha\Xi_1.$$

Now, by taking one-sided polars (cf. Lemma 7.1), with wk* $:= \sigma(X^* \oplus_1 X^*, X \oplus_\infty X)$, we obtain

$$\begin{aligned}\Sigma_1 &\subseteq \overline{\alpha\left(\overline{\mathbf{B}_{X\oplus_\infty X} + C \oplus_\infty (-C)}^{\text{wk}}\right)^{\odot} + \Sigma_0}^{\text{wk}^*}\\ &\subseteq \overline{\alpha(\mathbf{B}_{X\oplus_\infty X} + C \oplus_\infty (-C))^{\odot} + \Sigma_0}^{\text{wk}^*}\\ &= \overline{(\alpha\mathbf{B}_{X^*\oplus_1 X^*} \cap (-C^*) \oplus_1 C^*) + \Sigma_0}^{\text{wk}^*}.\end{aligned}$$

Since $\mathbf{B}_{X^*\oplus_1 X^*}$ is wk*-compact, the set $(\alpha\mathbf{B}_{X^*\oplus_1 X^*} \cap (-C^*) \oplus_1 C^*) + \Sigma_0$ can be seen to already be wk*-closed, so that taking the closure is redundant and we erase it. Hence $\Sigma_1 \subseteq (\alpha\mathbf{B}_{X^*\oplus_1 X^*} \cap (-C^*) \oplus_1 C^*) + \Sigma_0$, which is equivalent to (1)(b).

Assume (1)(b), which is equivalent to $\Sigma_1 \subseteq (\alpha\mathbf{B}_{X^*\oplus_1 X^*} \cap (-C^*) \oplus_1 C^*) + \Sigma_0$. Taking one-sided polars (cf. Lemma 7.1) with wk $:= \sigma(X \oplus_\infty X, X^* \oplus_1 X^*)$, we obtain

$$(\mathbf{B}_{X\oplus_\infty X} + C \oplus_\infty (-C)) \cap \Xi_\infty \subseteq \overline{\mathbf{B}_{X\oplus_\infty X} + C \oplus_\infty (-C)}^{\text{wk}} \cap \Xi_\infty \subseteq \alpha\Xi_1,$$

which is equivalent to (1)(a).

We adjust the definitions of $\Sigma_{(\cdot)}$ and $\Xi_{(\cdot)}$ appropriately, and prove (2).

Assume (2)(a). This is equivalent to, for every $\varepsilon > 0$, that $\Sigma_1 \subseteq (\alpha+\varepsilon)\mathbf{B}_{X\oplus_1 X} \cap C \oplus_1 (-C) + \Sigma_0$. Taking one-sided polars (cf. Lemma 7.1) with wk* $:= \sigma(X^* \oplus_\infty X^*, X \oplus_1 X)$, yields

$$\begin{aligned}(\mathbf{B}_{X^*\oplus_\infty X^*} + (-C^*) \oplus_\infty C^*) \cap \Xi_\infty &\subseteq \overline{(\mathbf{B}_{X^*\oplus_\infty X^*} + (-C^*) \oplus_\infty C^*)}^{\text{wk}^*} \cap \Xi_\infty\\ &\subseteq \bigcap_{\varepsilon>0} (\alpha+\varepsilon)\Xi_1 = \alpha\Xi_1,\end{aligned}$$

which is equivalent to (2)(b).

[3]Let $(x, x) \in \overline{(\mathbf{B}_{X\oplus_\infty X} + C \oplus_\infty (-C))}^{\|\cdot\|_\infty} \cap \Xi_\infty$. Then there exist sequences $((a_n, b_n)) \subseteq \mathbf{B}_{X\oplus_\infty X}$ and $((c_n, -d_n)) \subseteq C \oplus_\infty (-C)$ so that $(a_n, b_n) + (c_n, -d_n) \to (x, x)$ as $n \to \infty$. For every $n \in \mathbb{N}$, let $p_n := (a_n + c_n) - (b_n - d_n)$ and consider the sequence $S := ((a_n, b_n + p_n) + (c_n, -d_n)) = ((a_n + c_n, a_n + c_n)) \subseteq \Xi_\infty$. This sequence S converges to (x, x) and, since $p_n \to 0$ as $n \to \infty$, for every $\varepsilon > 0$ the tail of S eventually lies in $((1+\varepsilon/\alpha)\mathbf{B}_{X\oplus_\infty X} + C \oplus_\infty (-C)) \cap \Xi_\infty \subseteq (1+\varepsilon/\alpha)\alpha\Xi_1$, and hence $\|(x, x)\|_\infty \leq (\alpha+\varepsilon)$. But this holds for every $\varepsilon > 0$, so $\|(x, x)\|_\infty \leq \alpha$ and therefore $(x, x) \in \alpha\Xi_1$. Because the wk-closure and $\|\cdot\|_\infty$-closure of convex sets coincide, we obtain $\overline{(\mathbf{B}_{X\oplus_\infty X} + C \oplus_\infty (-C))}^{\text{wk}} \cap \Xi_\infty \subseteq \alpha\Xi_1$.

Conversely, assume (2)(b), which is equivalent to $(\mathbf{B}_{X^*\oplus_\infty X^*}+(-C^*)\oplus_\infty C^*)\cap \Xi_\infty \subseteq \alpha\Xi_1$. With $\text{wk}^* := \sigma(X^*\oplus_\infty X^*, X\oplus_1 X)$, the set $\mathbf{B}_{X^*\oplus_\infty X^*}$ is wk*-compact, so that $(\mathbf{B}_{X^*\oplus_\infty X^*} + (-C^*)\oplus_\infty C^*)$ is wk*-closed. Taking one-sided polars (cf. Lemma 7.1) with $\text{wk} := \sigma(X\oplus_1 X, X^*\oplus_\infty X^*)$, yields

$$\Sigma_1 \subseteq \overline{(\alpha\mathbf{B}_{X\oplus_1 X}\cap(C\oplus_1(-C))+\Sigma_0}^{\text{wk}}.$$

We note that since $\mathbf{B}_{X\oplus_1 X}$ is not necessarily wk-compact, the set $(\alpha\mathbf{B}_{X\oplus_1 X}\cap(C\oplus_1(-C))+\Sigma_0$ need not be wk-closed. By Lemma 7.3, $\mathbf{B}_{X\oplus_1 X}$ is norm-cs-compact, hence also wk-cs-compact. We note that the wk-closure and norm-closure of convex sets coincide, and that the sum $(\alpha\mathbf{B}_{X\oplus_1 X}\cap C\oplus_1(-C))+\Sigma_0$ is wk-cs-closed, being the sum of a wk-cs-compact set and a wk-cs-compact set. Applying Lemma 7.4 to the inclusion below,

$$\mathbf{B}_{X\oplus_1 X}\subseteq\Sigma_1\subseteq\overline{(\alpha\mathbf{B}_{X\oplus_1 X}\cap C\oplus_1(-C))+\Sigma_0}^{\text{wk}},$$

we obtain, for every $\varepsilon>0$ that $\Sigma_1\subseteq((\alpha+\varepsilon)\mathbf{B}_{X\oplus_1 X}\cap C\oplus_1(-C))+\Sigma_0$. This is equivalent to (2)(a). □

Theorem 7.2 and the duality $(F(M)\overline{\otimes}_\pi X)^* = B(F(M), X^*)\simeq \text{Lip}_0(M, X^*)$ observed in Sect. 6 suggest that it is possible to characterize the Lipschitz decomposition property in terms of the geometry of cones in projective tensor products with Free Lipschitz spaces. The remainder of the current section will be devoted to this.

Following Ryan [33] and Wittstock [38], we introduce the following terminology and notation for projective tensor products and projective tensor cones. For Banach spaces X and Y, we denote the projective tensor norm by π (cf. [33, Chapter 2]) and denote the completed projective tensor product of X and Y by $X\overline{\otimes}_\pi Y$. Let $C\subseteq X$ and $D\subseteq Y$ be norm-closed cones. We define the *projective tensor cone* $C\overline{\otimes}_\pi D$ as the norm closure of $\text{cspan}\{c\otimes d \mid (c,d)\in C\times D\}\subseteq X\overline{\otimes}_\pi Y$. We will view $B(X, Y^*)$ as the dual of $X\overline{\otimes}_\pi Y$ through the duality $\left\langle \sum_{i=1}^\infty x_i\otimes y_i \,\middle|\, T\right\rangle := \sum_{i=1}^\infty (Tx_i)(y_i)$ for $T\in B(X, Y^*)$ and $\sum_{i=1}^\infty x_i\otimes y_i\in X\overline{\otimes}_\pi Y$ (cf. [33, Section 2.2]). It can be seen that $(C\overline{\otimes}_\pi D)^* = \{T\in B(X,Y^*) \mid TC\subseteq D^*\}$.

The following is an immediate consequence of Theorem 7.2 and the observation $(X\overline{\otimes}_\pi Y)^* = B(X, Y^*)$.

Corollary 7.5 *Let $\alpha>0$ and X and Y be Banach spaces and $C\subseteq X$ and $D\subseteq Y$ norm-closed cones. The following are equivalent:*

(a) *For $v, w\in X\overline{\otimes}_\pi Y$, if $u\in(v+C\overline{\otimes}_\pi D)\cap(w-C\overline{\otimes}_\pi D))$, then $\pi(u)\leq \alpha\max\{\pi(v),\pi(w)\}$.*
(b) *For every $T\in B(X, Y^*)$, there exist $R, S\in\{T\in B(X,Y^*) \mid TC\subseteq D^*\}$ so that $T = R-S$ and $\|R\|+\|S\|\leq\alpha\|T\|$.*

Proof Since $(X\overline{\otimes}_\pi Y)^* = B(X, Y^*)$ (cf. [33, Section 2.2]) and since $(C\overline{\otimes}_\pi D)^* = \{T\in B(X,Y^*) \mid TC\subseteq D^*\}$, the equivalence follows from Theorem 7.2(1). □

Let M be any metric space and X a Banach space. In Sect. 6 we observed that $\mathrm{Lip}_0(M, X^*)$ is isometrically isomorphic to $B(F(M), X^*)$. Now applying Theorem 6.1 and Corollary 7.5, the Lipschitz decomposition problem for a dual space X^* (ordered by the dual cone C^*) can be transferred to statements on the geometry of the projective tensor cones $F(M)_+\overline{\otimes}_\pi C$ or $F(X)_+\overline{\otimes}_\pi C$.

Corollary 7.6 *Let X be a Banach space with $C \subseteq X$ a closed cone. The following are equivalent:*

(a) *The dual space X^* (ordered by the dual cone C^*) has the Lipschitz decomposition property.*
(b) *There exists some $\alpha > 0$, so that for every $w, v \in F(X)\overline{\otimes}_\pi X$, if $u \in (w + F(X)_+\overline{\otimes}_\pi C) \cap (v - F(X)_+\overline{\otimes}_\pi C)$, then $\pi(u) \le \alpha \max\{\pi(v), \pi(w)\}$.*
(c) *For every metric space M, there exists some $\alpha > 0$, so that for every $w, v \in F(M)\overline{\otimes}_\pi X$, if $u \in (w + F(M)_+\overline{\otimes}_\pi C) \cap (v - F(M)_+\overline{\otimes}_\pi C)$, then $\pi(u) \le \alpha \max\{\pi(v), \pi(w)\}$.*

Proof This follows from Theorems 6.1, 4.1, 6.2, and 7.5. □

We end this section by raising some final questions that are relevant to the current discussion.

Let X be a Banach space ordered by a closed cone $C \subseteq X$. Since $\mathrm{Lip}_0(X)_+$ is generating in $\mathrm{Lip}_0(X)$, by Theorems 4.1 and 7.2, there exists a constant $\alpha > 0$ so that, for $v, w \in F(X)$, if $u \in (v + F(X)_+) \cap (w - F(X)_+)$, then $\|u\| \le \alpha \max\{\|v\|, \|w\|\}$. This property is usually called *normality* in the literature. Similarly, by Theorems 4.1 and 7.2, if C^* is generating X^*, then X also has such a normality property. Therefore, using Corollary 7.6, the Lipschitz decomposition problem can be resolved for dual Banach spaces if one can answer the question: "If X is normal, is the projective tensor product $F(X)\overline{\otimes}_\pi X$ necessarily also normal, when ordered by the projective tensor cone $F(X)_+\overline{\otimes}_\pi C$?"; or more generally: "Do projective tensor products ordered by projective tensor cones always preserve normality of its tensor factors?"

By Theorem 6.2, the space $\mathrm{Lip}_0(X, X)$ is isometrically isomorphic to the space $B(F(X), X)$. If X is not a dual Banach space, Corollary 7.6 is not available. Certainly one may apply the general duality result, Theorem 7.2, to $B(F(X), X)$ and its dual $B(F(X), X)^*$, but this observation is not of use without knowledge of the structure of $B(F(X), X)^*$ that could be exploited to show that $B(F(X), X)^*$ is normal when ordered by the dual cone $\{T \in B(F(X), X) \mid TF(X)_+ \subseteq C\}^*$. Since $B(F(X), X) \subseteq B(F(X), X^{**}) = (F(X)\overline{\otimes}_\pi X^*)^*$, one is tempted to consider the dual pair $(B(F(X), X), F(X)\overline{\otimes}_\pi X^*)$ with the duality defined by restriction. However, the space $B(F(X), X)$ need not separate the points of $F(X)\overline{\otimes}_\pi X^*$ in general (cf. [18]).

Assume further that either X^* or $F(X)$ has the approximation property.[4] Then the space $B(F(X), X)$ does separate the points of $F(X)\overline{\otimes}_\pi X^*$ (cf. [12, Corollary 3 p. 65]), and $B(F(X), X)$ lies wk*-dense in $B(F(X), X^{**})$ (cf. [18, Proposition 2.2]). If C is generating in X, then so is C^{**} in X^{**} (cf. Theorems 4.1 and 7.2). Keeping Theorem 6.2 in mind brings us into the situation described in the previous paragraph with X^{**} being a dual Banach space, while introducing the further question: "When, if ever, does the cone $\{T \in B(F(X), X^{**}) \mid TF(X)_+ \subseteq C^{**}\}$ being generating in $B(F(X), X^{**})$, imply that $\{T \in B(F(X), X) \mid TF(X)_+ \subseteq C\}$ is generating in the wk*-dense subspace $B(F(X), X)$?"

Acknowledgements The author would like to express his thanks to the MathOverflow community, especially to Bill Johnson for bringing work described in Sect. 5 to the author's attention.

References

1. I. Aharoni, J. Lindenstrauss, Uniform equivalence between Banach spaces. Bull. Am. Math. Soc. **84**(2), 281–284 (1978)
2. F. Albiac, N.J. Kalton, *Topics in Banach Space Theory*, 2nd edn. (Springer, Berlin, 2016)
3. C.D. Aliprantis, K.C. Border, *Infinite Dimensional Analysis*, 3rd edn. (Springer, Berlin, 2006)
4. C.D. Aliprantis, R. Tourky, *Cones and Duality* (American Mathematical Society, Providence, 2007)
5. T. Andô, On fundamental properties of a Banach space with a cone. Pac. J. Math. **12**(4), 1163–1169 (1962)
6. L. Asimow, A.J. Ellis, *Convexity Theory and Its Applications in Functional Analysis* (Academic, London, 1980)
7. C.J.K. Batty, D.W. Robinson, Positive one-parameter semigroups on ordered Banach spaces. Acta Appl. Math. **2**(3–4), 221–296 (1984)
8. Y. Benyamini, J. Lindenstrauss, *Geometric Nonlinear Functional Analysis* (American Mathematical Society, Providence, 2000)
9. E.B. Davies, The structure and ideal theory of the predual of a Banach lattice. Trans. Am. Math. Soc. **131**, 544–555 (1968)
10. M. de Jeu, M. Messerschmidt, A strong open mapping theorem for surjections from cones onto Banach spaces. Adv. Math. **259**, 43–66 (2014)
11. K. de Leeuw, Banach spaces of Lipschitz functions. Stud. Math. **21**, 55–66 (1961/1962)
12. A. Defant, F. Floret, *Tensor Norms and Operator Ideals* (North-Holland, Amsterdam, 1993)
13. A.J. Ellis, The duality of partially ordered normed linear spaces. J. Lond. Math. Soc. **39**, 730–744 (1964)
14. M. Fabian, P. Habala, P. Hájek, V. Montesinos, V. Zizler, *Banach Space Theory* (Springer, New York, 2011)
15. G. Godefroy, N.J. Kalton, Lipschitz-free Banach spaces. Stud. Math. **159**(1), 121–141 (2003)
16. J. Grosberg, M. Krein, Sur la décomposition des fonctionnelles en composantes positives. C. R. (Dokl.) Acad. Sci. URSS (N.S.) **25**, 723–726 (1939)
17. J.B. Guerrero, G. López-Pérez, A. Rueda Zoca, Octahedrality in Lipschitz-free Banach spaces. Proc. Roy. Soc. Edinburgh Sect. A **148**(3), 447–460 (2018)

[4]Of some relevance here is Kalton and Godefroy's result [15, Theorem 5.3] showing that bounded approximation properties transfer between X and $F(X)$.

18. P. Hájek, R.J. Smith, Some duality relations in the theory of tensor products. Expo. Math. **30**(3), 239–249 (2012)
19. G. Jameson, *Ordered Linear Spaces* (Springer, Berlin, 1970)
20. W.B. Johnson, J. Lindenstrauss, G. Schechtman, Banach spaces determined by their uniform structures. Geom. Funct. Anal. **6**(3), 430–470 (1996)
21. N.J. Kalton, Spaces of Lipschitz and Hölder functions and their applications. Collect. Math. **55**(2), 171–217 (2004)
22. N.J. Kalton, Lipschitz and uniform embeddings into ℓ^∞. Fundam. Math. **212**(1), 53–69 (2011)
23. V.L. Klee, Boundedness and continuity of linear functionals. Duke Math. J. **22**(2), 263–269 (1955)
24. J. Lindenstrauss, Uniform embeddings, homeomorphisms and quotient maps between Banach spaces (a short survey). Topology Appl. **85**(1–3), 265–279 (1998). 8th Prague Topological Symposium on General Topology and Its Relations to Modern Analysis and Algebra (1996)
25. M. Messerschmidt, Geometric duality theory of cones in dual pairs of vector spaces. J. Funct. Anal. **269**(7), 2018–2044 (2015)
26. M. Messerschmidt, A pointwise Lipschitz selection theorem. Set-Valued Var. Anal. (2017). https://doi.org/10.1007/s11228-017-0455-2
27. M. Messerschmidt, Strong Klee–Andô theorems through an open mapping theorem for cone-valued multi-functions. J. Funct. Anal. **275**(12), 3325–3337 (2018)
28. K.F. Ng, The duality of partially ordered Banach spaces. Proc. Lond. Math. Soc. (3) **19**, 269–288 (1969)
29. K.F. Ng, On a theorem of Dixmier. Math. Scand. **29**(1971), 279–280 (1972)
30. K.F. Ng, C.K. Law, Monotonic norms in ordered Banach spaces. J. Austral. Math. Soc. Ser. A **45**(2), 217–219 (1988)
31. V.G. Pestov, Free Banach spaces and representations of topological groups. Funktsional. Anal. i Prilozhen. **20**(1), 81–82 (1986)
32. M. Ribe, Existence of separable uniformly homeomorphic nonisomorphic Banach spaces. Isr. J. Math. **48**(2–3), 139–147 (1984)
33. R.A. Ryan, *Introduction to Tensor Products of Banach Spaces* (Springer, London, 2002)
34. H.H. Schaefer, *Banach Lattices and Positive Operators* (Springer, Berlin, 1974)
35. N. Weaver, *Lipschitz Algebras* (World Scientific, River Edge, 1999)
36. N. Weaver, On the unique predual problem for Lipschitz spaces. Math. Proc. Camb. Philos. Soc. **165**(3), 467–473 (2018)
37. R. Whitley, Projecting m onto c_0. Am. Math. Mon. **73**(3), 285–286 (1966)
38. G. Wittstock, Ordered normed tensor products, in *Foundations of Quantum Mechanics and Ordered Linear Spaces*. Lecture Notes in Physics, vol. 29 (Springer, Berlin, 1974), pp. 67–84
39. D. Yost, There can be no Lipschitz version of Michael's selection theorem, in *Proceedings of the Analysis Conference*, Singapore 1986. North Holland Mathematics Studies, vol. 150 (North-Holland, Amsterdam, 1988), pp. 295–299

Classes of Localizable Measure Spaces

Susumu Okada and Werner J. Ricker

In honour of our good friend Ben de Pagter

Abstract The theory of localizable measures is an important part of measure theory in general (it includes all σ-finite measures, not to mention Haar measure on locally compact abelian groups!). Unfortunately, what is understood by the term "localizable measure" depends on whom one asks; it seems not to be an entirely fixed concept. Such a vagueness can make it precarious when applying the results from this area. The aim of this note is to make a careful investigation of the similarities and differences between four of the known and established classes of "localizable measures". It will be shown that there is a linear order between these four classes and that no pair of the classes within this order coincide. Moreover, some effort has been undertaken to present many relevant examples which illustrate the differences between these classes. Our hope is to be able to clarify and make more transparent the closely related (but, also different) notions of "localizability" that are available.

Keywords Localizable measure · Dedekind complete measure · Integration · Boolean algebra

1 Introduction

Localizable (also called Maharam) measures, which are intimately connected to measure algebras, form the most natural class of measures which properly contain the more traditional class of σ-finite measures; see, for example [2, 5–9, 19, 20, 23],

S. Okada
School of Mathematics and Physics, University of Tasmania, Hobart, TAS, Australia
e-mail: susbobby@grapevine.com.au

W. J. Ricker (✉)
Math.-Geogr. Fakultät, Katholische Universität Eichstätt-Ingolstadt, Eichstätt, Germany
e-mail: werner.ricker@ku.de

G. Buskes et al. (eds.), *Positivity and Noncommutative Analysis*,
Trends in Mathematics, https://doi.org/10.1007/978-3-030-10850-2_23

and the references therein. One of the many and diverse areas of applications of localizable measures, due to I. Kluvánek, is to the theory of conical and vector measures, [12, 14–16, 18]. Since the notion of localizability used in the work of I. Kluvánek differs from the traditional one mentioned above, let us call these K-localizable measures. A third notion of localizable measure, due to I. Segal, is also available, [22]; we call such measures S-localizable. And yes, in the classical monograph of A.C. Zaanen, [23], localizability is also treated, which we will term Z-localizability. Some inter-connections between these notions of localizable measures, which are closely connected to various versions of the Radon-Nikodym Theorem, occur in the recent article, [18]. The aim of this note is to further develop and clarify such inter-connections. It is time to be more precise.

Let Σ be a σ-algebra of subsets of a non-empty set Ω and $\mu : \Sigma \longrightarrow [0, \infty]$ be a σ-additive measure. Denote by $J_\mu : L^\infty(\mu) \longrightarrow (L^1(\mu))^*$ the canonical linear map which sends $\varphi \in L^\infty(\mu)$ to the continuous linear functional on $L^1(\mu)$ given by $f \longmapsto \int_\Omega \varphi f \, d\mu$; here $(L^1(\mu))^*$ is the dual Banach space of $L^1(\mu)$. Consider the following conditions:

(a) J_μ is surjective;
(b) J_μ is injective; and
(c) J_μ is bijective.

In [16, p. 9] the measure μ is called localizable if condition (a) is satisfied; for reasons of clarity, we will use the terminology *K-localizable*. Condition (b) is equivalent to μ being a *semifinite* measure, [8, Theorem 243G]. That is, if $E \in \Sigma$ satisfies $\mu(E) > 0$, then there exists $F \in \Sigma$ with $F \subseteq E$ such that $0 < \mu(F) < \infty$. Finally, μ is called *localizable*, [5, Section 64A], if it is semifinite and the quotient Boolean algebra (briefly, B.a.) $\Sigma/\mathcal{N}_0(\mu)$ is an order complete B.a., where $\mathcal{N}_0(\mu) := \{E \in \Sigma : \mu(E) = 0\}$ is the σ-ideal of all *μ-null sets*. This is equivalent to (c) being satisfied, [8, Theorem 243G]. Henceforth, a localizable measure μ *always means* that it satisfies condition (c). We also introduce the notion of a measure space (Ω, Σ, μ) being *Dedekind complete* namely, that the B.a. $\Sigma/\mathcal{N}_0(\mu)$ is order complete. In [16, p. 10] it is stated that the class of K-localizable measures coincides with the class of localizable measures introduced by I. Segal, [22, Definition 2.6], which we will refer to as *S-localizable*. This claim is incorrect; the confusion arises because the measure spaces (based on certain rings of sets) and the measurable sets considered in [22] (see Definitions 2.1 and 2.4 there) are different to those considered in [16].

The main aim of this note is to establish the following diagram for measure spaces.

Localizable ⇒ Dedekind complete ⇒ Z-localizable ⇒
K-localizable ⇒ S-localizable

The first implication (from the left) is per definition. The remaining implications are new results; see Theorem 3.4, Lemma 3.7(i), Corollary 3.9(i) and Theorem 4.3. Neither of the four implications is reversible. For instance, Example 3.5 exhibits two Dedekind complete measures which are not localizable and, just prior to

Proposition 3.8, a measure is presented which is Z-localizable but not Dedekind complete (see also Example 2.14). The measure of Example 5.1 is K-localizable but not Z-localizable. Finally, for an S-localizable measure which is not K-localizable see Example 2.14. We point out in Remark 2.15 that there also exist measures which are not S-localizable. Moreover, there is no connection between Dedekind completeness and semifiniteness.

The terminology used in various parts of the literature is not always the same and can even be inconsistent, so care needs to be taken. For instance, the semifiniteness of a measure (used in [2, 1.12.132, p. 97], [8, 211F], for example) is called the finite subset property in [19, Definition 2, p. 71], [23, Definition, p. 257]. The notion of a decomposable measure (also called strictly localizable) as given in [2, 1.12.131, p. 96], [8, Definition 211E], is *not* the same as a decomposable measure as specified in [11, Definition (19.25), p. 317]; see Proposition 2.4 and Example 5.1. As already noted previously, localizable measures as defined in [16] differ from those in [22] which in turn differ from those in [8]. Accordingly, we have attempted to provide detailed arguments and taken care to be rather precise concerning definitions. Several examples are included throughout the text in order to clarify and illustrate the differences between various results and concepts. There is a vast literature on this topic, of which we have made a somewhat personal selection. The encyclopedic works [7–9] have provided many of the results, techniques and methods that we required. The theory of integration presented in [7, 8], which is general and extensive, is perhaps not in the framework that is familiar to all. As to be expected, there is also a formulation of the theory available in the more traditional format, [8, Chapter 24]. For the sake of self-containment we have provided (in an Appendix) a brief explanation of the two approaches.

2 Preliminaries

Let (Ω, Σ, μ) be a measure space. Namely, Σ is a σ-algebra of subsets of a non-empty set Ω and $\mu : \Sigma \longrightarrow [0, \infty]$ is a measure (i.e., a σ-additive set function satisfying $\mu(\emptyset) = 0$). The measure space (Ω, Σ, μ) need not be complete. The following convention will be used, namely

$$\Sigma \cap F = \{E \cap F : E \in \Sigma\} = \{E \in \Sigma : E \subseteq F\}, \qquad F \in \Sigma .$$

A property which holds outside a μ-null set is said to hold *μ-almost everywhere*, (briefly μ-a.e.). Define

$$\Sigma^f := \{E \in \Sigma : \mu(E) < \infty\} \text{ and } \Sigma_0^f := \{E \in \Sigma : 0 < \mu(E) < \infty\}. \qquad (2.1)$$

Then μ, or (Ω, Σ, μ), is semifinite if, for every $E \in \Sigma$ with $\mu(E) = \infty$, there exists $F \in \Sigma_0^f$ satisfying $F \subseteq E$, [8, Definition 211F]. For a semifinite measure μ

we have

$$\mu(E) = \sup\{\mu(F) : F \subseteq E \text{ and } F \in \Sigma_0^f\}, \qquad E \in \Sigma; \tag{2.2}$$

see [8, Lemma 213A]. Of course, finite or σ-finite measure spaces are clearly semifinite. More generally, strictly localizable measure spaces, also called decomposable measure spaces (see [8, Definition 211E]), are semifinite, [8, Theorem 211L(d)]. It can happen that Σ_0^f is empty, equivalently $\mu(\Sigma) \subseteq \{0, \infty\}$. As the case $\mu(\Sigma) = \{0\}$ means exactly that μ is the zero measure, let us consider μ satisfying $\mu(\Sigma) = \{0, \infty\}$. Such a measure μ is called *totally infinite*, namely $\mu(E) = \infty$ for every non-μ-null set $E \in \Sigma$. For instance, given any measurable space (Ω, Σ) the measure μ defined by $\mu(E) = \infty$ for every $E \in \Sigma$ with $E \neq \emptyset$ and $\mu(\emptyset) = 0$ is totally infinite.

An important fact is that Σ is always a Dedekind σ-complete B.a. Indeed, Σ is a lattice with the empty set $\emptyset$ as zero and the whole set Ω as unit, in the order determined by set inclusion. The lattice Σ is distributive and complemented, where the complement E' of each $E \in \Sigma$ is its set-theoretic complement $E^c = \Omega \setminus E$. The B.a. operations $\wedge$ and $\vee$ are given by $E \wedge F := E \cap F$ and $E \vee F := E \cup F$ for $E, F \in \Sigma$. Clearly, Σ is Dedekind σ-complete. The subfamily $\mathcal{N}_0(\mu)$ of the B.a. Σ is a σ-ideal. So, the quotient B.a. $\Sigma/\mathcal{N}_0(\mu)$ is a Dedekind σ-complete B.a. To be precise, define an equivalence relation by $E \sim F$ for $E, F \in \Sigma$ if the symmetric difference $E \triangle F$ is μ-null, where $E \triangle F := (E \cup F) \setminus (E \cap F)$. Let $\pi_\mu(E)$ denote the quotient class $\{F \in \Sigma : E \sim F\}$ of E for each $E \in \Sigma$. The quotient

$$\Sigma/\mathcal{N}_0(\mu) := \{\pi_\mu(E) : E \in \Sigma\}$$

is then a B.a. with respect to the B.a. operations specified by

$$\pi_\mu(E) \wedge \pi_\mu(F) := \pi_\mu(E \cap F), \pi_\mu(E) \vee \pi_\mu(F) := \pi_\mu(E \cup F),$$
$$(\pi_\mu(E))' := \pi_\mu(\Omega \setminus E)$$

for $E, F \in \Sigma$. Since $\mathcal{N}_0(\mu)$ is a σ-ideal, the quotient B.a. $\Sigma/\mathcal{N}_0(\mu)$ is σ-complete and the so-defined quotient map $\pi_\mu : \Sigma \longrightarrow \Sigma/\mathcal{N}_0(\mu)$ is a B.a. σ-homomorphism.

According to [7, 112D], a set $A \subseteq \Omega$ is called *μ-negligible* if there exists $E \in \mathcal{N}_0(\mu)$ such that $A \subseteq E$. Let $\mathcal{N}$ denote the family of all μ-negligible sets. It is clear that a set $A \in \Sigma$ is μ-negligible if and only if $A \in \mathcal{N}_0(\mu)$. In other words, $\mathcal{N} \cap \Sigma = \mathcal{N}_0(\mu)$. So, we have

$$\Sigma/\mathcal{N}_0(\mu) = \Sigma/(\mathcal{N} \cap \Sigma),$$

that is, the quotient B.a. $\Sigma/\mathcal{N}_0(\mu)$ coincides with the quotient B.a. $\Sigma/(\mathcal{N} \cap \Sigma)$ considered in [9, Proposition 321H]. The same proposition defines the functional $\overline{\mu} : \Sigma/\mathcal{N}_0(\mu) \longrightarrow [0, \infty]$ by $\overline{\mu}(\pi_\mu(E)) := \mu(E)$ for $E \in \Sigma$ and verifies the well-definedness of $\overline{\mu}$ together with the properties that $\overline{\mu}(\mathbf{0}) = 0$, that $\overline{\mu}(\alpha) > 0$

whenever $\alpha \in \Sigma/\mathcal{N}_0(\mu)$ with $\alpha \neq \mathbf{0}$, and that $\overline{\mu}(\vee_{n=1}^{\infty}\alpha_n) = \sum_{n=1}^{\infty} \overline{\mu}(\alpha_n)$ for every pairwise disjoint sequence $\{\alpha_n\}_{n=1}^{\infty}$ in $\Sigma/\mathcal{N}_0(\mu)$. So, the pair $(\Sigma/\mathcal{N}_0(\mu), \overline{\mu})$ is a *measure algebra* (see [9, Definition 321A]) and is then called the measure algebra of the measure space (Ω, Σ, μ), [9, Definition 321I]. We say that $\overline{\mu}$, or $(\Sigma/\mathcal{N}_0(\mu), \overline{\mu})$, is semifinite if, given $\alpha \in \Sigma/\mathcal{N}_0(\mu)$ with $\overline{\mu}(\alpha) = \infty$, there is a non-zero $\beta \in \Sigma/\mathcal{N}_0(\mu)$ such that $\beta \leq \alpha$ and $\overline{\mu}(\beta) < \infty$.

Recall from Sect. 1 that the measure space (Ω, Σ, μ) is *Dedekind complete* if the quotient B.a. $\Sigma/\mathcal{N}_0(\mu)$ is complete. To determine the Dedekind completeness of an explicit measure can be rather complicated and many abstract results are available to assist in this regard. The following example shows that sometimes a direct argument can also be of interest! Let Ω be any uncountable set. Define $\Sigma \subsetneq 2^{\Omega}$ as the family of all sets $E \subseteq \Omega$ such that E is countable or cocountable (i.e., $\Omega \setminus E$ is countable). Then Σ is a σ-algebra, called the countable-cocountable σ-algebra. The set function $\mu : \Sigma \longrightarrow \{0, 1\}$ defined by $\mu(E) := 0$ for a countable set $E \in \Sigma$ and $\mu(E) := 1$ for a cocountable set $E \in \Sigma$ is a finite measure, called the *countable-cocountable measure*, [8, 211R]. To see directly that μ is Dedekind complete observe that $\mathcal{N}_0(\mu) = \{E \in \Sigma : E \text{ is countable}\}$. Given any cocountable set $E \in \Sigma$, say $E = \Omega \setminus M$ for some countable set $M \subseteq \Omega$, observe that

$$E \,\triangle\, \Omega := (E \setminus \Omega) \cup (\Omega \setminus E) = ((\Omega \setminus M) \setminus \Omega) \cup M = \emptyset \cup M = M.$$

Hence, $\mu(E \,\triangle\, \Omega) = 0$ showing that $\pi_\mu(\Omega) = \pi_\mu(E)$. So, the quotient B.a. $\Sigma/\mathcal{N}_0(\mu)$ is surely complete as it consists of only two equivalence classes, namely $\mathbf{0} := \mathcal{N}_0(\mu)$ and $\mathbf{1} := \pi_\mu(\Omega) = \{E \in \Sigma : E \text{ is cocountable}\}$.

Let (Ω, Σ, μ) be a measure space. According to [8, Definition 211G], an *essential supremum* of a subfamily $\mathcal{E} \subseteq \Sigma$ is defined as a set $H \in \Sigma$ satisfying the following two conditions:

(*E*-1) $E \setminus H \in \mathcal{N}_0(\mu)$ for all $E \in \mathcal{E}$, and
(*E*-2) if a set $G \in \Sigma$ satisfies $E \setminus G \in \mathcal{N}_0(\mu)$ for all $E \in \mathcal{E}$, then $H \setminus G \in \mathcal{N}_0(\mu)$.

Lemma 2.1 *The following statements hold for every measure space* (Ω, Σ, μ).

(i) *The measure μ is semifinite if and only if so is the corresponding functional $\overline{\mu}$ on $\Sigma/\mathcal{N}_0(\mu)$.*
(ii) *A set $H \in \Sigma$ is an essential supremum of a subfamily $\mathcal{E} \subseteq \Sigma$ if and only if $\pi_\mu(\mathcal{E})$ possesses a supremum in the quotient B.a. $\Sigma/\mathcal{N}_0(\mu)$ and $\pi_\mu(H) = \sup \pi_\mu(\mathcal{E})$.*
(iii) *The measure space (Ω, Σ, μ) is Dedekind complete if and only if every subfamily of Σ possesses an essential supremum.*

Proof

(i) See [9, Theorem 322B(d)].
(ii) See the proof of Theorem 322B(e) in [9].
(iii) This follows from (ii). □

Recall, given a measure space (Ω, Σ, μ) that μ, or (Ω, Σ, μ), is *localizable* (or *Maharam*) if μ is semifinite and (Ω, Σ, μ) is Dedekind complete.

Lemma 2.2 *The following conditions for a measure space (Ω, Σ, μ) are equivalent.*

(i) *The measure μ is localizable.*
(ii) *The measure μ is semifinite and every subfamily of Σ possesses an essential supremum.*
(iii) *The functional $\overline{\mu} : \Sigma/\mathcal{N}_0(\mu) \longrightarrow [0, \infty]$ is semifinite and the quotient B.a. $\Sigma/\mathcal{N}_0(\mu)$ is Dedekind complete.*

Proof

(i) $\Longleftrightarrow$ (ii). Apply Lemma 2.1(iii).
(i) $\Longleftrightarrow$ (iii). Apply Lemma 2.1(i),(ii). □

In [8, Definition 211G], the localizability of μ is defined by condition (ii) in Lemma 2.1. On the other hand, condition (iii) of Lemma 2.1 means exactly that the measure algebra $(\Sigma/\mathcal{N}_0(\mu), \overline{\mu})$ is localizable in the sense of [9, Definition 322A(e)].

Every σ-finite measure space is localizable, [8, Theorem 211L(c),(d)]. This fact can be extended to the class of strictly localizable measures, [8, Theorem 211L(d)]. As we shall present examples of such measures later, let us provide the precise definition now, following [2, 1.12.131, p. 96] and [8, Definition 211E]. A measure space (Ω, Σ, μ), or simply μ, is called *strictly localizable* (or *decomposable*) if there exists a partition $(\Omega_\gamma)_{\gamma\in\Gamma}$ of Ω into pairwise disjoint Σ-measurable sets satisfying the following three conditions:

$$(1\text{-}s)\quad \mu(\Omega_\gamma) < \infty \ \text{ for all } \ \gamma \in \Gamma;$$

$$(2\text{-}s)\quad \Sigma = \{E \subseteq \Omega : E \cap \Omega_\gamma \in \Sigma \ \text{ for all } \ \gamma \in \Gamma\}; \ \text{ and}$$

$$(3\text{-}s)\quad \mu(E) = \textstyle\sum_{\gamma\in\Gamma} \mu(E \cap \Omega_\gamma), \quad E \in \Sigma.$$

Here, the right-side of (3-s) is defined in the usual way as

$$\sup\{\textstyle\sum_{\gamma\in\Delta} \mu(E \cap \Omega_\gamma) : \Delta \subseteq \Gamma \text{ is a finite subset }\}.$$

On the other hand, consider the following less restrictive condition than (3-s) :
$$(3\text{-}s^*)\quad \mu(E) = \textstyle\sum_{\gamma\in\Gamma} \mu(E \cap \Omega_\gamma), \qquad E \in \Sigma^f.$$
We will say that μ, or (Ω, Σ, μ), is *weakly decomposable* if there exists a partition $(\Omega_\gamma)_{\gamma\in\Gamma}$ of Ω into pairwise disjoint Σ-measurable sets for which (1-s), (2-s) and (3-s^*) hold. These are the "decomposable measures" in the sense of [11, Definition (19.25)], a terminology which is somewhat unfortunate. Indeed, every strictly localizable measure is clearly weakly decomposable. However, the converse does not always hold. The measure μ in Example 5.1 provides a counterexample.

Remark 2.3 We point out that a decomposable measure space (Ω, Σ, μ) in the sense of [20, p. 243], with the *additional* assumption that the given decomposition $(\Omega_\gamma)_{\gamma\in\Gamma}$ of Ω satisfies (2-s), is strictly localizable; see [2, 1.12.131, p. 96]. □

The connection between a measure μ being weakly decomposable and strictly localizable is as follows. Further properties of weakly decomposable measures are presented in Corollary 4.8 and Proposition 5.3.

Proposition 2.4 *Let (Ω, Σ, μ) be a measure space. The following assertions are equivalent.*

(i) *μ is strictly localizable.*
(ii) *μ is semifinite and weakly decomposable.*

Proof

(i) $\Longrightarrow$ (ii). Let $(\Omega_\gamma)_{\gamma\in\Gamma}$ be a Σ-measurable partition of Ω satisfying (1-s), (2-s) and (3-s). Since (3-s) clearly implies (3-s^*) as $\Sigma^f \subseteq \Sigma$, it follows that μ is weakly decomposable. That μ is also semifinite follows from [8, Theorem 211L(d),(e)].
(ii) $\Longrightarrow$ (i). Let $(\Omega_\gamma)_{\gamma\in\Gamma}$ be a Σ-measurable partition of Ω satisfying (1-s), (2-s) and (3-s^*). Fix $E \in \Sigma$. It is required to verify that

$$\mu(E) = \Sigma_{\gamma\in\Gamma}\, \mu(E \cap \Omega_\gamma); \tag{2.3}$$

see (3-s). Let $\mathcal{F}(\Gamma)$ denote the family of all finite subsets of Γ. Since μ is semifinite, it follows from (2.2) that

$$\mu(E) = \sup\{\mu(E \cap F) : F \in \Sigma^f\}$$

as well as

$$\mu\left(E \cap \left(\bigcup_{\gamma\in\Delta} \Omega_\gamma\right)\right) = \sup\left\{\mu\left(E \cap F \cap \left(\bigcup_{\gamma\in\Delta} \Omega_\gamma\right)\right) : F \in \Sigma^f\right\}, \quad \Delta \in \mathcal{F}(\Gamma).$$

This, and (3-s^*) with $E \cap F \in \Sigma^f$ in place of E (for any $F \in \Sigma^f$), yield that

$$\begin{aligned}
\mu(E) &= \sup_{F\in\Sigma^f} \mu(E \cap F) = \sup_{F\in\Sigma^f} \Sigma_{\gamma\in\Gamma}\, \mu(E \cap F \cap \Omega_\gamma) \\
&= \sup_{F\in\Sigma^f} \sup_{\Delta\in\mathcal{F}(\Gamma)} \Sigma_{\gamma\in\Delta}\, \mu(E \cap F \cap \Omega_\gamma) \\
&= \sup_{F\in\Sigma^f} \sup_{\Delta\in\mathcal{F}(\Gamma)} \mu(E \cap F \cap (\bigcup_{\gamma\in\Delta} \Omega_\gamma)) \\
&= \sup_{\Delta\in\mathcal{F}(\Gamma)} \sup_{F\in\Sigma^f} \mu(E \cap F \cap (\bigcup_{\gamma\in\Delta} \Omega_\gamma)) \\
&= \sup_{\Delta\in\mathcal{F}(\Gamma)} \mu(E \cap (\bigcup_{\gamma\in\Delta} \Omega_\gamma)) \\
&= \sup_{\Delta\in\mathcal{F}(\Gamma)} \Sigma_{\gamma\in\Delta}\, \mu(E \cap \Omega_\gamma) \\
&= \Sigma_{\gamma\in\Gamma}\, \mu(E \cap \Omega_\gamma).
\end{aligned}$$

This is precisely (2.3) and (i) is thereby established. □

Example 2.5 We point out that *not* every localizable measure is strictly localizable; see [8, 216E]. Such a localizable measure, being semifinite (per definition), cannot then be weakly decomposable; see Proposition 2.4. □

Let (Ω, Σ, μ) be a measure space. A $\mathbb{C}$-valued, Σ-measurable function f on Ω is said to be *μ-integrable* if $\int_\Omega |f|\, d\mu < \infty$, [11, Sections 10–12], [21, p. 24]. Let $\mathcal{L}^1(\mu)$ denote the vector space of all $\mathbb{C}$-valued, μ-integrable functions. We say that a Σ-measurable function $f : \Omega \longrightarrow \mathbb{C}$ is *μ-null* if $f = 0$ pointwise μ-a.e. on Ω. Such a function f is necessarily μ-integrable and $\int_\Omega |f|\, d\mu = 0$. Let $\mathcal{N}(\mu)$ denote the vector space of all μ-null functions. The quotient vector space $L^1(\mu) := \mathcal{L}^1(\mu)/\mathcal{N}(\mu)$ is a Banach space with respect to the norm

$$\|\cdot\|_{L^1(\mu)} : f + \mathcal{N}(\mu) \longmapsto \int_\Omega |f|\, d\mu, \qquad f \in \mathcal{L}^1(\mu).$$

A Σ-measurable function $\varphi : \Omega \longrightarrow \mathbb{C}$ is called *μ-essentially bounded* if there exists $a > 0$ such that $\{\omega \in \Omega : |\varphi(\omega)| > a\} \in \mathcal{N}_0(\mu)$, so that the ess.sup $|\varphi| < \infty$ (cf. [21, pp. 64–66]). We write $\varphi \in \mathcal{L}^\infty(\mu)$. Observing that every $\varphi \in \mathcal{N}(\mu)$ belongs to $\mathcal{L}^\infty(\mu)$ and satisfies ess.sup $|\varphi| = 0$, we can also consider the quotient space $L^\infty(\mu) := \mathcal{L}^\infty(\mu)/\mathcal{N}(\mu)$, which is a Banach space with respect to the essential supremum norm

$$\|\cdot\|_{L^\infty(\mu)} : \varphi + \mathcal{N}(\mu) \longmapsto \text{ess.sup}\, |\varphi|, \qquad \varphi \in \mathcal{L}^\infty(\mu).$$

Except when precise arguments are required, we identify each function f in $\mathcal{L}^1(\mu)$ (resp. in $\mathcal{L}^\infty(\mu)$) with its corresponding equivalence class $(f + \mathcal{N}(\mu))$ in $L^1(\mu)$ (resp. in $L^\infty(\mu)$), that is, μ-a.e. equal functions will be identified. To avoid confusion we may, at times, speak of individual μ-integrable or μ-essentially bounded functions as opposed to their corresponding equivalence classes. In Sect. 6, we shall show that these above spaces $L^1(\mu)$ and $L^\infty(\mu)$ are isomorphic to $L^1(\mu)$ and $L^\infty(\mu)$ as defined in [7].

Recall the definition of the canonical map $J_\mu : L^\infty(\mu) \longrightarrow (L^1(\mu))^*$ from the Introduction and that a measure μ is K-localizable precisely when J_μ is *surjective*.

Lemma 2.6 *The following statements hold for any measure space* (Ω, Σ, μ).

(i) *The measure μ is semifinite if and only if J_μ is injective, in which case J_μ is a linear isometry from $L^\infty(\mu)$ into $(L^1(\mu))^*$.*
(ii) *The measure μ is localizable if and only if J_μ is bijective, in which case J_μ is a linear isometry from $L^\infty(\mu)$ onto $(L^1(\mu))^*$.*
(iii) *The measure μ is localizable if and only if μ is semifinite and K-localizable.*
(iv) *If μ is localizable, then it is also K-localizable.*
(v) *Suppose that μ is semifinite. Then (Ω, Σ, μ) is Dedekind complete if and only if μ is K-localizable.*

Proof

(i) and (ii). See [8, Theorem 243G].
(iii) This is immediate from (i) and the definition of K-localizability.
(iv) Clear from (iii).
(v) The direction ($\Longrightarrow$) follows from (iv). Conversely, the other direction ($\Longleftarrow$) is a consequence of (iii). □

Observe that a measure space (Ω, Σ, μ) satisfies $L^1(\mu) := \mathcal{L}^1(\mu)/\mathcal{N}(\mu) = \{0\}$ if and only if $\Sigma^f = \mathcal{N}_0(\mu)$ if and only if Σ_0^f is empty if and only if μ is either identically 0 or totally infinite. The following result is immediate from this observation.

Lemma 2.7 *If a measure is either identically zero or totally infinite, then it is K-localizable.*

The converse statement to Lemma 2.7 is not always valid. For instance, the measure in Example B.1(iv) (denoted there by ι_2) of [18] is not semifinite but it is K-localizable. Moreover, it has the properties that its corresponding L^1-space is isometrically isomorphic to the sequence space ℓ^1 and it is not Dedekind complete. In particular, the measure is neither zero nor totally infinite.

Let (Ω, Σ, μ) be a measure space and recall the definition of Σ^f from (2.1). Note that Σ^f is a δ-ring of sets from Σ. Define

$$\widetilde{\Sigma}^f := \{K \subseteq \Omega : K \cap E \in \Sigma^f \text{ for all } E \in \Sigma^f\}.$$

Then $\widetilde{\Sigma}^f$ is a σ*-algebra* of subsets in Ω and satisfies

$$\Sigma^f \subseteq \Sigma \subseteq \widetilde{\Sigma}^f.$$

The sets in $\widetilde{\Sigma}^f$ are called *locally measurable*, [2, 1.12.133, p. 97]; see also [1, p. 35] if $\mathscr{S}$ there is the *ring* of sets Σ^f. Let $\mu_f := \mu|_{\Sigma^f}$. We can extend μ_f from the ring Σ^f to the set function on the σ-algebra $\widetilde{\Sigma}^f$ defined by

$$\widetilde{\mu}_f(K) := \sup\{\mu_f(K \cap E) : E \in \Sigma^f\} = \sup\{\mu_f(E) : E \in \Sigma^f, E \subseteq K\} \tag{2.4}$$

$$= \sup\{\mu(E) : E \in \Sigma^f, E \subseteq K\}, \quad K \in \widetilde{\Sigma}^f.$$

Then $\widetilde{\mu}_f : \widetilde{\Sigma}^f \longrightarrow [0, \infty]$ is necessarily a semifinite measure, so that the triplet $(\Omega, \widetilde{\Sigma}^f, \widetilde{\mu}_f)$ is a semifinite measure space; see, for example, [19, pp. 72–73]. The σ-algebra $\widetilde{\Sigma}^f$ also occurs in [8, Definition 211H]. If $\Sigma = \widetilde{\Sigma}^f$, then μ is called *saturated*, [2, 1.12.133, p. 97]. Every weakly decomposable (hence, every strictly localizable) measure is saturated. Indeed, let (Ω, Σ, μ) be weakly decomposable and $(\Omega_\gamma)_{\gamma\in\Gamma}$ be a partition of Ω satisfying (1-s), (2-s) and (3-s^*). Let $K \in \widetilde{\Sigma}^f$, that is, $K \subseteq \Omega$ satisfies $K \cap E \in \Sigma^f$ for all $E \in \Sigma^f$. From (1-s) it is clear that $(\Omega_\gamma)_{\gamma\in\Gamma} \subseteq \Sigma^f$ and hence, $K \cap \Omega_\gamma \in \Sigma^f \subseteq \Sigma$ for all $\gamma \in \Gamma$. It follows from (2-s)

that $K \in \Sigma$. This shows that $\widetilde{\Sigma}^f \subseteq \Sigma$. Since $\Sigma \subseteq \widetilde{\Sigma}^f$ always holds, we conclude that $\Sigma = \widetilde{\Sigma}^f$, that is, μ is saturated.

A saturated measure which is semifinite is called *locally determined*, [8, Definition 211H], [2, 1.12.135, p. 98]. Clearly every σ-finite measure is locally determined. More generally, every strictly localizable measure is locally determined, [8, Theorem 211L(d)].

Remark 2.8 In the terminology of [22, Definition 2.1], the subfamily $\Sigma^f \subseteq \Sigma$ is a *conditional σ-ring* and the triplet $(\Omega, \Sigma^f, \mu_f)$ is a measure space. Note that $\widetilde{\Sigma}^f$ and $\widetilde{\mu}_f$ are defined in [22, Definition 2.1], where the elements of $\widetilde{\Sigma}^f$ are called the measurable sets, and [22, Theorem 2.1] asserts that $\widetilde{\mu}_f$ is a measure. Moreover, [1, p. 302] explicitly says that $(\Omega, \widetilde{\Sigma}^f, \widetilde{\mu}_f)$ is a measure space in our sense and briefly discusses different concepts of measure spaces by P.R. Halmos [10] and by I. Segal [22]. As in [1, p. 302] we also refer to [3] for further comparison of the Halmos and Segal approaches. □

A set $F \in \Sigma$ is said to be *locally μ-null* if $F \cap E \in \mathcal{N}_0(\mu)$ for all $E \in \Sigma^f$, [11, Definition (20.11)]. Of course, every μ-null set is locally μ-null. Denote the family of all locally μ-null sets by $\mathcal{N}_0^{\mathrm{loc}}(\mu)$. A Σ-measurable function g on Ω will be called *locally μ-essentially bounded* if there exists a real number $a > 0$ such that $\{\omega \in \Omega : |g(\omega)| > a\} \in \mathcal{N}_0^{\mathrm{loc}}(\mu)$. The linear space of all such functions is denoted by $\mathcal{L}^\infty_{\mathrm{loc}}(\mu)$. Identifying two functions when they coincide off a locally μ-null set generates the Banach space $L^\infty_{\mathrm{loc}}(\mu)$ (of equivalence classes) in the usual way. We refer to [11, pp. 346–347] for these concepts, with the *warning* that the notation $\mathcal{L}^\infty(\mu)$ used there (resp. the term "μ-essentially bounded" function) is what we denote by $\mathcal{L}^\infty_{\mathrm{loc}}(\mu)$ above (resp. call a "locally μ-essentially bounded" function).

Lemma 2.9 *Let (Ω, Σ, μ) be a measure space.*

(i) *The measure $\widetilde{\mu}_f : \widetilde{\Sigma}^f \longrightarrow [0, \infty]$ defined in (2.4) is semifinite and satisfies*

$$\widetilde{\mu}_f = \mu_f \ \text{on} \ \Sigma^f \ \text{with} \ \widetilde{\mu_f} \leq \mu \ \text{and} \ \Sigma \subseteq \widetilde{\Sigma}^f\,. \tag{2.5}$$

(ii) *The measure μ is semifinite if and only if $\widetilde{\mu}_f = \mu$ on Σ.*

(iii) *Suppose that μ is saturated, i.e., $\widetilde{\Sigma}^f = \Sigma$. Then $\widetilde{\mu}_f = \mu$ if and only if μ is semifinite.*

(iv) *A set $F \in \Sigma$ is locally μ-null if and only if $F \in \mathcal{N}_0(\widetilde{\mu}_f)$. In other words,*

$$\mathcal{N}_0^{\mathrm{loc}}(\mu) = \Sigma \cap \mathcal{N}_0(\widetilde{\mu}_f).$$

In particular, if μ is saturated, then $\mathcal{N}_0(\widetilde{\mu}_f) = \mathcal{N}_0^{\mathrm{loc}}(\mu)$.

(v) *We have the containments*

$$\mathcal{N}_0(\mu) \subseteq \mathcal{N}_0^{\mathrm{loc}}(\mu) = \Sigma \cap \mathcal{N}_0(\widetilde{\mu}_f) \subseteq \mathcal{N}_0(\widetilde{\mu}_f). \tag{2.6}$$

(vi) *The measure μ is semifinite if and only if $\mathcal{N}_0^{\mathrm{loc}}(\mu) = \mathcal{N}_0(\mu)$.*

(vii) *If $H \in \mathcal{N}_0^{\mathrm{loc}}(\mu)$ and $f \in \mathcal{L}^1(\mu)$, then $\int_\Omega |f|\, \chi_H \, d\mu = 0$.*

Proof Parts (i) and (ii) are clear from the relevant definitions and the fact that $\widetilde{\mu_f}$ is always semifinite. Part (iii), which follows from (ii), is in [1, p. 302]. The definition of locally μ-null sets implies that $\mathcal{N}_0^{\rm loc}(\mu) = \Sigma \cap \mathcal{N}_0(\widetilde{\mu}_f)$ in (iv). If μ is saturated, then $\widetilde{\Sigma}^f = \Sigma$ and so $\mathcal{N}_0(\widetilde{\mu}_f) \subseteq \widetilde{\Sigma}_f = \Sigma$, that is, $\mathcal{N}_0(\widetilde{\mu}_f) \cap \Sigma = \mathcal{N}_0(\widetilde{\mu}_f)$. Hence, in this case, $\mathcal{N}_0(\widetilde{\mu}_f) = \mathcal{N}_0^{\rm loc}(\mu)$. For part (v), the inequality $\widetilde{\mu_f} \leq \mu$ on Σ (see (2.5)) gives $\mathcal{N}_0(\mu) \subseteq \mathcal{N}_0(\widetilde{\mu}_f)$, by which the first inclusion in (2.6) holds. The second inclusion in (2.6) is clear. For part (vi), see the proof of Theorem 2 in [23, Section 34].

To verify (vii), note that the sets $E_n := \{\omega \in \Omega : |f(\omega)| \geq \frac{1}{n}\}$, for $n \in \mathbb{N}$, and the set $E := \bigcup_{n=1}^{\infty} E_n = \{\omega \in \Omega : |f(\omega)| > 0\}$ all belong to Σ. Since $\mu(E_n) \leq n \int_\Omega |f|\, d\mu$ for $n \in \mathbb{N}$, it follows that $\{E_n\}_{n=1}^{\infty} \subseteq \Sigma^f$ and hence, as $H \subset \mathcal{N}_0^{\rm loc}(\mu)$, that $\{H \cap E_n\}_{n=1}^{\infty} \subseteq \mathcal{N}_0(\mu)$. Then $f \in \mathcal{L}^1(\mu)$ implies that $\int_\Omega |f|\chi_{H\cap E_n}\, d\mu = 0$ for $n \in \mathbb{N}$. Moreover, $|f|\chi_{H\cap E_n} \uparrow |f|\chi_{H\cap E}$ pointwise on Ω and so Fatou's Lemma implies that $\int_\Omega |f|\chi_H\, d\mu = 0$. □

Combining Proposition 2.4 and Lemma 2.9(vi) yields the following result.

Corollary 2.10 *Let (Ω, Σ, μ) be a measure space. The following assertions are equivalent.*

(i) *μ is strictly localizable.*
(ii) *μ is weakly decomposable and semifinite.*
(iii) *μ is weakly decomposable and $\mathcal{N}_0^{\rm loc}(\mu) = \mathcal{N}_0(\mu)$.*

Clearly a measure space (Ω, Σ, μ) is locally determined if and only if $(\Omega, \Sigma, \mu) = (\Omega, \widetilde{\Sigma}^f, \widetilde{\mu}_f)$ as identical measure spaces (see Lemma 2.9(ii),(iii) above). Every strictly localizable measure space is locally determined, [8, Theorem 211L(d)]. The measure mentioned in Example 2.5 provides an example of a complete, locally determined, localizable measure space which is not strictly localizable.

Example 2.11 Let Ω be a set containing at least two distinct points and Σ be a σ-algebra satisfying $\Sigma \subsetneqq 2^\Omega$. Define a measure $\mu : \Sigma \longrightarrow [0, \infty]$ by $\mu(\emptyset) = 0$ and $\mu(E) = \infty$ for every non-empty $E \in \Sigma$. Then μ is *not* semifinite. Since $\Sigma^f = \{\emptyset\}$, we have $\mu_f = 0$ on Σ^f. Consequently, $\widetilde{\Sigma}^f = 2^\Omega$ and $\widetilde{\mu}_f$ is the zero measure on 2^Ω. In particular, μ is *not* saturated. The inclusions in (2.6) are both strict because $\mathcal{N}_0(\mu) = \{\emptyset\}$, $\Sigma \cap \mathcal{N}_0(\widetilde{\mu}_f) = \Sigma$ and $\mathcal{N}_0(\widetilde{\mu}_f) = 2^\Omega$. □

In view of Lemma 2.9(iv) above, we have that $\mathcal{N}_0(\mu) \neq \Sigma \cap \mathcal{N}_0(\widetilde{\mu}_f)$ if and only if μ admits a locally μ-null set in Σ which is not μ-null. Such a measure μ can be found, for example, in [4, Ex.5 and 6, p. 105]. See also Example 5.1 below.

Let (Ω, Σ, μ) be a measure space. We say that μ, or (Ω, Σ, μ), is *S-localizable* if the quotient B.a. $\widetilde{\Sigma}^f/\mathcal{N}_0(\widetilde{\mu}_f)$ is Dedekind complete.

Remark 2.12 In the terminology of [22, Definition 2.6], "the measure space" $(\Omega, \Sigma^f, \mu_f)$ is called localizable if the B.a. $\widetilde{\Sigma}^f/\mathcal{N}_0(\widetilde{\mu}_f)$ is Dedekind complete. □

The measure μ in Example 2.11 above is S-localizable because the B.a. $\widetilde{\Sigma}^f/\mathcal{N}_0(\widetilde{\mu}_f) = \{\mathbf{0}\}$ is Dedekind complete. On the other hand, the measure μ is *not* localizable (as it is not semifinite) whereas μ *is* K-localizable because $L^1(\mu) = \{0\}$; see Lemma 2.7.

As ascertained earlier, the measure $\widetilde{\mu}_f : \widetilde{\Sigma}^f \longrightarrow [0, \infty]$ is always semifinite. This leads to part (i) of the following result.

Lemma 2.13 *The following statements hold for a measure space* (Ω, Σ, μ).

(i) *The measure* μ *is S-localizable if and only if the measure* $\widetilde{\mu}_f : \widetilde{\Sigma}^f \longrightarrow [0, \infty]$ *is localizable.*
(ii) *Suppose that* (Ω, Σ, μ) *is locally determined. Then* μ *is S-localizable if and only if it is localizable.*

Proof

(i) Clear as $\widetilde{\mu}_f$ is semifinite.
(ii) This is immediate from the definitions of localizable and S-localizable measures, the fact that μ is locally determined if and only if $(\Omega, \Sigma, \mu) = (\Omega, \widetilde{\Sigma}^f, \widetilde{\mu}_f)$ as identical measure spaces, and part (i). □

Let us provide a measure space which is not K-localizable but is S-localizable.

Example 2.14 Let $\Omega := [0, 1]$ and Σ be the Borel σ-algebra of Ω. Let $\mu : \Sigma \longrightarrow [0, \infty]$ denote the counting measure, which *is* clearly semifinite. Since $\mathcal{N}_0(\mu) = \{\emptyset\}$, the B.a. $\Sigma/\mathcal{N}_0(\mu) \cong \Sigma$ is not complete, that is, μ is *not* Dedekind complete. It is shown in Example B.2.(i) of [18], where μ is denoted by ι_3, that μ is *not* K-localizable. Note, however, that μ *is* S-localizable. Indeed, it is routine to check that Σ^f is the collection of all finite subsets of Ω and hence, that $\widetilde{\Sigma}^f = 2^{\Omega}$. Moreover, $\widetilde{\mu}_f : \widetilde{\Sigma}^f \longrightarrow [0, \infty]$ is precisely the counting measure on 2^{Ω} and so $\mathcal{N}_0(\widetilde{\mu}_f) = \{\emptyset\}$. Hence, $\widetilde{\Sigma}^f/\mathcal{N}_0(\widetilde{\mu}_f) \cong 2^{\Omega}$ is a complete B.a., that is, μ *is* S-localizable. □

Remark 2.15 A complete, locally determined measure which is not localizable is presented in [8, 216D]. In view of Lemma 2.13(ii) above such a measure is *not* S-localizable. □

3 Dedekind Complete Measure Spaces

Let (Ω, Σ, μ) be a measure space. If μ is localizable, then it is K-localizable (see Lemma 2.6(iv)). Recall from Sect. 2 that the localizability of μ is defined via the two conditions that μ is semifinite and that (Ω, Σ, μ) is Dedekind complete. The main aim of this section is to obtain the fact that, if (Ω, Σ, μ) is Dedekind complete, then μ is K-localizable *without* assuming the semifiniteness of μ (see Theorem 3.4 below). A basic tool is Theorem 3.3 which asserts that μ admits a semifinite decomposition whenever (Ω, Σ, μ) is Dedekind complete.

Given $H \in \Sigma$, define $\mu_H : \Sigma \cap H \longrightarrow [0, \infty]$ by

$$\mu_H(F) := \mu(F), \quad F \in \Sigma \cap H,$$

in which case $(H, \Sigma \cap H, \mu_H)$ is a measure space. It is clear that

$$\mathcal{N}_0(\mu_H) := \{F \in \Sigma \cap H : \mu_H(F) = 0\} = \{F \in \Sigma : F \subseteq H,\ \mu(F) = 0\}.$$

Moreover, $E \cap H \in \mathcal{N}_0(\mu_H)$ whenever $E \in \mathcal{N}_0(\mu)$. If $H \in \mathcal{N}_0(\mu)$, then μ_H is the zero measure on $\Sigma \cap H$.

Lemma 3.1 *Let (Ω, Σ, μ) be a Dedekind complete measure space and $H \in \Sigma$ be a non-μ-null set. Then the measure space $(H, \Sigma \cap H, \mu_H)$ is also Dedekind complete.*

Proof Take an arbitrary subfamily $\mathcal{F} \subseteq \Sigma \cap H$, in which case also $\mathcal{F} \subseteq \Sigma$. Let $\mathcal{F}_1$ denote $\mathcal{F}$ when it is considered as a subfamily of Σ. Select $G \in \Sigma$ such that $\pi_\mu(G) = \sup \pi_\mu(\mathcal{F}_1)$ in the Dedekind complete B.a. $\Sigma/\mathcal{N}_0(\mu)$. Since π_μ is a B.a. homomorphism, the infinite distributive law in $\Sigma/\mathcal{N}_0(\mu)$, [9, Proposition 313B], gives

$$\begin{aligned}
\pi_\mu(G \cap H) &= \pi_\mu(G) \wedge \pi_\mu(H) = (\sup \pi_\mu(\mathcal{F}_1)) \wedge \pi_\mu(H) \\
&= \sup \left\{\pi_\mu(E) \wedge \pi_\mu(H) : E \in \mathcal{F}_1\right\} = \sup \left\{\pi_\mu(E \cap H) : E \in \mathcal{F}_1\right\} \\
&= \sup \left\{\pi_\mu(E) : E \in \mathcal{F}_1\right\},
\end{aligned}$$

where the last equality follows from $\mathcal{F} \subseteq \Sigma \cap H$. Thus, $G \cap H$ is an essential supremum of $\mathcal{F}_1$ in Σ; see Lemma 2.1(ii). In other words,

$$\mu(E \setminus (G \cap H)) = 0 \ \text{ for all } \ E \in \mathcal{F}_1; \ \text{ and} \tag{3.1}$$

$$\text{if } F \in \Sigma \text{ satisfies } \mu(E \setminus F) = 0 \text{ for all } E \in \mathcal{F}_1, \text{ then } \mu((G \cap H) \setminus F) = 0. \tag{3.2}$$

By (3.1) we have $\mu_H(E \setminus (G \cap H)) = \mu(E \setminus (G \cap H)) = 0$ for all $E \in \mathcal{F} = \mathcal{F}_1$. Moreover, if $K \in \Sigma \cap H$ satisfies $\mu_H(E \setminus K) = 0$ for all $E \in \mathcal{F}$, then $\mu_H((G \cap H) \setminus K) = \mu((G \cap H) \setminus K) = 0$ by (3.2) with K in place of F. So, $G \cap H$ is an essential supremum of $\mathcal{F}$ in $\Sigma \cap H$, which completes the proof. □

Let Ω be a non-empty set. Consider two measure spaces $(\Omega_1, \Sigma_1, \mu_1)$ and $(\Omega_2, \Sigma_2, \mu_2)$ such that Ω_1 and Ω_2 are disjoint subsets of Ω with $\Omega = \Omega_1 \cup \Omega_2$. The subfamily $\Sigma \subseteq 2^\Omega$ consisting of all sets $E \subseteq \Omega$ satisfying $E \cap \Omega_j \in \Sigma_j$ for $j = 1, 2$ is a σ-algebra and the set function $\mu : E \longmapsto \mu_1(E \cap \Omega_1) + \mu_2(E \cap \Omega_2)$ on Σ is a measure. In this case, we call (Ω, Σ, μ) the *direct sum* of $(\Omega_1, \Sigma_1, \mu_1)$ and $(\Omega_2, \Sigma_2, \mu_2)$; see [5, 61G].

We record the following result; its routine proof will be omitted.

Lemma 3.2 *Given are a measure space (Ω, Σ, μ) and a non-μ-null set $H \in \Sigma$ such that $\mu(\Omega \setminus H) > 0$. Then (Ω, Σ, μ) is the direct sum of the two measure spaces $(H, \Sigma \cap H, \mu_H)$ and $(\Omega \setminus H,\ \Sigma \cap (\Omega \setminus H),\ \mu_{\Omega \setminus H})$.*

Suppose that a measure space (Ω, Σ, μ) is either semifinite or totally infinite. Then it is not possible to express μ as the (direct) sum of a non-zero semifinite measure and a non-zero totally infinite measure based on two disjoint sets from Σ. However, for certain other measures μ such a direct sum decomposition is possible.

Theorem 3.3 *Let (Ω, Σ, μ) be a Dedekind complete measure space which is neither semifinite nor totally infinite. Then Σ_0^f is non-empty and an essential supremum $\Omega^s \in \Sigma$ of Σ_0^f exists in Σ satisfying the following conditions:*

(a) $\mu_{\Omega^s} : \Sigma \cap \Omega^s \longrightarrow [0, \infty]$ *is localizable;*
(b) $\mu_{\Omega \setminus \Omega^s} : \Sigma \cap (\Omega \setminus \Omega^s) \longrightarrow [0, \infty]$ *is totally infinite; and*
(c) (Ω, Σ, μ) *is the direct sum of the semifinite measure space* $(\Omega^s, \Sigma \cap \Omega^s, \mu_{\Omega^s})$ *and the totally infinite measure space* $(\Omega \setminus \Omega^s, \Sigma \cap (\Omega \setminus \Omega^s), \mu_{\Omega \setminus \Omega^s})$.

Moreover, both Ω^s and $\Omega \setminus \Omega^s$ are non-μ-null sets .

Proof Since μ is neither totally infinite nor non-zero, the subfamily $\Sigma_0^f \subseteq \Sigma$ is non-empty. The Dedekind completeness of $\Sigma/\mathcal{N}_0(\mu)$ guarantees that an essential supremum $\Omega^s \in \Sigma$ of Σ_0^f exists in Σ, that is, $\pi_\mu(\Omega^s) = \sup \pi_\mu(\Sigma_0^f)$ in $\Sigma/\mathcal{N}_0(\mu)$; see Lemma 2.1(ii),(iii). Select any $A \in \Sigma_0^f$ in which case $\mathbf{0} < \pi_\mu(A)$. Since $\pi_\mu(A) \le \pi_\mu(\Omega^s)$, it follows that $\mu(\Omega^s) > 0$. To prove that the measure $\mu_{\Omega^s} : \Sigma \cap \Omega^s \longrightarrow [0, \infty]$ is semifinite, take any set $F \in \Sigma$ with $F \subseteq \Omega^s$ satisfying $\mu(F) = \infty$. By the infinite distributive law in the B.a. $\Sigma/\mathcal{N}_0(\mu)$, we have

$$\begin{aligned}\pi_\mu(F) &= \pi_\mu(F \cap \Omega^s) = \pi_\mu(F) \wedge \pi_\mu(\Omega^s) = \pi_\mu(F) \wedge (\sup\{\pi_\mu(E) : E \in \Sigma_0^f\}) \\ &= \sup\{\pi_\mu(F) \wedge \pi_\mu(E) : E \in \Sigma_0^f\} = \sup\{\pi_\mu(F \cap E) : E \in \Sigma_0^f\}\end{aligned}$$

because π_μ is a B.a. homomorphism. Accordingly, there exists $E \in \Sigma_0^f$ such that $\pi_\mu(F \cap E) > \mathbf{0}$ because $\pi_\mu(F) > \mathbf{0}$ in the B.a. $\Sigma/\mathcal{N}_0(\mu)$. Thus

$$0 < \mu_{\Omega^s}(F \cap E) = \mu(F \cap E) \le \mu(E) < \infty,$$

which verifies that μ_{Ω^s} is semifinite. So, (a) holds because Lemma 3.1, with $H := \Omega^s$, gives that $(\Omega^s, \Sigma \cap \Omega^s, \mu_{\Omega^s})$ is Dedekind complete.

To verify (b), observe that $\mu(\Omega \setminus \Omega^s) > 0$ via (a) because μ is not semifinite. To prove that $\mu_{\Omega \setminus \Omega^s}$ is totally infinite assume, on the contrary, that $0 < \mu_{\Omega \setminus \Omega^s}(F) < \infty$, equivalently $F \in \Sigma_0^f$, for some $F \in \Sigma$ with $F \subseteq (\Omega \setminus \Omega^s)$. Then $F \in \Sigma_0^f$ implies that

$$\pi_\mu(F) \le \sup \pi_\mu(\Sigma_0^f) = \pi_\mu(\Omega^s),$$

from which it follows (via the definition of essential supremum) that $(F \setminus \Omega^s) \in \mathcal{N}_0(\mu)$. Hence, $\mu(F) = \mu(F \setminus \Omega^s) = 0$ because $F = F \setminus \Omega^s$. This contradicts the assumption that $F \in \Sigma_0^f$. Therefore $\mu_{\Omega \setminus \Omega^s}(F) \in \{0, \infty\}$ for every $F \in \Sigma \cap (\Omega \setminus \Omega^s)$ and so the non-zero measure $\mu_{\Omega \setminus \Omega^s}$ is totally infinite. We have thereby verified (b).

Finally, let us establish (c). It is clear that Ω is the disjoint union $\Omega^s \cup (\Omega \setminus \Omega^s)$. Moreover, a set $E \subseteq \Omega$ belongs to Σ if and only if $E \cap \Omega^s \in \Sigma \cap \Omega^s$ and $E \cap (\Omega \setminus \Omega^s) \in \Sigma \cap (\Omega \setminus \Omega^s)$. Since

$$\mu(E) = \mu(E \cap \Omega^s) + \mu(E \cap (\Omega \setminus \Omega^s)) = \mu_{\Omega^s}(E \cap \Omega^s) + \mu_{\Omega \setminus \Omega^s}(E \cap (\Omega \setminus \Omega^s)),$$

for all $E \in \Sigma$, condition (c) is satisfied, which completes the proof. □

Theorem 3.4 *If a measure space (Ω, Σ, μ) is Dedekind complete, then the canonical map $J_\mu : L^\infty(\mu) \longrightarrow (L^1(\mu))^*$ is surjective, that is, μ is K-localizable.*

Proof If μ is totally infinite, then μ is already K-localizable (see Lemma 2.7). In the case when μ is semifinite, Lemma 2.6(v) gives that μ is K-localizable. So, assume that μ is neither semifinite nor totally infinite. By Theorem 3.3 an essential supremum $\Omega^s \in \Sigma$ of Σ_0^f in Σ exists and satisfies conditions (a), (b) and (c) of that theorem. For simplicity of presentation, let us write $\mu_s := \mu_{\Omega^s}$. If $f \in L^1(\mu)$, then $f\chi_{\Omega \setminus \Omega^s} = 0$ (μ-a.e.) via (b). That is,

$$f = f\chi_{\Omega^s} \quad (\mu\text{-a.e.) for all } f \in L^1(\mu).$$

Accordingly, the restriction map $f \longmapsto f|_{\Omega^s}$ is a linear isometry from $L^1(\mu)$ onto $L^1(\mu_s)$. Its inverse $T : L^1(\mu_s) \longrightarrow L^1(\mu)$ maps each $g \in L^1(\mu_s)$ to the extension $(g + 0 \cdot \chi_{\Omega \setminus \Omega^s})$ of g by zero on $\Omega \setminus \Omega^s$ to the whole set Ω. Let $\xi \in (L^1(\mu))^*$. Then $\xi \circ T \in (L^1(\mu_s))^*$. By condition (a), the measure μ_s is localizable and hence, its corresponding canonical map $J_{\mu_s} : L^\infty(\mu_s) \longrightarrow (L^1(\mu_s))^*$ is bijective (see Lemma 2.6(ii) with μ_s in place of μ). So, there exists $\varphi \in L^\infty(\mu_s)$ such that

$$\langle g, \xi \circ T \rangle = \int_{\Omega^s} g\varphi \, d\mu_s, \qquad g \in L^1(\mu_s). \tag{3.3}$$

Write $\widetilde{\varphi} := \varphi + 0 \cdot \chi_{\Omega \setminus \Omega^s} \in L^\infty(\mu)$. Then

$$\langle f, \xi \rangle = \int_\Omega f\widetilde{\varphi} \, d\mu, \qquad f \in L^1(\mu). \tag{3.4}$$

In fact, since $T(f|_{\Omega^s}) = f$ in $L^1(\mu)$, we obtain (3.4) from (3.3) as

$$\langle f, \xi \rangle = \langle T(f|_{\Omega^s}), \xi \rangle = \langle f|_{\Omega^s}, \xi \circ T \rangle = \int_{\Omega^s} f\varphi \, d\mu_s = \int_\Omega f\widetilde{\varphi} \, d\mu.$$

So, J_μ is surjective. □

Let us provide some examples to illustrate Theorems 3.3 and 3.4.

Example 3.5

(i) Let $\Omega := \{1, 2, 3\}$ and $\Sigma := 2^{\Omega}$. Define a measure $\mu : \Sigma \longrightarrow [0, \infty]$ by $\mu(\emptyset) := 0$, $\mu(\{1\}) := 0$ and $\mu(E) := \infty$ for all $E \in \Sigma \setminus \{\emptyset, \{1\}\}$. Then $\mathcal{N}_0(\mu) = \Sigma^f = \{\emptyset, \{1\}\}$. Since Σ_0^f is empty, the measure μ is not semifinite but it is totally infinite. The measure space (Ω, Σ, μ) is clearly Dedekind complete and $L^1(\mu) = \{0\}$. The canonical map $J_\mu : L^\infty(\mu) \longrightarrow (L^1(\mu))^*$ is not injective because μ is not semifinite (via Lemma 2.6(i)). Of course, this can be seen directly because $J_\mu(\chi_{\{2\}}) = 0$ but $\chi_{\{2\}} \in L^\infty(\mu) \setminus \{0\}$. The equality $L^1(\mu) = \{0\}$ directly implies the K-localizability of μ whereas Theorem 3.4 above gives the same conclusion.

(ii) Let $\Omega_1 := [0, 1]$ and let μ_1 be Lebesgue measure on the σ-algebra Σ_1 of all Lebesgue measurable subsets in Ω_1. Define $\Omega_2 := \{2\}$ and $\Sigma_2 := \{\emptyset, \{2\}\}$. Let $\mu_2 : \Sigma_2 \longrightarrow [0, \infty]$ be the measure given by $\mu_2(E) := 0$ if $E = \emptyset$ and $\mu_2(E) := \infty$ if $E = \{2\}$. Let $\Omega := \Omega_1 \cup \Omega_2$,

$$\Sigma := \{E \subseteq \Omega : (E \cap \Omega_j) \in \Sigma_j \text{ for } j = 1, 2\}$$

and define a measure $\mu : \Sigma \longrightarrow [0, \infty]$ by

$$\mu(E) := \mu_1(E \cap \Omega_1) + \mu_2(E \cap \Omega_2), \qquad E \in \Sigma.$$

Thus, the measure space (Ω, Σ, μ) is the direct sum of $(\Omega_1, \Sigma_1, \mu_1)$ and $(\Omega_2, \Sigma_2, \mu_2)$, [5, 61G]. The measure μ is not semifinite because μ_2 is not semifinite.

The quotient B.a. $\Sigma_1/\mathcal{N}_0(\mu_1)$ is Dedekind complete as the *finite* measure μ_1 is localizable. On the other hand, the quotient B.a. $\Sigma_2/\mathcal{N}_0(\mu_2)$, which equals $\{\mathbf{0}, \mathbf{1}\}$, is Dedekind complete. Hence, (Ω, Σ, μ) is Dedekind complete as $\Sigma/\mathcal{N}_0(\mu)$ is isomorphic to the product B.a. $(\Sigma_1/\mathcal{N}_0(\mu_1)) \times (\Sigma_2/\mathcal{N}_0(\mu_2))$; see [5, 41K and 61H].

The set Ω^s obtained via Theorem 3.3 for this example equals $[0, 1] = \Omega_1$ and $\Omega \setminus \Omega^s = \{2\} = \Omega_2$ with $\mu(\Omega \setminus \Omega^s) = \infty$. Also note that $\Sigma_0^f = \{E \in \Sigma : 2 \notin E \text{ and } \mu(E) > 0\}$.

We note that $L^1(\mu)$, which is isomorphic to $L^1(\mu_1)$ as seen from the proof of Theorem 3.4, is equal to the vector space of all Σ-measurable functions f satisfying $f(2) = 0$ and $\int_0^1 |f| \, d\mu_1 < \infty$.

It is clear that μ is K-localizable as (Ω, Σ, μ) is Dedekind complete (see Theorem 3.4) and that J_μ is *not* injective as μ is not semifinite (see Lemma 2.6(i)). It is interesting to also verify this directly. The space $L^1(\mu)$ is described above. Fix any $\varphi \in L^\infty(\mu_1) = L^\infty([0, 1])$. For each $a \in \mathbb{C} \setminus \{0\}$ define $\varphi_a := \varphi + a\chi_{\{2\}}$ by extending φ from $[0, 1]$ to $\Omega = [0, 1] \cup \{2\}$ via $\varphi(2) := 0$. Then $\varphi_a - \varphi = a\chi_{\{2\}}$ is not μ-a.e. zero. But φ and φ_a induce the same element of $(L^1(\mu))^*$, namely

$$f \longmapsto \int_\Omega f\varphi \, d\mu = \int_\Omega f\varphi_a \, d\mu, \qquad f \in L^1(\mu),$$

because $f(2) = 0$ and so $f\varphi = f\varphi_a$ on Ω. In other words, $J_\mu(\varphi) = J_\mu(\varphi_a)$ whereas φ and φ_a are *different* elements of $L^\infty(\mu)$. □

Note that the converse statement of Theorem 3.4 above fails to hold, in general. In fact, the measures in Example B.1(i),(iv) of [18] are K-localizable but their corresponding measure spaces are not Dedekind complete. In short, Dedekind completeness is sufficient but not necessary for K-localizability.

We now make an improvement of Theorems 3.3 and 3.4 by obtaining the same conclusions under a weaker assumption.

Let (Ω, Σ, μ) be a measure space. Its completion is denoted by $(\Omega, \widehat{\Sigma}, \widehat{\mu})$ (see, [8, 212C], for example). We write $(\widehat{\Sigma})^f := \{E \in \widehat{\Sigma} : \widehat{\mu}(E) < \infty\}$ and $(\widehat{\Sigma})^f_0 := \{E \in \widehat{\Sigma} : 0 < \widehat{\mu}(E) < \infty\} = (\widehat{\Sigma})^f \setminus \mathcal{N}_0(\widehat{\mu})$. A.C. Zaanen has presented another notion of localizable measure in [23], where we point out that the measurable sets are always considered with respect to the completion of the measure space (see Sections 4–9 of Chapter 2). To be precise, we say that μ, or (Ω, Σ, μ), is *Z-localizable* if, for every subfamily $\mathcal{F} \subseteq (\widehat{\Sigma})^f$, the corresponding subset $\pi_{\widehat{\mu}}(\mathcal{F})$ possesses a supremum in the quotient B.a. $\widehat{\Sigma}/\mathcal{N}_0(\widehat{\mu})$; see [23, Definition, §35]. Here, take note that such a supremum may not belong to $\pi_{\widehat{\mu}}((\widehat{\Sigma})^f)$. Since it is simpler to deal with (Ω, Σ, μ) instead of its completion $(\Omega, \widehat{\Sigma}, \widehat{\mu})$, let us point out that equivalent conditions are available in terms of (Ω, Σ, μ).

Lemma 3.6 *The following conditions for a measure space (Ω, Σ, μ) are equivalent.*

(a) *μ is Z-localizable.*
(b) *For every subfamily $\mathcal{E} \subseteq \Sigma^f$, its corresponding subset $\pi_\mu(\mathcal{E})$ possesses a supremum in the quotient B.a. $\Sigma/\mathcal{N}_0(\mu)$.*
(c) *Every subfamily $\mathcal{E} \subseteq \Sigma^f$ possesses an essential supremum in Σ.*

It is routine to prove (a) $\Longleftrightarrow$ (b) by exploring the relationship between (Ω, Σ, μ) and $(\Omega, \widehat{\Sigma}, \widehat{\mu})$. So we omit the proof. The equivalence (b) $\Longleftrightarrow$ (c) is a consequence of Lemma 2.1(ii).

Thanks to the equivalence (a) $\Longleftrightarrow$ (c) in Lemma 3.6 we shall, henceforth, use (c) as an alternative definition of Z-localizable measure spaces.

Lemma 3.7 *The following statements hold for a measure space (Ω, Σ, μ).*

(i) *If (Ω, Σ, μ) is Dedekind complete, then it is Z-localizable.*
(ii) *If the measure space (Ω, Σ, μ) is semifinite and Z-localizable, then it is Dedekind complete. In other words, a semifinite measure space is Dedekind complete if and only if it is Z-localizable.*
(iii) *If μ is totally infinite, then it is Z-localizable.*

Proof

(i) This is clear from the respective definition.
(ii) See the argument in [19, p. 74].
(iii) This follows from the fact that μ satisfies $\Sigma^f = \mathcal{N}_0(\mu)$. □

The measure space given in [18, Example B.1(iv)] is neither semifinite, totally infinite nor Dedekind complete (as was discussed after Lemma 2.7). Moreover, it possesses a countable subset containing all sets from Σ_0^f, which implies its Z-localizability. Thus, the converse statements of parts (i) and (iii) are invalid in general.

The following result is an improvement of Theorem 3.3.

Proposition 3.8 *Let* (Ω, Σ, μ) *be a* Z*-localizable measure space, which is neither semifinite nor totally infinite. Then* Σ_0^f *is non-empty and possesses an essential supremum* $\Omega^s \in \Sigma$ *in* Σ *satisfying conditions* (a), (b) *and* (c) *in Theorem* 3.3.

Proof Since μ is neither identically zero nor totally infinite, it follows that Σ_0^f is non-empty. The Z-localizability of μ ensures that Σ_0^f possesses an essential supremum $\Omega^s \in \Sigma$ in Σ. We can verify that $\mu_{\Omega^s} : \Sigma \cap \Omega^s \longrightarrow [0, \infty]$ is semifinite exactly as in the proof of Theorem 3.3. The proof of Lemma 3.1 can be adapted to obtain the fact that μ_{Ω^s} is Z-localizable. So, $(\Omega^s, \Sigma \cap \Omega^s, \mu_{\Omega^s})$ is Dedekind complete by Lemma 3.7(ii), which establishes condition (a).

Conditions (b) and (c) can be verified as in the proof of Theorem 3.3. □

Corollary 3.9 *Let* (Ω, Σ, μ) *be any measure space. Then the following statements hold.*

(i) *If* (Ω, Σ, μ) *is* Z*-localizable, then* (Ω, Σ, μ) *is* K*-localizable.*
(ii) *Assume that* μ *is semifinite. Then the following conditions are equivalent.*

  (a) (Ω, Σ, μ) *is localizable.*
  (b) (Ω, Σ, μ) *is Dedekind complete.*
  (c) (Ω, Σ, μ) *is* Z*-localizable.*
  (d) (Ω, Σ, μ) *is* K*-localizable.*

Proof

(i) The proof of Theorem 3.4 remains valid if we apply Proposition 3.8 instead of Theorem 3.3.
(ii) By the definition of localizability we have (a) $\Longleftrightarrow$ (b). As formally stated in Lemma 3.7(ii), we have (b) $\Longleftrightarrow$ (c). The implication (c) $\Longrightarrow$ (d) is exactly part (i) above. Finally Lemma 2.6(v) asserts the equivalence (b) $\Longleftrightarrow$ (d). So, part (ii) is established. □

In Sect. 5, we shall provide an example of measure which is K-localizable but not Z-localizable. So, Z localizability is sufficient but not necessary for K-localizability.

Remark 3.10

(i) In Remark 2.15 we referred to the complete, locally determined, non-localizable measure μ due to D. Fremlin, [8, 216D]. Since μ is necessarily semifinite, it is not Z-localizable via Corollary 3.9(ii). Alternatively, that μ is not Z-localizable is also proved directly in part (f) of [8, 216D].

(ii) Let ν_1 and ν_2 be measures on the same measurable space (Ω, Σ). According to [13], we say that ν_2 is S-singular with respect to ν_1, denoted by $\nu_2 S \nu_1$, if for each $E \in \Sigma$, there exists a set $F \in \Sigma \cap E$ such that $\nu_2(E) = \nu_2(F)$ and $F \in \mathcal{N}_0(\nu_1)$.

Now, let (Ω, Σ, μ) be *any* measure space. It was shown by N.Y. Luther that there *always* exists a unique pair of measures $\mu_j : \Sigma \longrightarrow [0, \infty]$ for $j = 1, 2$, with μ_1 semifinite and $\mu_2(\Sigma) \subseteq \{0, \infty\}$, which satisfy $\mu_1 \, S \, \mu_2$ and $\mu_2 \, S \, \mu_1$ and such that $\mu = \mu_1 + \mu_2$, [17, Theorem 1]. It will be shown in Sect. 6 that μ_1 equals the measure μ_{sf} given in [8, 213X(c)].

Recall that Theorem 3.3 (or Proposition 3.8) provide the decomposition $\mu = \mu_{\Omega^s} + \mu_{\Omega \setminus \Omega^s}$, where μ_{Ω^s} and $\mu_{\Omega \setminus \Omega^s}$ are extended to Σ in the natural way. We can include the case when μ is semifinite and the case when μ is totally infinite in such a decomposition. It can be verified that $\mu_{\Omega^s} = \mu_1$ and $\mu_{\Omega \setminus \Omega^s} = \mu_2$. The K-localizability of μ is obtained via the localizability of μ_{Ω^s} (see the proof of Theorem 3.4). Luther's decomposition alone does *not* imply the K-localizability of μ. In fact, Example 2.14 provides a non-K-localizable, semifinite measure μ, that is, $\mu = \mu_1 + \mu_2$ with $\mu_2 = 0$. □

4 The Relationship Between K-localizable and S-localizable Measure Spaces

Given a measure space (Ω, Σ, μ), the main aim of this section is to establish the relationship between its K-localizability and its S-localizability. Since the two different measure spaces (Ω, Σ, μ) and $(\Omega, \widetilde{\Sigma}^f, \widetilde{\mu}_f)$ are involved, the definitions of K-localizability and S-localizability (see Sect. 2) do not provide any obvious relation between the two concepts. Our main tool is the fact that the Banach spaces $L^1(\mu)$ and $L^1(\widetilde{\mu}_f)$ are isometrically isomorphic, which is established in Lemma 4.2 below. This fact will lead us to Theorem 4.3 asserting that K-localizability necessarily implies S-localizability. The converse holds in the special case when μ is saturated, that is, when $\widetilde{\Sigma}^f = \Sigma$ (cf. Theorem 4.4) but not in general; see Example 2.14.

Since we are dealing with different measures μ and $\widetilde{\mu}_f$, we shall use the precise definitions (as explained in Sect. 2). Namely,

$$L^1(\mu) := \mathcal{L}^1(\mu)/\mathcal{N}(\mu) \text{ and } L^\infty(\mu) := \mathcal{L}^\infty(\mu)/\mathcal{N}(\mu),$$

as well as

$$L^1(\widetilde{\mu}_f) := \mathcal{L}^1(\widetilde{\mu}_f)/\mathcal{N}(\widetilde{\mu}_f) \text{ and } L^\infty(\widetilde{\mu}_f) := \mathcal{L}^\infty(\widetilde{\mu}_f)/\mathcal{N}(\widetilde{\mu}_f).$$

In other words, we distinguish between individual functions and their corresponding equivalence classes. Accordingly, the canonical maps J_μ and $J_{\widetilde{\mu}_f}$ are expressed as

$$\langle f + \mathcal{N}(\mu), J_\mu(\varphi + \mathcal{N}(\mu))\rangle := \int_\Omega f\varphi\, d\mu, \quad f \in \mathcal{L}^1(\mu),\ \varphi \in \mathcal{L}^\infty(\mu) \tag{4.1}$$

and

$$\langle h + \mathcal{N}(\widetilde{\mu}_f), J_{\widetilde{\mu}_f}(\psi + \mathcal{N}(\widetilde{\mu}_f))\rangle := \int_\Omega h\psi\, d\widetilde{\mu}_f, \quad h \in \mathcal{L}^1(\widetilde{\mu}_f),\ \psi \in \mathcal{L}^\infty(\widetilde{\mu}_f), \tag{4.2}$$

respectively.

Define a semifinite measure $\mu_{sf} : \Sigma \longrightarrow [0, \infty]$ by

$$\mu_{sf}(E) := \sup\{\mu(E \cap F) : F \in \Sigma^f\} = \sup\{\mu(F) : F \in \Sigma^f, F \subseteq E\}, \quad E \in \Sigma;$$

see [8, 213X(c)]. This definition and Lemma 2.9(i) imply (recall that $\Sigma^f \subseteq \Sigma \subseteq \widetilde{\Sigma}^f$)

$$\mu_{sf} = \widetilde{\mu}_f|_\Sigma \le \mu \quad \text{on } \Sigma. \tag{4.3}$$

Lemma 4.1 *The following statements hold for every measure space* (Ω, Σ, μ).

(i) *Every* μ*-integrable function* h *is* μ_{sf}*-integrable and*

$$\int_E h\, d\mu = \int_E h\, d\mu_{sf}, \quad E \in \Sigma,\ h \in \mathcal{L}^1(\mu). \tag{4.4}$$

(ii) *If* g *is a* μ_{sf}*-integrable function, then there is a* μ*-integrable function* h *such that* $(g - h) \in \mathcal{N}(\mu_{sf})$ *and*

$$\int_E g\, d\mu_{sf} = \int_E h\, d\mu, \quad E \in \Sigma. \tag{4.5}$$

Proof

(i) This is in [8, 213X(c)(ii)], which is a consequence of the facts that $\mu = \mu_{sf}$ on Σ^f and that h vanishes μ-a.e. off a μ-σ-finite set.
(ii) There exists $h \in \mathcal{L}^1(\mu)$ such that $(g - h) \in \mathcal{N}(\mu_{sf})$ because g vanishes μ_{sf}-a.e. off a μ_{sf}-σ-finite set and because $\mu = \mu_{sf}$ on Σ^f; see [8, 213X(c)(iv)]. So, $\int_E g\, d\mu_{sf} = \int_E h\, d\mu_{sf}$ for $E \in \Sigma$. This yields (4.5) via (4.4). □

Given a measure space (Ω, Σ, μ), let $h \in \mathcal{L}^1(\mu)$. Then $h \in \mathcal{L}^1(\mu_{sf})$ by Lemma 4.1(i) above. Recall from (4.3) that $\widetilde{\mu}_f$ is an extension of μ_{sf} from Σ to the larger σ-algebra $\widetilde{\Sigma}^f$. The Σ-measurable function h is also $\widetilde{\Sigma}^f$-measurable. When we consider h as $\widetilde{\Sigma}^f$-measurable, it will be denoted by $\widetilde{h}$. It is routine to verify that $\widetilde{h} \in \mathcal{L}^1(\widetilde{\mu}_f)$ and that

$$\int_E \widetilde{h}\, d\widetilde{\mu}_f = \int_E h\, d\mu_{sf}, \quad E \in \Sigma \subseteq \widetilde{\Sigma}^f. \tag{4.6}$$

Indeed, apply the Monotone Convergence Theorem to the special case when $h \geq 0$ and then express h as a (complex) linear combination of four positive functions for the general case. By (4.4) and (4.6) we have

$$\int_E \widetilde{h}\, d\widetilde{\mu}_f = \int_E h\, d\mu, \quad E \in \Sigma \subseteq \widetilde{\Sigma}^f. \tag{4.7}$$

Define a linear map $T : \mathcal{L}^1(\mu) \longrightarrow \mathcal{L}^1(\widetilde{\mu}_f)$ by $T(h) := \widetilde{h}$ for $h \in \mathcal{L}^1(\mu)$. Since $T(\mathcal{N}(\mu)) \subseteq \mathcal{N}(\widetilde{\mu}_f)$ by (4.7), we can define a linear map $\widehat{T} : L^1(\mu) \longrightarrow L^1(\widetilde{\mu}_f)$ associated with T via

$$\widehat{T}(h + \mathcal{N}(\mu)) := \widetilde{h} + \mathcal{N}(\widetilde{\mu}_f), \quad h \in \mathcal{L}^1(\mu). \tag{4.8}$$

Lemma 4.2 *Given any measure space (Ω, Σ, μ), the linear map $\widehat{T} : L^1(\mu) \longrightarrow L^1(\widetilde{\mu}_f)$ defined by* (4.8) *is a surjective linear isometry.*

Proof Let $h \in \mathcal{L}^1(\mu)$. Then, by (4.7) with $E := \Omega$ and with $|h|$ in place of h, we have that $\int_\Omega |\widetilde{h}|\, d\widetilde{\mu}_f = \int_\Omega |h|\, d\mu$. So,

$$\begin{aligned}\|\widehat{T}(h + \mathcal{N}(\mu))\|_{L^1(\widetilde{\mu}_f)} &= \|\widetilde{h} + \mathcal{N}(\widetilde{\mu}_f)\|_{L^1(\widetilde{\mu}_f)} \\ &= \int_\Omega |\widetilde{h}|\, d\widetilde{\mu}_f = \int_\Omega |h|\, d\mu = \|h + \mathcal{N}(\mu)\|_{L^1(\mu)},\end{aligned}$$

and hence, $\widehat{T}$ is a linear isometry.

To prove the surjectivity of $\widehat{T}$ we may assume that there exists a set $K \in \widetilde{\Sigma}^f$ for which $0 < \widetilde{\mu}_f(K) < \infty$. In fact, there exists no such set K if and only if $L^1(\mu) = \{0\}$ and $L^1(\widetilde{\mu}_f) = \{0\}$, in which case $\widehat{T}$ is zero and hence, is surjective.

So, take a set $K \in \widetilde{\Sigma}^f$ with $0 < \widetilde{\mu}_f(K) < \infty$. Via (2.4), select an increasing sequence $\{E_n\}_{n=1}^\infty$ in Σ^f such that $\widetilde{\mu}_f(K) = \sup_{n\in\mathbb{N}} \mu_f(K \cap E_n)$. With $F := \bigcup_{n=1}^\infty (K \cap E_n) \in \Sigma$, we have $F \subseteq K$ and $\widetilde{\mu}_f(K \setminus F) = 0$. As before, let $\widetilde{\chi}_F$ denote χ_F when it is considered to be $\widetilde{\Sigma}^f$-measurable. Then $\widetilde{\mu}_f(K \setminus F) = 0$ yields $(\chi_K - \widetilde{\chi}_F) \in \mathcal{N}(\widetilde{\mu}_F)$, so that

$$\widetilde{\chi}_F + \mathcal{N}(\widetilde{\mu}_f) = \chi_K + \mathcal{N}(\widetilde{\mu}_f) \text{ in } L^1(\widetilde{\mu}_f). \tag{4.9}$$

On the other hand, it follows from (4.8), with $h := \chi_F$, that

$$\widehat{T}(\chi_F + \mathcal{N}(\mu)) = \widetilde{\chi}_F + \mathcal{N}(\widetilde{\mu}_f). \tag{4.10}$$

With $\mathcal{R}(\widehat{T})$ denoting the range of $\widehat{T}$, it follows via (4.9) and (4.10) that

$$\chi_K + \mathcal{N}(\widetilde{\mu}_f) = \widehat{T}(\chi_F + \mathcal{N}(\mu)) \in \mathcal{R}(\widehat{T}).$$

Thus, we can conclude that

$$\operatorname{span}\{\chi_K + \mathcal{N}(\widetilde{\mu}_f) : K \in \widetilde{\Sigma}^f,\ 0 < \widetilde{\mu}_f(K) < \infty\} \subseteq \mathcal{R}(\widehat{T}). \tag{4.11}$$

This yields the identity $L^1(\widetilde{\mu}_f) = \mathcal{R}(\widehat{T})$ because the left-side of (4.11) is dense in the Banach space $L^1(\widetilde{\mu}_f)$ and because $\mathcal{R}(\widehat{T})$, being the range of the linear isometry $\widehat{T}$, is a *closed* linear subspace of $L^1(\widetilde{\mu}_f)$. So, $\widehat{T}$ is surjective. □

Theorem 4.3 *If (Ω, Σ, μ) is a K-localizable measure space, then it is also S-localizable.*

Proof Let $\eta \in (L^1(\widetilde{\mu}_f))^*$. Then $\eta \circ \widehat{T} \in (L^1(\widetilde{\mu}))^*$, so that $\eta \circ \widehat{T} = J_\mu(\varphi + \mathcal{N}(\mu))$ for some $\varphi \in \mathcal{L}^\infty(\mu)$ by the K-localizability assumption. In other words,

$$\langle h + \mathcal{N}(\mu), \eta \circ \widehat{T} \rangle = \int_\Omega h\varphi \, d\mu, \quad h \in \mathcal{L}^1(\mu). \tag{4.12}$$

Clearly, φ is $\widetilde{\Sigma}^f$-measurable as $\Sigma \subseteq \widetilde{\Sigma}^f$; let $\widetilde{\varphi}$ denote φ considered as being $\widetilde{\Sigma}^f$-measurable. Since $\mathcal{N}_0(\mu) \subseteq \mathcal{N}_0(\widetilde{\mu}_f)$ by (2.6), we have that

$$\widetilde{\mu}_f\text{-ess.sup}|\widetilde{\varphi}| \le \mu\text{-ess.sup}\,|\varphi| < \infty.$$

So, $\widetilde{\varphi} \in \mathcal{L}^\infty(\widetilde{\mu}_f)$.

Next, we claim that

$$\langle h + \mathcal{N}(\widetilde{\mu}_f), \eta \rangle = \int_\Omega h\widetilde{\varphi} \, d\widetilde{\mu}_f, \quad h \in \mathcal{L}^1(\widetilde{\mu}_f). \tag{4.13}$$

To see this fix $h \in \mathcal{L}^1(\widetilde{\mu}_f)$. Since $\widehat{T}$ is surjective, select $g \in \mathcal{L}^1(\mu)$ such that $(h + \mathcal{N}(\widetilde{\mu}_f)) = \widehat{T}(g + \mathcal{N}(\mu))$, that is, $(h - \widetilde{g}) \in \mathcal{N}(\widetilde{\mu}_f)$. Since the $\widetilde{\Sigma}^f$-measurable function $\widetilde{g}\widetilde{\varphi}$ is pointwise equal to the Σ-measurable function $g\varphi$ it follows, from (4.7) with $g\varphi$ in place of f and with $E := \Omega$, that $\int_\Omega \widetilde{g}\,\widetilde{\varphi}\, d\widetilde{\mu}_f = \int_\Omega g\varphi \, d\mu$. This yields $\int_\Omega h\widetilde{\varphi}\, d\widetilde{\mu}_f = \int_\Omega \widetilde{g}\,\widetilde{\varphi}\, d\widetilde{\mu}_f = \int_\Omega g\varphi d\mu$ via $(h - \widetilde{g}) \in \mathcal{N}(\widetilde{\mu}_f)$. So, (4.13) holds because (4.12) gives

$$\begin{aligned}\langle h + \mathcal{N}(\widetilde{\mu}_f), \eta \rangle &= \langle \widehat{T}(g + \mathcal{N}(\mu)), \eta \rangle = \langle g + \mathcal{N}(\mu), \eta \circ \widehat{T} \rangle \\ &= \int_\Omega g\varphi \, d\mu = \int_\Omega h\widetilde{\varphi} \, d\widetilde{\mu}_f.\end{aligned}$$

Recalling the definition of $J_{\widetilde{\mu}_f}$ from (4.2), with $\psi := \widetilde{\varphi}$ there, we have via (4.13) that

$$\langle h + \mathcal{N}(\widetilde{\mu}_f), \eta \rangle = \int_\Omega h\widetilde{\varphi} \, d\widetilde{\mu}_f = \langle h + \mathcal{N}(\widetilde{\mu}_f), J_{\widetilde{\mu}_f}(\widetilde{\varphi} + \mathcal{N}(\widetilde{\mu}_f)) \rangle.$$

Since $h \in \mathcal{L}^1(\widetilde{\mu}_f)$ is arbitrary, this means that $\eta = J_{\widetilde{\mu}_f}(\widetilde{\varphi} + \mathcal{N}(\widetilde{\mu}_f))$. That is, $J_{\widetilde{\mu}_f} : L^\infty(\widetilde{\mu}_f) \longrightarrow (L^1(\widetilde{\mu}_f))^*$ is surjective. Therefore $J_{\widetilde{\mu}_f}$ is bijective because the semifiniteness of $\widetilde{\mu}_f$ already gives that $J_{\widetilde{\mu}_f}$ is injective (by Lemma 2.6(i) with $\widetilde{\mu}_f$ in place of μ). So, $\widetilde{\mu}_f$ is localizable by Lemma 2.6(ii) (with $\widetilde{\mu}_f$ in place of μ) and hence, Lemma 2.13(i) ensures that μ is S-localizable. □

Theorem 4.4 *If a measure space (Ω, Σ, μ) is both saturated and S-localizable, then it is K-localizable.*

Proof Since $\Sigma = \widetilde{\Sigma}^f$ by assumption, we do not need to distinguish between a Σ-measurable function h and its corresponding $\widetilde{\Sigma}^f$-measurable function $\widetilde{h}$ (unlike in the proof of Theorem 4.3). Accordingly, the linear isometry $\widehat{T} : L^1(\mu) \longrightarrow L^1(\widetilde{\mu}_f)$ (see (4.8) and Lemma 4.2) can be expressed as

$$\widehat{T}(h + \mathcal{N}(\mu)) = h + \mathcal{N}(\widetilde{\mu}_f), \quad f \in \mathcal{L}^1(\mu).$$

Observe also that $\mu_{sf} = \widetilde{\mu}_f$ on $\Sigma = \widetilde{\Sigma}^f$ and that $\widetilde{\mu}_f = \mu_{sf} \leq \mu$ on Σ (see (4.3)). Moreover, Lemma 2.9(iv) implies that $\mathcal{N}_0(\widetilde{\mu}_f) = \mathcal{N}_0^{\text{loc}}(\mu)$.

Fix $\xi \in (L^1(\mu))^*$. Then $\xi \circ (\widehat{T})^{-1} \in (L^1(\widetilde{\mu}_f))^*$. Since $\widetilde{\mu}_f$ is localizable (by Lemma 2.13(i)) we have, from Lemma 2.6(ii) (with $\widetilde{\mu}_f$ in place of μ), that the canonical map $J_{\widetilde{\mu}_f} : L^\infty(\widetilde{\mu}_f) \longrightarrow (L^1(\widetilde{\mu}_f))^*$ is bijective. So, there exists $\psi \in \mathcal{L}^\infty(\widetilde{\mu}_f) = \mathcal{L}^\infty_{\text{loc}}(\mu)$ satisfying

$$\xi \circ (\widehat{T})^{-1} = J_{\widetilde{\mu}_f}(\psi + \mathcal{N}(\widetilde{\mu}_f)) \text{ as elements of } (L^1(\widetilde{\mu}_f))^*. \tag{4.14}$$

Choose $a > 0$ such that the subset $H := \{\omega \in \Omega : |\psi(\omega)| > a\} \in \widetilde{\Sigma}^f = \Sigma$ is $\widetilde{\mu}_f$-null, i.e., $H \in \mathcal{N}_0(\widetilde{\mu}_f) = \mathcal{N}_0^{\text{loc}}(\mu)$. The function $\varphi := \psi\chi_{\Omega\setminus H}$ is bounded pointwise everywhere on Ω, that is, $\varphi \in \mathcal{L}^\infty(\mu)$. Note that H may not be a μ-null set. To establish the identity

$$J_\mu(\varphi + \mathcal{N}(\mu)) = \xi, \tag{4.15}$$

fix $g \in \mathcal{L}^1(\mu)$. Then also $g \in \mathcal{L}^1(\mu_{sf}) = \mathcal{L}^1(\widetilde{\mu}_f)$ by Lemma 4.1(i). So, $\int_\Omega g\varphi\, d\widetilde{\mu}_f = \int_\Omega g\varphi\, d\mu_{sf} = \int_\Omega g\varphi\, d\mu$, again by Lemma 4.1(i) with $g\varphi \in \mathcal{L}^1(\mu)$ in place of h. Since $H \in \mathcal{N}_0(\widetilde{\mu}_f)$ and $g\psi \in \mathcal{L}^1(\widetilde{\mu}_f)$, we have that

$$\int_\Omega g\psi\, d\widetilde{\mu}_f = \int_\Omega g\psi\chi_{\Omega\setminus H}\, d\widetilde{\mu}_f + \int_\Omega g\psi\chi_H\, d\widetilde{\mu}_f = \int_\Omega g\psi\chi_{\Omega\setminus H}\, d\widetilde{\mu}_f.$$

This and (4.14), via the definitions (4.1) and (4.2), yield

$$\begin{aligned}
&\langle g + \mathcal{N}(\mu), \xi\rangle = \langle(\widehat{T})^{-1}(g + \mathcal{N}(\widetilde{\mu}_f)), \xi\rangle \\
&= \langle g + \mathcal{N}(\widetilde{\mu}_f), \xi \circ (\widehat{T})^{-1}\rangle = \langle g + \mathcal{N}(\widetilde{\mu}_f), J_{\widetilde{\mu}_f}(\psi + \mathcal{N}(\widetilde{\mu}_f))\rangle \\
&= \int_\Omega g\psi\, d\widetilde{\mu}_f = \int_\Omega g\psi\chi_{\Omega\setminus H}\, d\widetilde{\mu}_f = \int_\Omega g\varphi\, d\widetilde{\mu}_f = \int_\Omega g\varphi\, d\mu \\
&= \langle g + \mathcal{N}(\mu), J_\mu(\varphi + \mathcal{N}(\mu))\rangle.
\end{aligned}$$

This establishes (4.15) as $g \in \mathcal{L}^1(\mu)$ is arbitrary. So, J_μ is surjective and hence, μ is K-localizable. □

According to Theorem 4.4, a measure space which is S-localizable but not K-localizable cannot be saturated. Example 2.14 provides such a measure space.

The following result is immediate from Theorems 4.3 and 4.4.

Corollary 4.5 *A saturated measure space is K-localizable if and only if it is S-localizable.*

Let (Ω, Σ, μ) be a measure space. For the rest of this section we use individual functions instead of their corresponding equivalence classes. Define the canonical linear map $J_\mu^{\text{loc}} : L_{\text{loc}}^\infty(\mu) \longrightarrow (L^1(\mu))^*$ which sends $\varphi \in L_{\text{loc}}^\infty(\mu)$ to the continuous linear functional $J_\mu^{\text{loc}}(\varphi)$ on $L^1(\mu)$ given by $f \longmapsto \int_\Omega \varphi f \, d\mu$. That $J_\mu^{\text{loc}}(\varphi) \in (L^1(\mu))^*$ follows from [11, Theorem (20.13)], as does the continuity of J_μ^{loc}. Actually, J_μ^{loc} is an *isometry* of $L_{\text{loc}}^\infty(\mu)$ into $(L^1(\mu))^*$, [11, Theorem (20.16)].

Proposition 4.6 *Let (Ω, Σ, μ) be a measure space.*

(i) *The ranges of J_μ^{loc} and J_μ satisfy $\mathcal{R}(J_\mu^{\text{loc}}) = \mathcal{R}(J_\mu)$ and are a closed subspace of $(L^1(\mu))^*$.*

(ii) *The following statements are equivalent.*

(a) *μ is K-localizable.*
(b) *J_μ is surjective.*
(c) *J_μ^{loc} is surjective.*

Proof

(i) Since the canonical inclusion $L^\infty(\mu) \subseteq L_{\text{loc}}^\infty(\mu)$ always holds, it is clear that $\mathcal{R}(J_\mu) \subseteq \mathcal{R}(J_\mu^{\text{loc}})$.

Conversely, fix $\ell \in \mathcal{R}(J_\mu^{\text{loc}})$, that is, there exists $\psi \in L_{\text{loc}}^\infty(\mu)$ such that

$$\langle f, \ell \rangle = \int_\Omega f\psi \, d\mu, \quad f \in L^1(\mu).$$

That $f\psi \in L^1(\mu)$ for all $f \in L^1(\mu)$ follows from [11, Theorem (20.13)(ii)]. Choose any $a > 0$ such that the set $H := \{\omega \in \Omega : |\psi(\omega)| > a\}$ is locally μ-null. Observe that $\varphi := \psi\chi_{\Omega\setminus H}$ is a bounded, Σ-measurable function, that is, $\varphi \in L^\infty(\mu)$. Moreover, given any $f \in L^1(\mu)$ we know, via Lemma 2.9(vii), that $\int_\Omega \psi f \chi_H \, d\mu = 0$ (as $\psi f \in L^1(\mu)$). Hence,

$$\langle f, \ell \rangle = \int_\Omega f\psi \, d\mu = \int_\Omega f\psi\chi_H \, d\mu + \int_\Omega f\varphi \, d\mu = \langle f, J_\mu(\varphi) \rangle.$$

Accordingly, $\ell = J_\mu(\varphi) \in \mathcal{R}(J_\mu)$, that is, $\mathcal{R}(J_\mu^{\text{loc}}) \subseteq \mathcal{R}(J_\mu)$. So, we have established that $\mathcal{R}(J_\mu) = \mathcal{R}(J_\mu^{\text{loc}})$.

As already noted, J_μ^{loc} is an isometry and hence, it has closed range in $(L^1(\mu))^*$.

(ii) By the definition of K-localizability we have (a) $\Longleftrightarrow$ (b). Part (i) implies that (b) $\Longleftrightarrow$ (c). □

Remark 4.7 The function χ_H in the proof of Proposition 4.6(i) is the zero element of $L^\infty(\mu)$ if and only if the locally μ-null set H is actually μ-null. □

An immediate consequence of Proposition 4.6 is the following result.

Corollary 4.8 *Let (Ω, Σ, μ) be a weakly decomposable measure space. Then μ is K-localizable.*

Proof Fix any $\ell \in (L^1(\mu))^*$. By Hewitt and Stromberg [11, Theorem (20.19)] there exists $\psi \in L^\infty_{\mathrm{loc}}(\mu)$ such that $\ell = J_\mu^{\mathrm{loc}}(\psi)$, that is, $\ell \in \mathcal{R}(J_\mu^{\mathrm{loc}})$. Then Proposition 4.6(i) implies that $\ell \in \mathcal{R}(J_\mu)$. Accordingly, J_μ is surjective which means precisely that μ is K-localizable. □

Remark 4.9

(i) The measure in Example 5.1 below is weakly decomposable but not Z-localizable. The measure of Example 5.6 below is not K-localizable and hence, cannot be weakly decomposable by Corollary 4.8.
(ii) Let μ be any totally infinite measure space such that $\mathcal{N}_0(\mu) = \{\emptyset\}$. Then μ is K-localizable by Lemma 2.7. However, it is shown in Example 5.4(ii) below that μ is *not* weakly decomposable. Thus, the converse of Corollary 4.8 fails to hold. □

5 Examples

We present three examples to help us to understand localizable measures. First, let us introduce the notation to be used throughout this section. Define $I := [0, 1]$ and let $\mathcal{B}(I)$ denote the Borel σ-algebra of I. The Lebesgue measure on $\mathcal{B}(I)$ is denoted by λ. Consider the unit square $\Omega := I \times I$, whose points are indicated by (x, y), as usual. Given a subset $E \subseteq \Omega$ define, for each $x \in I$, its x-section by

$$E_x := \{y \in I : (x, y) \in E\} \subseteq I$$

and, for each $y \in I$, its y-section by $E^y := \{x \in I : (x, y) \in E\} \subseteq I$. We shall also use the notation

$$\Omega(x) := \{x\} \times I \subseteq \Omega, \quad x \in I.$$

The following example provides a measure which is *not* semifinite, *not* Dedekind complete and *not* Z-localizable but, it *is* K-localizable, S-localizable and weakly decomposable.

Example 5.1 The measure μ presented here is a slightly modified version of that in [4, Ex.6, p. 105]. Let Σ denote the family of all subsets $E \subseteq \Omega$ such that $E_x \in \mathcal{B}(I)$ for each $x \in I$. Then each set $E \in \Sigma$ equals $\bigcup_{x\in I}(\{x\} \times E_x)$ with $E_x \in \mathcal{B}(I)$ for $x \in I$. It follows from [1, §36 Theorem 2] that Σ is a σ-algebra in Ω. Given $E \in \Sigma$, let $s(E) \subseteq I$ be the set

$$s(E) := \{x \in I : E_x \neq \emptyset\}.$$

Define a set function $\mu : \Sigma \longrightarrow [0, \infty]$ by

$$\mu(E) := \begin{cases} \sum_{x \in I} \lambda(E_x) & \text{if } s(E) \text{ is countable,} \\ \infty & \text{otherwise,} \end{cases}$$

for each $E \in \Sigma$. Via the properties of x-sections presented in [1, §36, Theorem 2] and the σ-additivity of λ, it follows that μ is a measure. It is routine to check that both of the identities

$$\Sigma^f = \{E \in \Sigma : s(E) \text{ is countable and } \textstyle\sum_{x \in s(E)} \lambda(E_x) < \infty\}, \text{ and} \tag{5.1}$$

$$\mathcal{N}_0(\mu) = \{E \in \Sigma^f : E_x \in \mathcal{N}_0(\lambda) \text{ for all } x \in s(E)\}$$

are valid, where we recall that $\Sigma^f := \{E \in \Sigma : \mu(E) < \infty\}$.

Fact 1 The measure space (Ω, Σ, μ) is *not* complete.

To see this, take a set $A \in \mathcal{B}(I)$ such that $\lambda(A) = 0$ and A possesses a subset B which is not Borel measurable. Let $E := (\{0\} \times A) \in \Sigma^f$. Then $s(E) = \{0\}$ and (5.1) implies that $E \in \mathcal{N}_0(\mu)$. But, its subset $F := \{0\} \times B$ does not belong to Σ. So, the measure space (Ω, Σ, μ) is not complete.

Fact 2 $\mathcal{N}_0^{\text{loc}}(\mu) \neq \mathcal{N}_0(\mu)$. In other words, there exists a locally μ-null set which is not μ-null.

We shall establish Fact 2 by showing that the subset $C := (I \times \{0\}) \in \Sigma$ is locally μ-null but not μ-null. Observe first that $C_x = \{0\} \neq \emptyset$ for all $x \in I$, and so $s(C) = I$ is uncountable. Thus $\mu(C) = \infty$ and hence, C is surely not μ-null. To show that C is locally μ-null, fix $E \in \Sigma^f$. Then $s(E)$ is countable (see (5.1)) and $(C \cap E)_x = C_x \cap E_x = \{0\} \cap E_x = \emptyset$ whenever $x \notin s(E)$, from which it follows that

$$C \cap E = \textstyle\bigcup_{x \in I} (\{x\} \times (C \cap E)_x) = \bigcup_{x \in s(E)} (\{x\} \times (\{0\} \cap E_x)).$$

Hence, the set $(C \cap E) \in \mathcal{N}_0(\mu)$ because

$$\begin{aligned} \mu(C \cap E) &\leq \mu\left(\textstyle\bigcup_{x \in s(E)} (\{x\} \times (\{0\} \cap E_x))\right) \\ &\leq \textstyle\sum_{x \in s(E)} \mu(\{x\} \times (\{0\} \cap E_x)) = \sum_{x \in s(E)} \lambda(\{0\} \cap E_x) = 0. \end{aligned}$$

Accordingly, C is locally μ-null. Fact 2 is thereby established.

Fact 2 implies that μ is not semifinite because of Lemma 2.9(vi). It is not without interest to also show this directly.

Fact 3 The measure μ is *not* semifinite.

To verify this, take the set C above used in the proof of Fact 2. Our aim is to show that, given any set $E \in \Sigma$ with $E \subseteq C$, either $\mu(E) = 0$ or $\mu(E) = \infty$. If $x \in I$, then $E_x \subseteq C_x = \{0\}$ and hence, $E_x = \{0\}$ whenever $x \in s(E)$. Moreover, $E_x = \emptyset$ when $x \notin s(E)$. In the case when $s(E)$ is countable, we have

$$\mu(E) = \sum_{x \in s(E)} \lambda(E_x) = \sum_{x \in s(E)} \lambda(\{0\}) = 0.$$

Consider the remaining case when $s(E)$ is uncountable. Then $E_x = \{0\} \neq \emptyset$ for uncountably many $x \in I$, so that $\mu(E) = \infty$ by definition. Combining these two cases yields $\mu(E) \in \{0, \infty\}$. This provides a direct proof of Fact 3.

Fact 4 The measure space (Ω, Σ, μ) is *not* Z-localizable. Consequently, it is *not* Dedekind complete either.

To establish this, observe first that $\Omega(x) \in \Sigma^f$ for each $x \in I$ because $\mu(\Omega(x)) = \lambda((\Omega(x))_x) = \lambda(I) = 1$. Define $\mathcal{E} := \{\Omega(x) : x \in I\} \subseteq \Sigma^f$. To prove that $\mathcal{E}$ does *not* possess an essential supremum in Σ assume, on the contrary, that $\mathcal{E}$ does have an essential supremum $H \in \Sigma$. Fix $x \in I$ for the moment. Then $\lambda(H_x) = 1$ because

$$\lambda(I \setminus H_x) = \mu\left(\{x\} \times (I \setminus H_x)\right) = \mu(\Omega(x) \setminus H) = 0$$

by $(E\text{-}1)$ in Sect. 2 with $E := \Omega(x)$. So, choose any point y_x in the non-empty set H_x, in which case $\lambda(H_x \setminus \{y_x\}) = 1$. Observe that the subset

$$G := \bigcup_{x \in I} (H_x \setminus \{y_x\}) \subseteq \Omega$$

belongs to Σ. Moreover, $s(H \setminus G)$ equals the uncountable set I because $(H \setminus G)_x = \{y_x\} \neq \emptyset$ for every $x \in I$. Thus $\mu(H \setminus G) = \infty$. On the other hand,

$$\mu(\Omega(x) \setminus G) = \mu(\{x\} \times (I \setminus G_x)) = \lambda(I \setminus G_x) = \lambda(I \setminus H_x) = 0, \quad x \in I.$$

This, together with $(E\text{-}2)$ in Sect. 2 (for $E := \Omega(x)$) implies that $\mu(H \setminus G) = 0$, which contradicts $\mu(H \setminus G) = \infty$. So $\mathcal{E}$ cannot possess an essential supremum in Σ and hence, (Ω, Σ, μ) is *not* Z-localizable.

Lemma 3.7(i) implies that (Ω, Σ, μ) is *not* Dedekind complete.

Fact 5 The measure space (Ω, Σ, μ) *is S-localizable.*

Our argument to establish this fact requires four steps.

Step 1. The measure μ *is saturated*, that is, $\widetilde{\Sigma}^f = \Sigma$.

To verify this step, it suffices to prove the inclusion $\widetilde{\Sigma}^f \subseteq \Sigma$ as the reverse inclusion is always valid. Observe first that the equality

$$\Sigma = \{K \subseteq \Omega : K \cap \Omega(x) \in \Sigma \text{ for every } x \in I\} \tag{5.2}$$

holds because of the fact that

$$K_x = K_x \cap I = (K \cap \Omega(x))_x, \quad x \in I,\ K \subseteq \Omega. \tag{5.3}$$

Since $\Omega(x) \in \Sigma^f$ for every x (see the proof of Fact 4), every set $K \in \widetilde{\Sigma}^f$ necessarily belongs to the right-side of (5.2) and hence, to Σ. So, we have obtained the inclusion $\widetilde{\Sigma}^f \subseteq \Sigma$. Thus, $\widetilde{\Sigma}^f = \Sigma$ and so μ is saturated.

Next, in the following Step 2, we exhibit a strictly localizable measure ι defined on Σ and then show (in Step 3) that ι equals $\widetilde{\mu}_f$.

Step 2. The set function $\iota : \Sigma \longrightarrow [0, \infty]$ defined by

$$\iota(E) := \sum_{x \in I} \lambda(E_x), \quad E \in \Sigma, \tag{5.4}$$

is a *strictly localizable* measure.

First, as for the case of μ, the set function ι can be shown to be σ-additive via the properties of x-sections given in [1, §36, Theorem 2] and the σ-additivity of λ. Let us verify the identity

$$\iota(E) = \sum_{x \in I} \iota(E \cap \Omega(x)), \quad E \in \Sigma. \tag{5.5}$$

Given $x \in I$, we have both $E_x = (E \cap \Omega(x))_x$ (by (5.3) with $K := E$) and $(E \cap \Omega(x))_\xi = E_\xi \cap \emptyset = \emptyset$ for each $\xi \in I \setminus \{x\}$. This yields

$$\iota(E \cap \Omega(x)) = \sum_{\xi \in I} \lambda((E \cap \Omega(x))_\xi) = \lambda(E_x),$$

which implies (5.5) by recalling (5.4).

Now, it is clear that the sets $(\Omega(x))_{x \in I}$ form a partition of Ω into pairwise disjoint, Σ-measurable sets of *finite* measure. This, (5.2) and (5.5) assure us that ι is a strictly localizable measure (for the definition of strictly localizable measures see Sect. 2).

Step 3. The measures $\widetilde{\mu}_f$ and ι are identical on Σ.

To prove this, we first show that $\mu = \iota$ on Σ^f. So, let $E \in \Sigma^f$. By (5.1), the subset $s(E) \subseteq I$ is countable and hence, $\mu(E) = \Sigma_{x \in s(E)} \lambda(E_x)$. On the other hand, $\lambda(E_x) = \lambda(\emptyset) = 0$ for all $x \in I \setminus s(E)$. Thus

$$\begin{aligned}
\iota(E) &= \sum_{x \in I} \lambda(E_x) = \sum_{x \in s(E)} \lambda(E_x) + \sum_{x \in I \setminus s(E)} \lambda(E_x) \\
&= \sum_{x \in s(E)} \lambda(E_x) = \mu(E).
\end{aligned}$$

So, we have shown that $\mu = \iota$ on Σ^f.

Next, for a given set $F \in \Sigma$, observe that

$$\iota(F) \geq \sup\{\iota(E) : E \subseteq F,\ E \in \Sigma^f\} = \sup\{\mu(E) : E \subseteq F,\ E \in \Sigma^f\} \quad (5.6)$$

because $\mu = \iota$ on Σ^f. Let $\mathcal{F}(I) \subseteq 2^I$ denote the subfamily of all finite subsets of I. Then

$$\begin{aligned}\iota(F) &:= \sup_{J\in\mathcal{F}(I)} \textstyle\sum_{x\in J} \lambda(F_x) = \sup_{J\in\mathcal{F}(I)} \textstyle\sum_{x\in J} \mu(\{x\} \times F_x)\\ &= \sup_{J\in\mathcal{F}(I)} \mu\left(\textstyle\bigcup_{x\in J}(\{x\} \times F_x)\right) \leq \sup\{\mu(E) : E \subseteq F,\ E \in \Sigma^f\}\end{aligned}$$

because the subset $\bigcup_{x\in J}(\{x\} \times F_x) \subseteq F$ belongs to Σ^f for each $J \in \mathcal{F}(I)$. This and (5.6) yield

$$\iota(F) = \sup\{\mu(E) : E \subseteq F,\ E \in \Sigma^f\}.$$

In other words, $\iota(F) = \widetilde{\mu}_f(F)$ via (2.4) with $K := F$. As $F \in \Sigma$ is arbitrary, we can conclude that $\iota = \widetilde{\mu}_f$ on Σ.

The following and final step will establish Fact 5.

Step 4. The measure space (Ω, Σ, μ) *is* S-localizable.

To see this, Steps 2 and 3 yield that $\widetilde{\mu}_f$ equals the strictly localizable measure ι. In particular, $\widetilde{\mu}_f$ is localizable and hence, μ is S-localizable by Lemma 2.13(i).

At this stage it is worthwhile to point out that Σ *properly* contains the σ-algebra of all Borel subsets of Ω. Indeed, the subset B^* presented in Example 5.6 below belongs to Σ but it is *not* a Borel subset of Ω.

Fact 6 The measure μ *is weakly decomposable.*

To verify this fact, recall the identity $\mu = \iota$ on Σ^f (from the proof of Step 3 in Fact 5 above).This and (5.5) yield

$$\mu(E) = \iota(E) = \Sigma_{x\in I}\, \iota(E \cap \Omega(x)) = \Sigma_{x\in I}\, \mu(E \cap \Omega(x)), \quad E \in \Sigma^f.$$

It follows that μ is weakly decomposable because of (5.2) and because $(\Omega(x))_{x\in I}$ is a partition of Ω into pairwise disjoint, Σ-measurable sets of *finite* measure. Fact 6 is thereby established.

The following observation follows from Fact 6 and Corollary 4.8.

Fact 7 The measure μ *is K-localizable.*

It is clear that Fact 5 above also follows from Fact 7 and Theorem 4.3. We have chosen to prove Fact 5 directly as it reveals other features of μ (see Step 1 above) and because parts of the proof will be referred to in Example 5.7 below.

The measure μ provides an example of a measure which is weakly decomposable but not strictly localizable, as it is not semifinite by Fact 3; see Proposition 2.4. □

The following two results collect some further facts about weakly decomposable measures. Let (Ω, Σ, μ) be a measure space. Since we will consider two measures simultaneously, let us use (just for the remainder of this section) the more precise notation

$$\Sigma^f(\mu) := \{E \in \Sigma : \mu(E) < \infty\}.$$

Recall from Sect. 4 that the set function $\mu_{sf} : \Sigma \longrightarrow [0, \infty]$ is given by

$$\mu_{sf}(E) := \sup\{\mu(E \cap F) : F \in \Sigma^f(\mu)\}, \quad E \in \Sigma. \tag{5.7}$$

Lemma 5.2 *Let (Ω, Σ, μ) be a measure space.*

(i) *The inequality $\mu_{sf} \leq \mu$ is valid on Σ.*
(ii) *μ_{sf} is a semifinite measure on Σ.*
(iii) *The measure μ is semifinite if and only if $\mu = \mu_{sf}$ on Σ.*
(iv) *The following containment is valid:*

$$\Sigma^f(\mu) \subseteq \Sigma^f(\mu_{sf}) := \{E \in \Sigma : \mu_{sf}(E) < \infty\}.$$

(v) *$\mu = \mu_{sf}$ on $\Sigma^f(\mu)$.*

Proof (i) is clear from (5.7) as $\mu(E \cap F) \leq \mu(E)$ for all $E, F \in \Sigma$. For (ii) and (iii) we refer to [8, 213X(c)(i)]. Parts (iv) and (v) are a special case of [8, 213X(c)(ii)]. □

Proposition 5.3 *Let (Ω, Σ, μ) be a measure space.*

(i) *If μ is weakly decomposable, then μ_{sf} is strictly localizable.*
(ii) *Assume that $\Sigma^f(\mu) = \Sigma^f(\mu_{sf})$.*

 (a) *If μ_{sf} is strictly localizable, then μ is weakly decomposable.*
 (b) *The measure μ is weakly decomposable if and only if μ_{sf} is strictly localizable.*

Proof

(i) Let $(\Omega_\gamma)_{\gamma \in \Gamma}$ be a Σ-partition of Ω satisfying (1-s), (2-s) and (3-s^*); see Sect. 2. To verify the condition (3-s) for μ_{sf} (in place of μ) means to establish the validity of

$$\mu_{sf}(E) = \textstyle\sum_{\gamma \in \Gamma} \mu_{sf}(E \cap \Omega_\gamma), \quad E \in \Sigma. \tag{5.8}$$

Let $\mathcal{F}(\Gamma)$ be as in the proof of Proposition 2.4. Fix $E \in \Sigma$. Observe that

$$\mu_{sf}\left(E \cap \left(\textstyle\bigcup_{\gamma \in \Delta} \Omega_\gamma\right)\right) = \sup_{F \in \Sigma^f(\mu)} \mu\left(E \cap F \cap \left(\textstyle\bigcup_{\gamma \in \Delta} \Omega_\gamma\right)\right), \quad \Delta \in \mathcal{F}(\Gamma), \tag{5.9}$$

by (5.7) with $E \cap \left(\bigcup_{\gamma\in\Delta}\Omega_\gamma\right)$ in place of E. Next, by (3-s^*) with $E \cap F$ in place of E and the weak decomposability of μ, we have that

$$\mu(E\cap F) = \Sigma_{\gamma\in\Gamma}\,\mu(E\cap F\cap\Omega_\gamma), \quad F\in\Sigma^f(\mu). \tag{5.10}$$

Thus, by an analogous argument as for (ii) $\Longrightarrow$ (i) in the proof of Proposition 2.4, we see that (5.8) follows from (5.9) and (5.10) because

$$\begin{aligned}
\mu_{sf}(E) &:= \sup_{F\in\Sigma^f(\mu)}\mu(E\cap F) = \sup_{F\in\Sigma^f(\mu)}\sum_{\gamma\in\Gamma}\mu(E\cap F\cap\Omega_\gamma)\\
&= \sup_{F\in\Sigma^f(\mu)}\sup_{\Delta\in\mathcal{F}(\Gamma)}\sum_{\gamma\in\Delta}\mu(E\cap F\cap\Omega_\gamma)\\
&= \sup_{F\in\Sigma^f(\mu)}\sup_{\Delta\in\mathcal{F}(\Gamma)}\mu\left(E\cap F\cap\left(\bigcup_{\gamma\in\Delta}\Omega_\gamma\right)\right)\\
&= \sup_{\Delta\in\mathcal{F}(\Gamma)}\sup_{F\in\Sigma^f(\mu)}\mu\left(E\cap F\cap\left(\bigcup_{\gamma\in\Delta}\Omega_\gamma\right)\right)\\
&= \sup_{\Delta\in\mathcal{F}(\Gamma)}\mu_{sf}\left(E\cap\left(\bigcup_{\gamma\in\Delta}\Omega_\gamma\right)\right)\\
&= \sup_{\Delta\in\mathcal{F}(\Gamma)}\sum_{\gamma\in\Delta}\mu_{sf}(E\cap\Omega_\gamma) = \sum_{\gamma\in\Gamma}\mu_{sf}(E\cap\Omega_\gamma).
\end{aligned}$$

Moreover, $\Omega_\gamma \in \Sigma^f(\mu) \subseteq \Sigma^f(\mu_{sf})$ for all $\gamma \in \Gamma$; see (1-s) and Lemma 5.2(iv). So, (1-s) holds for μ_{sf} (in place of μ). Clearly condition (2-s) is valid for μ_{sf} (in place of μ). That (3-s) holds for μ_{sf} (in place of μ) is precisely (5.8). Accordingly, μ_{sf} is strictly localizable.

(ii) (a) By the assumption on μ_{sf} and Lemma 5.2(iv),(v) we have that

$$\mu(E) = \mu_{sf}(E) = \Sigma_{\gamma\in\Gamma}\,\mu_{sf}(E\cap\Omega_\gamma) = \Sigma_{\gamma\in\Gamma}\,\mu(E\cap\Omega_\gamma), \tag{5.11}$$

for every $E \in \Sigma^f(\mu) = \Sigma^f(\mu_{sf})$, because $(E\cap\Omega_\gamma) \subseteq \Omega_\gamma$ belongs to $\Sigma^f(\mu) = \Sigma^f(\mu_{sf})$ as $\mu(\Omega_\gamma) = \mu_{sf}(\Omega_\gamma) < \infty$, for $\gamma \in \Gamma$ (which also shows that (1-s) holds for μ). Again (2-s) is a condition on Σ and hence, holds for both μ, μ_{sf}. Since (5.11) is exactly (3-s^*) for μ, we conclude that μ is weakly decomposable.

(b) Apply parts (i) and (ii)(a). □

Example 5.4

(i) Let $\mu : \Sigma \longrightarrow [0,\infty]$ be any totally infinite measure. Then $\Sigma^f(\mu) = \mathcal{N}_0(\mu)$ and hence, $\mu_{sf} \equiv 0$ on Σ; see (5.7). Accordingly,

$$\Sigma^f(\mu) = \mathcal{N}_0(\mu) \subsetneqq \Sigma = \mathcal{N}_0(\mu_{sf}) = \Sigma^f(\mu_{sf}).$$

That is, the containment in Lemma 5.2(iv) can be *strict.*

(ii) Consider any totally infinite measure $\mu : \Sigma \longrightarrow [0, \infty]$ such that $\mathcal{N}_0(\mu) = \{\emptyset\}$; see Example 2.11, for instance. By part (i) we have that $\Sigma^f(\mu) \subsetneqq \Sigma^f(\mu_{sf})$ and $\mu_{sf} \equiv 0$ on Σ. In particular, μ_{sf} is strictly localizable. However, μ is *not* weakly decomposable as every member Ω_γ of any possible Σ-partition $(\Omega_\gamma)_{\gamma\in\Gamma}$ of Ω must satisfy $\Omega_\gamma = \emptyset$ and hence, $\Omega = \bigcup_{\gamma\in\Gamma} \Omega_\gamma = \emptyset$; contradiction. □

Example 5.4(ii) shows that we cannot omit the requirement that $\Sigma^f(\mu) = \Sigma^f(\mu_{sf})$ in Proposition 5.3(ii)(a).

Remark 5.5 The measure μ of Example 5.1 is weakly decomposable but not strictly localizable (as it is not semifinite). However, μ_{sf} is strictly localizable (as $\mu_{sf} = \widetilde{\mu}_f = \iota$ is strictly localizable) and satisfies $\Sigma^f(\mu) = \Sigma^f(\mu_{sf})$. □

The measure in the following example (from [11, Example (20.17)]) *is* semifinite but *not* Dedekind complete and *not* K-localizable. Hence, it is *neither* localizable *nor* Z-localizable.

Example 5.6 Let Σ be the Borel σ-algebra of $\Omega = I \times I$. Define measures ν_1 and ν_2 on Σ by

$$\nu_1(E) := \sum_{x\in I} \lambda(E_x) \text{ and } \nu_2(E) := \sum_{y\in I} \lambda(E^y), \quad E \in \Sigma.$$

Consider the measure

$$\mu := \nu_1 + \nu_2.$$

To see that μ *is semifinite* on Σ, take a set $E \in \Sigma$ with $\mu(E) = \infty$. Then, either $\nu_1(E) = \infty$ or $\nu_2(E) = \infty$. Assume first that $\nu_1(E) = \infty$. Then $\lambda(E_\xi) > 0$ for some $\xi \in I$. With $F := (\{\xi\} \times E_\xi) \in \Sigma \cap E$ we have both that $F_\xi = E_\xi$ and $F_x = \emptyset$ for each $x \in I \setminus \{\xi\}$, and that F^y equals either $\{y\}$ or $\emptyset$ for every $y \in I$. Therefore, $\nu_1(F) = \lambda(E_\xi)$ and $\nu_2(F) = 0$, which yields that $\mu(F) = \lambda(E_\xi) \in (0, 1]$. Similarly, when $\nu_2(E) = \infty$, we can find a set $G \in \Sigma \cap E$ satisfying $\mu(G) \in (0, 1]$. So, μ is semifinite.

Whereas it is possible to show that the measure space (Ω, Σ, μ) is *not* complete as in the proof of Fact 1 in Example 5.1, let us provide an alternate argument. To this end observe that the diagonal $\Delta := \{(x, x) : x \in I\}$ is a closed subset of Ω and hence, $\Delta \in \Sigma$. It is easy to see that $\mu(\Delta) = 0$, that is, $\Delta \in \mathcal{N}_0(\mu)$. Now, let $\mathcal{B}(\Delta)$ denote the Borel σ-algebra of Δ, in which case

$$\mathcal{B}(\Delta) = \Sigma \cap \Delta; \tag{5.12}$$

see, for example, [4, Lemma 7.2.2]. There exists a subset B^* of Δ with $B^* \notin \mathcal{B}(\Delta)$ because $\mathcal{B}(\Delta) \subsetneqq 2^\Delta$; see [11, Corollary (10.25)]. It follows from (5.12) that $B^* \notin \Sigma$ whereas B^* is contained in the μ-null set Δ, which implies that (Ω, Σ, μ) is *not* complete.

The measure μ is *not* K-localizable. In fact, consider the continuous linear functional

$$\ell : f \longmapsto \int_\Omega f \, d\nu_1, \quad f \in L^1(\mu). \tag{5.13}$$

As shown in [11, pp. 349–350], there is *no* $\varphi \in \mathcal{L}^\infty_{\text{loc}}(\mu)$ such that $\ell(f) = \int_\Omega f\varphi \, d\mu$ for $f \in L^1(\mu)$. Accordingly, $\ell \notin \mathcal{R}(J_\mu)$; see Proposition 4.6(i). That is, μ is *not* K-localizable. According to Theorem 3.4 and Corollary 3.9(i) the measure μ is *not* Dedekind complete and *not* Z-localizable. Moreover, Corollary 4.8 shows that μ is *not* weakly decomposable.

We are unable to decide whether μ is S-localizable or not. Of course, if μ were saturated, then it would be S-localizable by Theorem 4.4. However, it is undetermined whether or not μ is saturated. Not much seems to be known about the larger σ-algebra $\widetilde{\Sigma}^f \supseteq \Sigma$, except for the fact that B^* given above does not belong to $\widetilde{\Sigma}^f$ (because $\Delta \in \Sigma^f$ and $(B^* \cap \Delta) \notin \Sigma$). □

In the previous Example 5.6 the measure μ was shown to be non-K-localizable. It is natural to seek an extension of μ to a larger σ-algebra and try to determine whether or not the extended measure is K-localizable. The following example below shows that this can be done, modulo the continuum hypothesis.

Example 5.7 We continue with the notation of Example 5.6. Let $\mathscr{S}$ denote the subfamily of all subsets $E \subseteq \Omega$ such that $E_x \in \mathcal{B}(I)$ for all $x \in I$ and $E^y \in \mathcal{B}(I)$ for all $y \in I$. Again by Berberian [1, §36, Theorem 2], the subfamily $\mathscr{S}$ is a σ-algebra in Ω. The inclusion $\Sigma \subseteq \mathscr{S}$ is clear. Moreover, it is a *strict* inclusion as the subset B^* in Example 5.6 satisfies $B^* \in \mathscr{S} \setminus \Sigma$. Define two measures $\check{\nu}_1$ and $\check{\nu}_2$ on $\mathscr{S}$ by

$$\check{\nu}_1(E) := \sum_{x \in I} \lambda(E_x) \text{ and } \check{\nu}_2(E) := \sum_{y \in I} \lambda(E^y), \quad E \in \mathscr{S}, \tag{5.14}$$

respectively. The measures $\check{\nu}_1$ and $\check{\nu}_2$ are natural extensions of ν_1 and ν_2 from Σ to $\mathscr{S}$, respectively. So, the measure $\check{\mu} := \check{\nu}_1 + \check{\nu}_2$ is a natural extension of $\mu = \nu_1 + \nu_2$ from Σ to $\mathscr{S}$.

Fact 1 The measure space $(\Omega, \mathscr{S}, \check{\mu})$ is *not* complete.

This can be proved as in Fact 1 of Example 5.1.

Fact 2 The measure $\check{\mu}$ *is semifinite.*

Use the same arguments as for the proof of the semifiniteness of μ in Example 5.6.

Fact 3 The measure $\check{\mu}$ *is saturated.*

We shall establish this by adapting the proof of Step 1 in Fact 5 of Example 5.1. First observe that

$$\check{\mu}(\Omega(x)) = \check{\nu}_1(\Omega(x)) = \lambda(I) = 1, \quad x \in I,$$

because

$$\check{\nu}_2(\Omega(x)) = \Sigma_{y\in I}\lambda((\Omega(x))^y) = \Sigma_{y\in I}\lambda(\{y\}) = 0.$$

Let $K \in \widetilde{\mathscr{S}}^f$. Then $(K \cap \Omega(x)) \in \mathscr{S}$ and hence, $K_x = (K \cap \Omega(x))_x \in \mathcal{B}(I)$ (see (5.3)) for every $x \in I$. By using $I \times \{y\}$ instead of $\Omega(x)$ we can deduce by a similar argument that $K^y \in \mathcal{B}(I)$ for every $y \in I$. Thus, $K \in \mathscr{S}$, which yields the inclusion $\widetilde{\mathscr{S}}^f \subseteq \mathscr{S}$. So, Fact 3 is established as the reverse inclusion always holds.

Fact 4 Assuming the continuum hypothesis, the measure $\check{\mu}$ *is strictly localizable.*

It is the continuum hypothesis which, via [11, Ex.(21.26)], ensures the existence of a subset $H \subseteq \Omega$ such that H_x is countable for every $x \in I$ and $I \setminus H^y$ is countable for every $y \in I$. Note that $H \in \mathscr{S}$. Moreover,

$$\lambda(H_x) = \lambda(I \setminus H^y) = 0, \quad x, y \in I. \tag{5.15}$$

Thus, $\check{\nu}_1(H) = 0$ and $\check{\nu}_2(G) = 0$, with $G := \Omega \setminus H$. Also $G \in \mathscr{S}$. Therefore, $\check{\mu}(E \cap G) = \check{\nu}_1(E \cap G)$ and $\check{\mu}(E \cap H) = \check{\nu}_2(E \cap H)$ for $E \in \mathscr{S}$. In short,

$$\check{\mu} = \check{\nu}_1 \text{ on } \mathscr{S} \cap G \text{ and } \check{\mu} = \check{\nu}_2 \text{ on } \mathscr{S} \cap H. \tag{5.16}$$

The arguments used to verify Step 2 in the proof of Fact 5 in Example 5.1 can also be applied to verify that

$$\check{\nu}_1(E \cap G) = \textstyle\sum_{x\in I} \check{\nu}_1(E \cap G \cap (\Omega(x))), \quad E \in \mathscr{S}. \tag{5.17}$$

Similarly,

$$\check{\nu}_2(E \cap H) = \textstyle\sum_{y\in I} \check{\nu}_2(E \cap H \cap (I \times \{y\})), \quad E \in \mathscr{S}. \tag{5.18}$$

Let $G(x) := (\{x\} \times G_x) \subseteq G$ for $x \in I$ and $\widetilde{H}(y) := (H^y \times \{y\}) \subseteq H$ for $y \in I$. Then $G \cap H = \emptyset$ implies that $\{G(x), \widetilde{H}(y) : x, y \in I\}$ is a partition of Ω into pairwise disjoint, $\mathscr{S}$-measurable sets. By observing that $G(x) = G \cap (\Omega(x))$ for $x \in I$ and $\widetilde{H}(y) = H \cap (I \times \{y\})$ for $y \in I$, we can rewrite (5.17) and (5.18) as

$$\begin{aligned} &\check{\nu}_1(E \cap G) = \sum\nolimits_{x\in I} \check{\nu}_1(E \cap G(x)) \text{ and} \\ &\check{\nu}_2(E \cap H) = \sum\nolimits_{y\in I} \check{\nu}_2(E \cap \widetilde{H}(y)), \quad E \in \mathscr{S}, \end{aligned} \tag{5.19}$$

respectively. It follows from (5.14), (5.16) and (5.19) that

$$\begin{aligned} \check{\mu}(E) &= \check{\mu}(E \cap G) + \check{\mu}(E \cap H) = \check{\nu}_1(E \cap G) + \check{\nu}_2(E \cap H) \\ &= \textstyle\sum_{x\in I} \check{\nu}_1(E \cap G(x)) + \sum_{y\in I} \check{\nu}_2(E \cap \widetilde{H}(y)) \\ &= \textstyle\sum_{x\in I} \mu(E \cap G(x)) + \sum_{y\in I} \mu(E \cap \widetilde{H}(y)), \; E \in \mathscr{S}. \end{aligned} \tag{5.20}$$

Next, we claim that

$$\mathscr{S} = \{E \subseteq \Omega : E \cap G(x) \in \mathscr{S} \text{ for all } x \in I \text{ and } E \cap \widetilde{H}(y) \in \mathscr{S} \text{ for all } y \in I\}. \tag{5.21}$$

Indeed, it is clear that the left-side of (5.21) is contained in the right-side of (5.21). So, take a set E from the right-side of (5.21). Then, for every $x \in I$, we have

$$E_x = (E_x \cap G_x) \cup (E_x \cap H_x) = (E \cap G(x))_x \cup (E_x \cap H_x) \in \mathcal{B}(I)$$

because the assumption $E \cap G(x) \in \mathscr{S}$ yields $(E \cap G(x))_x \in \mathcal{B}(I)$ and because $(E_x \cap H_x) \subseteq H_x$ is countable. On the other hand, given $y \in I$,

$$E^y = (E^y \cap G^y) \cup (E^y \cap H^y) = (E^y \cap G^y) \cup (E \cap \widetilde{H}(y))_y \in \mathcal{B}(I)$$

because the assumption $E \cap \widetilde{H}(y) \in \mathscr{S}$ gives $(E \cap \widetilde{H}(y))^y \in \mathcal{B}(I)$ and because $(E^y \cap G^y) \subseteq G^y$ is countable. Thus, $E \in \mathscr{S}$ and hence, (5.21) holds.

Now, let $x \in I$. Since $(G(x))_\xi = G_x$ if $\xi = x$ and $(G(x))_\xi = \emptyset$ if $\xi \in I \setminus \{x\}$, we have from (5.14), (5.15) and (5.16) that

$$\check{\mu}(G(x)) = \check{\nu}_1(G(x)) = \textstyle\sum_{\xi \in I} \lambda((G(x))_\xi) = \lambda(G_x) = \lambda(I \setminus H_x) = 1.$$

Similarly, we have $\check{\mu}(\widetilde{H}(y)) = 1$, again by (5.14), (5.15), (5.16). So, the combined family of sets $G(x)$ with x varying through I and $\widetilde{H}(y)$ with y varying through I forms a partition of Ω into pairwise disjoint $\mathscr{S}$-measurable sets of *finite* measure. This, together with (5.20) and (5.21), enables us to conclude that $\check{\mu}$ is strictly localizable.

Under the continuum hypothesis, the measure $\check{\mu}$ is strictly localizable as ascertained above and hence, $\check{\mu}$ is also localizable. So, the canonical map $J_{\check{\mu}} : L^\infty(\check{\mu}) \longrightarrow (L^1(\check{\mu}))^*$ is bijective (see Lemma 2.6(ii)). This guarantees the existence of $\varphi \in L^\infty(\check{\mu})$ such that $\int_\Omega f \, d\check{\nu}_1 = \int_\Omega f\varphi \, d\check{\mu}$ for all $f \in L^1(\check{\mu})$ as the linear functional $\check{\ell} : f \longmapsto \int_\Omega f \, d\nu_1$ on $L^1(\check{\mu})$ is continuous. Accordingly, $J_{\check{\mu}}(\varphi) = \check{\ell}$. Compare this with the continuous linear functional ℓ on $(L^1(\mu))^*$ (see (5.13)) which has been shown to satisfy $\ell \in (L^1(\mu))^* \setminus J_\mu(L^\infty(\mu))$; see Example 5.6. Clearly $\check{\ell}$ is a continuous linear extension of ℓ. Moreover, $\check{\ell}$ lies in the range of $J_{\check{\mu}}$ (thanks to the continuum hypothesis) with $\check{\mu}$ being defined on the enlarged σ-algebra $\mathscr{S} \supseteq \Sigma$.

It is possible to exhibit an explicit element $\varphi \in L^\infty(\check{\mu})$ satisfying $J_{\check{\mu}}(\varphi) = \check{\ell}$. Indeed, we know that $\check{\nu}_1(\mathscr{S} \cap H) = \{0\}$ and that $\check{\mu} = \check{\nu}_1$ on $\mathscr{S} \cap G$; see the proof of Fact 4. This implies that $J_{\check{\mu}}(\chi_G) = \check{\ell}$ because

$$\langle f, \check{\ell} \rangle = \textstyle\int_\Omega f \, d\check{\nu}_1 = \int_G f \, d\check{\nu}_1 = \int_\Omega f \chi_G \, d\check{\mu}, \quad f \in L^1(\check{\mu}).$$

Fact 5 The measure $\breve{\mu}$ *is* both *K-localizable* and *S-localizable* (assuming the continuum hypothesis).

For this, recall that $\breve{\mu}$ is strictly localizable from Fact 4 and hence, is also localizable. So, $\breve{\mu}$ is K-localizable by Lemma 2.6(iv). This yields the S-localizability of $\breve{\mu}$ by Theorem 4.3. Alternatively, note that $\breve{\mu}$ is locally determined by Facts 2 and 3. So, Lemma 2.13(ii) assures us that $\breve{\mu}$ is S-localizable and hence, K-localizable via Theorem 4.4 above. □

6 Appendix

6.1 Luther's Decomposition via a Semifinite Measure

Let (Ω, Σ, μ) be a measure space. Our aim is to show that μ_{sf} equals the semifinite measure μ_1 given by [17, Theorem 1(i)], as announced earlier in Remark 3.10(ii). To be precise, recall from Sect. 4 (prior to (4.3)) that the semifinite measure μ_{sf} : $\Sigma \longrightarrow [0, \infty]$ is given by $\mu_{sf}(E) := \sup\{\mu(E \cap F) : F \in \Sigma^f\}$ for $E \in \Sigma$ (see [8, 213X(c)]). Its relevance to our theme is via the relationship $\mu_{sf} = \widetilde{\mu}_f|_\Sigma \leq \mu$ on Σ as given in (4.3).

N.Y. Luther has presented a semifinite measure μ_1 (see below) with certain properties as explained in Remark 3.10(ii). In order to give the precise definition of μ_1, for every $F \in \Sigma$ we recall from Sect. 3 the measure $\mu_F : \Sigma \longrightarrow [0, \infty]$ defined by

$$\mu_F(E) := \mu(F \cap E), \quad E \in \Sigma.$$

Moreover, let

$$\mathcal{M} := \{M \in \Sigma : M \text{ has } \mu\text{-}\sigma\text{-finite measure}\}.$$

Then the set function $\mu_1 : \Sigma \longrightarrow [0, \infty]$ defined by

$$\mu_1(E) := \sup\{\mu_M(E) : M \in \mathcal{M}\}, \quad E \in \Sigma,$$

is a measure, [1, §10, Theorem 1], and it is semifinite. Let μ_2 be the measure defined by

$$\mu_2(E) := \sup\{\mu_F(E) : F \in \mathcal{N}_0(\mu_1)\}, \quad E \in \Sigma.$$

Then μ_1, μ_2 are the unique measures constructed in the proof of Theorem 1(i) in [17], where it is also shown that they satisfy $\mu_2(\Sigma) \subseteq \{0, \infty\}$ and $\mu = \mu_1 + \mu_2$.

Proposition 6.1 *The measures μ_{sf} and μ_1 are identical on Σ.*

Proof Since $\Sigma^f \subseteq \mathcal{M}$, we have that

$$\mu_{sf}(E) := \sup\{\mu(F \cap E) : F \in \Sigma^f\} = \sup\{\mu_F(E) : F \in \Sigma^f\} \le \mu_1(E), \quad E \in \Sigma .$$

On the other hand, given any $M \in \mathcal{M}$, we can write $M = \bigcup_{n=1}^{\infty} F(n)$ for some increasing sequence $\{F(n)\}_{n=1}^{\infty}$ in Σ^f, so that

$$\begin{aligned} \mu_M(E) &= \mu(M \cap E) = \mu(\bigcup_{n=1}^{\infty}(F(n) \cap E)) \\ &= \sup_{n\in\mathbb{N}} \mu(F(n) \cap E) \le \mu_{sf}(E), \quad E \in \Sigma. \end{aligned}$$

So, $\mu_1(E) = \sup\{\mu_M(E) : M \in \mathcal{M}\} \le \mu_{sf}(E)$, for all $E \in \Sigma$. Thus, $\mu_1 = \mu_{sf}$ as measures on Σ. □

In general, there need not exist a set $\Omega^s \in \Sigma$ such that $\mu_1 = \mu_{sf}$ has the form μ_{Ω^s}; see Theorem 3.3.

6.2 D.H. Fremlin's Integration

Throughout this appendix let (Ω, Σ, μ) denote an arbitrary measure space. Our intention is to provide a *brief* explanation and summary of D.H. Fremlin's integration when it is reformulated in our notation (which follows the traditional process). The aim is to keep this note as self-contained as possible. We have used various results from [7–9] in order to discuss certain aspects of L^1-spaces and L^∞-spaces. Most relevant are 242Y(g) and 243X(b) in [8]; the L^1-space and L^∞-space given there can be identified with the real parts of the traditional spaces $L^1(\mu)$ and $L^\infty(\mu)$, respectively.

We have used the standard integration theory, as presented in [11, Sections 10–12], [21, Ch.1], for example. In particular, the μ-integrable and μ-essentially bounded functions are always considered to be defined on the *whole* set Ω; the corresponding spaces of (individual) functions are written as $\mathcal{L}^1(\mu)$, resp. $\mathcal{L}^\infty(\mu)$. In contrast with this, D.H. Fremlin allows such functions to be defined on smaller subsets of Ω.

Recall from Sect. 3 that $(\Omega, \widehat{\Sigma}, \widehat{\mu})$ denotes the completion of (Ω, Σ, μ). A subset of Ω is called *μ-negligible* if it is contained in a μ-null set, i.e., if it is $\widehat{\mu}$-null. Moreover, a subset $A \subseteq \Omega$ is called *μ-conegligible* if $\Omega \setminus A$ is μ-negligible; namely, there exists $E \in \mathcal{N}_0(\mu)$ such that $(\Omega \setminus A) \subseteq E$, [7, 112D]. A property holding off a μ-negligible set is phrased simply as "a.e." in [7, 112D(d)]; our notation for this is "$\widehat{\mu}$-a.e.".

Given a subset $D \subseteq \Omega$, not necessarily an element of Σ, define

$$\Sigma_D := \{E \cap D : E \in \Sigma\}.$$

Then Σ_D is a σ-algebra of subsets of D. Namely, (D, Σ_D) is a measurable space, [7, Lemma 121A]. In the special case when $D \in \Sigma$, we have the equality $\Sigma_D = \Sigma \cap D$ as defined in Sect. 2.

A function f with values in $\mathbb{R}$ and defined on a subset $A \subseteq \Omega$ is said to have domain A and we denote A by dom f. It is not assumed that $A \in \Sigma$. If $B \subseteq \Omega$ is any subset, then $f \upharpoonright B$ denotes the $\mathbb{R}$-valued function defined on dom $(f \upharpoonright B) := B \cap \mathrm{dom}\, f$ via $(f \upharpoonright B)(x) := f(x)$ for all $x \in \mathrm{dom}\,(f \upharpoonright B)$, [7, 121E(h)].

According to [7, Definition 121C], a function $f : \mathrm{dom}\, f \longrightarrow \mathbb{R}$ is said to be measurable if it is measurable with respect to the relative measurable space $(\mathrm{dom}\, f,\ \Sigma_{\mathrm{dom}\, f})$, that is, if $f^{-1}((-\infty, a)) \in \Sigma_{\mathrm{dom}\, f}$ for all $a \in \mathbb{R}$. In this case we say that f is $\Sigma_{\mathrm{dom}\, f}$-measurable in order to indicate explicitly the σ-algebra $\Sigma_{\mathrm{dom} f}$ of subsets of dom f. Moreover, a function $f : \mathrm{dom}\, f \longrightarrow \mathbb{R}$ is called *μ-virtually measurable* if dom f is *μ-conegligible* and if there is a μ-conegligible subset $E \subseteq \Omega$ such that $f \upharpoonright E : E \cap \mathrm{dom}\, f \longrightarrow \mathbb{R}$ is $\Sigma_{E \cap \mathrm{dom}\, f}$-measurable, [7, Remark 122Q].

Let sim Σ^f denote the $\mathbb{R}$-linear span of $\{\chi_E : E \in \Sigma^f\}$. Functions in sim Σ^f are called *simple functions* in [7, Definition 122A]. Note that they are often called μ-integrable simple functions in the literature as the functions in sim Σ^f are necessarily μ-integrable. Let $\mathrm{sim}^+\, \Sigma^f$ denote the positive cone of sim Σ^f, that is,

$$\mathrm{sim}^+\, \Sigma^f := \{s \in \mathrm{sim}\, \Sigma^f : s(\omega) \geq 0 \text{ for all } \omega \in \Omega\}.$$

By U we denote the set of all functions f with values in $[0, \infty)$ such that

(a) dom f is μ-conegligible, and
(b) there is an increasing sequence $\{s_n\}_{n=1}^{\infty}$ in $\mathrm{sim}^+\, \Sigma^f$ satisfying both $\lim_{n\to\infty} s_n(\omega) = f(\omega)$ for $\widehat{\mu}$-a.e. $\omega \in \Omega$ and $\sup_{n\in\mathbb{N}} \int_\Omega s_n\, d\mu < \infty$;

see [7, Definition 122H]. For a function $f \in U$ satisfying (a) and (b) we have that

$$\sup_{n\in\mathbb{N}} \int_\Omega s_n\, d\mu = \sup\left\{\int_\Omega s\, d\mu : s \in \mathrm{sim}^+\, \Sigma,\ s \leq f\ (\widehat{\mu}\text{-a.e.})\right\}, \tag{6.1}$$

[7, Lemma 122I]. This enables one to define

$$\int f\, d\mu := \sup\left\{\int_\Omega s\, d\mu : s \in \mathrm{sim}^+\, \Sigma,\ s \leq f\ (\widehat{\mu}\text{-a.e.})\right\}. \tag{6.2}$$

We have deliberately used the notation $\int f\, d\mu$ (i.e., omitted the subscript Ω) because we can only write $\int_\Omega f\, d\mu$ when $(\mathrm{dom}\, f) = \Omega$ and $f \in \mathcal{L}^1(\mu)$; note that this integral $\int f\, d\mu$ is denoted simply by $\int f$ in [7, Definition 122K]. It is routine to deduce that each $f \in U$ is μ-virtually measurable by using conditions (a) and (b) above.

Lemma 6.2 *Let $f \in \mathcal{L}^1(\mu)$ take its values in $[0, \infty)$. Then $f \in U$ and $\int_\Omega f\, d\mu = \int f\, d\mu$.*

Proof See [21, Definition 1.23]. □

Given $\mathbb{R}$-valued functions f, g whose domains lie within Ω, their sum $(f+g)$ is the $\mathbb{R}$-valued function with $\operatorname{dom}(f+g) := (\operatorname{dom} f) \cap (\operatorname{dom} g)$ and defined by $(f+g)(\omega) := f(\omega) + g(\omega)$ for all $\omega \in \operatorname{dom}(f+g)$. When $a \in \mathbb{R}$, the scalar multiplication af is the $\mathbb{R}$-valued function with $\operatorname{dom}(af) := \operatorname{dom} f$ and defined by $(af)(\omega) := af(\omega)$ for all $\omega \in \operatorname{dom}(af)$.

We say that a function $f : \operatorname{dom} f \longrightarrow \mathbb{R}$ is μ-integrable in the sense of Fremlin or simply *(F)-μ-integrable* if $f = f_1 - f_2$ for some $f_1, f_2 \in U$ and define its (F)-μ-integral over Ω by

$$\int f \, d\mu := \int f_1 \, d\mu - \int f_2 \, d\mu;$$

see [7, Definition 122M]. The integral $\int f \, d\mu$ does not depend on the choice of such $f_1, f_2 \in U$ satisfying $f = f_1 - f_2$; see [7, Remark 122N]. Moreover, every (F)-μ-integrable function is necessarily μ-virtually measurable because so are the functions in U. By $\widetilde{\mathcal{L}}^1(\mu)$ we denote the set of all (F)-μ-integrable functions. It turns out to be a real vector space, [7, Theorem 122O].

Let $\mathcal{L}^1(\mu)_{\mathbb{R}}$ denote the real part of $\mathcal{L}^1(\mu)$, namely

$$\mathcal{L}^1(\mu)_{\mathbb{R}} := \{f \in \mathcal{L}^1(\mu) : f \text{ is } \mathbb{R}\text{-valued}\}.$$

Given $\mathbb{R}$-valued functions f, g defined on subsets of Ω, we write $f = g$ ($\widehat{\mu}$-a.e.) if the subset $\{\omega \in (\operatorname{dom} f) \cap (\operatorname{dom} g) : f(\omega) \neq g(\omega)\}$ of Ω is $\widehat{\mu}$-null or equivalently, μ-negligible.

Lemma 6.3 *The following statements hold.*

(i) *A μ-virtually measurable function f is (F)-μ-integrable if and only if $|f| \in U$.*
(ii) *Let $f \in \widetilde{\mathcal{L}}^1(\mu)$. If g is a μ-virtually measurable function such that $f = g$ ($\widehat{\mu}$-a.e.), then $g \in \widetilde{\mathcal{L}}^1(\mu)$ and $\int f \, d\mu = \int g \, d\mu$.*
(iii) *The inclusion $\mathcal{L}^1(\mu)_{\mathbb{R}} \subseteq \widetilde{\mathcal{L}}^1(\mu)$ holds and $\int_\Omega f \, d\mu = \int f \, d\mu$ for every $f \in \mathcal{L}^1(\mu)_{\mathbb{R}}$.*
(iv) *Let $f : \Omega \longrightarrow \mathbb{R}$ be a Σ-measurable function. Then f is μ-integrable if and only if f is (F)-μ-integrable.*
(v) *If $f \in \widetilde{\mathcal{L}}^1(\mu)$, then there exists $g \in \mathcal{L}^1(\mu)_{\mathbb{R}}$ such that $f = g$ ($\widehat{\mu}$-a.e.) and $\int f \, d\mu = \int_\Omega g \, d\mu$.*

Proof

(i) See [7, Theorem 122P].
(ii) See [7, Corollary 122R(b)]
(iii) Let $f \in \mathcal{L}^1(\mu)_{\mathbb{R}}$. Write $f^+ := f \vee 0$ and $f^- := (-f) \vee 0$, so that $f^+, f^- \in U$ and $\int_\Omega f^+ d\mu = \int f^+ d\mu$ and $\int_\Omega f^- d\mu = \int f^- d\mu$; see Lemma 6.2 with f^+, f^- in place of f. Thus, $f = f^+ - f^- \in \widetilde{\mathcal{L}}(\mu)$. This fact can also be seen via part (i) above. Indeed, f is μ-virtually measurable and $|f|$ is μ-integrable. Hence, f belongs to U by Lemma 6.2 with $|f|$ in place of f. So, part (i) gives $f \in \widetilde{\mathcal{L}}^1(\mu)$.

Moreover, we have that

$$\int_\Omega f\,d\mu = \int_\Omega f^+\,d\mu - \int_\Omega f^-\,d\mu = \int f^+\,d\mu - \int f^-\,d\mu = \int f\,d\mu.$$

(iv) The 'only if' portion has been established in part (iii) above. All we need is to verify the 'if' portion. So, assume that f is (F)-μ-integrable. Then $|f| \in U$ by part (i) because the Σ-measurable function f is surely μ-virtually measurable. Choose an increasing sequence $\{s_n\}_{n=1}^\infty$ in $\text{sim}^+\,\Sigma^f$ such that $\lim_{n\to\infty} s_n(\omega) = |f(\omega)|$ for every $\omega \in \Omega$. The Monotone Convergence Theorem yields

$$\int_\Omega |f|\,d\mu = \lim_{n\to\infty} \int_\Omega s_n\,d\mu = \int |f|\,d\mu < \infty$$

by (6.1) and (6.2) with $|f|$ in place of f. Thus, f is μ-integrable.

(v) Since f is μ-virtually measurable, there is a μ-conegligible set E such that the function $f \restriction E : (\text{dom } f) \cap E \longrightarrow \mathbb{R}$ is $\Sigma_{(\text{dom } f)\cap E}$-measurable. As $(\text{dom } f) \cap E$ is μ-conegligible, choose any set $F \in \Sigma$ such that $\mu(\Omega \setminus F) = 0$ and $F \subseteq ((\text{dom } f) \cap E)$. Define $g(\omega) := f(\omega)$ for every $\omega \in F$ and $g(\omega) = 0$ for every $\omega \in \Omega \setminus F$. Then $g = f$ ($\widehat{\mu}$-a.e.). Moreover, $g \in \widetilde{\mathcal{L}}^1(\mu)$ and $\int g\,d\mu = \int f\,d\mu$ by part (ii) above. By applying part (iv), with g in place of f, we have that $g \in \mathcal{L}^1(\mu)$. Next part (iii), with g in place of f, yields $\int_\Omega g\,d\mu = \int g\,d\mu$. This establishes part (v). □

In view of Lemma 6.3(iv) above, the space $\mathcal{L}^1(\mu)_\mathbb{R}$ is exactly the space $\mathcal{L}^1_{\text{strict}}$ given in [8, 242Y(g)].

The $\widehat{\mu}$-a.e. equality specifies an equivalence relation in $\widetilde{\mathcal{L}}^1(\mu)$. Indeed, $f \sim g$ for $f, g \in \widetilde{\mathcal{L}}^1(\mu)$ is defined by requiring that $f = g$ ($\widehat{\mu}$-a.e.). Let $\widetilde{L}^1(\mu)$ denote the quotient vector space $\widetilde{\mathcal{L}}^1(\mu)/\!\sim$; see [8, 241C and 242A(a)]. Given $f \in \widetilde{\mathcal{L}}^1(\mu)$, its corresponding equivalence class is denoted by

$$f^\bullet := \{g \in \widetilde{\mathcal{L}}^1(\mu) : f \sim g\}.$$

For each $u \in \widetilde{L}^1(\mu)$, select any $f \in \widetilde{\mathcal{L}}^1(\mu)$ satisfying $u = f^\bullet$. Define the norm

$$\|u\|_1 := \int |f|\,d\mu,$$

which does not depend on the choice of such an f. Then $\widetilde{L}^1(\mu)$ is a real Banach space for the norm $\|\cdot\|$, [8, Theorem 242F].

Let $\mathcal{N}(\mu)_\mathbb{R}$ denote the set of all $\mathbb{R}$-valued, μ-null functions. Consider the quotient space $L^1(\mu)_\mathbb{R} := \mathcal{L}^1(\mu)_\mathbb{R}/\mathcal{N}(\mu)_\mathbb{R}$. Define a norm on $L^1(\mu)_\mathbb{R}$ by

$$\|f + \mathcal{N}(\mu)_\mathbb{R}\|_{L^1(\mu)_\mathbb{R}} := \int_\Omega |f|\,d\mu, \quad f \in \mathcal{L}^1(\mu)_\mathbb{R}.$$

This is well defined because if $f + \mathcal{N}(\mu)_\mathbb{R} = g + \mathcal{N}(\mu)_\mathbb{R}$ in $L^1(\mu)_\mathbb{R}$ for some $f, g \in \mathcal{L}^1(\mu)_\mathbb{R}$, then $\int_\Omega |f|\,d\mu = \int_\Omega |g|\,d\mu$. Of course, $L^1(\mu)_\mathbb{R}$ is the usual real-L^1-space for μ, so that it is a real Banach space. We refer to [21, pp. 64–66].

The following result from [8, 242Y(g)] allows us to identify $\mathcal{L}^1(\mu)_\mathbb{R}/\mathcal{N}(\mu)_\mathbb{R}$ with $\widetilde{L}^1(\mu)$.

Proposition 6.4 *The real Banach spaces $\mathcal{L}^1(\mu)_\mathbb{R}/\mathcal{N}(\mu)_\mathbb{R}$ and $\widetilde{L}^1(\mu)$ are linearly isometric.*

Proof Let $T : \mathcal{L}^1(\mu)_\mathbb{R} \longrightarrow \widetilde{\mathcal{L}}^1(\mu)$ denote the natural injection (see Lemma 6.3(iii)). The claim is that

$$(T(\varphi))^\bullet = 0 \quad \text{in } \widetilde{L}^1(\mu), \quad \varphi \in \mathcal{N}(\mu)_\mathbb{R}. \tag{6.3}$$

In fact, each $\varphi \in \mathcal{N}(\mu)_\mathbb{R}$ vanishes off a μ-null set. Thus, $\varphi(\omega) = 0$ for μ-a.e. $\omega \in \Omega$ and hence, $\varphi = 0$ ($\widehat{\mu}$-a.e.). Accordingly, $(T(\varphi))^\bullet = \varphi^\bullet = 0$ so that (6.3) holds. This enables us to define a linear map $\widehat{T} : L^1(\mu)_\mathbb{R} \longrightarrow \widetilde{L}^1(\mu)$ by

$$\widehat{T}(f + \mathcal{N}(\mu)_\mathbb{R}) := (T(f))^\bullet, \quad f \in \mathcal{L}^1(\mu)_\mathbb{R}.$$

To prove that $\widehat{T}$ is injective, let $f \in \mathcal{L}^1(\mu)_\mathbb{R}$ satisfy $(T(f))^\bullet = f^\bullet = 0$. By definition, $f(\omega) = 0$ for $\widehat{\mu}$-a.e. $\omega \in \Omega$ and hence, $f(\omega) = 0$ for μ-a.e. $\omega \in \Omega$. Thus, $f \in \mathcal{N}(\mu)_\mathbb{R}$. This implies that $\widehat{T}$ is injective.

For establishing the surjectivity of $\widehat{T}$, let $f \in \widetilde{\mathcal{L}}^1(\mu)$. Select $g \in \mathcal{L}^1(\mu)_\mathbb{R}$ such that $f = g$ ($\widehat{\mu}$-a.e.); see Lemma 6.3(v). Thus

$$\widehat{T}(g + \mathcal{N}(\mu)_\mathbb{R}) = (T(g))^\bullet = g^\bullet = f^\bullet,$$

which yields the surjectivity of $\widehat{T}$.

Given $f \in \mathcal{L}^1(\mu)_\mathbb{R} \subseteq \widetilde{\mathcal{L}}^1(\mu)$ it follows, from the definitions of the norms $\|\cdot\|_1$ and $\|\cdot\|_{L^1(\mu)_\mathbb{R}}$ together with Lemma 6.3(iii) (with $|f|$ in place of f), that

$$\begin{aligned}\|\widehat{T}(f + \mathcal{N}(\mu)_\mathbb{R})\|_1 &= \|(T(f))^\bullet\|_1 = \|f^\bullet\|_1 = \int |f|\, d\mu \\ &= \int_\Omega |f|\, d\mu = \|f + \mathcal{N}(\mu)_\mathbb{R}\|_{L^1(\mu)_\mathbb{R}}.\end{aligned}$$

This shows that $\widehat{T}$ is a linear isometry and thereby completes the proof. □

Remark 6.5 The complex Banach space $L^1(\mu) = \mathcal{L}^1(\mu)/\mathcal{N}(\mu)$ can be identified with the complexification of $L^1(\mu)_\mathbb{R} := \mathcal{L}^1(\mu)_\mathbb{R}/\mathcal{N}(\mu)_\mathbb{R}$. For the complex version of $\widetilde{L}^1(\mu)$, we refer to [8, 242P], where it is denoted by $L^1_\mathbb{C}$. Here, we will denote it by $\widetilde{L}^1(\mu)_\mathbb{C}$. So, $\widetilde{L}^1(\mu)_\mathbb{C}$ is the complexification of $\widetilde{L}^1(\mu)$; it is a complex Banach space for the norm induced by $\widetilde{L}^1(\mu)$. Moreover, we can extend $\widehat{T}$ to a surjective linear isometry $\widehat{T}_\mathbb{C} : L^1(\mu) \longrightarrow \widetilde{L}^1(\mu)_\mathbb{C}$. □

Next, let us turn our attention to $\mathcal{L}^\infty$-functions. An $\mathbb{R}$-valued, μ-virtually measurable function ψ is said to be μ-essentially bounded in the sense of Fremlin, or simply (F)-*essentially bounded* if there exists $a \geq 0$ such that $\{\omega \in \operatorname{dom} \psi :$

$|\psi(\omega)| \leq a\} \subseteq \Omega$ is a μ-conegligible set. The set $\widetilde{\mathcal{L}}^{\infty}(\mu)$ of all (F)-essentially bounded functions is a real vector space. Similar to the case of $\widetilde{\mathcal{L}}^{1}(\mu)$, the $\widehat{\mu}$-a.e. equality defines an equivalence relation in $\widetilde{\mathcal{L}}^{\infty}(\mu)$. Let $\widetilde{L}^{\infty}(\mu)$ denote the corresponding quotient space, which is also a real vector space. The equivalence class containing a function $\psi \in \widetilde{\mathcal{L}}^{\infty}(\mu)$ is denoted by $\psi^{\bullet}$. These facts have been formulated in [8, Definition 243A].

Given $\psi \in \widetilde{\mathcal{L}}^{\infty}(\mu)$, define the (F)-*essential supremum* of the function ψ, denoted by (F)-ess.supp $|\psi|$, as the infimum of those $a \geq 0$ such that $\{\omega \in \operatorname{dom}\psi : |\psi(\omega)| \leq a\}$ is a μ-conegligible set. For each $u \in \widetilde{L}^{\infty}(\mu)$, its (F)-essential supremum norm is defined by

$$\|u\|_{\infty} := (F)\text{-ess.sup}\,|\psi|, \qquad \text{whenever } \psi^{\bullet} = u.$$

The norm $\|u\|_{\infty}$ does not depend on the choice of $\psi \in \widetilde{\mathcal{L}}^{\infty}(\mu)$ satisfying $\psi^{\bullet} = u$. Then $\widetilde{L}^{\infty}(\mu)$ is a real Banach space for the norm $\|\cdot\|_{\infty}$; see [8, 243D] for the details.

Let

$$\mathcal{L}^{\infty}(\mu)_{\mathbb{R}} := \{\varphi \in \mathcal{L}^{\infty}(\mu) : \varphi \text{ is } \mathbb{R}\text{-valued}\}.$$

Lemma 6.6 *The following statements hold.*

(i) *If* $\varphi \in \mathcal{L}^{\infty}(\mu)_{\mathbb{R}}$, *then* $\varphi \in \widetilde{\mathcal{L}}^{\infty}(\mu)$ *and ess.sup* $|\varphi| = (F)$*-ess.sup* $|\varphi|$.
(ii) *A* Σ*-measurable function* $\varphi : \Omega \longrightarrow \mathbb{R}$ *belongs to* $\mathcal{L}^{\infty}(\mu)_{\mathbb{R}}$ *if and only if it is* (F)*-essentially bounded.*
(iii) *Given* $\psi \in \widetilde{\mathcal{L}}^{\infty}(\mu)$, *there exists* $\varphi \in \mathcal{L}^{\infty}(\mu)_{\mathbb{R}}$ *such that* $\psi = \varphi$ ($\widehat{\mu}$*-a.e.*).

Proof

(i) Given $a \geq 0$, the set $\{\omega \in \Omega : |\varphi(\omega)| \leq a\}$ belongs to Σ; we point out that $\operatorname{dom}\varphi = \Omega$. Hence, we have the equivalence that $\{\omega \in \Omega : |\varphi(\omega)| \leq a\}$ is μ-conegligible if and only if $\{\omega \in \Omega : |\varphi(\omega)| > a\}$ is μ-null. It is now routine to establish part (i) via this equivalence.
(ii) This also follows from the equivalence presented in the proof of part (i) above.
(iii) As ψ is μ-virtually measurable, we can find a μ-conegligible set $E \subseteq \Omega$ such that $\psi\restriction E : (\operatorname{dom}\psi) \cap E \longrightarrow \mathbb{R}$ is $\Sigma_{(\operatorname{dom}\psi)\cap E}$-measurable. As $(\operatorname{dom}\psi) \cap E$ is μ-conegligible, there exists $F \in \Sigma$ satisfying both $\mu(\Omega \setminus F) = 0$ and $F \subseteq (\operatorname{dom}\psi) \cap E$. Define a Σ-measurable function $\varphi : \Omega \longrightarrow \mathbb{R}$ by $\varphi(\omega) := \psi(\omega)$ for every $\omega \in F$ and $\varphi(\omega) := 0$ for every $\omega \in \Omega \setminus F$. Then $\varphi = \psi$ ($\widehat{\mu}$-a.e.) and hence, $\varphi \in \widetilde{\mathcal{L}}^{\infty}(\mu)$. By part (ii) we have $\varphi \in \mathcal{L}^{\infty}(\mu)_{\mathbb{R}}$. □

Lemma 6.6 implies that $\mathcal{L}^{\infty}(\mu)_{\mathbb{R}}$ equals the space $\mathcal{L}^{\infty}_{\text{strict}}$ consisting of all $\mathbb{R}$-valued, (F)-essentially bounded, Σ-measurable functions *on* Ω (see [8, 243X(b)]).

Recall the set $\mathcal{N}(\mu)_{\mathbb{R}}$ consisting of all $\mathbb{R}$-valued, μ-null functions. Then $\mathcal{N}(\mu)_{\mathbb{R}} = \mathcal{L}^{\infty}(\mu)_{\mathbb{R}} \cap \mathcal{N}(\mu)$ and a function $\varphi \in \mathcal{L}^{\infty}(\mu)_{\mathbb{R}}$ satisfies ess.sup $|\varphi| = 0$ if and only if $\varphi \in \mathcal{N}(\mu)_{\mathbb{R}}$. The quotient space $L^{\infty}(\mu)_{\mathbb{R}} := \mathcal{L}^{\infty}(\mu)_{\mathbb{R}}/\mathcal{N}(\mu)_{\mathbb{R}}$ is a real Banach space with respect to the norm

$$\|\varphi + \mathcal{N}(\mu)_{\mathbb{R}}\|_{L^{\infty}(\mu)_{\mathbb{R}}} := \text{ess.sup}\,|\varphi|, \quad \varphi \in \mathcal{L}^{\infty}(\mu)_{\mathbb{R}}.$$

The norm is well defined as $\varphi_1 + \mathcal{N}(\mu)_\mathbb{R} = \varphi_2 + \mathcal{N}(\mu)_\mathbb{R}$ for $\varphi_1, \varphi_2 \in \mathcal{L}^\infty(\mu)_\mathbb{R}$ implies that ess.sup $|\varphi_1| =$ ess.sup $|\varphi_2|$. The following result shows that we can identify the real Banach spaces $L^\infty(\mu)_\mathbb{R}$ and $\widetilde{L}^\infty(\mu)$ as in [8, 243X(b)].

Proposition 6.7 *The real Banach spaces $L^\infty(\mu)_\mathbb{R}$ and $\widetilde{L}^\infty(\mu)$ are linearly isometric.*

Proof Let $S : \mathcal{L}^\infty(\mu)_\mathbb{R} \longrightarrow \widetilde{\mathcal{L}}^\infty(\mu)$ denote the natural injection, which is well defined in view of Lemma 6.6(i). Each $\varphi \in \mathcal{N}(\mu)_\mathbb{R}$ satisfies (F)-ess.sup $|\varphi| =$ ess.sup $|\varphi| = 0$ (see Lemma 6.6(i)) and hence, $(S(\varphi))^\bullet = \varphi^\bullet = 0$ as elements of $\widetilde{L}^\infty(\mu)$. Because of this fact we can define a linear map $\widehat{S} : L^\infty(\mu)_\mathbb{R} \longrightarrow \widetilde{L}^\infty(\mu)$ by

$$\widehat{S}(\varphi + \mathcal{N}(\mu)_\mathbb{R}) := (S(\varphi))^\bullet, \quad \varphi \in \mathcal{L}^\infty(\mu)_\mathbb{R}.$$

To see that $\widehat{S}$ is injective, let $\varphi \in \mathcal{L}^\infty(\mu)_\mathbb{R}$ be a function satisfying $(S(\varphi))^\bullet = \varphi^\bullet = 0$. Then $\varphi = 0$ ($\widehat{\mu}$-a.e.) and hence, $\varphi \in \mathcal{N}(\mu)_\mathbb{R}$ as $\varphi(\omega) = 0$ for μ-a.e. $\omega \in \Omega$. So, $\widehat{S}$ is injective.

To verify the surjectivity of $\widehat{S}$, let $\psi \in \widetilde{\mathcal{L}}^\infty(\mu)$. Select $\varphi \in \mathcal{L}^\infty(\mu)_\mathbb{R} \subseteq \widetilde{\mathcal{L}}^\infty(\mu)$ such that $\psi = \varphi$ ($\widehat{\mu}$-a.e.); see Lemma 6.6(i),(iii). Since $\psi^\bullet = \varphi^\bullet$, we have that

$$\widehat{S}(\varphi + \mathcal{N}(\mu)_\mathbb{R}) = (S(\varphi))^\bullet = \varphi^\bullet = \psi^\bullet,$$

by which $\widehat{S}$ is surjective.

Given $\varphi \in \mathcal{L}^\infty(\mu)_\mathbb{R}$, it follows that

$$\begin{aligned}\|\widehat{S}(\varphi + \mathcal{N}(\mu)_\mathbb{R})\|_\infty &= \|(S(\varphi))^\bullet\|_\infty = \|\varphi^\bullet\|_\infty = (F)\text{-ess.sup } |\varphi| \\ &= \text{ess.sup } |\varphi| = \|\varphi + \mathcal{N}(\mu)_\mathbb{R}\|_{L^\infty(\mu)_\mathbb{R}},\end{aligned}$$

where we have used the definitions of $\|\cdot\|_\infty$ and $\|\cdot\|_{L^\infty(\mu)_\mathbb{R}}$ as well as Lemma 6.6(i). So, $\widehat{S}$ is a linear isometry and Proposition 6.7 is established. □

Remark 6.8 The complex Banach space $L^\infty(\mu) := \mathcal{L}^\infty(\mu)/\mathcal{N}(\mu)$ is the complexification of the real Banach space $L^\infty(\mu)_\mathbb{R} := \mathcal{L}^\infty(\mu)_\mathbb{R}/\mathcal{N}(\mu)_\mathbb{R}$. Regarding Fremlin's space $\widetilde{L}^\infty(\mu)$, its complexification $\widetilde{L}^\infty_\mathbb{C}(\mu)$ is a complex Banach space, [8, 243K]. So, $\widehat{S}$ has a natural extension to a surjective linear isometry $\widehat{S}_\mathbb{C} : L^\infty(\mu) \longrightarrow \widetilde{L}^\infty_\mathbb{C}(\mu)$. □

We define the canonical linear map $\widetilde{J}_\mu$ from $\widetilde{L}^\infty(\mu)$ into the dual Banach space $(\widetilde{L}^1(\mu))^*$ of $\widetilde{L}^1(\mu)$ by

$$\langle u, \widetilde{J}_\mu(v)\rangle = \int f\varphi\, d\mu, \quad u \in \widetilde{L}^1(\mu),\ v \in \widetilde{L}^\infty(\mu),$$

whenever $f \in \widetilde{\mathcal{L}}^1(\mu)$ and $\varphi \in \widetilde{\mathcal{L}}^\infty(\mu)$ satisfy $f^\bullet = u$ and $\varphi^\bullet = v$; see [8, 243F], where it is also shown that $\widetilde{J}_\mu$ is well defined. This map $\widetilde{J}_\mu$ can then be extended to the canonical map $(\widetilde{J}_\mu)_{\mathbb{C}} : \widetilde{L}^\infty_{\mathbb{C}}(\mu) \longrightarrow (\widetilde{L}^1_{\mathbb{C}}(\mu))^*$ such that $\widetilde{J}_\mu(\mathrm{Re}\, v) = \mathrm{Re}((\widetilde{J}_\mu)_{\mathbb{C}}(v))$ and $\widetilde{J}_\mu(\mathrm{Im}\, v) = \mathrm{Im}((\widetilde{J}\mu)_{\mathbb{C}}(v))$ for every $v \in \widetilde{L}^\infty_{\mathbb{C}}(\mu)$; see [8, 243K].

Parts (i) and (ii) of the following result are from [8, Theorem 243G] and parts (iii) and (iv) from [8, 243K].

Proposition 6.9 *The following statements hold.*

(i) *The canonical map $\widetilde{J}_\mu$ is injective if and only if the measure μ is semifinite, in which case $\widetilde{J}_\mu$ is a linear isometry from $\widetilde{L}^\infty(\mu)$ into $(\widetilde{L}^1(\mu))^*$.*
(ii) *The canonical map $\widetilde{J}_\mu$ is bijective if and only if the measure μ is localizable, in which case $\widetilde{J}_\mu$ is a linear isometry from $\widetilde{L}^\infty(\mu)$ onto $(\widetilde{L}^1(\mu))^*$.*
(iii) *The canonical map $(\widetilde{J}_\mu)_{\mathbb{C}}$ is injective if and only if the measure μ is semifinite, in which case $(\widetilde{J}_\mu)_{\mathbb{C}}$ is a linear isometry from the complex Banach space $\widetilde{L}^\infty_{\mathbb{C}}(\mu)$ into the complex dual Banach space $(\widetilde{L}^1_{\mathbb{C}}(\mu))^*$.*
(iv) *The canonical map $(\widetilde{J}_\mu)_{\mathbb{C}}$ is bijective if and only if the measure μ is localizable, in which case $(\widetilde{J}_\mu)_{\mathbb{C}}$ is a linear isometry from $\widetilde{L}^\infty_{\mathbb{C}}(\mu)$ onto $(\widetilde{L}^1_{\mathbb{C}}(\mu))^*$.*

Recall the canonical map $J_\mu : L^\infty(\mu) \longrightarrow (L^1(\mu))^*$ from the Introduction and Sect. 2. The following result is exactly parts (i) and (ii) of Lemma 2.6; we now clarify the reference quoted there.

Corollary 6.10 *The following statements hold.*

(i) *The measure μ is semifinite if and only if J_μ is injective, in which case J_μ is a linear isometry from $L^\infty(\mu)$ into $(L^1(\mu))^*$.*
(ii) *The measure μ is localizable if and only if J_μ is bijective, in which case J_μ is a linear isometry from $L^\infty(\mu)$ onto $(L^1(\mu))^*$.*

Proof Let $f \in \mathcal{L}^1(\mu)$. Then $\mathrm{Re}\, f$, $\mathrm{Im}\, f$ belong to $\mathcal{L}^1(\mu)_{\mathbb{R}} \subseteq \widetilde{\mathcal{L}}^1(\mu)$; see Lemma 6.3(iii). So, $f^\bullet := (\mathrm{Re}\, f)^\bullet + i(\mathrm{Im}\, f)^\bullet \in \widetilde{L}^1_{\mathbb{C}}(\mu)$ (see [8, 242P]. Next, let $\varphi \in \mathcal{L}^\infty(\mu)$. Then both $\mathrm{Re}\,\varphi$, $\mathrm{Im}\,\varphi$ belong to $\mathcal{L}^\infty(\mu)_{\mathbb{R}} \subseteq \widetilde{\mathcal{L}}^\infty(\mu)$; see Lemma 6.6(i). By D.H. Fremlin [8, 243K], we have $\varphi^\bullet := (\mathrm{Re}\,\varphi)^\bullet + i(\mathrm{Im}\,\varphi)^\bullet \in \widetilde{L}^\infty_{\mathbb{C}}(\mu)$. Note that $\widehat{T}_{\mathbb{C}}(f + \mathcal{N}(\mu)) = f^\bullet$ and $\widehat{S}_{\mathbb{C}}(\varphi + \mathcal{N}(\mu)) = \varphi^\bullet$; for the definition of $\widehat{T}_{\mathbb{C}}$ (resp. $\widehat{S}_{\mathbb{C}}$) see the proof of Proposition 6.4 and Remark 6.5 (resp. Proposition 6.7 and Remark 6.8). Thus, we have that

$$\begin{aligned}
&\langle f + \mathcal{N}(\mu), J_\mu(\varphi + \mathcal{N}(\mu))\rangle = \int_\Omega f\varphi\, d\mu \\
&= \langle f^\bullet, (\widetilde{J}_\mu)_{\mathbb{C}}(\varphi^\bullet)\rangle = \langle \widehat{T}_{\mathbb{C}}(f + \mathcal{N}(\mu)), (\widetilde{J}_\mu)_{\mathbb{C}} \circ \widehat{S}_{\mathbb{C}}(\varphi + \mathcal{N}(\mu))\rangle \\
&= \langle f + \mathcal{N}(\mu), (\widehat{T}_{\mathbb{C}})^* \circ (\widetilde{J}_\mu)_{\mathbb{C}} \circ \widehat{S}_{\mathbb{C}}(\varphi + \mathcal{N}(\mu))\rangle,
\end{aligned}$$

where $(\widehat{T}_{\mathbb{C}})^* : (\widetilde{L}^1_{\mathbb{C}}(\mu))^* \longrightarrow (L^1(\mu))^*$ is the dual operator of $\widehat{T}_{\mathbb{C}} : L^1(\mu) \longrightarrow \widetilde{L}^1_{\mathbb{C}}(\mu)$. So, we have the equality

$$J_\mu = (\widehat{T}_{\mathbb{C}})^* \circ (\widetilde{J}_\mu)_{\mathbb{C}} \circ \widehat{S}_{\mathbb{C}} \tag{6.4}$$

because $f \in \mathcal{L}^1(\mu)$ and $\varphi \in \mathcal{L}^\infty(\mu)$ are arbitrary.

Now, parts (i) and (ii) follow from Proposition 6.9(iii),(iv), via (6.4), respectively, because $(\widehat{T_{\mathbb{C}}})^*$, $(\widetilde{J_\mu})_{\mathbb{C}}$ and $\widehat{S_{\mathbb{C}}}$ are all surjective linear isometries. □

Acknowledgements The author Susumu Okada thanks the Mathematics Department of the Katholische Universität Eichstätt-Ingolstadt (Germany) for its hospitality and financial support during a 1 month research visit in October/November 2018. Both authors wish to thank Professor D.H. Fremlin for some relevant discussions.

References

1. S.K. Berberian, *Measure and Integration* (Chelsea Publishing, New York, 1965)
2. V.I. Bogachev, *Measure Theory*, vol. I (Springer, Berlin, 2007)
3. A. Brown, On the Lebesgue convergence theorem. Math. Nachr. **23**, 141–148 (1961)
4. D.L. Cohn, *Measure Theory* (Birkhäuser/Quinn-Woodbine, Woodbine, 1997)
5. D.H. Fremlin, *Topological Riesz Spaces and Measure Theory* (Cambridge University Press, Cambridge, 1974)
6. D.H. Fremlin, Measure Algebras, *Handbook of Boolean Algebras*, vol. 3 (North Holland, Amsterdam, 1989), pp. 877–980
7. D.H. Fremlin, *Measure Theory*, vol. 1 (Torres Fremlin, Colchester 2001). Corrected 2nd printing
8. D.H. Fremlin, *Measure Theory*, vol. 2 (Torres Fremlin, Colchester, 2003). Corrected 2nd printing
9. D.H. Fremlin, *Measure Theory*, vol. 3 (Torres Fremlin, Colchester, 2004). Corrected 2nd printing
10. P.R. Halmos, *Measure Theory* (van Nostrand, New York, 1950)
11. E. Hewitt, K. Stromberg, *Real and Abstract Analysis* (Springer, New York, 1969). 2nd corrected printing
12. B. Jefferies, *Evolution Processes and the Feynman-Kac Formula*, vol. 353 (Kluwer Academic Publishers, Dordrecht, 1996)
13. R.A. Johnson, On the Lebesgue decomposition theorem. Proc. Am. Math. Soc. **18**, 628–632 (1967)
14. I. Kluvánek, Characterization of the closed convex hull of the range of a vector-valued measure. J. Funct. Anal. **21**, 316–329 (1976)
15. I. Kluvánek, Conical measures and vector measures. Ann. Inst. Fourier (Grenoble) **27**, 83–105 (1977)
16. I. Kluvánek, G. Knowles, *Vector Measures and Control Systems* (North Holland, Amsterdam, 1976)
17. N.Y. Luther, A decomposition of measures. Can. J. Math. **20**, 953–959 (1968)
18. S. Okada, W.J. Ricker, Conical measures and closed vector measures. Funct. Approx. Comment. Math. **59**, 191–230 (2018)
19. M.M. Rao, *Measure Theory and Integration*, 2nd rev. edn. (Marcel Dekker, New York, 2004)
20. H.L. Royden, *Real Analysis*, 2nd edn. (Macmillan, New York, 1968)
21. W. Rudin, *Real and Complex Analysis*, 3rd edn. (McGraw-Hill, New York, 1987)
22. I.E. Segal, Equivalence of measure spaces. Am. J. Math. **73**, 275–313 (1951)
23. A.C. Zaanen, *Integration*, 2nd rev. edn. (North Holland, Amsterdam, 1967)

A Residue Formula for Locally Compact Noncommutative Manifolds

Denis Potapov, Fedor Sukochev, Dominic Vella, and Dmitriy Zanin

Dedicated to Ben de Pagter on the occasion of his 65th birthday

Abstract The most significant result in the integration theory for locally compact noncommutative spaces was proven in Carey et al. (J Funct Anal 263(2):383–414, 2012). We review and improve on this result. For bounded, positive operators A, B on a separable Hilbert space $\mathcal{H}$, if $AB \in \mathcal{M}_{1,\infty}$, $[A^{\frac{1}{2}}, B] \in \mathcal{L}_1$ and the limit $\lim_{p\downarrow 1}(p-1)\operatorname{Tr}(B^p A^p)$ exists, then AB is Dixmier measurable and, for any dilation-invariant extended limit ω, we have

$$\operatorname{Tr}_\omega(AB) = \lim_{p\downarrow 1}(p-1)\operatorname{Tr}(B^p A^p).$$

The proof of this result is facilitated by the efficient use of recent advances in the theory of singular traces and double operator integrals. The Dixmier traces of pseudo-differential operators of the form $M_f(1-\Delta)^{-\frac{d}{2}}$ on $\mathbb{R}^d$, where f is a Schwartz function on $\mathbb{R}^d$ and Δ denotes the Laplacian on $L_2(\mathbb{R}^d)$, are swiftly recovered using this result. An analogous result for the noncommutative plane is also obtained.

Keywords Dixmier trace · Connes integration formula · Noncommutative plane

D. Potapov · F. Sukochev (✉) · D. Vella · D. Zanin
School of Mathematics and Statistics, University of New South Wales, Sydney, Australia
e-mail: d.potapov@unsw.edu.au; f.sukochev@unsw.edu.au; d.vella@student.unsw.edu.au; d.zanin@unsw.edu.au

G. Buskes et al. (eds.), *Positivity and Noncommutative Analysis*,
Trends in Mathematics, https://doi.org/10.1007/978-3-030-10850-2_24

1 Introduction

Suppose $\mathcal{H}$ is a separable Hilbert space. We let $\mathcal{K}(\mathcal{H})$ denote the space of compact operators on $\mathcal{H}$, and $\mathcal{L}_1$ the ideal of trace-class operators on $\mathcal{H}$. The dual of the Macaev ideal, $\mathcal{M}_{1,\infty}$, is defined by

$$\mathcal{M}_{1,\infty} := \Big\{ A \in \mathcal{K}(\mathcal{H}) \, : \, \frac{1}{\log(2+n)} \sum_{j=0}^{n} \mu(j, A) = \mathcal{O}(1) \text{ as } n \uparrow \infty \Big\},$$

where $\{\mu(n, A)\}_{n=0}^{\infty}$ is the singular value sequence of $A \in \mathcal{K}(\mathcal{H})$ (see, e.g., [41]). A Dixmier trace is then defined on $\mathcal{M}_{1,\infty}$ by

$$\mathrm{Tr}_{\omega}(A) := \omega\bigg(\Big\{\frac{1}{\log(2+n)} \sum_{j=0}^{n} \mu(j, A)\Big\}_{n=0}^{\infty}\bigg), \quad 0 \le A \in \mathcal{M}_{1,\infty}. \tag{1.1}$$

Here, ω is a linear functional on the space of all bounded sequences $\ell_{\infty}(\mathbb{N})$ which vanishes on the subspace $c_0(\mathbb{N})$ of all sequences converging to zero; such functionals are called extended limits. In [21] (see also [13, §IV.2.β]), it is required that ω be invariant for the dilation semigroup on $\ell_{\infty}(\mathbb{N})$ for Tr_{ω} to be additive on $\mathcal{M}_{1,\infty}$. On the sub-ideal

$$\mathcal{L}_{1,\infty} := \Big\{ A \in \mathcal{K}(\mathcal{H}) \, : \, \mu(n, A) = \mathcal{O}\Big(\frac{1}{1+n}\Big) \text{ as } n \uparrow \infty \Big\},$$

of weak trace-class operators in $\mathcal{M}_{1,\infty}$, this dilation-invariance is, in fact, unnecessary [41, §9.7]. From the definition of Tr_{ω}, it can be seen that the Dixmier trace vanishes on $\mathcal{L}_1$; this is its most essential feature.

An operator $A \in \mathcal{M}_{1,\infty}$ is said to be Dixmier measurable if its Dixmier trace is independent of the choice of dilation-invariant extended limit ω. It is known that there exist Dixmier non-measurable operators in $\mathcal{L}_{1,\infty}$ (see [35, Theorem 1.4] wherein such an operator is constructed).

We may now state our main result.

Theorem 1.1 *Let* $0 \le A, B \in \mathcal{B}(\mathcal{H})$ *such that* $[A^{\frac{1}{2}}, B] \in \mathcal{L}_1$. *Let* $C_{AB} > 0$.

(a) If $AB \in \mathcal{M}_{1,\infty}$, *then the following are equivalent:*

(i) AB *is Dixmier measurable, and* $\mathrm{Tr}_{\omega}(AB) = C_{AB}$ *for all dilation-invariant extended limits* ω.

(ii) $\lim_{p \downarrow 1}(p-1)\,\mathrm{Tr}(B^p A^p) = C_{AB}$.

(b) If $AB \in \mathcal{L}_{1,\infty}$, *then the above also holds for any extended limit* ω.

In the following, we shall explain the components of this result, the primary technical ingredients of its proof, and its connections with the substantial body of the preceding work in this area.

1.1 Singular Traces in the Compact Setting

In 1966, Jacques Dixmier constructed an example of a non-normal trace [21] on a certain ideal of the compact operators. This functional was the Dixmier trace, which we defined above in (1.1).

Alain Connes in 1989 [12] applied the Dixmier trace in the setting of his noncommutative geometry. We state below the more recent version of this result from [35].

Theorem 1.2 (Connes Trace Theorem, [12, Theorem 1], [35, Corollary 7.22]) *If $\mathcal{H} = L_2(M)$, where M is a compact d-dimensional Riemannian manifold, and $B : C^\infty(M) \to C^\infty(M)$ is a classical pseudo-differential operator of order $-d$ with Wodzicki residue $\mathrm{Res}_W(B)$, then (its extension) $B \in \mathcal{L}_{1,\infty}$ and, for any extended limit ω, we have the equality*

$$\mathrm{Tr}_\omega(B) = \frac{1}{d(2\pi)^d} \mathrm{Res}_W(B).$$

It follows from Theorem 1.2 that the notion of integration of L_2-functions on compact manifolds is recovered by the Dixmier trace (see [40, Theorem 2.5], [35, Corollary 7.24]. Theorem 1.2 saw an immediate application [12] in the recovery of the Yang–Mills action functional in the setting of noncommutative differential geometry for compact manifolds.

In 1994 [13], Connes discovered a connection between Dixmier measurability and the residue of the ζ-function at its leading singularity using the Karamata theorem [33]:

Theorem 1.3 ([41, Theorem 9.3.1], [39, Corollary 6.8]) *Suppose $0 \leq B \in \mathcal{M}_{1,\infty}$, and $C_B \geq 0$. The following are equivalent:*

(*i*) *B is Dixmier measurable, and $\mathrm{Tr}_\omega(B) = C_B$ for all dilation-invariant extended limits ω.*
(*ii*) *$\lim_{p\downarrow 1}(p-1)\,\mathrm{Tr}(B^p) = C_B$, where Tr denotes the standard trace.*

Following a succession of results ([8, Theorem 3.8], [9, Theorem 4.11], [39, Theorem 6.6], [50, Corollary 16]), in 2017, this theorem was generalised in [51], where it was shown that a variant of Theorem 1.3 continues to hold for non-positive operators in $\mathcal{L}_{1,\infty}$.

Theorem 1.4 ([51, Theorem 1.2]) *Suppose $0 \leq B \in \mathcal{L}_{1,\infty}$, $V \in \mathcal{B}(\mathcal{H})$, and $C_{VB} \in \mathbb{C}$. The following are equivalent:*

(*i*) *VB is Dixmier measurable, and $\mathrm{Tr}_\omega(VB) = C_{VB}$ for all dilation-invariant extended limits ω.*
(*ii*) *$\lim_{p\downarrow 1}(p-1)\,\mathrm{Tr}(VB^p) = C_{VB}$.*

1.2 Singular Traces in the Locally Compact Setting

The previous results require that $B \in \mathcal{K}(\mathcal{H})$. However, it can occur that both A and B are not compact operators, but their product AB belongs to $\mathcal{L}_{1,\infty}$. For example, if $f \in \mathcal{S}(\mathbb{R}^d)$ is a Schwartz function on $\mathbb{R}^d$, and if $\Delta = \sum_{j=1}^{d} \frac{\partial^2}{\partial x_j^2}$ denotes the Laplace operator on $L_2(\mathbb{R}^d)$, then neither the multiplication operator by f on the Hilbert space $L_2(\mathbb{R}^d)$, denoted M_f, nor the resolvent operator $(1-\Delta)^{-\frac{d}{2}}$ acting on $L_2(\mathbb{R}^d)$ are compact operators, but their product $M_f(1-\Delta)^{-\frac{d}{2}}$ resides in $\mathcal{L}_{1,\infty}$ ([5, 5.7 (p. 103)], [35, Theorem 1.5]). There are analogous results to Theorems 1.3 and 1.4 in this case.

In 2012, Carey et al. [10, Theorem 4.13] established necessary conditions in the general setting of semifinite von Neumann algebras $\mathcal{M}$. We state below the restriction of their result to the case $\mathcal{M} = \mathcal{B}(\mathcal{H})$. In general, if $(\mathcal{E}, \|\cdot\|_{\mathcal{E}})$ is a symmetric operator space, the closure in the $\mathcal{E}$-norm of the subspace of all finite-rank operators is the separable part of $\mathcal{E}$ and is denoted $\mathcal{E}_0$. Note that an extended limit on $\ell_\infty(\mathbb{N})$ is called exponentiation-invariant if it is invariant for the exponentiation semigroup on $\ell_\infty(\mathbb{N})$ (more precisely, the restriction of the exponentiation semigroup acting on $L_\infty(\mathbb{R}_+)$ onto $\ell_\infty(\mathbb{N})$; see, e.g., [41, §2.3]).

Theorem 1.5 ([10, Theorem 4.13]) *Suppose $0 \le A, B \in \mathcal{B}(\mathcal{H})$. If there exists some $\varepsilon > 0$ such that $[A^{\frac{1}{2}-\varepsilon}, B] \in (\mathcal{M}_{1,\infty})_0$ and*

$$\sup_{1\le p\le 2} (p-1)\,\mathrm{Tr}(A^{\frac{1}{2}-\varepsilon} B^p A^{\frac{1}{2}-\varepsilon}) < \infty, \tag{1.2}$$

then $AB \in \mathcal{M}_{1,\infty}$ and, if $\lim_{p\downarrow 1}(p-1)\,\mathrm{Tr}(A^{\frac{1}{2}} B^p A^{\frac{1}{2}})$ exists, then for any dilation- and exponentiation-invariant extended limit ω,

$$\mathrm{Tr}_\omega(AB) = \lim_{p\downarrow 1}(p-1)\,\mathrm{Tr}(A^{\frac{1}{2}} B^p A^{\frac{1}{2}}). \tag{1.3}$$

The idea of using extended limits ω with additional invariance properties, in addition to dilation invariance, is due to Carey et al. [8]. There was also simultaneous work led by Dodds et al. [23, 24], who identified and established the existence of certain classes of symmetric functionals/singular traces which are defined with the help of Banach limits equipped with additional invariance properties.

The most significant advantage of Theorem 1.1 over Theorem 1.5 is that the requirement of exponentiation-invariant extended limits ω is dropped in the former, thus yielding Dixmier measurability. The necessary conditions in Theorem 1.1 are also easier to check in applications. The boundedness condition (1.2) in Theorem 1.5 is also modified in Theorem 1.1, and the required control on the commutator $[A^{\frac{1}{2}-\varepsilon}, B] \in (\mathcal{M}_{1,\infty})_0$ in the hypotheses of Theorem 1.5 is replaced with the trace-class condition $[A^{\frac{1}{2}}, B] \in \mathcal{L}_1$, which is simpler to verify. To illustrate the usefulness of Theorem 1.1, we apply it to two concrete examples of noncompact manifolds; one commutative and one noncommutative.

Since the relationship between pseudo-differential operators on (initially only compact) manifolds and the Dixmier trace originates from the Connes trace theorem, we apply Theorem 1.1 to the pseudo-differential operators $M_f(1-\Delta)^{-\frac{d}{2}}$ on $\mathbb{R}^d$ and recover the classical formula

$$\mathrm{Tr}_\omega\left(M_f(1-\Delta)^{-\frac{d}{2}}\right) = \frac{\mathrm{Vol}(\mathbb{S}^{d-1})}{d(2\pi)^d}\int_{\mathbb{R}^d} f(\mathbf{x})\,d\mathbf{x},$$

for Sobolev functions $f \in W_{d,1}(\mathbb{R}^d)$, where $\mathrm{Vol}(\mathbb{S}^{d-1})$ denotes the volume of the unit sphere $\mathbb{S}^{d-1}$ (see, e.g., [46, Corollary 14]). It is not immediate that the assumption required in Theorem 1.1, $\left[M_{f^{\frac{1}{2}}}, (1-\Delta)^{-\frac{d}{2}}\right] \in \mathcal{L}_1$, holds. Indeed, even the space of Schwartz functions is not closed under taking positive square roots (consider, e.g., $f(x) = x^2e^{-x^2}$, for which $(f^{\frac{1}{2}})'$ has a jump-discontinuity at $x = 0$), and this simple observation can complicate the analysis considerably. Fortunately, the assumption $\nabla(f^{\frac{1}{2}}) \in \ell_1(L_2)(\mathbb{R}^d)$ (which we show holds for all the nonnegative Schwartz functions f on $\mathbb{R}^d$) is all that is necessary to apply the Cwikel estimates from [48] and ensure that this commutator is indeed trace-class when $0 \leq f \in \mathcal{S}(\mathbb{R}^d)$; the result for the rest of $W_{d,1}(\mathbb{R}^d)$ follows from a standard density argument.

Using Theorem 1.1, we also recover the analogous result for the noncommutative (or Moyal) Euclidean plane recently shown in [52, Theorem 1.1] using a longer, totally different method which placed a strong emphasis on noncommutative Cwikel estimates found in [37]. We give a brief description of this noncommutative manifold. Suppose the convolution of functions on $\mathbb{R}^2$ is 'twisted' by a real skew-symmetric matrix Θ as follows: for $f, g \in \mathcal{S}(\mathbb{R}^2)$,

$$(f \diamond_\Theta g)(\mathbf{t}) := \int_{\mathbb{R}^2} f(\mathbf{t})g(\mathbf{s}-\mathbf{t})e^{i\mathbf{t}\cdot\Theta\mathbf{s}}\,d\mathbf{t}, \quad \mathbf{t} \in \mathbb{R}^2, \tag{1.4}$$

where $\Theta := \begin{pmatrix} 0 & \theta \\ -\theta & 0 \end{pmatrix}$, $\theta > 0$. This operation is the Fourier dual of the so-called Moyal product as studied in [27, 30, 31, 47].

Let $\mathrm{Op}_\Theta(f)$ correspond to left $\diamond_\Theta$-multiplication by a Schwartz function $f \in \mathcal{S}(\mathbb{R}^2)$; this provides an action of $\mathcal{S}(\mathbb{R}^2)$ onto itself which may be extended to a bounded linear operator on $L_2(\mathbb{R}^2)$.

In applying Theorem 1.1, we require that the commutator $\left[\mathrm{Op}_\Theta(f)^{\frac{1}{2}}, (1-\Delta_\Theta)^{-1}\right]$ belongs to $\mathcal{L}_1$, where the "Laplace-type" operator Δ_Θ is defined in [37] as the multiplication operator given by $\Delta_\Theta f(\mathbf{x}) = |\mathbf{x}|^2 f(\mathbf{x})$ defined on the weighted L_2-space, $L_2(\mathbb{R}^2, |\mathbf{x}|^4\,d\mathbf{x})$. Then, appealing to the double-sequence representation of the Fréchet $*$-algebra $\left(\mathcal{S}(\mathbb{R}^2), \diamond_\Theta\right)$ in [30], one may deduce that this algebra has a nontrivial positive cone.

The operator $\mathrm{Op}_\Theta(f)^{\frac{1}{2}}$, for $f \in \mathcal{S}(\mathbb{R}^2)$ such that $\mathrm{Op}_\Theta(f) \geq 0$, corresponds to multiplication by some $\diamond_\Theta$-square root, g, whose precise construction is generally unclear. However, we show (in Corollary 6.13 below) that $g \in \mathcal{S}(\mathbb{R}^2)$. This invariance also makes the trace-class condition on $\left[\mathrm{Op}_\Theta(f)^{\frac{1}{2}}, (1-\Delta_\Theta)^{-1}\right]$ follow from the fact that $\left(\mathcal{S}(\mathbb{R}^2), \diamond_\Theta\right)$ is naturally embedded in the noncommutative Sobolev space $W_{2,1}(\mathbb{R}^2_\Theta)$ defined in [37] (see, e.g., [52, Lemma 3.3]).

In [52, Theorem 1.1], it was shown that the Dixmier trace of $X(1-\Delta_\Theta)^{-1}$, for any operator $X \in W_{2,1}(\mathbb{R}^2_\Theta)$ on $L_2(\mathbb{R}^2)$, is given by the scalar multiple of $\tau(X)$, where τ is a faithful normal semifinite trace on the von Neumann algebra $L_\infty(\mathbb{R}^2_\Theta)$ (see Definition 6.1 below). For a Schwartz function $f \in \mathcal{S}(\mathbb{R}^2)$, we observe that

$$\tau\left(\mathrm{Op}_\Theta(f)\right) = \frac{\pi}{\theta} f(\mathbf{0}), \quad f \in \mathcal{S}(\mathbb{R}^2).$$

Then, choosing such Schwartz functions f on $\mathbb{R}^2$ that the operator $\mathrm{Op}_\Theta(f)$ is positive, we obtain the expression (see Proposition 6.17 below)

$$\mathrm{Tr}_\omega\left(\mathrm{Op}_\Theta(f)(1-\Delta_\Theta)^{-1}\right) = \pi f(\mathbf{0}),$$

which agrees with [52, Proposition 4.5]. It then follows from a density argument that

$$\mathrm{Tr}_\omega\left(X(1-\Delta_\Theta)^{-1}\right) = \theta\tau(X), \quad X \in W_{2,1}(\mathbb{R}^2_\Theta).$$

However, the reader is advised that the constant on the right-hand side of the above equation differs from the one in [52] due to a difference in normalisation conventions.

1.3 Double Operator Integration Theory

The proof of Theorem 1.1 depends strongly upon double operator integration techniques. The double operator integral first appeared in the literature in the work of Daleckiĭ and Kreĭn [15, 16], who worked with self-adjoint operators. These double operator integrals were then given a rigorous treatment in the 1960s by Birman and Solomyak in the setting of compact operators on separable Hilbert space [1–3], who utilised their value in investigating notions of differentiability in the setting of perturbation theory. The reader is referred to Birman and Solomyak's survey [6] for further details on their work.

For a sufficiently well-behaved Borel function h on $\mathbb{R}$, and self-adjoint operators A, B, one may ask how the difference $h(B)-h(A)$ depends on $B-A$. By [6, §1.4], integrating with respect to spectral measures yields the expression

$$h(B) - h(A) = \int_{\sigma(A)} \int_{\sigma(B)} h^{[1]}(\lambda, \mu)\, \mathrm{d}\mathsf{E}_A(\lambda)\,(B - A)\, \mathrm{d}\mathsf{E}_B(\mu),$$

where $\sigma(A)$ (resp. $\sigma(B)$) and E_A (resp. E_B) denote the spectrum and spectral measure of A (resp. B), and the function $h^{[1]}$ denotes the divided difference of h,

$$h^{[1]}(\lambda, \mu) = \frac{h(\lambda) - h(\mu)}{\lambda - \mu}, \quad \lambda \neq \mu.$$

Double operator integrals were treated once again in 2002 by de Pagter, Witvliet and one of the present authors in [18], this time in the Banach space setting, which required a more technical approach. Specifically, they were defined in terms of continuous Schur multipliers of the form

$$\mathcal{J}_\varphi := \int_{\mathbb{R}^2} \varphi \, \mathrm{d}(\mathsf{P} \otimes \mathsf{Q}),$$

where P, Q are spectral measures in a UMD-space X such that $\mathsf{P} \otimes \mathsf{Q}$ is finitely additive, which were constructed in [17], which in turn followed from their earlier work with Clément et al. [11] on discrete Schauder decompositions over UMD-spaces.

Suppose $\mathcal{M}$ is a semifinite von Neumann algebra, and suppose A is a self-adjoint operator affiliated with $\mathcal{M}$. Let $\sigma(A)$ denote the spectrum of A. In [19, Theorem 5.16], de Pagter et al. proceeded to show that if h is a function such that its divided difference $h^{[1]}$ admits a decomposition

$$h^{[1]}(\lambda, \mu) = \int_\Omega h_1(\lambda, s) h_2(\mu, s)\, \mathrm{d}\eta(s), \quad \lambda, \mu \in \sigma(A),$$

where (Ω, η) is some σ-finite measure space, and h_1, h_2 are bounded functions on $\sigma(A) \times \Omega$ continuous in the first argument satisfying the boundedness property

$$\int_\Omega \Big(\sup_{\lambda \in \sigma(A)} \big| h_1(\lambda, t) \big| \Big) \Big(\sup_{\mu \in \sigma(A)} \big| h_2(\mu, t) \big| \Big)\, \mathrm{d}\eta(t) < \infty, \tag{1.5}$$

then $h(A)$ has Gâteaux derivative given by the double operator integral $\mathcal{J}_{h^{[1]}}$. Thus, an analogy for the familiar Daletskiĭ–Kreĭn formula for $\mathcal{B}(\mathcal{H})$ (see [3]) holds also for the semifinite von Neumann algebra setting. In so doing, they showed that the operator norm of a double operator integral $\mathcal{J}_\varphi$ is controlled by the behaviour of φ. They then proceeded to show in [20, Corollary 7.5] that commutator estimates of the form

$$\big\| [A, h(B)] \big\|_{E(\mathcal{M})} \leq \mathrm{const}_h \cdot \big\| [A, B] \big\|_{E(\mathcal{M})} \tag{1.6}$$

may be obtained using double operator integral techniques for rearrangement-invariant Banach function spaces E.

Following the direction of obtaining Lipschitz and commutator estimates, two of the present authors in [43] showed that for certain h, if $A \in \mathcal{M}$ and B_0, B_1 self-adjoint operators such that $B_0A - AB_1 \in \mathcal{B}(\mathcal{H})$, then one has the difference formula

$$h(B_0)A - Ah(B_1) = \mathcal{J}_{h^{[1]}}(B_0A - AB_1).$$

Of particular interest to them was the case when $h(t) = t(1+t^2)^{-\frac{1}{2}}$, for $t \in \mathbb{R}$, is a smoothed signum function, which allows one to pass from unbounded to bounded spectral triples via the transformation $D \mapsto h(D)$ (see [44]). In this paper, we consider the case when $B = B_0 = B_1$, so that when h is chosen such that $h^{[1]}$ admits an integral decomposition satisfying (1.5), the difference formula above yields a commutator estimate analogous to (1.6): for $0 \leq A, B \in \mathcal{B}(\mathcal{H})$ such that $[A, B] \in \mathcal{L}_p$, then

$$\left\| \left[A, h(B)\right] \right\|_p \leq \mathrm{const}_h \cdot \left\| [A, B] \right\|_p.$$

This is the first of two key double operator integral tricks which play an instrumental role in this text (see, e.g., the proof of Proposition 4.3 below). The second trick is described as follows; for $0 \leq A, B \in \mathcal{B}(\mathcal{H})$, Connes et al. [14, Lemma 5.1] showed that one may use double operator integrals to characterise the difference $A^p - B^p$, for $p > 1$, as a weak integral of the form

$$\begin{aligned} A^p - B^p &= A^{p-1}(A-B) + (A-B)B^{p-1} \\ &\quad - \int_{\mathbb{R}} \hat{g}_p(t) A^{it}\big(A^{p-1}(A-B) + (A-B)B^{p-1}\big)B^{-it}\,\mathrm{d}t \end{aligned}$$

where g_p is some unique Schwartz function. Using this decomposition, it was shown that if $B \in \mathcal{L}_{p,\infty}$, where $\mathcal{L}_{p,\infty}$ is the pth weak Schatten class, such that $[A^{\frac{1}{2}}, B] \in (\mathcal{L}_{p,\infty})_0$, then the difference $B^pA^p - (A^{\frac{1}{2}}BA^{\frac{1}{2}})^p \in (\mathcal{L}_{1,\infty})_0$—that is, the Dixmier trace vanishes on this difference, which allows us to bridge the limit in Theorem 1.1(ii) with the limit $\lim_{p\downarrow 1}(p-1)\,\mathrm{Tr}\left((A^{\frac{1}{2}}BA^{\frac{1}{2}})^p\right)$ which was thoroughly treated in [41, Theorem 9.3.1].

1.4 Arrangement of This Paper

The contents of this paper are organised as follows. In the next section, the basic concepts and notations are established, including the precise definition of the Dixmier trace and measurability, as well as a short exposition on vector-valued integration. Section 3 gives the definition of the double operator integrals used in this paper, and proves estimates used in the proof of Theorem 1.5. The main result of the present paper, Theorem 1.1, is proved in Sect. 4. Sections 5 and 6 are dedicated to the examples.

2 Preliminaries

2.1 General Notation

We adopt the convention $\mathbb{N} = \{0, 1, 2, \ldots\}$. For $p \in [1, \infty)$, let $\ell_p(\mathbb{N})$ denote the space of p-summable sequences. We denote by $\ell_\infty(\mathbb{N})$ the space of all bounded sequences.

Likewise, for $d \geq 1$, denote by $L_p(\mathbb{R}^d)$ the L_p-space of Lebesgue p-integrable (complex-valued) functions on $\mathbb{R}^d$, and by $L_\infty(\mathbb{R}^d)$ the space of all essentially bounded functions. We let $\mathcal{S}(\mathbb{R}^d)$ denote the space of Schwartz functions f on $\mathbb{R}^d$. We use the unitary version of the Fourier transform $\mathcal{F}$; that is,

$$(\mathcal{F}f)(\mathbf{x}) = \hat{f}(\mathbf{x}) := \frac{1}{(2\pi)^{\frac{d}{2}}} \int_{\mathbb{R}^d} f(\mathbf{t}) e^{-i\mathbf{t}\cdot\mathbf{x}}\, d\mathbf{t}, \quad f \in \mathcal{S}(\mathbb{R}^d),\ \mathbf{x} \in \mathbb{R}^d,$$

with the inverse Fourier transform given by

$$(\mathcal{F}^{-1}f)(\mathbf{x}) = \check{f}(\mathbf{x}) := \frac{1}{(2\pi)^{\frac{d}{2}}} \int_{\mathbb{R}^d} f(\mathbf{t}) e^{i\mathbf{t}\cdot\mathbf{x}}\, d\mathbf{t}, \quad f \in \mathcal{S}(\mathbb{R}^d),\ \mathbf{x} \in \mathbb{R}^d.$$

Recall the Fourier inversion theorem (see, e.g., [45, Theorem IX.1]), which states that $\mathcal{F}^{-1}$ is the inverse operator of $\mathcal{F}$ as operators on $\mathcal{S}(\mathbb{R}^d)$. Recall also the Plancherel theorem (see, e.g., [45, Theorem IX.6]), which states that $\mathcal{F}$ and $\mathcal{F}^{-1}$ may be extended to mutually adjoint unitary operators on $L_2(\mathbb{R}^d)$; this implies that

$$\mathcal{F}^{-1}\mathcal{F}f = \mathcal{F}\mathcal{F}^{-1}f = f, \quad \forall f \in L_2(\mathbb{R}^d). \tag{2.1}$$

For $m \in \mathbb{N}$, we say that a function is C_b^m if it is C^m and all of its derivatives up to and including the mth degree are bounded, and we denote the space of C_b^m-functions (and C^m-functions) on $\mathbb{R}^d$ by $C_b^m(\mathbb{R}^d)$ (resp. $C^m(\mathbb{R}^d)$). Furthermore, for $p \in [1, \infty]$, we denote the (m, p)th Sobolev space on $\mathbb{R}^d$ by $W_{m,p}(\mathbb{R}^d)$ equipped with the corresponding Sobolev-norm

$$\|f\|_{W_{m,p}} := \sum_{\alpha : |\alpha| \leq m} \left\| \frac{\partial^{\alpha_1}}{\partial t_1^{\alpha_1}} \cdots \frac{\partial^{\alpha_d}}{\partial t_d^{\alpha_d}}(f) \right\|_p, \quad f \in W_{m,p}(\mathbb{R}^d),$$

where the summation above runs over all multi-indices $\alpha = (\alpha_1, \ldots, \alpha_d) \in \mathbb{N}^d$ such that $|\alpha| = \alpha_1 + \cdots + \alpha_d \leq m$. More generally, for $s \geq 0$, we denote the (s, p)th inhomogeneous Sobolev space on $\mathbb{R}^d$ by $W_{s,p}(\mathbb{R}^d)$ equipped with the corresponding norm $\|f\|_{W_{s,p}} = \left\|(1-\Delta)^{\frac{s}{2}} f\right\|_p$, where Δ is the d-dimensional Laplacian operator. The reader is advised that, since the two definitions of Sobolev spaces stated above are equivalent when $s \in \mathbb{N}$, there is no ambiguity in notations. For further details on these Sobolev spaces, the reader is referred to [32, §6.2].

We denote by $\mathcal{H}$ a separable infinite-dimensional Hilbert space. Furthermore, we let $\mathcal{B}(\mathcal{H})$ denote the $*$-algebra of all bounded linear operators on $\mathcal{H}$. This becomes a C^*-algebra when equipped with the uniform operator norm (which we shall denote by $\|\cdot\|_\infty$). For any operator A on $\mathcal{H}$, we let $\sigma(A)$ denote its spectrum.

For a compact operator $A \in \mathcal{K}(\mathcal{H})$, we let $\mu(A) := \{\mu(n, A)\}_{n \in \mathbb{N}}$ denote the singular value sequence of A—that is, the decreasing rearrangement of the eigenvalue sequence of $|A|$. We denote the standard trace on $\mathcal{B}(\mathcal{H})$ by Tr. For $p \in [1, \infty)$, we let

$$\mathcal{L}_p(\mathcal{H}) := \left\{A \in \mathcal{B}(\mathcal{H}) \,:\, \operatorname{Tr}\left(|A|^p\right) < \infty\right\}$$

denote the pth Schatten ideal of $\mathcal{B}(\mathcal{H})$, with associated norm

$$\|A\|_p := \operatorname{Tr}\left(|A|^p\right)^{\frac{1}{p}}, \quad A \in \mathcal{L}_p(\mathcal{H}).$$

In particular, the class $\mathcal{L}_1(\mathcal{H})$ denotes the ideal of trace-class operators. Likewise, we denote the pth weak Schatten ideal of $\mathcal{B}(\mathcal{H})$ by

$$\mathcal{L}_{p,\infty}(\mathcal{H}) = \left\{A \in \mathcal{B}(\mathcal{H}) \,:\, \mu(n, A) = \mathcal{O}\left((1+n)^{-\frac{1}{p}}\right)\right\},$$

equipped with the natural quasinorm

$$\|A\|_{p,\infty} := \sup_{n \in \mathbb{N}} (1+n)^{\frac{1}{p}} \mu(n, A), \quad A \in \mathcal{L}_{p,\infty}(\mathcal{H}).$$

By Sukochev [49], we note that for $0 < \alpha < 1$, if $A, B \in \mathcal{B}(\mathcal{H})$ such that $AB \in \mathcal{L}_1(\mathcal{H})$, then $B^\alpha A B^{1-\alpha} \in \mathcal{L}_1(\mathcal{H})$ and

$$\|B^\alpha A B^{1-\alpha}\|_1 \leq \|AB\|_1. \tag{2.2}$$

For two operators $C_1, C_2 \in \mathcal{B}(\mathcal{H})$, we let $[C_1, C_2] := C_1C_2 - C_2C_1$ denote its commutator. For $n \geq 1$, we have the identity

$$[C_1, C_2^n] = \sum_{k=0}^{n-1} C_2^k [C_1, C_2] C_2^{n-k-1}, \tag{2.3}$$

For an operator $A \in \mathcal{B}(\mathcal{H})$, we let E_A denote the spectral measure associated with A. Let $\mathcal{H} = L_2(\mathbb{R}^d)$, for $d \in \mathbb{Z}_+$, and let $A \in \mathcal{B}\big(L_2(\mathbb{R}^d)\big)$. If there exists a function $K \in L_2(\mathbb{R}^d \times \mathbb{R}^d)$ such that for every $f \in L_2(\mathbb{R}^d)$,

$$(Af)(\mathbf{x}) = \int_{\mathbb{R}^d} K(\mathbf{x}, \mathbf{y}) f(\mathbf{y})\, \mathrm{d}\mathbf{y}, \quad \mathbf{x} \in \mathbb{R}^d,$$

then A is called an integral operator with integral kernel $K \in L_2(\mathbb{R}^d \times \mathbb{R}^d)$. The following result for trace-class integral operators has become folklore:

Proposition 2.1 ([25, Theorem V.3.1.1], [7, Theorem 3.1]) *Suppose $A \in \mathcal{L}_1\big(L_2(\mathbb{R}^d)\big)$ is an integral operator with kernel $K \in L_2(\mathbb{R}^d \times \mathbb{R}^d)$. If K is continuous on $\mathbb{R}^d \times \mathbb{R}^d$, then $\int_{\mathbb{R}^d} K(\mathbf{x}, \mathbf{x})\, \mathrm{d}\mathbf{x} < \infty$ and*

$$\mathrm{Tr}(A) = \int_{\mathbb{R}^d} K(\mathbf{x}, \mathbf{x})\, \mathrm{d}\mathbf{x}.$$

2.2 *Dixmier Traces*

The interested reader is referred to [41] for a thorough exposition on the theory of singular traces, including Dixmier traces.

Definition 2.2 For $k \geq 1$, let σ_k be the mapping on $\ell_\infty(\mathbb{N})$ defined by

$$\sigma_k(x_0, x_1, \ldots) := (\underbrace{x_0, \ldots, x_0}_{k \text{ times}}, \underbrace{x_1, \ldots, x_1}_{k \text{ times}}, \ldots).$$

The semigroup generated by these maps is called the dilation semigroup. We call a state on $\ell_\infty(\mathbb{N})$ invariant under σ_k, for all $k \in \mathbb{Z}_+$, a *dilation-invariant extended limit*.

Recall the Dixmier–Macaev ideal of $\mathcal{B}(\mathcal{H})$ (the dual of the Macaev ideal) is defined by

$$\mathcal{M}_{1,\infty}(\mathcal{H}) := \Big\{ A \in \mathcal{B}(\mathcal{H}) \,:\, \sup_{n \in \mathbb{N}} \frac{1}{\log(2+n)} \sum_{j=0}^{n} \mu(j, A) < \infty \Big\},$$

and that $\mathcal{L}_{1,\infty}(\mathcal{H}) \subset \mathcal{M}_{1,\infty}(\mathcal{H})$.

Definition 2.3 Let ω be a dilation-invariant extended limit. The functional $\mathrm{Tr}_\omega :$ $\mathcal{M}^+_{1,\infty}(\mathcal{H}) \to \mathbb{C}$ defined by setting

$$\mathrm{Tr}_\omega(A) = \omega\bigg(\Big\{ \frac{1}{\log(2+n)} \sum_{j=0}^{n} \mu(j, A) \Big\}_{n \in \mathbb{N}} \bigg), \quad 0 \leq A \in \mathcal{M}_{1,\infty}(\mathcal{H})$$

is positive and additive, and extends to a linear trace on $\mathcal{M}_{1,\infty}(\mathcal{H})$ ([13, §IV.2.β], [22, Example 2.5], [41, Theorem 6.3.6]). We call such a trace Tr_ω a *Dixmier trace*. Additionally, an operator $A \in \mathcal{M}_{1,\infty}(\mathcal{H})$ is said to be *Dixmier measurable* if $\mathrm{Tr}_{\omega_1}(A) = \mathrm{Tr}_{\omega_2}(A)$, for all dilation-invariant extended limits ω_1, ω_2.

Fixing some dilation-invariant limit ω, we note several properties of the Dixmier trace. Firstly, Tr_ω is unitarily invariant, and therefore satisfies the tracial cyclic property,

$$\mathrm{Tr}_\omega(AB) = \mathrm{Tr}_\omega(BA), \quad \text{for all } A \in \mathcal{M}_{1,\infty}(\mathcal{H}),\ B \in \mathcal{B}(\mathcal{H}).$$

Secondly, Tr_ω is singular, in that it vanishes on $(\mathcal{L}_{1,\infty})_0$, the closure of $\mathcal{L}_1(\mathcal{H})$ in $\mathcal{L}_{1,\infty}(\mathcal{H})$.

2.3 Weak Integration in $\mathcal{B}(\mathcal{H})$

The exposition on weak integration below closely follows Section 2.7 of [14].

Definition 2.4 Let $f : \mathbb{R} \to \mathcal{B}(\mathcal{H})$ be a function. Such a function is *measurable in the weak operator topology* if, for all $\xi, \eta \in \mathcal{H}$, the map

$$t \mapsto \langle f(t)\xi, \eta\rangle_{\mathcal{H}}, \quad t \in \mathbb{R},$$

is measurable.

Note that real-valued mapping

$$t \mapsto \sup_{\|\xi\|, \|\eta\| \le 1} \left|\langle f(t)\xi, \eta\rangle_{\mathcal{H}}\right| = \|f(t)\|_\infty, \quad t \in \mathbb{R},$$

is measurable. For such functions, there is a notion of a weak integral.

Definition 2.5 Suppose $f : \mathbb{R} \to \mathcal{B}(\mathcal{H})$ is measurable in the weak operator topology. We say that f is *integrable in the weak operator topology* if

$$\int_{\mathbb{R}} \|f(t)\|_\infty \, \mathrm{d}t < \infty. \tag{2.4}$$

Now, define a sesquilinear form

$$(\xi, \eta)_f := \int_{\mathbb{R}} \langle f(s)\xi, \eta\rangle_{\mathcal{H}} \, \mathrm{d}s, \quad \text{for } \xi, \eta \in \mathcal{H},$$

from which we immediately see that

$$\left|(\xi,\eta)_f\right| \le \|\xi\| \Big(\int_{\mathbb{R}} \left\|f(t)\right\|_\infty \mathrm{d}t \Big) \|\eta\|, \quad \text{for } \xi,\eta \in \mathcal{H}.$$

Hence, fixing $\xi \in \mathcal{H}$, the map $\eta \mapsto (\xi,\eta)_f$ defines a bounded, anti-linear functional on $\mathcal{H}$. Hence, by the Riesz lemma (i.e., the description of the Hilbert space dual), there exists an element $x_\xi \in \mathcal{H}$ such that

$$\langle x_\xi, \eta\rangle_{\mathcal{H}} = (\xi,\eta)_f, \quad \text{for all } \eta \in \mathcal{H}.$$

The operator which maps $\xi \mapsto x_\xi$ is called the *weak integral* of f, and is denoted $\int_{\mathbb{R}} f(s)\,\mathrm{d}s$.

Lemma 2.6 ([53, Lemma 2.3.2]) *Suppose $f : \mathbb{R} \to \mathcal{B}(\mathcal{H})$ is continuous in the weak operator topology. If $f(t) \in \mathcal{L}_1(\mathcal{H})$, for all $t \in \mathbb{R}$, and if*

$$\int_{\mathbb{R}} \left\|f(s)\right\|_1 \mathrm{d}s < \infty,$$

then f is integrable in the weak operator topology, $\int_{\mathbb{R}} f(s)\,\mathrm{d}s \in \mathcal{L}_1(\mathcal{H})$ and

$$\left\| \int_{\mathbb{R}} f(s)\,\mathrm{d}s \right\|_1 \le \int_{\mathbb{R}} \left\|f(s)\right\|_1 \mathrm{d}s. \tag{2.5}$$

3 Double Operator Integrals

Suppose X, Y are self-adjoint operators on a separable Hilbert space $\mathcal{H}$, and f is a bounded, Borel-measurable function on $\sigma(X) \times \sigma(Y) \subset \mathbb{R}^2$. Intuitively speaking, the double operator integral $\mathcal{J}_f^{X,Y}$ is then defined as an operator in $\mathcal{B}(\mathcal{L}_2)$ expressed in terms of the product of the spectral measures of X, Y by

$$\mathcal{J}_f^{X,Y} = \int_{\sigma(Y)} \int_{\sigma(X)} f(\lambda,\mu)\,\mathrm{d}(\mathsf{E}_X \otimes \mathsf{E}_Y)(\lambda,\mu). \tag{3.1}$$

To ensure that this object defines a bounded operator on the other Schatten–von Neumann classes, we require that the function f belong to the integral projective tensor product (see [42] and also formula (1.5) above).

Definition 3.1 Let X, Y be self-adjoint operators on $\mathcal{H}$. Suppose f is a bounded, Borel-measurable function on $\sigma(X) \times \sigma(Y)$ such that there exists a σ-finite measure space (Ω,η) and functions f_1, f_2 on $\sigma(X) \times \Omega$, $\sigma(Y) \times \Omega$, respectively, satisfying the condition

$$\int_\Omega \Big(\sup_{\lambda\in\sigma(X)} \big|f_1(\lambda,t)\big|\Big)\Big(\sup_{\mu\in\sigma(Y)} \big|f_2(\mu,t)\big|\Big)\,\mathrm{d}\eta(t) < \infty, \tag{3.2}$$

such that

$$f(\lambda,\mu) = \int_\Omega f_1(\lambda,t) f_2(\mu,t)\,\mathrm{d}\eta(t), \quad \text{for } \lambda\in\sigma(X),\ \mu\in\sigma(Y). \tag{3.3}$$

We define the *double operator integral* $\mathcal{J}_f^{X,Y}$ by the expression

$$\mathcal{J}_f^{X,Y}(A) := \int_\Omega f_1(X,t) A f_2(Y,t)\,\mathrm{d}\eta(t), \quad \text{for } A\in\mathcal{B}(\mathcal{H}),$$

where the above integral is understood in the weak operator topology. The class of functions f with decomposition f_1, f_2 satisfying (3.2), (3.3) shall be denoted $\mathfrak{A}_0$, and is a Banach algebra under the norm

$$\|f\|_{\mathfrak{A}_0} := \inf_{f_1, f_2} \int_\Omega \Big(\sup_{\lambda\in\sigma(X)} \big|f_1(\lambda,t)\big|\Big)\Big(\sup_{\mu\in\sigma(Y)} \big|f_2(\mu,t)\big|\Big)\,\mathrm{d}\eta(t), \quad \text{for } f\in\mathfrak{A}_0,$$

where the above infimum runs over all possible f_1, f_2 satisfying (3.2), (3.3) (see [19] for details).

Remark 3.2 ([19, Proposition 4.7], [44, Corollary 2]) If X, Y are self-adjoint operators on $\mathcal{H}$, $f\in\mathfrak{A}_0$, and $1\le p\le\infty$, then $\mathcal{J}_f^{X,Y} : \mathcal{L}_p(\mathcal{H}) \to \mathcal{L}_p(\mathcal{H})$, and

$$\|\mathcal{J}_f^{X,Y}\|_{p\to p} \le \|f\|_{\mathfrak{A}_0}. \tag{3.4}$$

3.1 Connection to Commutators

Suppose h is some Borel function on $\mathbb{R}$, $A\in\mathcal{B}(\mathcal{H})$, and B is a self-adjoint operator on $\mathcal{H}$. The following results control the p-norm of the commutator $\big[A, h(B)\big]$ in terms of $[A, B]$.

Lemma 3.3 ([43, Theorem 3.1]) *Suppose B is a self-adjoint operator on $\mathcal{H}$, and suppose $A\in\mathcal{B}(\mathcal{H})$. Suppose h is a C_b^1-function on $\mathbb{R}$, and let $h^{[1]}$ be the function on $\mathbb{R}^2$ defined by*

$$h^{[1]}(x,y) = \begin{cases} \dfrac{h(x)-h(y)}{x-y}, & \text{if } x\ne y,\\ h'(x), & \text{if } x = y. \end{cases}$$

If $[A, B] \in \mathcal{B}(\mathcal{H})$ *and* $h^{[1]} \in \mathfrak{A}_0$, *then* $[A, h(B)] \in \mathcal{B}(\mathcal{H})$, *and*

$$[A, h(B)] = \mathcal{J}_{h^{[1]}}^{B,B}([A, B]). \tag{3.5}$$

Observe that, by [44, Theorem 4], if h is a C_b^2-function, then $h^{[1]} \in \mathfrak{A}_0$ and

$$\|h^{[1]}\|_{\mathfrak{A}_0} \leq \text{const} \cdot (\|h\|_\infty + \|h'\|_\infty + \|h''\|_\infty).$$

Corollary 3.4 ([43, Theorem 5.3]) *Let* $p \geq 1$. *Suppose* B *is a self-adjoint operator on* $\mathcal{H}$, *and suppose* $A \in \mathcal{B}(\mathcal{H})$ *such that* $[A, B] \in \mathcal{L}_p(\mathcal{H})$. *If* h *is a* C_b^2*-function on* $\mathbb{R}$, *then* $[A, h(B)] \in \mathcal{L}_p(\mathcal{H})$ *and*

$$\left\|[A, h(B)]\right\|_p \leq \text{const} \cdot (\|h\|_\infty + \|h'\|_\infty + \|h''\|_\infty)\left\|[A, B]\right\|_p. \tag{3.6}$$

The class of C_b^2-functions will not be sufficient for our purposes. However, if h is not bounded, we may still place controls on its derivatives to guarantee that $h^{[1]}$ belongs to $\mathfrak{A}_0$.

Lemma 3.5 ([55, Lemma 2.8]) *Let* $p \geq 1$. *Suppose* $A \in \mathcal{B}(\mathcal{H})$ *and* B *is a self-adjoint operator on* $\mathcal{H}$ *such that* $[A, B] \in \mathcal{L}_p(\mathcal{H})$. *If* h *is a* C^2*-function on* $\mathbb{R}$ *such that its derivative* $h' \in W_{1,2}(\mathbb{R})$ *is an absolutely continuous function, then*

$$\left\|[A, h(B)]\right\|_p \leq \text{const} \cdot (\|h'\|_2 + \|h''\|_2)\left\|[A, B]\right\|_p.$$

Proof By Fourier inversion (2.1), the divided difference $h^{[1]}$ of h may be expressed as

$$h^{[1]}(x, y) = \int_0^1 h'(sx + (1-s)y)\,\mathrm{d}s = \frac{1}{\sqrt{2\pi}} \int_0^1 \int_{\mathbb{R}} \widehat{(h')}(t) e^{it(sx+(1-s)y)}\,\mathrm{d}t\,\mathrm{d}s.$$

Observe that we may construct functions $h_1, h_2 : \mathbb{R} \times ([0, 1] \times \mathbb{R}) \to \mathbb{C}$ defined by

$$h_1(x, (s, t)) := e^{itsx}, \quad h_2(y, (s, t)) := e^{it(1-s)y},$$

and a measure η on $[0, 1] \times \mathbb{R}$ given by $\mathrm{d}\eta(s, t) := \widehat{(h')}(t)\,\mathrm{d}t\,\mathrm{d}s$ such that the decomposition

$$h^{[1]}(x, y) = \frac{1}{\sqrt{2\pi}} \int_{[0,1]\times\mathbb{R}} h_1(x, (s, t)) h_2(y, (s, t))\,\mathrm{d}\eta(s, t). \tag{3.7}$$

Therefore, appealing to [44, Lemma 7], since h' is absolutely continuous and $h', h'' \in L_2(\mathbb{R})$ by assumption, we have that $\widehat{(h')} \in L_1(\mathbb{R})$ and

$$\big\|\widehat{(h')}\big\|_1 \leq \sqrt{2}\big(\|h'\|_2 + \|h''\|_2\big).$$

Hence, the decomposition (3.7) satisfies (3.2) and (3.3), and $h^{[1]} \in \mathfrak{A}_0$, with $\|h^{[1]}\|_{\mathfrak{A}_0} \leq \big\|\widehat{(h')}\big\|_1$. This yields the estimate

$$\begin{aligned}\Big\|[A, h(B)]\Big\|_p &\overset{(3.5)}{=} \Big\|\mathcal{J}^{B,B}_{h^{[1]}}([A, B])\Big\|_p \\ &\leq \|\mathcal{J}^{B,B}_{h^{[1]}}\|_{p\to p}\big\|[A, B]\big\|_p \\ &\overset{(3.4)}{\leq} \|h^{[1]}\|_{\mathfrak{A}_0}\big\|[A, B]\big\|_p \leq \sqrt{2}\big(\|h'\|_2 + \|h''\|_2\big)\big\|[A, B]\big\|_p. \quad \square\end{aligned}$$

4 Main Result

In [13], Connes conjectured the following:

Conjecture 4.1 ([13, Lemma 11 (§IV.3.α)]) Let $p > 1$, and let $\mathcal{H} = L_2(\mathbb{S}^1)$. If $0 \leq f \in L_\infty(\mathbb{S}^1)$, and $0 \leq B \in \mathcal{L}_{p,\infty}(\mathcal{H})$ such that $[M_f, B] \in \big(\mathcal{L}_{p,\infty}(\mathcal{H})\big)_0$, where M_f denotes the multiplication operator of f, then

$$M_f^{\frac{p}{2}} B^p M_f^{\frac{p}{2}} - (M_f^{\frac{1}{2}} B M_f^{\frac{1}{2}})^p \in \big(\mathcal{L}_{1,\infty}(\mathcal{H})\big)_0.$$

A variant of this result was recently proved in [14] by Connes et al.

Proposition 4.2 ([14, Lemma 5.3]) *Suppose* $0 \leq A \in \mathcal{B}(\mathcal{H})$ *and* $0 \leq B \in \mathcal{L}_{p,\infty}$, *for some* $1 < p < \infty$. *If* $[A^{\frac{1}{2}}, B] \in (\mathcal{L}_{p,\infty})_0$, *then*

$$B^p A^p - (A^{\frac{1}{2}} B A^{\frac{1}{2}})^p \in (\mathcal{L}_{1,\infty})_0.$$

To prove our main result, we require a trace-class variant of Proposition 4.2.

Proposition 4.3 *If* $0 \leq A, B \in \mathcal{B}(\mathcal{H})$ *such that* $[A^{\frac{1}{2}}, B] \in \mathcal{L}_1(\mathcal{H})$, *then*

$$\lim_{p\downarrow 1}(p-1)\operatorname{Tr}\big(B^p A^p - (A^{\frac{1}{2}} B A^{\frac{1}{2}})^p\big) = 0. \tag{4.1}$$

The proof of Proposition 4.2 in [14] used double operator integrals to obtain a weak integral representation of the difference $B^p A^p - (A^{\frac{1}{2}} B A^{\frac{1}{2}})^p$. We use the same approach.

For $1 < p < \infty$, we define a function g_p on $\mathbb{R}$ by setting

$$g_p(t) := \begin{cases} \frac{1}{2}\Big(1 - \coth\Big(\frac{t}{2}\Big)\tanh\Big(\frac{(p-1)t}{2}\Big)\Big), & \text{if } t \neq 0, \\ 1 - \frac{p}{2}, & \text{if } t = 0. \end{cases}$$

It was shown in [53, Remark 5.2.2] that g_p is a Schwartz function. Proposition 4.2 is proved in [14] using the following decomposition lemma.

Lemma 4.4 ([53, Theorem 5.2.1]) *Let $0 \leq A, B \in \mathcal{B}(\mathcal{H})$. For brevity, let*

$$Y = Y(A, B) := A^{\frac{1}{2}} B A^{\frac{1}{2}}. \tag{4.2}$$

Additionally, define the family of operators

$$\begin{aligned} T_0 &:= B^{p-1}[B, A^p] + B^{p-1}A^{p-\frac{1}{2}}[A^{\frac{1}{2}}, B] \\ &\quad + [B, A]Y^{p-1} + A^{\frac{1}{2}}[A^{\frac{1}{2}}, B]Y^{p-1}, \end{aligned} \tag{4.3}$$

$$\begin{aligned} T_s &:= B^{p-1+is}[B, A^{p+is}]Y^{-is} + B^{p-1+is}A^{p-\frac{1}{2}+is}[A^{\frac{1}{2}}, B]Y^{-is} \\ &\quad + B^{is}[B, A^{1+is}]Y^{p-1-is} + B^{is}A^{\frac{1}{2}+is}[A^{\frac{1}{2}}, B]Y^{p-1-is}, \quad s > 0. \end{aligned} \tag{4.4}$$

If $1 < p < \infty$, then

$$B^p A^p - Y^p = T_0 - \frac{1}{\sqrt{2\pi}} \int_{\mathbb{R}} \hat{g}_p(s) T_s \, ds, \tag{4.5}$$

where the integral is understood in the weak sense of Definition 2.5.

We shall use this decomposition to show that the trace of the difference $(p-1)\big(B^p A^p - (A^{\frac{1}{2}} B A^{\frac{1}{2}})^p\big)$ is $o(1)$ as $p \downarrow 1$, whenever $[A^{\frac{1}{2}}, B]$ is trace-class.

Proof of Proposition 4.3 Firstly, we define the following operators for brevity:

$$X_1 := \int_{\mathbb{R}} \hat{g}_p(s) B^{is}[B, A^{p+is}]Y^{-is} \, ds, \quad X_2 := \int_{\mathbb{R}} \hat{g}_p(s) B^{is} A^{p-\frac{1}{2}+is}[A^{\frac{1}{2}}, B]Y^{-is} \, ds,$$

$$X_3 := \int_{\mathbb{R}} \hat{g}_p(s) B^{is}[B, A^{1+is}]Y^{-is} \, ds, \quad X_4 := \int_{\mathbb{R}} \hat{g}_p(s) B^{is} A^{\frac{1}{2}+is}[A^{\frac{1}{2}}, B]Y^{-is} \, ds.$$

Then, appealing to Lemma 4.4, we have the decomposition

$$B^p A^p - Y^p \overset{(4.5)}{=} T_0 - \frac{1}{\sqrt{2\pi}} \int_{\mathbb{R}} \hat{g}_p(s) T_s \, \mathrm{d}s$$
$$= T_0 - (2\pi)^{-\frac{1}{2}} B^{p-1}(X_1 + X_2) - (2\pi)^{-\frac{1}{2}}(X_3 + X_4) Y^{p-1}. \tag{4.6}$$

We treat only the term $B^{p-1}X_1$; the other terms may be considered using similar arguments. Consider the function $q_{p,s}$ defined by the expression

$$q_{p,s}(x) := \begin{cases} x^{2(p+is)} \psi_A(x), & \text{if } x \in \mathbb{R} \setminus \{0\}, \\ 0, & \text{if } x = 0. \end{cases}$$

where ψ_A is any compactly supported C_b^2-function such that $\|\psi_A\|_\infty, \|\psi_A'\|_\infty, \|\psi_A''\|_\infty \leq 1$ and $\psi_A(x) = 1$, for all $x \in \left[0, \|A\|_\infty\right]$. Then, since $p > 1$, $q_{p,s}$ is also a C_b^2-function. By Corollary 3.4, we have that

$$\left\|[B, A^{p+is}]\right\|_1 \overset{(3.6)}{\leq} \mathrm{const} \cdot \left(\|q_{p,s}\|_\infty + \|q_{p,s}'\|_\infty + \|q_{p,s}''\|_\infty\right) \left\|[B, A^{\frac{1}{2}}]\right\|_1$$
$$\leq \begin{cases} \mathcal{O}(1) \cdot \left\|[B, A^{\frac{1}{2}}]\right\|_1, & \text{if } |s| \leq 1 \\ \mathcal{O}(s^2) \cdot \left\|[B, A^{\frac{1}{2}}]\right\|_1, & \text{if } |s| > 1. \end{cases}$$

However, this gives us the estimate

$$\int_{\mathbb{R}} |\hat{g}_p(s)| \left\|B^{is}[B, A^{p+is}]Y^{-is}\right\|_1 \mathrm{d}s \leq \int_{\mathbb{R}} |\hat{g}_p(s)| \left\|[B, A^{p+is}]\right\|_1 \mathrm{d}s$$
$$\leq \mathrm{const} \cdot \left(\|\hat{g}_p\|_1 + \|\widehat{g_p''}\|_1\right) \left\|[B, A^{\frac{1}{2}}]\right\|_1$$
$$\leq \mathrm{const} \cdot \left(\|g_p\|_2 + \|g_p'\|_2 + \|g_p''\|_2 + \|g_p'''\|_2\right) \left\|[B, A^{\frac{1}{2}}]\right\|_1, \tag{4.7}$$

where the last line follows from [44, Lemma 7]. Note that the condition of [44, Lemma 7] is that g_p is absolutely continuous, which is satisfied since g_p is Schwartz (see, e.g., [36, Theorem 5, §33]). Therefore, appealing to Lemma 2.6, we obtain

$$\|X_1\|_1 \overset{(2.5)}{\leq} \int_{\mathbb{R}} |\hat{g}_p(s)| \left\|B^{is}[B, A^{p+is}]Y^{-is}\right\|_1 \mathrm{d}s$$
$$\overset{(4.7)}{\leq} \mathrm{const} \cdot \left(\|g_p\|_2 + \|g_p'\|_2 + \|g_p''\|_2 + \|g_p'''\|_2\right) \left\|[B, A^{\frac{1}{2}}]\right\|_1.$$

Hence, by Lemma 3 (see Appendix below), we have the estimate

$$\left|(p-1)\,\mathrm{Tr}(B^{p-1}X_1)\right| \leq (p-1)\|B^{p-1}\|_\infty \|X_1\|_1 \leq \mathcal{O}\big((p-1)^{\frac{1}{2}}\big), \quad p \downarrow 1.$$

Repeating this argument for the remaining terms on the right-hand side of (4.6), we obtain the estimate

$$\left|(p-1)\operatorname{Tr}(B^pA^p - Y^p)\right| \le \mathcal{O}\big((p-1)^{\frac{1}{2}}\big). \qquad \square$$

Appealing to Proposition 4.3, we obtain part (a) of Theorem 1.1.

Theorem 4.5 *Suppose* $0 \le A, B \in \mathcal{B}(\mathcal{H})$ *are such that* $AB \in \mathcal{M}_{1,\infty}(\mathcal{H})$. *Suppose that* $[A^{\frac{1}{2}}, B] \in \mathcal{L}_1(\mathcal{H})$. *Let* $C_{AB} > 0$. *The following are equivalent:*

(i) AB *is Dixmier measurable, and* $\operatorname{Tr}_\omega(AB) = C_{AB}$ *for all dilation-invariant extended limits* ω.
(ii) $\lim_{p\downarrow 1}(p-1)\operatorname{Tr}(B^pA^p) = C_{AB}$.

Proof Firstly, since $[A^{\frac{1}{2}}, B] \in \mathcal{L}_1(\mathcal{H})$ and $AB \in \mathcal{M}_{1,\infty}(\mathcal{H})$ by assumption, we have that $A^{\frac{1}{2}}BA^{\frac{1}{2}} = BA + [A^{\frac{1}{2}}, B]A^{\frac{1}{2}} \in \mathcal{M}_{1,\infty}(\mathcal{H})$ and

$$\operatorname{Tr}_\omega(AB) = \operatorname{Tr}_\omega(A^{\frac{1}{2}}BA^{\frac{1}{2}}) + \operatorname{Tr}_\omega\left([A^{\frac{1}{2}}, B]A^{\frac{1}{2}}\right) = \operatorname{Tr}_\omega(A^{\frac{1}{2}}BA^{\frac{1}{2}}).$$

First, we show that $(ii) \Rightarrow (i)$. Assume that the limit $\lim_{p\downarrow 1}(p-1)\operatorname{Tr}(B^pA^p)$ exists. Then, we have by Proposition 4.3 that the limit $\lim_{p\downarrow 1}(p-1)\operatorname{Tr}\left((A^{\frac{1}{2}}BA^{\frac{1}{2}})^p\right)$ exists and agrees with the former. Hence, appealing to [41, Theorem 9.3.1],

$$\operatorname{Tr}_\omega(A^{\frac{1}{2}}BA^{\frac{1}{2}}) = \lim_{p\downarrow 1}(p-1)\operatorname{Tr}\left((A^{\frac{1}{2}}BA^{\frac{1}{2}})^p\right) = \lim_{p\downarrow 1}(p-1)\operatorname{Tr}(B^pA^p).$$

Next, we show that $(i) \Rightarrow (ii)$. Assume that AB is Dixmier measurable. Then, again appealing to [41, Theorem 9.3.1], we have that the limit $\lim_{p\downarrow 1}(p-1)\operatorname{Tr}\left((A^{\frac{1}{2}}BA^{\frac{1}{2}})^p\right)$ exists and agrees with $\operatorname{Tr}_\omega(A^{\frac{1}{2}}BA^{\frac{1}{2}})$. Hence, by Proposition 4.3, the limit $\lim_{p\downarrow 1}(p-1)\operatorname{Tr}(B^pA^p)$ exists and

$$\lim_{p\downarrow 1}(p-1)\operatorname{Tr}(B^pA^p) = \lim_{p\downarrow 1}(p-1)\operatorname{Tr}\left((A^{\frac{1}{2}}BA^{\frac{1}{2}})^p\right) = \operatorname{Tr}_\omega(A^{\frac{1}{2}}BA^{\frac{1}{2}}). \qquad \square$$

From this, (b) of Theorem 1.1 follows easily:

Corollary 4.6 *Suppose* $0 \le A, B \in \mathcal{B}(\mathcal{H})$ *are such that* $AB \in \mathcal{L}_{1,\infty}(\mathcal{H})$. *Suppose that* $[A^{\frac{1}{2}}, B] \in \mathcal{L}_1(\mathcal{H})$. *Let* $C_{AB} > 0$. *The following are equivalent:*

(i) AB *is Dixmier measurable, and* $\operatorname{Tr}_\omega(AB) = C_{AB}$ *for all extended limits* ω.
(ii) $\lim_{p\downarrow 1}(p-1)\operatorname{Tr}(B^pA^p) = C_{AB}$.

Proof This follows from Theorem 4.5 and [41, Lemma 9.7.4], which implies that, for every extended limit ω, there exists a dilation-invariant extended limit ω_0 such that, if $AB \in \mathcal{L}_{1,\infty}$, then $\operatorname{Tr}_\omega(A) = \operatorname{Tr}_{\omega_0}(A)$. $\square$

5 Application to the Euclidean Plane

Let $d \geq 1$. For each $k = 1, \dots, d$, we define an (unbounded) closable operator on $L_2(\mathbb{R}^d)$ by

$$\partial_k := -i\frac{\partial}{\partial t_k}, \qquad \mathrm{dom}(\partial_k) = W_{1,2}(\mathbb{R}^d),$$

which, on the dense subspace $C^1(\mathbb{R}^d) \subset W_{1,2}(\mathbb{R}^d)$, corresponds to partial differentiation with respect to the k-th argument. Additionally, we define the Laplace operator on $L_2(\mathbb{R}^d)$ by

$$\Delta := -\sum_{k=1}^{d} \partial_k^2, \quad \mathrm{dom}(\Delta) = W_{2,2}(\mathbb{R}^d).$$

We shall adopt the notation $N_d := 2^{\lceil \frac{d}{2} \rceil}$. Denote by $\mathbb{I} := \mathbb{I}_{N_d}$ the $N_d \times N_d$ identity matrix, and for $j = 1, \dots, d$, let γ_j be *d-dimensional symmetric gamma matrices*; that is, a collection of $N_d \times N_d$ matrices which satisfy

$$(i)\ \gamma_j = \gamma_j^*,\ \gamma_j^2 = \mathbb{I},\ \text{for all } j = 1, \dots, d,$$

$$(ii)\ \gamma_j\gamma_k = -\gamma_k\gamma_j,\ \text{whenever } j \neq k.$$

We define the Dirac operator on $\mathbb{C}^{N_d} \otimes L_2(\mathbb{R}^d)$ (see, e.g., [54]) by

$$\mathcal{D} = \sum_{k=1}^{d} \gamma_k \otimes \partial_k, \qquad \mathrm{dom}(\mathcal{D}) = \mathbb{C}^{N_d} \otimes W_{1,2}(\mathbb{R}^d), \tag{5.1}$$

an unbounded (densely defined), self-adjoint operator. We observe that $\mathcal{D}^2 = \mathbb{I} \otimes (-\Delta)$.

For a bounded function $f \in L_\infty(\mathbb{R}^d)$, we let M_f denote the multiplication operator of f, which is the bounded operator on $L_2(\mathbb{R}^d)$ defined by

$$(M_f g)(\mathbf{x}) := f(\mathbf{x})g(\mathbf{x}), \quad g \in L_2(\mathbb{R}^d),\ \mathbf{x} \in \mathbb{R}^d.$$

We recall the following function spaces in the style of Birman and Solomyak (see [4]). Let $\mathcal{Q}_\mathbf{n}$ denote the unit cube centred at $\mathbf{n} \in \mathbb{Z}^d$. For a region $\Omega \subset \mathbb{R}^d$, denote by χ_Ω the characteristic function of Ω. Then, for $1 \leq p < 2$, $1 \leq q \leq \infty$, let

$$\ell_p(L_q)(\mathbb{R}^d) := \left\{ f \text{ measurable on } \mathbb{R}^d \ :\ \sum_{\mathbf{n} \in \mathbb{Z}^d} \|f\chi_{\mathcal{Q}_\mathbf{n}}\|_q^p < \infty \right\}$$

denote the (p,q)th Birman–Solomyak space, with the corresponding norm

$$\|f\|_{\ell_p(L_q)} := \Big(\sum_{\mathbf{n}\in\mathbb{Z}^d} \|f\chi_{Q_{\mathbf{n}}}\|_q^p\Big)^{\frac{1}{p}}, \qquad f \in \ell_p(L_q)(\mathbb{R}^d). \tag{5.2}$$

Theorem 5.1 *Suppose $f \in \ell_1(L_2)(\mathbb{R}^d)$, and suppose $\alpha \geq \frac{d}{2}$.*

(*i*) [48, Theorem 4.5] *If $\alpha > \frac{d}{2}$, then $M_f(1-\Delta)^{-\alpha} \in \mathcal{L}_1\big(L_2(\mathbb{R}^d)\big)$, and*

$$\big\|M_f(1-\Delta)^{-\alpha}\big\|_1 \leq \mathrm{const}_\alpha \cdot \|f\|_{\ell_1(L_2)}.$$

(*ii*) [5, 5.7 (p. 103)], *If $\alpha = \frac{d}{2}$, then $M_f(1-\Delta)^{-\frac{d}{2}} \in \mathcal{L}_{1,\infty}\big(L_2(\mathbb{R}^d)\big)$, and*

$$\big\|M_f(1-\Delta)^{-\frac{d}{2}}\big\|_{1,\infty} \leq \mathrm{const} \cdot \|f\|_{\ell_1(L_2)}.$$

It is immediate from Theorem 5.1(*ii*) that for a Schwartz function f, the operator $(\mathbb{I}\otimes M_f)(1+\mathcal{D}^2)^{-\frac{d}{2}} \in \mathcal{L}_{1,\infty}\big(\mathbb{C}^{N_d}\otimes L_2(\mathbb{R}^d)\big)$.

5.1 Square Roots of Smooth Functions

To show that the commutator $[A^{\frac{1}{2}}, B]$ is trace class, we will require control over the smoothness and decay of the *nonnegative* function $f^{\frac{1}{2}} = \big|\sqrt{f}\big|$, for nonnegative $f \in \mathcal{S}(\mathbb{R}^d)$. Since $f^{\frac{1}{2}}$ may not be differentiable at the zeros of f, we make the following observations, starting with the Malgrange lemma for *strictly positive* f, whose proof supplied in [28] is given below for the convenience of the reader.

Lemma 5.2 ([28, Lemma 1]) *If f is a strictly positive C_b^2-function on $\mathbb{R}$, then*

$$\big|(f^{\frac{1}{2}})'(x)\big| < \frac{\|f''\|_\infty^{\frac{1}{2}}}{\sqrt{2}}, \quad \textit{for every } x \in \mathbb{R}.$$

Proof Firstly, fix some $x \in \mathbb{R}$ and choose some $h > 0$. By Taylor's theorem, there exists a constant $c \in (x, x+h)$ such that

$$0 < f(x+h) = f(x) + hf'(x) + \frac{h^2}{2}f''(c) \leq f(x) + hf'(x) + \frac{h^2}{2}\|f''\|_\infty.$$

Observe that the expression on the right-hand side is a strictly positive quadratic in h. Particularly, it has no real solutions, so it has negative discriminant—that is,

$$f'(x)^2 < 2f(x)\|f''\|_\infty, \quad x \in \mathbb{R}.$$

Taking the absolute value followed by the square root of both sides of this inequality, and dividing through both sides by $2f^{\frac{1}{2}}(x)$, yields the result. □

The Malgrange lemma offers us the following Lipschitz condition on the derivative of $f^{\frac{1}{2}}$, for a *nonnegative* C_b^2-function f.

Corollary 5.3 *If f is a nonnegative C_b^2-function on $\mathbb{R}$, then $f^{\frac{1}{2}}$ is Lipschitz and $\|(f^{\frac{1}{2}})'\|_\infty \leq \|f''\|_\infty^{\frac{1}{2}}$.*

Proof For every $n \geq 1$, define the function $f_n(t) = f(t) + \frac{1}{n}$, $t \in \mathbb{R}$. Since f_n is strictly positive, we may immediately apply the Malgrange lemma above to bound the Lipschitz constant of $f_n^{\frac{1}{2}}$ by

$$\|(f_n^{\frac{1}{2}})'\|_\infty \leq \|f_n''\|_\infty^{\frac{1}{2}} = \|f''\|_\infty^{\frac{1}{2}}.$$

In particular, we have the expression

$$\left|f_n^{\frac{1}{2}}(x) - f_n^{\frac{1}{2}}(y)\right| \leq \|f''\|_\infty^{\frac{1}{2}} \cdot |x - y|, \quad \text{for all } x, y \in \mathbb{R}.$$

Taking the pointwise limit of the above as $n \to \infty$ yields the result. □

Remark 5.4 Suppose $f \geq 0$ is a C_b^2-function on $\mathbb{R}^d$. For each $j = 1, \ldots, d$, by fixing all variables x_k, for $k \neq j$, taking the partial derivative $\partial_j(f^{\frac{1}{2}})$ is the same as taking the derivative of a univariate function. Hence, by Corollary 5.3, we have that

$$\|\partial_j(f^{\frac{1}{2}})\|_\infty \leq \|\partial_j^2 f\|_\infty^{\frac{1}{2}}. \tag{5.3}$$

An immediate consequence of this remark is that $f^{\frac{1}{2}} \in W_{1,\infty}(\mathbb{R}^d)$ and, by the Leibniz rule and the Hölder inequality, defines a multiplication operator

$$M_{f^{\frac{1}{2}}} : W_{s,2}(\mathbb{R}^d) \to W_{s,2}(\mathbb{R}^d), \quad \text{for all } 0 \leq s \leq 1.$$

Lemma 5.5 *If $0 \leq f \in \mathcal{S}(\mathbb{R}^d)$, then, for every $j = 1, \ldots, d$, $\partial_j(f^{\frac{1}{2}}) \in \ell_1(L_2)(\mathbb{R}^d)$.*

Proof Define a function g by the expression

$$g(\mathbf{t}) = f(\mathbf{t}) \cdot \left(1 + |\mathbf{t}|^2\right)^{2d}, \quad \mathbf{t} \in \mathbb{R}^d.$$

Observe that g is also Schwartz, since f is rapidly decreasing. Then, for each $j = 1, \ldots, d$, the Leibniz rule yields

$$\partial_j(f^{\frac{1}{2}})(\mathbf{t}) = \Big(\partial_j(g^{\frac{1}{2}})(\mathbf{t}) - \frac{2dt_j(g^{\frac{1}{2}})(\mathbf{t})}{1+|\mathbf{t}|^2}\Big)(1+|\mathbf{t}|^2)^{-d}.$$

However, it follows from Remark 5.4 that the first factor on the right-hand side of the above defines a bounded function, while the function $(1+|\cdot|^2)^{-d} \in \ell_1(L_2)(\mathbb{R}^d)$ (see, e.g., [48, Remark (c) (p. 39)]). This concludes the proof. □

5.2 Calculating the Dixmier Trace of $(\mathbb{I} \otimes M_f)(1+\mathcal{D}^2)^{-\frac{d}{2}}$

Now that we have control over the derivatives of the square root of a Schwartz function, we may verify the following.

Lemma 5.6 *If* $0 \le f \in \mathcal{S}(\mathbb{R}^d)$*, then*

$$\big[(\mathbb{I} \otimes M_f)^{\frac{1}{2}}, (1+\mathcal{D}^2)^{-\frac{d}{2}}\big] \in \mathcal{L}_1\big(\mathbb{C}^{N_d} \otimes L_2(\mathbb{R}^d)\big).$$

Proof We begin by decomposing the commutator into treatable terms. Firstly, by the identity (2.3) (see Sect. 2 above), we obtain the expression

$$\big[(\mathbb{I}\otimes M_f)^{\frac{1}{2}}, (1+\mathcal{D}^2)^{-\frac{d}{2}}\big] = \sum_{k=0}^{3d-1} (1+\mathcal{D}^2)^{-\frac{k}{6}}\big[\mathbb{I} \otimes M_{f^{\frac{1}{2}}}, (1+\mathcal{D}^2)^{-\frac{1}{6}}\big](1+\mathcal{D}^2)^{\frac{k+1}{6}-\frac{d}{2}}.$$

Now, appealing to Remark 5.4, we have that $f^{\frac{1}{2}} \in W_{1,\infty}(\mathbb{R}^d) \subset W_{\frac{1}{3},\infty}(\mathbb{R}^d)$, so that the multiplication operator $M_{f^{\frac{1}{2}}}$ is bounded on both $L_2(\mathbb{R}^d)$ and $W_{\frac{1}{3},2}(\mathbb{R}^d)$. Therefore, since $(1+\mathcal{D}^2)^{\frac{1}{6}}$ is well-defined in the domain $\mathbb{C}^{N_d} \otimes W_{\frac{1}{3},2}(\mathbb{R}^d)$, and since $(1+\mathcal{D}^2)^{-\frac{1}{6}}$ maps $\mathbb{C}^{N_d} \otimes L_2(\mathbb{R}^d)$ into $\mathbb{C}^{N_d} \otimes W_{\frac{1}{3},2}(\mathbb{R}^d)$, the expression

$$\big[\mathbb{I} \otimes M_{f^{\frac{1}{2}}}, (1+\mathcal{D}^2)^{-\frac{1}{6}}\big] = -(1+\mathcal{D}^2)^{-\frac{1}{6}}\big[\mathbb{I} \otimes M_{f^{\frac{1}{2}}}, (1+\mathcal{D}^2)^{\frac{1}{6}}\big](1+\mathcal{D}^2)^{-\frac{1}{6}}$$

is well-defined on all of $\mathbb{C}^{N_d} \otimes L_2(\mathbb{R}^d)$. Therefore,

$$\begin{aligned}\big[\mathbb{I}\otimes M_{f^{\frac{1}{2}}}, (1+\mathcal{D}^2)^{-\frac{d}{2}}\big] &= \sum_{k=0}^{3d-1} (1+\mathcal{D}^2)^{-\frac{k+1}{6}}\big[\mathbb{I} \otimes M_{f^{\frac{1}{2}}}, (1+\mathcal{D}^2)^{\frac{1}{6}}\big](1+\mathcal{D}^2)^{\frac{k}{6}-\frac{d}{2}} \\ &= -\sum_{k=0}^{3d-1} \big[(1+\mathcal{D}^2)^{-\frac{k+1}{6}}(\mathbb{I} \otimes M_{f^{\frac{1}{2}}})(1+\mathcal{D}^2)^{\frac{k}{6}-\frac{d}{2}}, (1+\mathcal{D}^2)^{\frac{1}{6}}\big].\end{aligned}$$

Now, let $h(t) = (1+t^2)^{\frac{1}{6}}$. We have that

$$h'(t) = \frac{1}{3} t (1+t^2)^{-\frac{5}{6}}, \qquad h''(t) = \frac{1}{9}(3-2t^2)(1+t^2)^{-\frac{11}{6}}, \tag{5.4}$$

so $h', h'' \in L_2(\mathbb{R})$. Hence, by Lemma 3.5, it suffices to check that

$$\big[(1+\mathcal{D}^2)^{-\frac{k+1}{6}}(\mathbb{I} \otimes M_{f^{\frac{1}{2}}})(1+\mathcal{D}^2)^{\frac{k}{6}-\frac{d}{2}}, \mathcal{D}\big] \in \mathcal{L}_1\big(\mathbb{C}^{N_d} \otimes L_2(\mathbb{R}^d)\big),$$

for each $k = 0, \ldots, 3d-1$. Appealing to the fact that, for any function $g \in W_{1,\infty}(\mathbb{R}^d)$, the Leibniz rule produces the commutator identity $[\mathbb{I} \otimes M_g, \mathcal{D}] = \sum_{j=1}^d \gamma_j \otimes M_{\partial_j g}$, we obtain the expression

$$\begin{aligned}
&\big[(1+\mathcal{D}^2)^{-\frac{k+1}{6}}(\mathbb{I} \otimes M_{f^{\frac{1}{2}}})(1+\mathcal{D}^2)^{\frac{k}{6}-\frac{d}{2}}, \mathcal{D}\big] \\
&\qquad = (1+\mathcal{D}^2)^{-\frac{k+1}{6}}[\mathbb{I} \otimes M_{f^{\frac{1}{2}}}, \mathcal{D}](1+\mathcal{D}^2)^{\frac{k}{6}-\frac{d}{2}} \\
&\qquad = \sum_{j=1}^{d} (1+\mathcal{D}^2)^{-\frac{k+1}{6}}(\gamma_j \otimes M_{\partial_j(f^{\frac{1}{2}})})(1+\mathcal{D}^2)^{\frac{k}{6}-\frac{d}{2}}
\end{aligned}$$

on the dense domain $\mathbb{C}^{N_d} \otimes W_{1,2}(\mathbb{R}^d)$. Since the expression on the right-hand side defines a bounded operator on $\mathbb{C}^{N_d} \otimes L_2(\mathbb{R}^d)$, the operator on the left-hand side may be extended to a bounded operator. Furthermore, by Theorem 5.1(*i*) and Lemma 5.5, we have for each $j = 1, \ldots, d$ that

$$\begin{aligned}
&\big\|(1+\mathcal{D}^2)^{-\frac{k+1}{6}}(\gamma_j \otimes M_{\partial_j(f^{\frac{1}{2}})})(1+\mathcal{D}^2)^{\frac{k}{6}-\frac{d}{2}}\big\|_1 \\
&\qquad = \big\|(1-\Delta)^{-\frac{k+1}{6}} M_{\partial_j(f^{\frac{1}{2}})}(1-\Delta)^{\frac{k}{6}-\frac{d}{2}}\big\|_1 \\
&\qquad \le \big\|M_{\partial_j(f^{\frac{1}{2}})}(1-\Delta)^{-\frac{d}{2}-\frac{1}{6}}\big\|_1 \\
&\qquad \le \text{const}_d \cdot \big\|\partial_j(f^{\frac{1}{2}})\big\|_{\ell_1(L_2)} < \infty,
\end{aligned}$$

where the second-last inequality follows from (2.2) in Sect. 2. This concludes the proof. □

Finally, to explicitly calculate the Dixmier trace of $(\mathbb{I} \otimes M_f)(1+\mathcal{D}^2)^{-\frac{d}{2}}$ using Theorem 1.1, we need to establish existence of the relevant limit.

Proposition 5.7 *If $0 \le f \in \mathcal{S}(\mathbb{R}^d)$, then the limit*

$$\lim_{p \downarrow 1} (p-1)\,\mathrm{Tr}\big((1+\mathcal{D}^2)^{-\frac{dp}{2}}(\mathbb{I} \otimes M_f)^p\big) \text{ exists.}$$

In particular, $(\mathbb{I} \otimes M_f)(1 + \mathcal{D}^2)^{-\frac{d}{2}}$ *is a Dixmier measurable operator and, for any extended limit* ω,

$$\mathrm{Tr}_\omega \left((\mathbb{I} \otimes M_f)(1 + \mathcal{D}^2)^{-\frac{d}{2}} \right) = \frac{2^{\lceil \frac{d}{2} \rceil} \mathrm{Vol}(\mathbb{S}^{d-1})}{d(2\pi)^d} \int_{\mathbb{R}^d} f(\mathbf{x})\, d\mathbf{x},$$

where $\mathrm{Vol}(\mathbb{S}^{d-1})$ *denotes the volume of the unit sphere* $\mathbb{S}^{d-1}$.

Proof **Step 1** We verify that, for every $1 < p < 2$, the operator $B^p A^p$ is trace class. Since $(1 + \mathcal{D}^2)^{-\frac{dp}{2}} = \mathbb{I} \otimes (1 - \Delta)^{-\frac{dp}{2}}$, we observe that

$$B^p A^p = \mathbb{I} \otimes \left((1 - \Delta)^{-\frac{dp}{2}} M_f^p \right). \tag{5.5}$$

However, the trace on $\mathcal{L}_1\left(\mathbb{C}^{N_d} \otimes L_2(\mathbb{R}^d)\right) \cong M_{N_d}(\mathbb{C}) \otimes \mathcal{L}_1\left(L_2(\mathbb{R}^d)\right)$ is given by $\mathrm{tr} \otimes \mathrm{Tr}$, where tr is the matrix trace on $M_{N_d}(\mathbb{C})$ and Tr is the classical trace on $\mathcal{L}_1\left(L_2(\mathbb{R}^d)\right)$. That is, if either of the relevant norms exists, then we have

$$\|B^p A^p\|_1 = N_d \left\| (1 - \Delta)^{-\frac{dp}{2}} M_f^p \right\|_1.$$

It follows from [48, Remark (c) (p. 39)] that $f^p \in \ell_1(L_2)(\mathbb{R}^d)$ for every $1 < p < 2$. Hence, by Theorem 5.1(i), $(1 - \Delta)^{-\frac{dp}{2}} M_f^p$ is trace class for all $1 < p < 2$.

Step 2 We now calculate $\mathrm{Tr}\left((1 + \mathcal{D}^2)^{-\frac{dp}{2}} (\mathbb{I} \otimes M_f)^p \right)$ for $p > 1$. We do so by considering the continuity of its integral kernel. Observe that

$$\begin{aligned} \left((1 - \Delta)^{-\frac{dp}{2}} \phi \right)(\mathbf{x}) &= \left(\mathcal{F}\left[\left(1 + |\cdot|^2\right)^{-\frac{dp}{2}} \right] * \phi \right)(\mathbf{x}) \\ &= \int_{\mathbb{R}^d} \mathcal{F}\left[\left(1 + |\cdot|^2\right)^{-\frac{dp}{2}} \right](\mathbf{y} - \mathbf{x}) \phi(\mathbf{x})\, d\mathbf{y}, \end{aligned}$$

where $*$ denotes the convolution product. Hence, $(1 - \Delta)^{-\frac{dp}{2}} M_f^p$ has an integral kernel given by the expression

$$K(\mathbf{x}, \mathbf{y}) := f(\mathbf{y})^p \cdot \mathcal{F}\left[\left(1 + |\cdot|^2\right)^{-\frac{dp}{2}} \right](\mathbf{y} - \mathbf{x}),$$

which is continuous and, by the Fubini–Tonelli theorem, belongs to $L_2(\mathbb{R}^d \times \mathbb{R}^d)$. Appealing to Proposition 2.1 then yields

$$\begin{aligned}\mathrm{Tr}(B^p A^p) &\overset{(5.5)}{=} N_d \,\mathrm{Tr}\left((1-\Delta)^{-\frac{dp}{2}} M_f^p\right) = N_d \int_{\mathbb{R}^d} K(\mathbf{x},\mathbf{x})\,\mathrm{d}\mathbf{x} \\ &= \frac{N_d}{(2\pi)^d} \int_{\mathbb{R}^d} (1+|\mathbf{s}|^2)^{-\frac{dp}{2}}\,\mathrm{d}\mathbf{s} \cdot \int_{\mathbb{R}^d} f(\mathbf{x})^p\,\mathrm{d}\mathbf{x} \\ &= \frac{N_d \,\mathrm{Vol}(\mathbb{S}^{d-1})\Gamma(\frac{d}{2})}{2(2\pi)^d} \cdot \frac{\Gamma\left(\frac{d}{2}(p-1)\right)}{\Gamma(\frac{dp}{2})} \cdot \int_{\mathbb{R}^d} f(\mathbf{x})^p\,\mathrm{d}\mathbf{x}. \qquad (5.6)\end{aligned}$$

Therefore, since $f^p \to f$ pointwise as $p \downarrow 1$, the dominated convergence theorem yields

$$\lim_{p\downarrow 1} \int_{\mathbb{R}^d} f(\mathbf{x})^p\,\mathrm{d}\mathbf{x} = \int_{\mathbb{R}^d} f(\mathbf{x})\,\mathrm{d}\mathbf{x}.$$

Hence, the limit in Theorem 1.1(ii) exists and is given by the expression

$$\begin{aligned}&\lim_{p\downarrow 1}(p-1)\,\mathrm{Tr}\left((1+\mathcal{D}^2)^{-\frac{dp}{2}}(\mathbb{I}\otimes M_f)^p\right) \\ &\overset{(5.6)}{=} \frac{N_d \,\mathrm{Vol}(\mathbb{S}^{d-1})\Gamma(\frac{d}{2})}{2(2\pi)^d} \lim_{p\downarrow 1}\left(\frac{(p-1)\Gamma\left(\frac{d}{2}(p-1)\right)}{\Gamma(\frac{dp}{2})} \int_{\mathbb{R}^d} f(\mathbf{x})^p\,\mathrm{d}\mathbf{x}\right) \\ &= \frac{N_d \,\mathrm{Vol}(\mathbb{S}^{d-1})}{d(2\pi)^d} \int_{\mathbb{R}^d} f(\mathbf{x})\,\mathrm{d}\mathbf{x}.\end{aligned}$$

Finally, appealing to Lemma 5.6, the conditions of Theorem 1.1 (b) are satisfied. Therefore, $(\mathbb{I}\otimes M_f)(1+\mathcal{D}^2)^{-\frac{d}{2}}$ is Dixmier measurable, and we have that the Dixmier trace $\mathrm{Tr}_\omega\left((\mathbb{I}\otimes M_f)(1+\mathcal{D}^2)^{-\frac{d}{2}}\right)$ agrees with the above, for any extended limit ω. □

Corollary 5.8 *If $f \in W_{d,1}(\mathbb{R}^d)$, then $(\mathbb{I}\otimes M_f)(1+\mathcal{D}^2)^{-\frac{d}{2}}$ is a Dixmier measurable operator and, for any extended limit ω,*

$$\mathrm{Tr}_\omega\left((\mathbb{I}\otimes M_f)(1+\mathcal{D}^2)^{-\frac{d}{2}}\right) = \frac{2^{\lceil\frac{d}{2}\rceil}\,\mathrm{Vol}(\mathbb{S}^{d-1})}{d(2\pi)^d} \int_{\mathbb{R}^d} f(\mathbf{x})\,\mathrm{d}\mathbf{x},$$

where $\mathrm{Vol}(\mathbb{S}^{d-1})$ *denotes the volume of the unit sphere* $\mathbb{S}^{d-1}$.

Proof Without loss of generality, assume f is nonnegative. Since $\mathcal{S}(\mathbb{R}^d)$ is dense in $W_{d,1}(\mathbb{R}^d)$, one may construct a sequence $\{f_n\}_{n\in\mathbb{N}}$ of nonnegative Schwartz functions on $\mathbb{R}^d$ such that $f_n \to f$ in $W_{d,1}(\mathbb{R}^d)$. Then, by Theorem 5.1(ii), we have

$$\begin{aligned}\left\| M_f(1-\Delta)^{-\frac{d}{2}} - M_{f_n}(1-\Delta)^{-\frac{d}{2}} \right\|_{1,\infty} &\le \mathrm{const} \cdot \|f - f_n\|_{\ell_1(L_2)} \\ &\le \mathrm{const} \cdot \|f - f_n\|_{W_{d,1}} \to 0,\end{aligned}$$

where in the last line we used [38, Proposition 2.2]. Hence, the sequence of operators $\{M_{f_n}(1-\Delta)^{-\frac{d}{2}}\}_{n\in\mathbb{N}}$ converges to $M_f(1-\Delta)^{-\frac{d}{2}}$ in the $\mathcal{L}_{1,\infty}$-norm. Therefore, since Tr_ω is continuous in $\mathcal{L}_{1,\infty}$,

$$\begin{aligned}\mathrm{Tr}_\omega\left((\mathbb{I}\otimes M_f)(1+\mathcal{D}^2)^{-\frac{d}{2}}\right) &= \lim_{n\to\infty} \mathrm{Tr}_\omega\left((\mathbb{I}\otimes M_{f_n})(1+\mathcal{D}^2)^{-\frac{d}{2}}\right) \\ &= \frac{N_d \,\mathrm{Vol}(\mathbb{S}^{d-1})}{d(2\pi)^d} \lim_{n\to\infty} \int_{\mathbb{R}^d} f_n(\mathbf{x})\,\mathbf{dx} \\ &= \frac{N_d \,\mathrm{Vol}(\mathbb{S}^{d-1})}{d(2\pi)^d} \int_{\mathbb{R}^d} f(\mathbf{x})\,\mathbf{dx},\end{aligned}$$

where in the second equality we appealed to Proposition 5.7. □

6 Application to the Moyal Plane

The reader is advised that the definitions in this section are derived from [27, 30, 31, 37]. For the simplicity of the exposition, and in light of [37, Corollary 6.4], we need only consider the two-dimensional noncommutative Euclidean space, also known as Moyal plane.

6.1 Definition of the Noncommutative Plane

Let $\theta \in \mathbb{R}$ and let

$$\Theta = \begin{pmatrix} 0 & \theta \\ -\theta & 0 \end{pmatrix}.$$

Define a bilinear form on $\mathbb{R}^2$ by

$$\langle \mathbf{x}, \mathbf{y}\rangle_\Theta := \mathbf{x}\cdot\Theta\mathbf{y}, \quad \mathbf{x}, \mathbf{y} \in \mathbb{R}^2,$$

where skew-symmetry of Θ implies the skew-symmetry of $\langle\cdot,\cdot\rangle_\Theta$;

$$\langle \mathbf{x}, \mathbf{y}\rangle_\Theta = -\langle \mathbf{y}, \mathbf{x}\rangle_\Theta, \text{ for all } \mathbf{x}, \mathbf{y} \in \mathbb{R}^2. \tag{6.1}$$

Next, we consider the strongly continuous family of unitaries $\{U_{\mathbf{t}}^{\Theta}\}_{\mathbf{t}\in\mathbb{R}^2}$ on $L_2(\mathbb{R}^2)$ defined for each $\mathbf{t} \in \mathbb{R}^2$ by

$$(U_{\mathbf{t}}^{\Theta} f)(\mathbf{x}) := e^{-i\langle \mathbf{t},\mathbf{x}\rangle_{\Theta}} f(\mathbf{x}-\mathbf{t}), \quad f \in L_2(\mathbb{R}^2),\ \mathbf{x} \in \mathbb{R}^2.$$

Observe that these operators satisfy a version of the Weyl relations;

$$U_{\mathbf{t}}^{\Theta} U_{\mathbf{s}}^{\Theta} = e^{-i\langle \mathbf{t},\mathbf{s}\rangle_{\Theta}} U_{\mathbf{t}+\mathbf{s}}^{\Theta}, \quad \mathbf{s},\mathbf{t} \in \mathbb{R}^2. \tag{6.2}$$

Definition 6.1 The von Neumann algebra on $L_2(\mathbb{R}^2)$ generated by the group $\{U_{\mathbf{t}}^{\Theta}\}_{\mathbf{t}\in\mathbb{R}^2}$ is called the *Moyal plane*, and is denoted by $L_\infty(\mathbb{R}^2_\Theta)$.

It is well-known that (for $\theta \neq 0$) this von Neumann algebra is $*$-isomorphic to $\mathcal{B}\big(L_2(\mathbb{R}^2)\big)$ (see, e.g., [37, Theorem 6.5]). Therefore, it is equipped with a canonical (faithful, normal, semifinite) trace τ.

In this space, we define a class of "multiplication operators" corresponding to the Schwartz functions.

Definition 6.2 For a Schwartz function $f \in \mathcal{S}(\mathbb{R}^2)$, define the operator $\mathrm{Op}_\Theta(f) \in L_\infty(\mathbb{R}^2_\Theta)$ by

$$\mathrm{Op}_\Theta(f) := \int_{\mathbb{R}^2} f(\mathbf{t}) U_{\mathbf{t}}^{\Theta}\, d\mathbf{t}, \tag{6.3}$$

where the integral on the right-hand side may be understood in the weak sense of Definition 2.5 above. The family of such operators,

$$\mathcal{S}(\mathbb{R}^2_\Theta) := \big\{\mathrm{Op}_\Theta(f) : f \in \mathcal{S}(\mathbb{R}^2)\big\},$$

is called *noncommutative Schwartz space*.

Remark 6.3 When f and g are Schwartz functions, we have

$$\mathrm{Op}_\Theta(f) + \mathrm{Op}_\Theta(g) = \mathrm{Op}_\Theta(f+g), \quad \mathrm{Op}_\Theta(f)\cdot \mathrm{Op}_\Theta(g) = \mathrm{Op}_\Theta(f \diamond_\Theta g),$$

where $\diamond_\Theta$ denotes the twisted convolution

$$(f \diamond_\Theta g)(\mathbf{s}) = \int_{\mathbb{R}^2} f(\mathbf{t}) g(\mathbf{s}-\mathbf{t}) e^{-i\langle \mathbf{t},\mathbf{s}\rangle_{\Theta}} dt.$$

The Schwartz functions equipped with $\diamond_\Theta$ forms a Fréchet algebra [30]. The operation $\diamond_\Theta$ is a noncommutative analogue of the convolution.

One may consider elements of $\mathcal{S}(\mathbb{R}^2_\Theta)$ as operators from the Fourier dual picture of that treated in [27]. So, our differentiation operators are actually multiplication ones.

For $k = 1, 2$, we define the "differentiation" operators

$$(D_k f)(\mathbf{x}) := x_k f(\mathbf{x}), \quad f \in L_2\big(\mathbb{R}^2, |\mathbf{x}|^2 \, d\mathbf{x}\big), \ \mathbf{x} \in \mathbb{R}^2.$$

Observe that we have the commutator

$$[D_k, U_{\mathbf{t}}^{\Theta}] = t_k U_{\mathbf{t}}^{\Theta}, \quad \mathbf{t} \in \mathbb{R}^2, \tag{6.4}$$

which extends to a bounded operator on $L_2(\mathbb{R}^2)$ Recalling the gamma matrices from the previous section, we shall define the "twisted" Dirac operator by

$$\mathcal{D}_\Theta := \gamma_1 \otimes D_1 + \gamma_2 \otimes D_2, \quad \operatorname{dom}(\mathcal{D}_\Theta) = \mathbb{C}^2 \otimes L_2\big(\mathbb{R}^2, |\mathbf{x}|^2 \, d\mathbf{x}\big).$$

Note that this Dirac operator does not actually depend on Θ; this is merely a notational convention. In turn, let $-\Delta_\Theta := D_1^2 + D_2^2$ denote the corresponding Laplacian on $L_2\big(\mathbb{R}^2, |\mathbf{x}|^4 \, d\mathbf{x}\big)$.

Definition 6.4 Let $L_1(\mathbb{R}^2_\Theta)$ denote the trace class of $L_\infty(\mathbb{R}^2_\Theta)$ with respect to the canonical trace τ, equipped with the corresponding norm

$$\|X\|_1 := \tau\big(|X|\big), \quad X \in L_1(\mathbb{R}^2_\Theta).$$

For $X \in L_1(\mathbb{R}^2_\Theta)$, denote by $\eth_k X := [D_k, X]$ the commutator of D_k with X, for each $k = 1, 2$. For $m \in \mathbb{N}$, we define the *noncommutative Sobolev space* $W_{m,1}(\mathbb{R}^2_\Theta)$ by

$$W_{m,1}(\mathbb{R}^2_\Theta) := \big\{X \in L_1(\mathbb{R}^2_\Theta) : \eth_1^{\alpha_1} \eth_2^{\alpha_2} X \in L_1(\mathbb{R}^2_\Theta), \\ \text{for all } \alpha_1, \alpha_2 \in \mathbb{N} \text{ s.t. } \alpha_1 + \alpha_2 \leq m\big\},$$

and equip this space with the norm

$$\|X\|_{W_{m,1}} := \sum_{\alpha_1 + \alpha_2 \leq m} \|\eth_1^{\alpha_1} \eth_2^{\alpha_2} X\|_1, \quad \text{for } X \in W_{m,1}(\mathbb{R}^2_\Theta).$$

Remark 6.5 By [52, Lemma 3.3], the subspace $\mathcal{S}(\mathbb{R}^2_\Theta)$ is dense in $W_{2,1}(\mathbb{R}^2_\Theta)$.

We state a noncommutative analogue of Cwikel's estimates, whose proof may be found in [37].

Theorem 6.6 ([37, Theorem 7.6]) *If* $X \in W_{2,1}(\mathbb{R}^2_\Theta)$, *then* $X(1 - \Delta_\Theta)^{-1} \in \mathcal{L}_{1,\infty}\big(L_2(\mathbb{R}^2)\big)$ *and*

$$\big\|X(1 - \Delta_\Theta)^{-1}\big\|_{1,\infty} \leq \text{const} \cdot \|X\|_{W_{2,1}}.$$

6.2 The Algebra of Rapidly Decreasing Double-Sequences

In this section, we shall apply Theorem 1.1 to the operators $A := \mathrm{Op}_\Theta(f) \geq 0$, for some $f \in \mathcal{S}(\mathbb{R}^2)$, and $B := (1 - \Delta_\Theta)^{-1}$. To verify that these operators satisfy the conditions of Theorem 1.1, it is necessary to obtain some understanding of the behaviour of the fractional powers of A.

To this end, we shall investigate the algebraic structure of the Fréchet algebra $\big(\mathcal{S}(\mathbb{R}^2), \diamond_\Theta\big)$. This is done by recalling the algebra of rapidly decreasing double-sequences of Gracia-Bondía–Várilly [30].

Definition 6.7 We say that a square-summable double sequence $c \in \ell_2(\mathbb{N}^2)$ is *rapidly decreasing* if, for every $k \in \mathbb{N}$,

$$r_k(c) := \Big(\sum_{m,n\in\mathbb{N}} (m+1)^{2k}(n+1)^{2k}|c_{m,n}|^2\Big)^{\frac{1}{2}} < \infty.$$

The space of rapidly decreasing double sequences, denoted by $\mathbf{S} \subset \ell_2(\mathbb{N}^2)$, equipped with the family of seminorms $\{r_k\}_{k\in\mathbb{N}}$ forms a Fréchet space [30]. In addition, we equip this space with the *matrix product*, which we define by the expression

$$c \cdot d := \Big(\sum_{j\in\mathbb{N}} c_{m,j} d_{j,n}\Big)_{m,n\in\mathbb{N}}, \qquad c, d \in \mathbf{S}.$$

Remark 6.8 We have

$$r_k(c \cdot d) \leq r_k(c) r_k(d).$$

In particular, $\mathbf{S}$ equipped with the matrix product is a Fréchet algebra.

Proof Suppose $c, d \in \mathbf{S}$, and let $k \in \mathbb{N}$. By the triangle inequality and the Hölder inequality, we observe that

$$\Big|\sum_{j\in\mathbb{N}} c_{m,j} d_{j,n}\Big| \leq \Big(\sum_{j\in\mathbb{N}} |c_{m,j}|^2\Big)^{\frac{1}{2}} \Big(\sum_{j\in\mathbb{N}} |d_{j,n}|^2\Big)^{\frac{1}{2}},$$

so that

$$r_k(c \cdot d)^2 \leq \Big(\sum_{m,j\in\mathbb{N}} (m+1)^{2k}|c_{m,j}|^2\Big)\Big(\sum_{j,n\in\mathbb{N}} (n+1)^{2k}|d_{j,n}|^2\Big).$$

Hence, $r_k(c \cdot d) \leq r_k(c) r_k(d)$, □

Lemma 6.9 *If* $0 \leq c \in \mathbf{S}$, *then* $c^p \in \mathbf{S}$ *for all* $p > 0$.

Proof Firstly, we prove the assertion for $p = \frac{1}{2}$. Let $d = c^{\frac{1}{2}}$. Since d is self-adjoint, we observe that

$$|d_{m,n}|^2 = \big|\langle d\mathbf{e}_m, \mathbf{e}_n\rangle\big|\big|\langle \mathbf{e}_m, d\mathbf{e}_n\rangle\big|$$
$$\leq \|d\mathbf{e}_m\|\|d\mathbf{e}_n\| = \big|\langle d^2\mathbf{e}_m, \mathbf{e}_m\rangle\big|^{\frac{1}{2}}\big|\langle d^2\mathbf{e}_n, \mathbf{e}_n\rangle\big|^{\frac{1}{2}} = |c_{m,m}|^{\frac{1}{2}}|c_{n,n}|^{\frac{1}{2}}.$$

where the vectors $\{\mathbf{e}_j\}_{j=1}^{\infty}$ are the standard basis vectors for $\ell_2(\mathbb{N})$. Consequently, for $k \in \mathbb{N}$, we have that

$$r_k(d)^2 = \sum_{m,n=0}^{\infty} (m+1)^{2k}(n+1)^{2k}|d_{m,n}|^2$$
$$\leq \sum_{m,n=0}^{\infty} (m+1)^{2k}(n+1)^{2k}|c_{m,m}|^{\frac{1}{2}}|c_{n,n}|^{\frac{1}{2}} = \left(\sum_{m=0}^{\infty}(m+1)^{2k}|c_{m,m}|^{\frac{1}{2}}\right)^2.$$

By applying the Hölder inequality, we obtain

$$\sum_{m=0}^{\infty}(m+1)^{2k}|c_{m,m}|^{\frac{1}{2}} \leq \left(\sum_{m=0}^{\infty}\frac{1}{(m+1)^{\frac{8}{3}}}\right)^{\frac{3}{4}}\left(\sum_{m=0}^{\infty}(m+1)^{8k+8}|c_{m,m}|^2\right)^{\frac{1}{4}}.$$

Therefore, we have

$$r_k(d)^2 \leq \text{const} \cdot r_{2k+2}(c).$$

In particular, $r_k(d)$ is finite for every $k \in \mathbb{N}$. This proves the assertion for $p = \frac{1}{2}$.

By induction, the assertion holds for $p = 2^{-n}$, $n \in \mathbb{N}$.

Let $p > \frac{1}{2}$. Since c, c^p are self-adjoint, we have that

$$\big|(c^p)_{m,n}\big|^2 \leq \|c^p\mathbf{e}_m\|\|c^p\mathbf{e}_n\| \leq \|c\|_\infty^{2p-1}\cdot\|c^{\frac{1}{2}}\mathbf{e}_m\|\|c^{\frac{1}{2}}\mathbf{e}_n\| \leq \|c\|_\infty^{2p-1}\cdot|c_{m,m}|^{\frac{1}{2}}|c_{n,n}|^{\frac{1}{2}}.$$

Therefore,

$$r_k(c^p) \leq \|c\|_\infty^{2p-1}\sum_{m,n\in\mathbb{N}}(m+1)^{2k}(n+1)^{2k}|c_{m,m}|^{\frac{1}{2}}\cdot|c_{n,n}|^{\frac{1}{2}}$$
$$= \|c\|_\infty^{2p-1}\left(\sum_{m=0}^{\infty}(m+1)^{2k}|c_{m,m}|^{\frac{1}{2}}\right)^2.$$

Therefore,

$$r_k(c^p) \leq \text{const}\cdot\|c\|_\infty^{2p-1}r_{2k+2}(c)^{\frac{1}{2}}.$$

This proves the assertion for $p > \frac{1}{2}$. By considering $c^p = (c^{2^{-n}})^{2^n p}$, where $2^n p > 1$, we conclude the argument for $p > 0$. □

Remark 6.10 If $c \in \mathbf{S}$, then $c^2 \in \mathbf{S}$ and, by Lemma 6.9, $|c| = (c^2)^{\frac{1}{2}} \in \mathbf{S}$, but $c = |c| - \big(|c| - c\big)$. Hence, $\mathbf{S}$ is spanned by $\mathbf{S}_+$.

The algebra $\mathbf{S}$ is essential for us due to the following result of [30].

Theorem 6.11 *There exists a double sequence $(f_{m,n})_{m,n\in\mathbb{N}}$ of Schwartz functions such that*

(i) $f_{m,n} \diamond_\Theta f_{k,l} = \delta_{n,k} f_{m,l}$ *and* $f_{m,n} = \overline{f_{n,m}}$ *for* $m, n, k, l \in \mathbb{N}$.
(ii) *the mapping*

$$c \mapsto \sum_{m,n\in\mathbb{N}} c_{m,n} f_{m,n}, \quad c \in \mathbf{S},$$

is a $*$-isomorphism of Fréchet algebras.

We do not provide a proof here, but only mention that

$$f_{m,m}(\mathbf{x}) = \frac{\theta}{\pi} L_m\big(\theta|\mathbf{x}|^2\big) \exp\Big(-\frac{\theta}{2}|\mathbf{x}|^2\Big), \quad \mathbf{x} \in \mathbb{R}^2,$$

where L_m is the m-th Laguerre polynomial. A similar but much more complicated expression is available for $f_{m,n}$ with $m \neq n$. The reader is referred to [27] for further details.

Remark 6.12 Denote the origin of $\mathbb{R}^2$ by $\mathbf{0}$. Since τ is the normal trace on $L_\infty(\mathbb{R}^2_\Theta)$, we have that $\tau\big(\mathrm{Op}_\Theta(f_{m,m})\big) = 1$, since the operators $\mathrm{Op}_\Theta(f_{m,m})$ are atoms. In general, the decomposition in Theorem 6.11 yields the expression

$$\tau\big(\mathrm{Op}_\Theta(f)\big) = \frac{\pi}{\theta} f(\mathbf{0}). \tag{6.5}$$

Corollary 6.13 *If f is a Schwartz function such that $\mathrm{Op}_\Theta(f)$ is positive, then $\big(\mathrm{Op}_\Theta(f)\big)^p$ belongs to noncommutative Schwartz space for every $p > 0$.*

6.3 *Calculating the Dixmier Trace of* $\mathbf{Op}_\Theta(f)(1 - \Delta_\Theta)^{-1}$

It now remains to verify that the conditions of Theorem 1.1 are satisfied by the operators $A := \mathrm{Op}_\Theta(f)$, for $f \in \mathcal{S}(\mathbb{R}^2)$, and $B := (1 - \Delta_\Theta)^{-1}$, and calculate the value of the trace $\mathrm{Tr}(B^p A^p)$, for $p > 1$.

Lemma 6.14 *If $f \in \mathcal{S}(\mathbb{R}^2)$, then $\mathrm{Op}_\Theta(f)(1 - \Delta_\Theta)^{-p} \in \mathcal{L}_1\big(L_2(\mathbb{R}^2)\big)$ for every $p > 1$.*

Proof Recall from the previous section that $\mathcal{Q}_{\mathbf{n}}$ denotes the unit cube in $\mathbb{R}^2$ centred at $\mathbf{n} \in \mathbb{R}^2$. Since $\sum_{\mathbf{n}\in\mathbb{Z}^2} M_{\chi_{\mathcal{Q}_{\mathbf{n}}}} = 1$ (strongly), we have

$$\mathrm{Op}_\Theta(f)(1-\Delta_\Theta)^{-p} = \sum_{\mathbf{n}\in\mathbb{Z}^2} \mathrm{Op}_\Theta(f)(1-\Delta_\Theta)^{-p} M_{\chi_{\mathcal{Q}_{\mathbf{n}}}}.$$

By the triangle inequality we have

$$\begin{aligned}\left\|\mathrm{Op}_\Theta(f)(1-\Delta_\Theta)^{-p}\right\|_1 &\le \sum_{\mathbf{n}\in\mathbb{Z}^2} \left\|\mathrm{Op}_\Theta(f)(1-\Delta_\Theta)^{-p} M_{\chi_{\mathcal{Q}_{\mathbf{n}}}}\right\|_1 \\ &\le \sum_{\mathbf{n}\in\mathbb{Z}^2}\Big(\sup_{\mathbf{t}\in\mathcal{Q}_{\mathbf{n}}} (1+|\mathbf{t}|^2)^{-p}\cdot \left\|\mathrm{Op}_\Theta(f)M_{\chi_{\mathcal{Q}_{\mathbf{n}}}}\right\|_1\Big) \\ &\le \Big(\sum_{\mathbf{n}\in\mathbb{Z}^2}\sup_{\mathbf{t}\in\mathcal{Q}_{\mathbf{n}}} (1+|\mathbf{t}|^2)^{-p}\Big)\cdot\Big(\sup_{\mathbf{n}\in\mathbb{Z}^2}\left\|\mathrm{Op}_\Theta(f)M_{\chi_{\mathcal{Q}_{\mathbf{n}}}}\right\|_1\Big).\end{aligned}$$

Let $h_{\mathbf{n}}(\mathbf{t}) := e^{i\langle \mathbf{t},\mathbf{n}\rangle_\Theta}$ for $\mathbf{t}\in\mathbb{R}^2$. Let $T_{\mathbf{n}}$ denote the shift operator on $L_2(\mathbb{R}^2)$, then set $V_{\mathbf{n}} := M_{h_{\mathbf{n}}}T_{\mathbf{n}}$. We have

$$M_{\chi_{\mathcal{Q}_{\mathbf{n}}}} = V_{\mathbf{n}} M_{\chi_{\mathcal{Q}_{\mathbf{0}}}} V_{-\mathbf{n}}, \quad V_{-\mathbf{n}}\mathrm{Op}_\Theta(f)V_{\mathbf{n}} = \mathrm{Op}_\Theta(f).$$

Hence, it follows that

$$\left\|\mathrm{Op}_\Theta(f)(1-\Delta_\Theta)^{-p}\right\|_1 \le \Big(\sum_{\mathbf{n}\in\mathbb{Z}^2}\sup_{\mathbf{t}\in\mathcal{Q}_{\mathbf{n}}} (1+|\mathbf{t}|^2)^{-p}\Big)\cdot\left\|\mathrm{Op}_\Theta(f)M_{\chi_{\mathcal{Q}_{\mathbf{0}}}}\right\|_1.$$

Letting $g(\mathbf{t}) := f(-\mathbf{t})$, $\mathbf{t}\in\mathbb{R}^2$, and letting $k = g \diamond_\Theta f$, we have

$$\left|\mathrm{Op}_\Theta(f)M_{\chi_{\mathcal{Q}_{\mathbf{0}}}}\right|^2 = M_{\chi_{\mathcal{Q}_{\mathbf{0}}}}\mathrm{Op}_\Theta(k)M_{\chi_{\mathcal{Q}_{\mathbf{0}}}}.$$

Therefore, since $M_{\chi_{\mathcal{Q}_{\mathbf{0}}}}$ projects $L_2(\mathbb{R}^2)$ onto $L_2(\mathcal{Q}_{\mathbf{0}})$, it suffices to verify that

$$M_{\chi_{\mathcal{Q}_{\mathbf{0}}}}\mathrm{Op}_\Theta(k)M_{\chi_{\mathcal{Q}_{\mathbf{0}}}} \in \mathcal{L}_{\frac{1}{2}}\big(L_2(\mathcal{Q}_{\mathbf{0}})\big).$$

Appealing to (6.3), the operator on the right-hand side is an integral operator on $L_2(\mathcal{Q}_{\mathbf{0}})$ with a smooth integral kernel supported on $\mathcal{Q}_{\mathbf{0}}$; in fact,

$$\big(M_{\chi_{\mathcal{Q}_{\mathbf{0}}}}\mathrm{Op}_\Theta(k)M_{\chi_{\mathcal{Q}_{\mathbf{0}}}}\phi\big)(\mathbf{t}) = \int_{\mathcal{Q}_{\mathbf{0}}} k(\mathbf{t}-\mathbf{s})e^{i\langle \mathbf{s},\mathbf{t}\rangle_\Theta}\phi(\mathbf{s})\,d\mathbf{s}, \quad \phi\in L_2(\mathcal{Q}_{\mathbf{0}}),\ \mathbf{t}\in\mathcal{Q}_{\mathbf{0}}.$$

Therefore, by [29, Ch.III §10, Result 4], this operator belongs to $\mathcal{L}_p\big(L_2(\mathcal{Q}_{\mathbf{0}})\big)$ for every $p > 0$. In particular, it belongs to $\mathcal{L}_{\frac{1}{2}}\big(L_2(\mathcal{Q}_{\mathbf{0}})\big)$. This concludes the proof. □

We may now verify that the commutator $[A^{\frac{1}{2}}, B] \in \mathcal{L}_1$.

Lemma 6.15 *If $f \in \mathcal{S}(\mathbb{R}^2)$, then*

$$\big[\mathrm{Op}_\Theta(f), (1-\Delta_\Theta)^{-1}\big] \in \mathcal{L}_1\big(L_2(\mathbb{R}^2)\big).$$

Proof By the resolvent identity (see, e.g., [34, Theorem 4.8.2]), we have

$$\big[\mathrm{Op}_\Theta(f), (1-\Delta_\Theta)^{-1}\big] = (1-\Delta_\Theta)^{-1}\big[\mathrm{Op}_\Theta(f), \Delta_\Theta\big](1-\Delta_\Theta)^{-1}.$$

By (6.4), we have for $k = 1, 2$ that

$$\big[D_k, \mathrm{Op}_\Theta(f)\big] \overset{(6.3)}{=} \int_{\mathbb{R}^2} f(\mathbf{t})[D_k, U_{\mathbf{t}}^\Theta]\,\mathrm{d}\mathbf{t} \overset{(6.4)}{=} \int_{\mathbb{R}^2} t_k f(\mathbf{t}) U_{\mathbf{t}}^\Theta\,\mathrm{d}\mathbf{t} = \int_{\mathbb{R}^2} (D_k f)(\mathbf{t}) U_{\mathbf{t}}^\Theta\,\mathrm{d}\mathbf{t}.$$

Hence,

$$\begin{aligned}
&\big[\mathrm{Op}_\Theta(f), \Delta_\Theta\big] \\
&\quad= D_1\big[D_1, \mathrm{Op}_\Theta(f)\big] + D_2\big[D_2, \mathrm{Op}_\Theta(f)\big] + \big[D_1, \mathrm{Op}_\Theta(f)\big]D_1 + \big[D_2, \mathrm{Op}_\Theta(f)\big]D_2 \\
&\quad= D_1\mathrm{Op}_\Theta(D_1 f) + D_2\mathrm{Op}_\Theta(D_2 f) + \mathrm{Op}_\Theta(D_1 f)D_1 + \mathrm{Op}_\Theta(D_2 f)D_2.
\end{aligned}$$

Therefore, we have

$$\begin{aligned}
&\Big\|\big[\mathrm{Op}_\Theta(f), (1-\Delta_\Theta)^{-1}\big]\Big\|_1 \\
&\le \big\|(1-\Delta_\Theta)^{-\frac{1}{2}}\mathrm{Op}_\Theta(D_1 f)(1-\Delta_\Theta)^{-1}\big\|_1 + \big\|(1-\Delta_\Theta)^{-\frac{1}{2}}\mathrm{Op}_\Theta(D_2 f)(1-\Delta_\Theta)^{-1}\big\|_1 \\
&\qquad + \big\|(1-\Delta_\Theta)^{-1}\mathrm{Op}_\Theta(D_1 f)(1-\Delta_\Theta)^{-\frac{1}{2}}\big\|_1 + \big\|(1-\Delta_\Theta)^{-1}\mathrm{Op}_\Theta(D_2 f)(1-\Delta_\Theta)^{-\frac{1}{2}}\big\|_1 \\
&\overset{(2.2)}{\le} 2\big\|\mathrm{Op}_\Theta(D_1 f)(1-\Delta_\Theta)^{-\frac{3}{2}}\big\|_1 + 2\big\|\mathrm{Op}_\Theta(D_2 f)(1-\Delta_\Theta)^{-\frac{3}{2}}\big\|_1.
\end{aligned}$$

where in the last line we used the inequality (2.2) (see Sect. 2 above). The assertion now follows from Lemma 6.14. □

Furthermore, we obtain the following expression for the classical trace of $A^p B^p$.

Lemma 6.16 *If $f \in \mathcal{S}(\mathbb{R}^2)$ and if $p > 1$, then*

$$\mathrm{Tr}\big(\mathrm{Op}_\Theta(f)(1-\Delta_\Theta)^{-p}\big) = \frac{\pi}{p-1} f(\mathbf{0}).$$

Proof Appealing to (6.3), $\mathrm{Op}_\Theta(f)(1-\Delta_\Theta)^{-p}$ is an integral operator on $L_2(\mathbb{R}^2)$ with a continuous integral kernel defined by the expression

$$K(\mathbf{t}, \mathbf{s}) = f(\mathbf{t}-\mathbf{s})e^{i\langle \mathbf{s}, \mathbf{t}\rangle_\Theta}\big(1+|\mathbf{s}|^2\big)^{-p}, \quad \mathbf{t}, \mathbf{s} \in \mathbb{R}^2.$$

By Lemma 6.14, this operator belongs to $\mathcal{L}_1(L_2(\mathbb{R}^2))$. Since $K \in (L_1 \cap L_2)(\mathbb{R}^2 \times \mathbb{R}^2)$, Proposition 2.1 implies that

$$\begin{aligned}\operatorname{Tr}\left(\operatorname{Op}_\Theta(f)(1-\Delta_\Theta)^{-p}\right) &= \int_{\mathbb{R}^2} f(\mathbf{t}-\mathbf{t})e^{i\langle \mathbf{t},\mathbf{t}\rangle_\Theta}\left(1+|\mathbf{t}|^2\right)^{-p}\,d\mathbf{t} \\ &= f(\mathbf{0})\cdot\int_{\mathbb{R}^2}\left(1+|\mathbf{t}|^2\right)^{-p}\,d\mathbf{t}.\end{aligned}$$ □

Using these results, we arrive at the main result of this section.

Proposition 6.17 *If $f \in \mathcal{S}(\mathbb{R}^2)$, then $\operatorname{Op}_\Theta(f)(1-\Delta_\Theta)^{-1}$ is a Dixmier measurable operator and, for any extended limit ω,*

$$\operatorname{Tr}_\omega\left(\operatorname{Op}_\Theta(f)(1-\Delta_\Theta)^{-1}\right) = \pi f(\mathbf{0}).$$

Proof Firstly, denote $A := \operatorname{Op}_\Theta(f)$. Recalling Remark 6.10, we may assume without loss of generality that $A \geq 0$. Then, by Lemma 6.13, there exists some $g \in \mathcal{S}(\mathbb{R}^2)$ such that $A^{\frac{1}{2}} = \operatorname{Op}_\Theta(g)$. Setting $B := (1-\Delta_\Theta)^{-1}$, we infer from Lemma 6.15 that $[A^{\frac{1}{2}}, B] \in \mathcal{L}_1(L_2(\mathbb{R}^2))$ and, since A, B are positive operators, $\operatorname{Tr}(A^pB^p) = \operatorname{Tr}(B^pA^p)$.

Furthermore, by Lemma 6.13, there exists some Schwartz function $f_p \in \mathcal{S}(\mathbb{R}^2)$ such that $A^p = \operatorname{Op}_\Theta(f_p)$. By Remark 6.12 and Lemma 6.16, we have

$$(p-1)\operatorname{Tr}(A^pB^p) = \pi f_p(\mathbf{0}) = \theta\tau(\operatorname{Op}_\Theta(f_p)) = \theta\tau(A^p).$$

Hence, by Fack and Kosaki [26, Theorem 3.6], we have the limit

$$\lim_{p\downarrow 1}(p-1)\operatorname{Tr}(A^pB^p) = \theta\lim_{p\downarrow 1}\tau(A^p) = \theta\tau(A) = \pi f(\mathbf{0}).$$

Therefore, since $[A^{\frac{1}{2}}, B] \in \mathcal{L}_1(L_2(\mathbb{R}^2))$, and since $AB \in \mathcal{L}_{1,\infty}(L_2(\mathbb{R}^2))$ by Theorem 6.6, it follows from Lemma 1.1 (*b*) that

$$\operatorname{Tr}_\omega(AB) = \lim_{p\downarrow 1}(p-1)\operatorname{Tr}(B^pA^p) = \pi f(\mathbf{0}).$$ □

Using the noncommutative Cwikel estimate (see Theorem 6.6 above), this result may be easily extended to noncommutative Sobolev space.

Corollary 6.18 *If $X \in W_{2,1}(\mathbb{R}^2_\Theta)$, then $X(1-\Delta_\Theta)^{-1}$ is a Dixmier measurable operator and, for any extended limit ω,*

$$\operatorname{Tr}_\omega\left(X(1-\Delta_\Theta)^{-1}\right) = \theta\tau(X).$$

Proof By Remark 6.5, one may construct a sequence of functions $\{f_n\}_{n\in\mathbb{N}} \subset \mathcal{S}(\mathbb{R}^2)$ such that $\mathrm{Op}_\Theta(f_n) \to X$ as $n \to \infty$ in $W_{2,1}(\mathbb{R}^2_\Theta)$. In particular, since $\Theta(f_n) \to X$ as $n \to \infty$ in the L_1-norm, we also have that $\tau\big(\mathrm{Op}_\Theta(f_n)\big) \to \tau(X)$ as $n \to \infty$. By Theorem 6.6, we have

$$\left\| X(1-\Delta_\Theta)^{-1} - \mathrm{Op}_\Theta(f_n)(1-\Delta_\Theta)^{-1} \right\|_{1,\infty} \le \mathrm{const} \cdot \left\| X - \mathrm{Op}_\Theta(f_n) \right\|_{W_{2,1}} \to 0.$$

Hence, the sequence of operators $\left\{\mathrm{Op}_\Theta(f_n)(1-\Delta_\Theta)^{-1}\right\}_{n\in\mathbb{N}} \subset \mathcal{L}_{1,\infty}\big(L_2(\mathbb{R}^2)\big)$ converges to $X(1-\Delta_\Theta)^{-1}$ in the $\mathcal{L}_{1,\infty}$-norm. Therefore, since Tr_ω is continuous in $\mathcal{L}_{1,\infty}$,

$$\begin{aligned} \mathrm{Tr}_\omega\left(X(1-\Delta_\Theta)^{-1}\right) &= \lim_{n\to\infty} \mathrm{Tr}_\omega\left(\mathrm{Op}_\Theta(f_n)(1-\Delta_\Theta)^{-1}\right) \\ &= \lim_{n\to\infty} \pi f_n(\mathbf{0}) \overset{(6.5)}{=} \theta \lim_{n\to\infty} \tau\big(\mathrm{Op}_\Theta(f_n)\big) = \theta\tau(X) \end{aligned}$$

where in the second equality we appealed to Proposition 6.17. □

Acknowledgements The authors express their gratitude to Steven Lord for the extraordinary help he contributed related to the presentation of this text and the verification of its results. Denis Potapov, Fedor Sukochev and Dmitriy Zanin gratefully acknowledge the financial support from the Australian Research Council. Dominic Vella gratefully acknowledges the support received through the Australian Government Research Training Program Scholarship.

Appendix: Proof of Lemma 3

Let $n \in \mathbb{N}$ and $p \ge 1$. In the following, for $-\infty \le a < b \le \infty$, the Sobolev space of p-integrable functions on the interval (a, b) may be defined by

$$W_{n,p}(a,b) := \left\{ f \in L_p(a,b) \ : \ \|\partial^\alpha f\|_{L_p(a,b)} < \infty, \ \text{for multi-indices s.t. } |\alpha| \le n \right\},$$

with corresponding norm given by

$$\|f\|_{W_{n,p}(a,b)} := \sum_{\alpha:|\alpha|\le m} \left\| \frac{\partial^{\alpha_1}}{\partial t_1^{\alpha_1}} \cdots \frac{\partial^{\alpha_d}}{\partial t_d^{\alpha_d}}(f) \right\|_{L_p(a,b)}, \quad f \in W_{n,p}(a,b).$$

Lemma 1 *For $n \in \mathbb{N}$, we have*

$$\|g_p\|_{W_{n,2}(0,1)} = \mathcal{O}(1), \quad p \downarrow 1.$$

Proof Let $(\delta_u f)(t) = f(ut)$. We write

$$g_p = h \cdot \big(f - (p-1)(\delta_{p-1} f)\big),$$

where

$$h(t) = \frac{t}{2}\coth\Big(\frac{t}{2}\Big), \quad f(t) = \frac{1}{t}\tanh\Big(\frac{t}{2}\Big), \quad t \in \mathbb{R}.$$

By Leibniz rule, we have

$$\begin{aligned}\|g_p\|_{W_{n,2}(0,1)} &\le \Big\|h \cdot \big(f - (p-1)(\delta_{p-1}f)\big)\Big\|_{W_{n,\infty}(0,1)} \\ &\le \|h\|_{W_{n,\infty}(0,1)} \big\|f - (p-1)(\delta_{p-1}f)\big\|_{W_{n,\infty}(0,1)}.\end{aligned}$$

For $0 \le k \le n$, we have

$$\big(f - (p-1)(\delta_{p-1}f)\big)^{(k)} = f^{(k)} - (p-1)^{k+1}\delta_{p-1}(f^{(k)}).$$

Hence,

$$\Big\|\big(f - (p-1)(\delta_{p-1}f)\big)^{(k)}\Big\|_\infty \le \big(1 + (p-1)^{k+1}\big)\|f^{(k)}\|_\infty. \qquad \square$$

Lemma 2 *For $n \in \mathbb{N}$, we have*

$$\|g_p\|_{W_{n,2}(1,\infty)} = \mathcal{O}\big((p-1)^{-\frac{1}{2}}\big), \quad p \downarrow 1.$$

Proof Let $(\delta_u f)(t) = f(ut)$. We write

$$g_p = h \cdot (f - \delta_{p-1}f),$$

where

$$h(t) = \frac{1}{2}\coth\Big(\frac{t}{2}\Big), \quad f(t) = \tanh\Big(\frac{t}{2}\Big), \quad t \in \mathbb{R}.$$

By Leibniz rule, we have

$$\|g_p\|_{W_{n,2}(1,\infty)} \le \|h\|_{W_{n,\infty}(1,\infty)}\|f - \delta_{p-1}f\|_{W_{n,2}(1,\infty)}.$$

For $0 \le k \le n$, we have

$$(f - \delta_{p-1}f)^{(k)} = f^{(k)} - (p-1)^k\delta_{p-1}(f^{(k)}).$$

Hence,

$$\begin{aligned}\left\|(f-\delta_{p-1}f)^{(k)}\right\|_{L_2(1,\infty)} &\le \|f^{(k)}\|_{L_2(1,\infty)}+(p-1)^k\left\|\delta_{p-1}(f^{(k)})\right\|_{L_2(1,\infty)}\\ &\le \left(1+(p-1)^{k-\frac{1}{2}}\right)\cdot\|f^{(k)}\|_{L_2(0,\infty)}.\end{aligned}$$

□

Lemma 3 *For g_p defined as above, $\|g_p\|_{W_{n,2}}=\mathcal{O}\big((p-1)^{-\frac{1}{2}}\big)$ as $p\downarrow 1$.*

Proof As g_p is even, we have

$$\|g_p\|_{W_{n,2}}\le 2\big(\|g_p\|_{W_{n,2}(0,1)}+\|g_p\|_{W_{n,2}(1,\infty)}\big).$$

The assertion follows from the preceding lemmas. □

References

1. M. Birman, M. Solomyak, Double Stieltjes operator integrals, in *Spectral Theory and Wave Processes (Russian)*. Problems in Mathematical Physics, vol. I (Leningrad University, Leningrad, 1966), pp. 33–67
2. M. Birman, M. Solomyak, Double Stieltjes operator integrals II, in *Spectral Theory, Diffraction Problems (Russian)*. Problems of Mathematical Physics, vol. 2 (Leningrad University, Leningrad, 1967), pp. 26–60
3. M. Birman, M. Solomyak, Double Stieltjes operator integrals III (Russian). Prob. Math. Phys. **6**, 27–53 (1973)
4. M. Birman, M. Solomyak, Estimates for the singular numbers of integral operators (Russian). Usp. Mat. Nauk **32**(1), 17–84 (1977)
5. M. Birman, G. Karadzhov, M. Solomyak, Boundedness conditions and spectrum estimates for the operators $b(X)a(D)$ and their analogs, in *Estimates and Asymptotics for Discrete Spectra of Integral and Differential Equations (Leningrad, 1989–1990)*. Advances in Soviet Mathematics, vol. 7 (American Mathematical Society, Providence, 1991), pp. 85–106
6. M. Birman, M. Solomyak, Double operator integrals in a Hilbert space. Integr. Equ. Oper. Theory **47**(2), 131–168 (2003)
7. C. Brislawn, Kernels of trace class operators. Proc. Am. Math. Soc. **104**(4), 1181–1190 (1988)
8. A. Carey, J. Phillips, F. Sukochev, Spectral flow and Dixmier traces. Adv. Math. **173**(1), 68–113 (2003)
9. A. Carey, A. Rennie, A. Sedaev, F. Sukochev, The Dixmier trace and asymptotics of zeta functions. J. Funct. Anal. **249**(2), 253–283 (2007)
10. A. Carey, V. Gayral, A. Rennie, F. Sukochev, Integration on locally compact noncommutative spaces. J. Funct. Anal. **263**(2), 383–414 (2012)
11. P. Clément, B. de Pagter, F. Sukochev, H. Witvliet, Schauder decomposition and multiplier theorems. Stud. Math. **138**(2), 135–163 (2000)
12. A. Connes, The action functional in non-commutative geometry. Commun. Math. Phys. **117**, 673–683 (1988)
13. A. Connes, *Noncommutative Geometry* (Academic, San Diego, 1994)
14. A. Connes, F. Sukochev, D. Zanin, Trace theorem for quasi-Fuchsian groups. Sb. Math. **208**(10), 1473–1502 (2017)
15. Y. Daleckiĭ, S. Kreĭn, Formulas of differentiation according to a parameter of functions of Hermitian operators (Russian). Dokl. Akad. Nauk SSSR **76**, 13–16 (1951)

16. Y. Daleckiĭ, S. Kreĭn, Integration and differentiation of functions of Hermitian operators and applications to the theory of perturbations (Russian). Voronež. Gos. Univ. Trudy Sem. Funkcional. Anal. **1956**(1), 81–105 (1956)
17. B. de Pagter, F. Sukochev, H. Witvliet, Unconditional decompositions and Schur-type multipliers, in *Recent Advances in Operator Theory (Groningen, 1998)*. Operator Theory: Advances and Applications, vol. 124 (Birkhäuser, Basel, 2001), pp. 505–525
18. B. de Pagter, F. Sukochev, H. Witvliet, Double operator integrals. J. Funct. Anal. **192**(1), 52–111 (2002)
19. B. de Pagter, F. Sukochev, Differentiation of operator functions in non-commutative L_p-spaces. J. Funct. Anal. **212**(1), 28–75 (2004)
20. B. de Pagter, F. Sukochev, Commutator estimates and $\mathbb{R}$-flows in non-commutative operator spaces. Proc. Edinb. Math. Soc. (2) **50**(2), 293–324 (2007)
21. J. Dixmier, Existence de traces non normales (French). C. R. Acad. Sci. Paris Sér. A-B **262**, A1107–A1108 (1966)
22. P. Dodds, B. de Pagter, E. Semenov, F. Sukochev, Symmetric functionals and singular traces. Positivity **2**(1), 47–75 (1998)
23. P. Dodds, B. de Pagter, A. Sedaev, E. Semenov, F. Sukochev, Singular symmetric functionals and Banach limits with additional invariance properties (Russian). Izv. Ross. Akad. Nauk Ser. Mat. **67**(6), 111–136 (2003). English translation in: Izv. Math. **67**(6), 1187–1212 (2003)
24. P. Dodds, B. de Pagter, A. Sedaev, E. Semenov, F. Sukochev, Singular symmetric functionals (Russian). Zap. Nauchn. Sem. S.-Peterburg. Otdel. Mat. Inst. Steklov. (POMI) **290** (2002), Issled. po Lineĭn. Oper. i Teor. Funkts. **30**, 42–71, 178; English translation in: J. Math. Sci. (N.Y.) **124**(2), 4867–4885 (2004)
25. M. Duflo, Généralités sur les représentations induites (French), in *Représentations des Groupes de Lie Résolubles*. Monographies de la Société Mathématique de France, vol. 4 (Dunod, Paris, 1972), pp. 93–119
26. T. Fack, H. Kosaki, Generalised s-numbers of τ-measurable operators. Pac. J. Math. **123**(2), 269–300 (1986)
27. V. Gayral, J. Gracia-Bondía, B. Iochum, T. Schücker, J. Várilly, Moyal planes are spectral triples. Commun. Math. Phys. **246**(3), 569–623 (2004)
28. G. Glaeser, Racine carré d'une fonction différentiable (French). Ann. Inst. Fourier (Grenoble) **13**(2), 203–210 (1963)
29. I. Gohberg, M. Kreĭn, in *Introduction to the Theory of Linear Nonselfadjoint Operators*. Translations of Mathematical Monographs, vol. 18 (American Mathematical Society, Providence, 1969)
30. J. Gracia-Bondía, J. Várilly, Algebras of distributions suitable for phase-space quantum mechanics I. J. Math. Phys. **29**(4), 869–879 (1988)
31. J. Gracia-Bondía, J. Várilly, Algebras of distributions suitable for phase-space quantum mechanics II.: topologies on the Moyal algebra. J. Math. Phys. **29**(4), 880–887 (1988)
32. L. Grafakos, *Modern Fourier Analysis*, 2nd edn. Graduate Texts in Mathematics, vol. 250. (Springer, New York, 2009)
33. G. Hardy, *Divergent Series* (reprint of the revised (1963) edition) (Éditions Jacques Gabay, Sceaux, 1992)
34. E. Hille, R. Phillips, *Functional Analysis and Semi-groups* (3rd printing of the revised edition of 1957), American Mathematical Society Colloquium Publications, vol. 31 (American Mathematical Society, Providence, 1974)
35. N. Kalton, S. Lord, D. Potapov, F. Sukochev, Traces of compact operators and the noncommutative residue. Adv. Math. **235**, 1–55 (2013)
36. A. Kolmogorov, S. Fomin, *Introductory Real Analysis*. Translated from the second Russian edition and edited by R. Silverman (Dover Publications, New York, 1975)
37. G. Levitina, F. Sukochev, D. Zanin, Cwikel estimates revisited (2017). arXiv preprint arXiv:1703.04254
38. G. Levitina, F. Sukochev, D. Vella, D. Zanin, Schatten class estimates for the Riesz map of massless Dirac operators. Integr. Equ. Oper. Theory **90**(2), 19 (2018)

39. S. Lord, A. Sedaev, F. Sukochev, Dixmier traces as singular symmetric functionals and applications to measurable operators. J. Funct. Anal. **224**(1), 72–106 (2005)
40. S. Lord, D. Potapov, F. Sukochev, Measures from Dixmier traces and zeta functions. J. Funct. Anal. **259**, 1915–1949 (2010)
41. S. Lord, F. Sukochev, D. Zanin, *Singular Traces: Theory and Applications*. De Gruyter Studies in Mathematics, vol. 46 (Walter de Gruyter, Berlin, 2013)
42. V. Peller, Multiple operator integrals and higher operator derivatives. J. Funct. Anal. **233**(2), 515–544 (2006)
43. D. Potapov, F. Sukochev, Lipschitz and commutator estimates in symmetric operator spaces. J. Oper. Theory **59**(1), 211–234 (2008)
44. D. Potapov, F. Sukochev, Unbounded Fredholm modules and double operator integrals. J. Reine Angew. Math. **626**, 159–185 (2009)
45. M. Reed, B. Simon, *Fourier Analysis, Self-adjointness*. Methods of Modern Mathematical Physics, vol. II (Academic, New York, 1975)
46. A. Rennie, Summability for nonunital spectral triples. K-Theory **31**, 71–100 (2004)
47. M. Rieffel, Deformation quantization for actions of $\mathbb{R}^d$. Mem. Am. Math. Soc. **106**(506), x+93 (1993)
48. B. Simon, *Trace Ideals and Their Applications*. Mathematical Surveys and Monographs, 2nd edn., vol. 120 (American Mathematical Society, Providence, 2005)
49. F. Sukochev, On a conjecture of A. Bikchentaev, in *Spectral Analysis, Differential Equations and Mathematical Physics: A Festschrift in Honor of Fritz Gesztesy's 60th birthday*. Proceedings of Symposia in Pure Mathematics, vol. 87 (American Mathematical Society, Providence, 2013)
50. F. Sukochev, D. Zanin, ζ-Function and heat kernel formulae. J. Funct. Anal. **260**(8), 2451–2482 (2011)
51. F. Sukochev, A. Usachev, D. Zanin, Singular traces and residues of the ζ-function. Indiana Univ. Math. J. **66**(4), 1107–1144 (2017)
52. F. Sukochev, D. Zanin, Connes integration formula for the noncommutative plane. Commun. Math. Phys. **359**(2), 449–466 (2018)
53. F. Sukochev, D. Zanin, The Connes character formula for locally compact spectral triples. arXiv preprint arXiv:1803.01551 (2018)
54. B. Thaller, *The Dirac Equation* (Springer, Berlin, 1992)
55. A. Tomskova, Multiple operator integrals: development and applications. Ph.D. thesis, UNSW Sydney, 2017

Regular States and the Regular Algebra Numerical Range

Anton R. Schep and James Sweeney

Dedicated to Ben de Pagter on the occasion of his retirement

Abstract Let E be a Dedekind complete complex Banach lattice and let $\mathcal{L}_r(E)$ denote the Banach lattice algebra of regular operators on E. Then a bounded linear functional $\Phi : \mathcal{L}_r(E) \to \mathbb{C}$ is called a regular state if $\Phi(I) = \|\Phi\| = 1$. Basic properties of regular states are derived and a detailed description of order continuous regular states is obtained for ℓ_p-spaces. The regular states define the regular numerical algebra range $V(\mathcal{L}_r(E), T) = \{\Phi(T) : \Phi \text{ a regular state}\}$. We describe the regular algebra numerical range for operators T in the center $Z(E)$ of E and for operators T disjoint with the center. Using the description of regular states on $\ell_p(n)$, we characterize the regular numerical range preserving linear operators on $\mathcal{L}_r(\ell_p(n))$.

Keywords Regular operator · States · Numerical range · Banach lattice

1 Introduction

Originally the numerical range was defined as $W(T) = \{< Tx, x >: \|x\| = 1\}$ for a bounded linear operator T on a Hilbert space H. One of the main results is the Toeplitz–Hausdorff theorem, which states that $W(T)$ is convex for all bounded linear operators T on H. The natural extension of this numerical range is the *spatial numerical range* for a linear operator T on a Banach space E, which is defined as $V(T) = \{f(Tx) : x \in E, f \in E^*, \|x\| = \|f\| = 1 = f(x)\}$.

A. R. Schep (✉)
Department of Mathematics, University of South Carolina, Columbia, SC, USA
e-mail: schep@math.sc.edu

J. Sweeney
Coker College, Hartsville, SC, USA
e-mail: jsweeney@coker.edu

G. Buskes et al. (eds.), *Positivity and Noncommutative Analysis*,
Trends in Mathematics, https://doi.org/10.1007/978-3-030-10850-2_25

It was soon discovered that such numerical ranges no longer need to be convex, when the Banach space E is no longer a Hilbert space. For this reason the algebra numerical range was introduced. For a Banach algebra $\mathcal{A}$ with unit e the *algebra numerical range* for an element a in a Banach algebra $\mathcal{A}$ is defined as $V(\mathcal{A}, a) = \{\Phi(a) : \Phi \in \mathcal{A}^*, \Phi(e) = 1 = \|\Phi\|\}$. It is not difficult to see that this numerical is always a compact convex set. By taking $\mathcal{A}$ the set $\mathcal{L}(E)$ of bounded operators on a Banach space we get the algebra numerical range of a bounded linear operator. One can find an extensive discussion of this numerical range in the two books by Bonsall and Duncan [4, 5]. In this paper we will study the algebra numerical range for $\mathcal{A} = \mathcal{L}_r(E)$, the Banach lattice algebra of regular operators on a Dedekind complete complex Banach lattice E. Recently Radl [15, 16] considered the spatial numerical range of a positive operator on a Hilbert or Banach lattice, but did not consider either algebra numerical range. We start our study of the regular algebra numerical range in Sect. 2 by a systematic study of the regular states $\Phi \in (\mathcal{L}_r(E))^*$, i.e., $\Phi \in (\mathcal{L}_r(E))^*$ with $\Phi(I) = \|\Phi\| = 1$. Besides vector states we consider diagonal states and pure states. On finite dimensional Banach lattices we show that the pure states are those states Φ for which the modulus $|\Phi|$ is a pure vector state. In the infinite dimensional case we describe the order continuous regular states on the spaces $\mathcal{L}_r(\ell_p)$, where $1 \leq p \leq \infty$. In Sect. 3 we then use the obtained properties of regular states to describe the regular algebra numerical range. In particular we find the numerical range for T in the center $Z(E)$ of E and for $T \perp I$. Then we study some subsets and numerical quantities associated with the numerical ranges, by restricting ourselves to positive operators T and positive states Φ, where the state Φ is called positive if $\Phi(T) \geq 0$ for all $T \geq 0$. In Sect. 4 we study a linear preserving problem on $\mathcal{L}_r(E)$. The study of linear preserver problems is an active research area in operator theory, especially in matrix theory. For a general survey we refer to a paper by Li and Pierce [9] and for spectrum preserving operators on $\mathcal{L}_r(E)$ we refer to [19]. Of special interest for this paper is the paper [11] by Li and Sourour, who determined the general form of algebra numerical range preserving linear maps on the algebra of bounded operators on a finite dimensional symmetrically normed space. That paper motivated us in this section to consider regular numerical range preserving linear operators. In particular for $E = \ell_1(n)$ and $E = \ell_\infty(n)$ their results apply to our setting as the regular norm is identical to the operator norm on $\mathcal{L}_r(\ell_1(n))$ and $\mathcal{L}_r(\ell_\infty(n))$. We provide a necessary form for regular algebra numerical range preserving linear maps on the algebra of regular operators on a finite dimensional symmetrically normed space. For the case $E = \ell_p(n)$ we can then precisely characterize such operators. As an application of these results we obtain information about unital surjective isometries on the space of regular operators. For general background material on Banach lattices and operators on them, we refer to the books [2] and [20].

2 Regular States

Recall that for a Banach algebra $\mathcal{A}$ with unit e, a state Φ is defined as a linear functional on $\mathcal{A}$ such that $\Phi(e) = 1 = \|\Phi\|$. The collection of such states is denoted by $\mathcal{S}(\mathcal{A})$. In the literature, especially for the case that $\mathcal{A}$ is a C^*-algebra, such states are often called positive states. In our context this would be confusing and we will not use that terminology for arbitrary states. The following fact is well-known and easy to prove.

Proposition 2.1 *$\mathcal{S}(\mathcal{A})$ is a weak*compact convex subset of the dual space $\mathcal{A}^*$.*

Although we will not use it, we mention that $\mathcal{A}^*$ is the complex linear span of $\mathcal{S}(\mathcal{A})$ [5, Theorem 1 of section 31]. A consequence of the above proposition and the Krein–Milman theorem is that $\mathcal{S}(\mathcal{A})$ has extreme points. These extreme points are usually called pure states. From now on we will denote by E a complex Dedekind complete Banach lattice. Then we have several choices of Banach algebras $\mathcal{A}$ of bounded operators on E. If we consider the Banach algebra $\mathcal{A} = \mathcal{L}(E)$ of bounded operators on E, we get what we will call bounded states on E and use the notation $\mathcal{S}$ for $\mathcal{S}(\mathcal{L}(E))$. In this paper we will focus on the choice of $\mathcal{A} = \mathcal{L}_r(E)$, the Banach lattice algebra of regular operators on E. In that case we call the elements of $\mathcal{S}(\mathcal{L}_r(E))$ regular states and use the notation $\mathcal{S}_r$ for $\mathcal{S}(\mathcal{L}_r(E))$. We start with a simple observation.

Proposition 2.2 *Every bounded state is a regular state, i.e., $\mathcal{S} \subset \mathcal{S}_r$. Moreover if $\Phi \in \mathcal{S}_r$, then $|\Phi| \in \mathcal{S}_r$.*

Proof Let $\Phi \in \mathcal{S}$, i.e., $\Phi \in \mathcal{L}(E)^*$ such that $\|\Phi\| = 1 = \Phi(I)$. Let $\Phi_r = \Phi\,|_{\mathcal{L}_r(E)}$. Since $\|\Phi\| = 1$ we have that for all $S \in \mathcal{L}_r(E)$, $|\Phi_r(S)| = |\Phi(S)| \le \|S\| \le \|S\|_r$. Hence $||\Phi_r||_r \le 1$, where $||\Phi_r||_r$ denotes the norm of Φ_r in $\mathcal{L}_r(E)^*$. However, we have that $1 = \Phi(I) = |\Phi_r(I)| \le ||\Phi_r||_r||I||_r = ||\Phi_r||_r$ so $||\Phi_r||_r = 1$, i.e., $\Phi \in \mathcal{S}_r$. Let now $\Phi \in \mathcal{S}_r$, Then $\||\Phi|\| = \|\Phi\| = 1$ (as the norm on $\mathcal{L}_r(E)^*$ is a lattice norm), so it remains to show that also $|\Phi|(I) = 1$. This follows directly from

$$1 = \Phi(I) = |\Phi(I)| \le |\Phi|(I) \le \||\phi|\|\|I\| = 1.$$

□

The above proposition shows that for any regular state Φ we have $\Phi(I) = |\Phi|(I)$. We will now show that this is in fact true for all operators T in the center $Z(E) = \{T : |T| \le \lambda I \text{ for some } \lambda\}$ of E. First we state a well-known lemma, whose proof we include for the reader's convenience.

Lemma 2.3 *Let μ be a complex regular Borel measure on a compact Hausdorff space K such that $\mu(K) = 1 = |\mu|(K)$, where $|\mu|$ is the total variation of μ. Then μ is a positive measure.*

Proof Assume that there exists some measurable set $E \subset K$ such that $\mu(E) = c$ for some $c \in \mathbb{C} \setminus [0, 1]$. By additivity we have $\mu(E^c) = 1 - c$. Then $|c| + |1 - c| > 1$, since $c \notin [0, 1]$. Hence,

$$|\mu|(K) \geq |\mu(E)| + |\mu(E^c)| > 1.$$

This yields a contradiction, so μ must be a positive measure. □

Proposition 2.4 *Let $\Phi \in \mathcal{S}_r$ be a regular state. Then $\Phi \mid_{Z(E)} \geq 0$.*

Proof By the Kakutani's theorem, there exists a compact Hausdorff space K such that $Z(E)$ is lattice isomorphic to $C(K)$. Hence $\Phi \mid_{Z(E)}$ can be identified with some $\Psi \in C(K)^*$. By the Riesz representation theorem every functional on $C(K)$ can be represented by a regular complex Borel measure μ on K such that $\Psi(f) = \int_K f d\mu$ and $||\mu|| = |\mu|(K) = ||\Psi||$. Thus we have that $\mu(K) = 1 = |\mu|(K)$ and by the above lemma this means that μ must be a positive measure, which implies that Ψ must be a positive functional, and by the lattice isomorphism implies that $\Phi \mid_{Z(E)} \geq 0$. □

2.1 Diagonal States

Let $\mathcal{P} : \mathcal{L}_r(E) \to \mathcal{L}_r(E)$ denote the band projection from $\mathcal{L}_r(E)$ onto the center $Z(E)$. Then $\mathcal{P}^*$ is a band projection on $\mathcal{L}_r(E)^*$. It is clear that $\mathcal{P}^*(\Phi) = \Phi \mid_{Z(E)}$ on $Z(E)$. Moreover we have that $\Phi \mid_{Z(E)} \in \mathcal{S}(Z(E))$ for all $\Phi \in \mathcal{S}_r$ and that for every $\Psi \in \mathcal{S}(Z(E))$ there exists a $\Phi \in \mathcal{S}_r$ such that $\mathcal{P}^*(\Phi) \mid_{Z(E)} = \Psi$ (by defining, e.g., $\Phi = 0$ on $Z(E)^d$). We summarize these facts in the following proposition.

Proposition 2.5 *Let $\mathcal{P}^*$ be the adjoint of the band projection $\mathcal{P}$ from $L_r(E)$ onto $Z(E)$. Then $\mathcal{P}^*(\mathcal{S}_r) \subset \mathcal{S}_r$ and $\{\mathcal{P}^*(\Phi) \mid_{Z(E)} : \Phi \in \mathcal{S}_r\} = \mathcal{S}(Z(E)) = \mathcal{S}_r(Z(E)) = \{\Psi \in Z(E)^* : \Psi \geq 0, \|\Psi\| = 1\}$.*

We will call $\mathcal{P}^*(\Psi)$ the diagonal part of Ψ and call the elements of $\mathcal{P}^*(\mathcal{S}_r)$ diagonal states.

2.2 Vector States

Let $f \in E^*$ and $x \in E$ such that $1 = \|x\| = \|f\| = f(x)$. Then the linear functional $\Phi = f \otimes x$, defined by $\Phi(T) = f(Tx)$, is clearly a bounded state and is called a vector state. We will denote by $\mathcal{V}$ the set of all vector states. From Proposition 2.2 we know that if $\Phi = f \otimes x \in \mathcal{V}$, then $|\Phi|$ is a regular state. It is also clear that $|f| \otimes |x|$ is a vector state and that $|\Phi| \leq |f| \otimes |x|$. That in fact we have equality here is not completely obvious and therefore proved here.

Proposition 2.6 *Let $\Phi = f \otimes x \in \mathcal{V}$. Then $|\Phi| = |f| \otimes |x|$.*

Proof Let $0 \leq T \in \mathcal{L}_r(E)$. Then

$$|f|(T|x|) = \sup\{|f(y)| : |y| \leq T|x|\}.$$

Let $|y| \leq T|x|$. Then there exists $\pi \in Z(E)$ with $|\pi| = I$ such that $|x| = \pi x$ and there also exists $\sigma \in Z(E)$ with $|\sigma| \leq I$ such that $\sigma T|x| = y$. Define now $S = \sigma T \pi$. Then $|S| \leq T$ and $Sx = y$. This shows that

$$\begin{aligned} \sup\{|f(y)| : |y| \leq T|x|\} &= \sup\{|f(Sx)| : |S| \leq T\} \\ &= \sup\{|\Phi(S)| : |S| \leq T\} = |\Phi|(T). \end{aligned}$$

Hence $|\Phi|(T) = |f|(T|x|) = |f| \otimes |x|(T)$. □

In the above proof we used a formula for the modulus of a complex linear functional on a complex Banach lattice. We refer to [20, Theorem 36.4], for a detailed proof of this result (for order bounded complex linear operators). Using vector states one can easily construct examples of regular states, which are not bounded states.

Example Let $E = \ell_2(2)$. Then we can identify $\mathcal{L}_r(E)$ with the space $M_2(\mathcal{C})$ of two by two complex matrices, where the modulus is obtained by taking the pointwise modulus of the entries. Similarly we can represent $\Phi \in \mathcal{L}_r(E)^*$ by a two by two matrix, acting coordinate-wise on it. Let $\Phi = \begin{bmatrix} \frac{1}{2} & \frac{1}{2} \\ 0 & \frac{1}{2} \end{bmatrix}$. Note $0 \leq \Phi \leq f \otimes x$, where $f = (\frac{1}{2}\sqrt{2}, \frac{1}{2}\sqrt{2}) = x$ and $\Phi(I) = 1$. We see thus that the norm $\|\Phi\|_r = 1$. On the other hand, if we take $T = \begin{bmatrix} \frac{1}{2}\sqrt{2} & \frac{1}{2}\sqrt{2} \\ -\frac{1}{2}\sqrt{2} & \frac{1}{2}\sqrt{2} \end{bmatrix}$, then $\|T\| = 1$ and thus $\|\Phi\| \geq |\Phi(T)| = \frac{3}{4}\sqrt{2} > 1$, i.e., $\Phi \in \mathcal{S}_r \setminus \mathcal{S}$. Note that in this example the diagonal part of Φ is $\mathcal{P}^*(\Phi) = \begin{bmatrix} \frac{1}{2} & 0 \\ 0 & \frac{1}{2} \end{bmatrix}$.

Note that in [3, Theorem 4.4], an explicit description is given of all bounded states on $\mathcal{L}(\ell_2(2))$. In particular, if $\Phi = [\phi_{i,j}]$ is a bounded state on $\mathcal{L}(\ell_2(2))$, then $\phi_{1,2} = \overline{\phi_{2,1}}$. This shows immediately that the above regular state Φ is not a bounded state. From the Hahn-Banach theorem it is clear that the vector states $\mathcal{V}$ separate the points of $\mathcal{L}_r(E)$. This raises the question whether every diagonal state is the diagonal of a vector state. In general this is not true, as if E has order continuous norm, then every vector state is order continuous (or normal), while $Z(E)$ will have non-order continuous states when E is infinite dimensional. We provide a partial answer to the question.

Theorem 2.7 *Let E be a Banach function space with the Fatou property over a σ-finite measure space (X, Σ, μ). Then every order continuous diagonal state is the diagonal of an order continuous vector state.*

Proof It is well-known that $Z(E)$ is lattice isometric to $L_\infty(X,\Lambda,\mu)$, acting on E as multiplication operators. Hence, if $\phi \in Z(E)^*$ is an order continuous state, then there exists $0 \le h \in L_1(X,\Lambda,\mu)$ with $\int_X h(x)\,d\mu = 1$ such that $\phi(f) = \int_X f(x)h(x)\,d\mu$. Now by Lozanovskii's factorization theorem ([13] and [6]) we can factor $h = h_1h_2$ such that $\|h_1\|_E = \|h_2\|_{E^*} = 1$. Define now $\Phi = h_2 \otimes h_1$, where we identify h_2 with the order continuous functional it defines. It is now straightforward to verify that Φ is an order continuous vector state and that $\mathcal{P}^*(\Phi)\mid_{Z(E)}= \phi$, i.e., ϕ is the diagonal of the order continuous vector state Φ. □

2.3 Pure Regular States

First we recall some notations. For a Banach space F we denote by $B(F)$ the closed unit ball of F and for a convex set A we denote by $\mathcal{E}(A)$ the collection of all extreme points of A. For a Banach space F an extreme point of the set $\mathcal{S}$ of bounded states is called a pure state. Therefore we call an extreme point of the set $\mathcal{S}_r$ of regular states on a Dedekind complete complex Banach lattice E a pure regular state. We first state a simple fact about extreme points of faces of a unit ball.

Proposition 2.8 *Let E be a Dedekind complete complex Banach lattice. Then*

$$\mathcal{E}(\mathcal{S}_r) = \mathcal{E}(B(\mathcal{L}_r(E))^*) \cap \{\Phi \in B(\mathcal{L}_r(E))^* : \Phi(I) = 1\}.$$

In light of this proposition we first prove some properties of extreme points of unit balls of Banach lattices.

Theorem 2.9 *Let E be a Dedekind complete complex Banach lattice. Let $x \in B(E)$. Then the following holds.*

(i) *If $0 \le x \in \mathcal{E}(B(E))$, then x is maximal in $B(E)$, i.e., if $x \le y$ and $\|y\| = 1$, then $x = y$.*
(ii) *$x \in \mathcal{E}(B(E))$ if and only if $|x| \in \mathcal{E}(B(E))$.*

Proof Part (i) is an immediate consequence of Theorem 1 of [7]. To prove (ii), assume first that $x \in \mathcal{E}(B(E))$ and that $|x| = \frac{1}{2}(x_1 + x_2)$ with $x_1, x_2 \in B(E)$. Then we can find $\pi \in Z(E)$ with $|\pi| = I$ such that $\pi|x| = x$. It follows that $x = \frac{1}{2}(\pi x_1 + \pi x_2)$ with $\|\pi x_1\| = \|\pi x_2\| = 1$. This implies that $x = \pi x_1 = \pi x_2$. hence $x_1 = x_2 = \pi^{-1}x = |x|$ and it follows that $|x| \in \mathcal{E}(B(E))$. Now assume that $|x| \in \mathcal{E}(B(E))$ and that $x = \frac{1}{2}(x_1 + x_2)$ with $\|x_1\| = \|x_2\| = 1$. Let as above $\pi \in Z(E)$ with $|\pi| = I$ such that $\pi|x| = x$. It follows that $|x| = \frac{1}{2}(\pi^{-1}x_1 + \pi^{-1}x_2)$ with $\|\pi^{-1}x_1\| = \|\pi^{-1}x_2\| = 1$. This implies that $|x| = \pi^{-1}x_1 = \pi^{-1}x_2$. Hence $x_1 = x_2 = \pi|x| = x$ and it follows that $x \in \mathcal{E}(B(E))$. □

Corollary 2.10 *Let E be a Dedekind complete complex Banach lattice and $\Phi \in \mathcal{S}_r$. Then Φ is a pure regular state if and only if $|\Phi|$ is a pure regular state.*

Next we will discuss between pure vector states and pure regular states.

Lemma 2.11 *Let E be a Dedekind complete complex Banach lattice. Let $\pi \in \mathcal{Z}(E)$. Then $|\pi^*| = |\pi|^*$.*

Proof Let $\sigma \in Z(E)$ such that $\pi = \sigma|\pi|$ and $|\sigma| = I$. Via the Stone-Weierstrass theorem there exists $\sigma_k \in \mathcal{Z}(E)$ with $\sigma_k = \sum_1^{n_k} \alpha_{i,k} P_{i,k}$, where $\sum_i^{n_k} P_{i,k} = I$ and $|\alpha_{i,k}| = 1$ for all i such that

$$|\sigma_k - \sigma| \leq \epsilon_k I \text{ with } \epsilon_k \to 0.$$

By taking adjoints and using the fact that $|\sigma^*| \leq |\sigma|^*$ we have

$$|\sigma_k^* - \sigma^*| \leq |\sigma_k - \sigma|^* \leq \epsilon_k I^* = \epsilon_k I.$$

It follows that

$$|I - |\sigma^*|| \leq |\sigma_k^* - \sigma^*| \leq \epsilon_k I.$$

This gives us that $|\sigma^*| = I$. Now $\pi^* = |\pi|^*\sigma^*$ implies that $|\pi^*| = |\pi|^*|\sigma^*| = |\pi^*|$. □

We now use the above lemma in order to prove a result that a positive functional is related to another functional by a complex rotation.

Lemma 2.12 *Let E be a Dedekind complete complex Banach lattice and let $0 \leq f \in E^*$. Then for all $z \in E$ there exists $g \in E^*$ with $|g| = f$ such that $|g(z)| = g(z) = f(|z|)$.*

Proof There exists an $\pi \in Z(E)$ such that $|z| = \pi z$ and $|\pi| = I$. Let $g = \pi^* f$. Then

$$|g| = |\pi^* f| = |\pi^*|(|f|) = |f|,$$

where the final equality is due to the previous lemma. We also have

$$g(z) = \pi^* f(z) = f(\pi z) = f(|z|),$$

as desired. □

We now can state a formula for the regular norm, which gives information on the extreme points of $B(\mathcal{L}_r(E)^*)$.

Theorem 2.13 *Let E be a Dedekind complete complex Banach lattice. Then*

$$||T||_r = \sup\{|\Phi(T)| : \Phi \in \mathcal{L}_r(E)^*, |\Phi| = f \otimes x, ||x|| = ||f|| = 1\}.$$

Proof Using the previous lemma, we can move the modulus of a regular operator into the modulus of a vector state and obtain the following expression of the regular norm,

$$\begin{aligned} ||T||_r &= \sup\{f(|T|x) : ||f|| = ||x|| = 1, f, x \geq 0\} \\ &= \sup\{(f \otimes x)(|T|) : ||f|| = ||x|| = 1, f, x \geq 0\} \\ &= \sup\{|\Phi(T)| : \Phi \in \mathcal{L}_r(E)^* : |\Phi| = f \otimes x, ||f|| = ||x|| = 1, f, x \geq 0\}. \end{aligned}$$

□

Using now standard arguments from convexity theory (i.e., the Hahn-Banach theorem, the Krein-Milman theorem, and Milman's theorem, see [18] for details) we obtain the following corollary.

Corollary 2.14 *Let E be a Dedekind complete complex Banach lattice. Then the unit ball of $\mathcal{L}_r(E)^*$ is equal to the weak* closed convex hull of functionals whose modulus is of rank one. In particular,*

$$\mathcal{E}(B(\mathcal{L}_r(E)^*)) \subseteq \overline{\{\Phi \in \mathcal{L}_r(E)^* : |\Phi| = f \otimes x, f, x \geq 0, ||f|| = ||x|| = 1\}}, \tag{2.1}$$

where the closure is in the weak topology.*

Now we restrict ourselves to finite dimensions in order to remove the closure condition from the previous corollary and restrict ourselves to extreme points of $B(E)^*$, respectively, $B(E)$.

Lemma 2.15 *If E is a finite dimensional complex Banach lattice, then*

$$\mathcal{E}(B(\mathcal{L}_r(E)^*)) \subseteq \{\Phi \in \mathcal{L}_r(E)^* : |\Phi| = f \otimes x, f \in \mathcal{E}(B(E^*))^+, x \in \mathcal{E}(B(E))^+\}.$$

Proof Now that we are in finite dimensions, the right-hand side of equation in Lemma (2.1) is closed in the norm of $\mathcal{L}_r(E)^*$. Now let $\Phi \in \mathcal{E}(B(\mathcal{L}_r(E)^*))$. It is clear from the description in Eq. (2.1) that $|\Phi| \in \mathcal{E}(B(\mathcal{L}_r(E)^*))$ and $|\Phi| = f \otimes x$. If f is not extreme, then $f = \frac{1}{2}(f_1 + f_2)$ for some $f_1, f_2 \in B(\mathcal{L}_r E)^*$ and $|\Phi| = f \otimes x = \frac{1}{2}(f_1 \otimes x + f_2 \otimes x)$ which is a contradiction of $|\Phi|$ being extreme. If x is not extreme, then $x = \frac{1}{2}(x_1 + x_2)$ for some $x_1, x_2 \in B(E)$ and $|\Phi| = f \otimes x = \frac{1}{2}(f \otimes x_1 + f \otimes x_2)$ which is again a contradiction of $|\Phi|$ being extreme. Hence both f and x must be extreme as desired. □

We now prove that in fact we have equality in the above lemma.

Theorem 2.16 *Let E be a finite dimensional complex Banach lattice. Then,*

$$\mathcal{E}(B(\mathcal{L}_r(E)^*)) = \{\Phi \in \mathcal{L}_r(E)^* : |\Phi| = f \otimes x, f \in \mathcal{E}(B(E^*))^+, x \in \mathcal{E}(B(E))^+\}$$

Proof From the above lemma and previous results about extreme points of unit balls of Banach lattices it remains to show that if $\Phi = f \otimes x$ with $f, x \geq 0$, $f \in \mathcal{E}(B(E^*))$, and $x \in \mathcal{E}(B(E))$, then $\Phi \in \mathcal{E}(B(\mathcal{L}_r(E)^*))$. Assume such a Φ is not an extreme point of $B(\mathcal{L}_r(E)^*)$. Then we can write

$$f \otimes x = \sum_1^N \lambda_i \Phi_i$$

for some $N \in \mathbb{N}$ and $\Phi_i \in \mathcal{E}(B(\mathcal{L}_r(E)^*))$. By the above lemma $|\Phi_i| = f_i \otimes x_i$, where $f_i \in \mathcal{E}(B(E^*))$, and $x_i \in \mathcal{E}(B(E))$. Hence, $f \otimes x$ is dominated by a convex combination of other f_i and x_i such that

$$f \otimes x \leq \lambda_1 f_1 \otimes x_1 + \cdots \lambda_n f_n \otimes x_n, \quad \lambda_1, \ldots, \lambda_n > 0, \sum_{i=1}^n \lambda_i = 1.$$

Let $0 \leq u \in B(E)$ such that $f(u) = 1$ and let $f_i(u) = u_i$ for $i = 1, \ldots, n$. Note that $0 \leq u_j \leq 1$. Now considering the tensor products as operators on E, we have

$$x = (f \otimes x)(u) \leq \sum_{i=1}^n \lambda_i (f_i \otimes x_i)(u) = \sum_{i=1}^n \lambda_i u_i x_i.$$

Since x is assumed to be an extreme point we have from

$$1 \leq || \sum_{i=1}^n \lambda_i u_i x_i || \leq || \sum_{i=1}^n \lambda_i x_i || = 1,$$

by the theorem of Grząślewicz and Schaefer [7], that $x = \sum_{i=1}^n \lambda_i u_i x_i$ which implies that $u_i = 1$ and $x = x_i$ for all i. A similar argument shows that $f = f_i$ for all i. This shows that $f \otimes x = |\Phi_i|$ is extreme, which is a contradiction with our assumption. □

As a consequence of the above theorem we get immediately the following description of the pure regular states on a finite dimensional Banach lattice.

Corollary 2.17 *Let E be a finite dimensional complex Banach lattice. Then,*

$$\mathcal{E}(S_r(\mathcal{L}_r(E))) =$$
$$\{\Phi \in \mathcal{L}_r(E)^* : \Phi(I) = 1, |\Phi| = f \otimes x, 0 \leq f \in \mathcal{E}(B(E^*)), 0 \leq x \in \mathcal{E}(B(E))\}.$$

Note that, if Φ is as in the corollary, then also $|\Phi|(I) = 1$, so that $f(x) = 1$, i.e., $|\Phi|$ is a vector state. Thus the extreme points of the regular states on a finite dimensional Banach lattice are those Φ for which $|\Phi|$ is an extreme vector state. This shows the difference with the extreme points of the bounded states on a finite dimensional

Banach space, which are given by the extreme vector states (see Li-Schneider [10] for this results and Lima-Olsen [12] for an infinite dimensional version of this). In connection with the above corollary it is worthwhile to observe the following proposition, which shows that the modulus of a pure vector state is determined by its diagonal.

Proposition 2.18 *Let E be a finite dimensional complex Banach lattice such that both E and E^* have strictly monotone norms. Let $\Phi = f \otimes x$ and $\Psi = g \otimes y$ be pure regular states with the same diagonal, i.e., with $\mathcal{P}^*(\Phi) = \mathcal{P}^*(\Psi)$. Then $|\Phi| = |\Psi|$, i.e., $|f| = |g|$ and $|x| = |y|$. Moreover, every diagonal state has a unique extension to a positive pure regular state.*

Proof As observed earlier $|\Phi| = |f| \otimes |x|$ and $|\Psi| = |g| \otimes |x|$. Then $\mathcal{P}^*(\Phi)$ is the pointwise product of $|f|$ and $|x|$, respectively $|g|$ and $|y|$. The monotonicity conditions now imply that $\mathcal{P}^*(\Phi)$, $|f|$, $|g|$, $|x|$, and $|y|$ all have the same support. It follows now from the uniqueness part of Lozanovskii's factorization theorem (see Gillespie [6]) that $|f| = |g|$ and $|x| = |y|$. The existence of an extension as a vector state follows from Lozanovskii's factorization theorem and the monotonicity conditions imply that every vector state is a pure regular state. □

2.4 Order Continuous States on ℓ_p

In this section we will consider regular states on $\mathcal{L}_r(\ell_p)$. Our goal is to get again a characterization of pure regular states as the states with modulus equal to a vector state. We start with an example to show that in the infinite dimensional case we have singular states.

Example Let $1 < p < \infty$ and let e_n denote the standard basis of ℓ_p and e_n^* the standard dual basis. Let $\mathcal{U}$ denote a free ultrafilter on $\mathbb{N}$. Define $\Phi(T) = \lim_{\mathcal{U}} x_n^* \otimes x_n(T)$. Then $\Phi \in \mathcal{S}_r(\mathcal{L}_r(\ell_p))$. The diagonal part $\mathcal{P}^*(\Phi)$ corresponds now to a point evaluation at a point in $\beta(\mathbb{N}) \setminus \mathbb{N}$. It is easy to see that Φ is zero on the order dense ideal of compact regular operators (i.e., the closure of the finite rank operators in the regular norm). Therefore Φ is a singular functional on $(\mathcal{L}_r(\ell_p))^*$ and can thus not be dominated by any vector state, which are order continuous. For this reason we restrict ourselves to order continuous (or normal) states.

We start with a lemma, which shows that we can use the order continuous states of $\mathcal{L}_r(\ell_1)$ and $\mathcal{L}_r^n(\ell_\infty)$, where $\mathcal{L}_r^n(\ell_\infty)$ denotes the order continuous operators on ℓ_∞, to get order continuous states on $\mathcal{L}_r(\ell_p)$. For $1 < p < \infty$ every regular operator can be represented by an infinite matrix, so we can consider $\mathcal{L}_r(\ell_p)$ as a Banach function space on $\mathbb{N} \times \mathbb{N}$. Therefore the order continuous dual of $\mathcal{L}_r(\ell_p)$ can be represented by infinite matrices, acting coordinate-wise on the matrix, with $\Phi(T)$ then being the sum over all the products $\phi_{i,j} t_{i,j}$ where $T = [t_{i,j}]$ and $\Phi = [\phi_{in}]$.

Lemma 2.19 *Let* $0 \leq \Phi_1 = [a_{ij}] \in \mathcal{S}_r(\mathcal{L}_r(\ell_1))$ *and* $0 \leq \Phi_2 = [b_{ij}] \in \mathcal{S}_r(\mathcal{L}_r(\ell_\infty))$. *Let* $1 < p < \infty$ *and* $\frac{1}{p} + \frac{1}{p'} = 1$. *If for all* $1 \leq i < \infty$ *we have* $a_{ii} = b_{ii}$ *(i.e., the diagonals* $\mathcal{P}^*(\Phi)$ *and* $\mathcal{P}^*(\Psi)$ *are equal), then* $\Phi := \Phi_1^{1/p}\Phi_2^{1/p'}$ *is in* $\mathcal{S}_r(\mathcal{L}_r(\ell_p))$, *where the exponent and multiplication are done entry-wise.*

Proof Let $\Phi = [\phi_{ij}]$

$$\Phi(I) = \sum_{i=1}^{\infty} \phi_{ii} = \sum_{i=1}^{\infty} a_{ii}^{1/p} b_{ii}^{1/p'} = \sum_{i=1}^{\infty} a_{ii}^{1/p} a_{ii}^{1/p'} = \sum_{i=1}^{\infty} a_{ii} = \Phi_1(I) = 1.$$

Thus we have that $\Phi(I) = 1$. Now via Hölder's inequality we have

$$\begin{aligned} |\Phi(T)| &\leq \sum_{i,j=1}^{\infty} a_{i,j}^{\frac{1}{p}} |t_{i,j}|^{\frac{1}{p}} b_{i,j}^{\frac{1}{p'}} |t_{i,j}|^{\frac{1}{p'}} \\ &\leq \Phi_1(|T|)^{\frac{1}{p}} \Phi_2(|T|)^{\frac{1}{p'}}, \end{aligned}$$

for $T = [t_{i,j}]$ in $\mathcal{L}_r(\ell_1) \cap \mathcal{L}_r^n(\ell_\infty)$. Hence

$$||\Phi||_r \leq ||\Phi_1||_r^{1/p} ||\Phi_2||_r^{1/p'} = 1^{1/p} 1^{1/p'} = 1.$$

However we already know that $\Phi(I) = 1$, so we must have that $||\Phi||_r = 1$. This shows that Φ is a state on $\mathcal{L}_r(\ell_p)$. □

To use the above lemma, we get an explicit description of the order continuous states on $\mathcal{L}_r(\ell_1)$ and the order continuous states on the order continuous operators in $\mathcal{L}_r(\ell_\infty)$. We first recall that if $T = [t_{i,j}]$ is in $\mathcal{L}_r(\ell_1)$, then it is well-known that $\|T\| = \|T\|_r = \sup_j \sum_{i=1}^{\infty} |t_{i,j}|$ and if $T = [t_{i,j}]$ is in $\mathcal{L}_r(\ell_\infty)$, then $\|T\| = \|T\|_r = \sup_i \sum_{j=1}^{\infty} |t_{i,j}|$. This explicit description of the (regular) norm leads directly to an explicit description of the order continuous functionals given by an infinite matrix.

Lemma 2.20 *Let* $\Phi = [\phi_{i,j}]$. *Then the following holds.*

(i) $\Phi \in \mathcal{L}_r(\ell_1)^*$ *if and only if* $\|\Phi\| = \sum_{j=1}^{\infty} \sup_i |\phi_{i,j}| < \infty$.
(ii) $\Phi \in \mathcal{L}_r(\ell_\infty)^*$ *if and only if* $\|\Phi\| = \sum_{i=1}^{\infty} \sup_j |\phi_{i,j}| < \infty$.

From the lemma we get the following description of order continuous regular states.

Corollary 2.21 *Let* $\Phi = [\phi_{i,j}]$. *Then the following holds.*

(i) $\Phi \in \mathcal{S}_r(\mathcal{L}_r(\ell_1))$ *if and only if* $0 \leq \phi_{i,i} \leq 1$ *for all* $i \geq 1$, $\sum_{i=1}^{\infty} \phi_{i,i} = 1$, *and* $|\phi_{i,j}| \leq \phi_{j,j}$ *for all* $i \geq 1$.
(ii) $\Phi \in \mathcal{S}_r(\mathcal{L}_r(\ell_\infty))$ *if and only if* $0 \leq \phi_{i,i} \leq 1$ *for all* $i \geq 1$, $\sum_{i=1}^{\infty} \phi_{i,i} = 1$, *and* $|\phi_{i,j}| \leq \phi_{i,i}$ *for all* $j \geq 1$.

To prove the converse of the above lemma we use a theorem proven in [17].

Theorem 2.22 *Let $1 < p < \infty$. Then the order continuous dual of the band of integral operators $\mathcal{K}_r(L^p)$ is equal to $(L_{1,\infty})^{1/p'}(L^t_{1,\infty})^{1/p} = L_{p',\infty}L^t_{p,\infty}$.*

Note that in the discrete case every regular operator on ℓ_p, $1 < p < \infty$, is an integral operator.

Proposition 2.23 *Let $1 < p < \infty$, $\frac{1}{p} + \frac{1}{p'} = 1$ and $0 \leq \Phi \in \mathcal{S}_r(\mathcal{L}_r(\ell_p))$ be an order continuous state. Let Φ have matrix representation $\Phi = [\phi_{ij}]$. There exist order continuous $0 \leq \Phi_1 \in \mathcal{S}_r(\mathcal{L}_r(\ell_1))$ and $0 \leq \Phi_2 \in \mathcal{S}_r(\mathcal{L}_r(\ell_\infty))$ with $\Phi_1 = [a_{ij}]$, $\Phi_2 = [b_{ij}]$ and for all $1 \leq i < \infty$, $a_{ii} = b_{ii}$ such that $\Phi = \Phi_1^{1/p}\Phi_2^{1/p'}$, where the exponent and multiplication are done entry-wise.*

Proof By the above theorem there exist order continuous $0 \leq \Phi_1 = [a_{ij}] \in \mathcal{L}_r(\ell_1)^*$ and $0 \leq \Phi_2 = [b_{i,j}] \in \mathcal{L}_r(\ell_\infty)^*$ with $||\Phi_1|| = ||\Phi_2|| = 1$ such that $\Phi = \Phi_1^{1/p}\Phi_2^{1/p'}$. Thus we need to show that the diagonals of the matrices are equal and that they are in fact states. To show that the diagonal of the matrices Φ_1 and Φ_2 are equal, observe

$$
\begin{aligned}
1 = \Phi(I) = \sum_{i=1}^{\infty} \phi_{ii} &= \sum_{i=1}^{\infty} a_{ii}^{1/p} b_{ii}^{1/p'} \\
&\leq \left(\sum_{i=1}^{\infty} a_{ii}\right)^{1/p} \left(\sum_{i=1}^{\infty} b_{ii}\right)^{1/p'} \leq ||\Phi_1||^{1/p}||\Phi_2||^{1/p'} = 1.
\end{aligned}
$$

Hence we must have equality throughout and by a corollary of the Holder's inequality the a_{ii} and the b_{ii} must be linearly dependent. Thus we must have that $a_{ii} = \mu b_{ii}$.

$$
1 = \sum_{i=1}^{\infty} \phi_{ii} = \sum_{i=1}^{\infty} (\mu b_{ii})^{1/p} b_{ii}^{1/p'} = \mu^{1/p} \sum_{i=1}^{\infty} b_{ii}.
$$

Since $||\Phi_2|| = 1$ we must have that $\Phi_2(I) = \sum_{i=1}^{\infty} b_{ii} \leq 1$. This in turn means that $\mu^{1/p} \geq 1$, i.e., $\mu \geq 1$. Similarly, considering $b_{ii} = \frac{1}{\mu}a_{ii}$, we find $\frac{1}{\mu} \geq 1$. Combining the inequalities involving μ gives us that $\mu = 1$ and thus $a_{ii} = b_{ii}$ and the diagonals of Φ_1 and Φ_2 are equal. Since the diagonals are equal we now have that

$$
\phi_{ii} = a_{ii}^{1/p} b_{ii}^{1/p'} = a_{ii}^{1/p} a_{ii}^{1/p'} = a_{ii}.
$$

Thus,

$$
\Phi_1(I) = \sum_{i=1}^{\infty} a_{ii} = \sum_{i=1}^{\infty} \phi_{ii} = 1.
$$

The same argument can be used to show that $\Phi_2(I) = 1$. Since we already have that $||\Phi_1|| = ||\Phi_2|| = 1$, we have that both are states. □

Combining the above results we have

Theorem 2.24 *Let $1 < p < \infty$, $\frac{1}{p} + \frac{1}{p'} = 1$ and $\Phi = [\phi_{i,j}]$. Then $\Phi \in \mathcal{S}_r(\mathcal{L}_r(\ell_p))$ if and only if $0 \leq \phi_{i,i} \leq 1$ for all $i \geq 1$, $\sum_{i=1}^{\infty} \phi_{i,i} = 1$, and $|\phi_{i,j}| \leq \phi_{i,i}^{\frac{1}{p'}} \phi_{j,j}^{\frac{1}{p}}$ for all $i, j \geq 1$. In particular, there exist $0 \leq f \in \ell_{p'}$, $0 \leq x \in \ell_p$ with $||f||_{p'} = ||x||_p = 1 = f(x)$ such that $|\Phi| \leq f \otimes x$.*

Proof The first part of the theorem is a direct consequence of the factorization theorem and the characterization of order continuous states in $\mathcal{S}_r(\mathcal{L}_r(\ell_1))$, respectively $\mathcal{S}_r(\mathcal{L}_r(\ell_\infty))$. The second part follows if we take $f = (\phi_{i,i}^{\frac{1}{p'}})$ and $x = (\phi_{j,j}^{\frac{1}{p}})$. □

As in Corollary 2.17 we can now characterize the pure order continuous states. We will denote by $\mathcal{S}_r(\mathcal{L}_r(\ell_p))_n$ the order continuous (or normal) states on $\mathcal{L}_r(\ell_p)$.

Theorem 2.25 *Let $1 < p < \infty$ and $\frac{1}{p} + \frac{1}{p'} = 1$. Then,*

$$\mathcal{E}(\mathcal{S}_r(\mathcal{L}_r(\ell_p))_n) = \left\{\Phi \in \mathcal{S}_r(\mathcal{L}_r(\ell_p)) : |\Phi| = f \otimes x,\ f, x \geq 0,\ ||f||_{p'} = ||x||_p = 1 = f(x)\right\}.$$

Proof Let first $\Phi = [\phi_{i,j}] \in \mathcal{E}(\mathcal{S}_r(\mathcal{L}_r(\ell_p))_n)$. Then $|\Phi| = [|\phi_{i,j}|]$ is a pure order continuous state. By the above theorem there exist $0 \leq f \in \ell_{p'}$, $0 \leq x \in \ell_p$ with $||f||_{p'} = ||x||_p = 1 = f(x)$ such that $|\Phi| \leq f \otimes x$. By the maximality property of extreme points it now follows that $|\Phi| = f \otimes x$. Let now $\Phi = [\phi_{i,j}] \in \mathcal{S}_r(\mathcal{L}_r(\ell_p))$ with $|\Phi| = f \otimes x$, where $0 \leq f \in \ell_{p'}$, $0 \leq x \in \ell_p$ with $||f||_{p'} = ||x||_p = 1 = f(x)$. It suffices to show that $f \otimes x$ is a pure state. Assume $f \otimes x = \frac{1}{2}(\Phi_1 + \Phi_2)$, where $\Phi_1, \Phi_2 \in \mathcal{S}_r(\mathcal{L}_r(\ell_p))_n$. Then by the above theorem there exists $0 \leq f_1, f_2 \in \ell_{p'}$, $0 \leq x_1, x_2 \in \ell_p$ with $||f_i||_{p'} = ||x_i||_p = 1 = f_i(x_i)$ for $i = 1, 2$ such that $|\Phi_i| \leq f_i \otimes x_i$ for $i = 1, 2$. Hence we have

$$f \otimes x \leq \frac{1}{2}(f_1 \otimes x_1 + f_2 \otimes x_2).$$

As in the proof of Theorem 2.16 it follows now that $f = f_1 = f_2$ and $x = x_1 = x_2$ and thus $f \otimes x$ is a pure state. □

3 The Regular Algebra Numerical Range

We now use the regular states to define the numerical range associated with the Banach algebra $\mathcal{L}_r(E)$ of regular operators on a Dedekind complete complex Banach lattice.

Definition 3.1 The *regular algebra numerical range* of a regular operator T on a Banach lattice E is defined to be

$$V(\mathcal{L}_r(E), T) = \{\Phi(T) : \Phi \in \mathcal{S}_r\}.$$

Recall that the standard algebra numerical range for a bounded operator $T \in \mathcal{L}(E)$ is defined as $V(\mathcal{L}(E), T) = \{\Phi(T) : \Phi \in \mathcal{S}\}$. Since $\mathcal{S} \subset \mathcal{S}_r$, it is clear that $V(\mathcal{L}(E), T) \subset V(\mathcal{L}_r(E), T)$ for all $T \in \mathcal{L}_r(E)$. Also, since $\mathcal{S}_r$ is convex and compact in the weak*-topology, we have that $V(\mathcal{L}_r(E), T)$ is a compact convex subset of $\mathbb{C}$. From the general theory of algebra numerical ranges we get the following description

$$V(\mathcal{L}_r(E), T) = \bigcap_{z \in \mathbb{C}} \{\lambda : |z - \lambda| \leq ||T - zI||_r\}.$$

Theorem 3.2 *Let E be a Dedekind complete complex Banach lattice and let $T \in \mathcal{L}_r(E)$ with $T \perp I$. Then $V(\mathcal{L}_r(E), T)$ is a disk centered around $z = 0$, i.e. if $\lambda \in V(\mathcal{L}_r(E), T)$, then $e^{i\theta}\mu \in V(\mathcal{L}_r(E), T)$ for all $0 \leq \theta \leq 2\pi$, and for all $|\mu| \leq |\lambda|$.*

Proof Assume $T \perp I$ and let $\lambda \in V(\mathcal{L}_r(E), T)$ and $\theta \in [0, 2\pi)$. Since $\lambda \in V(\mathcal{L}_r(E), T)$,by the above observation, we have that $|\lambda - \omega| \leq ||(T - \omega I)||_r$ for all $\omega \in \mathbb{C}$. Now we have

$$|e^{i\theta}\lambda - \omega| = |\lambda - e^{-i\theta}\omega| \leq ||(T - e^{-i\theta}\omega I)||_r = ||\,|T| + |\omega|I||_r = ||T - \omega I||_r$$

for all $\omega \in \mathbb{C}$. By the above description of $V(\mathcal{L}_r(E), T)$, we have that $e^{i\theta}\lambda \in V(\mathcal{L}_r(E), T)$. □

Remark 3.3 Recall the definition of the spatial numerical range $V(T)$ of a bounded linear operator,

$$V(T) = \{(f \otimes x)(T) = f(Tx) : f \otimes x \in \mathcal{V}\}.$$

Then the numerical radius for a bounded linear operator T is given by $v(T) := \sup\{|\lambda| : \lambda \in V(T)\}$. Now $\frac{1}{e} \leq v(T) \leq \|T\|$ for $T \in \mathcal{L}(E)$. In particular, if $(f \otimes x)(T) = 0$ for all $f \otimes x \in \mathcal{V}$, then $T = 0$ (note this uses crucially that we work over the complex scalars). This shows that if $T \neq 0$ in the above theorem, then $V(\mathcal{L}_r(E), T)$ is a non-degenerate disk centered around $z = 0$

We now give an example where $V(\mathcal{L}(E), T) \neq V(\mathcal{L}_r(E), T)$.

Example Let $T = \begin{bmatrix} 0 & 1 \\ 1 & 0 \end{bmatrix}$ over $E = \ell_2(\mathcal{C})$. Since E is a Hilbert space we can use the fact that the standard spatial numerical range $W(T) \subseteq \mathbb{R}$, since T is a Hermitian matrix. Therefore $V(\mathcal{L}(E), T) = \overline{\text{co}}\, W(T) \subseteq \mathbb{R}$, in fact one can easily see that $V(\mathcal{L}(E), T) = [-1, 1]$. However, as $T \perp I$, $V(\mathcal{L}_r(E), T)$ must be a disk and is thus not a subset of $\mathbb{R}$. It is easy to see that $V(\mathcal{L}_r(E), T)$ is the unit disk around the origin.

Theorem 3.4 *Let E be a Dedekind complete complex Banach lattice and $T \in \mathcal{Z}(E)$. Then $V(\mathcal{L}_r(E), T) = co(\sigma(T))$, where $co(\sigma(T))$ denotes the convex hull of the spectrum $\sigma(T)$ of T.*

Proof By Proposition 2.5 we have that $V(\mathcal{L}_r(E), T) = V(\mathcal{Z}(E), T)$. By Kakutani's theorem we have that $\mathcal{Z}(E)$ is lattice isomorphic to $C(K)$ for some compact Hausdorff space K. We can identify T with some $f \in C(K)$ such that $V(\mathcal{Z}(E), T) = V(C(K), f)$. Now it is known that $V(C(K), f) = \text{co}(f(K))$. It is also known that $\sigma(T) = f(K)$. Thus we have $V(\mathcal{L}_r(E), T) = \text{co}(\sigma(T))$. □

Throughout the study of numerical ranges people have studied Hermitian operators, i.e., those operators whose numerical range is a subset of the real line. For this reason we consider the operators whose regular algebra numerical range is a subset of the real line.

Theorem 3.5 *Let E be a Dedekind complete complex Banach lattice and $T \in \mathcal{L}_r(E)$. Then $V(\mathcal{L}_r(E), T) \subset \mathbb{R}_+$ if and only if $T \geq 0$ and $T \in \mathcal{Z}(E)$*

Proof Assume first that $0 \leq T \in \mathcal{Z}(E)$. Then by Proposition 2.5 we have that $\Phi \mid_{Z(E)} \geq 0$ for all $\Phi \in \mathcal{S}_r$. Hence $V(\mathcal{L}_r(E), T) \subset \mathbb{R}_+$. Assume now that $V(\mathcal{L}_r(E), T) \subset \mathbb{R}_+$. Then $T = T_1 + T_2$ where $T_1 = \mathcal{P}(T) \in \mathcal{Z}(E)$ and $T_2 \perp I$. Then $\mathcal{P}^*(\Phi)(T) = \Phi(T_1) \geq 0$ for $\Phi \in S_r$. This implies that $T_1 \geq 0$. Since $\Phi(T_1) \in \mathbb{R}$ for all regular states, we must have that $\Phi(T_2) \in \mathbb{R}$ for all regular states. If $T_2 \neq 0$ then by the remark following Theorem 3.2 there exists a regular state Ψ such that $\Psi(T_2) \in \mathbb{C}\backslash\mathbb{R}$ which is in contradiction with $V(\mathcal{L}_r(E), T) \subseteq \mathbb{R}_+$. Hence $T_2 = 0$ and thus $0 \leq T = T_1 \in \mathcal{Z}(E)$. □

Corollary 3.6 *Let E be a Dedekind complete complex Banach lattice and $T \in \mathcal{L}_r(E)$. Then $V(\mathcal{L}_r(E), T) \subset \mathbb{R}$ if and only if $T \in \mathcal{Z}_{\mathbb{R}}(E)$.*

Proof Observe that $V(\mathcal{L}_r(E), T + cI) = V(\mathcal{L}_r(E), T) + c$, so that by adding a large enough real multiple of I to T we can reduce the corollary to the theorem. □

Thus the Hermitian operators for the regular algebra numerical range are the real central operators. In particular it follows that $\|e^{itT}\|_r = 1$ for all $t \in \mathbb{R}$ if and only if $T \in \mathcal{Z}_{\mathbb{R}}(E)$. In [8] Nigel Kalton proved that the linear span of the Hermitian operators for the algebra numerical range is a C^*-algebra. It follows that for E a complex AL- or Dedekind complete AM-space this C^*-algebra equals the center $Z(E)$ of E.

3.1 Positive Numerical Ranges

Let E be a complex Dedekind complete Banach lattice. The *positive spatial numerical range* of a linear operator $T \in \mathcal{L}_r(E)$ is defined to be

$$V_+(T) = \{f(Tx) : 0 \leq x \in E, 0 \leq f \in E^*, ||x|| = ||f|| = 1 = f(x)\}$$

and the *positive algebra numerical range* of a linear operator $T \in \mathcal{L}_r(E)$ is defined to be

$$V_+(\mathcal{L}_r(E), T) = \{\Phi(T) : 0 \leq \Phi \in \mathcal{S}_r(E)\}.$$

Obviously $V_+(T) \subset V_+(\mathcal{L}_r(E), T)$ and $V_+(\mathcal{L}_r(E), T)$ is weak*-compact convex subset of $\mathbb{R}$, so $V_+(\mathcal{L}_r(E), T)$ is a closed bounded interval. The *positive vector numerical radius* of a linear operator T on a Banach lattice E is defined to be

$$v_+(T) = \sup\{|\lambda| : \lambda \in V_+(T)\},$$

while the *positive regular algebra numerical radius* of a linear operator T on a Banach lattice E is defined to be

$$v_+^o(T) := \sup\{|\lambda| : \lambda \in V_+(\mathcal{L}_r(E), T)\}.$$

Since $|(f \otimes x)(T)| \leq (|f| \otimes |x|)(T)$ for $T \geq 0$, we have $0 \leq v(T) = v_+(T) \leq v_+^o(T)$ for $T \geq 0$. In general it is not clear whether $v_+(T) = v_+^o(T)$, but we can show this in case E is finite dimensional.

Proposition 3.7 *Let E be a finite dimensional complex Banach lattice. Let $0 \leq T \in \mathcal{L}_r(E)$. Then $v_+^o(T) = v_+(T)$.*

Proof Clearly

$$v_+(T) \leq v_+^o(T). \tag{3.1}$$

Let $0 \leq \Phi \in \mathcal{S}_r$. Then by Theorem 2.16 there exists a convex combination of positive pure vector states $\sum_{i=1}^m \lambda_i f_i \otimes x_i$ such that for $T \geq 0$ we have

$$\Phi(T) \leq \sum_{i=1}^m \lambda_i f_i \otimes x_i(T) \leq \sum_{i=1}^m v_+(T) = v_+(T).$$

Hence $v_+^o(T) \leq v_+(T)$ and thus equality holds. □

Before continuing we recall a special case of Problem 3.3.10. of [2]. An outline of the proof of the exercise can be found in [1]. We note that the exercise as stated in [2] is not correct, but the statement and outlined proof are correct for the special case used here.

Proposition 3.8 *Let E be a Dedekind complete Banach lattice and let $T : E \to E$ such that $T \perp I$. For each $0 < x \in E$ and each $0 < \epsilon < 1$ there exists a non-zero component a of x (i.e., $a \wedge (x - a) = 0$) such that $P_a(|T|a) \leq \epsilon a$.*

Lemma 3.9 *Let $0 \leq T \in \mathcal{L}_r(E)$ with $T \perp I$. Then we have $0 \in \overline{V_+(T)}$.*

Proof Let $0 < \epsilon < 1$ be given. Let $0 < x \in E$. By the above proposition there exists a component of x, a, such that $P_a(Ta) \leq \epsilon a$, where P_a is the band projection of E onto the band generated by a. By normalizing, we can assume that $||a|| = 1$. Now consider a norm-attainer $0 \leq g \in E^*$ such that $||g|| = ||a|| = 1 = g(a)$. Now note that $||g|_{P_a}|| = 1 = g|_{P_a}(a)$ so $g|_{P_a}$ is also a norm-attainer. We have

$$|g|_{P_a}(Ta)| = |g|_{P_a}(P_a(Ta))| \leq |g(\epsilon a)| = \epsilon.$$

Thus every ϵ-neighborhood of 0 contains an element of $V_+(T)$, which implies that $0 \in \overline{V_+(T)}$. □

It is clear from the definitions that $V_+(T - \delta I) = \{\lambda - \delta : \lambda \in V_+(T)\}$ for all $\delta \in \mathbb{R}$. Similarly to $v_+(T)$, which is the supremum of the positive spatial numerical range, we also want to define the infimum of the positive spatial numerical range. Define $\nu_-(T) := \inf\{|\lambda| : \lambda \in V_+(T)\}$.

Theorem 3.10 *Let $0 \leq T \in \mathcal{L}_r(E)$. Then we have that $\nu_-(T) = \sup\{c : T \geq cI\}$.*

Proof Let $\delta := \sup\{c : T \geq cI\}$. By definition, $T - \delta I \geq 0$ however $T - (\delta + \epsilon)I \ngeq 0$ for all $\epsilon > 0$. Let $\mathcal{P}$, as before, be the projection of $\mathcal{L}_r(E)$ onto the center $\mathcal{Z}(E)$. This projection is a positive map, so we have

$$\mathcal{P}(T - \delta I) = \mathcal{P}(T) - \delta I \geq 0.$$

Also,

$$\mathcal{P}(T - (\delta + \epsilon)I) = \mathcal{P}(T) - (\delta + \epsilon)I \ngeq 0$$

for all $\epsilon > 0$. We may assume that $\delta = 0$, for if not, we can consider $T' = T - \delta I \geq 0$. Thus we have $(\mathcal{P}(T) - \epsilon I)^- \gneq 0$. Hence there must exist some $0 \neq y \in E$ such that $\mathcal{P}(T)y < \epsilon y$. Now consider P_y the band projection onto B_y the band generated by y. For each $x \in B_y$ we must also have $\mathcal{P}(T)x \leq \epsilon x$ for if not then $0 \leq (\mathcal{P}(T) - \epsilon I)x \leq (\mathcal{P}(T) - \epsilon I)\lambda y$ where $0 < \lambda \in \mathbb{R}$. This is a contradiction as it implies that $\mathcal{P}(T)y \geq \epsilon y$.

Now consider $T = \mathcal{P}(T) + T_1$ where $T_1 \perp I$. Thus there exists a non-zero component of y, y_0 such that $P_{y_0}(T_1 y_0) \leq \epsilon y_0$. We can normalize y_0 to ensure that $||y_0|| = 1$. Consider a norm-attainer, g_0, such that $||g_0|| = ||y_0|| = 1 = g_0(y_0)$. Now considering the original T we have

$$\begin{aligned} g_0(Ty_0) &= g_0((\mathcal{P}(T) + T_1)y_0) = g_0(\mathcal{P}(T)y_0) + g_0(T_1 y_0) \\ &< g_0(\epsilon y_0) + g_0(T_1 y_0) < 2\epsilon. \end{aligned}$$

Thus every ϵ-neighborhood of 0 contains an element of $V_+(T)$. Since T was assumed positive, which implies that $V_+(T) \subset [0, \infty)$, we must have that $\nu_-(T) = 0$. □

Corollary 3.11 *Let $0 \leq T \in \mathcal{L}_r(E)$. Then we have that $\nu_-(T) = \inf\{\lambda : \lambda \in V_+(\mathcal{L}_r(E), T)\}$.*

Proof Assume that $\inf\{\lambda : \lambda \in V_+(\mathcal{L}_r(E), T)\} < \nu^-(T)$. Consider $T - \nu^-(T)I$. Then we have that $(T - \nu^-(T)I) \geq 0$ and that $0 \in V_+(T - \nu^-(T))I$. However, since $\inf\{\lambda : \lambda \in V_+(\mathcal{L}_r(E), T)\} < \nu^-(T)$, we must have that $V_+(\mathcal{L}_r(E), T - \nu^-(T)I) \not\subseteq [0, \infty)$. This is a contradiction because of the positivity of $T - \nu^-(T)I$. □

We now relate the two positive numerical ranges. As with the traditional numerical ranges these two numerical ranges are related by a convex closure.

Theorem 3.12 *Let E be a finite dimensional complex Banach lattice. Let $T \in \mathcal{L}_r(E)$. If $T \geq 0$, then $\overline{co(V_+(T))} = V_+(\mathcal{L}_r(E), T)$.*

Proof Based on the above results, we already have that $v_+(T) = v_+^o(T) = \sup\{\lambda : \lambda \in V_+(\mathcal{L}_r(E), T)\}$, that $V_+(\mathcal{L}_r(E), T)$ is an interval on the positive real axis, and $V_+(T) \subseteq V_+(\mathcal{L}_r(E), T)$. Moreover $\nu_-(T) = \inf\{\lambda : \lambda \in V_+(\mathcal{L}_r(E), T)\}$ by the above corollary and the result follows. □

4 Regular Numerical Range Preserving Maps

We start by recalling a classical result by Pellegrini [14].

Theorem 4.1 *Let $\mathcal{A}$ be a unital Banach algebra, $a \in \mathcal{A}$ and $\mathcal{F}$ a bounded linear operator on $\mathcal{A}$. Then the following are equivalent:*

(i) $V(\mathcal{A}, \mathcal{F}(a)) \subset V(\mathcal{A}, a)$
(ii) $\mathcal{F}^*(S(\mathcal{A})) \subset S(\mathcal{A})$

A bounded linear map $\mathcal{F} : \mathcal{L}_r(E) \to \mathcal{L}_r(E)$ is called *regular numerical range preserving*, if $V(\mathcal{L}_r(E), \mathcal{F}(T)) = V(\mathcal{L}_r(E), T)$ for all $T \in \mathcal{L}_r(E)$. As $V(\mathcal{L}_r(E), T) = \{0\}$ if and only $T = 0$, any regular numerical range preserving linear map is always injective. It follows now from Pellegrini's Theorem that a surjective linear map $\mathcal{F}$ is regular numerical range preserving if and only if $\mathcal{F}^*(\mathcal{S}_r) = \mathcal{S}_r$. Note that in case E is finite dimensional, a regular numerical range preserving linear map is automatically surjective, as it is injective.

Lemma 4.2 *Let $\mathcal{F} : \mathcal{L}_r(E) \to \mathcal{L}_r(E)$ be a surjective regular numerical range preserving map. Then the restriction of $\mathcal{F}$ to the center of $Z(E)$ is numerical range preserving map. Hence $\mathcal{F}\mid_{Z(E)}$ is a positive surjective isometry with $\mathcal{F}(I) = I$.*

Proof Let $0 \leq T \in Z(E)$. Then $V(\mathcal{L}_r(E), T) \subset \mathbb{R}^+$ and thus we have that $V(\mathcal{L}_r(E), \mathcal{F}(T)) \subset \mathbb{R}^+$. This implies $0 \leq \mathcal{F}(T) \in Z(E)$. Hence $\mathcal{F}\mid_{Z(E)}$ is numerical range preserving. As the same argument applies to $\mathcal{F}^{-1}$ we conclude that $\mathcal{F}\mid_{Z(E)}$ is a positive surjective isometry with $\mathcal{F}(I) = I$. □

We use the above lemma to modify the map $\mathcal{F}$, so that the modified map fixes the center $Z(E)$. To do so, we restrict ourselves to the finite dimensional case, as there are obstacles to do this in general.

Lemma 4.3 *Let E be a finite dimensional complex Banach lattice and let $\mathcal{F} : \mathcal{L}_r(E) \to \mathcal{L}_r(E)$ be a regular numerical range preserving map. Then there exists a permutation matrix P such that $\tilde{\mathcal{F}}(T) = P^{-1}\mathcal{F}(T)P$ satisfies $\tilde{\mathcal{F}}(T) = T$ for all $T \in Z(E)$.*

Proof As E is finite dimensional, we can identify $Z(E)$ lattice isometrically with $\ell_\infty(n)$, acting as multiplication operators on E. For $f = (f_i) \in \ell_\infty(n)$ we will denote by M_f the corresponding multiplication operator on E. By the above lemma and the known structure of isometries on spaces $C(K)$ we conclude that there exists a permutation σ of $\{1, \cdots, n\}$ such that $\mathcal{F}(M_f) = M_{f\circ\sigma}$, where $(f \circ \sigma)_i = f_{\sigma(i)}$. Define now the permutation matrix by $Px = x \circ \sigma$. Then a straightforward calculation shows that $\tilde{\mathcal{F}}(T) = P^{-1}\mathcal{F}(T)P$ satisfies $\tilde{\mathcal{F}}(M_f) = M_f$ for all $f \in \ell_\infty$. □

Recall now that a norm on a finite dimensional Banach lattice E is called *symmetric*, if for all permutation matrices P and all $x \in E$ we have that $\|Px\| = \|x\|$, in other words E has symmetric norm if E is a finite dimensional rearrangement invariant Banach function space.

Lemma 4.4 *Let E be a finite dimensional complex Banach lattice with symmetric norm and let $\mathcal{F} : \mathcal{L}_r(E) \to \mathcal{L}_r(E)$ be a regular numerical range preserving map. Let $\tilde{\mathcal{F}}(T) = P^{-1}\mathcal{F}(T)P$, where P is a permutation matrix. Then $\tilde{\mathcal{F}}$ is also a regular numerical range preserving map.*

Proof Let $T \in \mathcal{L}_r(E)$ and P be a permutation matrix. Then it suffices to show that $V(\mathcal{L}_r(E), P^{-1}TP) = V(\mathcal{L}_r(E), T)$. Let $\Phi \in \mathcal{S}_r$ and define $\Phi_P(T) = \Phi(P^{-1}TP)$. All we need to show now is that $\Phi_P \in \mathcal{S}_r$. This follows however immediately from $\Phi_P(I) = 1$ and $|\Phi_P(T)| \leq \|P^{-1}\|_r\|T\|_r\|P\|_r = \|T\|_r$. □

Summarizing the above lemmas we have:

Proposition 4.5 *Let E be a finite dimensional complex Banach lattice with symmetric norm and let $\mathcal{F} : \mathcal{L}_r(E) \to \mathcal{L}_r(E)$ be a regular numerical range preserving map. Then there exists a permutation matrix P such that $\tilde{\mathcal{F}}(T) = P^{-1}\mathcal{F}(T)P$ satisfies $\tilde{\mathcal{F}}(T) = T$ for all $T \in Z(E)$ and $\tilde{\mathcal{F}}$ is also a regular numerical range preserving map.*

Now that we know how to utilize the diagonal action of a regular numerical range preserving map we turn to the off-diagonal behavior.

Lemma 4.6 *Let E be a finite dimensional complex Banach lattice and let $\mathcal{F} : \mathcal{L}_r(E) \to \mathcal{L}_r(E)$ be a regular numerical range preserving map, such that $\mathcal{F}$ fixes the center $Z(E)$. Then $\mathcal{F}(\{I\}^d) = \{I\}^d$, i.e., $\mathcal{F}$ maps zero diagonal operators to zero diagonal operators.*

Proof This is immediate, as if $T : E \to E$ with $\mathcal{P}(T) = 0$, then $\mathcal{P}(\mathcal{F}(T)) = 0$ as $\mathcal{F}$ and $\mathcal{F}^{-1}$ fixes the diagonal. Thus $\mathcal{F}$ maps zero diagonal operators to zero diagonal operators. □

An immediate consequence of the above results is:

Corollary 4.7 *Let E be a finite dimensional complex Banach lattice and let $\mathcal{F} : \mathcal{L}_r(E) \to \mathcal{L}_r(E)$ be a regular numerical range preserving map, which fixes the center $Z(E)$. Then $\mathcal{F}^*$ fixes the diagonal of any state $\Phi \in \mathcal{S}_r$ and maps zero diagonal $\Phi \in \mathcal{L}_r(E)^*$ to zero diagonal functionals.*

Lemma 4.8 *Let E be a finite dimensional complex Banach lattice and let $\mathcal{F} : \mathcal{L}_r(E) \to \mathcal{L}_r(E)$ be a regular numerical range preserving map, which fixes the center $Z(E)$. Let $0 \le f \in \mathcal{E}(B(E^*))$ and $0 \le x \in \mathcal{E}(B(E))$ with $f(x) = 1$. Assume $|\Phi| \le f \otimes x$ is a state with the same diagonal as $f \otimes x$. Then $|\mathcal{F}^*(\Phi)| \le f \otimes x$.*

Proof Let $|\Phi| \le f \otimes x$ be a state. Then by Corollary 2.17 there exists Φ_i with $|\Phi_i| = f \otimes x$ and $\lambda_i \ge 0$ with $\sum_{i=1}^m \lambda_i = 1$ such that $\Phi = \sum_{i=1}^m \lambda_i \Phi_i$. Now $|\mathcal{F}^*(\Phi_i)| = f \otimes x$ (as $\mathcal{F}^*$ maps extreme points to extreme points) implies that $|\mathcal{F}^*(\Phi)| \le |\sum_{i=1}^m \lambda_i \mathcal{F}^*(\Phi_i)|| \le f \otimes x$. □

Proposition 4.9 *Let E be a finite dimensional complex Banach lattice of dimension n and let $\mathcal{F} : \mathcal{L}_r(E) \to \mathcal{L}_r(E)$ be a regular numerical range preserving map, which fixes the center $Z(E)$. Then there exist a permutation σ of $\{(i, j) : 1 \le i, j \le n, \}$ with $\sigma(i, i) = (i, i)$ for all i and a matrix $\mathcal{U} = [u_{i,j}]$ with $u_{i,i} = 1$ for all i and $|u_{i,j}| = 1$ for all $i \ne j$ such that for any $\Phi \in \mathcal{L}_r(E)^*$ with $\Phi = [\phi_{i,j}]$ we have that $\mathcal{F}^*(\Phi) = [u_{i,j}\phi_{\sigma(i,j)}]$.*

Proof Consider the diagonal state $(\frac{1}{n})$ on $\mathcal{L}_r(E)$. Via Lozanovskii's factorization theorem we can find $0 \le f \in B(E^*)$ and $0 \le x \in B(E)$ with $f(x) = 1$ such that the state $f \otimes x$ has diagonal $(\frac{1}{n})$. From the extreme point characterization of $\mathcal{S}_r$ we obtain Φ_i with $|\Phi_i| = f_i \otimes x_i$, $0 \le f_i \in \text{extr}(B(E^*))$ and $0 \le x_i \in \text{extr}(B(E))$ with $f_i(x_i) = 1$, and $\lambda_i \ge 0$ with $\sum_{i=1}^m \lambda_i = 1$ such that

$$f \otimes x = \sum_{i=1}^{m} \lambda_i \Phi_i.$$

Observe that this implies that the diagonal of $\sum_{i=1}^m \lambda_i f_i \otimes x_i$ is still $(\frac{1}{n})$. Let now Φ be a regular state with diagonal $(\frac{1}{n})$ such that $|\Phi| \le \sum_{i=1}^m \lambda_i f_i \otimes x_i$. The $\Phi = \sum_{i=1}^m \lambda_i \Psi_i$, where $|\Psi_i| \le f_i \otimes x_i$. Since Φ is a regular state, it follows that also the Ψ_i are states and $|\Psi_i| \le f_i \otimes x_i$. This implies by the above Lemma that $|\mathcal{F}^*(\Psi_i)| \le f_i \otimes x_i$ for each index i. Hence $|\mathcal{F}^*(\Phi)| \le \sum_{i=1}^m \lambda_i f_i \otimes x_i$. This shows that $\mathcal{F}^*$ is a contraction with respect to the AM-norm on the ideal generated by the off-diagonal of $\sum_{i=1}^m \lambda_i f_i \otimes x_i$. As the same holds now for $\mathcal{F}^{*-1}$ it follows that $\mathcal{F}^*$ is an isometry on $(I - \mathcal{P})\mathcal{L}_r(E)^*$ with respect to the AM-norm defined by the off-diagonal of $\sum_{i=1}^m \lambda_i f_i \otimes x_i$. From the well-known characterization of isometries on AM-spaces with unit follows now the existence of $\mathcal{U}$ and the permutation σ of the off-diagonal indices so that $\mathcal{F}^*$ has the required representation on zero diagonal matrices. By extending $\mathcal{U}$ and σ on the diagonal as indicated, the result follows directly. □

Theorem 4.10 *Let E be a finite dimensional complex Banach lattice of dimension n with symmetric norm and let $\mathcal{F} : \mathcal{L}_r(E) \to \mathcal{L}_r(E)$ be a regular numerical range preserving map. Then there exist a permutation τ of $\{(i, j) : 1 \le i, j \le n, \}$ with $\tau = \tau_1\tau_2$, where τ_1 is a permutation of the diagonal $\{(i, i); 1 \le i \le n\}$, and a matrix $\mathcal{U} = [u_{i,j}]$ with $u_{i,i} = 1$ for all i and $|u_{i,j}| = 1$ for all $i \ne j$ such that for any regular operator $T = [t_{i,j}]$ we have that $\mathcal{F}(T) = [u_{i,j}t_{\tau(i,j)}]$. In particular, $\mathcal{F}$ is a lattice isomorphism of $\mathcal{L}_r(E)$.*

Proof By Proposition 4.5 there exists a permutation matrix P such that $\tilde{\mathcal{F}}(T) = P^{-1}\mathcal{F}(T)P$ fixes $Z(E)$ and $\tilde{\mathcal{F}}$ is also a regular numerical range preserving map. By the above proposition there exist a permutation σ of $\{(i, j) : 1 \le i, j \le n, \}$ with $\sigma(i, i) = (i, i)$ for all i and a matrix $\tilde{\mathcal{U}} = [\tilde{u}_{i,j}]$ with $\tilde{u}_{i,i} = 1$ for all i and $|\tilde{u}_{i,j}| = 1$ for all $i \ne j$ such that for any $\Phi \in \mathcal{L}_r(E)^*$ with $\Phi = [\phi_{i,j}]$ we have that $\tilde{\mathcal{F}}^*(\Phi) = [\tilde{u}_{i,j}\phi_{\sigma(i,j)}]$. Assume $Px = x \circ \phi$. Then $\mathcal{F}(T) = P\tilde{\mathcal{F}}(T)P^{-1}$ and as can be seen by means of a direct calculation we have for $T = [t_{i,j}]$ that $\mathcal{F}(T) = [\tilde{u}_{\sigma^{-1}(\phi^{-1}(i),\phi^{-1}(j))}t_{\sigma^{-1}(\phi^{-1}(i),\phi^{-1}(j))}]$. Define now $u_{i,j} = \tilde{u}_{\sigma^{-1}(\phi^{-1}(i),\phi^{-1}(j))}$ and $\tau(i, j) = \sigma^{-1}(\phi^{-1}(i), \phi^{-1}(j))$. Then $\mathcal{F}(T) = [u_{i,j}t_{\tau(i,j)}]$ and we can split τ by defining $\tau_1(i, i) = \sigma^{-1}(\phi^{-1}(i), \phi^{-1}(i)) = (\phi^{-1}(i), \phi^{-1}(i))$, as σ^{-1} fixes the diagonal. We see thus that τ_1 defines a permutation of the diagonal. □

Remark 4.11

1. In the above theorem we don't know of any E for which every τ will define a regular numerical range preserving map. In fact as we will see below for $1 < p \ne 2 < \infty$ only the case $\tau_2 = id$ will define a regular numerical range preserving map.
2. From the above theorem (or its proof more accurately) we can also describe $\mathcal{F}^*$ as follows. If $\mathcal{F}$ is as in the above theorem, then $\mathcal{F}^*(\Phi) = [u_{\tau^{-1}(i,j)}\phi_{\tau^{-1}(i,j)}]$ for all $\Phi \in \mathcal{L}_r(E)^*$. In describing the regular numerical range preserving mappings on $\ell_p(n)$ we will in effect describe which τ_2 lead to regular state preserving maps $\mathcal{F}^*$.

In case $E = \ell_p(n)$ with $1 \le p \le \infty$, we can now completely characterize which τ are allowed in the above theorem.

Theorem 4.12 *Let $E = \ell_p(n)$ with $1 \le p \le \infty$ and suppose $\mathcal{F}(T) = [u_{i,j}t_{\tau(i,j)}]$ for $T = [t_{i,j}]$, where τ is a permutation of $\{(i, j) : 1 \le i, j \le n, \}$ with $\tau = \tau_1\tau_2$, where τ_1 is a permutation of the diagonal $\{(i, i); 1 \le i \le n\}$ given by $\tau_1(i, i) = (\phi^{-1}(i), \phi^{-1}(i))$, and a matrix $\mathcal{U} = [u_{i,j}]$ with $u_{i,i} = 1$ for all i and $|u_{i,j}| = 1$ for all $i \ne j$. Then*

(i) *$\mathcal{F}$ is a regular numerical range preserving map on $\mathcal{L}_r(\ell_1(n))$ if and only τ maps the i-th column onto the $\phi^{-1}(i)$-th column.*

(ii) *$\mathcal{F}$ is a regular numerical range preserving map on $\mathcal{L}_r(\ell_\infty(n))$ if and only τ maps the j-th row onto the $\phi^{-1}(j)$-th row.*
(iii) *$\mathcal{F}$ is a regular numerical range preserving map on $\mathcal{L}_r(\ell_2(n))$ if and only τ_2 is a permutation of a partial transpose, i.e. there exists $A \subset \{(i,j) : 1 \le i < j \le n\}$ such that $\tau_2(t_{i,j}) = t_{\phi^{-1}(j),\phi^{-1}(i)}$ for all $(i,j) \in A$ or $(j,i) \in A$, and $\tau_2(t_{i,j}) = t_{\phi^{-1}(i),\phi^{-1}(j)}$ otherwise.*
(iv) *$\mathcal{F}$ is a regular numerical range preserving map on $\mathcal{L}(\ell_p(n))$ for $1 < p \ne 2 < \infty$ if and only τ_2 is the identity permutation.*

Proof By considering, as before, $\tilde{\mathcal{F}}(T) = P^{-1}\mathcal{F}(T)P$ we can assume that ϕ (and thus τ_1) is the identity permutation. Now the sufficiency conditions in (i) and (ii) are immediate consequences of Corollary 2.21. To prove necessity in (i), assume that $\tau^{-1}(i,j) = (k,l)$ for $j \ne l$. Then there exists a state $\Psi = [\psi_{ij}]$ such that $\operatorname{diag}(\Psi) = (\kappa_1, \ldots, \kappa_n)$ and $\kappa_j > \kappa_l$. Furthermore, let $\psi_{ij} = \kappa_j$ for all $1 \le i, j \le n$. Then by Corollary 2.21, $\mathcal{F}^*(\Psi)$ is no longer a state, which yields a contradiction. Hence we must have that τ only permutes within each column vector. The necessity in case (ii), i.e. $p = \infty$, follows similarly. To show the sufficiency of the condition in (iii) we show that such a representation will preserve the regular states. To that end let $\lambda_1 + \cdots + \lambda_n = 1$ with $\lambda_i \ge 0$ for every $1 \le i \le n$. Consider the pure state $\Phi = [\phi_{ij}]$ where $|\phi_{ij}| = \lambda_i^{1/2}\lambda_j^{1/2}$ as in Theorem 2.24. Since $\lambda_i^{1/2}\lambda_j^{1/2} = \lambda_j^{1/2}\lambda_i^{1/2}$, partial transposes will preserve the pure regular states and thus all regular states. Now assume there is (i,j) such that $\tau((i,j)) = (j,l)$ with $i \ne l$. Then there exists a state $\Psi = [\psi_{ij}]$ such that $\operatorname{diag}(\Psi) = (\kappa_1, \ldots, \kappa_n)$ and $\kappa_i > \kappa_l$. Furthermore, let $\psi_{ij} = \kappa_i^{1/2}\kappa_j^{1/2}$ for all $1 \le i, j \le n$. Clearly $\kappa_i^{1/2}\kappa_j^{1/2} > \kappa_j^{1/2}\kappa_l^{1/2}$ and thus $\mathcal{F}^*(\Psi)$ is no longer a regular state, which yields a contradiction. The same argument as above will show that there cannot be an (i,j) such that $\tau((i,j)) = (k,i)$ where $k \ne j$. Together these arguments show that τ must be a partial transpose. Hence (iii) holds. The sufficiency in (iv) has been observed before. Now assume there exists an (i,j) such that $\tau((i,j)) = (k,j)$, where $i \ne k$. There exists a state $\Psi = [\psi_{ij}]$ such that $\operatorname{diag}(\Psi) = (\kappa_1, \ldots, \kappa_n)$ and $\kappa_i > \kappa_k$. Furthermore, let $\psi_{ij} = \kappa_i^{1/p'}\kappa_j^{1/p}$ for all $1 \le i, j \le n$. Clearly $\kappa_i^{1/p'}\kappa_j^{1/p} > \kappa_k^{1/p'}\kappa_j^{1/p}$ and thus $\mathcal{F}^*(\Psi)$ is no longer a regular state, which yields a contradiction. The same argument as above will show that there cannot be an (i,j) such that $\tau((i,j)) = (i,l)$ where $j \ne l$. Together these arguments show that τ must be the identity. □

The cases $p = 1$ and $p = \infty$ were already obtained, by a different method, in [11].

5 Unital Isometries

Let $\mathcal{A}$ be a Banach algebra with unit e. Then a linear isometry $\mathcal{F} : \mathcal{A} \to \mathcal{A}$ is called *unital* if $\mathcal{F}(e) = e$. The following proposition shows that there is a relation between surjective unital linear isometries and numerical range preserving maps.

Proposition 5.1 *Let $\mathcal{A}$ be a Banach algebra with unit e. Then a surjective unital linear isometry $\mathcal{F} : \mathcal{A} \to \mathcal{A}$ is numerical range preserving, i.e., $V(\mathcal{A}, \mathcal{F}(a)) = V(\mathcal{A}, a)$ for all $a \in \mathcal{A}$.*

Proof Let $\Phi \in \mathcal{S}(\mathcal{A})$. Then $\|\mathcal{F}^*(\Phi)\| = 1$ and $\mathcal{F}^*(\Phi)(e) = 1$, since $\mathcal{F}$ is unital. Hence, by Theorem 4.1, $V(\mathcal{A}, \mathcal{F}(a)) \subset V(\mathcal{A}, a)$. As the same argument applies to $\mathcal{F}^{-1}$ it follows that $\mathcal{F}$ is numerical range preserving. □

Corollary 5.2 *Let E be a finite dimensional complex Banach lattice of dimension n with symmetric norm and let $\mathcal{F} : \mathcal{L}_r(E) \to \mathcal{L}_r(E)$ be a surjective unital linear isometry. Then $\mathcal{F}$ is a lattice isomorphism of $\mathcal{L}_r(E)$.*

Proof This is an immediate consequence of the above proposition and Theorem 4.10. □

One can ask now whether in the above proposition the converse implication is true. Using the characterization of regular numerical range preserving operators from the previous section we will see now that this is sometimes, but not always, true.

Theorem 5.3 *Let $1 \le p \le \infty$ and $\mathcal{F} : \mathcal{L}_r(\ell_p(n)) \to \mathcal{L}_r(\ell_p(n))$ a regular numerical range preserving operator. Then for $p \neq 2$ the operator $\mathcal{F}$ is a unital isometry. For $p = 2$ and $n = 2$ the operator $\mathcal{F}$ is also a unital isometry, but for $n \ge 3$ this is no longer necessarily true.*

Proof For any permutation P we have that $|PTP^{-1}| = P|T|P^{-1}$, which implies that $\|PTP^{-1}\|_r = \|T\|_r$ for any permutation P and $T \in \mathcal{L}_r(\ell_p(n))$. This implies that we can assume, as in the proof of Theorem 4.12 that $\mathcal{F}$ fixes the centrum. Then for $p \neq 2$ we see from the representation of $\mathcal{F}$ that $\mathcal{F}$ is a unital isometry. For $p = 2$ and $n = 2$ we get in the proof of Theorem 4.12 for τ only the full transpose $T \mapsto T^t$ as the only other option than the identity, and this map is also an isometry. For $n > 2$ it suffices to consider the case $n = 3$. In this case define $\mathcal{F}([t_{i,j}]) = [s_{i,j}]$, where $s_{i,j} = t_{i,j}$ when $(i, j) \notin \{(1, 2), (2, 1)\}$, $s_{1,2} = t_{2,1}$, and $s_{2,1} = t_{1,2}$. Then by Theorem 4.12 the operator $\mathcal{F}$ is a regular numerical range preserving operator on $\mathcal{L}_r(\ell_2(3))$. On the other hand, it is easy to see that $\mathcal{F}$ is not an isometry, e.g., let

$$T = \begin{bmatrix} 1 & 1 & 1 \\ 0 & 1 & 0 \\ 0 & 0 & 1 \end{bmatrix}.$$

Then $\|T\| = \sqrt{2 + \sqrt{3}} \approx 1.93$, while $\|\mathcal{F}(T)\| \approx 1.8$. Hence $\mathcal{F}$ is not an isometry. □

6 Possible Extensions and Open Problems

To extend the results about regular numerical range preserving maps to infinite dimensional Banach lattices we face several difficulties. First of all we can try to extend Proposition 4.5 to, say, rearrangement invariant Banach function spaces. However, in general the permutation operator induced from the action on the center will not be an isometry on the underlying Banach lattice. This difficulty can be overcome, if we restrict ourselves to atomic rearrangement invariant Banach function spaces and if we assume that the map $\mathbb{F}$ and its inverse are order continuous. Even assuming this, we run into the problem of producing a strong unit, like in the proof of Theorem 4.10. This problem goes away if we restrict ourselves to ℓ_p-spaces, as in that case each regular state is dominated by a unique positive vector state with the same support set for the diagonals. Starting with a diagonal like $(\frac{1}{2^n})$ we can use the associated vector state as AM-unit. Therefore we have the following infinite dimensional version of Theorem 4.12.

Theorem 6.1 *Let $E = \ell_p$ with $1 \leq p \leq \infty$ and suppose $\mathcal{F}$ is a surjective regular numerical range preserving map such that both $\mathcal{F}$ and $\mathcal{F}^{-1}$ are regular order continuous maps. Then $\mathcal{F}(T) = [u_{i,j} t_{\tau(i,j)}]$ for $T = [t_{i,j}]$, where τ is a permutation of $\{(i,j) : 1 \leq i, j < \infty, \}$ with $\tau = \tau_1\tau_2$, where τ_1 is a permutation of the diagonal $\{(i,i); 1 \leq i < \infty\}$ given by $\tau_1(i,i) = (\phi^{-1}(i), \phi^{-1}(i))$, and a matrix $\mathcal{U} = [u_{i,j}]$ with $u_{i,i} = 1$ for all i and $|u_{i,j}| = 1$ for all $i \neq j$. Then we have*

(i) *if $p = 1$, then τ maps the i-th column onto the $\phi^{-1}(i)$-th column.*
(ii) *if $p = \infty$, then τ maps the j-th row onto the $\phi^{-1}(j)$-th row.*
(iii) *If $p = 2$, then τ_2 is a permutation of a partial transpose, i.e. there exists $A \subset \{(i,j) : 1 \leq i < j < \infty\}$ such that $\tau_2(t_{i,j}) = t_{\phi^{-1}(j),\phi^{-1}(i)}$ for all $(i,j) \in A$ or $(j,i) \in A$, and $\tau_2(t_{i,j}) = t_{\phi^{-1}(i),\phi^{-1}(j)}$ otherwise.*
(iv) *If $1 < p \neq 2 < \infty$, then τ_2 is the identity permutation.*

Moreover, in each of these case, each mapping $\mathcal{U}$ and permutation τ, as described, defines a regular numerical range preserving map on $\mathcal{L}_r(\ell_p)$ for the indicated p..

Open Problems

1. Can we describe $\mathcal{E}(\mathcal{S}_r^n)$ in terms of Φ with $|\Phi| = f \otimes x$, where $f \in \mathcal{E}(B(E^*))$ and $x \in \mathcal{E}(B(E))$ for infinite dimensional E?
2. Is it true for infinite dimensional E that for every regular state Φ, there exists a bounded state Ψ such that $|\Phi| \leq |\Psi|$? For finite dimensional spaces E and for order continuous states on $\mathcal{L}_r(\ell_p)$, the results of this paper show that the answer is affirmative.
3. Is it true that $v_+^o(T) = v_+(T)$ for infinite dimensional E?
4. Can we describe order continuous surjective regular numerical range preserving maps for rearrangement invariant Banach sequence spaces?

References

1. Y.A. Abramovich, C.D. Aliprantis, *Problems in Operator Theory*. Graduate Studies in Mathematics, vol. 51 (American Mathematical Society, Providence, 2002)
2. Y.A. Abramovitch, C.D. Aliprantis, *An Invitation to Operator Theory*. Graduate Studies in Mathematics, vol. 50 (American Mathematical Society, Providence, 2002)
3. E.M. Alfsen, F.W. Shultz, *State Spaces of Operator Algebras*. Mathematics: Theory & Applications (Birkhäuser Boston, Inc., Boston, 2001)
4. F.F. Bonsall, J. Duncan, *Numerical Ranges of Operators on Normed Spaces and of Elements of Normed Algebras*. London Mathematical Society Lecture Note Series, vol. 1 (Cambridge University Press, London/New York, 1971)
5. F.F. Bonsall, J. Duncan, *Numerical Ranges. II*. London Mathematical Society Lecture Notes Series, vol. 10 (Cambridge University Press, New York/London, 1973)
6. T. Gillespie, Factorization in Banach function spaces. Indag. Math. **84**, 287–300 (1981)
7. R. Grząślewicz, H.H. Schaefer, Extreme points of the positive part of the unit ball and strictly monotone norm. Math. Z. **207**(3), 481–483 (1991)
8. N.J. Kalton, Hermitian operators on complex Banach lattices and a problem of Garth Dales. J. Lond. Math. Soc. (2) **86**(3), 641–656 (2012)
9. C-K. Li, S. Pierce, Linear preserver problems. Am. Math. Mon. **108**(7), 591–605 (2001)
10. C-K. Li, H. Schneider, Orthogonality of matrices. Linear Algebra Appl. **347**, 115–122 (2002)
11. C-K. Li, A.R. Sourour, Linear operators on matrix algebras that preserve the numerical range, numerical radius or the states. Can. J. Math. **56**(1), 134–167 (2004)
12. A. Lima, G. Olsen, Extreme points in duals of complex operator spaces. Proc. Am. Math. Soc. **94**(3), 427–440 (1985)
13. G.Ya. Lozanovskii, On some Banach lattices. Siberian Math J. **10**, 419–431 (1969)
14. V.J. Pellegrini, Numerical range preserving operators on a Banach algebra. Studia Math. **54**(2), 143–147 (1975)
15. A. Radl, The numerical range of positive operators on Hilbert lattices. Integr. Equ. Oper. Theory **75**(4), 459–472 (2013)
16. A. Radl, The numerical range of positive operators on Banach lattices, Positivity **19**(3), 603–623 (2015)
17. A.R. Schep, The order continuous dual of the regular integral operators on L^p. Vladikavk. Math. J. **11**, 45–48 (2009)
18. B. Simon, *Convexity*. Cambridge Tracts in Mathematics, vol. 187 (Cambridge University Press, Cambridge, 2011)
19. A.R. Sourour, Spectrum-preserving linear maps on the algebra of regular operators, in *Aspects of Positivity in Functional Analysis (Tübingen, 1985)*. North-Holland Mathematics Studies, vol. 122 (North-Holland, Amsterdam, 1986), pp. 255–259
20. A.C. Zaanen, *Introduction to Operator Theory in Riesz Spaces* (Springer, Berlin, 1997)

Bilaplace Eigenfunctions Compared with Laplace Eigenfunctions in Some Special Cases

Guido Sweers

To Ben, incomparable and always positive

Abstract On any bounded domain the Dirichlet or Neumann–Laplace operator has a first eigenfunction which is the unique one of fixed sign. For the bilaplace operator, as for any fourth order operator, there is no direct maximum principle and hence no obvious argument for a first eigenfunction of one sign. We will address some examples where fourth order eigenvalue problems have an unexpected behaviour concerning positivity of an eigenfunction.

Keywords Fundamental eigenfunction · Dirichlet and Navier bilaplace · Positivity preserving

1 Introduction

The eigenvalue problems for which we will show some peculiar features are described as follows:

- The Dirichlet-bilaplace problem

$$\begin{cases} \Delta^2\phi = \lambda\phi & \text{in } \Omega, \\ \phi = \partial_\nu\phi = 0 & \text{on } \partial\Omega, \end{cases} \text{ with } R_4(\phi) = \frac{\int_\Omega |\Delta\phi|^2\,dx}{\int_\Omega \phi^2\,dx}. \tag{1}$$

- The Navier-bilaplace problem

$$\begin{cases} \Delta^2\phi = \lambda\phi & \text{in } \Omega, \\ \phi = \Delta\phi = 0 & \text{on } \partial\Omega, \end{cases} \text{ with } R_4(\phi) = \frac{\int_\Omega |\Delta\phi|^2\,dx}{\int_\Omega \phi^2\,dx}. \tag{2}$$

G. Sweers (✉)
Mathematisches Institut, Universität zu Köln, Köln, Germany
e-mail: gsweers@math.uni-koeln.de

G. Buskes et al. (eds.), *Positivity and Noncommutative Analysis*,
Trends in Mathematics, https://doi.org/10.1007/978-3-030-10850-2_26

We will compare these with the eigenvalue problem in the well studied Dirichlet–Laplace case:

$$\begin{cases} -\Delta\phi = \lambda\phi & \text{in } \Omega, \\ \phi = 0 & \text{on } \partial\Omega, \end{cases} \quad \text{with } R_2(\phi) = \frac{\int_\Omega |\nabla\phi|^2\, dx}{\int_\Omega \phi^2\, dx}. \tag{3}$$

In our notation Ω is a domain, meaning an open and connected subset of $\mathbb{R}^n$, which we moreover assume to be bounded.

Interest in the spectrum of fourth problems boundary value problems has a long history. The equation $\Delta^2 u = f$ in $\Omega \subset \mathbb{R}^2$ is used in the model for the deviation u of thin horizontal plate under a vertical force density f. The boundary conditions in (1) represent a plate clamped at its boundary, see [14] by Friedrichs. For the corresponding eigenvalue problems, see, for example, Courant's paper from 1922 [8], Weinstein in 1937 [32], Szegö in 1950 [29] or Weinberger [31]. The last authors also considered so-called buckling eigenvalues in the clamped setting: $\Delta^2\phi = \lambda\Delta\phi$ in Ω with $\phi = \partial_\nu\phi = 0$ on $\partial\Omega$, which we will not address. A recent survey on variational methods for eigenvalues, with nonlinear applications, can be found in [3, 30].

The cases (2) and (1) are obviously different but even here, when considering $\Omega = B$, a disk, the eigenfunctions do not seem to be very different except near the boundary. Indeed, for $\Omega = B \subset \mathbb{R}^2$ one finds a sketch in Fig. 1 of the first four eigenfunctions of (1) and the first four of (3) and (2). Note that when $\partial\Omega$ is smooth, the eigenfunctions for (3) are smooth, and are hence also (all) eigenfunctions for (2).

Our aim is to show and recall some unexpected features for the first eigenfunctions of (1) and (2). Indeed, we will show that in both cases the second eigenvalue may be the positive one. For (1) we recall a result from Duffin and coauthors [6, 7, 13], where $\Omega \subset \mathbb{R}^2$ is an annulus with a small hole. For (2) we will show that on some extreme sector also the second one is the positive one. Note that if the

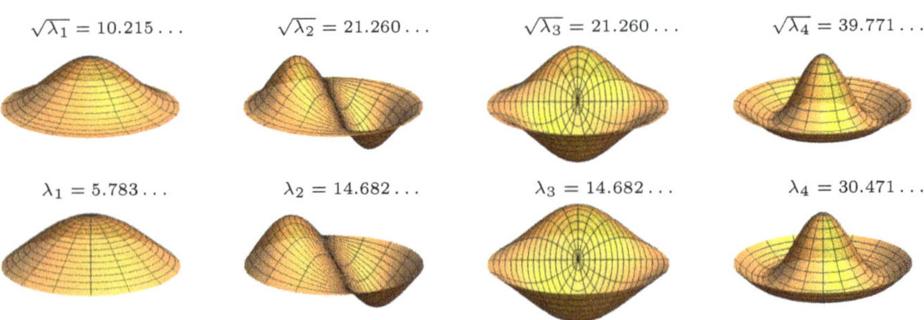

Fig. 1 For $\Omega = \{x \in \mathbb{R}^2;\, |x| < 1\}$ in the top row the first four eigenfunctions and eigenvalues for the Dirichlet-bilaplace case (1); on the bottom the first four eigenfunctions/values for the Dirichlet–Laplace case (3). Since $\lambda_{k,(1)} = \mathcal{O}(k^2)$ and $\lambda_{k,(3)} = \mathcal{O}(k)$ we compare $\sqrt{\lambda_{k,(1)}}$ with $\lambda_{k,(3)}$. Note that the bottom row also represents the first four eigenfunctions/values for (2) when we replace λ_k by $\sqrt{\lambda_k}$. Indeed $\sqrt{\lambda_{k,(2)}} = \lambda_{k,(3)}$

domain has a reentrant corner, the eigenfunctions of (3) and (2) do not coincide in the most natural settings for the quotient R_4, which is in sharp contrast to the case that $\partial\Omega$ is smooth.

We will start by recalling the relation between the domain regularity and the setting for the eigenfunctions. The peculiarities of fourth order eigenfunctions we will discuss by examples, where also the relation between the 'fundamental' eigenvalue and a positivity preserving property is discussed. We will also recall that a positive eigenfunction may lead to a maximum principle, which has recently been proven in [25].

2 Recalling Existence and Domain Regularity

In this section we recall the arguments that guarantee the existence of eigenfunctions in an appropriate setting. The arguments are standard but it is hard to find a condense reference with explicit assumptions for the regularity of the boundary. Moreover, since the two examples use nonsmooth boundary points, respectively, the convergence of the boundary to a single point, it should be helpful to see the results in relation with the regularity of the domain. For general elliptic problems on domains with corners see [17].

2.1 The Dirichlet–Laplace Case

For general domains one finds countably many eigenvalues for (3). Indeed, without further regularity assumptions on $\partial\Omega$, the Riesz representation theorem yields that

$$\begin{cases} -\Delta u = f & \text{in } \Omega \\ u = 0 & \text{on } \partial\Omega \end{cases} \tag{4}$$

has a well-defined (weak) solution operator $\mathcal{G}_1 : f \mapsto u : L_2(\Omega) \to W_0^{1,2}(\Omega)$, i.e., $u = \mathcal{G}_1 f$ satisfies

$$u \in W_0^{1,2}(\Omega) \text{ and } \int_\Omega (\nabla u \cdot \nabla \psi - f\,\psi)\,dx = 0 \text{ for all } \psi \in W_0^{1,2}(\Omega). \tag{5}$$

Indeed, by using the Poincaré–Friedrichs inequality

$$\int_\Omega u^2 dx \le \operatorname{diam}(\Omega)^2 \int_\Omega |\nabla u|^2\,dx \quad \text{for all } u \in W_0^{1,2}(\Omega), \tag{6}$$

one finds that $\|\nabla u\|_{L_2(\Omega)}$ and $\|u\|_{W^{1,2}(\Omega)}$ are equivalent norms on $W_0^{1,2}(\Omega)$ and one uses (5) and $\|\nabla u\|_{L_2(\Omega)}$ as a norm on $W_0^{1,2}(\Omega)$. Having a unique weak solution u, results concerning interior regularity show that $-\Delta u = f$ holds locally in $L_2(\Omega)$-sense and so $\Delta u \in L_2(\Omega)$. Hence we have a solution operator for (4) as follows:

$$\mathcal{G}_1 : L_2(\Omega) \to \mathcal{W}_0^1(\Omega) := \left\{ u \in W_0^{1,2}(\Omega) ; \Delta u \in L_2(\Omega) \right\}. \tag{7}$$

Remark 1 Note that in general an individual second derivative like u_{xy} does not lie in $L_2(\Omega)$, unless the boundary has some regularity such as $\partial\Omega \in C^2$ with the possible exception of convex corners [20], see, for example, [23].

By Rellich's theorem (see [1]) $W_0^{1,2}(\Omega)$ is compactly imbedded in $L_2(\Omega)$, and $\mathcal{G}_1$ followed by the imbedding is compact on $L_2(\Omega)$. So (3) has at most countable many eigenvalues λ_k, which cannot accumulate except at ∞. These eigenvalues lie in $\mathbb{R}$ by the symmetry of the quadratic form in R_2. Since (6) holds, all eigenvalues satisfy $\lambda_k \geq c_\Omega$.

The famous Weyl's law [33] even tells us that for second order problems $\lambda_k \sim k^{2/n}$, more precisely with N the counting function:

$$N(\lambda_k < t) = \frac{\omega_n}{(2\pi)^n} t^{n/2} \operatorname{vol}(\Omega) + \mathcal{O}\left(t^{(n-1)/2}\right).$$

A special property for second order elliptic problems is the maximum principle. It shows us that solution operator in (7) preserves positivity and even shows us positivity in a strong sense, namely, that

$$f \gneqq 0 \text{ in } \Omega \implies u > 0 \text{ in } \Omega.$$

Here $f \gneqq 0$ in Ω means $f \geq 0$ and $f \not\equiv 0$. Then the Krein–Rutman theorem [21], simplified in its use on general domains through the result of de Pagter ([12], see also [26]), tells us that the first eigenvalue $\lambda_1 > 0$ is simple and is the only one with a positive eigenfunction. One also finds for $\lambda < \lambda_1$ that

$$\begin{cases} -\Delta u = \lambda u + f & \text{in } \Omega \\ u = 0 & \text{on } \partial\Omega \end{cases} \tag{8}$$

is positivity preserving. Indeed, for $\lambda < 0$ and assuming $u \not\geq 0$ one uses the maximum principle on $\Omega \cap [u < 0]$ to find contradiction. For $\lambda \in [0, \lambda_1)$ the solution of (8) is positive for f, since with $\nu(\mathcal{G}_1) = \lambda_1^{-1}$

$$u = \sum_{k=0}^{\infty} (\lambda \mathcal{G}_1)^k \mathcal{G}_1 f \tag{9}$$

is a convergent series of positive operators.

Let us summarize:

Theorem 2 *Suppose that $\Omega \subset \mathbb{R}^n$ is a bounded domain. Then the Dirichlet–Laplace eigenvalue problem (3)*

- *has countable many eigenvalues* $0 < \lambda_1 < \lambda_2 \leq \lambda_3 \leq \dots$,
- *with* $\lim_{k\to\infty} \lambda_k = \infty$, *even* $\lambda_k \sim k^{2/n}$,
- *with* λ_1 *simple and the corresponding* ϕ_1 *of fixed sign,*
- *for all* $k \geq 2$ *'the' eigenfunction* ϕ_k *is sign-changing and*
- *for all* $\lambda < \lambda_1$ *one finds that:* $f \gneq 0$ *implies* $u > 0$ *in (8).*

2.2 The Dirichlet-Bilaplace Case

In two dimensions the boundary value problem

$$\begin{cases} \Delta^2 u = f & \text{in } \Omega \\ u = \partial_\nu u = 0 & \text{on } \partial\Omega \end{cases} \tag{10}$$

is known as the clamped plate problem. In that model f is the force density and u the resulting deviation from the equilibrium at 0. Fixing both the position and the angle of the elastic plate to be 0 at the boundary of Ω, with Ω representing the shape of the plate, one derives, respectively, $u = 0$ and $\nabla u = 0$ at $\partial\Omega$. When the exterior normal ν is defined on $\partial\Omega$, it is equivalent to $u = \partial_\nu u = 0$ on $\partial\Omega$. Also for (10) one finds by Riesz for each $f \in L_2(\Omega)$ a weak solution u, that is,

$$u \in W_0^{2,2}(\Omega) \text{ with } \int_\Omega (\Delta u\, \Delta\psi - f\,\psi)\, dx = 0 \text{ for all } \psi \in W_0^{2,2}(\Omega). \tag{11}$$

Here $W_0^{2,2}(\Omega) = \overline{C_c^\infty(\Omega)}$ is the closure with respect to $\|\cdot\|_{W^{2,2}}$ of $C_c^\infty(\Omega)$, the C^∞-functions on Ω with compact support in Ω. So indeed, one may show through approximation and integration by parts for $u \in C_c^\infty(\Omega)$ that

$$\int_\Omega \sum_{1\leq i\leq j\leq n} \left(\partial_i\partial_j u\right)^2 dx = \int_\Omega \sum_{1\leq i\leq j\leq n} \partial_i^2 u\, \partial_j^2 u\, dx = \int_\Omega (\Delta u)^2\, dx$$

and with (6) for u and $\partial_i u$ that $\|\Delta u\|_{L_2(\Omega)}$ is a norm equivalent to $\|u\|_{W^{2,2}(\Omega)}$ on $W_0^{2,2}(\Omega)$. As for the Dirichlet–Laplace problem there is no condition for any smoothness of $\partial\Omega$ involved. We have a solution operator for (10) as follows:

$$\mathcal{G}_{2,D} : L_2(\Omega) \to \mathcal{W}_{00}^2(\Omega) := \left\{u \in W_0^{2,2}(\Omega);\, \Delta^2 u \in L_2(\Omega)\right\}. \tag{12}$$

Similar as before, a compact imbedding and symmetry implies that the spectrum consists of eigenvalues $\{\lambda_k\}_{k\in\mathbb{N}} \subset \mathbb{R}^+$ with no accumulation point besides ∞.

In [15] and [2] one finds a Weyl result for general elliptic operators implying that $\lambda_k \sim k^{4/n}$. For (1) the estimate was already obtained by Courant in 1922 [9]. The question concerning the fixed sign of (first) eigenfunctions, however, cannot be answered in general by lack of a positivity preserving property of $\mathcal{G}_{2,D}$.

When we compare with the theorem above we find:

Theorem 3 *Suppose that $\Omega \subset \mathbb{R}^n$ is a bounded domain. Then the Dirichlet-bilaplace eigenvalue problem (1)*

- *has countable many eigenvalues* $0 < \lambda_1 \leq \lambda_2 \leq \lambda_3 \leq \ldots,$
- *with* $\lim_{k\to\infty} \lambda_k = \infty$, *even* $\lambda_k \sim k^{4/n}$,

- *but no general argument exists that λ_1 is simple or that the corresponding ϕ_1 has a fixed sign and*
- *there is no general argument that eigenfunctions ϕ_k with $k \geq 2$ are sign-changing.*

Remark 4 By 'general argument' we mean an argument that is not depending on any special property of the domain. Boggio [4] gave an explicit positive formula for the Green function for (1) in case of $\Omega = B$ a ball. So in that case we may use the Krein–Rutman theorem to conclude that λ_1 is simple, ϕ_1 has a fixed sign and all other eigenfunctions are sign-changing. On the other hand, Brown and allies [5] gave numerical evidence that on some dumbbell domains there is no positive eigenfunction or, depending on shape of the connecting part that the positive eigenfunction is the second one, see also [27] and [16]. However, all examples up to now seem to indicate that no matter which domain, there is always an almost positive eigenfunction ϕ_i in the sense that

$$\int_\Omega \phi_i^+ dx \gg \int_\Omega \phi_i^- dx. \tag{13}$$

Here $u^+ = \max(u, 0)$ and $u^- = \max(-u, 0)$.

Below we recall the example of Duffin and coauthors, in which the third eigenfunction has a fixed sign and the first two are sign-changing.

2.3 The Navier-Bilaplace Case

The problem

$$\begin{cases} \Delta^2 u = f & \text{in } \Omega \\ u = \Delta u = 0 & \text{on } \partial\Omega \end{cases} \tag{14}$$

for $\Omega \subset \mathbb{R}^2$ is sometimes used as a model for a plate that is hinged or fixed at its boundary. Erroneously sometimes even called supported. By fixing just the position through $u = 0$, and not u_ν like for the clamped situation, the second condition comes out as a natural boundary condition. Weinstein in [32] calls (14) the model for 'une plaque immobilisé'.

Remark 5 The boundary values in (14) do not seem to be an appropriate model for a thin plate fixed at its boundaries. A more realistic energy formulation, see [14], which includes torsional rigidities, uses the functional $E : \mathcal{W}_0^2(\Omega) \to \mathbb{R}$, defined by

$$E(u) := \int_\Omega \left(\tfrac{1}{2} (\Delta u)^2 + (1-\sigma) \left(u_{xy}^2 - u_{xx}u_{yy} \right) - fu \right) dxdy,$$

where

$$\mathcal{W}_0^2(\Omega) := W^{2,2}(\Omega) \cap W_0^{1,2}(\Omega). \tag{15}$$

The Hessian type term in E is a null-Lagrangian which has no influence on the differential equation but through the corresponding Euler–Lagrange equation yields, besides $u = 0$, a second boundary condition depending on σ, namely,

$$-\Delta u + (1-\sigma)\kappa \partial_\nu u = 0$$

with $\partial_\nu u$ the exterior normal derivative. The function κ is the signed curvature of the boundary, which is positive on convex boundary parts, and $\sigma \in \left(-1, \frac{1}{2}\right)$ is the Poisson ratio, a material depending constant usually near 0.3. So only on straight boundary parts where $\kappa = 0$ the condition $\Delta u = 0$ appears.

For the supported plate one finds a unilateral boundary condition [24].

Coming back to (14), the formulation for the weak solution, with $\mathcal{W}_0^2(\Omega)$ as in (15), is

$$u \in \mathcal{W}_0^2(\Omega) \text{ with } \int_\Omega (\Delta u\, \Delta\psi - f\,\psi)\, dx = 0 \text{ for all } \psi \in \mathcal{W}_0^2(\Omega). \tag{16}$$

Remark 6 This formulation fixes the boundary condition $u = 0$ in a trace sense since $u \in W_0^{1,2}(\Omega)$. The second boundary condition follows as a natural boundary condition for functions u with sufficient regularity through the integration by parts of (16) and using $\psi = 0$ on $\partial\Omega$. Indeed, then (16) implies for $\psi \in \mathcal{W}_0^2(\Omega)$ that

$$0 = \int_\Omega (\Delta u\, \Delta\psi - f\,\psi)\, dx = \int_{\partial\Omega} \Delta u\, \partial_\nu \psi\, ds + \int_\Omega \left(\Delta^2 u - f\right) \psi dx. \tag{17}$$

By taking appropriate test functions one finds $\Delta^2 u - f = 0$ in Ω and $\Delta u = 0$ on $\partial\Omega$.

In the present setting it is not true that Riesz' theorem directly yields a weak solution. It is a priori missing that $\|\Delta u\|_{L_2(\Omega)}$ and $\|u\|_{W^{2,2}(\Omega)}$ are equivalent norms on $\mathcal{W}_0^2(\Omega)$. For domains with a C^2-boundary one may solve (3) not only in $\mathcal{W}_0^1(\Omega)$ but even in $\mathcal{W}_0^2(\Omega)$, see [22] or [20]. So for $f \in L_2(\Omega)$ it follows that $\mathcal{G}_1 f \in \mathcal{W}_0^2(\Omega)$ and even that there is a constant $c > 0$ such that the following estimate holds for all $f \in L_2(\Omega)$:

$$\tfrac{1}{2n}\|f\|_{L_2(\Omega)} \leq \|\mathcal{G}_1 f\|_{W^{2,2}(\Omega)} \leq c\,\|f\|_{L_2(\Omega)}, \tag{18}$$

which implies for $u \in \mathcal{W}_0^2(\Omega)$:

$$\tfrac{1}{2n}\|\Delta u\|_{L_2(\Omega)} \leq \|u\|_{W^{2,2}(\Omega)} \leq c\,\|\Delta u\|_{L_2(\Omega)}.$$

Then both expressions $\|\Delta u\|_{L_2(\Omega)}$ and $\|u\|_{W^{2,2}(\Omega)}$ make equivalent norms. Even for domains with a $C^{2,1}$-boundary except for some convex corners [20], one finds that (18) holds. So $u := (\mathcal{G}_1)^2 f$ for $f \in L_2(\Omega)$, meaning $u := \mathcal{G}_1 v \in \mathcal{W}_0^1(\Omega)$ with $v := \mathcal{G}_1 f \in \mathcal{W}_0^1(\Omega)$, implies $u, v \in \mathcal{W}_0^2(\Omega)$ and hence $\Delta^2 u = f$ in Ω. Arguing backwards from (17) we indeed have a weak solution as in (16). Even if those corners are not convex but 'nice' enough, these norms are equivalent on $\mathcal{W}_0^2(\Omega)$, see [23] or [11].

For

$$\begin{cases} \Delta^2 u = \lambda u + f & \text{in } \Omega, \\ u = \Delta u = 0 & \text{on } \partial\Omega, \end{cases} \tag{19}$$

we find that positivity is preserved for $\lambda \in [0, \lambda_1)$ by a similar argument as in (9). There is no maximum principle on subdomains but by using the so-called 3G-theorem it can be shown that (19) is also positivity preserving on $(\lambda_c, 0)$ for some $\lambda_c < 0$, see [16].

So we find:

Theorem 7 *Suppose that $\Omega \subset \mathbb{R}^n$ is a bounded domain. If $\partial\Omega \in C^2$ except for at most finitely many convex conical points, then problem (2)*

- *inherits its eigenfunctions from problem (3), and*
- *there exists $\lambda_c < 0$ such that for all $\lambda \in (\lambda_c, \lambda_1)$ one finds that: $f \gneq 0$ implies $u > 0$ in (19).*

Remark 8 By 'inherit' we mean that an eigenfunction–eigenvalue pair (ϕ_i, λ_i) for (3) leads to an eigenfunction pair (ϕ_i, λ_i^2) for (2) and all eigenfunctions for (2) one finds this way.

For two-dimensional domains with reentrant corners $\{a_k\}_{k=1}^m \subset \partial\Omega$ the function $\mathcal{G}_1 f$ lies in general not in $\mathcal{W}_0^2(\Omega)$ and $\mathcal{G}_1^2 f$ is no weak solution as defined in (16). In [23] one finds however that for each of those corners a_k there exists a special function $\zeta_k \in \mathcal{W}_0^1(\Omega)$ with $\zeta_k \in C^\infty(\Omega)$, $\|\zeta_k\|_{L_2(\Omega)} = 1$ and $\Delta^2 \zeta_k = 0$, and such that there is a (weak) solution operator $\mathcal{G}_{2,N} : L_2(\Omega) \to \mathcal{W}_0^2(\Omega)$ for (14), with N for Navier, satisfying

$$\mathcal{G}_{2,N} = \mathcal{G}_1^2 - \sum_{1 \le k \le m} \mathcal{G}_1 \mathcal{P}_{\zeta_k} \mathcal{G}_1. \tag{20}$$

Here $\mathcal{P}_{\zeta_k}$ is the $L_2(\Omega)$-projection on ζ_k, $\mathcal{G}_1$ as in (7). The function $u := \mathcal{G}_{2,N} f$ gives for each $f \in L_2(\Omega)$ a weak solution of (14) in the sense of (16). Similar results one finds for three-dimensional domains with some nonconvex conical points. From (20) there is no obvious positivity preservation. Since $\mathcal{G}_{2,N}$ is still a bounded linear operator and $\mathcal{W}_0^2(\Omega)$ is compactly imbedded in $L_2(\Omega)$ one finds:

Theorem 9 *Suppose that $\Omega \subset \mathbb{R}^2$ is a bounded domain. If $\partial\Omega \in C^2$ except for at most finitely many conical points some of which are nonconvex, then problem (2):*

- *has countable many eigenvalues $0 < \lambda_1 \le \lambda_2 \le \lambda_3 \le \ldots$, which are not necessarily inherited from problem (2),*
- *with $\lim_{k\to\infty} \lambda_k = \infty$, even $\lambda_k \sim k^{4/n}$ with $n = 2$,*

- *has no general argument that λ_1 is simple or that the corresponding ϕ_1 has a fixed sign and*
- *there is no general argument that eigenfunctions ϕ_k with $k \ge 2$ are sign-changing.*

Remark 10 In two dimensions a domain with finitely many corners seems quite natural. In higher dimensions the variety of nonsmooth boundary points is too diverse for a general statement. Only domains that are smooth except for finitely many isolated conical points can be dealt with by Nazarov and Sweers [23]. In $\mathbb{R}^4$ such domains inherit the eigenfunctions from (2) even if these cones are not convex.

In Sect. 4 we will reconsider the last two items of Theorem 9 on a nonconvex sector.

3 Dirichlet-Bilaplace on an Annulus

Hadamard mentions in [19] without proof that on domains with small holes the clamped plate problem (10) is not positivity preserving. Hence the Boggio-Hadamard from around the same became '*On convex domains Ω problem (10) is positivity preserving*', until again Duffin found a counterexample.

The interest concerning the behaviour of the first eigenfunction near small holes reappears in the thesis by Weinstein [32], a student of Hadamard, which was noticed by Szegö in [29], which in turn motivated Duffin and Shaffer. They considered the sign of the first eigenvalue on an annulus with a small hole. Indeed, in 1952 in a

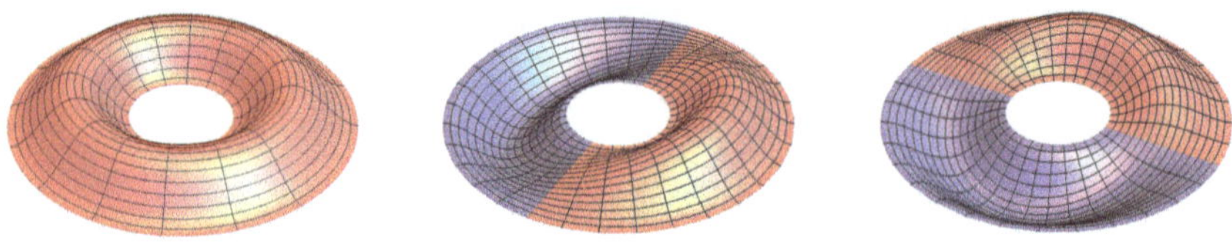

Fig. 2 Graphs of the first three eigenfunctions ϕ_1, ϕ_2 and ϕ_3 for (22). Here $\varepsilon = 0.3$

note consisting of just a few lines [13] it was announced that for $\Omega = A_\varepsilon$, where

$$A_\varepsilon = \left\{x \in \mathbb{R}^2; \varepsilon < |x| < 1\right\}, \tag{21}$$

there is a positive eigenfunction for all $\varepsilon \in (0, 1)$ of

$$\begin{cases} \Delta^2\phi = \lambda\phi & \text{in } A_\varepsilon, \\ \phi = \partial_\nu\phi = 0 \text{ on } \partial A_\varepsilon, \end{cases} \tag{22}$$

but for ε small enough it is no longer the first. Using Bessel functions and modified Bessel functions of the first and second kind, see [7] or [6] one can find almost explicit formulations for the solutions of (22). One finds that for all $\varepsilon > 0$ the first three eigenfunctions are as in Fig. 2.

Throughout this section the eigenfunction with a fixed sign we call ϕ_1 and the two with the x_1, respectively, x_2-axes as a nodal line, ϕ_2 and ϕ_3, independent of the actual order of the eigenvalues in $\mathbb{R}$. With this notation we can describe the result of Coffman, Duffin and Shaffer as follows:

Theorem 11 (Coffman, Duffin and Shaffer) *Let $A_\varepsilon \subset \mathbb{R}^2$ be as in (21) with ϕ_1, ϕ_2 and ϕ_3 as above in Fig. 2. Then there is $\varepsilon^* > 0$ such that:*

- *If $\varepsilon \in (\varepsilon^*, 1)$, then $\lambda_1 < \lambda_2 = \lambda_3$.*
- *If $\varepsilon = \varepsilon^*$, then $\lambda_1 = \lambda_2 = \lambda_3$.*
- *If $\varepsilon \in (0, \varepsilon^*)$, then $\lambda_2 = \lambda_3 < \lambda_1$.*

Remark 12 A numerical approximation shows $\varepsilon^* = 0.00131\ldots$. Moreover, all other eigenvalues are strictly larger: $\lambda_i > \max(\lambda_1, \lambda_2)$ for $i \notin \{1, 2, 3\}$, see Fig. 3 for the relation between ε and $\lambda_1, \lambda_2 = \lambda_3$.

For $\Omega \subset \mathbb{R}^2$ the space $W_0^{2,2}(\Omega)$ is imbedded in $C\left(\overline{\Omega}\right)$ but not in $C^1\left(\overline{\Omega}\right)$, so pointwise derivatives are not well-defined. As Coffman and Duffin remarked in [6], one obtains for $\varepsilon \downarrow 0$ the limit problem on $B := \left\{x \in \mathbb{R}^2; |x| < 1\right\}$:

$$\begin{cases} \Delta^2\phi = \lambda\phi & \text{in } B \setminus \{0\}, \\ \phi = \partial_\nu\phi = 0 & \text{on } \partial B, \\ \phi(0) = 0. \end{cases} \tag{23}$$

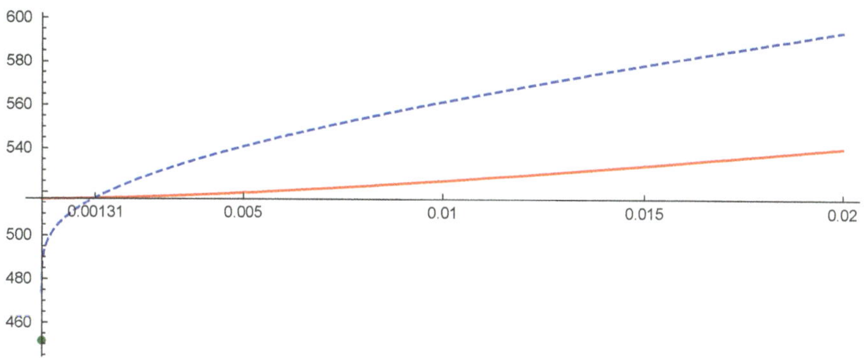

Fig. 3 Numerical approximations of the graphs $\varepsilon \mapsto \lambda_i(\varepsilon)$ of the first three eigenvalues λ_1 and $\lambda_2 = \lambda_3$ for (22) in dependence of $\varepsilon \in (0, 0.02)$. The dashed one represents $\lambda_2 = \lambda_3$. And indeed, for $\varepsilon < \varepsilon^* = 0.00131\ldots$ one finds $\lambda_2 < \lambda_1$. Moreover, the point appearing on the left axis corresponds with the second eigenvalue $(21.260\ldots)^2$ on B, as depicted in the bottom row of Fig. 1, where according to [6] the dashed line has its limit for $\varepsilon \downarrow 0$

Moreover, writing $\phi_{i,\varepsilon}$ for an eigenfunction of (22), see Fig. 4, and extending by 0 on

$$B_\varepsilon = \left\{x \in \mathbb{R}^2; |x| < \varepsilon\right\},$$

the extended eigenfunction lies in $W_0^{2,2}(B)$. In general however

$$\phi_{i,\varepsilon|_{r=\varepsilon}} = \partial_r \phi_{i,\varepsilon|_{r=\varepsilon}} = 0$$

and letting $\varepsilon \downarrow 0$ does not lead to $\nabla\phi_{i,0}(0)$ being 0. Nevertheless, for the radially symmetric eigenfunctions $\phi_{i,\varepsilon}$ the radial symmetry implies that $\nabla\phi(0) = 0$. Indeed, writing $r = |x|$ one finds through the explicit solution

$$\phi_{1,0}(r) = c_1\left(I_0(\nu r) - J_0(\nu r)\right) + c_2\left(\frac{2}{\pi}K_0(\nu r) + Y_0(\nu r)\right)$$

that

$$\phi_{1,0}(r) = -c_2\frac{\nu^2}{\pi}\log(r)\,r^2 + \mathcal{O}\left(r^2\right) \text{ for } r \downarrow 0.$$

Here J_s, I_s, Y_s and K_s are the (modified) Bessel functions of the first and second kind (Fig. 4):

$$\begin{aligned} J_s(t) &= \sum_{m=0}^{\infty} \tfrac{(-1)^m}{m!\Gamma(m+s+1)}\left(\tfrac{1}{2}t\right)^{2m+s} \quad \text{and } I_s(t) = \sum_{m=0}^{\infty} \tfrac{1}{m!\Gamma(m+s+1)}\left(\tfrac{1}{2}t\right)^{2m+s}, \\ Y_0(r) &= \lim_{s\downarrow 0}\frac{J_s(r) - J_{-s}(r)}{\pi s} \quad \text{and } K_0(r) = \lim_{s\downarrow 0}\frac{I_{-s}(t) - I_s(t)}{2s}. \end{aligned} \tag{24}$$

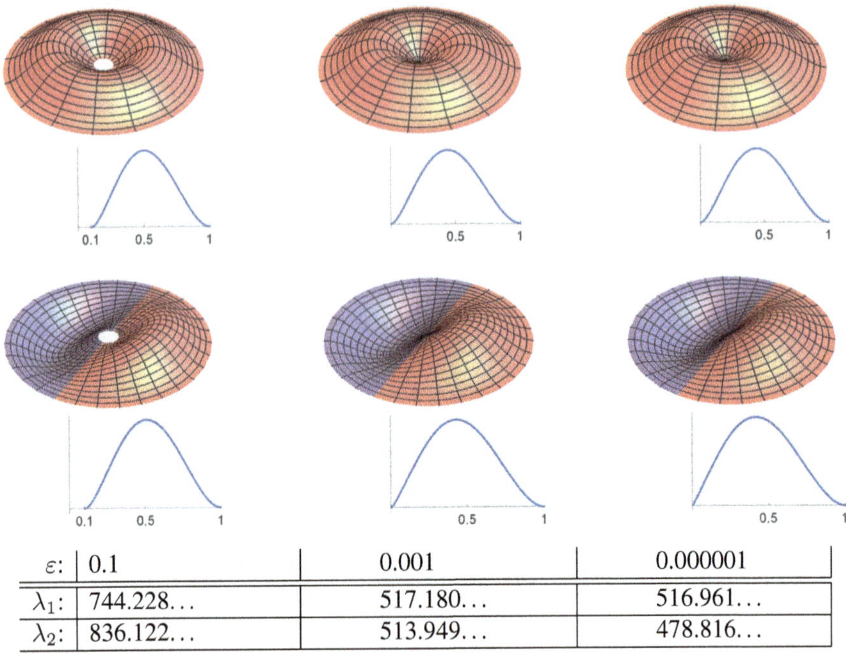

ε:	0.1	0.001	0.000001
λ_1:	744.228...	517.180...	516.961...
λ_2:	836.122...	513.949...	478.816...

Fig. 4 Graphs of $\phi_{1,\varepsilon}$ and $\phi_{2,\varepsilon}$ with ε and λ as in the table. Only $\phi_{2,\varepsilon}$ 'loses' the second boundary condition at $r = \varepsilon$ for $\varepsilon \downarrow 0$

For the eigenfunctions $\phi_{i,0}(r,\varphi) = \hat{\phi}_{i,0}(r)\cos\varphi$, with a nodal line through 0, no condition for $\hat{\phi}'(0)$ remains and

$$\hat{\phi}_{i,0}(r) = c\,r + \mathcal{O}\left(r^3\right).$$

Concerning (10) we cannot say if there is A_ε for which one finds positivity preservation in the sense that $f \geq 0$ implies $u \geq 0$. However, if we add a term λu and consider for $\Omega = A_\varepsilon$

$$\begin{cases} \Delta^2 u = \lambda u + f & \text{in } \Omega, \\ u = \partial_\nu u = 0 & \text{on } \partial\Omega, \end{cases} \tag{25}$$

then through a recent result from [25] one finds:

Theorem 13 *Set $\Omega = A_\varepsilon$. Let $\lambda_{1,\varepsilon}$ and ε^* be as in Theorem 11. Then for all $\varepsilon \in (0,\varepsilon^*) \cup (\varepsilon^*, 1)$ there exists $\delta_\varepsilon > 0$, only depending on ε, such that for all $\lambda \in \left(\lambda_{1,\varepsilon} - \delta_\varepsilon, \lambda_{1,\varepsilon}\right)$ one finds for $f \in L_2(A_\varepsilon)$ and u_λ the solution of (25) that:*

$$f \gneq 0 \implies u_\lambda > 0.$$

Remark 14 From the existence of a simple strictly positive eigenfunction with eigenvalue λ_1 one concludes a positivity preserving property on $(\lambda_1 - \delta, \lambda_1)$. As such one might call this result a converse to the Krein–Rutman theorem [21].

Remark 15 This theorem is a special case of a more general result stated in [25]. Its proof is based on an estimate from below for the Green function for (25), which in turn is based on [18], and a Weyl type result for the growth rate of the eigenvalues.

For the solution on $\Omega = A_0$ however, no result as in Theorem 13 holds. As Coffman remarked in [6], one loses $u'(0) = 0$ and the (strong) formulation would be

$$\begin{cases} \Delta^2 u = \lambda u + f \text{ on } A_0 = \left\{x \in \mathbb{R}^2; 0 < |x| < 1\right\}, \\ u = \partial_\nu u = 0 \quad \text{on } \left\{x \in \mathbb{R}^2; |x| = 1\right\}, \\ u(0) = 0. \end{cases} \tag{26}$$

The more appropriate weak formulation is, again with $\Omega = A_0$,

$$u \in W_0^{2,2}(\Omega) \text{ with } \int_\Omega (\Delta u \, \Delta\psi - (\lambda u + f)\, \psi)\, dx = 0 \text{ for all } \psi \in W_0^{2,2}(\Omega). \tag{27}$$

By Riesz there is for $f \in L_2(A_0)$, which identifies with $L_2(B)$, and $\lambda \neq \lambda_{i,0}$, a unique weak solution u of (27). By the imbedding of $W^{2,2}(B)$ in $C\left(\overline{B}\right)$ one finds $u(0) = 0$.

Lemma 16 *Set $\Omega = A_0$. For all $\lambda \in \mathbb{R} \setminus \left\{\lambda_{1,0}\right\}$ there is $f \gneq 0$ such that a (weak) solution u of (26) satisfies $u(x) < 0$ for some $x \in A_0$.*

Proof

- For $\lambda > \lambda_{1,0}$ one takes $f = \phi_{1,0} > 0$ to find $u = \left(\lambda_{1,0} - \lambda\right)^{-1} \phi_{1,0} < 0$.
- For $\lambda = 0$ we may use Boggio's formula for the disk to find a Green function for

$$\begin{cases} \Delta^2 u = f \quad \text{in } D = \left\{x \in \mathbb{R}^2; |x| < 1 \text{ and } x_1 > 0\right\}, \\ u = \partial_\nu u = 0 \text{ on } \left\{x \in \mathbb{R}^2; |x| = 1 \text{ and } x_1 > 0\right\}, \\ u = \Delta u = 0 \text{ on } \left\{x \in \mathbb{R}^2; |x| < 1 \text{ and } x_1 = 0\right\}. \end{cases} \tag{28}$$

Indeed, (28) is solved by

$$u(x) = \int_D G_{2,D}(x, y)\, f(y)\, dy \tag{29}$$

for $y^* = (-y_1, y_2)$ and through $G_{2,D}(x,y) = G_{\text{Boggio}}(x,y) - G_{\text{Boggio}}(x,y^*)$, with G_{Boggio} from [4]

$$G_{2,D}(x,y) = \tfrac{1}{16\pi}\left(\left|x-y^*\right|^2 \log\left(1+\tfrac{(1-|x|^2)(1-|y|^2)}{|x-y^*|^2}\right)\right.$$
$$\left. - |x-y|^2 \log\left(1+\tfrac{(1-|x|^2)(1-|y|^2)}{|x-y|^2}\right)\right).$$

Since $t \mapsto t\log\left(1+\frac{1}{t}\right)$ is increasing and $|x-y^*| > |x-y|$ for $x, y \in D$ one finds $G_{2,D}(x,y) > 0$ for $x, y \in D$. So taking any function $f_a \in C_0^\infty(D)$ with $f_a \geq 0$ and extending antisymmetrically by $f_a(-x_1, x_2) = -f_a(x_1, x_2)$, we find a solution u_a for $f = f_a$ of

$$\begin{cases} \Delta^2 u = f \quad \text{in } B = \left\{x \in \mathbb{R}^2; |x| < 1\right\}, \\ u = \partial_\nu u = 0 \text{ on } \partial B, \end{cases} \tag{30}$$

which is positive on D, and satisfies $u_a(x_1, x_2) = -u_a(-x_1, x_2)$. One finds moreover that $\partial_{x_1} u_a(0) > 0$. Next we take $f_b = c\phi_{1,0}$ with c such that $f_b \geq f_a$ on D. Then $f = f_b - f_a \geq 0$ on A_0, while the sign of the solution $u = \lambda_{1,0}^{-1} f_b - u_a$ of (26) near 0 is determined by u_a since $\phi_{1,0}(r) = \mathcal{O}\left(r^2 \log r\right)$.

- For $\lambda \in \left(0, \min\left(\lambda_{1,0}, \lambda_{2,0}\right)\right)$ one takes the same right-hand side. One finds

$$u_\lambda = \frac{1}{\lambda_{1,0} - \lambda} f_b - \sum_{k=0}^{\infty} \lambda^k \mathcal{G}_{2,D}^{k+1} f_a$$

with the same arguments concerning the sign near 0. Here $\mathcal{G}_{2,D}$ is the positive operator determined by (29), and the series converges for $\lambda \leq \lambda_{2,0}$.

- When $\lambda_{2,0} < \lambda_{1,0}$ we have to consider $\lambda \in \left(\lambda_{2,0}, \lambda_{1,0}\right)$. One of the eigenfunctions for $\lambda_{2,0}$ is $\phi_2(r,\varphi) = \hat{\phi}_2(r)\cos\varphi$ and the other one is $\phi_3(r,\varphi) = \hat{\phi}_2(r)\sin\varphi$. All other eigenfunctions have eigenvalues larger than $\lambda_{1,0}$. We consider

$$f_{a,\varepsilon}(r,\varphi) = \psi_\varepsilon(r)\,\phi_k(r,\varphi), \tag{31}$$

where $\phi_k(r,\varphi) = \hat{\phi}_k(r)\cos\varphi$ is the second eigenfunction in the class $\{\hat{\phi}(r)\cos\varphi\}$. The functions $\psi_\varepsilon \in C^\infty[0,1]$ are taken such that

$$\text{support}(\psi_\varepsilon) \subset [\varepsilon, 1], \qquad \lim_{\varepsilon \downarrow 0} \|\psi_\varepsilon - 1\|_{L_2((0,1),rdr)} = 0$$

$$\text{and} \quad \left\langle \hat{\phi}_2, \psi_\varepsilon \hat{\phi}_k \right\rangle_{L_2((0,1),rdr)} = 0.$$

Then $\langle \phi_2, f_{a,\varepsilon} \rangle_{L_2(A_0)} = \langle \phi_3, f_{a,\varepsilon} \rangle_{L_2(A_0)} = 0$. Hence with these assumptions we find that for $f = f_{a,\varepsilon}$ from (31) there is a unique solution $u_{\lambda,a,\varepsilon}$ of (27) for all $\lambda < \lambda_{1,0}$. Since for right-hand sides in the class $\{\hat{\phi}(r) \cos \varphi\}$ the solution of (25) for $\Omega = B$ and (26) coincides, we find that $u_{\lambda,a,\varepsilon} \in C^\infty(\overline{B})$ and $\left\| u_{\lambda,a,\varepsilon} - \frac{1}{\lambda - \lambda_k} \phi_k \right\|_{W^{4,2}(B)} \to 0$ for $\varepsilon \downarrow 0$. Since for $n = 2$ the space $W^{4,2}(B)$ is imbedded in $C^1(\overline{B})$, it follows for some $\varepsilon_0 > 0$ that $\partial_{x_1} u_{\lambda,a,\varepsilon}(0)$ and $\partial_{x_1} \phi_k(0) \neq 0$ have the same sign. Fixing $f_a = f_{a,\varepsilon_0}$ we may proceed as for the case $\lambda < \lambda_1$ with $f = c\phi_1 - f_{a,\varepsilon_0} \geq 0$ to find that the solution takes the signs of ϕ_k near 0.

- For $\lambda < 0$ one argues by contradiction. If there is a positive solution operator $\mathcal{G}_{A_0,\lambda}$ for (27), then

$$\mathcal{G}_{A_0,0} = \sum_{k=0}^{\infty} |\lambda|^k \mathcal{G}_{A_0,\lambda}^{k+1} \geq \mathcal{G}_{A_0,\lambda} \geq 0,$$

contradicting the case $\lambda = 0$. □

4 Navier-Bilaplace on a Nonconvex Sector

In this section we will consider for nonconvex sectors

$$C_\alpha = \left\{ (r \cos \varphi, r \sin \varphi) ; 0 < r < 1 \text{ and } |\varphi| < \tfrac{1}{2}\alpha \right\} \tag{32}$$

with $\alpha \in (\pi, 2\pi)$ the Navier-bilaplace eigenvalue problem

$$\begin{cases} \Delta^2 \phi = \lambda \phi & \text{in } C_\alpha, \\ \phi = \Delta \phi = 0 \text{ on } \partial C_\alpha. \end{cases} \tag{33}$$

If one just considers this pointwise definition for eigenvalues, there is a problem in $x = 0$. Taking ϕ_1 from (3) one finds

$$\phi_1 \in C_0\left(\overline{C_\alpha}\right) \cap C^\infty(C_\alpha) \cap W_0^{1,2}(\Omega) \text{ but } \phi_1 \notin C^2\left(\overline{C_\alpha}\right) \cup \mathcal{W}_0^2(\Omega),$$

and $\Delta\phi_1$ in 0 can only be defined through extending $-\Delta\phi_1 = \lambda_{1,(3)}\phi_1$ up to the boundary and using $\phi_1(0) = 0$.

So should we assume $\Delta\phi = 0$ only on $\partial C_\alpha \setminus \{0\}$? Then, however, we have at least two positive eigenfunctions for (33), see Fig. 5.

In other words, one finds that the pointwise description in (33) is not precise. To find a well-defined setting one should fix the boundary values through the function spaces. Indeed, the eigenfunction on the left in Fig. 5 is the unique minimizer of R_2 on $W_0^{1,2}(C_{3\pi/2})$, which means for the Navier-bilaplace eigenvalue problem and

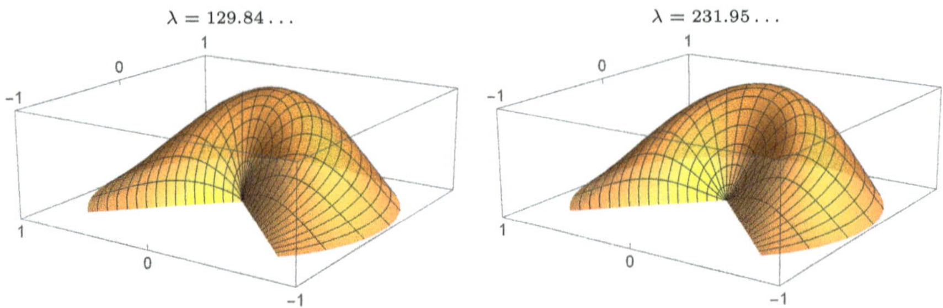

Fig. 5 Two positive eigenfunctions for (33) with $\alpha = \frac{3}{2}\pi$. On the left the one for (34); on the right for (35)

$\alpha = \frac{3}{2}\pi$ that

$$\begin{cases} \Delta^2\phi = \lambda\phi & \text{in } C_\alpha, \\ \phi = \Delta\phi = 0 & \text{on } \partial C_\alpha \setminus \{0\}, \\ \phi, \Delta\phi \in W_0^{1,2}(C_\alpha). \end{cases} \tag{34}$$

The one on the right is the unique minimizer of R_4 on $\mathcal{W}_0^2(C_{3\pi/2})$, i.e., for $\alpha = \frac{3}{2}\pi$ it satisfies:

$$\begin{cases} \Delta^2\phi = \lambda\phi & \text{in } C_\alpha, \\ \phi = \Delta\phi = 0 & \text{on } \partial C_\alpha \setminus \{0\}, \\ \phi \in \mathcal{W}_0^2(C_\alpha). \end{cases} \tag{35}$$

4.1 The $W^{1,2} \times W^{1,2}$-System Setting

This case is described in (34) and the eigenfunctions coincide with those of the second order Dirichlet–Laplace problem. The explicit eigenfunctions one obtains by a separation-ansatz for (3). Indeed we find the corresponding eigenfunctions in polar coordinates through

$$\phi_{n,k}(r,\varphi) = J_{n\frac{\pi}{\alpha}}\left(\nu_{n\frac{\pi}{\alpha},k}\, r\right)\sin\left(n\frac{\pi}{\alpha}\left(\varphi + \tfrac{1}{2}\alpha\right)\right), \tag{36}$$

where $J_{n\frac{\pi}{\alpha}}$ is the standard Bessel function as in (24), and $\nu_{s,k}$ is the kth positive zero of J_s. The sinus functions make a complete orthogonal system for $L_2\left(-\frac{1}{2}\alpha, \frac{1}{2}\alpha\right)$ and we are left to find a complete system in the radial direction for each $n \in \mathbb{N}^+$. This is where the Bessel functions $J_{n\frac{\pi}{\alpha}}$ come in.

Since

$$\int_0^1 \left(t^{s-1}\right)^2 t\, dt < \infty \Leftrightarrow s > 0,$$

and with $s = n\frac{\pi}{\alpha}$ these eigenfunctions $\phi_{n,k}$ lie in $W_0^{1,2}(C_\alpha)$ for $n \in \mathbb{N}^+$. One indeed finds

$$-\Delta\phi_{n,k}(r,\varphi) = \nu^2_{n\frac{\pi}{\alpha},k}\phi_{n,k}(r,\varphi).$$

Setting $c_{n,k} = \left\|\phi_{n,k}\right\|^{-1}_{L_2(C_\alpha)}$ and

$$\Phi_{\alpha,n,k} = c_{n,k}\phi_{n,k} \tag{37}$$

we obtain:

Lemma 17 *For C_α as in (32) with $\alpha \in (0, 2\pi)$ the set*

$$\left\{\Phi_{\alpha,n,k}\right\}_{n,k\in\mathbb{N}^+} \subset W_0^{1,2}(C_\alpha)$$

is a complete orthonormal system in $L_2(C_\alpha)$ of eigenfunctions for (34) with $\lambda = \lambda_{\alpha,n,k} = \nu^4_{n\frac{\pi}{\alpha},k}$. The smallest eigenvalue is $\lambda_{\alpha,1,1}$.

The problem with a right-hand side $f \in L_2(C_\alpha)$ corresponding to the setting of (34) as in (19) is

$$\begin{cases} \Delta^2 u = \lambda u + f & \text{in } C_\alpha, \\ u, \Delta u \in W_0^{1,2}(C_\alpha). \end{cases} \tag{38}$$

Corollary 18 *For each $f \in L_2(C_\alpha)$ the solution u_λ of (38) satisfies*

$$u = \sum_{n,k=1}^{\infty} \frac{1}{\lambda_{\alpha,n,k} - \lambda} \left\langle \Phi_{\alpha,n,k}, f\right\rangle \Phi_{\alpha,n,k}.$$

Remark 19 Note that this solution is not in $\mathcal{W}_0^2(C_\alpha)$ generically. Since

$$\int_0^1 \left(t^{s-2}\right)^2 t\, dt < \infty \Leftrightarrow s > 1 \tag{39}$$

one finds that $\Phi_{\alpha,n,k} \in \mathcal{W}_0^2(C_\alpha)$ if and only if $n\frac{\pi}{\alpha} > 1$. For $\alpha \in (\pi, 2\pi)$ this inequality is satisfied if and only if $n \geq 2$.

Concerning positivity preservation for (38) we have, see also [28]:

Lemma 20 *Let C_α be as in (32) with $\alpha \in (0, 2\pi)$ and set $\lambda_{\alpha,1,1}$ as in Lemma 17. Then there exists $\lambda_{\alpha,c} < 0$ such that for all $\lambda \in \left(\lambda_{\alpha,c}, \lambda_{\alpha,1,1}\right)$ the following holds for $f \in L_2(C_\alpha)$ and u_λ the corresponding solution of (38):*

$$f \gneq 0 \Longrightarrow u_\lambda > 0.$$

Proof Let us write $(-\Delta + \mu)_0^{-1} : L_2(C_\alpha) \to W_0^{1,2}(C_\alpha)$ for the solution operator of

$$\begin{cases} -\Delta u + \mu u = f & \text{in } C_\alpha, \\ \quad u \in W_0^{1,2}(C_\alpha). \end{cases}$$

For $\lambda \in \left[0, \lambda_{\alpha,1,1}\right)$ the positivity preserving property follows from the maximum principle for $(-\Delta + \mu)_0^{-1}$, whenever $\mu < \mu_1$, with here $\mu_1 = \sqrt{\lambda_{\alpha,1,1}}$, the first eigenvalue for the Dirichlet Laplace. Indeed, one uses

$$u_\lambda = \left(-\Delta - \sqrt{\lambda}\right)_0^{-1} \left(-\Delta + \sqrt{\lambda}\right)_0^{-1} f.$$

For $\lambda \in \left(-\lambda_{\alpha,1,1}, 0\right)$ and setting $\mathcal{G} = (-\Delta)_0^{-1}$ one finds

$$u_\lambda = \sum_{k=0}^{\infty} \left(\lambda \mathcal{G}^2\right)^k \mathcal{G} f = \sum_{k=0}^{\infty} \left(\lambda^2 \mathcal{G}^4\right)^k \left(\mathcal{I} + \sqrt{-\lambda}\mathcal{G}\right)\left(\mathcal{I} - \sqrt{-\lambda}\mathcal{G}\right) \mathcal{G} f. \tag{40}$$

By the 3G-theorem from Cranston et al. [10], which holds on bounded Lipschitz domains, one finds that $\left(\mathcal{I} - \sqrt{-\lambda}\mathcal{G}\right)\mathcal{G}$ is a positivity preserving operator for $\sqrt{-\lambda}$ small enough. The remaining terms in (40) are positive and the series converges when $|\lambda| < \lambda_{\alpha,1,1}$. □

4.2 The $W^{2,2} \times L_2$-Equation Setting

This case refers to (35) and again we start with the almost explicit eigenfunctions.

Of the eigenfunctions that we found for (34) only the ones with $n \geq 2$ lie in $\mathcal{W}_0^2(C_\alpha)$. In [23] the solution operator for solutions in $\mathcal{W}_0^2(C_\alpha)$ is given by (20), which means for the sector C_α with $\alpha \in (\pi, 2\pi)$ by

$$\mathcal{G}_{2,\text{Navier}} = \mathcal{G}_1^2 - \mathcal{G}_1 \mathcal{P}_\zeta \mathcal{G}_1 \tag{41}$$

with $\mathcal{P}_\zeta$ the projection in L_2 on ζ, defined by

$$\zeta_\alpha(r,\varphi) = c_\alpha\left(r^{-\pi/\alpha} - r^{\pi/\alpha}\right)\cos\left(\frac{\pi}{\alpha}\varphi\right) \tag{42}$$

and $c_\alpha = \|\zeta\|^{-1}_{L_2(C_\alpha)}$. Also the operator $\mathcal{G}_{2,\text{Navier}}$ is symmetric and compact, moreover

$$\left\langle \mathcal{G}_{2,\text{Navier}} f, g\right\rangle_{L_2(C_\alpha)} = \left\langle \left(\mathcal{I} - \mathcal{P}_\zeta\right)\mathcal{G}_1 f, \left(\mathcal{I} - \mathcal{P}_{\zeta_\alpha}\right)\mathcal{G}_1 g\right\rangle_{L_2(C_\alpha)}$$

implies $\left\langle \mathcal{G}_{2,\text{Navier}} f, f\right\rangle_{L_2(C_\alpha)} \geq 0$, so the spectrum consists of nonnegative eigenvalues. Moreover, 0 is not an eigenvalue since $\zeta_\alpha \notin W_0^{1,2}(C_\alpha)$.

Since $\left\langle \zeta_\alpha, \Phi_{\alpha,n,k}\right\rangle = 0$ for $n \geq 2$, indeed $\left\{\Phi_{\alpha,n,k}\right\}_{n,k\in\mathbb{N}^+, n\geq 2}$ is an orthonormal system in $L_2(C_\alpha)$ of eigenfunctions for $\mathcal{G}_{2,\text{Navier}}$. We may split $L_2(C_\alpha)$ by

$$L_2(C_\alpha) = \mathcal{L}_I \oplus \mathcal{L}_{II}$$

with $\mathcal{L}_I = \text{Span}\left(\Phi_{\alpha,1,k}; k \geq 1\right)$ and $\mathcal{L}_{II} = \text{Span}\left(\Phi_{\alpha,n,k}; n \geq 2, k \geq 1\right)$. It remains to find a complete orthonormal system for $\mathcal{L}_I$ of eigenfunctions for $\mathcal{G}_{2,\text{Navier}}$. Since $\Delta\zeta_\alpha = 0$ we find that

$$\lambda\mathcal{G}_{2,\text{Navier}}\left(R(r)\cos\left(\frac{\pi}{\alpha}\varphi\right)\right) = R(r)\cos\left(\frac{\pi}{\alpha}\varphi\right)$$

turns into

$$\left(\partial_r^2 + \frac{1}{r}\partial r - \left(\frac{\pi}{\alpha r}\right)^2\right)^2 R(r) = \lambda R(r). \tag{43}$$

Besides the Bessel functions $J_{\frac{\pi}{\alpha}}(\nu r)$ and $J_{-\frac{\pi}{\alpha}}(\nu r)$ one obtains the modified Bessel functions $I_{\frac{\pi}{\alpha}}(\nu r)$ and $I_{-\frac{\pi}{\alpha}}(\nu r)$ for the solutions of (43), see (24). Note that $\pi/\alpha \in \left(\frac{1}{2}, 1\right)$ is noninteger and hence these four solutions are independent. With $s = \frac{\pi}{\alpha} \in \left(\frac{1}{2}, 1\right)$ and the condition that the solution lies in $\mathcal{W}_0^2(C_\alpha)$ the first coefficient in the series has to be larger than 1 as in (39). That leaves the combinations of two functions:

$$R(r) = c_1\left(I_{\frac{\pi}{\alpha}}(\nu r) - J_{\frac{\pi}{\alpha}}(\nu r)\right) + c_2\left(I_{-\frac{\pi}{\alpha}}(\nu r) - J_{-\frac{\pi}{\alpha}}(\nu r)\right).$$

Indeed

$$I_{\frac{\pi}{\alpha}}(\nu r) - J_{\frac{\pi}{\alpha}}(\nu r) = \frac{2}{\Gamma(2+\frac{\pi}{\alpha})}(\nu r)^{2+\frac{\pi}{\alpha}} \text{ and}$$

$$I_{-\frac{\pi}{\alpha}}(\nu r) - J_{-\frac{\pi}{\alpha}}(\nu r) = \frac{2}{\Gamma(2+\frac{\pi}{\alpha})}(\nu r)^{2-\frac{\pi}{\alpha}}.$$

One finds

$$-\Delta\left(J_{\pm\frac{\pi}{\alpha}}(\nu r)\cos\left(\frac{\pi}{\alpha}\varphi\right)\right)=\nu^2 J_{\pm\frac{\pi}{\alpha}}(\nu r)\cos\left(\frac{\pi}{\alpha}\varphi\right),$$
$$-\Delta\left(I_{\pm\frac{\pi}{\alpha}}(\nu r)\cos\left(\frac{\pi}{\alpha}\varphi\right)\right)=-\nu^2 I_{\pm\frac{\pi}{\alpha}}(\nu r)\cos\left(\frac{\pi}{\alpha}\varphi\right),$$

and in order to satisfy the boundary conditions for $r=1$ one needs

$$\det\begin{pmatrix} I_{\frac{\pi}{\alpha}}(\nu)-J_{\frac{\pi}{\alpha}}(\nu) & I_{-\frac{\pi}{\alpha}}(\nu)-J_{-\frac{\pi}{\alpha}}(\nu) \\ I_{\frac{\pi}{\alpha}}(\nu)+J_{\frac{\pi}{\alpha}}(\nu) & I_{-\frac{\pi}{\alpha}}(\nu)+J_{-\frac{\pi}{\alpha}}(\nu)\end{pmatrix}=0,$$

which simplifies to

$$f_\alpha(\nu):=J_{\frac{\pi}{\alpha}}(\nu)-\frac{I_{\pi/\alpha}(\nu)}{I_{-\pi/\alpha}(\nu)}J_{-\frac{\pi}{\alpha}}(\nu)=0. \tag{44}$$

The function $I_{\pi/\alpha}(\nu)\,/I_{-\pi/\alpha}(\nu)$ is strictly positive on $\mathbb{R}^+$ and converges to 1 for $\nu\to\infty$. Since

$$J_{\pm\frac{\pi}{\alpha}}(\nu)=\sqrt{\frac{2}{\pi\nu}}\left(\cos\left(\nu\mp\frac{\pi^2}{2\alpha}-\frac{\pi}{4}\right)+\mathcal{O}\left(\frac{1}{\nu}\right)\right)\text{ for }\nu\to\infty,$$

the function f_α from (44) has countably many zeroes $\left\{\tilde{\nu}_{\frac{\pi}{\alpha},k}\right\}_{k=1}^{\infty}$ and

$$\tilde{\nu}_{\frac{\pi}{\alpha},k}=\left(k+\tfrac{1}{4}\right)\pi+\mathcal{O}\left(\tfrac{1}{k}\right)\text{ for }k\to\infty.$$

We define for $k\in\mathbb{N}^+$ and $\nu=\tilde{\nu}_{\frac{\pi}{\alpha},k}$ the function

$$\begin{aligned}\tilde{\phi}_{1,k}(r,\varphi)=\Big(&\left(I_{-\frac{\pi}{\alpha}}(\nu)-J_{-\frac{\pi}{\alpha}}(\nu)\right)\left(I_{\frac{\pi}{\alpha}}(\nu r)-J_{\frac{\pi}{\alpha}}(\nu r)\right)\\ &-\left(I_{\frac{\pi}{\alpha}}(\nu)-J_{\frac{\pi}{\alpha}}(\nu)\right)\left(I_{-\frac{\pi}{\alpha}}(\nu r)-J_{-\frac{\pi}{\alpha}}(\nu r)\right)\Big)\cos\left(\frac{\pi}{\alpha}\varphi\right),\end{aligned}$$

the constant $c_{1,k}=\left\|\tilde{\phi}_{1,k}\right\|_{L_2(C_\alpha)}^{-1}$ and set

$$\tilde{\Phi}_{\alpha,1,k}(r,\varphi)=c_{1,k}\tilde{\phi}_{1,k}(r,\varphi). \tag{45}$$

The corresponding eigenvalue is $\tilde{\lambda}_{\alpha,1,k}=\left(\tilde{\nu}_{\frac{\pi}{\alpha},k}\right)^4$.

For $n \geq 2$ we set $\tilde{\Phi}_{\alpha,n,k}(r,\varphi) = \Phi_{\alpha,n,k}(r,\varphi)$ and $\tilde{\nu}_{n\frac{\pi}{\alpha},k} = \nu_{n\frac{\pi}{\alpha},k}$ as before in (37). Then we may state:

Lemma 21 *For C_α as in (32) with $\alpha \in (\pi, 2\pi)$ the set*

$$\left\{\tilde{\Phi}_{\alpha,n,k}\right\}_{n,k\in\mathbb{N}^+} \subset \mathcal{W}_0^2(C_\alpha)$$

is a complete orthonormal system in $L_2(C_\alpha)$ of eigenfunctions for (35) with eigenvalues $\tilde{\lambda}_{\alpha,n,k} = \left(\tilde{\nu}_{n\frac{\pi}{\alpha},k}\right)^4$.

Corollary 22 *For each $f \in L_2(C_\alpha)$ the solution $u = \mathcal{G}_{2,Navier}f$ of (25) satisfies*

$$u = \sum_{n,k=1}^{\infty} \frac{1}{\tilde{\lambda}_{\alpha,n,k} - \lambda} \left\langle \tilde{\Phi}_{\alpha,n,k}, f \right\rangle \tilde{\Phi}_{\alpha,n,k}.$$

Remark 23 For all $\alpha \in (\pi, 2\pi)$ the only eigenvalue in $\{\tilde{\lambda}_{\alpha,n,k}\}_{n,k\in\mathbb{N}^+}$ with a positive eigenfunction is $\tilde{\lambda}_{\alpha,1,1}$. By numerical computations one finds that there is a constant $c^* \in (1,2)$ such that (Figs. 6 and 7):

- for $\alpha \in (\pi, c^*\pi)$ the unique smallest eigenvalue is $\tilde{\lambda}_{\alpha,1,1}$;
- for $\alpha = c^*\pi$ one finds that $\tilde{\lambda}_{\alpha,1,1} = \tilde{\lambda}_{\alpha,2,1}$;
- for $\alpha \in (c^*\pi, 2\pi)$ one finds that $\tilde{\lambda}_{\alpha,2,1} < \tilde{\lambda}_{\alpha,1,1}$.

A numerical approximation yields $c^* = 1.87369319\ldots$

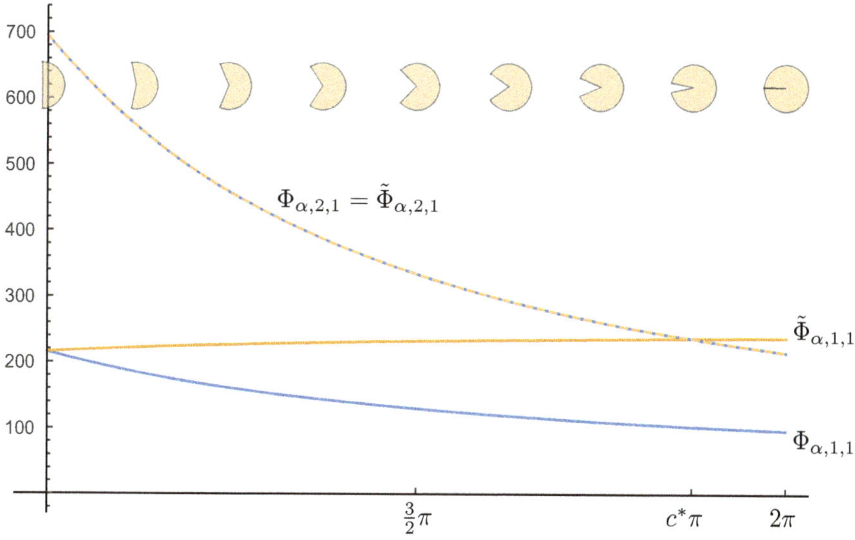

Fig. 6 The first eigenvalues on C_α both for (34) and for (35) as a function of $\alpha \in (\pi, 2\pi)$

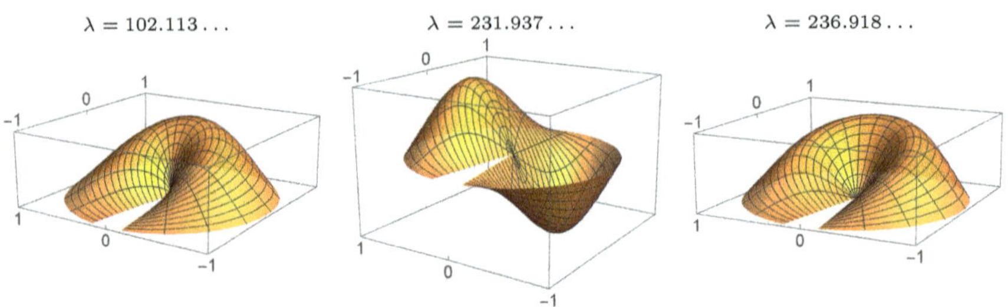

Fig. 7 Here $\alpha = 1.9\,\pi$. The left two are graphs of the eigenfunctions for (34) that correspond with the first and second eigenvalue. The right two are graphs of the eigenfunctions for (35) that correspond with the first and second eigenvalue. Indeed, the middle one is the second for (34) and the first for (35)

Concerning a positivity preserving property for

$$\begin{cases} \Delta^2 u = \lambda u + f \quad \text{in } C_\alpha, \\ u = \Delta u = 0 \text{ on } \partial C_\alpha \setminus \{0\}, \\ u \in \mathcal{W}_0^2(C_\alpha), \end{cases} \tag{46}$$

one should compare the behaviour of the positive eigenfunction near 0 and the lowest one with the x_1-axis as a nodal line. For these two eigenfunctions one finds, with $c_1, c_2 \neq 0$:

$$\tilde{\Phi}_{\alpha,1,1}(r,\varphi) = \left(c_1\, r^{2-\frac{\pi}{\alpha}} + \mathcal{O}\left(r^{2+\frac{\pi}{\alpha}}\right)\right)\cos\left(\frac{\pi}{\alpha}\varphi\right), \tag{47}$$

$$\tilde{\Phi}_{\alpha,2,1}(r,\varphi) = \left(c_2\, r^{2\frac{\pi}{\alpha}} + \mathcal{O}\left(r^{2+2\frac{\pi}{\alpha}}\right)\right)\sin\left(2\frac{\pi}{\alpha}\varphi\right). \tag{48}$$

Lemma 24 *Let C_a be as in (32) with $\left(\frac{3}{2}\pi, \pi\right)$. Then for each $\lambda \neq \tilde{\lambda}_{\alpha,1,1}$ there exists $f \in L_2(C_\alpha)$ with $f > 0$ and u_λ a solution of (46) satisfying $u_\lambda(x) < 0$ for some $x \in C_\alpha$.*

Proof This can be proven like for the annulus with $\lambda \in (\lambda_2, \lambda_1)$. We consider for either $k = 1$ or $k = 2$:

$$f_\varepsilon(r,\varphi) = \psi_\varepsilon(r)\,\tilde{\Phi}_{\alpha,2,k}(r,\varphi) + C_\varepsilon\,\tilde{\Phi}_{\alpha,1,1}(r,\varphi) \geq 0, \tag{49}$$

where $\psi_\varepsilon \in C^\infty(\mathbb{R})$ with $\psi_\varepsilon(r) = 0$ for $r < \varepsilon$, $\left\langle \psi_\varepsilon \tilde{\Phi}_{\alpha,2,1}, \tilde{\Phi}_{\alpha,2,2}\right\rangle_{L_2(C_\alpha)} = 0$ and

$$\|\psi_\varepsilon - 1\|_{L_2((0,r),rdr)} \to 0 \text{ for } \varepsilon \downarrow 0.$$

For $k = 1$ one has to avoid $\tilde{\lambda}_{\alpha,2,1}$ and for $k = 2$ the value $\tilde{\lambda}_{\alpha,2,2}$. Since $\tilde{\lambda}_{\alpha,2,1} < \tilde{\lambda}_{\alpha,2,2}$ this covers $\mathbb{R} \setminus \{\tilde{\lambda}_{\alpha,1,1}\}$. The solution for the first part of (49) converges to a multiple of $\tilde{\Phi}_{\alpha,2,k}$, while the solution for the second part is a multiple of $\tilde{\Phi}_{\alpha,1,1}$. Using the growth rates near 0 from (47) and (48), which also holds for $\tilde{\Phi}_{\alpha,2,2}$, and remarking that

$$2\frac{\pi}{\alpha} < 2 - \frac{\pi}{\alpha} \text{ for } \alpha > \frac{3}{2}\pi,$$

one finds that the sign of u_λ near 0 is determined by $\tilde{\Phi}_{\alpha,2,k}$. Hence u_λ changes sign. □

Onc is left with the case $\alpha \in \left(\pi, \frac{3}{2}\pi\right)$. Although numerical evidence [23] supports the following conjecture for $\alpha^* \approx 1.2\pi$, a proof is still missing.

Conjecture 25 *Let C_a be as in (32). Then there exists $\alpha^* \in \left(\pi, \frac{3}{2}\pi\right)$, such that for each $\alpha \in (\pi, \alpha^*)$ and for each $\lambda \in \left[0, \tilde{\lambda}_{\alpha,1,1}\right)$ one finds*

$$f \gneq 0 \Longrightarrow u_\lambda > 0$$

for $f \in L_2(C_\alpha)$ and u_λ the corresponding solution of (46).

In order to prove this conjecture one has to show that $\mathcal{G}_{2,\text{Navier}} \geq 0$ and one knows by (41) that this coincides with

$$\mathcal{G}_{C_\alpha}^2 - \mathcal{G}_{C_\alpha}\mathcal{P}_{\zeta_\alpha}\mathcal{G}_{C_\alpha} \geq 0. \tag{50}$$

Here $\mathcal{G}_{C_\alpha}$ is the Dirichlet–Laplace Green function for $\Omega = C_\alpha$ and ζ_α the function in (42). Using the kernel expression G_{C_α}, one finds that (50) coincides with: for all $x, y \in C_\alpha$

$$\int_{C_\alpha} G_{C_\alpha}(x,z)\, G_{C_\alpha}(z,y)\, dz \geq \int_{C_\alpha} G_{C_\alpha}(x,z)\, \zeta_\alpha(z)\, dz \int_{C_\alpha} G_{C_\alpha}(z,y)\, \zeta_\alpha(z)\, dz. \tag{51}$$

For the right-hand side of (51) we have, again switching from z to (r, φ),

$$\zeta_\alpha(r,\varphi) = \frac{\sqrt{\alpha^2-\pi^2}}{\sqrt{\alpha/2\pi}}\left(r^{-\pi/\alpha} - r^{\pi/\alpha}\right)\cos\left(\frac{\pi}{\alpha}\varphi\right),$$

$$\left(\mathcal{G}_{C_\alpha}\zeta_\alpha\right)(r,\varphi) = \frac{\sqrt{\alpha/2}}{\sqrt{\alpha^2-\pi^2}}\left(r^{\pi/\alpha} - \tfrac{\alpha+\pi}{2\pi}r^{2-\pi/\alpha} + \tfrac{\alpha-\pi}{2\pi}r^{2+\pi/\alpha}\right)\cos\left(\frac{\pi}{\alpha}\varphi\right).$$

By the transformation $z \mapsto z^{\pi/\alpha}$ from C_α to $D := \{x \in B; x_1 > 0\}$, one finds that (51) transforms into

$$\int_D G_D(x,z)\, G_D(z,y)\left(\frac{\alpha}{\pi}\right)^2 |z|^{2\alpha/\pi-2}\, dz \geq g_\alpha(r_x,\varphi_x)\, g_\alpha\left(r_y,\varphi_y\right) \tag{52}$$

for all $x, y \in D$, where for $y^* = (-y_1, y_2)$

$$G_D(x, y) = \frac{1}{4\pi} \log \left(\frac{1 + \left(1 - |x|^2\right)\left(1 - |y|^2\right) |x - y|^{-2}}{1 + \left(1 - |x|^2\right)\left(1 - |y|^2\right) |x - y^*|^{-2}} \right)$$

and with $(x_1, x_2) = (r_x \cos(\varphi_x), r_x \sin(\varphi_x))$ and similarly for y

$$g_\alpha(r_x, \varphi_x) = \left(\mathcal{G}_{C_\alpha} \zeta_\alpha\right) \left(r_x^{\alpha/\pi}, \tfrac{\alpha}{\pi}\varphi_x\right)$$

$$= \frac{\sqrt{\alpha/2}}{\sqrt{\alpha^2 - \pi^2}} \left(r_x - \tfrac{\alpha+\pi}{2\pi} r_x^{2\alpha/\pi - 1} + \tfrac{\alpha-\pi}{2\pi} r_x^{2\alpha/\pi + 1}\right) \cos(\varphi_x).$$

Whether or not and for which α (52) holds true has not been proven at the time this manuscript was submitted.

References

1. R.A. Adams, J. Fournier, *Sobolev Spaces*, 2nd edn. (Academic, New York, 2003)
2. S. Agmon, Asymptotic formulas with remainder estimates for eigenvalues of elliptic operators. Arch. Rational Mech. Anal. **28**, 165–183 (1967/1968)
3. G. Auchmuty, Variational principles for eigenvalues of compact operators. SIAM J. Math. Anal. **20**, 1321–1335 (1989)
4. T. Boggio, Sulle funzioni di Green d'ordine m. Rendiconti del Circolo Matematico di Palermo **20**, 97–135 (1905)
5. B.M. Brown, E.B. Davies, P.K. Jimack, M.D. Mihajlović, A numerical investigation of the solution of a class of fourth-order eigenvalue problems. R. Soc. Lond. Proc. Ser. A Math. Phys. Eng. Sci. **456**(1998), 1505–1521 (2000)
6. C.V. Coffman, R.J. Duffin, On the fundamental eigenfunctions of a clamped punctured disk. Adv. Appl. Math. **13**, 142–151 (1992)
7. C.V. Coffman, R.J. Duffin, D.H. Shaffer, The fundamental mode of vibration of a clamped annular plate is not of one sign, in *Constructive Approaches to Mathematical Models* (Proceeding Conference in Honor of R. J. Duffin, Pittsburgh, PA, 1978) (Academic, New York/London/Toronto, 1979), pp. 267–277
8. R. Courant, Über die Eigenwerte bei den Differentialgleichungen der mathematischen Physik. Math. Z. **7**, 1–57 (1920)
9. R. Courant, Über die Schwingungen eingespannter Platten. Math. Zeitschr. **15**, 195–200 (1922)
10. M. Cranston, E. Fabes, Z. Zhao, Conditional gauge and potential theory for the Schrödinger operator. Trans. Am. Math. Soc. **307**, 171–194 (1988)
11. C. De Coster, S. Nicaise, G. Sweers, Comparing variational methods for the hinged Kirchhoff plate with corners (submitted)
12. B. de Pagter, Irreducible compact operators. Math. Z. **192**(1), 149–153 (1986)
13. R.J. Duffin, D.H. Shaffer, On the modes of vibration of a ring-shaped plate. Bull. A.M.S. **58**, 652 (1952)
14. K. Friedrichs, Die Randwert- und Eigenwertprobleme aus der Theorie der elastischen Platte. Math. Ann. **98**, 205–247 (1928)
15. L. Gårding, On the asymptotic distribution of the eigenvalues and eigenfunctions of elliptic differential operators. Math. Scand. **1**, 237–255 (1953)

16. F. Gazzola, H.-Ch. Grunau, G. Sweers, *Polyharmonic Boundary Value Problems*. Springer Lecture Notes Series, vol. 1991 (Springer, Berlin, 2010)
17. P. Grisvard, *Elliptic Problems in Nonsmooth Domains*. Monographs and Studies in Mathematics, vol. 24 (Pitman, Boston, 1985)
18. H.-Ch. Grunau, F. Robert, G. Sweers, Optimal estimates from below for biharmonic Green functions. Proc. Am. Math. Soc. **139**, 2151–2161 (2011)
19. J. Hadamard, Mémoire sur le problème d'analyse relatif à l'équilibre des plaques élastiques encastrées, in *Memoires presents par divers savants* à l'Acad*émie des Sciences*, vol. 33 (1908), pp. 1–128
20. J. Kadlec, The regularity of the solution of the Poisson problem in a domain which boundary is similar to that of a convex domain. Czechoslovak Math. J. **14**, 89, 386–393 (1964)
21. M.G. Krein, M.A. Rutman, Linear operators leaving invariant a cone in a Banach space, Uspchi Matem. Nauk (N.S.) 3 (1948) **1**(23), 3–95 (translation in: American Mathematical Society Translation 26 (1950))
22. O.A. Ladyzhenskaya, N.N. Ural'tseva, *Linear and Quasilinear Elliptic Equations*. Translated from the Russian, Academic, New York/London 1968
23. S.A. Nazarov, G. Sweers, A hinged plate equation and iterated Dirichlet Laplace operator on domains with concave corners. J. Differ. Equ. **233**, 151–180 (2007)
24. S. Nazarov, A. Stylianou, G. Sweers, Hinged and supported plates with corners. ZAMP **63**, 929–960 (2012)
25. I. Schnieders, G. Sweers, A biharmonic converse to Krein-Rutman: a maximum principle near a positive eigenfunction (submitted)
26. G. Sweers, Strong positivity in $C(\overline{\Omega})$ for elliptic systems. Math. Z. **209**, 251–271 (1992)
27. G. Sweers, When is the first eigenfunction for the clamped plate equation of fixed sign? in *Electron. J. Diff. Eqns. Conf. 06* (2001), pp. 285–296
28. G. Sweers, Positivity for the Navier bilaplace, an anti-eigenvalue and an expected lifetime. DCDS Series S **7**(4), 839–855 (2014)
29. G. Szegö, On membranes and plates. Proc. Natl. Acad. Sci. **36**, 210–216 (1950)
30. H. Voss, Variational principles for eigenvalues of nonlinear eigenproblems, in *Numerical Mathematics and Advanced Applications, ENUMATH 2013*. Lecture Notes of Computer Science & Engineering, vol. 103 (Springer, Cham, 2015), pp. 305–313
31. H.F. Weinberger, Variational methods for eigenvalue approximation, in *Conference Board of the Mathematical Sciences Regional Conference Series in Applied Mathematics*, vol. 15 (SIAM, Philadelphia, 1974)
32. A. Weinstein, Étude des spectres des équations aux derivées partielles de la théorie des plaques élastiques. Mem. Sci. Math. **88**, 62 pp. (1937)
33. H. Weyl, Die asymptotische Verteilungsgesetze der Eigenwerte linearer partieller Differentialgleichungen. Math. Ann. **71**, 441–479 (1912)

Representations of the Dedekind Completions of Spaces of Continuous Functions

Jan Harm van der Walt

Dedicated to Prof. Ben de Pagter, on the occasion of his 65th birthday

Abstract Various representations of the Dedekind completions of Riesz spaces of continuous functions are known in the literature. The aim of this review paper is to collect together some of these results and to connect the various constructions to each other.

Keywords Continuous functions · Riesz spaces · Dedekind completion

1 Introduction

Among the classical examples of Riesz spaces, the space $C(X)$ of continuous, real-valued functions on a topological space X stands apart. It lacks many of the 'good properties' with which its fairer relatives are so richly endowed. It is neither Dedekind complete nor does it satisfy the projection property, or even the principle projection property, see, for instance, [15, Example 23.3 (ii)] and [15, Proof of Theorem 25.1]. Those topological spaces X for which $C(X)$ is a 'nice' Riesz space seem at first to be rather pathological. Indeed, if X is a compact Hausdorff space, $C(X)$ is Dedekind complete if and only if X is extremely disconnected; that is, the closure of every open in set in X is open [15, Theorem 43.11]. However, in some sense, spaces of continuous (extended) real-valued functions, and the Riesz subspaces of these, in particular on extremely disconnected topological spaces, are in fact the only examples of Riesz spaces. This is due to the various representation theorems for Riesz spaces, see, for instance, [25].

J. H. van der Walt (✉)
Department of Mathematics and Applied Mathematics, University of Pretoria, Pretoria, South Africa
e-mail: janharm.vanderwalt@up.ac.za

G. Buskes et al. (eds.), *Positivity and Noncommutative Analysis*,
Trends in Mathematics, https://doi.org/10.1007/978-3-030-10850-2_27

In view of the fact that, in contrast to, say, spaces of measurable functions, $C(X)$ is not Dedekind complete, it is natural to ask what its Dedekind completion $C(X)^\delta$ looks like. Many answers to this question have been given, among others by Dillworth [11], Horn [13], Nakano and Shimogaki [18], Maxey [17], see also [14], Ercan and Onal [12], Anguelov [2], and Dăneţ [8], to name but a few. The aim of this paper is to provide an overview of some of the known representations of $C(X)^\delta$.

When searching for representations of $C(X)^\delta$, a good starting point is a result of Veksler [24].

Theorem 1.1 *Let L and K be Archimedean Riesz spaces with K Dedekind complete, L^δ the Dedekind completion of L and $\Gamma : L \to K$ a Riesz isomorphism onto $\Gamma[L]$. Then there exists a Dedekind complete Riesz subspace $\hat{L}$ of K and a Riesz isomorphism $\hat{\Gamma} : L^\delta \to \hat{L}$ onto $\hat{L}$ so that $\Gamma[u] = \hat{\Gamma}[u]$ for all $u \in L$.*

Veksler's theorem indicates that a first step towards obtaining a concrete realisation of $C(X)^\delta$ is to identify a Dedekind complete Riesz space which contains $C(X)$ as a Riesz subspace. Two candidates immediately spring to mind. The Riesz space $B_{\ell oc}(X)$ of locally bounded, real-valued functions on X, and the order bidual $C(X)^{\sim\sim}$ of $C(X)$. These two possibilities lead to two classes of representations for $C(X)^\delta$. The latter possibility has been investigated by Kaplan [14] and Maxey [17]. A third possibility is to represent $C(X)^\delta$ as a space of set-valued maps on X. As we will see, such representations arise naturally from the construction of $C(X)^\delta$ by cuts. As we show, and as is alluded to by Dăneţ [8], such representations of $C(X)^\delta$ lead to interpretations of certain set-valued maps as equivalence classes of real-valued functions.

The representation theory for Archimedean Riesz spaces leads to a fourth representation of $C(X)^\delta$. As mentioned, every Archimedean Riesz space can be represented as a Riesz subspace of the space $C^\infty(S)$ of continuous extended real-valued functions on some extremely disconnected compact Hausdorff space, the latter being a Dedekind complete Riesz space.

In this paper, we concern ourselves with representations of $C(X)^\delta$ in terms of functions and set-valued maps on X. The paper is organised as follows. In Sect. 2 we recall the construction of the Dedekind completion of an Archimedean Riesz space by cuts, essentially due to MacNeille [16], see also [15, Section 32]. We present the general construction in the context of $C(X)$, to make clearer the connection with the concrete representations dealt with in subsequent sections. The construction of $C(X)^\delta$ by cuts leads naturally to various representations in terms of semi-continuous functions, which is discussed in Sect. 3. We turn, in Sect. 4, to the representation of $C(X)^\delta$ as a space of set-valued maps on X. Here we consider Anguelov's construction [2] using interval-valued functions, as well as a representation in terms of minimal upper semi-continuous compact-valued (musco) maps [21]. Pointwise discontinuous functions are considered in Sect. 5, while Sect. 6 deals with Borel measurable functions. Finally, we pose some open questions in Sect. 7.

Before proceeding any further, we fix some notation. By X we denote a topological space which, unless otherwise stated, is only assumed to be completely regular and Hausdorff. The set of open neighbourhoods of a point $x \in X$ is denoted

$\mathcal{V}_x$. We recall that X is a Baire space if the union of any countable collection of closed nowhere dense subsets of X has empty interior; equivalently, the intersection of countably many open and dense subsets of X is dense in X. By $B_{\ell oc}(X)$ we denote the set of locally bounded real-valued functions on X; that is, $u : X \to \mathbb{R}$ belongs to $B_{\ell oc}(X)$ if for every $x \in X$ there exists $V \in \mathcal{V}_x$ and $K > 0$ so that $|u(y)| \leq K$ for every $y \in V$. The function on X which is identically 0 is denoted $\mathbf{0}$, while $\mathbf{1}$ is the function identically 1 on X.

2 Constructing $C(X)^\delta$ via Dedekind Cuts

In this section we describe $C(X)^\delta$ in terms of the abstract construction of the Dedekind completion of an Archimedean Riesz space, as presented, for instance, in [15].[1]

Definition 2.1 A cut in $C(X)$ is a pair (A, B) of subsets of $C(X)$ which satisfies the following properties.

(i) $A = \{u \in C(X) \ : \ u \leq v, \ v \in B\}$.
(ii) $B = \{v \in C(X) \ : \ u \leq v, \ u \in A\}$.

The pairs $(C(X), \emptyset)$ and $(\emptyset, C(X))$ are cuts in $C(X)$; if $w \in C(X)$, then $K(w) = (A(w), B(w))$ is the cut in $C(X)$, with

$$A(w) = \{u \in C(X) \ : \ u \leq w\}, \ B(w) = \{v \in C(X) \ : \ w \leq v\}. \tag{2.1}$$

In particular,

$$K(\mathbf{0}) = (-C(X)^+, C(X)^+).$$

Denote by $C(X)^\kappa$ the set of cuts in $C(X)$, and order this set through set inclusion as follows. For $K_0 = (A_0, B_0)$ and $K_1 = (A_1, B_1)$ in $C(X)^\kappa$ we set

$$K_0 \leq K_1 \text{ if and only if } A_0 \subseteq A_1.$$

Observe that

$$K_0 \leq K_1 \text{ if and only if } B_1 \subseteq B_0.$$

With respect to this ordering, $C(X)^\kappa$ is a complete lattice, with least element $(\emptyset, C(X)$ and greatest element $(C(X), \emptyset)$; it is the MacNeille completion of $C(X)$ [16].

[1]The general construction, which applies to an arbitrary partially ordered set, is due to MacNeille [16].

However, $C(X)^\kappa$ is not a Riesz space. We remedy the situation by deleting the least and greatest elements from $C(X)^\kappa$; that is, we set

$$C(X)^\delta = C(X)^\kappa \setminus \{(\emptyset, C(X), (C(X), \emptyset))\}.$$

$C(X)^\delta$ becomes a Dedekind complete Riesz space when equipped with addition defined as

$$K_0 + K_1 = (A_0 + A_1, B_0 + B_1),$$

and scalar multiplication defined as

$$\alpha K_0 = \begin{cases} (\alpha A_0, \alpha B_0) \text{ if } \alpha > 0 \\ K(\mathbf{0}) \qquad\quad \text{if } \alpha = 0 \\ (\alpha B_0, \alpha A_0) \text{ if } \alpha < 0 \end{cases}$$

for $K_0 = (A_, B_0)$, $K_1 = (A_1, B_1) \in C(X)^\delta$. The cut $K(\mathbf{0})$ is the additive identity in $C(X)^\delta$. Hence

$$C(X)^{\delta +} = \{K = (A, B) \ : \ -C(X)^+ \subseteq A, \ B \subseteq C(X)^+\}.$$

The mapping

$$T : C(X) \ni w \mapsto K(w) \in C(X)^\delta$$

is a Riesz isomorphism onto a Riesz subspace of $C(X)^\delta$. Furthermore, for each nonzero cut $K = (A, B) \in C(X)^{\delta +}$, $u \in A$ and $v \in B$,

$$A \subseteq A(v), \ B \subseteq B(u)$$

so that

$$Tu \leq K \leq Tv.$$

In particular, if $u \in (A \cap C(X)^+) \setminus \{\mathbf{0}\}$, then

$$K(\mathbf{0}) < Tu \leq K \leq Tv.$$

Therefore $T[C(X)]$ is order dense in $C(X)^\delta$, and the ideal generated by $T[C(X)]$ in $C(X)^\delta$ is all of $C(X)^\delta$. All this is summarised in the following.

Theorem 2.2 *$C(X)^\delta$ is, up to Riesz isomorphism, the Dedekind completion of $C(X)$.*

3 Semi-continuous Functions

As is well known, a function $u \in \mathcal{B}_{\ell oc}(X)$ is upper semi-continuous if for every $x \in X$ and each $u(x) < \alpha$ there exists $V \in \mathcal{V}_x$ so that

$$u(y) < \alpha, \ y \in V.$$

Likewise, u is lower semi-continuous if for every $x \in X$ and each $\alpha < u(x)$ there exists $V \in \mathcal{V}_x$ so that

$$\alpha < u(y), \ y \in V.$$

We denote by $U(X)$ the set of real-valued upper semi-continuous functions in $\mathcal{B}_{\ell oc}(X)$, and by $L(X)$ the set of real-valued lower semi-continuous functions in $\mathcal{B}_{\ell oc}(X)$.

It will frequently be convenient to express semi-continuity in terms of the so-called Baire operators.[2] The lower Baire operator $I : u \in \mathcal{B}_{\ell oc}(X) \to \mathcal{B}_{\ell oc}(X)$ is defined by setting

$$I[u] : X \ni x \mapsto \sup\{\inf\{u(y) \ : \ y \in V\} \ : \ V \in \mathcal{V}_x\} \in \mathbb{R} \tag{3.1}$$

for each $u \in \mathcal{B}_{\ell oc}(X)$, while the upper Baire operator $S : \mathcal{B}_{\ell oc}(X) \to \mathcal{B}_{\ell oc}(X)$ is given by

$$S[u] : X \ni x \mapsto \inf\{\sup\{u(y) \ : \ y \in V\} \ : \ V \in \mathcal{V}_x\} \in \mathbb{R}. \tag{3.2}$$

A moment of reflection will convince the reader of the following, see, for instance, [5, Chapter IV, Section 2] and [2].

Proposition 3.1 *Let $u \in \mathcal{B}_{\ell oc}(X)$. Then the following statements are true:*

(i) *u is lower semi-continuous if and only if $u = I[u]$.*
(ii) *u is upper semi-continuous if and only if $u = S[u]$.*

Proposition 3.2 *The operators I, S, $I \circ S$ and $S \circ I$ are monotone and idempotent, and $I[u] \leq u \leq S[u]$ for all $u \in \mathcal{B}_{\ell oc}(X)$.*

The connection between semi-continuous functions on X and cuts in $C(X)$ is based on the following two results, see, for instance, [5, Chapter IV, Theorem 4] and [6, Chapter IX, Proposition 5], respectively.

[2]The operators I and S were first introduced by Baire [4] for functions of one real variable. Subsequently, Sendov [19] extended the definition to interval functions, see Sect. 4, and Anguelov [2] considered functions defined on an arbitrary topological space.

Theorem 3.3 *Let A and B be a nonempty sets of continuous functions on X. Then the following statements are true:*

(i) *If A is bounded above, then the function*

$$\lambda_A : X \ni x \mapsto \sup\{u(x) \ : \ u \in A\} \in \mathbb{R}$$

is lower semi-continuous.

(ii) *If B is bounded below, then the function*

$$\mu_A : X \ni x \mapsto \inf\{v(x) \ : \ v \in A\} \in \mathbb{R}$$

is upper semi-continuous.

Theorem 3.4 *Let $\lambda \in L(X)$ and $\mu \in S(X)$. Then the following statements are true:*

(i) *If there exists a continuous function u on X so that $u \leq \lambda$, then*

$$\lambda(x) = \sup\{w(x) \ : w \in C(X), \ w \leq \lambda\}, \ x \in X.$$

(ii) *If there exists a continuous function v on X so that $\mu \leq v$, then*

$$\mu(x) = \inf\{w(x) \ : \ w \in C(X), \ \mu \leq w\}, \ x \in X.$$

Through Theorem 3.3 we associate with each cut $K = (A, B) \in C(X)^\delta$ a pair of functions in $L(X) \times U(X)$. In particular, we have a mapping

$$C(X)^\delta \ni K = (A, B) \mapsto (\lambda_A, \mu_B) \in L(X) \times U(X)$$

where, as in Theorem 3.3,

$$\lambda_A : X \ni x \mapsto \sup\{u(x) \ : \ u \in A\} \in \mathbb{R}$$

and

$$\mu_B : X \ni x \mapsto \inf\{v(x) \ : \ v \in B\} \in \mathbb{R}.$$

Not every pair $(\lambda, \mu) \in L(X) \times U(X)$ arises in this way, as shown in the following. In particular, the pair (λ_A, μ_B) corresponding to a cut $K = (A, B)$ in $C(X)^\delta$ satisfy a certain minimality property.

Theorem 3.5 *Let $(\lambda, \mu) \in L(X) \times U(X)$. Assume that there exist $u, v \in C(X)$ so that $u \leq \lambda \leq \mu \leq v$. The following statements are equivalent:*

(i) *There exists a cut $K = (A, B)$ in $C(X)$ so that $(\lambda, \mu) = (\lambda_A, \mu_B)$.*
(ii) $\{(\lambda_0, \mu_0) \in L(X) \times U(X) \ : \ \lambda \leq \lambda_0 \leq \mu_0 \leq \mu\} = \{(\lambda, \mu)\}$.
(iii) $S[\lambda] = \mu$, $I[\mu] = \lambda$.

Proof Consider a pair $(\lambda, \mu) \in L(X) \times U(X)$. Set

$$A_\lambda = \{u \in C(X) : u \leq \lambda\}, \ B_\mu = \{v \in C(X) : \mu \leq v\}.$$

We first prove the equivalence of (i) and (ii). Assume that (ii) holds. We claim that (A_λ, B_μ) is a cut in $C(X)$. By assumption, A_λ and B_μ are nonempty. By virtue of the definitions of A_λ and B_μ, respectively, $u \leq \lambda \leq \mu \leq v$ for all $u \in A_\lambda$ and $v \in B_\mu$. It therefore remains to show that every upper bound for A_λ is a member for B_μ, and every lower bound for B_μ is an element of A_λ.

Consider $v \in C(X)$ so that $u \leq v$ for all $u \in A_\lambda$. According to Theorem 3.4,

$$\lambda(x) = \sup\{u(x) : u \in A_\lambda\} \leq v(x), \ x \in X.$$

The pointwise infimum of finitely many upper semi-continuous functions is again upper semi-continuous, so the function μ_0 given by

$$\mu_0(x) = v(x) \wedge \mu(x), \ x \in X$$

is upper semi-continuous and satisfies $\lambda \leq \mu_0 \leq \mu$. According to (ii), $\mu_0 = \mu$, which shows that $\mu \leq v$. Hence $v \in B_\mu$.

In the same way we see that $u \in A_\lambda$ whenever u is a lower bound for B_μ in $C(X)$. Therefore the pair (A_λ, B_μ) is a cut in $C(X)$. We invoke Theorem 3.4 to conclude that $(\lambda, \mu) = (\lambda_{A_\lambda}, \mu_{B_\mu})$.

Let $K = (A, B)$ be a cut in $C(X)$. Clearly, $A_{\lambda_A}, B_{\mu_B} \neq \emptyset$. Suppose that $(\lambda_0, \mu_0) \in L(X) \times U(X)$ satisfies $\lambda_A \leq \lambda_0 \leq \mu_0 \leq \mu_A$. Since $u \leq \lambda_0 \leq \mu_0 \leq v$ for all $u \in A$ and $v \in B$, it follows that $A \subseteq A_{\lambda_0}$ and $B \subseteq B_{\mu_0}$. Let $u \in A_{\lambda_0}$. Then $u \leq \lambda_0 \leq \mu \leq v$ for all $v \in B$. According to the definition of a cut in $C(X)$, Definition 2.1, $u \in A$ so that $A = A_{\lambda_0}$. In the same way, $B = B_{\mu_0}$. Theorem 3.4 shows that $\lambda_0 = \lambda_A$ and $\mu_0 = \mu_B$. This shows that (i) implies (ii).

We now turn to the equivalence of (ii) and (iii). Assume that (ii) is true. According to Proposition 3.2, $\lambda \leq I[\mu] \leq \mu$ and $\lambda \leq S[\lambda] \leq \mu$. Then $\lambda = I[\mu]$ and $\mu = S[\lambda]$.

Conversely, assume that (iii) holds, and consider $(\lambda_0, \mu_0) \in L(X) \times S(X)$ so that $\lambda \leq \lambda_0 \leq \mu_0 \leq \mu$. Then Propositions 3.1 and 3.2 yield

$$\mu = S[\lambda] \leq S[\mu_0] = \mu_0 \leq \mu$$

and

$$\lambda \leq \lambda_0 = I[\lambda_0] \leq I[\mu] = \lambda$$

so that $\lambda_0 = \lambda$ and $\mu_0 = \mu$. □

Let $LU_m^{\Delta+}(X)$ be the subset of $L(X) \times U(X)$ consisting of all pairs (λ, μ) such that $u \leq \lambda \leq \mu \leq v$ for some $u, v \in C(X)$, and

$$\{(\lambda_0, \mu_0) \in L(X) \times U(X) : \lambda \leq \lambda_0 \leq \mu_0 \leq \mu\} = \{(\lambda, \mu)\}.$$

We observe that $C(X)$ is embedded in $LU_m^{\Delta+}(X)$ in a natural way. In particular,

$$C(X) \ni u \mapsto (u, u) \in LU_m^{\Delta+}(X) \tag{3.3}$$

is an injection.

Theorem 3.5 gives rise to the bijections

$$LU_m^{\Delta+}(X) \ni (\lambda, \mu) \mapsto (A_\lambda, B_\mu) \in C(X)^\delta \tag{3.4}$$

where

$$A_\lambda = \{u \in C(X) \ : \ u \leq \lambda_0\}, \ B_\mu = \{v \in C(X) \ : \ \mu \leq v\},$$

and

$$C(X)^\delta \ni (A, B) \mapsto (\lambda_A, \mu_B) \in LU_m^{\Delta+}(X), \tag{3.5}$$

where

$$\lambda_A : X \ni x \mapsto \sup\{u(x) \ : \ u \in A\} \in \mathbb{R}$$

and

$$\mu_B : X \ni x \mapsto \inf\{v(x) \ : \ v \in B\} \in \mathbb{R}.$$

The proof of Theorem 3.5 shows that, actually, the mappings (3.4) and (3.5) are inverses of each other. Moreover, (3.4) defines a Riesz isomorphism, when $LU_m^{\Delta+}(X)$ is equipped with a suitable linear space structure and partial order.

In this regard, we order $LU_m^{\Delta+}(X)$ componentwise; that is, for $(\lambda_0, \mu_0), (\lambda_1, \mu_1) \in LU_m^{\Delta+}(X)$ we set

$$(\lambda_0, \mu_0) \leq (\lambda_1, \mu_1) \text{ if and only if } \lambda_0 \leq \lambda_1 \text{ and } \mu_0 \leq \mu_1.$$

Scalar multiplication is defined as

$$\alpha(\lambda, \mu) = \begin{cases} (\alpha\lambda, \alpha\mu) & \text{if } \alpha > 0 \\ (\mathbf{0}, \mathbf{0}) & \text{if } \alpha = 0 \\ (\alpha\mu, \alpha\lambda) & \text{if } \alpha < 0. \end{cases}$$

Some care must be taken when introducing addition; if $(\lambda_0, \mu_0), (\lambda_1, \mu_1) \in LU_m^{\Delta+}(X)$ it is in general not true that $(\lambda_0 + \lambda_1, \mu_0 + \mu_1) \in LU_m^{\Delta+}(X)$.

Example Let X be the real line $\mathbb{R}$, and set

$$\lambda_0(x) = \begin{cases} 0 \text{ if } x \leq 0 \\ 1 \text{ if } x > 0, \end{cases} \quad \mu_0(x) = \begin{cases} 0 \text{ if } x < 0 \\ 1 \text{ if } x \geq 0 \end{cases}$$

and

$$\lambda_1(x) = \begin{cases} 0 & \text{if } x < 0 \\ -1 & \text{if } x \geq 0, \end{cases} \quad \mu_0(x) = \begin{cases} 0 & \text{if } x \leq 0 \\ -1 & \text{if } x > 0. \end{cases}$$

Then $(\lambda_0, \mu_0), (\lambda_1, \mu_1) \in LU_m^{\Delta+}(X)$. But

$$[\lambda_0 + \lambda_1](x) = \begin{cases} 0 & \text{if } x \neq 0 \\ -1 & \text{if } x = 0, \end{cases} \quad [\mu_0 + \mu_1](x) = \begin{cases} 0 \text{ if } x \neq 0 \\ 1 \text{ if } x = 0. \end{cases}$$

Therefore the pair $(\lambda_0 + \lambda_1, \mu_0 + \mu_1)$ fails the minimality condition imposed on members of $LU_m^{\Delta+}(X)$. Indeed, $\lambda_0 + \lambda_1 < \mathbf{0} < \mu_0 + \mu_1$.

The situation is remedied at the hand of the following.

Proposition 3.6 *Consider a pair $(\lambda, \mu) \in L(X) \times S(X)$ so that $\lambda \leq \mu$. For $\epsilon > 0$ let $D_\epsilon(\lambda, \mu) = \{x \in X \,:\, \mu(x) - \lambda(x) < \epsilon\}$.*

(i) *If $(\lambda, \mu) \in LU_m^{\Delta+}(X)$, then $D_\epsilon(\lambda, \mu)$ is open and dense in X for every $\epsilon > 0$.*
(ii) *If $D_\epsilon(\lambda, \mu)$ is open and dense in X for every $\epsilon > 0$, then there exists a unique $(\lambda_0, \mu_0) \in LU_m^{\Delta+}(X)$ so that $\lambda \leq \lambda_0 \leq \mu_0 \leq \mu$.*

Proof of (i) Assume that $(\lambda, \mu) \in LU_m^{\Delta+}(X)$. Fix $\epsilon > 0$. Since $\mu - \lambda$ is upper semi-continuous, $D_\epsilon(\lambda, \mu)$ is open in X. We show that $D_\epsilon(\lambda, \mu)$ is dense in X. Consider any $x \in X$ and a neighbourhood V of x. With a view towards obtaining a contradiction, suppose that

$$\mu(y) \geq \lambda(y) + \epsilon, \ y \in V.$$

Since λ is lower semi-continuous there exists a neighbourhood $W \subseteq V$ of x so that

$$\lambda(x) + \frac{\epsilon}{2} < \lambda(y) + \epsilon \leq \mu(y), \ y \in W.$$

It follows from the definition (3.1) of I and Propositions 3.2 and 3.1 that

$$\lambda(x) < \lambda(x) + \frac{\epsilon}{2} = I[\lambda](x) + \frac{\epsilon}{2} \leq I[\mu](x).$$

But according to Theorem 3.5, $I[\mu] = \lambda$. We thus arrive at a contradiction and conclude that, indeed, $D_\epsilon(\mu, \lambda)$ is dense in X.

Proof of (ii) Assume that $D_\epsilon(\lambda, \mu)$ is open and dense in X for every $\epsilon > 0$. We claim that

$$I \circ S[\lambda] = I[\mu], \ S \circ I[\mu] = S[\lambda]. \tag{3.6}$$

It follows from Propositions 3.1 and 3.2 that

$$I \circ S[\lambda] \leq I \circ S[\mu] = I[\mu].$$

Suppose that $I \circ S[\lambda](x_0) < I[\mu](x_0)$ for some $x_0 \in X$. Then there exists $\alpha \in \mathbb{R}$ and $\epsilon > 0$ so that

$$I \circ S[\lambda](x_0) < \alpha < \alpha + \epsilon < I[\mu](x_0).$$

Since $I[\mu]$ is lower semi-continuous, there exists $V \in \mathcal{V}_{x_0}$ so that

$$\alpha + \epsilon < I[\mu](x) \leq \mu(x), \ x \in V.$$

The inequality $I \circ S[\lambda](x_0) < \alpha$, together with the definition (3.1) of I, implies that $S[\lambda](x') < \alpha$ for some $x' \in V$. According to the definition (3.2) of S, there exists an open neighbourhood $W \subseteq V$ of x' so that $\lambda(x) < \alpha, \ x \in W$. Then $\mu(x) - \lambda(x) \geq \epsilon, \ x \in W$ so that $W \cap D_\epsilon(\lambda, \mu) = \emptyset$, which contradicts the fact that $D_\epsilon(\lambda, \mu)$ is dense in X. In the same way, it follows that $S \circ I[\mu] = S[\lambda]$.

Now consider $(\lambda_0, \mu_0) \in LU_m^{\Delta+}(X)$ so that $\lambda \leq \lambda_0 \leq \mu_0 \leq \mu$. By the monotonicity of the Baire operators, see Proposition 3.2, the upper semi-continuity of μ, see Proposition 3.1, and (3.6), it follows that

$$I[\mu] = I \circ S[\lambda] \leq I \circ S[\lambda_0] \leq I \circ S[\mu] = I[\mu].$$

But by Theorem 3.5, $I \circ S[\lambda_0] = I[\mu_0] = \lambda_0$ so that $\lambda_0 = I[\mu]$. Likewise, we see that $\mu_0 = S[\lambda]$. Therefore there exists at most one pair $(\lambda_0, \mu_0) \in LU_m^{\Delta+}(X)$ so that $\lambda \leq \lambda_0 \leq \mu_0 \leq \mu$. That there exists at least one such pair follows from Zorn's Lemma once we recall that membership of $LU_m^{\Delta+}(X)$ is nothing but a minimality condition. □

In view of Proposition 3.6, there exists for all $(\lambda_0, \mu_0), (\lambda_1, \mu_1) \in LU_m^{\Delta+}(X)$ a unique $(\lambda, \mu) \in LU_m^{\Delta+}(X)$ so that $\lambda_0(x) + \lambda_1(x) \leq \lambda(x) \leq \mu(x) \leq \mu_0(x) + \mu_1(x)$ for all $x \in X$. We thus define the sum of $(\lambda_0, \mu_0), (\lambda_1, \mu_1) \in LU_m^{\Delta+}(X)$ as the unique pair $(\lambda, \mu) \in LU_m^{\Delta+}(X)$ so that

$$\lambda_0(x) + \lambda_1(x) \leq \lambda(x) \leq \mu(x) \leq \mu_0(x) + \mu_1(x), \ x \in X.$$

The following is easily verified using Proposition 3.6.

Theorem 3.7 *$LU_m^{\Delta+}(X)$ is a Riesz space, (3.3) is a Riesz isomorphism into $LU_m^{\Delta+}(X)$ and the mapping (3.4) is a Riesz isomorphism which leaves $C(X)$ invariant.*

3.1 Normal Semi-continuous Functions

Theorem 3.7 gives rise, immediatcly, to two distinct representations of $C(X)^{\delta}$ in $B_{\ell oc}(X)$. In 1950, Dilworth [11] characterised $C(X)^{\delta}$ in terms of so-called normal semi-continuous functions. While Dilworth dealt only with the case in which X is a compact Hausdorff space, his result was extended to the general case considered here by Horn [13].

Definition 3.8 A function $u \in B_{\ell oc}(X)$ is

(i) normal lower semi-continuous if $I \circ S[u] = u$.
(ii) normal upper semi-continuous if $S \circ I[u] = u$.

We denote by $NL(X)$ the set of locally bounded, normal lower semi-continuous functions, and by $NU(X)$ the set of locally bounded, normal upper semi-continuous functions. As is shown in [23],[3] see also [7], $NL(X)$ is a Dedekind complete Riesz space, containing $C(X)$ as a Riesz subspace. While $NL(X)$ is a subset of $B_{\ell oc}(X)$, it is not a Riesz subspace of $B_{\ell oc}(X)$. Indeed, with respect to the pointwise operations on $B_{\ell oc}(X)$, $NL(X)$ is neither a vector space nor a lattice, as can be seen at the hand of elementary examples, see, for instance, [23, Example 7]. The algebraic and lattice operations on $NL(X)$ are defined as

$$\begin{array}{c} u + v = I \circ S[u \oplus v], \quad \alpha u = I \circ S[\alpha \odot u] \\ u \vee v = I \circ S[\max\{u, v\}], \quad u \wedge v = I \circ S[\min\{u, v\}], \end{array} \tag{3.7}$$

where $\oplus$ and $\odot$ denote the usual pointwise operations on functions in $B_{\ell oc}(X)$, and $\min\{u, v\}$, $\max\{u, v\}$ are the pointwise minimum and maximum of $u, v \in B_{\ell oc}(X)$, respectively. Likewise, in $NU(X)$ we define

$$u+v = S\circ I[u\oplus v], \ \alpha u = S\circ I[\alpha\odot u], \ u\vee v = S\circ I[\max\{u, v\}], \ u\wedge v = S\circ I[\min\{u, v\}].$$

The ideals generated by $C(X)$ in $NL(X)$ and $NU(X)$, respectively, are denoted by $NL_c(X)$ and $NU_c(X)$. As an immediate consequence of Proposition 3.5, we have the following.

[3]The relevant results in [23] deal with nearly finite normal semi-continuous functions, but it is evident that these results apply equally in the locally bounded case.

Proposition 3.9 *Let $\lambda, \mu \in B_{\ell oc}(X)$. Then the following statements are true:*

(i) *If $(\lambda, \mu) \in LU_m^{\Delta+}(X)$, then $\lambda \in NL_c(X)$ and $\mu \in NU_c(X)$.*
(ii) *If $\lambda \in NL_c(X)$ then $S[\lambda] \in NU_c(X)$, and $(\lambda, S[\lambda]) \in LU_m^{\Delta+}(X)$.*
(iii) *If $\mu \in NU_c(X)$, then $I[\mu] \in NL_c(X)$, and $(I[\mu], \mu) \in LU_m^{\Delta+}(X)$.*

We therefore have the following.

Theorem 3.10 *The mapping $LU_m^{\Delta+}(X) \ni (\lambda, \mu) \mapsto \lambda \in NL_c(X)$ is a Riesz isomorphism leaving $C(X)$ invariant.*

Theorem 3.11 *The mapping $LU_m^{\Delta+}(X) \ni (\lambda, \mu) \mapsto \mu \in NU_c(X)$ is a Riesz isomorphism leaving $C(X)$ invariant.*

Theorem 3.12 *$S : NL_c(X) \to NU_c(X)$ is a Riesz isomorphism which leaves $C(X)$ invariant, and $I : NU_c(X) \to NL_c(X)$ is its inverse.*

4 Set-Valued Maps

In [2], Anguelov characterised $C(X)^\delta$ in terms of so-called Hausdorff continuous (H-continuous) interval-valued functions. In this section, we show how Anguelov's construction follows as a natural consequence of the representation of $C(X)^\delta$ as $LU_m^{\Delta+}(X)$. We also discuss the closely related representation of $C(X)^\delta$ in terms of upper semi-continuous compact-valued maps.

4.1 Hausdorff Continuous Functions

Denote by $\mathbb{IR}$ the set of nonempty, compact intervals in $\mathbb{R}$. An interval function on X is a function $\mathbf{u} : X \to \mathbb{IR}$. With every interval function $\mathbf{u}$ we associate the real-valued functions

$$\underline{u} : X \ni x \mapsto \inf \mathbf{u}(x) \in \mathbb{R},\ \overline{u} : X \ni x \mapsto \sup \mathbf{u}(x) \in \mathbb{R}.$$

Then $\mathbf{u}(x) = [\underline{u}(x), \overline{u}(x)],\ x \in X$. We call an interval function $\mathbf{u}$ locally bounded if $\underline{u}, \overline{u} \in B_{\ell oc}(X)$. An interval-valued function $\mathbf{u}$ is called S-continuous [19] if $\underline{u}$ is lower semi-continuous and $\overline{u}$ is upper semi-continuous.

Definition 4.1 An S-continuous function $\mathbf{u} : X \to \mathbb{IR}$ is called Hausdorff continuous (H-continuous) if $\mathbf{v} = \mathbf{u}$ for every S-continuous function $\mathbf{v} : X \to \mathbb{IR}$ such that $\mathbf{v}(x) \subseteq \mathbf{u}(x), x \in X$. The set of all H-continuous functions on X is denoted $\mathbb{H}(X)$.

Among the S-continuous functions, H-continuous functions are characterised as follows, see, for instance, [2].

Proposition 4.2 *An S-continuous function* $\mathbf{u}$ *on* X *is H-continuous if and only if* $S[\underline{u}] = \overline{u}$ *and* $I[\underline{u}] = \overline{u}$.

In [20] it is shown that, for an arbitrary topological space X, $\mathbb{H}(X)$ is a Dedekind complete Riesz space. The vector space operations are defined as follows. Given $\mathbf{u}, \mathbf{v} \in \mathbb{H}(X)$ and $\alpha \in \mathbb{R}$, there exist unique H-continuous functions $\mathbf{u} + \mathbf{v}$ and $\alpha\mathbf{u}$ so that

$$[\mathbf{u} + \mathbf{v}](x) \subseteq [\underline{u}(x) + \underline{v}(x), \overline{u}(x) + \overline{v}(x)], \ x \in X$$

and

$$[\alpha\mathbf{u}](x) \subseteq [\min\{\alpha\underline{u}(x), \alpha\overline{u}(x)\}, \max\{\alpha\underline{u}(x), \alpha\overline{u}(x)\}], \ x \in X.$$

The functions $\mathbf{u} + \mathbf{v}, \alpha\mathbf{u} \in \mathbb{H}(X)$ are precisely the sum and scalar product, respectively, in $\mathbb{H}(X)$. An elementary computation, utilising Propositions 3.2 and 4.2, yields

$$\mathbf{u} + \mathbf{v} = [I \circ S[\underline{u} \oplus \underline{v}], S \circ I[\overline{u} \oplus \overline{v}]]$$

and

$$\alpha\mathbf{u} = [I \circ S[\alpha \odot \underline{u}], S \circ I[\alpha \odot \overline{u}]]$$

for all $\mathbf{u}, \mathbf{v} \in \mathbb{H}(X)$. Likewise,

$$\mathbf{u} \vee \mathbf{v} = [I \circ S[\max\{\underline{u}, \underline{v}\}], S \circ I[\max\{\overline{u}, \overline{v}\}]]$$

and

$$\mathbf{u} \wedge \mathbf{v} = [I \circ S[\min\{\underline{u}, \underline{v}\}], S \circ I[\min\{\overline{u}, \overline{v}\}]].$$

Note that $\mathbf{u} \leq \mathbf{v}$ if and only if $\underline{u} \leq \underline{v}$ and $\overline{u} \leq \overline{v}$. We denote by $\mathbb{H}_c(X)$ the ideal generated by $C(X)$ in $\mathbb{H}(X)$. Theorem 3.5 and Proposition 4.2 immediately yield the following.

Theorem 4.3 *The mapping*

$$LU_m^{\Delta+}(X) \ni (\lambda, \mu) \mapsto [\lambda, \mu] \in \mathbb{H}_c(X)$$

is a Riesz isomorphism leaving $C(X)$ *invariant, with inverse*

$$\mathbb{H}_c(X) \ni \mathbf{u} \mapsto (\underline{u}, \overline{u}) \in LU_m^{\Delta+}(X).$$

4.2 Minimal Upper Semi-continuous Compact-Valued Maps

A real set-valued map on X is a function $\mathbf{u} : X \to 2^{\mathbb{R}}$. Such a map $\mathbf{u}$ is called upper semi-continuous if for every $x \in X$, and every open set $W \supseteq \mathbf{u}(x)$ there exists $V \in \mathcal{V}_x$ so that $\mathbf{u}(y) \subseteq W$ for every $y \in V$. If $\mathbf{u}$ is upper semi-continuous and compact-valued, then it is called usco.

Definition 4.4 An usco map $\mathbf{u} : X \to 2^{\mathbb{R}}$ is called minimal, musco for short, if $\mathbf{u} = \mathbf{v}$ for every usco map $\mathbf{v} : X \to 2^{\mathbb{R}}$ such that $\mathbf{v}(x) \subseteq \mathbf{u}(x)$, $x \in X$.

We denote by $\mathcal{M}(X)$ the set of all musco maps from X into $\mathbb{R}$ so that $\mathbf{u}(x) \neq \emptyset$ for every $x \in X$. As is shown in [21], $\mathcal{M}(X)$ is a Dedekind complete Riesz space containing $C(X)$ as a Riesz subspace. As in the case of H-continuous interval functions, the definitions of the vector space operations on $\mathcal{M}(X)$ are based on the following fact [3]. For all $\mathbf{u}, \mathbf{v} \in \mathcal{M}(X)$ and $\alpha \in \mathbb{R}$, there exist unique musco maps $\mathbf{u} + \mathbf{v}$ and $\alpha\mathbf{u}$ so that

$$[\mathbf{u} + \mathbf{v}](x) \subseteq \{y + z \ : \ y \in \mathbf{u}(x),\ z \in \mathbf{v}(x)\},\ [\alpha\mathbf{u}](x) \subseteq \{\alpha y \ : \ y \in \mathbf{u}(x)\},\ x \in X.$$

These are precisely the sum and scalar product, respectively, in $\mathcal{M}(X)$. The positive cone in $\mathcal{M}(X)$ is $\mathcal{M}(X)^+ = \{\mathbf{u} \ : \ \mathbf{u}(x) \subseteq [0, \infty)\}$.

Anguelov and Kalenda [3] showed that for each $\mathbf{u} \in \mathcal{M}(X)$ there exists a unique $\mathbf{u}^* \in \mathbb{H}(X)$ so that $\mathbf{u}(x) \subseteq \mathbf{u}^*(x)$, $x \in X$. In particular, for $\mathbf{u} \in \mathcal{M}(X)$, $\mathbf{u}^*(x) = [\min \mathbf{u}(x), \max \mathbf{u}(x)]$, $x \in X$. Conversely, for every $\mathbf{u} \in \mathbb{H}(X)$ there exists a unique $\mathbf{u}_* \in \mathcal{M}(X)$ so that $\mathbf{u}_*(x) \subseteq \mathbf{u}(x)$, $x \in X$. The following is easily verified.

Theorem 4.5 *The mapping*

$$\mathcal{M}(X) \ni \mathbf{u} \mapsto \mathbf{u}^* \in \mathbb{H}(X)$$

is a Riesz isomorphism which leaves $C(X)$ invariant, and the mapping

$$\mathbb{H}(X) \ni \mathbf{u} \mapsto \mathbf{u}_* \in \mathcal{M}(X)$$

is its inverse.

Denoting by $\mathcal{M}_c(X)$ the ideal generated by $C(X)$ in $\mathcal{M}(X)$, we therefore have the following.

Corollary 4.6 *The mapping*

$$LU_m^{\Delta+}(X) \ni (\lambda, \mu) \mapsto [\lambda, \mu]_* \in \mathcal{M}_c(X)$$

is a Riesz isomorphism leaving $C(X)$ invariant, with inverse

$$\mathcal{M}_c(X) \ni \mathbf{u} \mapsto (\underline{u}^*, \overline{u}^*) \in LU_m^{\Delta+}(X).$$

5 Pointwise Discontinuous Functions

In this section we discuss the representation of $C(X)^\delta$ in terms of pointwise discontinuous functions, due to Dăneţ [8]. We derive this result as a consequence of a theorem of Maxey [17], see also [1, Theorem 2.18].

Theorem 5.1 *Let K be a Dedekind complete Riesz space, L a Riesz subspace of K and A an ideal in K, maximal with respect to the property $L \cap A = \{0\}$. Then the quotient map $Q_A : K \to K/A$ restricted to L is a Riesz isomorphism into K/A, and the ideal generated by $Q_A[L]$ in K/A is Riesz isomorphic to L^δ.*

Throughout this section, we assume that X is a Baire space. For a function $u \in B_{\ell oc}(X)$, denote by C_u the set of points of continuity of u; that is

$$C_u = \{x \in X \ : \ u \text{ is continuous at } x\}.$$

Then u is called pointwise discontinuous if C_u is dense in X. The set $C_d(X)$ of locally bounded, pointwise discontinuous functions on X is a Riesz subspace of $B_{\ell oc}(X)$.[4] Clearly, $C(X)$ is a Riesz subspace of $C_d(X)$. Denote by $C_{d,c}(X)$ the ideal generated by in $C(X)$ in $C_d(X)$.

In order to apply Maxey's Theorem 5.1 to $C(X)$ as a Riesz subspace of $C_{d,c}(X)$, we must identify the appropriate ideal in $C_{d,c}(X)$. This is the ideal of rare elements [8].

Definition 5.2 A function $u \in C_d(X)$ is called rare if $I \circ S[|u|] = \mathbf{0}$.

Denote by $R_c(X)$ the set of rare functions in $C_d(X)$. According to [8, Corollary 4.4], $R_c(X)$ is an ideal in $C_{d,c}(X)$. Furthermore, since every continuous function is both upper- and lower semi-continuous, it follows from Proposition 3.1 that $I \circ S[|u|] = |u|$ for every $u \in C(X)$. Therefore

$$C(X) \cap R_c(X) = \{\mathbf{0}\}.$$

Consider an ideal A in $C_{d,c}(X)$, and assume that A contains a non-rare member $u_0 > \mathbf{0}$. According to [8, Proposition 4.5], a function $u \in C_d(X)$ is rare if and only if $u(x) = 0$ for every $x \in C_u$. Since $u_0 \in A^+ \setminus \{\mathbf{0}\}$, there exists a point $x_0 \in C_{u_0}$ so that $u_0(x_0) > 0$. Since u_0 is continuous at x_0 there exists $V \in \mathcal{V}_{x_0}$ and $\epsilon > 0$ so that $u_0(x) > \epsilon$, $x \in V$. By the complete regularity of X there exists a continuous function $u : X \to [0, \epsilon)$ so that $0 < u(x_0)$ and $u(x) = 0$ if $x \in X \setminus V$. Then $\mathbf{0} < u \leq u_0$ so that, A being an ideal in $C_{d,c}(X)$, it follows that $u_0 \in A \cap C(X)$. This shows that $R_c(X)$ is a maximal, in fact the maximum, ideal of $C_{d,c}(X)$ which intersects $C(X)$ only in $\mathbf{0}$. We therefore have the following [8, Theorem 6.1].

[4] If X is not a Baire space, then $C_d(X)$ may fail to be a Riesz subspace of $B_{\ell oc}(X)$. In fact, it may not even be a linear subspace of $B_{\ell oc}(X)$.

Theorem 5.3 *The quotient mapping* $Q_{R_c(X)} : C_{d,c}(X) \to C_{d,c}(X)/R_c(X)$ *restricted to* $C(X)$ *is a Riesz isomorphism into* $C_{d,c}(X)/R_c(X)$*, and* $C_{d,c}(X)/R_c(X)$ *is the Dedekind completion of* $C(X)$.

We end this section, as we did the preceding ones, by giving an explicit description of the Riesz isomorphism from $C_{d,c}(X)/R_c(X)$ onto $C(X)^\delta$. Combining [8, Definition 5.6 & Theorem 5.7 (i)] and Theorem 3.5, we see that if $u \in C_{d,c}(X)$, then $(I \circ S[u], S \circ I[u]) \in LU_m^{\Delta+}(X)$, and $(I \circ S[u], S \circ I[u]) = (I \circ S[v], S \circ I[v])$ whenever $Q_{R_c(X)}u = Q_{R_c(X)}v$. Conversely, if $(\lambda, \mu) \in LU_m^{\Delta+}(X)$, then $\lambda, \mu \in C_{c,d}(X)$ and $Q_{R_c(X)}\lambda = Q_{R_c(X)}\mu$.

Theorem 5.4 *The mapping*

$$LU_m^{\Delta+}(X) \ni (\lambda, \mu) \mapsto Q_{R_c(X)}\mu \in C_{d,c}(X)/R_c(X)$$

is a Ries isomorphism which leaves $C(X)$ *invariant, with inverse*

$$C_{d,c}(X)/R_c(X) \ni Q_{R_c(X)}u \mapsto (I \circ S[u], S \circ I[u]) \in LU_m^{\Delta+}(X).$$

6 Borel Measurable Functions

Let X be a compact Hausdorff space. Denote by $Bo(X)$ the set of bounded, real-valued Boreal measurable functions on X. With respect to the pointwise operations, $Bo(X)$ is an Archimedean Riesz space containing $C(X)$ as a Riesz subspace. The set $N(X)$ of functions in $Bo(X)$ which vanish everywhere except possibly on a Borel set of first Baire category is an ideal in $Bo(X)$, see [9, Definition 14.7]. Since X, being a compact Hausdorff space, is also a Baire space, every set of first category has dense complement. Therefore

$$C(X) \cap N(X) = \{\mathbf{0}\}.$$

Hence the quotient map $Q_{N(X)} : Bo(X) \to Bo(X)/N(X)$, restricted to $C(X)$, is a Riesz isomorphism into $Bo(X)/N(X)$. We thus identify $C(X)$ with its image under $Q_{N(X)}$. Evidently, the ideal generated by $C(X)$ in $Bo(X)/N(X)$ is all of $Bo(X)/N(X)$. In order to see that $Bo(X)/N(X)$ is the Dedekind completion of $C(X)$, it remains to show that $Bo(X)/N(X)$ is Dedekind complete, and $C(X)$ is order dense in $Bo(X)/N(X)$. Both of these assertions follow, see [9, Theorem 14.9], from the remarkable fact that every $u \in Bo(X)$ is equal to a lower semi-continuous function everywhere except possibly on a set of first category [9, Theorem 14.6]. In fact, a closer examination of the proof of de Jonge and van Rooij's result shows that, in fact, we have the following.

Proposition 6.1 *Let* $u \in Bo(X)$*. Then there exists a bounded function* $v \in NL(X)$ *so that* $u - v \in N(X)$.

Theorem 6.2 *$Bo(X)/N(X)$ is the Dedekind completion of $C(X)$.*

Consider $(\lambda, \mu) \in LU_m^{\Delta+}(X)$. Because X is compact, hence a Baire space, $\lambda(x) = \mu(x)$ for every x outside some set of first Baire category. Hence $\mu - \lambda \in N(X)$ so that μ and λ generate the same equivalence class in $Bo(X)/N(X)$. By abuse of notation, we denote this equivalence class by $Q_{N(X)}(\lambda, \mu)$. By the definition of the vector space operations and partial order on $LU_m^{\Delta+}(X)$, it follows that

$$Q_{N(X)}(\alpha(\lambda_0, \mu_0) + \beta(\lambda_1, \mu_1)) = \alpha Q_{N(X)}(\lambda_0, \mu_0) + \beta Q_{N(X)}(\lambda_1, \mu_1)$$

and

$$Q_{N(X)}((\lambda_0, \mu_0)) \leq Q_{N(X)}((\lambda_1, \mu_1)) \text{ if and only if } (\lambda_0, \mu_0) \leq (\lambda_1, \mu_1)$$

for all $(\lambda_0, \mu_0), (\lambda_1, \mu_1) \in LU_m^{\Delta+}(X)$ and $\alpha, \beta \in \mathbb{R}$. We therefore have the following.

Theorem 6.3 *The mapping*

$$LU_m^{\Delta+}(X) \ni (\lambda, \mu) \mapsto Q_{N(X)}\lambda \in Bo(X)/N(X)$$

is a Riesz isomorphism which leaves $C(X)$ invariant.

7 Open Questions

As is demonstrated in the preceding sections, much is known concerning representations of $C(X)^\delta$. However, the following questions remain open, to the best of our knowledge.

(1) Is it possible to generalise Dăneţ's characterisation of $C(X)^\delta$ in terms of pointwise discontinuous functions, discussed in Sect. 5, to spaces X which are not Baire? This will require a suitable generalisation of the concept of a pointwise discontinuous function. Indeed, if X is not a Baire space, then $C_d(X)$ is not a Riesz space. In fact, it is not even a vector space with respect to pointwise addition and scalar multiplication.

(2) Can de Jonge and van Rooij's characterisation of $C(X)^\delta$ in terms of measurable functions be generalised to non-compact spaces X? This seems feasible, at least for completely regular Baire spaces.

(3) Does de Jonge and van Rooij's result, Theorem 6.2, follow from Maxey's Theorem 5.1? That is, is the ideal $N(X)$ maximal with respect to the property $I \cap C(X) = \{\mathbf{0}\}$, for ideals I in $Bo(X)$?

(4) The characterisation of $C(X, E)^\delta$, with E a Banach lattice, is investigated in [21] and [22]. Partial positive answers have been obtained for representations

of $C(X, E)^\delta$ in terms of H-continuous functions [22] and musco maps [21]. Can characterisations of $C(X, E)^\delta$ be obtained in terms of, for instance, measurable or pointwise discontinuous functions?

(5) As mentioned in the introduction, it is possible to represent $C(X)^\delta$, for X a compact Hausdorff space, in terms of its order bidual, see, for instance, [14]. The function spaces discussed in this paper can therefore be related to the order bidual of $C(X)$. It may be interesting to investigate the exact nature of these relationships.

References

1. C.D. Aliprantis, O. Burkinshaw, *Locally Solid Riesz Spaces* (Academic, New York, 1978)
2. R. Anguelov, Dedekind order completion of $C(X)$ by Hausdorff continuous functions. Quaest. Math. **27**(2), 153–169 (2004)
3. R. Anguelov, O.F.K. Kalenda, The convergence space of minimal USCO mappings. Czechoslovak Math. J. **59**(1), 101–128 (2009)
4. R. Baire, *Leçons sur les fonctions discontinues*, Les Grands Classiques Gauthier-Villars. [Gauthier-Villars Great Classics], Éditions Jacques Gabay, Sceaux, 1995, Reprint of the 1905 original.
5. N. Bourbaki, *Elements of Mathematics. General Topology. Part 1*, Hermann, Paris (Addison-Wesley, Reading, 1966)
6. N. Bourbaki, *Elements of Mathematics. General Topology. Part 2*, Hermann, Paris (Addison-Wesley, Reading, 1966)
7. N. Dăneţ, Riesz spaces of normal semicontinuous functions. Mediterr. J. Math. **12**(4), 1345–1355 (2015)
8. N. Dăneţ, The Dedekind completion of $C(X)$ with pointwise discontinuous functions, in ed. by M. de Jeu et al. [10] (Birkhäuser, Cham, 2016), pp. 111–125
9. E. de Jonge, A.C.M. van Rooij, *Introduction to Riesz Spaces* (Mathematisch Centrum, Amsterdam, 1977)
10. M. de Jeu, B. de Pagter, O. van Gaans, M. Veraar (eds.), *Ordered Structures and Applications*. Trends in Mathematics (Birkhäuser/Springer, Cham, 2016)
11. R.P. Dilworth, The normal completion of the lattice of continuous functions. Trans. Amer. Math. Soc. **68**, 427–438 (1950)
12. Z. Ercan, S. Onal, A new representation of the Dedekind completion of $C(K)$-spaces. Proc. Am. Math. Soc. **133**(11), 3317–3321 (2005)
13. A. Horn, The normal completion of a subset of a complete lattice and lattices of continuous functions. Pacific J. Math. **3**, 137–152 (1953)
14. S. Kaplan, *The Bidual of* $C(X)$. *I*. North-Holland Mathematics Studies, vol. 101 (North-Holland, Amsterdam, 1985)
15. W.A.J. Luxemburg, A.C. Zaanen, *Riesz Spaces*, vol. I (North-Holland, Amsterdam; American Elsevier Publishing Co., New York, 1971)
16. H.M. MacNeille, Partially ordered sets. Trans. Amer. Math. Soc. **42**(3), 416–460 (1937)
17. W. Maxey, The Dedekind completion of $c(x)$ and its second dual. Ph.D. thesis, Purdue University, West Lafayette, 1973
18. K. Nakano, T. Shimogaki, A note on the cut extension of C-spaces. Proc. Japan Acad. **38**, 473–477 (1962)
19. B. Sendov, *Hausdorff Approximations*. Mathematics and Its Applications (East European Series), vol. 50 (Kluwer Academic Publishers Group, Dordrecht, 1990). Translated and revised from the Russian

20. J.H. van der Walt, The linear space of Hausdorff continuous interval functions. BIOMATH **2**(2), 1311261, 6 (2013)
21. J.H. van der Walt, The Riesz space of minimal USCO maps, in ed. by M. de Jeu et al. [10] (Birkhäuser, Cham, 2016), pp. 481–502
22. J.H. van der Walt, Vector-valued interval functions and the Dedekind completion of $C(X, E)$. Positivity **21**(3), 1143–1159 (2017)
23. J.H. van der Walt, The Universal Completion of $C(X)$ and unbounded order convergence. J. Math. Anal. Appl. **460**(1), 76–97 (2018)
24. A.I. Veksler, A new construction of the Dedekind completion of vector lattices and l-groups with division. Sibirsk. Mat. Ž. **10**, 1206–1213 (1969)
25. T. Wickstead, Representations of archimedean RIESZ spaces by continuous functions. Acta Appl. Math. **27**, 123–133 (1992)

Joint Representation of a Riesz Space and Its Conjugate Space

A. C. M. van Rooij

To Ben de Pagter, on the occasion of his formal retirement, trusting that our factual cooperation will continue

Abstract Our main goal is to prove the following theorem. *Let E be a Riesz space such that $E_c^\sim$ separates the points of E. Then there exist a measure space $(X, \mathcal{A}, \mu)$ and injective σ-order continuous Riesz homomorphisms $x \mapsto \hat{x}$ and $f \mapsto \hat{f}$ of E and $E_c^\sim$, respectively, into $L^0(\mu)$ such that*

$$\textit{if } x \in E \textit{ and } f \in E_c^\sim, \textit{ then } \hat{f}\hat{x} \in L(\mu) \textit{ and } \int \hat{f}\hat{x}\, \mathrm{d}\mu = f(x).$$

Here $E_c^\sim$ is the Riesz space of all integrals on E, and $L^0(\mu)$ is the space of μ-equivalence classes of $\mathcal{A}$-measurable functions $X \to \mathbb{R}$. Our results are related to a theorem of A.W. Wickstead for which we refer to a postscript.

Keywords Riesz spaces · Conjugate space · Representation

1 Our terminology and notations are those of [2]. "Functions" are allowed to have the values $\pm\infty$. For $a, x_1, x_2, \ldots$ in a Riesz space E, by "$x_n \downarrow a$ in E" we mean that $x_1 \geq x_2 \geq \cdots$ and a is the greatest lower bound of $\{x_1, x_2, \ldots\}$ in E. For functions f, g on a set X we define $[f \leq g] = \{x \in X\colon\ f(x) \leq g(x)\}$; similarly, $[f = 0]$, etc.

The indicator of a set A is $\mathbb{1}_A$; the constant function with value 1 is $\mathbb{1}$.

2 We start with a closer look at C^∞. For details we refer to Chapter 7 in [2].

For an extremally disconnected compact Hausdorff space Z we denote by $C^\infty(Z)$ the (Dedekind complete) Riesz space of all continuous functions $f\colon Z \to$

A. C. M. van Rooij (✉)
Department of Mathematics, Radboud University, Nijmegen, The Netherlands
e-mail: maths@math.ru.nl

G. Buskes et al. (eds.), *Positivity and Noncommutative Analysis*,
Trends in Mathematics, https://doi.org/10.1007/978-3-030-10850-2_28

$[-\infty, \infty]$ for which the set $[|f| < \infty]$ is dense in Z. On $C^\infty(Z)$ there is a multiplication, indicated by juxtaposition, or, for emphasis, by a dot: $\bullet$. With the convention $0 \cdot \infty = \infty \cdot 0 = 0$, for $f, g \in C^\infty(Z)$ we have $(f \bullet g)(z) = f(z)g(z)$ for all z in a dense subset of Z.

There is a partially defined division:

Lemma 3 *Let Z be an extremally disconnected compact Hausdorff space. Let $f, g \in C^\infty(Z)$, $f = 0$ on $\operatorname{int}[g = 0]$. Then there exists a unique element f/g of $C^\infty(Z)$ satisfying*

$$(f/g)(z) = \frac{f(z)}{g(z)} \quad \textit{if} \quad z \in [|f| < \infty] \cap [0 < |g| < \infty],$$
$$(f/g)(z) = 0 \quad \textit{if} \quad z \in \operatorname{int}[g = 0].$$

We have

$$(f/g) \bullet g = f,$$
$$f/g \textit{ lies in the band generated by } f.$$

If $|f| \leq |g|$, then

$$f/g \in C(Z), \ |f/g| \leq \mathbb{1},$$
$$(f/g)(z)g(z) = f(z) \textit{ if } z \in [|g| \neq \infty].$$

Proof The proof is an application of the observation that a continuous $\mathbb{R}$-valued function on an open dense subset of Z extends uniquely to an element of $C^\infty(Z)$ (Theorem 7.25 in [1]). □

From here on, *X and Y are extremally disconnected compact Hausdorff spaces, $F \subset C^\infty(X)$ and $G \subset C^\infty(Y)$ are order dense Riesz ideals, and Ω is a Riesz isomorphism $F_c^\sim \to G$.*

4 Every $h \in C(X)$ induces a map $M_h\colon F \to F$ by $M_h f = hf$ $(f \in F)$. Every $j \in C(Y)$ induces $M_j\colon G \to G$ by $M_j g = jg$ $(g \in G)$.

5 Take h in $C(X)^+$. If $f_n \downarrow 0$ in F, then $M_h f_n = hf_n \downarrow 0$ in F, so that $\varphi(M_h f_n) \to 0$ for every φ in $F_c^\sim$. Thus, if $\varphi \in F_c^\sim$, then $\varphi \circ M_h \in F_c^\sim$, and h determines a positive linear map $\varphi \mapsto \varphi \circ M_h$ of $F_c^\sim$ into $F_c^\sim$.

Here we had $h \geq 0$. By linearity, every h in $C(X)$ yields an order bounded linear map $M_h^\sim\colon F_c^\sim \to F_c^\sim$:

$$M_h^\sim(\varphi) = \varphi \circ M_h \quad (\varphi \in F_c^\sim, \ h \in C(X)).$$

Lemma 6 *There is a continuous $\omega\colon Y \to X$ such that for all h in $C(X)$*

$$\Omega \circ M_h^{\sim} = M_{h \circ \omega} \circ \Omega \quad on\ F_c^{\sim}.$$

Proof If $h \in C(X)$ and $0 \leq h \leq \mathbb{1}$, then $\Omega \circ M_h^{\sim} \circ \Omega^{-1}$ lies in the center of F, hence is M_{h^*} for some h^* in $C(Y)$ with $0 \leq h^* \leq \mathbb{1}$. (See [3].)

Thus we obtain a Riesz homomorphism $h \mapsto h^*$ of $C(X)$ into $C(Y)$ such that $\Omega \circ M_h^{\sim} = M_{h^*} \circ \Omega$ ($h \in C(X)$). Since $\mathbb{1}^* = \mathbb{1}$, there exists a continuous $\omega\colon Y \to X$ with $h^* = h \circ \omega$ for all h in $C(X)$ (Theorem 2.34 in [1]). □

Lemma 7 *Let $h_n \downarrow 0$ in $C(X)$. Then $h_n \circ \omega \downarrow 0$ in $C(Y)$.*

Proof Let j be the infimum of $\{h_n \circ \omega\colon n \in \mathbb{N}\}$ in $C(Y)$.

Take φ in $F_c^{\sim}$, $\varphi \geq 0$. For all $f \in F^+$ we have $h_n f \downarrow 0$ in $C(X)$, hence in F, so $\varphi(h_n f) \downarrow 0$, i.e., $(M_{h_n}^{\sim}\varphi)(f) \downarrow 0$. Thus, $M_{h_n}^{\sim}\varphi \downarrow 0$ in $F_c^{\sim}$. Then $\Omega(M_{h_n}^{\sim}\varphi) \downarrow 0$ in G, hence in $C^\infty(Y)$ because G is an ideal in $C^\infty(Y)$. With Lemma 6 it follows that $M_{h_n \circ \omega}(\Omega\varphi) \downarrow 0$ in $C^\infty(Y)$.

The above says that $(h_n \circ \omega)g \downarrow 0$ in $C^\infty(Y)$ for every g in G^+. Therefore, $jg = 0$ whenever $g \in G^+$. It follows from the order denseness of G in $C^\infty(Y)$ that $j = 0$. □

Lemma 8 *Let $h_0, h \in C(X)$. If h lies in the band of $C(X)$ generated by h_0, then $h \circ \omega$ lies in the band of $C(Y)$ generated by $h_0 \circ \omega$.*

Proof We consider $h_0, h \geq 0$. Apply Lemma 7 with $h_n = h - h \wedge nh_0$. □

Lemma 9 *If $h \in C^\infty(X)$, then $h \circ \omega \in C^\infty(Y)$.*

Proof We consider $h \geq 0$. Of course, $h \circ \omega$ is continuous $Y \to [0, \infty]$; it remains to prove that $[h \circ \omega = \infty]$ has empty interior.

For $n \in \mathbb{N}$ let h_n be the indicator of clo$[h > n]$; then $h_n \in C(X)$ for every n. As $h \in C^\infty(X)$, we have $h_n \downarrow 0$ in $C(X)$, so that $h_n \circ \omega \downarrow 0$ in $C(Y)$, by Lemma 7. But $[h \circ \omega = \infty] \subset \bigcap_{n \in \mathbb{N}}[h_n \circ \omega = 1]$. Thus, int$[h \circ \omega = \infty] = \emptyset$. □

10 Definition:

$$FF_c^{\sim} = \{(f \circ \omega) \bullet \Omega\varphi\colon\ f \in F,\ \varphi \in F_c^{\sim}\}.$$

Theorem 11 *$FF_c^{\sim}$ is a Riesz ideal in $C^\infty(Y)$.*

Proof

(I) For f in F^+ put $A_f = \{(f \circ \omega)\Omega\varphi\colon\ \varphi \in F_c^{\sim}\} = \{(f \circ \omega)g\colon\ g \in G\}$. Every A_f is a Riesz subspace of $C^\infty(Y)$.

(II) Take f in F^+; we prove A_f to be an ideal in $C^\infty(Y)$. To that end, let $g \in G^+$, $j \in C^\infty(Y)^+$, $j \leq (f \circ \omega)g$; we are done if we can prove $j \in A_f$. As j and $f \circ \omega$ lie in $C^\infty(Y)$ and $j = 0$ on int$[f \circ \omega = 0]$, we can form

$$g' = j/(f \circ \omega).$$

Then $g' \in C^\infty(Y)^+$ and $j = g' \bullet (f \circ \omega)$. But then $(f \circ \omega)g' = j \le (f \circ \omega)g$ whereas $g' = 0$ on $\mathrm{int}[f \circ \omega = 0]$. Consequently, $g' \le g$, so $g' \in G$ and $j \in A_f$.

(III) If $f, f' \in F^+$ and $f \le f'$, then $A_f \subset A_{f'}$, as follows from (II). Hence, $FF_c^\sim$, being the union of all sets A_f, is a Riesz ideal of $C^\infty(Y)$. □

Lemma 12 *Let $N \in \mathbb{N}$ and let $f_1, \dots, f_N \in F$ and $\varphi_1, \dots, \varphi_N \in F_c^\sim$ be such that $\sum (f_n \circ \omega)\Omega\varphi_n \ge 0$. Then $\sum \varphi_n(f_n) \ge 0$.*

Proof Set $f = |f_1| + \cdots + |f_N|$ and $g_n = f_n/f$ $(n = 1, \dots, N)$ in $C(X)$. From Lemma 3 we obtain $f_n \circ \omega = (f \circ \omega)(g_n \circ \omega)$ pointwise on $[f \circ \omega \ne \infty]$, and by Lemma 9 the latter set is dense in Y; so $f_n \circ \omega = (f \circ \omega) \bullet (g_n \circ \omega)$ in $C^\infty(Y)$. Thus,

$$0 \le \sum (f_n \circ \omega)\Omega\varphi_n = (f \circ \omega) \bullet \sum (g_n \circ \omega)\Omega\varphi_n.$$

g_n lies in the band generated by f, so $g_n \circ \omega$ lies in the band of $C^\infty(Y)$ generated by $f \circ \omega$ (Lemma 8). Therefore (with Lemma 6)

$$0 \le \sum (g_n \circ \omega)\Omega\varphi_n = \Omega\left(\sum M_{g_n}^\sim \varphi_n\right).$$

Then $0 \le \sum M_{g_n}^\sim \varphi_n$ and in particular

$$0 \le \left(\sum M_{g_n}^\sim \varphi_n\right)(f) = \sum \varphi_n\left(M_{g_n} f\right) = \sum \varphi_n f_n.$$

□

Theorem 13 *The formula*

$$\gamma\left((f \circ \omega)\Omega\varphi\right) = \varphi(f) \qquad (f \in F,\ \varphi \in F_c^\sim)$$

defines a strictly positive integral γ on $FF_c^\sim$.

Proof It follows directly from Lemma 12 that the given formula defines a function γ on $FF_c^\sim$ and that this γ is linear and positive.

To prove that γ is strictly positive: Let $u \in FF_c^\sim$, $u > 0$. As $\Omega(F_c^\sim)$ (which is G) is order dense in $C^\infty(Y)$, there is a φ in $F_c^\sim$ with $0 < \Omega(\varphi) \le u$. Then $\varphi > 0$, so there is an f in F^+ with $\varphi(f) > 0$. For $n \in \mathbb{N}$ put $f_n = (n^{-1} f) \wedge \mathbb{1}$. Then $nf_n \uparrow f$, so (φ being an integral) $\varphi(f_n) > 0$ for some n. Now $0 \le (f_n \circ \omega)\Omega\varphi \le \Omega\varphi \le u$ whereas $\gamma\left((f_n \circ \omega)\Omega\varphi\right) = \varphi(f_n) > 0$.

To prove that γ is an integral: Let $u_n \downarrow 0$ in $FF_c^\sim$. In the language of the proof of Theorem 11, choose f in F^+ with $u_1 \in A_f$. Then $u_n \in A_f$ for every n, so there exist $\varphi_1, \varphi_2, \dots$ in $F_c^\sim$ with $u_n = (f \circ \omega)\Omega\varphi_n$ $(n \in \mathbb{N})$. Putting $\psi_n = |\varphi_1| \wedge \cdots \wedge |\varphi_n|$ we obtain $u_n = (f \circ \omega)\Omega\psi_n$ $(n \in \mathbb{N})$. With $\psi = \inf_n \psi_n$ we find $\psi \ge 0$ and $\psi_n \downarrow \psi$ in $F_c^\sim$. Consequently,

$$\psi_n(f) \downarrow \psi(f) \text{ in } \mathbb{R}$$

but also $\Omega\psi_n \downarrow \Omega\psi$ in $C^\infty(Y)$. The latter implies $u_n = (f \circ \omega)\Omega\psi_n \downarrow (f \circ \omega)\Omega\psi$ in $C^\infty(Y)$. Thus, $(f \circ \omega)\Omega\psi = 0$. Now $\gamma(u_n) = \gamma((f \circ \omega)\Omega\psi_n) = \psi_n(f) \downarrow \psi(f) = \gamma((f \circ \omega)\Omega\psi) = \gamma(0) = 0$. □

Theorem 14 *Let F be a Dedekind complete Riesz space. Choose (applying Theorem 7.29 in [2]) an extremally disconnected compact Hausdorff space Y and a Riesz isomorphism $\varphi \mapsto \hat{\varphi}$ of $F_c^\sim$ onto an order dense Riesz ideal of $C^\infty(Y)$. Then there exists a σ-order continuous Riesz homomorphism $f \mapsto \hat{f}$ of F into $C^\infty(Y)$ such that $A = \left\{\hat{\varphi}\hat{f} : \varphi \in F_c^\sim,\ f \in F\right\}$ is a Riesz ideal of $C^\infty(Y)$ and there exists a positive integral γ on A with*

$$\varphi(f) = \gamma\left(\hat{\varphi}\hat{f}\right) \qquad (\varphi \in F_c^\sim,\ f \in F).$$

15 It remains to obviate the Dedekind completeness condition and to reformulate the result in terms of integration. The first is easy:

Let E be a Riesz space and apply the above to $F = E_c^\sim$. For $x \in E$ let φ_x be the functional $f \mapsto f(x)$ $(f \in F)$. If $f_n \downarrow 0$ in F, then $g(x) = \lim_n f_n(x)$ exists for all $x \in E^+$, hence for all $x \in E$. Now $g \in E^{\sim+}$ and $g \leq f_n$, so $g = 0$. This means that for every $x \in E$ we have $\varphi_x(f_n) \to 0$. We see that $\varphi_x \in F_c^\sim$ for all x. Thus, we can define a Riesz homomorphism $x \mapsto \hat{x}$ of E into $C^\infty(Y)$ by: $\hat{x} = \hat{\varphi}_x$ $(x \in E)$.

The upshot is that for any Riesz space E we obtain an extremally disconnected compact Hausdorff space Y, σ-order continuous Riesz homomorphisms $x \mapsto \hat{x}$ and $f \mapsto \hat{f}$ of E and $E_c^\sim$ into $C^\infty(Y)$, and a positive integral γ on a Riesz ideal of $C^\infty(Y)$ such that

$$f(x) = \gamma\left(\hat{x}\hat{f}\right) \qquad (x \in E,\ f \in E_c^\sim).$$

Replacing the topology by integration theory is done in the following theorem, that is conceptually easy but presents some technical complications. (Little is new in the following passage.)

Theorem 16 *Let Y be an extremally disconnected compact Hausdorff space, A a Riesz ideal of $C^\infty(Y)$, and α a positive integral on A. Then there exist a measure space $(Y, \mathcal{W}, \mu)$ and a σ-order continuous multiplicative Riesz homomorphism $f \mapsto f_\mu$ of $C^\infty(Y)$ into $L^0(\mu)$ such that*

$$f \in A \quad \Longrightarrow \quad f_\mu \in L(\mu) \text{ and } \int f_\mu \,\mathrm{d}\mu = \alpha(f).$$

Proof The proof consists of six steps, (I)–(VI).

(I) For functions f, g on Y, set

$$f \leq g \text{ n.e. ("nearly everywhere")}$$

if the complement of the set $[f \leq g]$ is meager.

Similarly we use expressions such as "$f = g$ n.e.", "$f_n \to 0$ n.e.", "$U \subset V$ n.e." (where U and V are subsets of Y), etc.

Some simple observations:

(1a) If f, g are continuous and $f \le g$ n.e., then $f \le g$.
(1b) If U is an open set and $\overline{U}$ is its closure, then $U = \overline{U}$ n.e.
(1c) If for every n in $\mathbb{N}$ we have $U_n \subseteq V_n$ n.e., then $\bigcup_n U_n \subset \bigcup_n V_n$ n.e.
(1d) Let $f_1 \ge f_2 \ge \cdots$ in $C^\infty(Y)$. Then $f_n \downarrow 0$ in the sense of the Riesz space $C^\infty(Y)$ if and only if $f_n \downarrow 0$ n.e.

For a proof of (1d), assume $f_n \downarrow 0$ in $C^\infty(Y)$. For k in $\mathbb{N}$ the set $S_k = \{y \in Y\colon\ f_n(y) \ge k^{-1}$ for all $n\}$ is closed and has empty interior, so $\bigcup_k S_k$ is meager. But $f_n(y) \downarrow 0$ for every y outside $\bigcup_k S_k$. Hence $f_n \downarrow 0$ n.e.

(II) Define $\mathcal{W} = \{W \subset Y\colon\ W = U$ n.e. for some clopen $U \subset Y\}$. Then:

(2a) $\mathcal{W}$ contains all open sets. (Follows from (1b) and the fact that $\overline{U}$ is clopen.)
(2b) $\mathcal{W}$ contains all meager sets.
(2c) $\mathcal{W}$ is a σ-algebra. (If $W_1, W_2, \ldots \in \mathcal{W}$, then $\bigcup_n W_n \in \mathcal{W}$, thanks to (1c) and (2a).)
(2d) Every continuous function $Y \to \mathbb{R}$ is $\mathcal{W}$-measurable (by (2a)).

(III) Define $\mu\colon \mathcal{W} \to [0, \infty]$ by

$$\mu(W) = \sup\{\alpha(\mathbb{1}_U)\colon\ U \text{ clopen}, \mathbb{1}_U \in A,\ U \subset W \text{ n.e.}\}$$

Claim: μ is a measure; all meager sets are μ-negligible. Proof of the σ-additivity:

Let $W_1, W_2, \ldots$ be pairwise disjoint elements of $\mathcal{W}$. It is straightforward that $\mu(W_1) + \mu(W_2) \le \mu(W_1 \cup W_2)$ and, by extension, that $\sum_n \mu(W_n) \le \mu\left(\bigcup_n W_n\right)$. For the reverse inequality, let $U \subset Y$ be clopen, $\mathbb{1}_U \in A$, and $U \subset \bigcup_n W_n$ n.e. We are done if it follows that $\alpha(\mathbb{1}_U) \le \sum_n \mu(W_n)$. For each n there is a clopen set V_n with $V_n = W_n$ n.e. As A is an ideal in $C^\infty(Y)$ we have $\mathbb{1}_{U\cap V_v} \in A$. Now $\mathbb{1}_U = \sum_n \mathbb{1}_{U\cap W_n} = \sum_n \mathbb{1}_{U\cap V_n}$ n.e. It follows from (1d) that $\mathbb{1}_U = \sum_n \mathbb{1}_{U\cap V_n}$ in the sense of the Riesz space A. Then $\alpha(\mathbb{1}_U) = \sum_n \alpha\left(\mathbb{1}_{U\cap V_n}\right) \le \sum_n \mu(W_n)$ as $U \cap V_n \subset W_n$.

(IV) Take f in $C^\infty(Y)$ and form $f'\colon Y \to \mathbb{R}$ by setting $f'(y) = f(y)$ if $f(y)$ is finite, $f'(y) = 0$ otherwise. Claim: f' is $\mathcal{W}$-measurable and $f' = f$ n.e. Proof for $f \ge 0$:

Let $S = [f = \infty]$. Then S is meager, so $f' = f$ n.e. For $n \in \mathbb{N}$ the function $f \wedge n\mathbb{1}$ is continuous and finite-valued, hence $\mathcal{W}$-measurable (see 2d); then so is $\lim_n (f_n \wedge \mathbb{1})\mathbb{1}_{Y\setminus S}$, which is f'.

(V) For f in $C^\infty(Y)$, let f_μ be the collection of all $\mathcal{W}$-measurable functions that are equal to f μ-almost everywhere.

Then $f \mapsto f_\mu$ is a map $C^\infty(Y) \to L^0(\mu)$; it is a multiplicative Riesz homomorphism. To prove that it is also σ-order continuous, take a sequence $(f_n)_{n\in\mathbb{N}}$ in $C^\infty(Y)^+$ with $f_n \downarrow 0$ in the sense of $C^\infty(Y)$. Then $f_n \downarrow 0$ n.e.

(1d), so $f_n \downarrow 0$ μ-a.e. (by (3)). It follows that $f'_n \downarrow 0$ μ-a.e., whence $(f_n)_\mu \downarrow 0$ in $L^0(\mu)$, as f'_n is a representative of the class $(f_n)_\mu$.

(VI) Take f in A^+; we prove that f_μ is μ-integrable and $\int f_\mu \, d\mu = \alpha(f)$, or, equivalently (as f' is a positive $\mathcal{W}$-measurable function) that $\int f' \, d\mu = \alpha(f)$. First, assume there exist clopen sets $U_1, \ldots, U_N$ and positive numbers $t_1, \ldots, t_N$ such that $f = \sum_n t_n \mathbb{1}_{U_n}$. Then $\mathbb{1}_{U_n} \in A$ for each n, and

$$\alpha(f) = \sum_n t_n \alpha\left(\mathbb{1}_{U_n}\right) = \sum_n t_n \mu(U_n) = \int f \, d\mu \; (= \int f' \, d\mu).$$

Now the general case. Choose an enumeration $(q_n)_{n \in \mathbb{N}}$ of $\mathbb{Q} \cap (0, \infty)$. For each n in $\mathbb{N}$, let U_n be the (clopen) closure of $[f > q_n]$ and put

$$f_n = q_1 \mathbb{1}_{u_1} \vee \cdots \vee q_n \mathbb{1}_{U_n}.$$

By the above, $\alpha(f_n) = \int f'_n \, d\mu$.

Now $(f_n)_{n \in \mathbb{N}}$ is a sequence in A and $f_n \uparrow f$ pointwise. As α is an integral on A we find

$$\alpha(f) = \lim_n \alpha(f_n) = \lim_n \int f'_n \, d\mu = \int f' \, d\mu.$$

This concludes the proof of Theorem 16. □

Combining Theorem 16 with the conclusion in 15 yields:

Theorem 17 *Let E be a Riesz space. Then there exist a measure space $(Y, \mathcal{W}, \mu)$ and σ-order continuous Riesz homomorphisms $x \mapsto \hat{x}$ and $f \mapsto \hat{f}$ of E and $E_c^\sim$, respectively, into $L^0(\mu)$ such that*

$$\text{if } x \in E \text{ and } f \in E_c^\sim, \text{ then } \hat{f}\hat{x} \in L(\mu) \text{ and } \int \hat{f}\hat{x} \, d\mu = f(x).$$

Comments 18

(1) If $f \in E_c^\sim$ and $\hat{f} = 0$, then $f(x) = 0$ for all $x \in E$. Thus, the map $f \mapsto \hat{f}$ is always injective.

(2) Similarly, the map $x \mapsto \hat{x}$ is injective if the points of E are separated by the integrals. This gives us the theorem promised at the beginning of this article.

Postscript The above is closely related to a result of Wickstead [4], which (roughly speaking) states that there exist Riesz homomorphisms $\hat{}$ of E and $E^\sim$ into $L^0(\mu)$ such that $f(x) = \int \hat{f}\hat{x} \, d\mu$ for all $x \in E$ and $f \in E^\sim$. At first sight this may seem to be more general than our Theorem 17 in that $E^\sim$ is a larger space than $E_c^\sim$. However, in Wickstead's theorem the map $E \to L^0(\mu)$ cannot be σ-order continuous, except if $E^\sim$ equals $E_c^\sim$. (The map $E^\sim \to L^0(\mu)$ is order continuous, just like our map $E_c^\sim \to L^0(\mu)$.)

References

1. C.D. Aliprantis, O. Burkinshaw, *Positive Operators* (Academic, London, 1985)
2. C.D. Aliprantis, O. Burkinshaw, *Locally Solid Riesz Spaces with Applications to Economics*, 2nd edn. (A.M.S., Providence, 2003)
3. E. Chil, M. Meyer, On the centre of a vector lattice. Indag. Math. **23**, 167–183 (2012)
4. A.W. Wickstead, Representation and duality of multiplication operators on Archimedean Riesz spaces. Compos. Math. **35**, 225–238 (1977)

When Do the Regular Operators Between Two Banach Lattices Form a Lattice?

A. W. Wickstead

This work is dedicated to Ben de Pagter on the occasions of his 65th birthday and his impending retirement

Abstract We investigate when the regular operators from one Banach lattice into another form a Riesz space. We give complete results when the domain is either separable or has an order continuous norm. In these two settings, at least, the lattice operators are given by the Riesz-Kantorovich formulae, in contrast with Elliott's negative result for the general setting.

Keywords Banach lattice · Regular operator

1 Introduction

Ever since the initial studies of regular operators on Banach lattices, the order structure of the space of regular operators $\mathcal{L}^r(E, F)$ from a Banach lattice E into a Banach lattice F has been an important question. If the range space, F, is Dedekind complete, then $\mathcal{L}^r(E, F)$ is even a Dedekind complete Riesz space and the lattice operations are given by the Riesz-Kantorovich formulae.[1] At the other extreme, if E is atomic with an order continuous norm, then again $\mathcal{L}^r(E, F)$ is a Riesz space for any choice of F and again the lattice operations are given by the Riesz-Kantorovich

[1]This result, essentially, was proved independently around 1936 by each of Freudenthal [8], Kantorovich [9], and F. Riesz [12]. In spite of the historical justification for naming the Riesz-Kantorovich formulae after all three of these authors, we stick to what is by now the accepted terminology.

A. W. Wickstead (✉)
Mathematical Sciences Research Centre, Queen's University Belfast, Belfast, Northern Ireland
e-mail: A.Wickstead@qub.ac.uk

G. Buskes et al. (eds.), *Positivity and Noncommutative Analysis*,
Trends in Mathematics, https://doi.org/10.1007/978-3-030-10850-2_29

formulae.[2] After these early results, most authors have concentrated on the question of whether or not the modulus of a regular operator, if indeed it exists, must be given by the Riesz-Kantorovich formula. Since Elliott [7] has settled this question in the negative, I will discuss here the much more restricted question as to whether the lattice operations are given by the Riesz-Kantorovich formulae if *all* regular operators from E into F have a modulus, i.e. if $\mathcal{L}^r(E, F)$ is a Riesz space. The only obvious way to tackle this question seems to be to first describe exactly when $\mathcal{L}^r(E, F)$ is a Riesz space and then check whether or not the Riesz-Kantorovich formulae hold in these cases. Needless to say, we have not accomplished this task in its generality. We can, however, give a complete answer when the domain E is either separable or has an order continuous norm. In these cases it turns out that the lattice operations *are* all given by the Riesz-Kantorovich formulae.

The two cases listed above do not, of course, exhaust the cases when $\mathcal{L}^r(E, F)$ is known to be a Riesz space. In [17] there is a proof that if order intervals in E are separable and F is Dedekind σ-complete then $\mathcal{L}^r(E, F)$ is a lattice, although the theorem stated there actually assumes that the whole of E is separable. The proof indicated there implies that the Riesz-Kantorovich formulae hold for the lattice operations in this case. There is nothing special about the cardinal $\aleph_0$ that implicitly occurs in the conditions on both E and F here—a similar result is true for any infinite cardinal as we will note below.

Essentially, we suggest as a possibility (I am diffident about actually conjecturing this!) that the cases already listed are the only ones where $\mathcal{L}^r(E, F)$ is a Riesz space. We will make this statement precise later.

Our terminology is standard and all terms that we do not define below may be found in, for example, [1, 3, 4, 10, 11, 15] or [18]. Only sketch proofs of the results for the case of order continuous domain are given, and the interested reader must refer to Elliott's thesis [6], which I had the pleasure of supervising, for full details.

2 History

Recall that this topic of research essentially started with the result quoted above:

Theorem 2.1 *If E is any Banach lattice and F a Dedekind complete Banach lattice, then $\mathcal{L}^r(E, F)$ is a Dedekind complete Riesz space. Furthermore if $T \in \mathcal{L}^r(E, F)$ and $x \in E_+$, then*

$$|T|(x) = \sup\{Ty : -x \leq y \leq x\}.$$

[2] This actually characterizes atomic Banach lattices with an order continuous norm, which is due to van Rooij, see [13] and [14].

The formula for $|T|$ plus the similar ones for T^+ and T^- are commonly known as the *Riesz-Kantorovich formulae*. It has long been the subject of research to discover whether or not a regular operator between Banach lattices which has a modulus must have that modulus given by the Riesz-Kantorovich formula. We now know, thanks to Elliott's example [7], that this is not the case. However, the main interest in Theorem 2.1 is in the fact that $\mathcal{L}^r(E, F)$ is then a lattice. There are of course other cases when $\mathcal{L}^r(E, F)$ is a lattice. In 1993, a paper of myself and Yuri Abramovich contained a variant of this [2, Theorem 3.10], which assumed that E was separable and F Dedekind σ-complete, with the conclusion being that $\mathcal{L}^r(E, F)$ was a Dedekind σ-complete Riesz space in which the lattice operations were given by the Riesz-Kantorovich formulae. In fact, all that was really needed for E was that all order intervals are separable. There is nothing special about the cardinal $\aleph_0$ here and the same methods suffice to prove more, namely Theorem 2.3 below. We need a definition to state that result.

Definition 2.2 If $\mathfrak{a}$ is an infinite cardinal, then a Riesz space is *Dedekind $\mathfrak{a}$-complete* if every non-empty set of cardinality at most $\mathfrak{a}$, which is bounded above, has a supremum.

Theorem 2.3 *If $\mathfrak{a}$ is an infinite cardinal, E a Banach lattice in which every order interval has a dense subset of cardinality at most $\mathfrak{a}$ and F a Banach lattice which is Dedekind $\mathfrak{a}$-complete, then $\mathcal{L}^r(E, F)$ is a Dedekind $\mathfrak{a}$-complete Riesz space in which the lattice operations are given by the Riesz-Kantorovich formulae.*

On a different line, the following result is due to van Rooij [13, Theorem 10.2]. In fact van Rooij shows that this characterizes atomic Banach lattices with an order continuous norm.

Theorem 2.4 *If E is an atomic Banach lattice with an order continuous norm and F any Banach lattice, then $\mathcal{L}^r(E, F)$ is a Riesz space in which the lattice operations are given by the Riesz-Kantorovich formulae.*

3 When the Domain Is Separable

An important notion in these studies is the property identified by van Rooij in [13, Theorem 8.2].

Definition 3.1 A Banach lattice has *property* $(\star)$ if, for every sequence (f_n) in E^*_+ which converges $\sigma(E^*, E)$ to $f \in E^*_+$ as $n \to \infty$, we have $|f_n - f| \to 0$ for $\sigma(E^*, E)$ as $n \to \infty$.

He showed, in [13, Theorem 8.2], that E has property ($\star$) if and only if $\mathcal{L}^r(E, c)$ is a Riesz space. We also need [13, Theorem 8.12]:

Theorem 3.2 *If E is a Riesz space and F a uniformly complete Riesz space which is not Dedekind σ-complete and $\mathcal{L}^r(E, F)$ is a Riesz space, then $\mathcal{L}^r(E, c)$ is a Riesz space (and therefore E has property ($\star$).)*

Property ($\star$) is not very intuitive. Atomic Banach lattices with an order continuous norm have property ($\star$) and I know of no others.[3] [5, Theorem 3.1] states:

Theorem 3.3 *If a Banach lattice E has property ($\star$) and*

1. *E is separable,*
2. *$E = C_0(\Sigma)$ for some locally compact Hausdorff space Σ,*
3. *or E has an order continuous norm,*

then E is atomic with an order continuous norm.

Answering our original question when the domain is separable is now merely a question of piecing together known results.

Theorem 3.4 *If E is a separable Banach lattice and F any Banach lattice, then the following are equivalent:*

1. *Either E is atomic with an order continuous norm or F is Dedekind σ-complete.*
2. *$\mathcal{L}^r(E, F)$ is a Riesz space.*

In this case, the lattice operations in $\mathcal{L}^r(E, F)$ are given by the Riesz-Kantorovich formulae.

Proof That 1 implies 2 follows from the results of van Rooij, Theorem 2.4, and Abramovich-Wickstead [2, Theorem 3.10]. To see that 2 implies 1, van Rooij's result, Theorem 3.2, shows that either E has property ($\star$), and therefore is atomic with an order continuous norm as E is separable, or else F is Dedekind σ-complete. □

Let me emphasize that Theorem 3.3, and therefore the last result, does require that E be separable, rather than that all its order intervals are separable.

4 When the Domain Has an Order Continuous Norm

Full details of the results in this section may be found in Elliott's thesis [6]. In places we outline techniques that differ from his.

[3] After this paper was completed, Michael Elliott has informed the author that he can prove the existence of an atomic AM-space with property ($\star$) which does not have an order continuous norm.

As the results to be stated fundamentally involve Dedekind $\mathfrak{a}$-completeness, we must start by stating some relevant characterizations of the notion. First we need a generalization of Nakano's characterization of when $C(\Omega)$ is Dedekind σ-complete.

Definition 4.1 Let Ω be a topological space and $\mathfrak{a}$ be an infinite cardinal.

1. A subset Λ of Ω is an $F_{\mathfrak{a}}$-set if it is the union of at most $\mathfrak{a}$ closed subsets of Ω.
2. Ω is $\mathfrak{a}$-Stonean if the closure of every open $F_{\mathfrak{a}}$ subset is open.

From Elliott's thesis, [6, Theorem II.13], we need:

Theorem 4.2 *If Ω is a normal topological space and $\mathfrak{a}$ an infinite cardinal, then $C(\Omega)$ is Dedekind $\mathfrak{a}$-complete if and only if Ω is $\mathfrak{a}$-Stonean.*

We don't actually need this result directly, but it is an important tool in the proof of the following variant of a famous result of Veksler and Gejler [6, Theorem II.6].

Theorem 4.3 *If E is a uniformly complete Riesz space and $\mathfrak{a}$ an infinite cardinal, then the following are equivalent:*

1. *E is Dedekind $\mathfrak{a}$-complete.*
2. *Every pairwise disjoint order bounded set of cardinality at most $\mathfrak{a}$ has a supremum.*

Order continuity of the norm is also of obvious relevance here, so we need to look at the structure of Banach lattices which possess one. Luckily there is a good starting point for this in Maharam's characterization of abstract L-spaces. As stated in [16, Proposition 26.4.7], Maharam's representation involved products of Lebesgue measure of the unit interval, but the fact that its measure algebra is isometrically isomorphic to that of $\{-1,1\}^{\aleph_0}$ when equipped with the product of the measure γ on $\mathbf{2}=\{-1,1\}$ with $\gamma(\{-1\})=\gamma(\{1\})=\frac{1}{2}$, gives us:

Theorem 4.4 *If E is an abstract L-space, then there is a unique well-ordered family of cardinals, $(\mathfrak{a}_\sigma)_{-1\le\sigma<\tau}$, indexed by ordinals, such that:*

1. *for each $\sigma\ge 0$ each $\mathfrak{a}_\sigma$ is equal to 0 or to 1 or is uncountable,*
2. *$\{\sigma:\mathfrak{a}_\sigma\neq 0\}$ is cofinal in τ, and*
3. *E is isometrically order isomorphic to*

$$\ell_1(\mathfrak{a}_{-1})\oplus_1\ell_1(\mathfrak{a}_\sigma L_1(\mathbf{2}^{\aleph_\sigma},\gamma^{\aleph_\sigma});0\le\sigma<\tau).$$

In this result, $\mathfrak{a}F$ means an ℓ_1 direct sum of $\mathfrak{a}$ copies of F. To exploit this structure in more generality, recall from [11, Theorem 2.7.8]:

Theorem 4.5 *If E is a Banach lattice with an order continuous norm and a weak order unit, then there is a probability measure μ such that:*

1. *There is a linear order isomorphism $x\mapsto\hat{x}$ of E onto to a dense ideal in $L_1(\mu)$ containing $L_\infty(\mu)$.*
2. *There is a linear order isomorphism $f\mapsto\hat{f}$ of E^* onto to a dense ideal in $L_1(\mu)$ containing $L_\infty(\mu)$.*
3. *For all $x\in E$ and $f\in E^*$, $f(x)=\int\hat{f}\hat{x}\,d\mu$.*

Amemiya's theorem, [11, Theorem 2.4.8], is also needed:

Theorem 4.6 *If E, F, and G are Banach lattices with E being an ideal in both F and G and both F and G having order continuous norms, then on every order interval of E the norm topologies of F and G coincide as do their weak topologies.*

The consequence of this that we need is that when we embed a Banach lattice E with a weak order unit and an order continuous norm into $L_1(\mu)$ as above, a subset of an order interval in E is dense for the norm in E if and only if it is dense for the $L_1(\mu)$-norm.

This is relevant because in $L_1(\mathbf{2}^{\mathfrak{a}}, \gamma^{\mathfrak{a}})$ every order interval has density character precisely $\mathfrak{a}$, where the density character of a set is the cardinality of its smallest dense subset, see §26 of [16]. We call Banach lattices with this property *$\mathfrak{a}$-homogeneous.* It follows from Maharam's representation that every abstract L-space is a direct sum of an atomic band and $\mathfrak{a}$-homogenous bands for varying $\mathfrak{a}$. Theorem 4.5 and the consequence of Amemiya's theorem give the same decomposition for Banach lattices with an order continuous norm and a weak order unit and hence (decomposing into bands with weak order units) for every Banach lattice with an order continuous norm. The interested reader is referred to [6] for a representation-free proof of this decomposition.

We can now start proving something about the case when the domain has an order continuous norm.

Proposition 4.7 *Let $\mathfrak{a}$ be an infinite cardinal, E an $\mathfrak{a}$-homogeneous Banach lattice with an order continuous norm and a weak order unit, and F any Banach lattice. Then $\mathcal{L}^r(E, F)$ is a Riesz space if and only if F is Dedekind $\mathfrak{a}$-complete.*

Proof The "if" part of the claim follows from Theorem 2.3. For the converse, we suppose that E is actually an ideal in $L_1(\mathbf{2}^{\mathfrak{a}}, \gamma^{\mathfrak{a}})$ containing $L_\infty(\mathbf{2}^{\mathfrak{a}}, \gamma^{\mathfrak{a}})$. In order to prove that F is Dedekind $\mathfrak{a}$-complete, it suffices by Theorem 4.3 to prove that every disjoint family $(y_{\mathfrak{b}})_{\mathfrak{b}\in\mathfrak{a}}$ in F_+ with upper bound y has a supremum. For each $\mathfrak{b} \in \mathfrak{a}$ we may define a function on $\mathbf{2}^{\mathfrak{a}}$ taking the $\mathfrak{b}$'th component. Call this function $e_{\mathfrak{b}}$ when we regard it as a member of E and $\phi_{\mathfrak{b}}$ when considered as a member of E^*. Define also $\mathbf{1}$ to be the constantly one function.[4] Define operators $S, T : E \to F$ by $S(x) = \sum_{\mathfrak{b}\in\mathfrak{a}} \phi_{\mathfrak{b}}(x) y_{\mathfrak{b}}$ and $T(x) = \mathbf{1}(x)y$. It is routine to show that the series for S converges, and that $T \geq \pm S$. As we are assuming that $\mathcal{L}^r(E, F)$ is a lattice, $T \geq |S|$. For each $\mathfrak{b} \in \mathfrak{a}$, we have

$$|S|(\mathbf{1}) = |S|(|e_{\mathfrak{b}}|) = |S|(e_{\mathfrak{b}}^+ + e_{\mathfrak{b}}^-) \geq S(e_{\mathfrak{b}}^+) + (-S)(e_{\mathfrak{b}}^-) = S(e_{\mathfrak{b}}) = y_{\mathfrak{b}}.$$

Therefore $|S|(\mathbf{1})$ is an upper bound for the $(y_{\mathfrak{b}})_{\mathfrak{b}\in\mathfrak{a}}$. But also, $|S|(\mathbf{1}) \leq T(\mathbf{1}) = \mathbf{1}(\mathbf{1})y = y$. As y was any upper bound for $(y_{\mathfrak{b}})_{\mathfrak{b}\in\mathfrak{a}}$ we see that $|S|(\mathbf{1})$ is the supremum of $(y_{\mathfrak{b}})_{\mathfrak{b}\in\mathfrak{a}}$ and the proof is complete. □

[4] These functions are simply, apart from numbering, an uncountable analogue of the usual Rademacher functions.

Putting together this result, Theorem 2.4, and the observation that if E has an order continuous norm and is decomposed into an order direct sum $E = \bigoplus_{\alpha \in A} E_\alpha$ then $\mathcal{L}^r(E, F)$ is a Riesz space if and only if each $\mathcal{L}^r(E_\alpha, F)$ is a Riesz space gives us:

Theorem 4.8 *Let E be a Banach lattice with an order continuous norm and F be any Banach lattice. Let $\mathfrak{a}$ be the smallest cardinal that is strictly greater than the density character of every order interval in E. The following are equivalent:*

1. *Either E is atomic with an order continuous norm or every order bounded subset of F, of cardinality strictly less than $\mathfrak{a}$, which is bounded above has a supremum.*
2. *$\mathcal{L}^r(E, F)$ is a Riesz space.*

Furthermore, in this case the lattice operations in $\mathcal{L}^r(E, F)$ are given by the Riesz-Kantorovich formulae.

An extremely bold (and very unlikely) conjecture would be that order continuity of the norm could be dropped from the statement of this theorem.

5 Can We Simplify the Statement of the Last Result?

The last result uses a cardinality assumption that seems rather convoluted. Surely we must actually have the following result?

Conjecture 5.1 Let E be a Banach lattice with an order continuous norm with $\mathfrak{a}$ being the smallest cardinal that is greater than or equal to the density character of every order interval in E and let F be any Banach lattice. The following are equivalent:

1. Either E is atomic with an order continuous norm or F is Dedekind $\mathfrak{a}$-complete.
2. $\mathcal{L}^r(E, F)$ is a Riesz space.

Furthermore, in this case the lattice operations in $\mathcal{L}^r(E, F)$ are given by the Riesz-Kantorovich formulae.

Recall that a *weakly inaccessible* cardinal is a regular limit cardinal. Their existence cannot be proved in ZFC. Conjecture 5.1 can be proved in the expected manner if we assume the non-existence of weakly inaccessible cardinals. Surprisingly, the truth of the conjecture implies the non-existence of weakly inaccessible cardinals! Thus the conjecture is actually equivalent to the non-existence of inaccessible cardinals. The following result is due to Elliott [6, Example 5.3].

Theorem 5.2 *Suppose that $\mathfrak{a}$ is a weakly inaccessible cardinal. There exist an AL-space E and AM-space F such that*

1. *$\mathfrak{a}$ is the smallest cardinal such that every order interval in E is $\mathfrak{a}$-separable.*
2. *$\mathcal{L}^r(E, F)$ is a Riesz space, but*
3. *E is not atomic and F is not Dedekind $\mathfrak{a}$-complete.*

Proof Let τ be the ordinal such that $\mathfrak{a} = \aleph_\tau$. We take $E = \ell_1\big(L_1(\mathbf{2}^{\aleph_\sigma}, \gamma^{\aleph_\sigma})\big)_{\sigma\in\tau}$. Each $L_1(\mathbf{2}^{\aleph_\sigma}, \gamma^{\aleph_\sigma})$ is $\aleph_\sigma$-homogeneous. It follows easily that $\mathfrak{a}$ is the smallest cardinal such that every order interval in X is $\mathfrak{a}$-separable. In fact every element of X lies in the sum of only countably many of these factors, so any order interval lies within such a countable sum. As $\mathfrak{a}$ is uncountable and regular, the supremum ρ of any countable family $\{\sigma_n : n \in \mathbb{N}\}$ will be less than τ. It follows that any order interval has density character at most $\aleph_\rho < \aleph_\tau = \mathfrak{a}$. Thus every order interval in E has density character strictly less than $\mathfrak{a}$. Note that E is certainly not purely atomic.

Take F the space of all bounded functions $f : \mathfrak{a} \to \mathbb{R}$ such that there is a real number ℓ with the property that for all $\epsilon > 0$ the set $\{\mathfrak{b} \in \mathfrak{a} : |f(\mathfrak{b}) - \ell| > \epsilon\}$ has cardinality less than $\mathfrak{a}$. As $\mathfrak{a}$ is regular, F can be shown to be Dedekind $\mathfrak{b}$-complete for all $\mathfrak{b} < \mathfrak{a}$, and also not to be Dedekind $\mathfrak{a}$-complete. Theorem 4.8 above tells us that $\mathcal{L}^r(E, F)$ is a Riesz space. □

References

1. Y.A. Abramovich, C.D. Aliprantis, *An Invitation to Operator Theory*. Graduate Studies in Mathematics, vol. 50 (American Mathematical Society, Providence, 2002). MR1921782
2. Y.A. Abramovich, A.W. Wickstead, The regularity of order bounded operators into C(K). II. Quart. J. Math. Oxford Ser. (2) **44**(175), 257–270 (1993). https://doi.org/10.1093/qmath/44.3.257.MR1240470
3. C.D. Aliprantis, O. Burkinshaw, *Positive Operators*. Pure and Applied Mathematics, vol. 119 (Academic, Orlando, 1985). MR809372 (87h:47086)
4. C.D. Aliprantis, O. Burkinshaw, *Positive Operators* (Springer, Dordrecht, 2006). Reprint of the 1985 original, MR2262133
5. Z.L. Chen, A.W. Wickstead, Equalities involving the modulus of an operator. Math. Proc. R. Ir. Acad. **99A**(1), 85–92 (1999). MR1883067
6. M.E. Elliott, Abstract Rademacher systems in Banach lattices. Ph.D. thesis, Queen's University of Belfast, Belfast, 2001
7. M.E. Elliott, The Riesz-Kantorovich formulae (accepted)
8. H. Freudenthal, Teilweise geordnete Moduln. Nederl. Akad.Wetensch. Proc. Ser. A **39**, 641–651 (1936)
9. L.V. Kantorovich, Concerning the general theory of operations in partially ordered spaces. Dok. Akad. Nauk. SSSR **1**, 271–274 (1936) (Russian)
10. W.A.J. Luxemburg, A.C. Zaanen, *Riesz Spaces*, vol. I (North-Holland, Amsterdam; American Elsevier Publishing Co., New York, 1971). North-Holland Mathematical Library, MR0511676
11. P. Meyer-Nieberg, *Banach Lattices*. Universitext (Springer, Berlin, 1991). MR1128093 (93f:46025)
12. F. Riesz, Sur quelques notions fondamentals dans la theorie générale des opérations linéaires. Ann. Math. **41**, 174–206 (1940) [This work was originally published in Hungarian in 1936]
13. A.C.M. van Rooij, When do the regular operators between two Riesz spaces form a Riesz space? Technical Report 8410, Katholieke Universiteit, Nijmegen, 1984
14. A.C.M. van Rooij, On the space of all regular operators between two Riesz spaces. Nederl. Akad. Wetensch. Indag. Math. **47**(1), 95–98 (1985). MR783009 (86k:46011)
15. H.H. Schaefer, *Banach Lattices and Positive Operators* (Springer, New York, 1974). Die Grundlehren der mathematischen Wissenschaften, Band 215, MR0423039

16. Z. Semadeni, *Banach Spaces of Continuous Functions*, vol. I (PWN|Polish Scientific Publishers, Warsaw, 1971). Monografie Matematyczne, Tom 55, MR0296671
17. A.W. Wickstead, The regularity of order bounded operators into C(K). Quart. J. Math. Oxford Ser. (2) **41**(163), 359–368 (1990). MR1067490 (91k:47086)
18. A.C. Zaanen, *Riesz Spaces. II*. North-Holland Mathematical Library, vol. 30 (North-Holland, Amsterdam, 1983). MR704021

Lexicographic Cones and the Ordered Projective Tensor Product

Marten Wortel

Dedicated to Ben de Pagter on the occasion of his 65th birthday

Abstract We introduce lexicographic cones, a method of assigning an ordered vector space $\mathrm{Lex}(S)$ to a poset S, generalising the standard lexicographic cone. These lexicographic cones are then used to prove that the projective tensor cone of two arbitrary cones is a cone, and to find a new characterisation of finite-dimensional vector lattices.

Keywords Lexicographic cone · Finite-dimensional vector lattices · Ordered projective tensor product

1 Introduction

In the theory of Archimedean vector lattices, the Fremlin projective tensor product is an important tool with many applications. The Fremlin projective tensor product was introduced by Fremlin in [3], and it satisfies the usual universal property for Riesz bimorphisms. However, the construction of the Fremlin tensor product is fairly complicated and uses representation theory.

In [4], Grobler and Labuschagne gave an easier construction of the Fremlin projective tensor product. A crucial ingredient in their construction is the wedge generated by tensors of positive elements in the algebraic tensor product, called the projective cone (the projective cone of ordered vector spaces was introduced and investigated earlier, cf. [1, 2, 5–8]). Amongst other things, they show that the projective cone of two Archimedean ordered vector spaces with the Riesz decomposition property is actually a cone [4, Theorem 2.5]; despite its name, it

---

M. Wortel (✉)
Department of Mathematics and Applied Mathematics, University of Pretoria, Pretoria, South Africa
e-mail: marten.wortel@up.ac.za

G. Buskes et al. (eds.), *Positivity and Noncommutative Analysis*,
Trends in Mathematics, https://doi.org/10.1007/978-3-030-10850-2_30

is a priori not clear at all that the projective cone is a cone. In [10, Theorem 3.3], this result was extended by van Gaans and Kalauch to Archimedean ordered vector spaces, removing the Riesz decomposition requirement.

In this paper we go one step further and show that the projective tensor cone of two arbitrary ordered vector spaces is a cone, removing the Archimedean requirement. Our main tool is lexicographic cones, which is a method for assigning an ordered vector space $\mathrm{Lex}(S)$ to any poset S. In finite dimensions, by choosing S appropriately, this generates the standard lexicographic cone $\mathbb{R}^d_{lex}$, the standard cone $\mathbb{R}^d_+$, and many new intermediate ordered vector spaces. It turns out that the projective tensor product of these lexicographic cones has a very nice description, cf. Proposition 4.2, which allows us to prove the above-mentioned result.

In [9, Theorem 3.9], Schaefer gave a recursive characterisation of finite-dimensional vector lattices. It turns out that these can be reformulated in terms of $\mathrm{Lex}(S)$ for appropriate S, so the lexicographic cones also yield an alternative, direct characterisation of finite-dimensional vector lattices, cf. Theorem 3.5.

We briefly explain the structure of the paper. In Sect. 2, we start by introducing $\mathrm{Lex}(S)$ and proving some basic properties of these ordered vector spaces; we also investigate the dual cone of these spaces. We characterise when $\mathrm{Lex}(S)$ is a vector lattice in Sect. 3 and prove the characterisation of finite-dimensional vector lattices mentioned above. In Sect. 4, we investigate the projective tensor product of these lexicographic cones and prove the main result that this is a cone.

2 Lexicographic Cones

A *wedge* C in a vector space X is a convex subset satisfying $C + C \subset C$ and $\lambda C \subset C$ for all $\lambda \geq 0$. A wedge C is called a *cone* if $C \cap -C = \{0\}$. An *ordered vector space* is a vector space X equipped with a linear order, i.e., if $x, y \in X$ and $x \leq y$, then $x + z \leq y + z$ for all $z \in X$ and $\lambda x \leq \lambda y$ for all $\lambda \geq 0$. A linear order on X generates the cone of positive elements $C := \{x \in X : x \geq 0\}$, and conversely, every cone C generates a linear order defined by $x \leq y$ if and only if $y - x \in C$.

In this paper, S will always denote a poset (partially ordered set). For $s \in S$, we denote the set $\{t \in S : t < s\}$ by $\langle s)$ and the set $\{t \in S : t \leq s\}$ is denoted by $\langle s]$. The symbols $(s\rangle$ and $[s\rangle$ have similar meaning.

Let $F_0(S)$ be the vector space of finitely supported real-valued functions on S. Then $\mathrm{Lex}(S)$ is defined to be the vector space $F_0(S)$ equipped with the cone

$$\mathrm{Lex}(S)_+ := \{f \in F_0(S) : f(s) < 0 \Rightarrow \exists t < s \text{ with } f(t) > 0\}.$$

Lemma 2.1 $\mathrm{Lex}(S)_+$ *is a cone.*

Proof To show that $\mathrm{Lex}(S)_+$ is a wedge, let $f, g \in \mathrm{Lex}(S)_+$. If $s_0 \in I$ is such that $(f + g)(s_0) < 0$, then either $f(s_0) < 0$ or $g(s_0) < 0$; assume the first case. Then there is an $s_1 < s_0$ with $f(s_1) > 0$. If $g(s_1) \geq 0$ then $(f + g)(s_1) > 0$, and if $g(s_1) < 0$ then there is an $s_2 < s_1$ with $g(s_2) > 0$. If $f(s_2) \geq 0$ then $(f+g)(s_2) > 0$, and if $f(s_2) < 0$ then there is an $s_3 < s_2$ with $f(s_3) > 0$. Hence either $(f+g)(s_k) > 0$ for some $k \in \mathbb{N}$ or $f(s_{2n+1}) > 0$ and $g(s_n) > 0$ for all $n \in \mathbb{N}$, the latter case contradicting the fact that f and g are finitely supported. Since $s_k < s_0$, $f + g \in \mathrm{Lex}(S)_+$.

We now show that $\mathrm{Lex}(S)_+$ is a cone, so suppose $\pm f \in \mathrm{Lex}(S)_+$ and that $f(s_0) < 0$ for some $s_0 \in S$. Then $f(s_1) > 0$ for some $s_1 < s_0$, and so $-f(s_1) < 0$ which implies that $-f(s_2) > 0$ for some $s_2 < s_1$. Repeating this argument shows that f is supported on an infinite set, which contradicts $f \in F_0(S)$. Hence $\mathrm{Lex}(S)_+$ is a cone. □

Denote by e_s the function $t \mapsto \delta_{st}$, then $\{e_s\}_{s \in S}$ forms a basis for $\mathrm{Lex}(S)$. If $S = \{1, \ldots, d\}$ with the standard ordering, then $\mathrm{Lex}(S)$ is the usual lexicographic cone $\mathbb{R}^d_{lex}$, whereas if $S = \{1, \ldots, d\}$ with no elements comparable, then $\mathrm{Lex}(S)$ is the standard cone $\mathbb{R}^d_+$. Generalising the previous example, if S is an arbitrary disjoint union of posets S_k, then it is easy to see that $\mathrm{Lex}(S) \cong \bigoplus_k \mathrm{Lex}(S_k)$ (this direct sum is an order direct sum: an element is positive if and only if all its components are positive).

Let $F(S)$ be the vector space of real-valued functions on S. Under the natural duality $\langle f, g \rangle := \sum_{s \in S} f(s)g(s)$, the space $F(S)$ can be identified with the algebraic dual of $F_0(S)$. We now identify $\mathrm{Lex}(S)^*_+$, the dual cone of $\mathrm{Lex}(S)_+$. The set $F(S)_+$ denotes the functions $g \in F(S)$ for which $g(s) \geq 0$ for all $s \in S$.

Lemma 2.2 $\mathrm{Lex}(S)^*_+ \cong \{g \in F(S)_+ \colon \mathrm{supp}(g) \subset \{s \in S \colon s \text{ } \textit{is minimal}\}\}$.

Proof Let g be supported on the minimal elements of S. Any $f \in \mathrm{Lex}(S)_+$ is nonnegative on minimal elements of S, so $\langle f, g \rangle \geq 0$ for all $f \in \mathrm{Lex}(S)_+$, hence $g \in \mathrm{Lex}(S)^*_+$. Conversely, if $g(s) > 0$ for a nonminimal element s and $t < s$, then $f_n := e_t - ne_s \in \mathrm{Lex}(S)_+$ and $\langle f_n, g \rangle < 0$ for large enough n, so $g \notin \mathrm{Lex}(S)^*_+$. □

In particular, if S contains no minimal elements, then $\mathrm{Lex}(S)^*_+$ is trivial.

3 Lattices

First we will investigate when $\mathrm{Lex}(S)$ is a vector lattice. Let $\bigwedge := \{s, t, m\}$ where $s, t < m$ are the only nontrivial relations. Then $\mathrm{Lex}(\bigwedge) \cong (\mathbb{R}^3, C)$ where

$$C := \{(x, y, z) \colon x, y \geq 0, z \in \mathbb{R}\} \setminus \{(0, 0, z) \colon z < 0\}.$$

The set of upper bounds of the zero vector and $(1, -1, -1)$ equals

$$\{(x, y, z) \colon x \geq 1, y \geq 0, z \in \mathbb{R}\},$$

and if (x, y, z) is in this set, then $(x, y, z - 1)$ is a smaller element in this set, so it has no least element. Therefore the zero vector and $(1, -1, -1)$ have no supremum, and so $\mathrm{Lex}(\bigwedge)$ is not a vector lattice. We will show that this is in some sense the only possible counterexample.

Definition 3.1 A poset S is called a *forest* if for each $s \in S$, the set $\langle s)$ is totally ordered. A forest is called a *tree* if every two elements have a common lower bound. A *root* of a tree S is a minimal element of S.

The forests are precisely the disjoint unions of trees. Note that a tree may not have a root (e.g. $\mathbb{Z}$), but if it exists it is unique, and every finite tree has a root.

Remark 3.2 Note that a common definition of a tree in the literature requires the initial segments to be well-ordered, not just totally ordered. For finite sets, both definitions coincide.

If S is a forest, then for $m \in S$, the set $\langle m)$ is totally ordered, hence S contains no subposet isomorphic to $\bigwedge$. If S is not a forest, then for some $m \in S$, two predecessors are incomparable, hence S contains a subposet isomorphic to $\bigwedge$. Therefore S is a forest if and only if it contains no subposet isomorphic to $\bigwedge$.

Remark 3.3 In the next theorem, we will be concerned with possible suprema of a function $f \in \mathrm{Lex}(S)$ and 0. We claim that any upper bound g not supported on $\mathrm{supp}(f)$ is not a supremum. Indeed, if $s \notin \mathrm{supp}(f)$ and $g(s) > 0$, then $g - (g(s)/2)e_s$ is a lower upper bound of f and 0, and if $g(s) < 0$, then $g - e_s$ is a lower upper bound of f and 0. Hence it suffices to only consider functions supported on $\mathrm{supp}(f)$, and so we may assume that $S = \mathrm{supp}(f)$.

Theorem 3.4 $\mathrm{Lex}(S)$ *is a vector lattice if and only if* S *is a forest.*

Proof Suppose S is not a forest. Let $\{s, t, m\} \subset S$ be isomorphic to $\bigwedge$, and let $f := e_s - e_t - e_m$. By Remark 3.3 and the example at the beginning of this section, f and 0 have no supremum.

Conversely, suppose that S is a forest, and let $f \in \mathrm{Lex}(S)$; we will compute $f \vee 0$. By Remark 3.3 we may assume that $S = \mathrm{supp}(f)$ is a finite forest, which is a disjoint union of finite trees S_k. Then $\mathrm{Lex}(S)$ is a finite order direct sum of $\mathrm{Lex}(S_k)$, and since the order is coordinatewise, it suffices to consider a single $\mathrm{Lex}(S_k)$. Let $s \in \mathrm{supp}(f)$ be the root of S_k. If $f(s) > 0$, then $f > 0$, and so $f \vee 0 = f$, and if $f(s) < 0$, then $f < 0$, and so $f \vee 0 = 0$. □

The class $\mathrm{Lex}(S)$ where S is a forest without minimal element is a class of vector lattices with no nontrivial positive functionals (cf. Lemma 2.2); a particular example is $\mathrm{Lex}(\mathbb{Z})$. (Another example is $L^p(0, 1)$ for $0 < p < 1$; here one can show that positive functionals are automatically continuous and that there are no nontrivial continuous functionals.)

Each finite forest S yields a finite-dimensional vector lattice $\mathrm{Lex}(S)$, and we will show that those are the only finite-dimensional vector lattices. If X is a finite-dimensional vector lattice, then [9, Theorem 3.9] shows that X is a direct sum of $\mathbb{R} \circ M$'s, where $\circ$ denotes the lexicographic union and M is a maximal ideal in $\mathbb{R} \circ M$. Each maximal ideal is of the same form, so this yields a recursive characterisation of finite-dimensional vector lattices. Our result below yields an alternative, non-recursive characterisation.

Theorem 3.5 *An ordered vector space X is a finite-dimensional vector lattice if and only if it is isomorphic to* $\mathrm{Lex}(S)$ *for some finite forest S.*

Proof If S is a finite forest, then the above theorem shows that $\mathrm{Lex}(S)$ is a vector lattice. (Alternatively, one can prove this directly: let $s_1, \ldots, s_n$ be the roots of S. Let $M_k := \mathrm{Lex}(\{s \in S : s > s_k\})$, then $S \cong \oplus_{k=1}^n \mathbb{R} \circ M_k$. Now the ordered vector space M_k is of smaller dimension, so it follows by induction on the dimension.)

Conversely, suppose X is a finite dimensional vector lattice. By induction on $\dim(X)$ we will show that it is isomorphic to $\mathrm{Lex}(S)$ for some finite forest S. The result is obvious if $\dim(X) = 1$. Suppose it holds for all dimensions $1 \leq k \leq d$ and suppose $\dim(X) = d + 1$, then $X \cong \oplus_{k=1}^n \mathbb{R} \circ M_k$ by Schaefer [9, Theorem 3.9]. The induction hypothesis now implies that $M_k \cong \mathrm{Lex}(S_k)$ for some finite forest S_k. Let S_k' be the forest S_k adjoined with a new element that is below every element in S_k, then $\mathbb{R} \circ M_k \cong \mathrm{Lex}(S_k')$ by the definition of lexicographic union. Hence $X \cong \oplus_{k=1}^n \mathrm{Lex}(S_k') \cong \mathrm{Lex}(S)$ where S is the forest defined by the disjoint union of the trees S_k'. □

4 Projective Tensor Product

If X and Y are ordered vector spaces, then the *projective tensor product* of X and Y is the vector space $X \otimes Y$ equipped with the *projective cone*

$$K_p = K_p(X, Y) := \left\{ \sum_{i=1}^n x_i \otimes y_i : n \in \mathbb{N}, x_i \in X_+, \in Y_+ \right\}.$$

It is obvious that K_p is a wedge.

Our next goal is to show that K_p is actually a cone. Let X be an ordered vector space. A set $G \subset X_+$ is a *generating set* for X_+ if every $x \in X_+$ can be written as a positive linear combination of elements of G. A set of positive elements such that its positive linear span contains a generating set is obviously generating. Suppose G is a generating set for X_+ and H is a generating set for Y_+. Then if $u = \sum_{i=1}^n x_i \otimes y_i \in K_p \subset X \otimes Y$, then there exist $\lambda_{ij} \geq 0$, $g_{ij} \in G$ and $h_{ij} \in H$ such that $x_i = \sum_{j=1}^{n_i} \lambda_{ij} g_j$ and $y_i = \sum_{k=1}^{m_i} \mu_{ik} h_k$. Therefore

$$u = \sum_{i=1}^n x_i \otimes y_i = \sum_{i=1}^n \sum_{j=1}^{n_i} \sum_{k=1}^{m_i} \lambda_{ij} \mu_{ik} (g_{ij} \otimes h_{ik}),$$

and so $G \otimes H := \{g \otimes h : g \in G, h \in H\}$ is a generating set for $K_p(X, Y)$.

Lemma 4.1 *The set*

$$G := \{e_s - \lambda e_t : \lambda > 0, s < t\} \cup \{e_s : s \in S\}$$

is a generating set for $\mathrm{Lex}(S)_+$.

Proof We will show that every $f \in \mathrm{Lex}(S)_+$ can be written as a positive linear combination of elements of G by induction on $|\operatorname{supp}(f)|$. If $|\operatorname{supp}(f)| = 1$, then the result is obvious. Suppose it holds for every f with $1 \leq |\operatorname{supp}(f)| \leq n$, and take an $f \in \mathrm{Lex}(S)_+$ with $|\operatorname{supp}(f)| = n + 1$. Then there is an $s \in S$ with $f(s) > 0$. Write $f = f|_{[s\rangle} + f|_{[s\rangle^c}$. Clearly $f|_{[s\rangle}$ is positive since s is the smallest element in its support and $f(s) > 0$. To show that $f|_{[s\rangle^c}$ is positive, let $t \in [s\rangle^c$ with $f(t) < 0$, then there is an $r < t$ with $f(r) > 0$. If $r > s$, then $t > r > s$, contradicting $t \in [s\rangle^c$, so $r \in [s\rangle^c$, and therefore $f|_{[s\rangle^c} \geq 0$.

Let $T^- := \{t \in S : t > s, f(t) < 0\}$ and $T^+ := \{t \in S : t > s, f(t) > 0\}$. If $T^- = \emptyset$ then

$$f|_{[s\rangle} = \sum_{t \in T^+} f(t)e_t + f(s)e_s,$$

and if $T^- \neq \emptyset$ then

$$f|_{[s\rangle} = \sum_{t \in T^+} f(t)e_t + \sum_{t \in T^-} \left(\frac{f(s)}{|T^-|} e_s - |f(t)| e_t \right).$$

In both cases $f|_{[s\rangle}$ can be written as a positive linear combination of elements of G, and the same holds for $f|_{[s\rangle^c}$ by the induction hypothesis. Hence $f = f|_{[s\rangle} + f|_{[s\rangle^c}$ can be written as a positive linear combination of elements of G. □

Using this lemma we can show that the projective tensor product behaves well with respect to lexicographic cones.

Proposition 4.2 *Let S and T be posets. Then there is a natural linear isomorphism* $\mathrm{Lex}(S) \otimes \mathrm{Lex}(T) \cong \mathrm{Lex}(S \times T)$*, and under this isomorphism,* $K_p \subset \mathrm{Lex}(S) \otimes \mathrm{Lex}(T)$ *satisfies* $K_p \cong \mathrm{Lex}(S \times T)_+$.

Proof Let $\{e_s\}_{s \in I}$, $\{f_t\}_{t \in T}$ and $\{g_{s,t}\}_{(s,t) \in S \times T}$ be the natural bases of $\mathrm{Lex}(S)$, $\mathrm{Lex}(T)$ and $\mathrm{Lex}(S \times T)$; then $e_s \otimes f_t \mapsto g_{s,t}$ induces the natural linear isomorphism $\mathrm{Lex}(S) \otimes \mathrm{Lex}(T) \cong \mathrm{Lex}(S \times T)$.

By Lemma 4.1,

$$G = \{e_{s_1} - \lambda e_{s_2} : \lambda > 0, s_1 < s_2\} \cup \{e_s : s \in S\}$$

is a generating set for $\mathrm{Lex}(S)$,

$$H = \{f_{t_1} - \lambda f_{t_2} : \lambda > 0, t_1 < t_2\} \cup \{f_t : t \in T\}$$

is a generating set for Lex(T), and a generating set of Lex($S \times T$) is given by

$$\{g_{s_1,t_1} - \alpha g_{s_2,t_2} : \alpha > 0, (s_1, t_1) < (s_2, t_2)\} \cup \{g_{s,t} : (s, t) \in S \times T\}.$$

We will compute $G \otimes H$ in Lex($S \times T$). Let $s \in S$, $t \in T$, $s_1 < s_2$, $t_1 < t_2$ and $\lambda, \mu > 0$, then

$$\begin{aligned}(e_{s_1} - \lambda e_{s_2}) \otimes (f_{t_1} - \mu f_{t_2}) &\cong g_{s_1,t_1} - \lambda g_{s_2,t_1} - \mu g_{s_1,t_2} + \lambda\mu g_{s_2,t_2}\\ (e_{s_2} - \lambda e_{s_2}) \otimes f_t &\cong g_{s_1,t} - \lambda g_{s_2,t}\\ e_s \otimes (f_{t_1} - \mu f_{t_2}) &\cong g_{s,t_1} - \mu g_{s,t_2}\\ e_s \otimes f_t &\cong g_{s,t}.\end{aligned}$$

All elements on the right-hand side are positive in Lex($S \times T$). It now suffices to show that the positive span of these elements contains a generating set for Lex($S \times T$). For that, we have to show that for $s_1 < s_2$ and $t_1 < t_2$ (the case where $s_1 = s_2$ or $t_1 = t_2$ is already covered by the second and third line above) and $\alpha > 0$, $g_{s_1,t_1} - \alpha g_{s_2,t_2}$ is a positive linear combination of these elements. Indeed,

$$g_{s_1,t_1} - \alpha g_{s_2,t_2} = (g_{s_1,t_1} - g_{s_2,t_1}) + (g_{s_2,t_1} - \alpha g_{s_2,t_2}). \qquad \square$$

For the next theorem we need a proposition about maximal cones in finite-dimensional vector spaces, and for this we need some preparations.

Lemma 4.3 *Let C be a cone in a vector space X. Then C is contained in a maximal cone. Moreover, the following are equivalent:*

(i) *C is maximal;*
(ii) *$C \cup -C = X$;*
(iii) *(X, C) is totally ordered.*

Proof The first statement easily follows from Zorn's Lemma.

To prove $(i) \Rightarrow (ii)$, suppose $x \in X \setminus (C \cup -C)$. Towards showing that the wedge generated by C and x is a cone, let $\lambda, \mu \geq 0$ and $y, z \in C$ be such that $\lambda x + y = -\mu x - z$. Then $(\lambda + \mu)x = -y - z \in -C$, which is only possible if $\lambda = \mu = 0$ and $y = z = 0$. Hence the wedge generated by x and C is a cone, contradicting the maximality of C.

$(ii) \Rightarrow (i)$ is obvious.

To prove $(ii) \Leftrightarrow (iii)$, note that (X, C) being totally ordered means that every element of X is either positive or negative, i.e., $C \cup -C = X$. $\square$

Proposition 4.4 *Let X be an ordered vector space of dimension d. Then X_+ is contained in a maximal cone which is isomorphic to $\mathbb{R}^d_{lex}$.*

Proof By Lemma 4.3, X_+ is contained in a maximal cone C which induces a total order. A totally ordered vector space is obviously a vector lattice, so $C \cong \text{Lex}(S)_+$

by Theorem 3.5. The total order forces S to have no incomparable elements, hence S must be equal to $\{1, \ldots, d\}$ with the usual order, and so $\text{Lex}(S)_+ \cong \mathbb{R}^d_{lex}$. □

An alternative proof of Proposition 4.4, which does not rely on the results from Sect. 3, follows from the following results, whose easy proof is left to the reader.

Lemma 4.5 *If W is a wedge that is not dense in a locally convex space X, then there exists an $x^* \in X^*$ such that the closed half-space $\{x \in X : \langle x, x^* \rangle \geq 0\}$ contains W.*

Lemma 4.6 *If W is a wedge contained in a finite-dimensional space X, then W is dense in X if and only if $W = X$.*

Corollary 4.7 *If C is a cone in a finite-dimensional vector space X, then it is contained in a closed half-space.*

Note that it is in general not true that a cone is contained in a half-space, since the half-space would induce a positive functional and some cones (even lattice cones) do not admit any positive functionals, as shown before.

The alternative proof of Proposition 4.4 now follows easily by induction on the dimension and the above corollary.

Theorem 4.8 *Let X and Y be ordered vector spaces. Then $K_p(X, Y)$ is a cone.*

Proof Let $\pm u \in K_p(X, Y)$. Then $u = \sum_{i=1}^k x_i \otimes y_i$ and $u = -\sum_{i=k+1}^n x_i \otimes y_i$, for $x_i \in X_+$ and $y_i \in Y_+$. Hence $\pm u \in K_p(E, F)$, where $E = \text{Sp}\{x_i\}$ and $F = \text{Sp}\{y_i\}$ are finite dimensional. Therefore, to show the theorem, we may assume that X and Y are finite dimensional.

In this case, by Proposition 4.4, X_+ and Y_+ are contained in lexicographic cones, of which the tensor product is of the form $\text{Lex}(\{1, \ldots, d_X\} \times \{1, \ldots, d_Y\})_+$ by Proposition 4.2. This is a cone by Lemma 2.1, and so K_p, as a subwedge of this cone, is a cone. □

References

1. D.A. Birnbaum, Cones in the tensor product of locally convex lattices. Amer. J. Math. **98**(4), 1049–1058 (1976)
2. A.J. Ellis, Linear operators in partially ordered normed vector spaces. J. London Math. Soc. **41**, 323–332 (1966)
3. D.H. Fremlin, Tensor products of Archimedean vector lattices. Amer. J. Math. **94**, 777–798 (1972)
4. J.J. Grobler, C.C.A. Labuschagne, The tensor product of Archimedean ordered vector spaces. Math. Proc. Cambridge Philos. Soc. **104**(2), 331–345 (1988)
5. H. Merklen, Tensor product of ordered vector spaces. Univ. Nac. Ingen. Inst. Mat. Puras Apl. Notas Mat. **2**, 41–57 (1964)
6. H. Nakano, Product spaces of semi-ordered linear spaces. J. Fac. Sci. Hokkaido Univ. Ser. I **12**, 163–210 (1953)

7. A.L. Peressini, D.R. Sherbert, Ordered topological tensor products. Proc. London Math. Soc. (3) **19**, 177–190 (1969)
8. H. Schaefer, Halbgeordnete lokalkonvexe Vektorräume. II. Math. Ann. **138**, 259–286 (1959)
9. H.H. Schaefer, *Banach Lattices and Positive Operators* (Springer, New York, 1974). Die Grundlehren der mathematischen Wissenschaften, Band 215
10. O. van Gaans, A. Kalauch, Tensor products of Archimedean partially ordered vector spaces. Positivity **14**(4), 705–714 (2010)

www.ingramcontent.com/pod-product-compliance
Ingram Content Group UK Ltd.
Pitfield, Milton Keynes, MK11 3LW, UK
UKHW020141300726
14059UKWH00008B/91